Land Use, Climate Change and Biodiversity Modeling:

Perspectives and Applications

Yongyut Trisurat
Kasetsart University, Thailand

Rajendra P. Shrestha
Asian Institute of Technology, Thailand

Rob Alkemade
PBL Netherlands Environmental Assessment Agency & Wageningen University, The Netherlands

Information Science
REFERENCE

Senior Editorial Director: Kristin Klinger
Director of Book Publications: Julia Mosemann
Editorial Director: Lindsay Johnston
Acquisitions Editor: Erika Carter
Development Editor: Hannah Abelbeck
Production Editor: Sean Woznicki
Typesetters: Jennifer Romanchak, Mike Brehm, and Julia Mosemann
Print Coordinator: Jamie Snavely
Cover Design: Nick Newcomer

Published in the United States of America by
Information Science Reference (an imprint of IGI Global)
701 E. Chocolate Avenue
Hershey PA 17033
Tel: 717-533-8845
Fax: 717-533-8661
E-mail: cust@igi-global.com
Web site: http://www.igi-global.com/reference

Library of Congress Cataloging-in-Publication Data

Land use, climate change and biodiversity modeling: perspectives and applications / Yongyut Trisurat, Rajendra P. Shrestha and Rob Alkemade, editors.
p. cm.
Includes bibliographical references and index.
Summary: "This book combines state-of-the-art modeling approaches at various scales with case studies from across the world, discussing how to translate models into results and illustrate how pro-active implementation can mitigate biodiversity loss"--Provided by publisher.
ISBN 978-1-60960-619-0 (hbk.) -- ISBN 978-1-60960-620-6 (ebook) 1. Biodiversity conservation. 2. Biodiversity--Monitoring. 3. Landscape changes. 4. Land use--Environmental aspects. 5. Climatic changes--Environmental aspects. I. Trisurat, Yongyut, 1962- II. Shrestha, Rajendra Prasad. III. Alkemade, Rob, 1960-
QH75.L26 2011
333.95'16--dc22
2010043009

British Cataloguing in Publication Data
A Cataloguing in Publication record for this book is available from the British Library.

All work contributed to this book is new, previously-unpublished material. The views expressed in this book are those of the authors, but not necessarily of the publisher.

Table of Contents

Section 3
Methods: Land Use and Biodiversity Modeling

Section 4
Case Studies

Section 5
Conclusion

Detailed Table of Contents

Chapter 1 provides a coherent presentation of the essential concepts, key terminology, and historical background of land use informatics, deforestation and climate change, as the global threat to biodiversity. In addition, it also raises some key issues on consequences of these threats and discusses why biodiversity conservation practitioners have to think and map out integrated strategies to cope with these issues.

Chapter 2 reviews and describes the relationship between forest and climate, and forest ecosystem functions and biodiversity. Based on meta-analyses of peer-reviewed literature, this chapter also discusses in details the impacts of deforestation that will diminish population viability, and the predicted climate changes based on several development scenarios on plants and animals.

Chapter 3 explores identification and analysis of key natural, socio-economic and regulatory drivers for land use/land cover (LU/LC). Finally, it collates a number of LU/LC studies involving usage of Geo-informatics provide decision makers, land managers, stakeholders and researchers the scientific grounds for better management and formulation of conservation strategies and policies.

Chapter 4 discuses quantitative methodologies for the spatial identification and monitoring of European landscapes and their habitats. The developed methodology is now possible to model quantitatively the spatial extent of widespread habitats and landscapes on the basis of land cover information and to provide a synoptic overview of the European landscape.

Chapter 5 describes briefly the data and models used in IMAGE 2.4. It starts from basic driving forces like demographics and economic development, energy consumption and production, agricultural demand, trade and production. Then, this chapter provides the potential use of data and information derived from IMAGE to feed broader policy-exploring tools for global assessment of terrestrial biodiversity and climate mitigation.

Chapter 6 presents an integrated modeling approach for assessing land use changes and its effects for biodiversity. A modeling framework consisting of a macro-economic model, a land use change model (Dyna-CLUE) and biodiversity indicator models is described and illustrated with a scenario study for the European Union. The modeling framework can provide ex-ante assessments of policies and identify critical regions for biodiversity conservation and assist in targeting policies and incentives to protect biodiversity to vulnerable areas.

This chapter discusses various aspects of biodiversity parameters and landscape indices that can be estimated using remote sensing data. Moderate resolution satellite (MODIS) data was used to generate forest type map and to demonstrate the biodiversity characterization of ecoregion 29. The outcome states that remote sensing and geographic information systems can be used in combination to derive various parameters related to biodiversity surveillance at a regional scale.

This chapter introduces the GLOBIO3 model , which is one of the most advanced biodiversity pressure models. The model is built on simple cause–effect relationships between environmental drivers and biodiversity impacts, based on state-of-the-art knowledge. The mean species abundance of original species relative to their abundance in undisturbed ecosystems (MSA) is used as the indicator for biodiversity. Previously, GLOBIO3 described impacts on terrestrial ecosystems, but recently a separate GLOBIO aquatic model is developed based on a similar approach.

This chapter elaborates on the concepts of species distribution modeling and presents three popular techniques to generate species distribution: cartographic overlay (habitat suitability index), binary response (presence/absence), prediction model (logistic regression), and presence-only data model (maximum entropy method or MAXENT). The Asian elephant (Elephas maximus) was selected as a proxy species for this study. The study was conducted in Bun Tharik-Yod Mon, a proposed wildlife sanctuary in northeast Thailand.

This chapter investigates characteristics of floristic regions in Thailand and predict the impacts of future climate change (2050) on the recognized phytogeographical areas. Based on the MAXENT model results and clustering, the authors propose to reduce the existing seven phytogeographic regions as used in the Flora of Thailand to four regions. In addition, the future climate will strongly diminish the number of species in the northern and northeastern region. Peninsular Thailand appears to be stable, but high endemism shows that there is a decrease in suitable niche in this area, while far eastern Thailand and the Peninsular region will gain species.

This chapter indicates the history and the development of biodiversity modeling in Ukraine in order to support policy making and for providing information to e.g. the reporting to the UN Convention of Biological Diversity. It indicates that Ukrainian researchers have used extensive biodiversity modeling methods (e.g. species-based model and pressure biodiversity models) to predict the distributions of vascular plants, insects, amphibians, birds and mammals. Later, the researchers evaluate effects in habitats condition of selected species caused by land use change and climate change in 2050. This study suggests that expected climate change together with land-use change would provoke numerous non-simplified and unexpected habitat changes. In addition, scientists expect to find about 4% of new species by 2050 and approximately 13% of existing species would disappear. The model approaches and results were integrated in the education system and mass media for awareness raising.

This chapter evaluates the remaining biodiversity at the regional level and for three countries in the tropical Andes: Colombia, Ecuador and Peru in 2000 and under two scenarios in 2030: Market forces and Policy Reform using GLOBIO model. This research aims to identify the most vulnerable areas to biodiversity loss and the most important drivers of such losses. The results indicate that at the country level Ecuador would have the lowest values of remaining MSA for 2030, followed by Colombia and finally by Peru for both scenarios. In a comparison with the values of the year 2000, Ecuador also showed the highest losses of biodiversity and Peru is the second highest. The model results are used for policy formulation to maintain biodiversity in the Tropical Andean countries.

This chapter briefly describes the evolution of different approaches to modeling the potential impacts of climate change on biodiversity. The author looks in detail at the BIOCLIMA model, which simulates trends in a sample chosen to represent regional plant biodiversity and how climate change directly influences the processes determining a plant's response to climate change, i.e. reproductive rate, dispersal mechanisms and pre-adaptations to expected stresses and its application to Amazonia. It then discusses conservation planning applications of the three other contemporary paradigms, illustrated by examples from Amazonia and Kenya. This chapter also recommends authorized agencies to identify and establish optimal locations of protected areas when climate is changing, and to use protection to mitigate climate change.

This chapter first discusses the issues and status of land degradation and biodiversity in Southeast Asia and goes on to present two case studies. The first case study is a land degradation assessment in the Lower Mekong Basin demonstrating the use of spatial data and technologies and various land degradation indicators. The second case study specifically documents plant diversity and examines the relationship of plant diversity with biomass and soil erosion by making use of field surveyed primary data.

This chapter focuses on sustainable land use and watershed management. The first part covers some definitions and background on sustainable land use and watershed management. The second part describes the use of the Markov's Chain model to predict land use change in Dong Nai watershed, Vietnam and the Soil and Water Assessment Tool (SWAT) for modeling watershed hydrology and simulating the movement of sediment. Finally, the example of methodology development for sustainable land use and watershed management in response to climate change in Dong Nai watershed, Vietnam is presented.

This chapter uses Dyna-CLUE and GLOBIO3 models to predict the current and future state of biodiversity in seven countries in the Central America: Guatemala, Belize, Honduras, El Salvador, Nicaragua, Costa Rica and Panama and integrates the results into one regional assessment. The results show that in the current state, the region has a remaining MSA of 48%. The remaining MSA values are above 50% for Belize, Nicaragua, and Panama. The remaining countries experience less than 50% in the current situation. However, the future state of biodiversity is expected to be lower than 50% for all countries, especially under Baseline and Trade Liberalization scenarios.

This chapter uses logistic regression to find relationships between deforestation and biophysical and socio-economic factors in Java, Indonesia. Deforestation was detected from interpretation of MODIS satellite imageries between 2000 and 2008. Result of the study showed that impacts of population density, road density and number of households engaged in the agricultural sector are significant and they have negative impact on deforestation. Measures to reduce future deforestation are also included.

This chapter demonstrates the collaborative project between the Netherlands Environmental Assessment Agency (PBL) and the Environmental Operations Centre to integrate the results of biodiversity modeling into Strategic Environmental Assessment (SEA) in Vietnam, both at national and local levels. This collaborative project introduces an effective new indicator and biodiversity assessment method that is already endorsed by the Ministry of Environment to be embedded in the national policy process of Vietnam.

Section 5
Conclusion

Foreword

Biodiversity is declining and will decline in the near future due to ongoing land use change, climate change, increased consumption, pollution, the introduction of exotic species and the overexploitation of natural areas and natural resources. This loss of biodiversity is an issue of profound concern for its own sake and also because it underpins the functioning of ecosystems, which provide a wide range of services to human societies. This affects people in poor countries, who often depend directly on forest land and its resources, but it also affects societies in the west. Biodiversity loss and the depletion of natural resources ultimately threaten everyone's survival. These are important conclusions of the third Global biodiversity Outlook, and is the main reason why PBL - The Netherlands Environmental Assessment Agency developed a series of integrated models to support policy and decision making at global, regional and national levels. These models were successfully applied to evaluate scenarios in various global assessments like the Global Environmental Outlooks, the Global Biodiversity Outlooks and the Millennium Ecosystem Assessment.

PBL started to develop modeling tools for national use in 2005, together with a large group of modelers from various regions of world. Modelers from Meso- and South America, from eastern Europe and Southeast Asia and from Eastern and Southern Africa were brought together to actually develop tools to be used in their own countries, based on the principles of models developed at a global level. Aspects of models like IMAGE and GLOBIO and also the land use allocation model CLUE proved to be very useful at (sub-) national levels.

This book includes some of these experiences and gives the wider, methodological, context of these experiences. I hope that this book may inspire researches in many countries to set up science based policy support even if resources are limited or data are scarce. Biodiversity loss declines too quickly to wait for sufficient data. I appreciate the initiative of especially Yongyut Trisurat to assemble all these experiences and join them together with theory and background.

Maarten Hajer
PBL Netherlands Environmental Assessment Agency, The Netherlands

***Maarten Hajer** is Director of PBL Netherlands Environmental Assessment Agency since 2008 and responsible for the strategic assessments and policy evaluations to facilitate political deliberation and decision making, ranging from environment, nature, land use to water and transport. While primarily focused on the Dutch political decision making PBL is also active for international bodies, such as the European Commission, OECD and UNEP. He is professor of Public Policy at the University of Amsterdam since 1998, and continues his professorship part time. Hajer holds MA degrees in Urban and Regional Planning and Political Science from University of Amsterdam, as well as a D.Phil. in Politics from University of Oxford. Hajer is the author of over ten books and many articles and contributions to books. His most recent book is Authoritative Governance: Policy Making in the Age of Mediatization which appeared at Oxford University Press in 2009.*

Foreword

This book on "Land Use, Climate Change and Biodiversity Modeling: Perspectives and Applications" is a very useful reference material as it attempts to integrate three main topics (land use, climate change and biodiversity) using case studies in different parts of Asia and the world and state-of-the-art modelling approaches to effectively address global problems on the environment and use of natural resources.

Climate change, land degradation and loss of biodiversity are complex in nature and hence integrated solutions are imperative as we cannot deal with these issues independently. Thus, I believe that the UNFCCC, UNCCD, and CBD are doing their important role to address these concerns. The centre of the many critical issues being discussed in these conventions is the Forest.

Forest is the origin of life and the source of human existence. About 2/3 of the terrestrial species belongs to forest. It is essential as carbon sink and has immense contribution in mitigating climate change. It has a foremost role in the human economic activity.

According to the World Bank, the welfare of about 160 million people is being affected by the rapidly deteriorating global economic environment. Therefore, in this era of environmental dilemma, we should work together towards improving our environment by putting together realistic approaches.

This book serves as a source of sound information for many practitioners like investors, environmental advocacy groups, forestry professionals and educators, policy-makers, and the general public, which can be used to formulate recommendations for future policies and management strategies needed in support of sustainable development.

I would like to congratulate the authors and editors of this book that I know would be a worthwhile contribution to our society at large.

Don Koo Lee
International Union of Forest Research Organizations

***Don Koo Lee** is now the Minister of the Korean Forest Service, Korea. He was a professor of Silviculture and Restoration Ecology at Seoul National University (SNU), Korea. He received BS and MS degrees in forestry and forest genetics, respectively from SNU, and MS and Ph. D. degrees in forest biometry and silviculture, respectively from Iowa State University, USA. He was Dean for College of Agriculture and Life Sciences, SNU (1999-2001), a Board of Trustees member of the Center for International Forestry Research (1999-2004) in Indonesia, President of the Korean Forest Society (2004-2006), and President of the International Union of Forest Research Organizations or IUFRO (2006-2010). He was awarded an Honorary Doctoral degree from the Moscow State Forest University (2007) in Russia. He has been Project Leader of the ASEAN-Korea Environmental Cooperation Project since 2000, a member of the Royal Swedish Academy of Agriculture and Forestry, Sweden since 2003, and Co-Representative of Forest for Life, Korea since 2004. His research interests are: restoration of degraded forest ecosystems, eco-friendly management of forest ecosystem and development of silvicultural strategies for natural forests, growth and nitrogen fixation, and biomass production and nutrient cycling of forest ecosystems.*

Preface

This edited book focuses on discussing three interrelated issues namely land use, climate change and biodiversity. It particularly looks at the impacts of land-use change and climate change on biodiversity with reference to the state-of-the-art modeling approaches at various scales and through case studies representing various regions of the world. In addition, we hope it will help natural resource managers, scientists and decision makers in overcoming their fear of models and help them in translating the model results into pro-active implementation to mitigate biodiversity loss. This subject area is of high importance nowadays because it is not only of interest to individual scientists, but also to policy-makers who are committed to at least three global commitments, namely Convention on Biological Diversity (CBD), the United Nations Framework Convention on Climate Change (UNFCC) and the United Nations Convention to Combat Desertification (UNCCD). Most countries of the World have already agreed upon their implementation.

The idea to develop this book originated in 2006 when the Netherlands Environmental Assessment Agency organized an international meeting on biodiversity modeling in the Netherlands. Approximately 20 land use, climate and biodiversity modelers gathered in this meeting. One key message derived from the meeting was to document the concept and methodology currently practiced in various parts of the world and to disseminate it to a wide audience.

Why another biodiversity modeling book? While there already are a number of publications on the subject of a particular threat, like land-use change or climate change, to biodiversity, there is no single book volume at the moment that combines these important issues. Therefore, this book, "*Land Use, Climate Change and Biodiversity Modeling: Perspectives and Applications*", is unique and is distinguished from existing titles within the subject area. In addition, it responds to the remaining controversial issue on the effects of accumulative land-use and climate change on biodiversity.

This edited book contains 20 chapters divided into five parts. Part I Introduction provides a coherent presentation of the essential concepts, key terminology, and historical background of land use informatics, deforestation and climate change, in light of their global threat to biodiversity. Part II Setting the Scene starts off the overview of deforestation and climate impacts on biodiversity followed by the information on how to monitor and quantify these impacts. Part III Methods: Land Use and Biodiversity Modeling gives readers essential tools for land use studies and biodiversity modeling. These insights are put into practice in Part IV: Case Studies. Part V Conclusions and Recommendations summarizes and presents analytical views on the status, trend and way forward with regard to the issues of biodiversity and land use modeling and conservation in the context of climate change. More details about the content of each chapter can be found at the end of the Chapter 1.

The contributors to this book are university professors, scientists and conservation practitioners, who are internationally recognized and have published a number of scientific papers in international journals and at international conferences. Some are the architects of land use and biodiversity models currently used worldwide.

The edited book not only contains recent concepts and methods on land-use modeling and species modeling and accumulative impacts of land use and climate change on biodiversity, but also several case studies of practical importance from various parts of the world. In addition, contributors elaborate methods and processes integrated in a single volume using a simple language that is understandable to non-modelers and resource managers. This edited book is of value for resource managers, scientists, and graduate students. It has a potential for use as a text book and reference in various university classes. Further, researchers should be interested in our conclusion and recommendations on future research needs in order to fill existing gaps on land use and biodiversity modeling. Land use planners and protected area managers would be interested in finding optimal land allocation options and in prioritizing protected area network to effectively conserve biodiversity.

Yongyut Trisurat
Kasetsart University, Thailand

Rajendra Prasad Shrestha
Asian Institute of Technology, Thailand

Rob Alkemade
PBL Netherlands Environmental Assessment Agency & Wageningen University, The Netherlands

Acknowledgment

The editors would like to express their sincere appreciation to many scientists and organizations contributing to the development of this book. First, IGI Global was instrumental in formulating the content of the book by accepting the initial concept of the book proposal. Hannah Abelbeck, (Editorial Assistant) and Christine Bufton (Editorial Communications Coordinator) of IGI Global provided tremendous assistance in the development process for the edited book. We are very grateful to the Editorial Advisory Board, namely Nipon Tangtham at Kasetsart University, Olavi Luukkanen at Helsinki University, and Roger Kjelgren at Utah State University, as well as anonymous reviewers for their diligent and rigorous evaluation of the chapter manuscripts. We would like to thank all the contributing authors for submitting high quality manuscripts and promptness in responding to the subsequent alterations of reviews, despite being scattered all over the globe. We would especially like to thank our editor, Palle Havmoller, for his strong language editorial and proof reading during the final stage of producing the text. Last but not least, we would like to thank Maarten Hajer, Director of the Netherlands Environmental Assessment Agency (PBL) and Don K. Lee, President of International Union of Forest Research Organizations (IUFRO) and now the Minister of the Korean Forest Service for contributing for contributing the remarkable forewords for this book.

Our respective employers, namely the Faculty of Forestry at Kasetsart University, the School of Environment, Resources and Development at Asian Institute of Technology, and the Netherlands Environmental Assessment Agency (PBL), provided logistic and administration support for this undertaking. Our family and friends always encouraged and provided moral support through the long development process. Our research assistants, Suramongkon Siripon at Kasetsart University and Binaya Pasakhala at Asian Institute of Technology, patiently copyedited, formatted various permutations, and searched often obscure references of the manuscripts and deserve foremost gratitude for getting the job done.

Yongyut Trisurat
Kasetsart University, Thailand

Rajendra P. Shrestha
Asian Institute of Technology, Thailand

Rob Alkemade
PBL Netherlands Environmental Assessment Agency & Wageningen University, The Netherlands

Section 1
Introduction

Chapter 1
Linkage between Biodiversity, Land Use Informatics and Climate Change

Yongyut Trisurat
Kasetsart University, Thailand

Rajendra P. Shrestha
Asian Institute of Technology, Thailand

Rob Alkemade
PBL Netherlands Environmental Assessment Agency, The Netherlands

ABSTRACT

Biodiversity is the variety and variability among living organisms and ecological complexes in which they occur, and it can be divided into three levels – gene, species and ecosystems. Biodiversity is an essential component of human development and security in terms of proving ecosystem services, but also it is important for its own right to exist in the globe. Failure to conserve and use biological diversity in a sustainable manner would result in degrading environments, new and more rampant illnesses, deepening poverty and a continued pattern of inequitable and untenable growth. This chapter provides a coherent presentation of the essential concepts, key terminology, historical background of biodiversity, and drivers to biodiversity loss, especially land use/land cover and climate change. A number of land use change models and a general circulation model for prediction of future climate change and its effects on individuals, populations, species, and ecosystems are briefly described. The chapter also introduces the structure of the book including summaries of each chapter.

DOI: 10.4018/978-1-60960-619-0.ch001

1. INTRODUCTION

Humans have been using Earth's terrestrial land for various purposes from time immemorial. It has been increasingly recognized that conversion of pristine land into various forms of land uses and especially their misappropriate use affect ecosystems and cause biodiversity loss. Biodiversity is declining at an unprecedented rate and is expected to continue to decline (sCBD, 2010). The concern over rapid biodiversity decline has urged the international community to organize the Earth Summit in Rio de Janeiro in Brazil in 1992. This general concern is not limited to the decline of biodiversity itself but also includes the very notion that biodiversity is a key factor in the provision of a series of ecosystem goods and services on which humanity depends. Especially the rural poor depend largely on ecosystems that provide food, shelter and protection to natural hazards.

One of the outcomes of the 1992 Earth Summit was the adoption of the Convention on Biological Diversity (CBD), which has been ratified by more than 190 countries. The CBD focuses on conservation of biodiversity, sustainable uses and fair and equitable sharing of benefits arising out of the use of biodiversity. It is one of the most important international conventions and is implemented widely across the World.

Failure to conserve and use biological diversity in a sustainable manner would result in degrading environments, new and more rampant illnesses, deepening poverty and a continued pattern of inequitable and untenable growth. Therefore, in 2002 the Parties to the Convention committed themselves to achieve by 2010 a significant reduction of the current rate of biodiversity loss at the global, regional and national level as a contribution to poverty alleviation and to the benefit of all life on Earth. These targets were endorsed during the World Summit on Sustainable Development in Johannesburg (UN, 2002)

The so-called "Biodiversity 2010 Targets" were developed and indicators, measures and options were indentified to guide implementable activities. Targets are increasingly being used in various areas of public policy. Clear, long-term outcome-oriented targets that are adopted by the international community can help shape expectations and create the conditions in which all actors, whether governments, the private sector, or civil society, have the confidence to develop solutions to common problems. By establishing targets and indicators, progress can be assessed and appropriate actions taken. In addition to the 2010 Biodiversity Targets, the Convention has established other targets, such as the Global Strategy for Plant Conservation, and the Programme of Work on Protected Areas.

Concurrently, the Millennium Development Goals (MDGs) were formally established when the United Nationals General Assembly adopted the Millennium Declaration in 2002. MDGs address issues of poverty eradication and sustainable development through a set of targets and dates. One of the significant elements of the MDGs is Goal 7 which focuses on addressing challenges to biodiversity from climate change and pollution. Attempts are being made to maintain and enhance resilience to adapt to climate change, and to reduce pollution and its impacts on biodiversity. These measures are to mainstream biodiversity into not only Goal 7, but also across other MDGs, as achieving the targets of the MDGs will directly or indirectly impinge on the status and use of biodiversity (UN, 2005).

In general land use change has been the main driver of terrestrial biodiversity loss during the past century. Climate change will be a major driver in the near future. Other important factors are nutrient loading, overexploitation, fragmentation and the effects of invasive species (Leadley *et al.*, 2010).

Table 1. Definitions of biodiversity or biological diversity

Definition	Source
The amount of genetic variability within species and the number of species in a community of organisms.	Norse & McManus (1980)
The variety and variability among living organisms and ecological complexes in which they occur.	Office of Technology Assessment (1987)
The variety of and variability among living organisms and the ecological complexes of which they are part; this included diversity within species, between species and ecosystems.	United Nations Environment Programme (1991)
Biological diversity or biodiversity as the variety and variability among living organisms and ecological complexes in which they occur. Biodiversity occurs at three hierarchical categories – gene, species and ecosystems that describe different aspects of living organisms.	Office of Technology Assessment (1987)
"Biological diversity" means the variability among living organisms from all sources including, inter alia, terrestrial, marine and other aquatic ecosystems and the ecological complexes of which they are part; this includes diversity within species, between species and of ecosystems.	Convention on Biological Diversity (1992)
Biodiversity has to be thought of in a number of different ways over evolutionary time, as a characteristic of natural communities, globally, and collectively.	Lovejoy (1997)
Biodiversity is, in one sense, everything. Biodiversity is all hereditarily based variation at all levels of organization, from the genes within a single local population or species, to the species composing all or part of a local community, and finally to the communities themselves that compose the living parts of the multifarious ecosystems of the world.	Wilson (1997)
Biodiversity means the variability among living organisms from all sources, including terrestrial, marine and other aquatic ecosystems and the ecological complexes of which they are part.	Millennium Ecosystem Assessment (2005)

Modified from Baydack & Campa III (1999) and extended

Thus, policies to protect biodiversity or to reduce the rate of biodiversity loss are preferably targeted on lowering the effects of these drivers. Models may help inform policy makers on the current state of biodiversity and the drivers affecting it's future. Models may be used to guide the selection of effective and cost-efficient measures at global, regional, national, and local levels.

In chapter 1, we review the concept of biodiversity and briefly describe the elements of land use/land cover and climate change as the two dominant drivers of biodiversity loss Furthermore we present some general concepts on modeling of the effects of land-use and climate changes on biodiversity, and briefly describe the set-up of this volume. Many chapters report activities performed in the context of the PBL Netherlands Environmental Assessment Agency projects. Moreover, case studies, using the results of biodiversity modeling and using land use and climate change models focusing on policy support, are included.

2. BIODIVERSITY

2.1 What is Biodiversity?

The expression and concept of biodiversity or biological diversity that did not exist decades ago has now become one of the most commonly used expressions in the biological sciences, political sciences, economics and management planning. Biodiversity can mean different things to different people, so that anyone using the term needs to define or at least imply its definition to ensure that others are aware of the specific orientation under consideration. Numerous sources have provided detailed definitions of biodiversity as shown in Table 1.

Generally biodiversity can be divided into three hierarchical categories – gene, species and ecosystems that describe different aspects of living organisms:

Genetic diversity refers to the variation of genes within species. This covers distinct population of the same species or genetic variation within a population. Genetic diversity is an important aspect of a species and a population. The presence of different ecotypes permits a species to survive in a variety of physical and biotic environments, whereas genetic variation allows a population of a species to adapt to changing environmental conditions.

Species diversity refers to the variety of species within a region. It is the most commonly considered aspect of biological diversity. Basically species diversity can be measured in many ways, such as species richness and species diversity. Species richness is often used to refer to the number of species of a particular group found in a particular ecosystem, such as the number of bird species or the number of mammal species. However, species richness is an incomplete description of species biodiversity because it does not account for differences in the relative abundance of the different species in the community. Therefore, both species richness and species abundance are normally measured together to represent species diversity.

Ecosystem diversity refers to the diversity of a place at the level of ecosystem. The physical environment, especially the annual cycle of temperature and precipitation, affects the structure and characteristics of a biological community, determining whether a particular site will be forest, grassland or desert. In addition, biological factors can also alter the physical characteristics of an ecosystem. Conservation management of whole landscapes becomes an important consideration to ensure the survival of species that range widely across different ecosystems. Therefore, some biological scientists also propose *landscape diversity* as another level on the top of ecosystem diversity.

Another way of looking at biodiversity is a more conceptual one, with much more focus on the meaning it has for people instead of describing a biophysical entity. Gaston (1996) classified the meanings of biodiversity into three broad concepts: a theoretical concept; a measurable entity; and a social/political construct.

1. *Biodiversity as a concept*: The definition given by the US Congress Office of Technology Assessment (1987) is biological diversity or biodiversity as the variety and variability among living organisms and ecological complexes in which they occur. Generally, biodiversity can be divided into three levels of ascending scopes – gene, species and ecosystems that describe different aspects of living organisms. This definition, perhaps the most widely cited by scientists and publics (Gaston, 1996), strongly establishes biodiversity as a concept. Recently, landscape biodiversity is sometimes added to determine habitat configuration that affects species viability in a large landscape, which further reinforcing the interdisciplinary nature of biodiversity research.
2. *Biodiversity as a measurable entity*: Haila & Kouki (1994) put forward that the biodiversity is a measurable entity and not simply an abstract concept. The choice and derivation of a measure of biodiversity will depend fundamentally on the use to which it will be put. There are two kinds of measures, those which simply count entities and those which additionally attempt to incorporate some elements of their differences. Diversity measures in biology were derived from information theory, which combines number of species and the evenness or equality of their abundances, and several indices are available for usage.
3. *Biodiversity as a social/political construct*: There is a general acceptance in many communities that biodiversity is per se a good thing, that its loss is bad, and it should be maintained. Thus, in this concept, biodiversity is not a neutral scientific concept but it is perceived as a value or as having value (Bowman, 1993).

Today, nearly two decades since the Earth Summit in Rio de Janeiro in Brazil and eventual evolvement of biodiversity concept it is important to examine whether it has allowed us to progress ecological understanding and systems science any further, and more importantly, whether it has led to progress in environmental conservation, management and policy making.

2.2 Values of Biodiversity

Biodiversity is an essential component of human development and security. Demonstrating the value of biodiversity is a complex issue because value is determined by a variety of economic and ethical factors. McNeely (1998) classified the values of biological resources into two broad categories. *Direct* values are known as private goods which are assigned to those products harvested by people. *Indirect* values, known as public goods which are assigned to benefits provided by biological diversity, include water quality, soil protection, etc. Direct and indirect values of biodiversity are more recently included in the concept of ecosystem goods and services (MEA, 2005).

Generally, there are four categories of ecosystem services recognized as given below.

1. *Provisioning services*
   - food (including seafood and game), crops, wild foods, and spices
   - water
   - pharmaceuticals, biochemicals, and industrial products
   - energy (hydropower, biomass fuels)
2. *Regulating services*
   - carbon sequestration and climate regulation
   - waste decomposition and detoxification
   - purification of water and air
   - crop pollination
   - pest and disease control
3. *Supporting services*
   - nutrient dispersal and cycling
   - seed dispersal
   - Primary production
4. *Cultural services*
   - cultural, intellectual and spiritual inspiration
   - recreational experiences (including ecotourism)
   - scientific discovery

In this book we will focus on the species diversity aspect of biodiversity and will consider it mainly as a measurable entity. From that perspective we believe we can support policy making and also may be able to link biodiversity to ecosystems and ecosystem goods and services, as ecosystems constitute an important part of species and their interactions with the environment.

3. LAND USE, LAND COVER AND INFORMATICS

3.1 Land Use Informatics

Land cover and land use are distinct terms despite often being used interchangeably. The origins of the 'land cover / land use' couplet and the implications of their confusion are discussed in Fisher *et al.* (2005). FAO (1997) defined *land cover* as "the observed (bio)physical cover on the earth's surface". Strict consideration of land cover should be confined to describe the vegetation and the man-made features. Consequently, areas where the surface consists of bare rock or bare soil are describing land itself rather than land cover. Also water surfaces can be disputed as being real land cover. However, in practice the scientific community typically describes those aquatic aspects under the term land cover.

Land use is characterized by the arrangements, activities and inputs people undertake to produce, change or maintain a certain land cover type (FAO, 1997). Land use is a description of how people utilize the land and thus a socio-economic activity. Land use defined in this way establishes a direct link between a particular land cover and the actions of people interacting with that land cover. At any one point or place, there may be multiple and alternate land uses, the specification of which may have a political dimension. The major effect of land use on land cover since 1750 has been deforestation of temperate regions. More recent significant effects of land use include urban sprawl, soil erosion, soil degradation, salinization, and desertification. In addition, land-use change, and the use of fossil fuels are the major anthropogenic sources of carbon dioxide, a dominant greenhouse gas. Land use has also been defined as "the total of arrangements, activities, and inputs that people undertake in a certain land cover type" (FAO, 1997).

Land use directly affects peoples' livelihood, and also biogeochemical cycles and biodiversity through land surface processes. Predicting how land use changes affect land degradation, the feedback on livelihood strategies from land degradation, and the vulnerability of places requires a good understanding of the dynamic of human-environment interactions associated with land use changes (Kasperson *et al.*, 1999). This requires asking the following fundamental questions in specific respective geographical context (Lambin & Geist, 2006).

- How has land use and land cover been changed in the past?
- What are the causes and circumstances of land use change?
- How will change in land use affect land cover in immediate and distant future?
- How do human and biophysical dynamics affect the coupled human-environment system?
- How do climate variability and change affect land use and land cover, and what are the potential feedbacks of changes in land use and land cover to climate and vice versa?
- How do land uses and land covers affect the vulnerability of the coupled human-environment system?

Key to answering these questions is land use informatics, which is an emerging and growing discipline that combines sciences related to information and technology, land use, and other sciences, like socio-political. It strives to develop methods to organize knowledge on land use/land cover. Information system is generally defined in terms of databases, which focus on data requirements and the mechanism to store, organize, process and analyze data (Cruz, 2006) that would contribute to achieving organizational goals and objectives. Therefore, the goal of land use informatics is to update and unify information on land use and land cover change at all levels (local to global), and to pursue specific analysis from such data that would improve decision making by responding to the questions raised above.

3.2 Land Use Modeling

In order to structure our understanding of land uses, we often need to approximate the real world situation through land use modeling. These modeling approaches can range from a simple spatial prediction of land characteristics to predicting specific land use responses in the short term, such as yield modeling, to a long term integrated land use model for predicting future land use development. A range of models of land-use change have been developed to meet land use management needs, and to better assess and project the future role of land-use and land-cover change in the functioning of the earth system (Veldcamp & Lambin, 2001). Spatially explicit, integrated and multi-scale modeling is an important technique for the projection of alternative pathways into the future. Modeling allows for conducting experiments that test our understanding of key processes, and for describing the latter in quantitative terms (Lambin *et al.*, 2001).

Land-use change models represent part of the complexity of land-use systems. Models of land-use change can address two separate questions: where are land-use changes likely to take place (location of change) and at what rates are changes likely to progress (quantity of change). The first question requires the identification of the natural and cultural landscape attributes which are the spatial determinants of change. The rate or quantity of change is driven by demands for land-based commodities and are often captured using economic models accounting for demand-supply relations and (international) trade (Verburg *et al.*, 2008). Land-use change models range from simple system representations, including a few driving forces, to simulation systems based on profound understanding of situation-specific interactions among a large number of factors at different spatial and temporal scales, as well as environmental policies.

A number of land use models are being used. Verburg & Veldkamp (2004), Matthews *et al.* (2007) and Priess & Schaldach (2008) have provided reviews on different land use models. Parker *et al.* (2001) categorized six broad land use predictive models: mathematical equation-based, statistical, expert system, system dynamics, cellular, and hybrid. Statistical techniques are a common approach to modeling land-use/cover change given their power, wide acceptance, and relative ease of use (see chapter 17). System models are dynamic and represent stocks and flows of information, material, as sets of differential equations linked through intermediary functions and data structures. An example of system models is Integrated Modeling of Global Environmental Change (IMAGE), a model developed by the Netherlands Environmental Assessment Agency (see chapter 5 for more details). Hybrid models usually combine any of above-mentioned techniques, examples are DELTA (Southworth *et al.*, 1991), GEOMOD2 (Hall *et al.*, 1995) and Dyna-CLUE (Verburg & Veldkamp, 2004). Hybrid models are quite advanced modeling approaches for complex, dynamic and spatial problems and widely applied in the tropics (see chapter 6 for details on Dyna-CLUE).

3.3 Effects of Land-Use/Land-Cover Changes on Biodiversity

Deforestation has been given much attention in land use and landscape changes because of the high rate of forest change and the ecological importance forest ecosystems. Basic data on the rate and spatial distribution of deforestation is typically derived from remote sensing images and Geographic Information Systems (GIS) (Brannstrom *et al.*, 2008). For example, the Royal Forest Department in Thailand has monitored forest cover using satellite images to show that 1961 forest cover of 27.36 million ha (53.3% of the country area), has declined to approximately 12.97 million ha (25.2%) in 1998. The average annual loss was approximately 400,000 ha or 2.0%, while the total area of reforestation between 1961-2001 was approximately equivalent to one year of deforestation (Trisurat, 2007). At global level, the United Nations Food and Agriculture Organization (FAO, 2010) indicated that around 13 million ha of forests were converted to other uses or lost through natural causes each year between 2000 and 2010 as compared to around 16 million ha per year during the 1990s. Even though, the worldwide pace of deforestation has slowed down for the first time on record as a result of concerted efforts at both local and international levels, it still remains alarming in many countries. The highest annual losses were registered in South America, which lost four million hectares, and Africa, which lost 3.4 million ha.

Deforestation causes a number of consequent effects on the biological and physical environment, such as habitat loss, habitat fragmentation, species extinction, deterioration of soil properties, drought, flooding, etc. Habitat fragmentation is the process of dissecting large and contiguous areas of similar native vegetation types into smaller units separated by different vegetation types and/ or by areas of intensive human activity (Saunders *et al.*, 1991). Fragmentation occurs in conjunction with loss of area and includes changes in composition, shape and configuration of resulting patches (Rutledge, 2003). A number of landscape indices have been developed to measure the effects of fragmentation. In addition, the FRAGSTATS 3.0 software is usually used to assess landscape structure and fragmentation indices (McGarigal & Marks, 1995).

The popular landscape indices to explain forest fragmentation that may have a direct impact on biodiversity include area indices such as total area, number of patches, mean patch size and largest patch index, (the percentage of landscape area occupied by the largest patch of a particular land use class), edge indices, shape index, core area indices (mean core area and total core area) and neighbor index (Forman, 1995; Ochoa-Gaona, 2001).

Increased fragmentation often results in the subdivision of the natural environment into isolated patches of different size and shapes (Turner & Corlett, 1996). The effects of fragmentation include decreased species richness, increased habitat edges, favoring species adapted to edge habitats at the expense of species living in core areas (Yahner, 1988), diminished species distribution and gene flow (Raabova *et al.*, 2007). Donovan & Flather (2002) found that forest birds had lower reproductive rates in small patches than in large patch. If small patches occur in areas with less forest, the reduced reproductive rate may not be the result of patch size, but rather from larger populations of nest predators and brood parasites that occur in landscapes with more open habitat (Schmiegelow & Monkkinen, 2002). In addition, Zanette (2000) studied patch size and demography of an area-sensitive songbird and reported that many songbird species were absent from small forest patches and likely decline to extinction because reproductive success was too low to compensate for adult mortality. More examples are presented in chapters 10 to 19 of this volume.

Study on land use and land cover change is important specifically with regard to a number of aspects including their role on biodiversity, climate change and vice versa. As discussed above, land use is influenced by biophysical, socio-cultural, economic and political factors, and a number of different information inputs are required to study land use land cover change and its influence on biodiversity. Some high priority topics within the area of land use informatics are included below.:

- Historic and up to date land use and land cover data at various scales
- Drivers and underlying factors affecting land use change.
- People's perception of land and preference for type of land uses
- Future land use projection under different potential scenarios along with benefit cost analysis
- Strategic plans, including funding, for the collection, use, and availability of consistent and standardized land use data

Knowledge of the effect of land use change on biodiversity is mainly based on relatively simple assumptions on the suitability of habitat to a species. An assumption often made is simply to state that biodiversity is completely destroyed if land conversion has taken place. Examples are the models based on the species area relationship (SAR) (e.g. van Vuuren *et al.*, 2006). Other models allow for the existence of species in converted land (e.g. Alkemade *et al.*, 2009; Louette *et al.*, 2010). Modeling the effects of land use change on biodiversity requires knowledge on suitability of the different land use and land cover classes for occurrence, and abundance, of individual species. This information can be used for individual species (Louette *et al.*, 2010) or can be grouped into a general cause-effect relationship as demonstrated in GLOBIO3 (Alkemade *et al.*, 2009; Chapter 8).

4. CLIMATE CHANGE

Climate is always changing, as it is largely determined by sun activity and circulations in the atmosphere Recent climate change is at least partly human induced by the increased concentrations of green house gases (GHGs).GHGs are gases in an atmosphere that absorb and emit radiation within the thermal infrared range. The main greenhouse gases in the Earth's atmosphere are carbon dioxide, methane and nitrous oxide. In addition, other greenhouse gases include sulfur hexafluoride, hydrofluorocarbons and perfluorocarbons. The rapid increase in GHGs is expected to continue for several decades to come and greatly affect the temperature of the Earth.

4.1 Climate Change Scenarios

Future levels of GHGs emissions will be the product of very complex dynamic systems, determined by driving forces such as demographic development, socio-economic development and technological change. In 1996 the IPCC (Intergovernmental Panel on Climate Change) Special Reports on Emissions Scenarios (SRES) developed four different narrative scenarios to represent the range of driving forces and emissions in the scenario literature so as to reflect current understanding and knowledge about underlying uncertainties (IPCC, 2000).

Each storyline represents different demographic, social, economic, technological, and environmental developments, which may be viewed positively by some people and negatively by others. The SRES global scenarios are presented in Table 2, which includes the estimated population, CO_2 emission, economic growth and per capita income for each scenario. Data in columns 2 to 4 are taken from Nakicenovic *et al.* (2000). CO_2 concentrations were estimated by using the same model runs. Below is a description of those four storylines.

- *A1*: A future world of very rapid economic growth, global population that peaks in mid-century and declines thereafter, and rapid introduction of new and more effi-

Table 2. Some aspects of predicted population, CO_2 emission, economic growth and per capita income against SRES scenarios

Emissions Scenario	Global Population (billions)	Global GDP (10^{12} US$ a^{-1})	Per Capita Income Ratio	CO_2 Concentration (ppm)	Temperature change (°C at 2090-2099 relative to 1980-1999)		Sea level rise (m at 2090-2099 relative to 1980-1999)
					Best estimate	Likely range	Model-based range excluding rapid dynamic changes in ice flow
Constant year 2000 concentrations	6.1-6.2	25-28	12.3-14.2	367	0.6	0.3 – 0.9	NA
2100							
- A1FI scenario	7.1	525	1.5	976	4.0	2.4 – 6.4	0.26 – 0.59
- A1B scenario	7.1	529	1.6	711	2.8	1.7 – 4.4	0.21 – 0.48
- A1T scenario	7.1	550	1.6	569	2.4	1.4 – 3.8	0.20 – 0.45
- A2 scenario	15.1	243	4.2	857	3.4	2.0 – 5.4	0.23 – 0.51
- B1 scenario	7.0	328	1.8	538	1.8	1.1 – 2.9	0.18 – 0.38
- B2 scenario	10.4	235	3.0	615	2.4	1.4 – 3.8	0.20 - 0.43

Source: IPCC (2000)

cient technologies. Major underlying themes are economic and cultural convergence and capacity-building, with a substantial reduction in regional differences in per capita income. The A1 scenario family develops into three groups that describe alternative directions of technological change in the energy system: fossil-intensive (A1FI), nonfossil energy sources (A1T), and a balance across all sources (A1B).

A1FI represents "business-as-usual" - a world that still runs on coal and gas. It is here that predictions are most shocking: temperature gains of some 2.4 to 6.4 °C are within reach. The sea would rise some 26 to 59 centimeters until the end of the century, flooding large coastal cities and numerous islands. A1B, the most probable scenario given current trends, is also alarming. While fossil fuels are still widely used, they are part of a more balanced energy mix. Still, by the end of the century, temperatures will have risen some 1.7 to 4.4 °C, with the oceans gaining some 21 to 48 centimeters. Rainfall is likely to decrease by some 20 percent in the subtropics, while more rain will fall in the more northern and southern latitudes. The Gulf Stream will not stop, but it will lose about a quarter of its force. A1T is a world that has lived through a third industrial revolution - a widespread conversion to green energy sources. It is similar to B1 in the sense that temperatures and oceans will rise, but to an extent that is manageable.

- *A2*: A differentiated world in which the underlying theme is self-reliance and preservation of local identities. Fertility patterns across regions converge very slowly, resulting in continuously increasing population. Economic development is primarily regionally orientated, and per capita economic growth and technological change are more fragmented and slower than other storylines.

The predicted average temperature rise for the SRES A2 (relative to the 1980 to1999 average) is +3.4°C with a range of +2.0 to +5.4°C. It is expected that the global sea level will have risen some 23 to 51 centimeters in the 21st century. In addition, the average precipitation response, using the SRES A2 forcing for the 30-year average 2071 to 2100 compared with 1961 to 1990, is an increase of 3.9% with a range of 1.3 to 6.8%.

- *B1*: A convergent world with rapid change in economic structures toward a service and information economy, reductions in material intensity, and introduction of clean technologies. The emphasis is on global solutions to economic, social, and environmental sustainability, including improving equity, but without additional climate change policies.

The B1 scenario with rapid change in economic structures toward a service and information economy predicts that temperatures will rise by 1.8 ºC (likely range between 1.1 ºC and 2.9 ºC). The sea would rise the least compared to other SRES scenarios.

- *B2*: A world in which the emphasis is on local solutions to economic, social, and environmental sustainability. This is a world with continuously increasing global population at a lower rate than in scenario A2, intermediate levels of economic development, and less rapid and more diverse technological change than in the A1 and B1 storylines. Although this scenario also is orientated toward environmental protection and social equity, it focuses on the local and regional levels.

The B2 scenario produces a smaller warming, which is consistent with its lower positive radiative forcing at the end of the 21st century. Under this scenario, the average temperature will gain +2.4 °C with a range of +1.4 to 3.8 °C. The sea level would increase by 20 to 43 centimeters. The average precipitation response, using the SRES B2 scenarios, is an increase of 3.3% with a range of 1.2 to 6.1%. The lower precipitation increase values for the B2 scenario are consistent with less globally averaged warming for that scenario at the end of the 21st century compared with A2.

4.2 Climate Change Models

A general circulation model (GCM) is a numerical model that gives the analysis of atmosphere on an hourly basis in all three spatial dimensions based on conservation laws of momentum, energy and water vapor. GCMs are the most reliable and powerful tools used to enhance our understanding of the factors that influence climate and improve our ability to predict future climate patterns (Patta, 2004).

GCM resolutions have become finer with time due to the advances in computing technology and also with more recent models having spatial resolution of 250 km and about 20 vertical levels, compared to a resolution of about 1000 km and between 2 and 10 vertical levels in earlier GCMs. However, this resolution is quite coarse for assessing impacts of climate change on biodiversity because plants and animals are highly dependent on landscape features.

Regional climate and global atmospheric modeled, high resolution scenarios allow a more realistic representation of the response of climate to fine-scale topographic features. Numerous regionalization techniques have been employed to obtain high-resolution using medium-coarse resolution GCM outputs as a starting point. The most common methods for transferring GCM output to variable at the local scale include: (1) delta change/ratio methods which look at the percent or amount of change from present to future conditions; (2) dynamic downscaling, using Regional Circulation Models (RCMs) driven by GCMs; and (3) statistical downscaling.

Barron & Sorooshin (1997) developed the Regional Climate Models to perform essentially the same functions as global GCMs but over a restricted domain. The reduced area allows higher grid resolutions (~75 km), achieving more detailed results by incorporating additional small-scale physical phenomena and local knowledge of the topology/vegetation. The Regional Climate Model is then driven at its boundaries by periodic updates from the coarse GCM. In addition, Samuels *et al.* (2010) used the RCM models to downscale the global climate and assess its impacts on Jordan River flow. In addition, Trisurat (2009) downscaled the Tyndall Centre for Climate Change Research, School of Environmental Sciences, University of East Anglia (TYN SC 2.0) global climate generated at a spatial resolution of 0.5° (approximately 45 km) to a resolution of 500 m using spline interpolation method (ESRI, 1996) with latitude, longitude and digital elevation model (DEM) in the model to reduce statistical error (Hutchinson, 1995). The 500-m resolution was chosen as an appropriate size for regional assessment and an intermediate point between the high resolutions of digital elevation model. In addition, Hutchinson (2000) developed the ANUCLIM, commercial software package that enables the user to obtain estimates of monthly mean climate variables, bioclimatic parameters, and indices relating to crop growth. These models use mathematical descriptions to characterize change of climate variables across a region in order to estimate those climate variables at user specified points within the region. The independent surface variables include longitude, latitude, and elevation.

Much additional work has currently been produced using methods of statistical downscaling (SD) for climate scenario generation. Various SD techniques have been used in downscaling directly to (physically-based) impacts and to a greater variety of climate variables than previously, including variable extremes. While statistical downscaling has mostly been applied for single locations, Hewitson (2003) developed empirical downscaling for point-scale precipitation at numerous sites, and on a 0.1° resolution grid over Africa. Finally, the wider availability of statistical downscaling tools is being reflected in wider application. An example is the Statistical Downscaling Model (SDSM) tool of Wilby *et al.* (2002) used to produce scenarios for the Thames river basin (Wilby & Harris, 2006). Statistical downscaling does have some limitations, for example it cannot take account of small-scale processes with strong time-scale dependencies (e.g., land-cover change).

4.3 Effects of Climate Change on Biodiversity

Changes in climate have the potential to affect the geographic location of ecological systems, the mix of species that they contain, and their ability to provide the wide range of benefits on which societies rely for their continued existence. Ecological systems are intrinsically dynamic and are constantly influenced by climate variability. The primary influence of anthropogenic climate change on ecosystems is expected to be through the rate and magnitude of change in climate means and extremes—climate change is expected to occur at a rapid rate relative to the speed at which ecosystems can adapt and re-establish themselves—and through the direct effects of increased atmospheric CO_2 concentrations, which may increase the productivity and efficiency of water use in some plant species under situations where water or temperature are not limiting (Secretariat of the Convention on Biological Diversity, 2003; Korner, 2009). Secondary effects of climate change involve changes in soil characteristics and disturbance regimes (e.g., fires, pests and diseases), which would favor some species over others and thus change the species composition of ecosystems. Saipunkaew *et al.* (2007) found that the significant increase of dust significantly reduces lichen diversity in seven Northern provinces in Thailand.

The Secretariat of the Convention on Biological Diversity (2003) summarized the projected impacts of climate change on individuals, populations, species and ecosystems. Significant impacts with reliable evidences are for example:

- Extinction of wildlife populations may be hastened by increasing temporal variability in precipitation. Mclaughlin *et al.* (2002) revealed that changes in precipitation amplified the population of checkspot butterfly, leading to extinction. Miles *et al.* (2004) predicted that up to 43% of a sample of plant species in Amazonia could become non-viable by 2095. In addition, approximately 59% of plant and 37% of bird species in the Northern Tropical Andes will become extinct or classified as critically endangered species by the year 2080 as a result of A2 climate change scenario (regionally-oriented economic development) due to high rainfall intensity and long drought (Cuesta-Camacho *et al.*, 2006).
- Changes in phenology, hatching and immigration of insects, birds and mammals have been observed and are expected to continue. Over the mid-20th century, the theoretical distance a small mammal in northern Indiana must move to remain at the same temperature ranged from 0.40–2.07 km/year and appears to have been attainable. However, based on future temperature changes projected under the SRES higher (A1FI) and lower (B1) emissions scenarios, Francl *et al.* (in press) found that significantly larger increases in temperature-maintaining distance (TMDs) some greater than 4 km/year in some scenarios will ap-

pear less viable than those experienced in the past in this region.

- Plant communities are expected to be disrupted, as species that make up a community are unlikely to shift together. Trivedi *et al.* (2008) indicated that Arctic-alpine communities in protected areas could undergo substantial species turnover, even under the lower climate change scenario for the 2080s. For example, *Racomitrium-Carex* moss-heath, a distinctive community type of the British uplands, could lose suitable climate space as other communities spread uphill.
- Changes in rainfall and flooding patterns across large areas of arid land will adversely affect bird species in inland wetlands that rely on a network of wetlands and lakes that are alternatively or even episodically wet and fresh and dry and saline (Roshier *et al.*, 2001), or even affect a small number of wetlands, such as those used by the banded stilt, which breeds opportunistically in Australia's arid interior (Williams, 1998).
- Species and ecosystems are projected to be impacted by extreme climatic events. Attorre *et al.* (2007) indicated that the potential habitat for dragonblood (*Dracaena cinnabari)*, which is a spectacular relict of the Mio-Pliocene Laurasian subtropical forest in Socotra (Yemen), will be reduced approximately 45% by 2080 because of a predicted increased aridity. In addition, Trisurat *et al.* (2009) indicated that among 19 bioclimatic variables, e.g. minimum temperature of coldest month, precipitation of driest month and precipitation of coldest quarter are significant factors for future plant distribution in northern Thailand.
- Habitats of many species will move poleward or upward. The climatic zones suitable for temperate and boreal plant species may be displaced by 200-1,200 km poleward. Parolo & Rossi (2008) compared historical records (1954–1958) with results from recent plant surveys (2003–2005) from alpine to nival ecosystems in the Rhaetian Alps, N-Italy. An increase in species richness from 153 to 166 species was observed in higher altitude. Climate warming is therefore considered as a primary cause of the observed upward migration of high mountain plants.
- For lakes and streams, the effects of temperature-dependent changes would be least in the tropics, moderate at mid-latitude and pronounced in high latitudes. This latitudinal trend is projected to be due decreasing extent and duration of ice cover in some high latitude lakes, thus affecting biodiversity of species adapted to shorter ice cover (Christensen & Christensen, 2003).
- Climate change will have most pronounced effects on wetlands through altering the hydrological regime as most wetland species are water dependent. This is expected to affect biodiversity and the phenology of wetland species (van Dam *et al.*, 2002).
- Disturbance can both increase the rate of loss of species and create opportunities for the establishment of new species. Trisurat *et al.* (2009) used maximum entropy theory (MAXENT) to generate ecological niche models of forest plant species in the northern Thailand. The results showed high spatial configuration and turnover rate, especially for evergreen tree species. Ten plant species will lose, from 2-13%, ecological niches (suitable locations), while the remaining 12 species will gain substantial suitable habitats. The assemblages of evergreen species or species richness are likely to shift toward the north where lower temperature are anticipated for year 2050. In contrast, the deciduous species will expand their distribution ranges. However, the impact on the distribution of species richness

is not substantial in southern Thailand between year 2000 and year 2100 due to the peninsular effects (Trisurat, 2010).

More examples and many chapters describing the potential effects of climate change on biodiversity in different parts of the Globe are included in this book.

5. BOOK STRUCTURE AND CHAPTER SUMMARIES

With an increasing population pressure and resulting increase in demand for food and other basic needs, land use will continue to remain an issue on the global agenda. Land use ultimately impacts ecological services including biodiversity values and the rapidly emerging global concern over climate change, which has added a further challenge to existing problems. As presented in this chapter, a substantial technical knowledge base exists in terms of assessing the value of our resources, such as land use systems and biodiversity, yet our understanding on these issues is limited, particularly in the context of developing countries. Hence, it is of utmost importance for stocktaking of key ecological resources and phenomena to exercise sound planning and develop policy strategies for climate change adaption and biodiversity conservation in order to ultimately provide food security and improved livelihood to the people.

This book is an attempt to compile selected existing bodies of knowledge from different parts of the world on the assessment of these issues, i.e. land use, biodiversity, climate change, their Inter-relationship and to demonstrate the use of various tools and modeling techniques at different scales. The edited book contains 20 chapters divided into five parts.

The Introduction Chapter (Section 1 Introduction) already provides a coherent presentation of the essential concepts, key terminology, and historical background of land use informatics, deforestation and climate change, as the global threat to biodiversity. In addition, it also raises some key issues on consequences of these threats and discusses why biodiversity conservation practitioners have to think and map out integrated strategies to cope with these issues.

Section 2, Setting the Scene: This section off the overview of deforestation and climate impacts on biodiversity followed by information on how to monitor and quantify these impacts. Section 2 contains three chapters. Chapter 2 (Consequence of deforestation and climate change on biodiversity) reviews and describes the relationship between forest and climate, and forest ecosystem functions and biodiversity. Based on meta-analyses of peer-reviewed literature, the chapter then discusses in details the impacts of deforestation (e.g. habitat loss, habitat destruction and fragmentation) that will diminish population viability, and the predicted climate changes based on several development scenarios on plants and animals. The interlinkages among deforestation and climate change on biodiversity are also included.

Chapter 3 (The role of Geo-informatics for land use and biodiversity studies) includes concepts and terminology of Geo-informatics technologies, namely remote sensing, geographic information system, global positioning system, as well as spatial analysis methods as useful tools for land-use/land-cover (LU/LC) studies worldwide and their impacts on biodiversity. Chapter 3 then explores identification and analysis of key natural, socio-economic and regulatory drivers for LU/LC. Finally, it collates a number of LU/LC studies involving usage of Geo-informatics provide decision makers, land managers, stakeholders and researchers the scientific grounds for better management and formulation of conservation strategies and policies.

Chapter 4 (Monitoring biodiversity using remote sensing and field surveys) aims to develop quantitative methodologies for the spatial identification and monitoring of European landscapes and their habitats. This chapter concludes that, in combination with additional environmental data sets from field surveys, it is now possible to model quantitatively the spatial extent of wide-spread habitats and landscapes on the basis of land cover information derived from satellite imagery. Field surveys are always limited to relatively small areas and therefore the spatial modeling of habitats and landscapes with the help of remotely sensed information remains important to provide a synoptic overview of the European landscape.

Section 3, Land use and biodiversity modeling: The natural environment, such as land use and biodiversity, is very complex, thus simplification through abstraction is essential to communicate concepts and relationships concerning different components of the ecosystem and its environmental factors in order to decide effective conservation measures. This Part, through five chapters (chapters 5-9) provides essential tools for land use studies and biodiversity modeling.

Chapter 5 (Integrated modeling of global environmental change: IMAGE) describes briefly the data and models used in IMAGE 2.4. It starts from basic driving forces like demographics and economic development, energy consumption and production, agricultural demand, trade and production. Important elements in the bio-physical modeling of land-cover and land-use processes are addressed. Finally, the use of data and information from IMAGE to feed broader policy-exploring tools is presented, including global assessment of terrestrial biodiversity and climate mitigation. Chapter 6 presents a land use allocation model, called the Dyna-CLUE model, which is one of the most used land allocation models globally and is highly applicable for scenario analysis. In addition, it has been used in many case studies at local and continental scale by different institutions worldwide, including several cases studies in this book.

Chapter 7 (Landscape biodiversity characterization in Ecoregion 29 using MODIS) discusses various aspects of biodiversity parameters that can be estimated using remote sensing data. Moderate resolution satellite (MODIS) data was used to demonstrate the biodiversity characterization of ecoregion 29. A forest type map linked to density of the study area was also developed by MODIS data. The outcome states that remote sensing and geographic information systems can be used in combination to derive various parameters related to biodiversity surveillance at a regional scale.

This book also introduces two perspectives of biodiversity models: individual species, and species abundance relative to undisturbed ecosystem (pressure-based model) as a proxy of biodiversity. GLOBIO3 (chapter 8) is clearly one of the most advanced models, a biodiversity pressure model per se. GLOBIO3 is a quantitative model used in the assessment of policy options for reducing global biodiversity loss. The model is built on simple cause–effect relationships between environmental drivers and biodiversity impacts, based on state-of-the-art knowledge. The mean species abundance of original species relative to their abundance in undisturbed ecosystems (MSA) is used as the indicator for biodiversity. Previously, GLOBIO3 described impacts on terrestrial ecosystems, but recently a separate GLOBIO aquatic model is developed based on a similar approach. These two chapters are suitable readings for modelers to explore advantages and disadvantages of individual species and biodiversity modeling techniques as well as non-modelers who may regard modeling as a black box.

Meanwhile, Chapter 9 (Species distribution modeling) elaborates on the concepts of species modeling and presents three popular techniques to generate species distribution: cartographic overlay (habitat suitability index), binary response (presence/absence), prediction model (logistic regression), and presence-only data model (maximum entropy method or MAXENT). The latter approach is illustrated for Asian elephant in Bun Tharik-Yod Mon wildlife sanctuary in northeast Thailand.

Section 4, Case Studies: Following concepts and detailed methods discussed in previous chapters, Section 4 provides "real case studies" implemented in various regions (east to west, north to south) and multi-scale studies (global to local) across the globe. Six case studies from Southeast Asia, South Asia, Central America, South America, Africa, West Europe, West Europe and Africa are presented.

Chapter 10 (Modeling land-use and biodiversity in northern Thailand) presents an analysis in northern region where rapid deforestation has occurred over the last few decades and is expected to continue due to high land demand for rubber plantations and infrastructure and tourism development. This analysis suggests that deforestation would continue, as establishing only a fixed percentage of forests was not efficient in conserving biodiversity. Measures aimed at the conservation of locations with high biodiversity values, limited fragmentation and careful consideration of road expansion in pristine forest areas may be more efficient for achieving biodiversity conservation.

Chapter 11 (The current and future status of floristic provinces in Thailand) investigates characteristics of floristic regions in Thailand and predict the impacts of future climate change (2050) on the recognized phytogeographical areas. Based on the MAXENT model results and clustering it is proposed to reduce the existing seven phytogeographic regions as used in the Flora of Thailand to four regions. In addition, the future climate will strongly diminish the number of species in the Northern and North-eastern region. Peninsular Thailand appears to be stable, but high endemism shows that there is a decrease in suitable niche also in this area, while far eastern Thailand and the Peninsular region will gain species.

Besides studies on plant species, Chapter 12 (Biodiversity experiences in Ukraine) indicates that Ukrainian researchers have used extensive biodiversity modeling methods (e.g. species-based model and pressure biodiversity models) to predict the distributions of vascular plants, insects, amphibians, birds and mammals. Later, the Ukrainian researchers evaluate effects in habitats condition of selected species caused by land use change and climate change in 2050. This study suggests that expected climate change together with land-use change would provoke numerous non-simplified and unexpected habitat changes. The model approaches and results were integrated in the education system and mass media for awareness raising.

Chapter 13 (Regional scenarios of biodiversity states in the Tropical Andes) evaluates the remaining biodiversity for 2000 and for two 2030 scenarios: Market forces and Policy Reform at regional level and for three countries in the tropical Andes: Colombia, Ecuador and Peru. In addition, this research also aims to identify the most vulnerable areas to biodiversity loss and the most important drivers of such losses. At the country level Ecuador would have the lowest values of remaining MSA for 2030, followed by Colombia and finally by Peru for both scenarios. In a comparison with the values of the year 2000, Ecuador also showed the highest losses of biodiversity (5.7% for Market Forces scenario), Peru is the second highest, while Colombia would have a loss of 3.9% for the Market Forces scenario. The model results are used for policy formulation to maintain biodiversity in the Tropical Andean countries.

Chapter 14 (The influence of changing conservation paradigms on identification of priority protected area locations) briefly describes the evolution of six different approaches to modeling the potential impacts of climate change on biodiversity: i.e., biome models; dynamic global vegetation models; and climate envelope models. The author looks in detail at the BIOCLIMA model, which simulates trends in a sample chosen to represent regional plant biodiversity and how climate change directly influences the processes determining a plant's response to climate change, i.e. reproductive rate, dispersal mechanisms and pre-adaptations to expected stresses and its application to Amazonia. This chapter recommends authorized agencies to establish more protected areas and to include both lowland and montane forests or migration corridors between these in order to protect the best remaining lowland moist forest species and montane forest flora.

Chapter 15 (Land degradation and biodiversity loss in Southeast Asia) examines the general status of land degradation and biodiversity in Southeast Asia and goes on to present two case studies. The first case study is a land degradation assessment in the Lower Mekong Basin demonstrating the use of spatial data and technologies and various land degradation indicators. The study suggests that about one quarter of the Lower Mekong Basin is severely degraded and another three quarters are moderately, slightly or with no degradation with their area distribution in the LMB. The second case study specifically documents plant diversity and examines the relationship of plant diversity with biomass and soil erosion by making use of field surveyed primary data. The results revealed that the trend towards mono-cropping of shrubs, which can be expected to accelerate in Thailand due to the prioritization of export crops and more recently bio-fuels, will lead to a further reduction in plant diversity on a landscape level.

Chapter 16 (Sustainable land use and watershed management in response to climate change impacts: Overview and proposed research techniques) applies the Markov's Chain model to determine probability of land use change based on the land use evolution in Dong Nai watershed, Vietnam. The outputs were used in the Soil and Water Assessment Tool (SWAT) for modeling watershed hydrology and simulating the movement of sediment, and agricultural chemical yields in large complex basins with varying soil type, land use and management conditions over long periods of time. In this chapter, the author formulates sustainable land use and watershed management in response to future land use/land cover and climate change based on three scenarios: future trends, land allocation for maximizing economic, and land allocation for sustainable land use.

Similar to Chapter 13, Chapter 17 (Modeling of the current and future status of biodiversity in Central America using GLOBIO3 methodology) predicts the current and future state of biodiversity in seven countries in the Central America: Guatemala, Belize, Honduras, El Salvador, Nicaragua, Costa Rica and Panama and integrates the results into one regional assessment using Dyna-CLUE and GLOBIO3. Besides, this chapter also suggests a methodology to effectively downscale the existing models for national implementation. Results show that in the current state, the region has a remaining MSA of 48%. The main source of biodiversity loss identified was the land use driver followed by infrastructure, fragmentation and climate change, respectively. Individual country results show that remaining MSA values are above 50% for Belize, Nicaragua, and Panama. The remaining countries experience less than 50% in the current situation. However, the future state of biodiversity is expected to be lower than 50% for all countries, especially under Baseline and Trade Liberalization scenarios. Minimum loss is expected under the ALIDES policy options.

In Chapter 18 (Spatial model approach for deforestation: case study in Java island), logistic regression was used to find relationships between deforestation and biophysical and socioeconomic factors in Java, Indonesia which is under human pressure. Deforestation was detected from interpretation of MODIS satellite imagery between 2000-2008. Result of the study showed that impacts of population density, road density and number of households engaged in the agricultural sector are significant and they have negative impact on deforestation. Implication of the model is the recommendation to control population growth, promotion of alternative non agricultural jobs and be aware of road construction into remaining forests.

Chapter 19 (Embedding biodiversity modeling in the policy process) demonstrates the collaborative project between the Netherlands Environmental Assessment Agency (PBL) and the Environmental Operations Centre (EOC) to integrate the results of biodiversity modeling into Strategic Environmental Assessment (SEA) in Vietnam, both at national and local levels. In addition, four case studies on the linkages between biodiversity and poverty were selected each with a special theme: shifting cultivation, migration, hydro power, and construction of roads and infrastructure. For each case study the current state of both poverty and biodiversity was assessed and trends and linkages were analysed. This collaborative project introduces an effective new indicator and biodiversity assessment method that is already endorsed by the Ministry of Environment to be embedded in the national policy process of Vietnam.

Chapter 20 (Conclusions and recommendations) in Section 5 summarizes and presents analytical views on the status, trend and way forward with regard to the issues of biodiversity and land use modeling and conservation in the context of climate change. It provides researchers with a range of options to improve existing models with identified research needs for effective modeling and conservation of land and biodiversity.

REFERENCES

Alkemade, R., vanOorschot, M., Miles, L., Nellemann, C., Bakkenes, M., & ten Brink, B. (2009). GLOBIO3: A framework to investigate options for reducing global terrestrial biodiversity loss. *Ecosystems, 12*(3), 374–390. doi:10.1007/s10021-009-9229-5

Attorre, F., Francesconi, F., Taleb, N., Scholte, P., Saed, A., Alfo, M., & Bruno, F. (2007). Will dragonblood survive the next period of climate change? Current and future potential distribution of Dracaena cinnabari (Socotra, Yemen). *Biological Conservation, 138*, 430–439. doi:10.1016/j.biocon.2007.05.009

Barron, E., & Sorooshian, S. (1997). *Assessing the impacts of climate on regional water resources.* Paper presented at the Investigators Working Group (IWG) of the Mission to Planet Earth/Earth Observing System *(MTPE/EOS)*, San Diego, California.

Baydack, R. K., & Campa, H. III. (1999). Setting the context. In Baydack, R. K., Campa, H. III, & Haufler, J. B. (Eds.), *Practical approaches to the conservation of biological diversity*. Washington, DC: Island Press.

Bowman, D. M. J. S. (1993). Biodiversity much more than biology inventory. In Gaston, K. J. (Ed.), *Biodiversity: A Biology of numbers and difference*. Oxford, UK: Blackwell Science.

Christensen, J. H., & Christensen, O. B. (2003). Climate modeling Sever summertime flooding in Europe. *Nature, 421*, 805–806. doi:10.1038/421805a

Convention on Biological Diversity. (1992). *Text of the convention on biological diversity*. Retrieved April 15, 2010, from http://www.cbd.int/convention/convention.shtml

Cruz, R. B. (2006). Land use information system for local government: The case of Naga City, Philippines. *Systemics. Cybernetics and Informatics, 4*, 67–74.

Cuesta-Camacho, F., Ganzenmuller, A., Peralvo, M., Novoa, J., & Riofrio, G. (2006). *Predicting species' niche distribution shifts and biodiversity change within climate change scenarios: A regional assessment for bird and plant species in the Northern Tropical Andes*. Unpublished report, Biodiversity Monitoring Program (EcoCiencia), Peru.

Donovan, T. M., & Flather, C. H. (2002). Relationships among North American songbird trends, habitat fragmentation and landscape occupancy. *Ecological Applications, 12*, 364–374.

ESRI, Inc. (1996). *Using Arc View GIS*. Redlands, CA: ESRI, Inc.

FAO. (1997). *Africover land cover classification. Environment and Natural Resources Service (SDRN), Food and Agriculture Organization of the United Nations*. Rome: FAO.

FAO. (2010). *Global forest resources assessment. Food and Agriculture Organization of the United Nations*. Rome: FAO.

Fisher, P. F., Comber, A. J., & Wadsworth, R. A. (2005). Land use and land cover: Contradiction or complement. In Fisher, P., & Unwin, D. (Eds.), *Re-Presenting GIS* (pp. 85–98). Chichester, UK: Wiley Press.

Forman, R. T. T. (1995). *Land mosaics: The ecology of landscapes and regions*. Cambridge, UK: Cambridge University Press.

Francl, K. E., Hayhoe, K., Saunders, M., & Maurer, E. P. (in press). Ecosystem adaptation to climate change: Small mammal migration pathways in the Great Lakes states. *Journal of Great Lakes Research*. doi: 10.1018.jglr.2009.09.007.

Gaston, K. J. (Ed.). (1996). *Biodiversity: A biology of numbers and difference*. Oxford, UK: Blackwell Science.

Haila, Y., & Kouki, J. (1994). The phenomenon of biodiversity in conservation biology. In Gaston, K. J. (Ed.), *Biodiversity: A Biology of numbers and difference*. Oxford, UK: Blackwell Science.

Hall, C. A. S., Tian, H., Qi, Y., Pontius, G., & Cornell, J. (1995). Modelling spatial and temporal patterns of tropical land use change. *Journal of Biogeography*, *22*(4/5), 753–757. doi:10.2307/2845977

Hewitson, B. (2003). Developing perturbations for climate change impact assessments. *Eos, Transactions, American Geophysical Union*, *84*, 337–348. doi:10.1029/2003EO350001

Hutchinson, M. F. (1995). Interpolating mean rainfall using thin plate smoothing splines. *International Journal of GIS*, *9*, 305–403.

Hutchinson, M. F. (2000). *ANUSPLIN, ver 4.1. user guide. Centre for Resource and Environmental Studies*. Canberra, Australia: The Australian National University.

IPCC. (2000). *IPCC special report on emission scenarios: Summary for policymakers*. Nairobi: Intergovernmental Panel on Climate Change. United Nations Environment Programme and World Meteorological Organization.

Kasperson, R. E., Kasperson, J. X., & Turner, B. L. II. (1999). Risk and criticality: Trajectories of regional environmental degradation. *Ambio*, *28*(6), 562–568.

Korner, C. (2009). Response of humid tropical trees to rising CO_2. *Annual Review of Ecology Evolution and Systematics*, *40*, 61–79. doi:10.1146/annurev.ecolsys.110308.120217

Lambin, E. F., & Geist, H. (Eds.). (2006). *Land-use and land-cover change: Local processes and global impacts*. Berlin, Heidelberg, Germany: Springer-Verlag.

Lambin, E. F., Turner, B. L. II, Geist, H. J., Agbola, S. B., Angelsen, A., & Bruce, J. W. (2001). The causes of land-use and land-cover change – moving beyond the myths. *Global Environmental Change: Human and Policy Dimensions*, *11*, 261–269.

Leadley, P., Pereira, H. M., Alkemade, R., Fernandez-Manjarrés, J. F., Proenca, V., Scharlemann, J. P. W., & Walpole, M. J. (2010). *Biodiversity scenarios: Projections of 21st century change in biodiversity and associated ecosystem services, Technical series no. 50*. Montreal: Secretariat of the Convention on Biological Diversity.

Louette, G., Maes, D., Alkemade, J. R. M., Boitani, L., de Knegt, B., & Eggers, J. (2010). BioScore–cost-effective assessment of policy impact on biodiversity using species sensitivity scores. *Journal of Nature Conservation, 18*, 142–148. doi:10.1016/j.jnc.2009.08.002

Lovejoy, T. E. (1997). Biodiversity: What is it? In Reaka-Kudld, M. L., Wilson, D. E., & Wilson, E. O. (Eds.), *Biodiversity II: Understanding and protecting our biological resources* (pp. 7–14). Washington, DC: Joseph Henry Press.

Matthews, R., Gilbert, N., Roach, A., Polhill, J. G., & Gotts, N. M. (2007). Agent-based land-use models: A review of applications. *Landscape Ecology, 22*(10), 1447–1459. doi:10.1007/s10980-007-9135-1

McGarigal, K., & Marks, B. (1995). *FRAGSTATS: Spatial pattern analysis program for quantifying landscape structure*. (Gen. Tech. Rep. PNW-GTR-351). Portland.

McLaughlin, J. F., Hellmann, J. J., Boggs, C. L., & Ehrlich, P. R. (2002). Climate change hastens population extinctions. *Proceedings of the National Academy of Sciences of the United States of America, 99*, 6070–6074. doi:10.1073/pnas.052131199

McNeely, J. A. (1998). *Economics and biological diversity: Developing and using economic incentives to conserve biological resources*. Gland, Switzerland: IUCN.

Miles, L., Grainger, A., & Phillips, O. (2004). Impact of global climate change on tropical forest biodiversity in Amazonia. *Global Ecology and Biogeography, 13*, 553–565. doi:10.1111/j.1466-822X.2004.00105.x

Millennium Ecosystem Assessment (MEA). (2005). *Ecosystems and human well-being. synthesis*. Washington, D.C: Island Press.

Nakicenovic, N., Alcamo, J., Davis, G., De Vries, B., Fenhann, J., & Gaffin, S. (2000). *Special report on emissions scenarios. IPCC Special Reports*. Cambridge, UK: Cambridge University Press.

Norse, E. A., & McManus, R. E. (1980). *Ecology and living resources biological diversity.* In Council on Environmental Quality, 11th Annual Report, 31-80. Council on Environmental Quality, Washington, DC.

Ochoa-Gaona, S. (2001). Traditional land-use systems and patterns of forest fragmentation in the highlands of Chiapas, Mexico. *Environmental Management, 27*, 571–586. doi:10.1007/s002670010171

Office of Technology Assessment (OTA). (1987). *Technologies to maintain biological diversity. (OTA-F-330)*. Washington, DC: Government Printing Office.

Parker, D. C., Manson, S. M., Janssen, M. A., Hoffman, M., & Deadman, P. (2001). *Multi-agent systems for the simulation of land-use and land-cover change: A review*. Indiana University, Retrieved April 20, 2010, from http://www.csiss.org/events/other/agent-based/.../maslucc_overview.pdf

Parolo, G., & Rossi, G. (2008). Upward migration of vascular plants following a climate warming trend in the Alps. *Basic and Applied Ecology, 9*(2), 100–107. doi:10.1016/j.baae.2007.01.005

Patta, S. (2004). *Application of stochastic dowscaling techniques to global climate model data for regional climate prediction*. Unpublished M.Sc. thesis, Louisiana State University and Agricultural and Mechanical College, Louisiana.

Priess, J. A., & Schaldach, R. (2008). Integrated models of the land system: A review of modelling approaches on the regional to global scale. *Living Reviews in Landscape Research, 2*. Retrieved January 12, 2010, from http://www.livingreviews.org/lrlr-2008-1.

Raabová, J., Münzbergová, Z., & Fischer, M. (2007). Ecological rather than geographic or genetic distance affects local adaptation of the rare perennial herb, Aster amellus. *Biological Conservation, 139*, 348–357. doi:10.1016/j.biocon.2007.07.007

Roshier, D. A., Whetton, P., Allan, R. J., & Robertson, A. I. (2001). Distribution and persistence of temporary wetland habitats in arid Australia in relation to climate. *Austral Ecology, 26*, 371–384. doi:10.1046/j.1442-9993.2001.01122.x

Saipunkaew, W., Wolseley, P. A., Chimonides, P. J., & Boonpragob, K. (2007). Epiphytic macrolichens as indicators of environmental alteration in northern Thailand. *Environmental Pollution, 146*, 366–376. doi:10.1016/j.envpol.2006.03.044

Samuels, R., Rimmer, A., Hartmann, A., Krichak, S. & Alpert, P. (2010). *Change impacts on Jordan River flow: Downscaling application from a regional climate model.* American Meterological Society. doi: 10.1175/2010JHM1177.1.

Schmiegelow, F. K. A., & Monkkonen, M. (2002). Habitat loss and fragmentation in dynamic landscape: Avian perspectives from the boreal forest. *Ecological Applications, 12*, 375–389.

Secretariat of the Convention on Biological Diversity. (2003). *Interlinakages between biological diversity and climate change. Advise on the integration of biodiversity considerations into the implementation of the United Nations Framework Convention on Climate Change and Its Kyoto protocol* (CBD Technical Series no. 10). Montreal.

Secretariat of the Convention on Biological Diversity (sCBD). (2006). *Convention on biological diversity: Global biodiversity outlook 2.* Montreal.

Secretariat of the Convention on Biological Diversity (sCBD). (2010). *Global biodiversity outlook 3 – executive summary.* Montreal.

Southworth, F., Dale, V. H., & O'Neill, R. V. (1991). Contrasting patterns of land use in Rondonia, Brazil: Simulating the effects on carbon release. *International Social Science Journal, 130*, 681–698.

Trisurat, Y. (2007). Applying gap analysis and a comparison index to assess protected areas in Thailand. *Environmental Management, 39*, 235–245. doi:10.1007/s00267-005-0355-3

Trisurat, Y., Alkemade, R., & Arets, E. (2009). Projecting forest tree distributions and adaptation to climate change in northern Thailand. *Journal of Ecology and Natural Environment, 1*(3), 55–63.

Trisurat, Y., Suraphabmiatree, S., & Saengnil, S. (2010). *Plant species vulnerability to climate change during 2002-2100.* Unpublished report submitted to the National Research Council of Thailand, Bangkok.

Trivedi, M. R., Morecroft, M. D., Berry, P. M., & Dawson, T. P. (2008). Potential effects of climate change on plant communities in three Montane nature reserves in Scotland, UK. *Biological Conservation, 141*(6), 1665–1675. doi:10.1016/j.biocon.2008.04.008

Turner, I. M., & Corlett, R. T. (1996). The conservation value of small, isolated fragments of lowland tropical rain forest. *Trends in Ecology & Evolution, 11*, 330–333. doi:10.1016/0169-5347(96)10046-X

United Nations. (2002). (*UN*). New York: Report of the World Summit on Sustainable Development.

United Nations. (UN). (2005). *The millennium development goals report 2005.* New York.

United Nations Environmental Programme. (1991). *Fourth revised draft convention on biodiversity.* Nairobi, Kenya.

Urban, D. L., O'Neill, R. V., & Shugart, H. H. (1987). Landscape ecology. *Bioscience, 37*, 119–127. doi:10.2307/1310366

Van Dam, R., Gitay, H., & Finlayson, M. (2002). *Climate change and wetlands: Impact and mitigation*. Ramsar Draft, COP8 paper.

van Vuuren, D., Sala, O., & Pereira, H. M. (2006). The future of vascular plant diversity under four global scenarios. *Ecology and Society*, *11*, 25.

Veldkamp, A., & Lambin, E. F. (2001). Predicting land-use change. *Agriculture Ecosystems & Environment*, *85*, 1–6. doi:10.1016/S0167-8809(01)00199-2

Verburg, P., Eickhout, B., & van Meijl, H. (2008). A multi-scale, multi-model approach for analyzing the future dynamics of European land use. *The Annals of Regional Science*, *42*, 57–77. doi:10.1007/s00168-007-0136-4

Verburg, P. H., & Veldkamp, A. (2004). Projecting land use transitions at forest fringes in the Philippines at two spatial scales. *Landscape Ecology*, *19*, 77–98. doi:10.1023/B:LAND.0000018370.57457.58

Wilby, R. L., Dawson, R., & Barrow, E. M. (2002). SDSM: A decision support tool for the assessment of regional climate change assessments. *Environmental Modelling & Software*, *17*, 145–157. doi:10.1016/S1364-8152(01)00060-3

Wilby, R. L., & Harris, I. (2006). *A framework for assessing uncertainties in climate change impacts: Low-flow scenarios for the River Thames, UK.* doi: 10.1029/2005WR004065

Williams, W. D. (1998). Dryland wetlands. In McComb, A. J., & Davis, J. A. (Eds.), *Wetlands for the future*. Glen Osmond, Australia: Gleneagles Publishing.

Wilson, E. O. (1997). Introduction to Biodiversity. In Reaka-Kudld, M. L., Wilson, D. E., & Wilson, E. O. (Eds.), *Biodiversity II: Understanding and protecting our biological resources* (pp. 7–14). Washington, DC: Joseph Henry Press.

Yahner, R. H. (1988). Changes in wildlife communities near edges. *Conservation Biology*, *2*(4), 333–339..doi:10.1111/j.1523-1739.1988.tb00197.x

Zanette, L. (2000). Fragment size and the demography of an area-sensitive songbird. *Journal of Animal Ecology*, *69*, 458–470. doi:10.1046/j.1365-2656.2000.00408.x

Section 2
Setting the Scene

Chapter 2
Consequences of Deforestation and Climate Change on Biodiversity

Roland Cochard
Asian Institute of Technology, Thailand

ABSTRACT

Ever since their evolution, forests have been interacting with the Earth's climate. Species diversity is particularly high in forests of stable moist tropical climates, but patterns of diversity differ among various taxa. Species richness typically implies high ecosystem resilience to ecosystem disturbances; many species are present to fill in newly created niches and facilitate regeneration. Species loss, on the other hand, often entails environmental degradation and erosion of essential ecosystem services. Until now species extinction rates have been highest on tropical islands which are characterized by a high degree of species endemism but comparatively low species richness (and therefore high vulnerability to invasive species). Deforestation and forest degradation in many countries has lead to forest fragmentation with similar effects on increasingly insularized and vulnerable forest habitat patches. If forest fragments are becoming too small to support important keystone species, further extinctions may occur in cascading ways, and the vegetation structure and composition may eventually collapse. Until now relatively few reported cases of species extinctions can be directly attributed to climate change. However, climate change in combination with habitat destruction, degradation, and fragmentation may lead to new waves of species extinctions in the near future as species are set on the move but are unable

DOI: 10.4018/978-1-60960-619-0.ch002

to reach cooler refuges due to altered, obstructing landscapes. To mitigate the future risks of extinctions as well as climate change, major efforts should be undertaken to protect intact large areas of forests and restore wildlife corridors. Carbon sequestration may be seen as just one of many other environmental services of forest biodiversity that deserve economic valuation as alternatives to conversion to often unsustainable agricultural uses.

1. FORESTS AND CLIMATES

The world's climates and forests are intimately interlinked. Dense communities of tree species can only grow in environments with sufficient soil water. Where mean annual precipitation is less than about one meter, continuous forests are commonly replaced by smaller woody vegetation (e.g. scrub forests, dry thickets), grasslands (e.g. savannas, steppes) or deserts. In these dry regions tree stands and forests may only be found in topographic depressions where water accumulates and is stored well into the dry season, e.g. along river beds and in periodic floodplains (Whittaker, 1975). Likewise, primary productivity in forests is principally related to rainfall as well as temperature, ranging from averages of about 2200 g m^{-2} yr^{-1} in lush tropical rainforests (trees of more than 30 m height and mean woody biomass of around 45 kg m^{-2}) to 800 g m^{-2} yr^{-1} in the northern taiga forests (stunted trees of less than 15 m height and woody biomass of around 20 kg m^{-2}; Gurevitch *et al.*, 2006). Seasonal weather patterns also influence tree physiology and determine the distribution of forest biomes, e.g. temperate deciduous forests occur in regions characterized by cold winters, whereas dry deciduous forests are widespread in parts of South and Southeast Asia that are influenced by the monsoon. The growing season of these vast deciduous northern forests is reflected as a small seasonal decrease in atmospheric carbon dioxide measured at weather stations around the northern hemisphere.

Ever since their evolution, forests have been influencing the gas composition in the atmosphere, which in turn influenced temperatures and weather patterns on planet Earth (Zachos *et al.*, 2001, Sigman & Boyle, 2000). The accumulation of oxygen in the atmosphere and the absorption of carbon dioxide into the biosphere and earth crust began with the evolution of photosynthesis in algae around 3500 million years ago. After the 'great oxygenation event' around 2400 million years ago plants began to spread and diversify on land (Anbar *et al.*, 2007, Dole, 1965). Carbon sequestration further increased when woody vascular plants started reaching for the sun in the middle Devonian (ca. 385 million years ago), and during the Carboniferous (ca. 359 million years ago) biomass accumulation reached a first climax in tropical peat swamp forests of Pangaea, which lead to the formation of large coal deposits (Ghazoul & Sheil, 2010).

Forests are still sequestering carbon at significant rates which can offset emissions from deforestation to some degree (Lewis *et al.*, 2009; Bunker *et al.*, 2005). South East Asian peat soils up to >20 m deep constitute probably the largest carbon stores of any living ecosystem (Phillips, 1998). Poor drainage, permanent waterlogging, high rainfall and substrate acidification are conditions in which plant materials accumulate faster than they decay (Brady, 1997). The average rate of carbon accumulation in pristine peat swamps in Indonesia has been estimated at 0.8-1.9 t ha^{-1} y^{-1} (Immirzi & Maltby in Rieley *et al.*, 1997), respectively 0.4-1.1 t ha^{-1} y^{-1} (Sorensen, 1993). South East Asian peat soils developed in coastal floodplains as early as 30'000 BP (Whitten *et al.*, 1997; Page *et al.*, 2004). On average these swamp forests may comprise about 200 t C ha^{-1} in the standing tree biomass and more than 2500 t C ha^{-1} in the peat soil (typical median depth of 5 m); this is about nine times as much carbon as is stored in tropical rainforest standing on mineral soils (Diemont *et al.,* 1997).

As much as forests have been influencing the mix of gases in the atmosphere, as much have they been functional in shaping the global water cycle and stabilizing local weather patterns (Bonan, 2008). The colonization of land by arborescent plants and the associated evolution of deep organic soils and peatlands significantly increased the capacity of continents to absorb and retain water. Forest trees not only provided stability to soils and retained soil water via their roots, but via their canopies they also offered large transpiration surfaces that could return much of the water back to the atmosphere in a relatively short time period. Forests reduce temperature extremes and variations; in the tropics evaporation from forests cools down the air, whereas the low albedo of boreal forests absorbs much of the solar radiation, converting it to heat (Bonan, 2008). Fast transpiration rates lead to a fast buildup of atmospheric water vapor; as much 68% of rainfall water may thereby be returned to the atmosphere in Amazonian rainforest (Leopoldo *et al.,* 1995). Cooler air, loaded with water vapor over forests, implies fast rates of condensation and precipitation. Tropical rainforests therefore act like extensive water sponges, and the water reservoir is moved inland like a conveyor belt by the force of trade winds and via cycles of convective precipitation.

2. FORESTS, CLIMATES AND BIODIVERSITY

Biological diversity (biodiversity) - the variation of life forms - may be seen as a measure of health and stability in forests and other ecosystems. Biodiversity can be considered in various groupings and at different levels of variation: genetic diversity (from nucleotides to populations of organisms), organismal diversity (from individuals to kingdoms), and ecological diversity (from populations to biogeographic realms; Gaston, 2010). Most commonly, however, biodiversity refers to species diversity within a specified ecosystem or geographic area. Organisms of species are conventionally regarded as the basic building blocks of ecosystems; generally species are also easier to assess as units of diversity as compared to genetic diversity or ecological diversity. The simplest measure often used to describe biodiversity is ‘species richness’, i.e. the number of species found for a given area. This measure, however, disregards the common observation that species are not evenly distributed, but ecosystems are often dominated by a few highly competitive species and many rare species. Various diversity indices have therefore been developed that account for the importance of dominant and rare species, whereby the Shannon index and the Simpson index are among the most popularly used indices (Magurran, 2003).

Appraisal of the world’s biodiversity is far from concluded. Currently, the number of described extant eukaryote species is around 2 million with about another 13’000 species being described each year; this includes around 2000 plant species and many vertebrates, i.e. around 130-160 new fish species, 95 amphibians, 6-7 bird species, and 25-30 mammals (Prance *et al.,* 2000; Gaston, 2010). Regarding plant species many tropical regions are still poorly surveyed, e.g. an estimated 15-35% of species in Borneo are not yet described (Beaman & Burley, 2003). In recent years some spectacular new discoveries of large mammals and a monitor lizard still occurred in the tropical forests of the Annamite Ranges in Vietnam, on the island of New Guinea, and in the Sierra Madre Ranges in the Philippines (Ceballos & Ehrlich, 2009; Welton *et al.,* 2010). The total estimates of eukaryote species on earth range from 3.5 to over 108 million species, with major uncertainties of species accounts in poorly sampled environments such as forest canopies and soils (Gaston, 2010). Even the number of the currently described species remains vague inasmuch as some species have been described under more than one name (synonymy), whereas other species are sometimes found to encompass several cryptic species (Gaston, 2010). There are also biases as the described species mostly tend to be larger in size, more abundant and widespread, and are disproportionately from (northern) temperate regions. Particularly in the realms of microorganism there may still be much to discover; these are the oldest organisms on earth which are characterized by a much greater phyletic diversity as compared to macroorganisms (Gaston, 2010). Equally, only a fraction of the insect species is currently described; many of the species may perform crucial ecosystem services ranging from pollination to decomposition and nutrient cycling (Foottit & Adler, 2009; Ghazoul & Sheil, 2010).

Biodiversity is spread heterogeneously around the world; it is found at high densities in so-called 'hotspots', at low densities in 'coldspots' (commonly deserts and anthropogenic landscapes), and at intermediate densities in 'extensive plains' in between (Gaston, 2000). Species richness varies significantly along climatic gradients, in particular gradients of mean annual temperatures and precipitation, and along gradients of seasonality. Richness tends to increase from northern latitudes (seasonally cold climates) towards the tropics (stable warm climates), and it tends to decrease with altitude from the lowlands to mountain peaks. These increases are, however, not linear but they are commonly interrupted in parts, mostly reflecting patterns of precipitation which also correlate with the distribution of forests and other vegetation types. For example, Mediterranean ecoregions are generally more species rich than the adjacent more arid belts at lower latitudes (Groombridge & Jenkins, 2002). Equally, moist mid-altitude forests on mountain slopes are often more diverse than adjacent lowland forests; mist forests in particular are often characterized by a high diversity of epiphyte species and associated arthropod diversity (Nieder *et al.*, 2001). Diversity is typically highest in wet tropical regions close to the equator, where primary productivity is high, climates are fairly stable, but a multitude of niches are created in a richly structured forest with intermediate disturbances (Wright, 2002). The highest diversities of terrestrial vertebrates and vascular plant species are found in tropical forest ecoregions in South and Central America, Indochina and parts of Africa, whereby peaks of diversity are, however, often at some distance to the equator at latitudes of about 20-30° N or S (Olson *et al.*, 2001; Myers *et al.*, 2000; Gaston, 2000; Groombridge & Jenkins, 2002).

The general increases of species richness towards the wet tropics is manifested in virtually all groups of plant and animal species, however there are differences in climate sensitivity related to differences in physiological and life cycle characteristics of organisms of different plant and animal taxa. The diversity of amphibians, for example, is much more influenced by rainfall amounts than the diversity of reptiles, which decline primarily along temperature gradients (Currie, 1991). While species richness of both birds and butterflies decrease with distance away from the tropics, this decrease is much steeper for the butterflies: insects have a much shorter life cycle and turnover rate than vertebrates, and their capacities for migration are limited (Gaston, 2010; Groombridge & Jenkins, 2002). Tree species diversity is also highly influenced by rainfall amounts and associated productivity, but other plant growth forms such as grasses are more diverse in warm environments where certain nutrients are limited (Currie, 1991; Whittaker, 1975; Givnish, 1999).

Variability in the terrain and patchiness of soils and soil nutrients, the frequency and intensity of disturbances, and biogeographical history are also important determinants of biodiversity (Gurevitch *et al.*, 2006; Gaston, 2000). Diversity of mammal species correlates with a high diversity of habitats created by terrain variability (Simpson, 1964); equally bird diversity tends to increase with foliage height diversity in forests (McArthur &

McArthur, 1961). Due to a limited area the overall diversity may not be highest in forests on islands; however, richness in species endemism (i.e. the state of being unique to a particular geographic location) is particularly high on chains of islands, e.g. the Indonesian archipelago, the Philippines, the Galapagos Islands and various other oceanic island groups (Metcalfe *et al.*, 2002; Gaston, 2000; Myers *et al.*, 2000). It is such areas of high endemism, i.e. areas where species are unique to forests that are locally confined, where deforestation bears the highest threats to lead to the extinction of many species (Pimm & Jenkins, 2010; Brooks *et al.*, 1999; Sodhi *et al.*, 2004).

3. FOREST ECOSYSTEM FUNCTIONS AND BIODIVERSITY

Forests perform countless essential functions that are of immense value to human populations as 'ecosystem services'. Forest ecosystems mediate local and regional climates, retain and form soil resources thereby sequestering carbon, retain and purify water, regulate water flows and mitigate against floods and land slides, retain and sequester plant nutrients such as nitrogen, thereby preventing eutrophication of water bodies, mediate against human and animal diseases, and provide a vast diversity of fundamentally important products, ranging from valued timber to non-timber products (NTFPs) such as food products (fruits, seeds, bush meat, etc.), medicinal plants and pharmaceuticals, and building products (e.g. rattan and bamboo) (Sodhi *et al.*, 2007). In addition, forests themselves are a habitat that has shaped human societies and contributed to cultural diversity; still today's modern urbanised societies highly value forests as an environment for recreation (Sodhi *et al.*, 2007; Maffi & Woodley, 2010). Populations of plant and animal species are the building blocks of forests as well as other ecosystems. Each species occupies a specific ecological niche, and certain (keystone) species may be of particular importance for maintaining ecosystem functions.

The relationships between species diversity and ecosystem functions are vigorously debated in ecology. Biodiversity tends to be high in environments which have been relatively stable over long evolutionary time periods, respectively where the impacts of disturbances have been intermediate and variable but not catastrophic over large scales (Wright, 2002). Likewise, various studies indicate that ecosystems which are characterized by high species diversity are generally more robust and resilient towards disturbances (Naeem & Li, 1997; Hooper *et al.*, 2005). Plant and animal species play particular roles in forest ecosystems. Ecosystems are often dominated by a few species, which therefore account for most of the biomass and structure of the ecosystems, and the services they provide. However, various other species which may not directly appear to be important may actually serve as an insurance policy in the case of environmental change or disturbances; rare species may fill in gaps or even proliferate and largely replace dominant species (Sekercioglu, 2010). High biodiversity also means that generally more ecological niches are filled; this implies a higher primary productivity and resource use efficiency in plant communities (Tilman, 1997). Conversely, high biodiversity can only occur in complex and richly structured ecosystems which provide a wealth of ecological niches – such as in luxuriant tropical moist forests that are rich in epiphytes. A diversity of various plant and animal forms may therefore create further niches and facilitate the evolution of even more species.

Certain species may be crucial in keeping ecosystems in balance, even if they are comparatively rare. Such species – called 'keystone species' – are often unknown to perform important ecosystem roles until their loss eventually reveals their importance via dramatic ecosystem transformations (Power *et al.*, 1996). For example, large predators (e.g. wolves, bears, tigers, etc.) are well-known to keep populations of mammalian herbivores in check and to shape their spatial distributions, thereby facilitating establishment and growth of woody vegetation in certain areas (Ripple & Larson, 2000, Riginos & Grace, 2008; Terborgh *et al.,* 2001). Herbivores, and in particular large megaherbivores (e.g. elephants and giraffes), may in turn provide for habitat diversity via direct or indirect destruction of established vegetation and facilitation of woody regrowth (Pringle, 2008; Prins & van der Jeugd, 1993; Palmer *et al.*, 2008; Cochard & Agosti, 2008). While the role of large animals is often well-known and well-publicized for conservation efforts, smaller organisms - alone or in combination - may be equally important in any ecosystem. Determining systems of so-called 'functional groups' of species has therefore gained increasing attention for conservation biologists (e.g. Schwartz *et al.,* 2000; Lavorel *et al.,* 1997; Walker, 1995; Sekercioglu, 2006).

Other species again may be important as vectors of pollen and propagules as well as nutrients, biomass and energy, these species are called 'mobile links' (Sekercioglu, 2010). New habitat niches may be created by animals dislocating nutrients via behavioral activities or their faeces (O'Dowd & Lake, 2009; Tobler *et al.,* 2003). Ants and termites are particularly important vectors and keystone species that break down and redistribute biomass and nutrients within forest ecosystems (Brussaard, 1997). Bees and a vast diversity of other insects are well known to maintain gene flow between flowering plants. In some flowering plant species such as in the figs, some orchids etc. the relationship between plant species and their pollinators is often one of obligate mutualism. Plant-animal mutualisms are particularly common in tropical forests which are characterized by a high richness of mostly rare and widely dispersed tree and other plant species (Ghazoul & Sheil, 2010); up to 98% of tropical rainforest trees are pollinated by animals (Bawa, 1990). The inter-dependency of the plant with the respective vector species ensures pollen transfer between widely dispersed and rare trees; the loss of the co-specific insect vector, on the other hand, can potentially also incur the extinction of the plant species (Cox & Elmquist, 2000; Fontaine *et al.,* 2006). Seed dispersal by various types of vertebrate and invertebrate animals also ensures the spread and survival of many seed plants within their respective ecological niches; extinction of frugivorous megafauna (e.g. rhinoceros, elephants, great apes, etc.) may have lead to the decline of various fruit-bearing tree species (Ghazoul & Sheil, 2010; Guimarães *et al.,* 2008; Miller, 1994). Protection of pollinators, seed dispersers, nutrient depositors, scavengers, and predators must have highest priority in order to maintain the continuing stability and integrity of forest and other ecosystems (Sekercioglu, 2010).

4. CLIMATE CHANGE AND FORESTS

There is now little disagreement by leading scientists that global warming is occurring at rates unprecedented in human history, and that this is linked to the effects of anthropogenic greenhouse gas emissions (Anderegg *et al.,* 2010; Solomon *et al.,* 2009; IPCC, 2007). Mean temperatures are rising, with estimated increases over the last 40 years of up to 0.4° C in some parts the tropics (but significant regional differences) and more than 2° C in some polar regions, the global average being at about 0.75° C (IPCC, 2007; Hanson *et al.,* 2006; Malhi & Wright, 2004). Projections of average temperature increases until the end of the 21st century range from 2°-5° C in the tropics and up to more than 10° C in some polar regions (IPCC, 2007; Cramer *et al.,* 2004). Future changes in rainfall patterns are more uncertain. The hitherto observed changes in rainfall have been very variable geographically; for example, amounts of precipitation have been decreasing by more than 4% in parts of the Sahel zone of Africa, and parts of India; in contrast, rainfalls increased in some parts of Amazonia by up to 3% (Malhi & Wright, 2004). Overall, many uncertainties still exist in regards to the longer-term rates of global warming, the unfolding dynamics of weather patterns, feedbacks between climatic and other environmental changes and thresholds (cf. Colman & Power, 2010; Dallmeyer *et al.*, 2010; Zheng & Yoon, 2009; Bonan, 2008; Schuur *et al.*, 2008; IPCC, 2007; Arnell, 2000).

Climate change can affect forests by altering the frequency, intensity, duration, and timing of droughts and fires as well as storms and associated landslides. More indirectly (and difficult to assert via scientific research) introduced species, insect pests and pathogen outbreaks may be favored (Dale *et al.*, 2001). Under the most pessimistic climate change scenarios the earth's forest ecosystems may change dramatically. Some scenarios, for example, predict an alarming tree-dieback of the Amazon rainforest and eventually a fire-driven transformation to open vegetation (Cowling *et al.*, 2004; Cox *et al.*, 2004). Such transformations could imply further emissions of 72% carbon currently stored in the forests biomass, with further feedbacks on global warming (Cox *et al.*, 2004). However, even moderate scenarios suggest that most tropical forests will experience mean annual temperatures which are higher than were hitherto recorded in regions supporting tropical forest (Wright *et al.*, 2009). The exact consequences cannot be foreseen, but clearly, plant and animal species' phenologies are sensitive to changes in temperature, rainfall and humidity, and seasonal cycles; plant species may also react to the increased levels of CO_2 in the atmosphere, whereby some species (e.g. plants with the C3 photosynthetic pathway, pioneer species, etc.) may gain an advantageous competitive edge over other species (Ghazoul & Sheil, 2010).

Effects of climate change on forest ecosystems are increasingly becoming evident, even if the effects of logging and other types of disturbances and land use changes are currently still predominant. Changes and impacts are particularly evident in the polar regions and on high mountain ranges. Polar ice sheets and glaciers are in retreat. Glaciers on Mount Kilimanjaro – Africa's highest mountain – have been receding at a fast rate over the last decades and are expected to disappear entirely in less than 15 years (UNEP, 2007). Most probably this will lead to major changes in the water household and forest ecosystems of the mountain (Agrawala *et al.*, 2003). Similar fears also exist for other mountain regions, e.g. the Himalayas which feed huge river systems such as the Ganges, Brahmaputra and the Mekong, parts of the Andes and also the Alps (UNEP, 2007; Kundzewicz *et al.*, 2008; Böhner & Lehmkuhl, 2005). Other weather extremes appear to be increasingly related to global climate change and associated changes in marine circulation patterns. For example, the intense droughts and associated wildfires in 2008-2009 in Southern Australia and in 2005 in Amazonia affected forests over large areas (CSIRO, 2009; Malhi *et al.*, 2009); equally forests in California are feared to be affected by increasingly frequent and intense droughts (van Mantgem & Stephenson, 2007). The frequency and intensity of tropical cyclones is believed to have increased in the last decades, disproportionately affecting coastal communities and their environment (including coastal forests) in parts of South and Southeast Asia and the Caribbean (Webster *et al.*, 2005; Cochard *et al.*, 2008). In addition, rising sea levels are assumed to be increasingly stressing and threatening the remaining coastal forests, in particular mangroves (Alongi, 2008; Gilman *et al.*, 2008).

5. CLIMATE CHANGE AND FOREST BIODIVERSITY

Temperature increases and changes in rainfall patterns in terrestrial ecosystems are expected to occur increasingly at rates which outpace the capacities of many species populations (especially slow-breeding vertebrates) to adapt via processes of natural selection and evolution (Sodhi *et al.*, 2007). Species that are equipped with a high phenotypic plasticity may be able to survive in changing environments; some of these species may even thrive as other species of competitors and predators go extinct or emigrate from an area. However, those species (probably the majority) that have a small phenotypic plasticity and cannot adapt to climate change may only have one option for survival: the migration to regions that still reflect the range of climates within which the species has evolved (e.g. regions of higher latitudes or higher altitudes). Shifts in species dominance may generally be expected. Such shifts, however, are virtually impossible to predict because of the persisting uncertainties of climate predictions, the complex nature of most ecosystems, the very limited knowledge about the species' phenotypic plasticity and ecological competitiveness – in general, a lack of understanding of ecosystem dynamics under shifting climate parameters, whatever the climate predictions may be. As noted by Lovejoy (2010, p. 158), studies of pre-historic climate change suggest "that biological communities do not move as a unit, but rather it is the individual species that move each at its own rate and in its own direction. The consequence is that ecosystems, as we know them, will disassemble and the surviving species will assemble into new species configurations that largely defy the ability to foresee." Various projective computer models have been produced (see for example Lovejoy & Hannah, 2005); some models suggest extinction rates of 18-40% with a twofold increase in atmospheric carbon dioxide (Thomas *et al.*, 2004), or 20-30% with a temperature increase of about 3° C and considerably higher extinction rates at higher temperatures (IPCC, 2007; Lovejoy & Hannah, 2005).

The reliability of some of these models may soon be tested via comparison to real world observations. Currently there is scope for hope, as most forest ecosystems still appear to be fairly unaffected by climate change. However, even with the current increase of 0.75° C several consequences of climate change and associated environmental changes on forest biodiversity are becoming increasingly evident. Some low-lying isolated islands in the Pacific are feared to disappear entirely within the next decades; with them several endemic forest species they support would be lost (Yu *et al.*, 2006). Equally, in many isolated mountain regions many endemic species of plants and animals may be at significant risk of extinction from climate change (Thomas *et al.*, 2004; Foster, 2001; La Sorte & Jetz, 2010). In Switzerland the upper tree lines of alpine forests have been shifting upwards due to land use changes and supposedly due to climate change about 1-4 m per decade (Gehrig-Fasel *et al.*, 2007; Walther *et al.*, 2002); equally several bird species typical in coniferous forests (e.g. the Eurasian bullfinch, *Pyrrhula pyrrhula*, the common crossbill, *Loxia curvirostra*, and the spotted nutcracker, *Nucifraga caryocatactes*) have been declining in the lowlands and are still common only at higher altitudes as is obvious from the new edition of the Bird Atlas of the Canton of Zurich (see ZVS, 2010). Tree lines appear to have shifted in mountain regions of the temperate and high latitudes around the world; the recorded shifts vary from 2 m to more than 300 m for some species (Walther, 2003). Not much is known about tree-line shifts in the tropical regions, but upward species shifts seem to have occurred in forest bird communities in tropical montane regions of America and Southeast Asia (Pounds *et al.*, 1999; Peh, 2007). In addition, those regions have been the scene of a recent wave of amphibian extinctions which is well documented in tropical America and in the coastal rainforests of Australia; this is believed to be at least partially due to climate change (Pounds *et al.*, 2006; Laurance, 2008). The well-publicized loss of the golden toad, an endemic species of Monteverde Cloud Forest in Costa Rica, may be due to prolonged drought conditions that stem from decreased cloud contacts with the cloud forest habitat (Pounds *et al.*, 1999; Still *et al.*, 1999). It is hypothesized that prolonged warm and dry periods may cause stress conditions which weaken the amphibians and make them more susceptible to fungal infections and other parasite outbreaks. Regarding the extinctions of various *Atelopus* frogs in South America, Pounds *et al.* (2006) found that the percentage of species lost were highest at intermediate elevations of between 1000 and 2300 meters. This may best be explained by an interaction of climate change and parasite range expansion up to an elevation of 2300 m. Seen overall, the direct effects of climate change on species may currently still be minor. Many examples, however, show that climate change may already affect forest species populations indirectly by promoting the proliferation and range expansion of various diseases and parasites, e.g. fungal diseases and disease-carrying mosquitoes (Sodhi *et al.,* 2007; Walther *et al.,* 2002; Vors & Boyce, 2009) and plant insect pests, e.g. pine bark beetles and wood boring beetles (Lovejoy, 2010; Volney & Fleming, 2000).

Changes in animal and plant phenological patterns have also been observed. Small organisms with a high turnover rate, such as microorganisms, insects and annual plant species, are more sensitive and responsive to climate change over a short time period than larger organisms, such as vertebrates and trees. Equally, migratory species, being highly mobile and adaptable to various environmental situations, can respond to changes in climates and land uses by traveling to other more suitable regions. It is therefore no surprise that some of the best evidence for effects of ongoing climate change on species phenologies and distributions currently comes from studies about small species or from migratory species, or from studies using long-term data. In temperate northern latitudes (Europe, Northern America, Japan and China) plant species are now on average flowering about 2-5 days earlier in the spring than at the time of the first records some 100 years ago, and growing seasons have been extended to more than 10 days (Miller-Rushing & Primack, 2008; Walther *et al.,* 2002). In tropical forests the determinants of plant phenological patterns are still poorly understood; coupled factors of seasonal temperature, moisture and sunshine cues appear to play a significant role (Ghazoul & Sheil, 2010). There is as yet little evidence, but climate change has been implicated to cause failure of seed set of various tropical tree species (Chapman *et al.,* 2005). Range shifts northward and upward in altitude have been observed in some butterfly species in North America as well as in Europe (Parmesan, 2006; Parmesan *et al.,* 1999). In North America tree swallows (*Tachycineta bicolor*) have been observed nesting and laying their eggs earlier (Dunn & Winkler, 1999), and a hummingbird species (*Selasphorus rufus*) has entirely ceased to migrate (Parmesan, 2006). Observed shifts of bird species densities and distributions in the Canton of Zurich, Switzerland, may partially be attributable to climate change (see ZVS, 2010). There are two new breeding species typical of the Mediterranean (the European bee-eater, *Merops apiaster*, and the melodious warbler, *Hippolais polyglotta*). The Garden Warbler (*Sylvia borin*) has significantly declined in abundance (by about 51%) whereas its relative, the blackcap (*Sylvia atricapilla*), has slightly increased in dominance (by about 9%). The blackcap migrates to the Mediterranean in winter and is on average now returning several days earlier in spring; it may therefore be able to secure the best habitat patches, before the garden warbler, migrating farther South to Africa, arrives later in spring. Several other far migrating song birds have also shown significant population declines, the most noteworthy being the wood warbler (*Phylloscopus sibilatrix*) which has declined by almost 97% from an estimated population of 5500 breeding pairs in 1988 to only 180 pairs in 2008; in contrast, populations of its close relative the chiffchaff (*Phylloscopus collybita*), overwintering in the Mediterranean, were stable (slight declines of about 6%; ZVS, 2010). Other changes include population increases of all but one woodpecker species; this may reflect an increase of dead woody biomass in forests (and an associated abundance of insects) following several severe winter storms. In all of these examples climate change may play a role, but significant land use changes in the breeding ranges or along the migration routes of birds may be equally important.

6. DEFORESTATION AND CLIMATES

In the tropical regions the effects of global climate change have been much less evident than in the polar regions. However, even under the most optimistic scenarios, climate change may have significant detrimental effects in interaction with other anthropogenic ecosystem disturbances that continue unabated, and take their toll on forest environments. At regional and local scales climatic changes may be the result of changing land use patterns (Gash & Nobre, 1997). Deforestation implies that less solar radiation can be absorbed by foliage, and a lack of water storage and diminished transpiration surfaces disrupts the replenishment of water in the atmosphere, leading to more irregular rainfalls of lesser frequency and water volume (Malhi & Wright, 2004). Deforestation and forest burning may therefore lead to changes in air movements, increases in temperatures, and decreases in atmospheric moisture and cloud formation, whereby convection-related forest patterns are particularly sensitive (Berbet & Costa, 2003; Ghazoul & Sheil, 2010). In some regions such climatic changes may already have occurred early in human history; according to Miller *et al.* (2005) late Pleistocene fire-driven woodland destruction and desertification in Central Australia probably caused significant declines in monsoon rainfalls. In Southeast Asia deforestation has been predicted to result in a precipitation decline of 8%, with a decline of up to 17% for Indonesia (Hoffmann *et al.*, 2003).

Deforestation has been practiced by humans for thousands of years, first using primitive stone axes and applying fire. China's Loess Plateau was cleared of forests thousands of years ago; ever since then it has been eroding, evidenced by the colour of the sediment-loaded mighty Yellow River (Xiubin *et al.*, 2004). Equally, siltation of some coastal areas in Asia Minor and Syria is traceable to the deforestation and the introduction of agricultural practices in Ancient Greece (van Andel *et al.*, 1990). Much of the deforestation in Western Europe occurred between 1100 to 1800 AD, when human populations expanded, and the rising colonial powers started building large fleets of wooden sailing ships. With the industrialisation, the spread of colonial forestry and globalised trade, and the invention of the chain saw, deforestation sharply accelerated on a global scale – particularly in the tropical regions (Angelson & Kaimowitz, 2001). Within the last decade about half of the Earth's mature tropical forests (about 7.5-8 million km^2) have been cleared, and deforestation is continuing at a fast rate of about 130,000 km^2 a year (about 50 football fields a minute), particularly in the developing world (including a loss of about 55'000 km^2 of pristine tropical forests), but also in parts of the developed world (FAO, 2007; Asner *et al.*, 2009; Laurance, 2010). In South Asia, about 90% of the rainforests have been lost, much of it before the 1970ies. Deforestation rates in Southeast Asia and Africa increased dramatically after the 1970ies, with a loss of up to 90% in West Africa's coastal rainforests and 70-80% in countries such as China, Thailand and the Philippines (Sodhi *et al.*, 2007; Usher, 2009). Deforestation rates are currently highest in Nigeria (annual losses of 11.1%) and Southeast Asia, with annual losses of more than 2% in Indonesia, Cambodia, Vietnam and the Philippines; these losses are particularly concerning as the forests of these countries are characterised by high levels of endemism (FAO, 2007; Corlett, 2009). The largest block of rainforest (approximately 4 million square kilometres) remains in the Amazon basin, but even there already an area greater than the size of France (731'000 km^2) had been cleared by 2008, and deforestation was continuing at an ever faster pace (Ghazoul & Sheil, 2010). In absolute terms there had been more forest clearance in Brazil than in any other nation. Equally, other last intact forest blocks in the Congo Basin and in New Guinea are increasingly under pressure (Ghazoul & Sheil, 2010).

The current drivers of deforestation are manifold and often coupled. Industrial logging is often followed by conversion to agriculture (e.g. soybean cultivation and cattle ranching in the Amazon) and plantations (e.g. rubber and oil palm in Malaysia and Indonesia, and acacia and eucalypt in Vietnam) (Ghazoul & Sheil, 2010; FAO, 2007; Pearce, 2001). The local environmental consequences of deforestation – regional climatic changes, losses of soil resources, losses of biological and bio-cultural diversity, increasing threats from more frequent and hot fire spells, increasing frequencies of disasters such as floods and drought events, etc. – can severely affect countries' social systems and economies, and further increase the dependency of the world's poorest nations (e.g. Haiti, Bolivia, Madagascar, countries of the African Sahel, Afghanistan, etc.) to fuel imports (FAO, 2007; Pearce, 2001). Deforestation in the tropical regions may account for as much as 20-35% of global CO_2 emissions (IPCC, 2007); other calculations suggest emissions of 6-17% (excluding emissions from peatlands) (van der Werf, 2009). Deforestation in Southeast Asia alone is estimated to release approximately 465 million tonnes of carbon per year into the atmosphere, or 29% of the carbon release due to deforestation (Phat *et al.,* 2004). Similar amounts are being released from the Amazon Basin, spurred by fires during intermittent drought periods (Santilli *et al.,* 2005).

Deforestation in peat swamp forest lands - for conversion to rice fields (as in Borneo) or oil palm plantations (as in Aceh, Sumatra) - is particularly damaging. Cleared peat swamp forests become a net source of carbon, with oxidation being accelerated by active water drainage and fires. The massive peat and forest fires in 1997 in Borneo and Sumatra released an estimated 0.81-2.57 gigatonnes of carbon which was equivalent to 13-40% emissions from fossil fuels in that year (Page *et al.,* 2002). Total emissions are likely to be in the upper range; in a study in Malaysia Wösten *et al.* (1997) found that on average 7.2 t C ha^{-1} y^{-1} were released from drained peatlands even in the absence of fire, leading to land subsidence of 2 cm y^{-1}. In addition to carbon release, the destruction of peatlands and conversion to rice fields leads to significantly higher emissions of other greenhouse gases, such as methane and nitric oxide (Inubushi *et al.,* 2003; Takakai *et al.,* 2006; Hadi *et al.,* 2005; Sodhi *et al.,* 2007). The recurrent hazes emanating from peat fires also had serious impacts on people's health and on ecosystems in the region (Page *et al.,* 2002; Nichol, 1997; Abram *et al.,* 2003).

7. DEFORESTATION AND FOREST BIODIVERSITY

Deforestation in peat swamp forest lands is also disastrous for other reasons. The peat swamp forests in the lowlands of Borneo and Sumatra are centers of endemism and provide some of the last retreats for some of the most critically endangered Asian large mammal and bird species. For example, possibly more than 40% of the remaining populations of Sumatran orangutan (~7500 animals) live in the coastal swamp forests of Aceh (van Schaik *et al.,* 2001; Wich *et al.,* 2003). Other highly endangered animals found in the swamp forests include the Sumatran tiger (~250 animals in Sumatra), Asian tapir, otter civet, siamang, Storm's stork, masked finfoot, white-winged wood-duck, several hornbills, and the freshwater crocodile (Rijksen *et al.,* 1997). The swamps are also characterized by a high diversity and endemism of edible fishes, and socio-economic studies indicate that local community livelihoods may depend for over 80% on the peat swamp forest resources (Ng, 1994; Rijksen *et al.,* 1997; Kuniyasu, 2002). Peat swamp forests are one of the least researched vegetation types in South East Asia (Whitmore, 1995), yet over the last two decades some vast peat swamp areas have been destroyed at appalling rates (cf. Limin & Putir, 2000; Giesen, 2004; Page *et al.,* 2002). For example, since the beginning of the 1990ies about 70% of the Tripa swamps – originally the largest peat swamp forest area in Aceh – have been logged, partially drained and planted with oil palms by several palm estates of predominantly foreign ownership. Ironically, palm oil from Asia serves the world market of biofuels which are often heralded as a climate-friendly alternative to fossil fuels (Koh & Ghazoul, 2008; Fitzherbert *et al.,* 2008; Danielsen *et al.,* 2008). With continuing peat oxidation and subsidence, lowering of the water level and soil nutrient losses the palm plantations at Tripa are also becoming increasingly less viable; peat swamp areas are generally not suitable for conversion into croplands, especially where the peat is deeper than 1-2 m (Rieley *et al.,* 1997; Sorensen, 1993; Phillips, 1998). However, with the loss of the peat resources, these unique ecosystems may be permanently damaged or lost.

Long before the industrial age have forest species become extinct due to hunting and overuse by humans, by habitat modification (mostly the impact of fires), or the introduction of alien species. Large mammals, reptiles and many species of birds (particularly flightless birds, and endemic birds on islands) are known to have gone extinct in Australia, Java and other parts of Asia, in Madagascar, the Americas, and Islands in the Pacific and Indian Ocean (Pimm & Jenkins, 2010; Guimarães *et al.,* 2008; Ghazoul & Sheil, 2010). However, the rate of extinctions is now reaching unprecedented levels - currently more than 100 times higher than natural (background) rates of extinction - and is expected to sharply increase (Pimm & Jenkins, 2010). Of the world's avifauna 2% of the species have been lost in the last 500 years, and more than 12% of the known bird species are currently threatened with extinction (Pimm & Jenkins, 2010). Levels of threats for various other lesser known and lesser mobile taxa (e.g. amphibians, endemic freshwater fishes, small mammals and sensitive insects) are considerably higher (Pimm & Jenkins, 2010). For example, 35% of the world's amphibians are listed as endangered with many endemic tropical mountain forest species at high risk (Ghazoul & Sheil, 2010). Of the world's plant species 16% are deemed threatened, but given the large percentage of species (>25%) still to be described (of which many are rare and endemic species), up to 37% of the extant species (described and undescribed) may actually be threatened (Pimm & Jenkins, 2010).

The most important causes of extinction vary in different regions of the world. In the USA habitat destruction (or modification) is the most important threat to vertebrate biodiversity followed by the impact of alien species, pollution, overexploitation, and diseases, whereas in China the most important is overexploitation, followed by habitat destruction, pollution, alien species and diseases (Li & Wilcove, 2005). Until recently overexploitation may also have been the main driver of extinctions in Southeast Asia and tropical America, but habitat destruction and degradation is now becoming ever more important (Sodhi *et al.,* 2007; Sodhi & Brooks, 2006; Sodhi *et al.,* 2004). In contrast, in Australia, New Zealand and the Pacific and Indian Ocean Islands introduced exotic species which transform the environment and compete with native species are the most important causes of extinctions up to this date (Simberloff, 2010; Smith & Quin, 1996). In forest areas, and in the tropics in particular, all of these drivers are commonly interlinked in some ways with deforestation or forest degradation by logging activities. While deforestation is the 'conversion of forest to another land use', forest degradation has been defined as a 'reduction of the forest canopy cover or stocking within a forest' (FA0, 2000). In which way and to what degree these processes impact upon biodiversity overall depends on the amount of total impacted area, the intensity of these impacts per area and the spatial arrangement of the severely impacted areas (the anthropogenic matrix) in relation to the remaining non-impacted or lesser-impacted habitat fragments. These may all influence the types and dynamics of environmental changes which can then further exacerbate the losses of biodiversity by facilitating the invasion of alien species, promoting the spread and effect of deadly diseases, facilitating access of human hunters and collectors to previously remote and locked-up areas, and cause siltation of waterbodies. Ecosystem transformation may be further driven on by increasingly frequent and intense bush fires emanating from human settlements and by the losses of soil organic matter and nutrients.

Tropical forests are composed of a high diversity of comparatively rare and widely spread tree species; selectively logged species (e.g. large dipterocarp trees in Southeast Asia) may therefore become significantly rarer or even become locally extinct (Ghazoul & Sheil, 2010; Horne & Hickey, 1990). In tropical forests remaining populations of logged-over tree species may also further decline due to reduced pollination and recruitment success of the remaining trees, albeit there is as yet little empirical evidence (Ghazoul *et al.,* 1998; Curran *et al.,* 1999). Forest habitat changes due to logging may lead to various changes in the species composition. The drier microclimate in degraded forests may affect humidity-dependent species, such as amphibians, molluscs, ferns, bryophytes, fungi and tree epiphytes (e.g. Iskander, 1999; Padmawathe *et al.,* 2004). Reduced numbers of larger trees may lead to fewer suitable nesting sites, tree hollows, and certain feeding niches for various birds and mammals, and disjointed tree canopies can affect the movement and increase predation of canopy species such as monkeys, squirrels, possums and various birds (e.g. Johns, 1997; Johns & Skorupa, 1987; Johns, 1986; Yahner & Mahan, 1997). Equally, the movement of insects may be affected and dispersal patterns of pollen and seeds may be altered (e.g. Hill *et al.,* 1995; Ghazoul *et al.,* 1998; Johns, 1997). Increased plant density in the undergrowth may promote some ground-dwelling herbivorous species such as deer and smaller mammals which are better protected from predation; in contrast, understorey insects, such as ant communities, and insectivorous birds may be negatively affected (e.g. Floren & Linsenmair, 2001; Lambert & Collar, 2002). Despite such changes and the reduction of the target timber species, selective timber logging may leave forest habitats relatively intact overall provided that tree fall does not severely affect the remaining non-harvested trees (Meijard & Sheil, 2008). While shifts of species dominance occur in logged-over forests, overall species richness may remain high (e.g. Hill, 1999; Johns, 1997; Vallan *et al.,* 2004; Martin-Smith, 1998). Timber harvesting practices, however, differ significantly between countries and types of forests, and the impacts may be very variable. For example in the Amazon, commercial timber species comprise 2-10% of the woody volume in forests but up to 60% is destroyed during careless logging operations, including severe impacts on soil resources (Uhl & Viera, 1989). In the worst case entire forest plots are logged over and subsequently burnt either accidentally or on purpose, leading to permanent ecosystem transformation via further degradation and fragmentation processes. In the Amazon, the likelihood of complete deforestation is four times greater in degraded forest than in pristine forest (Asner *et al.,* 2006).

Logging trails open up forests and enable hunters and settlers to penetrate deeper into previously pristine areas. Settlers may clear further forest parts along trails, thereby widening the gaps fragmenting the forests and increasing the risks of fires. Fragmentation is a process whereby forest habitat blocks are subdivided into smaller and more isolated pieces of habitat (fragments); it encompasses an increase in the length of forest edges, i.e. edge effects overall become more important, affecting the forest integrity (Bennet & Saunders, 2010). Disturbance of forest structure and soils at the edges of fragments lead to increased light and other micro-climatic modifications which penetrate the fragments and decrease the core area (i.e. the least altered area at the centre of the fragments); this may allow the deeper penetration of alien species and disease vectors into the forests with further ensuing impacts (Green *et al.,* 2004; Lake & Leishman, 2004; Fensham *et al.,* 1994; Allan *et al.,* 2003; Gash & Nobre, 1997). The shrinking of habitats, the widening gaps between the habitat fragments (i.e. increased insularisation of fragments) and the decreasing habitat quality means that populations of species are becoming more vulnerable to extinction, particularly species that require large habitats for their survival and are unable to move between habitat fragments. Species of large mammals that require a large home range are typically vulnerable; in Southeast Asia some of the representative and most threatened megafauna, such as tigers, rhinoceroses, elephants and the great apes, clearly fall into this category. Overhunting in combination with forest fragmentation can soon lead to local extinctions of wildlife (e.g. Michalski & Peres, 2005; Sodhi & Brook, 2006; Poulsen *et al.,* 2009; Peres, 2010). Also other much smaller species may be affected and become extinct. Ant birds which require large undisturbed forest areas for foraging are often affected by fragmentation (Van Houtan *et al.,* 2007); this is also the case for other insectivorous forest birds that forage in mixed-species flocks and are unable to move across the surrounding matrix landscape (Stouffer & Bierregard, 1995; Sekercioglu *et al.,* 2002). Many insect taxa, in particular beetles and bees, may also show significant declines, whereas other taxa, e.g. butterflies, are somewhat more resilient to fragmentation (Didham *et al.,* 1998; Brown & Hutchings, 1997). The decline of important pollinators and seed dispersers, and their inability to move between fragments may decrease the pollination success and seed set, and the recruitment of seedlings of the dependent tree species (Ghazoul, 2005; Cramer *et al.,* 2007; Chapman *et al.,* 2003).

The loss of species generally render ecosystems - respectively the fragmented remains of ecosystems - less resilient to disturbances such as fires, and the impact of insect pests, diseases and invasive alien species. At the forest edges, evapotranspiration of trees is increased; this can result in dramatic dieback of drought sensitive trees up to 300 m and more from the edge into forest fragments (Laurance, 1997; Kapos, 1989). Edge effects in Southeast Asian peat swamp forests, such as the Tripa forests, can be even much greater as drainage canals in oil palm plantations significantly change the hydrology of adjacent forests (e.g. Rijksen *et al.,* 1997). The resulting dead biomass accumulation, increased light penetration, higher temperatures (also higher temperature variations), and reduced moisture depresses seedling establishment and survival and increases the susceptibility of the forest edges to fires, creating positive feedbacks that increase the penetration of edge effects (Ghazoul, 2005; Bruna, 1999; Laurance, 1997; Nascimiento & Laurance, 2004). Species composition typically changes to contain more pioneer species, lianas and other climbers (Laurance *et al.,* 2006); under optimal conditions these edge plant communities buffer the remaining core habitats against impacts emanating from the matrix, such as fires. Frequently, forest gaps and edges are, however, invaded by alien plant species which can further transform the forest fragments by facilitating hot fires, smothering other vegetation, transforming the soil chemistry and causing soil erosion and land slides (Asner *et al.,* 2008; Forseth & Innis, 2004; Jenkins & Pimm, 2003; Fensham *et al.,* 1994). Oceanic islands are known to be particularly vulnerable to exotic invasions because of their isolation, comparatively low species richness and simpler plant community structure (Pimm & Jenkins, 2010). As forests become increasingly fragmented, and fragments lose species and become structurally simpler, the threat of species invasions equally increases with potentially dramatic effects as demonstrated from some island cases (cf. Mortenson *et al.,* 2008; Drake & Hunt, 2009; O'Dowd *et al.,* 2003; Asner *et al.,* 2008). Meltdown of diversity in fragments may also occur due to the loss of species which are performing a keystone role in forests. The loss of large predators in small forest fragments (e.g. due to hunting) can lead to 10- to 100-fold increases in herbivore densities; this in turn can severely affect and transform the vegetation, leading to the extinction of plant species that are not well adapted to high pressures of herbivory (Terborgh *et al.,* 2001; Terborgh *et al.,* 2006).

8. CONCLUSION

At the time of writing this chapter the effects of deforestation and other types of anthropogenic habitat losses and degradation on the earth's biodiversity are still far more dramatic than the current effects of global climate change; until now only a few species are thought to have gone extinct as a direct result of climatic changes (with somewhat debatable evidence at hand). However, also at the time of writing this chapter, the specter of slowly rising temperatures and increasing frequencies of weather extremes is once again rising in human consciousness as communities around the world stand witness to new heat waves in Eurasia and North America, hazes stemming from blasting forest and smoldering peat fires of previously unseen extent in Russia, and the worst flooding disasters in Pakistan since more than 60 years (The Economist, 2010a, b).

Climate change science is a heated business (cf. Hulme, 2010; Kitcher, 2010). Much of it is still based on computer-aided projections that depend on the parameters of our current (limited) understanding and beliefs. Given the dimensions of what is at stake, scientists' climate change debates are exposed to open publicity and politics with a danger to overheat and jump the rails. In the rich countries climate change skepticism may stifle determined and swift action by politicians to promote carbon-free societies – a development that will have to come at any rate as fossil fuels are gradually being depleted and continue to be spilled in disastrous ways (cf. The Economist, 2010c). In poorer countries, on the other hand, the 'climate change hype' may distract attention away from a large array of more 'conventional' and much more immediately pressing risks that may be summarized under the caption 'population increases and unsustainable land use practices'.

The threats of regional and local deforestation may be comparatively easier to address than the problem of worldwide climate change which requires cooperation at a global level. Various countries in Europe and Asia have already passed through what has been termed a 'forest transition', i.e. episodes of deforestation coming to a standstill (e.g. by introducing a logging ban) and reversing to episodes of partial forest increases (e.g. by state-subsidized reforestation programs) (Rudel *et al.,* 2005; Usher, 2009; FAO, 2007). In some countries particularly in Asia (e.g. China, Vietnam, Malaysia and Indonesia) such 'transitions' and increases may be largely attributed to growing forest plantations for pulp, oil and rubber production, while natural forests and the associated biodiversity continue to be declining in many regions (FAO, 2007; Fitzherbert *et al.,* 2008; Kikkawa, 1999). North-South carbon trading schemes may now invigorate efforts to conserve the remaining primary forests and introduce sustainable practices in forestry. However, there are also several threats associated with such schemes, as they may affect the governance of forestry (cf. Phelps & Webb, 2010; Palmer, 2010). All efforts should be concentrated to address the risks of deforestation and climate change in a synergistic way. Rainforest timber is becoming ever more valuable and rainforest can store and sequester more carbon per unit area than any plantation monocultures of exotic species (Kikkawa, 2006; Zhou, 2008; Kanowski *et al.,* 2005; Wright *et al.,* 2000). Rainforest conservation and reforestation as well as introducing sustainable harvesting methods of rainforests should therefore be put much higher on national and international agendas. In addition, destruction of valuable peat swamp forests in the tropics and elsewhere must be halted; further senseless infringements should be internationally ostracized. Any failure to reverse the trends of deforestation and associated greenhouse emissions may exacerbate the biodiversity crisis as well as human crises.

Species loss will be eternal. However, the combination of forest fragmentation with climate change may bring about a new wave of extinctions, as insularized populations are barred from moving to cooler refuges. According to an assessment by Wright *et al.* (2009) such a synergistic risk of climate change and habitat fragmentation to biodiversity may be as high or even higher in the tropics than in temperate regions as the distances for species to migrate to cooler refuges is greatest for equatorial regions, including key tropical forest areas such as the Amazon and Congo River Basins, and upper elevations of many tropical mountain ranges. Climate change may increasingly cause food shortages and further increasing pressures and exploitative impacts by humans on wild plant and animal populations. Climate change may promote the spread of diseases and invasive species, accelerate oxidation and erosion of soils, and exacerbate the stresses on aquatic life caused by sediment runoff. By eroding the forests' resources also humans as a species may be increasingly set on the move in search of refuges. It is more humanly dignified to start investing now in native trees, peats and wildlife corridors.

REFERENCES

Abram, N. J., Gagan, M. K., McCulloch, M. T., Chappell, J., & Hantoro, W. S. (2003). Coral reef death during the 1997 Indian Ocean dipole linked to Indonesian wildfires. *Science*, *301*, 952–955. doi:10.1126/science.1083841

Agrawala, S., Moehner, A., Hemp, A., van Aalst, M., Hitz, S., & Smith, J. … Mwaipopo, O. U. (2003). *Development and climate change in Tanzania: Focus on Mount Kilimanjaro*. Organisation for Economic Co-operation and Development, Paris.

Allan, B. F., Keesing, F., & Ostfeld, R. S. (2003). Effect of forest fragmentation on Lyme disease risk. *Conservation Biology*, *17*, 267–272. doi:10.1046/j.1523-1739.2003.01260.x

Alongi, D. M. (2008). Mangrove forests: Resilience, protection from tsunamis, and responses to global climate change. *Estuarine, Coastal and Shelf Science*, *76*, 1–13. doi:10.1016/j.ecss.2007.08.024

Anbar, A., Duan, Y., Lyons, T., Arnold, G., Kendall, B., & Creaser, R. (1903–1906). Buick, R. (2007). A whiff of oxygen before the great oxidation event? *Science*, *317*(5846).

Anderegg, W. R. L., Prall, J. W., Harold, J., & Schneider, S. H. (2010). Expert credibility in climate change. *Proceedings of the National Academy of Sciences of the United States of America*, *107*(27), 12107–12109. doi:10.1073/pnas.1003187107

Angelson, A., & Kaimowitz, D. (Eds.). (2001). *Agricultural technologies and tropical deforestation*. New York, NY: CABI Publishing. doi:10.1079/9780851994512.0000

Arnell, N. W. (2000). Thresholds and response to climate change forcing: The water sector. *Climatic Change*, *46*, 305–316. doi:10.1023/A:1005699210660

Asner, G. P., Broadbent, E. N., Oliveira, P. J. C., Keller, M., Knapp, D. E., & Sylva, J. N. M. (2006). Condition and fate of logged forests in the Brazilian Amazon. *Proceedings of the National Academy of Sciences of the United States of America*, *103*, 12947–12950. doi:10.1073/pnas.0604093103

Asner, G. P., Jones, M. O., Martin, R. E., Knapp, D. E., & Hughes, R. F. (2008). Remote sensing of native and invasive species in Hawaiian forests. *Remote Sensing of Environment*, *112*(5), 1912–1926. doi:10.1016/j.rse.2007.02.043

Asner, G. P., Rudel, T. K., Aide, T. M., DeFries, R., & Emerson, R. (2009). A contemporary assessment of change in humid tropical forest. *Conservation Biology, 23*(6), 1386–1395. doi:10.1111/j.1523-1739.2009.01333.x

Bawa, K. S. (1990). Plant-pollinator interactions in tropical rainforests. *Annual Review of Ecology and Systematics, 21*, 399–422. doi:10.1146/annurev.es.21.110190.002151

Beaman, J. H., & Burley, J. S. (2003). Progress in the floristic inventory of Borneo. In Padoch, C., & Peluso, N. L. (Eds.), *Borneo in transition: People, forests, conservation and development* (pp. 93–113). Oxford, UK: Oxford University Press.

Bennet, A. F., & Saunders, D. A. (2010). Habitat fragmentation and landscape change. In Sodhi, N. S., & Ehrlich, P. R. (Eds.), *Conservation Biology for all* (pp. 88–106). Oxford, UK: Oxford University Press. doi:10.1093/acprof:oso/9780199554232.003.0006

Berbet, M. L. C., & Costa, M. H. (2003). Climate change after tropical deforestation: Seasonal variability of surface albedo and its effects on precipitation change. *Journal of Climate, 16*, 2099–2104. doi:10.1175/1520-0442(2003)016<2099:CCATDS>2.0.CO;2

Böhner, J., & Lehmkuhl, F. (2005). Environmental change modelling for Central and High Asia: Pleistocene, present and future scenarios. *Boreas, 34*(2), 220–231. doi:10.1080/03009480510012917

Bonan, G. B. (2008). Forests and climate change: Forcings, feedbacks, and the climate benefits of forests. *Science, 320*(5882), 1444–1449. doi:10.1126/science.1155121

Brady, M. A. (1997). Effects of vegetation changes on organic matter dynamics in three coastal peat deposits in Sumatra, Indonesia. In Rieley, J. O., & Page, S. E. (Eds.), *Biodiversity and sustainability of tropical peatlands* (pp. 113–134). Cardigan, UK: Samara Publishing Limited.

Brooks, T. M., Pimm, S. L., Kapos, V., & Ravilious, C. (1999). Threat from deforestation to montane and lowland birds and mammals in insular South-east Asia. *Journal of Animal Ecology, 68*(8), 1061–1078. doi:10.1046/j.1365-2656.1999.00353.x

Brown, K. S., & Hutchings, R. W. (1997). Disturbance, fragmentation and the dynamics of diversity in Amazonian forest butterflies: Implications for conservation. In Laurance, W. F., & Bierregaard, R. O. (Eds.), *Tropical forest remnants: Ecology, management and conservation of fragmented communities* (pp. 91–110). Chicago, IL: University of Chicago Press.

Bruna, E. (1999). Seed germination in rainforest fragments. *Nature, 402*, 139. doi:10.1038/45963

Brussaard, L. (1997). Biodiversity and ecosystem functioning in soils. *Ambio, 26*(8), 563–570.

Bunker, D. E., DeClerck, F., Bradford, J. C., Colwell, R. K., Perfecto, I., & Phillips, O. L. (2005). Species loss and aboveground carbon storage in a tropical forest. *Science, 310*, 1029–1031. doi:10.1126/science.1117682

Ceballos, G., & Ehrlich, P. R. (2009). Discoveries of new mammal species and their implications for conservation and ecosystem services. *Proceedings of the National Academy of Sciences of the United States of America, 105*, 11505–11511.

Chapman, C. A., Chapman, L. J., Struhsaker, T. T., Zanne, A. E., Clark, C. J., & Poulsen, J. R. (2005). A long-term evaluation of fruiting phenology: Importance of climate change. *Journal of Tropical Ecology, 21*, 31–45. doi:10.1017/S0266467404001993

Chapman, C. A., Chapman, L. J., Vulinec, K., Zanne, A., & Lawes, M. J. (2003). Fragmentation and alteration of seed dispersal processes: An initial evaluation of dung beetles, seed fate, and seedling diversity. *Biotropica, 35*, 382–393.

Cochard, R., & Agosti, D. (2008). Putting ant-acacia mutualisms to the fire. *Science*, *319*, 759–761. doi:10.1126/science.319.5871.1759d

Cochard, R., Ranamukharachchi, S. L., Shivakoti, G., Shipin, O., & Edwards, P. J. (2008). The 2004 tsunami in Aceh and Southern Thailand: A review on coastal ecosystems, wave hazards and vulnerability. *Perspectives in Plant Ecology, Evolution and Systematics*, *10*, 3–40. doi:10.1016/j.ppees.2007.11.001

Colman, R. A., & Power, S. B. (2010). Atmospheric radiative feedbacks associated with transient climate change and climate variability. *Climate Dynamics*, *34*(7-8), 919–933. doi:10.1007/s00382-009-0541-8

Corlett, R. T. (2009). *The ecology of tropical East Asia*. Oxford, UK: Oxford University Press.

Cowling, S. A., Betts, R. A., Cox, P. M., Ettwein, V. J., Jones, C. D., Maslin, M. A., & Spall, S. A. (2004). Contrasting simulated past and future responses of the Amazonian forest to atmospheric change. *Philosophical transactions of the Royal Society B*, *359*, 539-547.

Cox, P. A., & Elmquist, T. (2000). Pollinator extinction in the Pacific Islands. *Conservation Biology*, *14*(5), 1237–1239. doi:10.1046/j.1523-1739.2000.00017.x

Cox, P. M., Betts, R. A., Collins, M., Harris, P. P., Huntingford, C., & Jones, C. D. (2004). Amazonian forest dieback under climate-carbon cycle projections for the 21st century. *Theoretical and Applied Climatology*, *78*, 137–156. doi:10.1007/s00704-004-0049-4

Cramer, W., Bondeau, A., Schaphoff, S., Lucht, W., Smith, B., & Sitch, S. (2004). Tropical forests and the global carbon cycle: impacts of atmospheric carbon dioxide, climate change and rate of deforestation. *Philosophical Transactions of the Royal Society of London Series B*, *359*, 331–343. doi:10.1098/rstb.2003.1428

Cramer, W., Mesquita, R. C. G., & Williamson, G. B. (2007). Forest fragmentation differentially affects seed dispersal of large and small-seeded tropical trees. *Biological Conservation*, *137*, 415–423. doi:10.1016/j.biocon.2007.02.019

CSIRO. (2009). *Climate Change and the 2009 Bushfires*. ([Victorian Bushfires Royal Commission). Commonwealth Scientific and Industrial Research Organisation, Canberra.]. *CSIRO Submission*, *091345*, 2009.

Curran, L. M., Caniago, I., Paoli, G. D., Astianti, D., Kusneti, M., & Leighton, M. (1999). Impact of El Niño and logging on canopy tree recruitment in Borneo. *Science*, *286*, 2184–2188. doi:10.1126/science.286.5447.2184

Currie, D. J. (1991). Energy and large-scale patterns of animal and plant species richness. *American Naturalist*, *137*, 27–49. doi:10.1086/285144

Dale, V. H., Joyce, L. A., McNulty, S., & Neilson, R. P. (2001). Climate change and forest disturbances. *Bioscience*, *51*(9), 723–734. doi:10.1641/0006-3568(2001)051[0723:CCAFD]2.0.CO;2

Dallmeyer, A., Claussen, M., & Otto, J. (2010). Contribution of oceanic and vegetation feedbacks to Holocene climate change in monsoonal Asia. *Climate of the Past*, *6*(2), 195–218. doi:10.5194/cp-6-195-2010

Danielson, F., Beukema, H., Burgess, N. D., Parish, F., Brühl, C. A., & Donald, P. F. (2008). Biofuel plantations on forested lands: Double jeopardy for biodiversity and climate. *Conservation Biology*, *23*(2), 348–358. doi:10.1111/j.1523-1739.2008.01096.x

Didham, R. K., Hammond, P. M., Lawton, J. H., Eggleton, P., & Stork, N. E. (1998). Beetle species responses to tropical forest fragmentation. *Ecological Monographs*, *68*, 295–323. doi:10.1890/0012-9615(1998)068[0295:BSRTTF]2.0.CO;2

Diemont, W. H., Nabuurs, G. J., Rieley, J. O., & Rijksen, H. D. (1997). Climate change and management of tropical peatlands as a carbon reservoir. In Rieley, J. O., & Page, S. E. (Eds.), *Tropical peatlands* (pp. 363–368). Cardigan, UK: Samara Publishing Limited.

Dole, M. (1965). The natural history of oxygen. *The Journal of General Physiology*, *49*, 5–27. doi:10.1085/jgp.49.1.5

Drake, D. R., & Hunt, T. L. (2009). Invasive rodents on islands: Integrating historical and contemporary ecology. *Biological Invasions*, *11*(7), 1483–1487. doi:10.1007/s10530-008-9392-1

Dunn, P. O., & Winkler, D. W. (1999). Climatic change has affected breeding date of tree swallows throughout North America. *Proceedings. Biological Sciences*, *266*, 2487–2490. doi:10.1098/rspb.1999.0950

FAO. (2000). *On definitions of forest and forest change*. Rome, Italy: Food and Agriculture Organisation of the United Nations.

FAO. (2007). *State of the world's forests*. Rome, Italy: Food and Agriculture Organisation of the United Nations.

Fensham, R. J., Fairfax, R. J., & Cannell, R. J. (1994). The invasion of Lantana camara L. in Forty Mile Scrub National Park, North Queensland. *Austral Ecology*, *19*(3), 297–305. doi:10.1111/j.1442-9993.1994.tb00493.x

Fitzherbert, E. B., Struebig, M. J., Morel, A., Danielson, F., Brühl, C. A., Donald, P. F., & Phalan, B. (2008). How will oil palm expansion affect biodiversity? *Trends in Ecology & Evolution*, *23*(10), 538–545. doi:10.1016/j.tree.2008.06.012

Floren, A., & Linsenmair, E. (2001). The influence of anthropogenic disturbances on the structure of arboreal arthropod communities. *Plant Ecology*, *153*, 153–167. doi:10.1023/A:1017510312462

Fontaine, C., Dajoz, I., Meriguet, J., & Loreau, M. (2006). Functional diversity of plant—pollinator interaction webs enhances the persistence of plant communities. *PLoS Biology*, *4*(1), e1. doi:10.1371/journal.pbio.0040001

Foottit, R. G., & Adler, P. H. (Eds.). (2009). *Insect biodiversity*. Oxford, UK: Blackwell Publishing. doi:10.1002/9781444308211

Forseth, I. N. Jr, & Innis, A. F. (2004). Kudzu (Pueraria montana): History, physiology, and ecology combine to make a major ecosystem threat. *Critical Reviews in Plant Sciences*, *23*(5), 401–413. doi:10.1080/07352680490505150

Foster, P. (2001). The potential negative impacts of global climate change on tropical montane cloud forests. *Earth-Science Reviews*, *55*(1-2), 73–106. doi:10.1016/S0012-8252(01)00056-3

Gash, J. H. C., & Nobre, C. A. (1997). Climatic effects of Amazonian deforestation: Some results from ABRACOS. *Bulletin of the American Meteorological Society*, *78*, 823–830. doi:10.1175/1520-0477(1997)078<0823:CEOADS>2.0.CO;2

Gaston, K. J. (2000). Global patterns in biodiversity. *Nature*, *405*, 220–227. doi:10.1038/35012228

Gaston, K. J. (2010). Biodiversity. In Sodhi, N. S., & Ehrlich, P. R. (Eds.), *Conservation Biology for all* (pp. 27–44). Cambridge, UK: Oxford University Press. doi:10.1093/acprof:oso/9780199554232.003.0003

Gehrig-Fasel, J., Guisan, A., & Zimmermann, N. E. (2007). Tree line shifts in the Swiss Alps: Climate change or land abandonment? *Journal of Vegetation Science*, *18*, 571–582. doi:10.1111/j.1654-1103.2007.tb02571.x

Ghazoul, J. (2005). Pollen and seed dispersal among dispersed plants. *Biological Reviews of the Cambridge Philosophical Society*, *80*, 413–443. doi:10.1017/S1464793105006731

Ghazoul, J., Liston, K. A., & Boyle, T. J. B. (1998). Disturbance-induced density-dependent seed set in *Shorea siamensis* (Dipterocarpaceae), a tropical forest tree. *Journal of Ecology*, *86*, 462–473. doi:10.1046/j.1365-2745.1998.00270.x

Ghazoul, J., & Sheil, D. (2010). *Tropical rainforest: Ecology, diversity and conservation*. New York, NY: Oxford University Press.

Giesen, W. (2004). *Causes of peatswamp forest degradation in Berbak NP, Indonesia, and recommendations for restoration. International Agricultural Centre (IAC)*. Netherlands: Arcadis Euroconsult Arcadis Euroconsult, Wageningen University.

Gilman, E. L., Ellison, J., Duke, N. C., & Field, C. (2008). Threats to mangroves from climate change and adaptation options, a review. *Aquatic Botany*, *89*(2), 237–250. doi:10.1016/j.aquabot.2007.12.009

Givnish, T. J. (1999). On the causes of gradients in tropical tree diversity. *Journal of Ecology*, *87*, 193–210. doi:10.1046/j.1365-2745.1999.00333.x

Green, P. T., Lake, P. S., & O'Dowd, D. J. (2004). Resistance of island rainforest to invasion by alien plants: Influence of microhabitat and herbivory on seedling performance. *Biological Invasions*, *6*(1), 1–9. doi:10.1023/B:BINV.0000010144.12808.cb

Groombridge, B., & Jenkins, M. D. (2002). *World atlas of biodiversity. Earth's living resources in the 21st century*. Berkeley, CA: University of California Press.

Guimarães, P. R. Jr, Galetti, M., & Jordano, P. (2008). Seed dispersal anachronisms: Rethinking the fruits extinct megafauna ate. *PLoS ONE*, *3*(3), e1745.

Gurevitch, J., Scheiner, S. M., & Fox, G. A. (2006). *The ecology of plants*. Massachusetts: Sinauer Associates, Inc.

Hadi, A., Inubushi, K., & Furukawa, Y., Pur nomo, E., Rasmadi, M., & Tsurata, H. (2005). Greenhouse gas emissions from tropical peatlands of Kalimantan, Indonesia. *Nutrient Cycling in Agroecosystems*, *71*, 73–80. doi:10.1007/s10705-004-0380-2

Hanson, J., Sato, M., Ruedy, R., Lo, K., Lea, D. W., & Medina-Elizade, M. (2006). Global temperature change. *Proceedings of the National Academy of Sciences of the United States of America*, *103*(39), 14288–14293. doi:10.1073/pnas.0606291103

Hill, J. K. (1999). Butterfly spatial distribution and habitat requirements in a tropical forest: impacts of selective logging. *Journal of Applied Ecology*, *36*, 564–572. doi:10.1046/j.1365-2664.1999.00424.x

Hill, J. K., Hamer, K. C., Lace, L. A., & Banham, W. M. T. (1995). Effects of selective logging on tropical forest butterflies on Buru, Indonesia. *Journal of Applied Ecology*, *32*, 754–760. doi:10.2307/2404815

Hoffman, W. A., Schroeder, W., & Jackson, R. B. (2003). Regional feedbacks among fire, climate, and tropical deforestation. *Journal of Geophysical Research*, *108*, 4721. doi:10.1029/2003JD003494

Hooper, D. U., Chapin, F. S., Ewel, J. J., Hector, A., Inchausti, P., & Lavorel, S. (2005). Effects of biodiversity on ecosystem functioning: A consensus of current knowledge. *Ecological Monographs*, *75*, 3–5. doi:10.1890/04-0922

Horne, R., & Hickey, J. (2006). Ecological sensitivity of Australian rainforests to selective logging. *Austral Ecology*, *16*(1), 119–129. doi:10.1111/j.1442-9993.1991.tb01487.x

Hulme, M. (2010). *Why we disagree about climate change. Understanding controversy, inaction, and opportunity*. Cambridge, UK: Cambridge University Press.

Inubushi, K., Furukawa, Y., & Hadi, A., Pur nomo, E., & Tsurata, H. (2003). Seasonal changes of CO_2, CH_4 and N_2O fluxes in relation to land-use change in tropical peatlands located in coastal area of South Kalimantan. *Chemosphere, 52*, 603–608. doi:10.1016/S0045-6535(03)00242-X

IPCC. (2007). *Observations: Oceanic climate change and sea level. Working Group 1: The Physical Science Basis*. Geneva, Switzerland: Intergovernmental Panel on Climate Change.

Iskander, D. T. (1999). Amphibian declines monitoring in the Leuser Management Unit, Aceh, North Sumatra, Indonesia. *Froglog, 34*, 2.

Jenkins, C. N., & Pimm, S. L. (2003). How big is the global weed patch? *Annals of the Missouri Botanical Garden, 90*, 172–178. doi:10.2307/3298581

Johns, A. G. (1986). Effects of selective logging on the behavioural ecology of West Malaysian primates. *Ecology, 67*, 684–694. doi:10.2307/1937692

Johns, A. G. (1997). *Timber production & biodiversity: Conservation in tropical forests*. Cambridge, UK: Cambridge University Press. doi:10.1017/CBO9780511525827

Johns, A. G., & Skorupa, J. P. (1987). Responses of rainforest primates to habitat disturbances: A review. *International Journal of Primatology, 8*, 157–191. doi:10.1007/BF02735162

Kanowski, J., Catterall, C. P., & Wardell-Johnson, G. W. (2005). Consequences of broadscale timber plantations for biodiversity in cleared rainforest landscapes of tropical and subtropical Australia. *Forest Ecology and Management, 208*(1-3), 359–372. doi:10.1016/j.foreco.2005.01.018

Kapos, V. (1989). Effects of isolation on the water status of forest patches in the Brazilian Amazon. *Journal of Tropical Ecology, 5*, 173–185. doi:10.1017/S0266467400003448

Kikkawa, J. (2006). Reforestation and biodiversity in the Asia-Pacific region. In Suzuki, K., Ishii, K., Sakurai, S., & Sasaki, S. (Eds.), *Plantation technology in tropical forest science* (pp. 247–263). Tokyo, Japan: Springer. doi:10.1007/4-431-28054-5_25

Kitcher, P. (2010). The climate change debates. *Science, 328*(5983), 1230–1234. doi:10.1126/science.1189312

Koh, L. P., & Ghazoul, J. (2008). Biofuels, biodiversity, and people: Understanding the conflicts and finding opportunities. *Biological Conservation, 141*(10), 2450–2460. doi:10.1016/j.biocon.2008.08.005

Kundzewicz, Z. W., Mata, L. J., Arnell, N. W., Döll, P., Jimenez, B., & Miller, K. (2008). The implications of projected climate change for freshwater resources and their management. *Hydrological Sciences Journal, 53*, 3–10.

Kuniyasu, M. (2002). Environments and people of Sumatran peat swamp forest II: Distribution of villages interactions between people and forests. *Southeast Asian Studies, 40*(1), 87–108.

La Sorte, F. A., & Jetz, W. (2010). Avian distribution under climate change: Towards improved projections. *The Journal of Experimental Biology, 213*, 862–869. doi:10.1242/jeb.038356

Lake, J. C., & Leishman, M. R. (2004). Invasion success of exotic plants in natural ecosystems: The role of disturbance, plant attributes and freedom from herbivores. *Biological Conservation, 117*(2), 215–226. doi:10.1016/S0006-3207(03)00294-5

Lambert, F. R., & Collar, N. J. (2002). The future of Sundaic lowland forest birds: Long-term effects of commercial logging and fragmentation. *Forktail, 18*, 127–146.

Laurance, W. F. (1997). Biomass collapse in Amazonian forest fragments. *Science, 278*, 1117–1118. doi:10.1126/science.278.5340.1117

Laurance, W. F. (2008). Global warming and amphibian extinctions in eastern Australia. *Austral Ecology, 33*, 1–9. doi:10.1111/j.1442-9993.2007.01812.x

Laurance, W. F. (2010). Habitat destruction: Death by a thousand cuts. In Sodhi, N. S., & Ehrlich, P. R. (Eds.), *Conservation Biology for all* (pp. 73–87). New York, NY: Oxford University Press. doi:10.1093/acprof:oso/9780199554232.003.0005

Laurance, W. F., Nascimiento, H. E. M., Laurance, S. G., Andrade, A., Ribeiro, J. E. L. S., & Giraldo, J. P. … D'Angelo, S. (2006). Rapid decay of tree-community composition in Amazonian forest fragments. *Proceedings of the National Academy of Sciences of the United States of America, 103*(50), 19010-19014.

Lavorel, S., McIntyre, S., Landsberg, J., & Forbes, T. D. A. (1997). Plant functional classifications: From general groups to specific groups based on response to disturbance. *Trends in Ecology & Evolution, 12*(12), 474–478. doi:10.1016/S0169-5347(97)01219-6

Leopoldo, P. R., Franken, W. K., & Villa Nova, N. A. (1995). Real evapotranspiration and transpiration through a tropical rain forest in central Amazonia as estimated by the water balance method. *Forest Ecology and Management, 73*(1-3), 185–195. doi:10.1016/0378-1127(94)03487-H

Lewis, S. L., Lopez-Gonzalez, G., Sonke, B., Affum-Baffoe, K., Baker, T. R., & Ojo, L. O. (2009). Increasing carbon storage in intact African tropical forests. *Nature, 457*, 1003–U3. doi:10.1038/nature07771

Li, Y., & Wilcove, D. S. (2005). Threats to vertebrate species in China and the United States. *Bioscience, 55*, 147–153. doi:10.1641/0006-3568(2005)055[0147:TTVSIC]2.0.CO;2

Limin, S. H., & Putir, P. E. (2000). The massive exploitation of peat swamp forest potentially has not successfully increased the local people's prosperity in Central Kalimantan. In T. Iwakuma, T. Inoue, T. Kohyama, M. Osaki, H. Simbolon, H. Tachibana, H.... K. Yabe (Eds.), *Proceedings of the International Symposium on: Tropical Peatlands. Bogor, Indonesia, 22-24 November 1999* (pp. 491-498). Japan: Sapporo Editors.

Lovejoy, T. E. (2010). Climate change. In Sodhi, N. S., & Ehrlich, P. R. (Eds.), *Conservation Biology for all* (pp. 153–161). New York, NY: Oxford University Press. doi:10.1093/acprof:oso/9780199554232.003.0009

Lovejoy, T. E., & Hannah, L. (Eds.). (2005). *Climate change and biodiversity*. New Haven, CT: Yale University Press.

Maffi, L., & Woodley, E. (2010). *Biocultural diversity conservation. A global sourcebook*. London, UK: Earthscan Ltd.

Magurran, A. E. (2003). *Measuring biological diversity*. Oxford, UK: Blackwell Publishing.

Malhi, Y., Aragão, L. E. O. C., Galbraith, D., Huntingford, C., Fisher, R., & Zelazowski, P. … Meir, P. (2009). Exploring the likelihood and mechanism of a climate-change-induced dieback of the Amazon rainforest. *Proceedings of the National Academy of Sciences of the United States of America, 106*(49), 20610-20615.

Malhi, Y., & Wright, J. (2004). Spatial patterns and recent trends in the climate of tropical rainforest regions. *Philosophical Transactions of the Royal Society of London Series B, 359*, 311–329. doi:10.1098/rstb.2003.1433

Martin-Smith, K. M. (1998). Effects of disturbance caused by selective timber extraction on fish communities in Sabah, Malaysia. *Environmental Biology of Fishes, 53*, 155–167. doi:10.1023/A:1007496424730

McArthur, R. H., & McArthur, J. W. (1961). On bird species diversity. *Ecology, 42*, 594–598. doi:10.2307/1932254

Meijaard, E., & Sheil, D. (2008). A logged forest in Borneo is better than none at all. *Nature, 446*, 976.

Metcalfe, I., Smith, J. M. B., Morwood, M., & Davidson, I. (Eds.). (2002). *Faunal and floral migrations and evolution in SE Asia-Australasia*. New York, NY: Oceania Publication.

Michalski, F., & Peres, C. A. (2005). Anthropogenic determinants of primate and carnivore local extinctions in a fragmented forest landscape of southern Amazonia. *Biological Conservation, 124*, 383–396. doi:10.1016/j.biocon.2005.01.045

Miller, G. H., Mangan, J., Pollard, D., Thompson, S., Felzer, B., & Magee, J. (2005). Sensitivity of the Australian monsoon to insolation and vegetation: implications for human impact on continental moisture balance. *Geology, 33*, 65–68. doi:10.1130/G21033.1

Miller, M. (1994). Large African herbivores, bruchid beetles and their interactions with Acacia seeds. *Oecologia, 97*(2), 265–270. doi:10.1007/BF00323159

Miller-Rushing, A., & Primack, R. B. (2008). Global warming and flowering times in Thoreau's Concord: A community perspective. *Ecology, 89*, 332–341. doi:10.1890/07-0068.1

Mortenson, H. S., Dupont, Y. L., & Olesen, J. M. (2008). A snake paradise: Disturbance of plant reproduction following extirpation of bird flower-visitors on Guam. *Biological Conservation, 141*(8), 2146–2154. doi:10.1016/j.biocon.2008.06.014

Myers, N., Mittermeier, R. A., Mittermeier, C. G., da Fonseca, G. A. B., & Kent, J. (2000). Biodiversity hotspots for conservation priorities. *Nature, 403*, 853–858. doi:10.1038/35002501

Naeem, S., & Li, S. (1997). Biodiversity enhances ecosystem reliability. *Nature, 390*, 507–509. doi:10.1038/37348

Nascimiento, H. E. M., Laurance, W. F., Condit, R., Laurance, S., D'Angelo, S., & Andrade, A. C. (2005). Demographic and life-history correlates for Amazonian trees. *Journal of Vegetation Science, 16*, 625–634. doi:10.1111/j.1654-1103.2005.tb02405.x

Ng, P. K. L., Tay, J. B., & Lim, K. K. P. (1994). Diversity and conservation of blackwater fishes in Peninsular Malaysia, particularly in the North Selangor peat swamp forest. *Hydrobiologia, 285*, 203–218. doi:10.1007/BF00005667

Nichol, J. (1997). Bioclimatic impacts of the 1994 smoke haze event in Southeast Asia. *Atmospheric Environment, 31*(8), 1209–1219. doi:10.1016/S1352-2310(96)00260-9

Nieder, J., Prospera, J., & Michaloud, G. (2001). Epiphytes and their contribution to canopy diversity. *Plant Ecology, 153*, 51–63. doi:10.1023/A:1017517119305

O'Dowd, D. J., Green, P. T., & Lake, P. S. (2003). Invasional meltdown on an oceanic island. *Ecology Letters, 6*(9), 812–817. doi:10.1046/j.1461-0248.2003.00512.x

O'Dowd, D. J., & Lake, P. S. (2009). Red crabs in rain forest, Christmas Island: Removal and relocation of leaf-fall. *Journal of Tropical Ecology, 5*, 337–348. doi:10.1017/S0266467400003746

Olson, D. M., Dinerstein, E., Wickramanayake, E. D., Burgess, N. D., Powell, G. V. N., & Underwood, E. C. (2001). Terrestrial ecoregions of the world: A new map of life on earth. *Bioscience, 51*, 933–938. doi:10.1641/0006-3568(2001)051[0933:TEOTWA]2.0.CO;2

Padmawathe, R., Qureshi, Q., & Rawat, G. S. (2004). Effects of selective logging on vascular epiphyte diversity in a moist lowland forest of Eastern Himalaya, India. *Biological Conservation, 119*(1), 81–92. doi:10.1016/j.biocon.2003.10.024

Page, S. E., Siegert, F., Rieley, J. O., Boehm, H. V., Jaya, A., & Limin, S. (2002). The amount of carbon released from peat and forest fires in Indonesia during 1997. *Nature*, *420*, 61–65. doi:10.1038/nature01131

Page, S. E., Wüst, R. A. J., Weiss, D., Rieley, J. O., Shotyk, W., & Limin, S. (2004). A record of late Pleistocene and Holocene carbon accumulation and climate change from an equatorial peat bog (Kalimantan, Indonesia): Implications for past, present and future carbon dynamics. *Journal of Quaternary Science*, *19*, 625–635. doi:10.1002/jqs.884

Palmer, C. (2010). REDD+: Property rights and liability. *Science*, *328*, 1105. doi:10.1126/science.328.5982.1105-a

Palmer, T. M., Stanton, M. L., Young, T. P., Goheen, J. R., Pringle, R. M., & Karban, R. (2008). Breakdown of an ant-plant mutualism follows the loss of large herbivores from an African savanna. *Science*, *319*, 192–195. doi:10.1126/science.1151579

Parmesan, C. (2006). Ecological and evolutionary responses to recent climate change. *Annual Review of Ecology Evolution and Systematics*, *37*, 637–669. doi:10.1146/annurev.ecolsys.37.091305.110100

Parmesan, C., Ryrholm, N., Stefanescu, C., Hill, J. K., Thomas, C. D., & Descimon, H. (1999). Poleward shifts in geographic ranges of butterfly species associated with regional warming. *Nature*, *399*, 579–583. doi:10.1038/21181

Pearce, D. W. (2001). The economic value of forest ecosystems. *Ecosystem Health*, *7*(4), 284–296. doi:10.1046/j.1526-0992.2001.01037.x

Peh, K. S. H. (2007). Potential effects of climate change on altitudinal distributions in tropical birds. *The Condor*, *109*, 437–440. doi:10.1650/0010-5422(2007)109[437:PEOCCO]2.0.CO;2

Peres, C. A. (2010). Overexploitation. In Sodhi, N. S., & Ehrlich, P. R. (Eds.), *Conservation Biology for all* (pp. 107–130). Oxford, UK: Oxford University Press. doi:10.1093/acprof:oso/9780199554232.003.0007

Phat, N. K., Knorr, W., & Kim, S. (2004). Appropriate measures for conservation of terrestrial carbon stocks – analysis of trends of forest management in Southeast Asia. *Forest Ecology and Management*, *191*, 283–299. doi:10.1016/j.foreco.2003.12.019

Phelps, J., Webb, E. L., & Agrawal, A. (2010). Does REDD+ threaten to recentralize forest governance? *Science*, *328*, 312–313. doi:10.1126/science.1187774

Phillips, V. D. (1998). Peatswamp ecology and sustainable development in Borneo. *Biodiversity and Conservation*, *7*, 651–671. doi:10.1023/A:1008808519096

Pimm, S. L., & Jenkins, C. N. (2010). Extinctions and the practice of preventing them. In Sodhi, N. S., & Ehrlich, P. R. (Eds.), *Conservation Biology for all* (pp. 181–198). Oxford, UK: Oxford University Press. doi:10.1093/acprof:oso/9780199554232.003.0011

Poulsen, J. R., Clark, C. J., Mavah, G., & Elkan, P. W. (2009). Bushmeat supply and consumption in a tropical logging concession in Northern Congo. *Conservation Biology*, *23*(6), 1597–1608. doi:10.1111/j.1523-1739.2009.01251.x

Pounds, J. A., Bustamante, M. R., Coloma, L. A., Consuegra, J. A., Fogden, M. P. L., & Foster, P. N. (2006). Widespread amphibian extinctions from epidemic disease driven by global warming. *Nature*, *439*, 161–167. doi:10.1038/nature04246

Pounds, J. A., Fodgen, M. P. L., & Campbell, J. H. (1999). Biological response to climate change on a tropical mountain. *Nature*, *398*, 611–615. doi:10.1038/19297

Power, M. E., Tilman, D., Estes, J. A., Menge, B. A., Bond, W. J., & Mills, L. S. (1996). Challenges in the quest for keystones. *Bioscience*, *46*, 609–620. doi:10.2307/1312990

Prance, G. T., Beentje, H. J. D., & Johns, R. (2000). The tropical flora remains undercollected. *Annals of the Missouri Botanical Garden*, *87*, 67–71. doi:10.2307/2666209

Pringle, R. M. (2008). Elephants as agents of habitat creation for small vertebrates at the patch scale. *Ecology*, *89*(1), 26–33. doi:10.1890/07-0776.1

Prins, H. T. P., & van der Jeugd, H. (1993). Herbivore population crashes and woodland structure in East Africa. *Journal of Ecology*, *81*, 305–314. doi:10.2307/2261500

Rieley, J. O., Page, S. E., & Shepherd, P. A. (1997). Tropical bog forests of South East Asia. In Stoneman, L. P. R. E., & Ingram, H. A. P. (Eds.), *Conserving peatlands* (pp. 35–41). Wallingford, UK: CAB International.

Riginos, C., & Grace, J. B. (2008). Savanna tree density, herbivores, and the herbaceous community: Bottom-up versus top-down effects. *Ecology*, *89*, 2228–2238. doi:10.1890/07-1250.1

Ripple, W. J., & Larson, E. J. (2000). Historic aspen recruitment, elk, and wolves in northern Yellowstone National Park, USA. *Biological Conservation*, *95*(3), 361–370. doi:10.1016/S0006-3207(00)00014-8

Rudel, T. K., Coomes, O. T., Moran, E., Achard, F., Angelsen, A., Xu, J. C., & Lambin, E. (2005). Forest transitions: Towards a global understanding of land use change. *Global Environmental Change*, *15*, 23–31. doi:10.1016/j.gloenvcha.2004.11.001

Santilli, M., Moutinho, P., Schwartzman, S., Nepstad, D., Curran, L., & Nobre, C. (2005). Tropical deforestation and the Kyoto Protocol. *Climatic Change*, *71*, 267–276. doi:10.1007/s10584-005-8074-6

Schuur, E. A. G., Bockheim, J., Canadell, J. G., Euskirchen, E., Field, C. B., & Goryachkin, S. V. (2008). Vulnerability of permafrost carbon to climate change: Implications for the global carbon cycle. *Bioscience*, *58*(8), 701–714. doi:10.1641/B580807

Schwartz, M. W., Brigham, C. A., Hoeksema, J. D., Lyons, K. G., Mills, H., & van Mantgem, P. J. (2000). Linking biodiversity to ecosystem function: Implications for conservation biology. *Oecologia*, *122*(3), 297–305. doi:10.1007/s004420050035

Sekercioglu, C. H. (2006). Conservation ecology: Area trumps mobility in fragment bird extinctions. *Current Biology*, *17*, R283–R286. doi:10.1016/j.cub.2007.02.019

Sekercioglu, C. H. (2010). Ecosystem functions and services. In Sodhi, N. S., & Ehrlich, P. R. (Eds.), *Conservation Biology for all* (pp. 45–72). Oxford, UK: Oxford University Press. doi:10.1093/acprof:oso/9780199554232.003.0004

Sekercioglu, C. H., Ehrlich, P. R., Daily, G. C., & Aygen, D. Goehring, D., & Sandi, R. F. (2002). Disappearance of insectivorous birds from tropical fragments. *Proceedings of the National Academy of Sciences of the United States of America, 99*, 263-267.

Sigman, D. M., & Boyle, E. A. (2000). Glacial/interglacial variations in atmospheric carbon dioxide. *Nature*, *407*, 859–869. doi:10.1038/35038000

Simberloff, D. (2010). Invasive species. In Sodhi, N. S., & Ehrlich, P. R. (Eds.), *Conservation Biology for all* (pp. 131–148). Oxford, UK: Oxford University Press. doi:10.1093/acprof:oso/9780199554232.003.0008

Simpson, G. G. (1964). Species density of North American recent mammals. *Systematic Zoology*, *13*, 57–73. doi:10.2307/2411825

Smith, A. P., & Quin, D. G. (1996). Patterns and causes of extinction and decline in Australian conilurine rodents. *Biological Conservation, 77*, 243–267. doi:10.1016/0006-3207(96)00002-X

Sodhi, N. S., & Brook, B. W. (2006). *Southeast Asian biodiversity in crisis*. Cambridge, UK: Cambridge University Press.

Sodhi, N. S., Brook, B. W., & Bradshaw, C. J. A. (2007). *Tropical conservation biology*. Singapore: Blackwell Publishing.

Sodhi, N. S., Koh, L. P., Brook, B. W., & Ng, P. L. K. (2004). Southeast Asian biodiversity: An impending disaster. *Trends in Ecology & Evolution, 19*(12), 654–660. doi:10.1016/j.tree.2004.09.006

Solomon, S., Plattner, G., Knutti, R., & Friedlingstein, P. (2009). Irreversible climate change due to carbon dioxide emissions. *Proceedings of the National Academy of Sciences of the United States, 106*(6), 1704–1709. doi:10.1073/pnas.0812721106

Sorensen, K. W. (1993). Indonesian peat swamp forests, and their role as a carbon sink. *Chemosphere, 27*(6), 1065–1082. doi:10.1016/0045-6535(93)90068-G

Still, C. J., Foster, P. N., & Schneider, S. H. (1999). Simulating the effects of climate change on tropical montane cloud forests. *Nature, 398*, 608–610. doi:10.1038/19293

Stouffer, P. C., & Bierregaard, R. O. (1995). Use of Amazonian forest fragments by understorey insectivorous birds. *Ecology, 76*, 2429–2445. doi:10.2307/2265818

Takakai, F., Morishita, T., Hashidoko, Y., Darung, U., Kuramochi, K., & Dohong, S. (2006). Effects of agricultural land-use change and forest fire on N_2O emission from tropical peatlands, Central Kalimantan, Indonesia. *Soil Science and Plant Nutrition, 52*, 662–674. doi:10.1111/j.1747-0765.2006.00084.x

Terborgh, J., Feeley, K., Silman, M., Nunez, P., & Balukjian, B. (2006). Vegetation dynamics of predator-free land-bridge islands. *Journal of Ecology, 94*, 253–263. doi:10.1111/j.1365-2745.2006.01106.x

Terborgh, J., Lopez, L., Nunez, P., Rao, M., Shahabuddin, G., & Orihuela, G. (1923-1926). Balbas, L. (2001). Ecological meltdown in predator-free forest fragments. *Science*, ▪▪▪, 294.

The Economist. (2010a). *Russia's heat wave. A hazy shade of summer. A potent mix of heat, haze, alcohol and corruption.* The Economist, July 29th 2010. Retrieved August 3rd, 2010 from http://www.economist.com/node/16703362

The Economist. (2010b). *Flood of miseries.* The Economist, August 4th 2010. Retrieved August 5th, 2010, from http://www.economist.com/node/21009205

The Economist. (2010c). *The gulf oil spill: No end in sight. High tempers and low expectations in the Gulf of Mexico.* The Economist, June 10th 2010. Retrieved August 2nd, 2010, from http://www.economist.com/node/16322752

Thomas, C. D., Cameron, A., Green, R. E., Bakkenes, M., Beaumont, L. J., & Collingham, Y. C. (2004). Extinction risk from climate change. *Nature, 427*, 145–148. doi:10.1038/nature02121

Tilman, D. (1997). Biodiversity and ecosystem functioning. In Daily, G. C. (Ed.), *Nature's services: Societal dependence on natural ecosystems* (pp. 93–112). Washington, DC: Island Press.

Tobler, M. W., Cochard, R., & Edwards, P. J. (2003). The impact of cattle ranching on large-scale vegetation patterns in a coastal savanna in Tanzania. *Journal of Applied Ecology, 40*, 430–444. doi:10.1046/j.1365-2664.2003.00816.x

Uhl, C., & Viera, I. C. G. (1989). Ecological impacts of selective logging in the Brazilian Amazon: A case study from the Paragominas region of the state of Para. *Biotropica, 21*, 98–106. doi:10.2307/2388700

UNEP. (2007). *Global outlook for ice and snow*. Nairobi, Kenya: United Nations Environment Programme.

Usher, A. D. (2009). *Thai forestry. A critical history*. Chiang Mai, Thailand: Silkworm Books.

Vallan, D., Andreone, F., Raherisoa, V. H., & Dolch, R. (2004). Does selective wood exploitation affect amphibian diversity? The case of An'Ala, a tropical rainforest in eastern Madagascar. *Oryx, 38*, 410–417. doi:10.1017/S003060530400078X

van Andel, T. H., Zangger, E., & Demitrack, A. (1990). Land use and soil erosion in prehistoric and historical Greece. *Journal of Field Archaeology, 17*(4), 379–396. doi:10.2307/530002

van der Werf, G. R., Morton, D. C., DeFries, R. S., Olivier, J. G. J., Kasibhatla, P. S., & Jackson, R. B. (2009). CO_2 emissions from forest loss. *Nature Geoscience, 2*, 737–738. doi:10.1038/ngeo671

van Houtan, K. S., Pimm, S. L., Halley, J. M., Bierregaard, R. O., & Lovejoy, T. E. (2007). Dispersal of Amazonian birds in continuous and fragmented forest. *Ecology Letters, 10*(3), 219–229. doi:10.1111/j.1461-0248.2007.01004.x

van Mantgem, P. J., & Stephenson, N. L. (2007). Apparent climatically induced increase of tree mortality rates in a temperate forest. *Ecology Letters, 10*(10), 909–916. doi:10.1111/j.1461-0248.2007.01080.x

van Schaik, C. P., Monk, K. A., & Robertson, J. M. Y. (2001). Dramatic decline in orangutan numbers in the Leuser Ecosystem, Northern Sumatra. *Oryx, 35*, 14–25.

Volney, W. J. A., & Fleming, R. A. (2000). Climate change impacts of boreal forest insects. *Agriculture Ecosystems & Environment, 82*, 283–294. doi:10.1016/S0167-8809(00)00232-2

Vors, L. S., & Boyce, M. S. (2009). Global declines of caribou and reindeer. *Global Change Biology, 15*, 2626–2633. doi:10.1111/j.1365-2486.2009.01974.x

Walker, B. (1995). Conserving biological diversity through ecosystem resilience. *Conservation Biology, 9*(4), 747–752. doi:10.1046/j.1523-1739.1995.09040747.x

Walther, G. (2003). Plants in a warmer world. *Perspectives in Plant Ecology, Evolution and Systematics, 6*(3), 169–185. doi:10.1078/1433-8319-00076

Walther, G., Post, E., Convey, P., Menzel, A., Parmesan, C., & Beebee, T. J. C. (2002). Ecological responses to recent climate change. *Nature, 416*, 389–395. doi:10.1038/416389a

Webster, P. J., Holland, G. J., Curry, J. A., & Chang, H.-R. (2005). Changes in tropical cyclone number, duration, and intensity in a warming environment. *Science, 309*(5742), 1844–1846. doi:10.1126/science.1116448

Welton, L. J., Siler, C. D., Bennet, D., Diesmos, A., Duya, M. R., & Dugay, R.... Brown, R. M. (2010). A spectacular new Philippine monitor lizard reveals a hidden biogeographic boundary and a novel flagship species for conservation. *Biology Letters*. doi:.doi:10.1098/rsbl.2010.0119

Whitmore, T. C. (1995). Comparing South East Asian and other tropical rainforests. In Primack, R. E., & Lovejoy, T. E. (Eds.), *Ecology, conservation and management of South East Asian rainforests* (pp. 5–15). New Haven, CT: Yale University Press.

Whittaker, R. H. (1975). *Communities and ecosystems*. New York, NY: Macmillan.

Whitten, T., Damanik, S. J., Anwar, J., & Hisyam, N. (1997). *The ecology of Sumatra*. Jakarta, Indonesia: Periplus.

Wich, S. A., Singleton, I., Utami-Atmoko, S. S., Geurts, M. L., Rijksen, H. D., & van Schaik, C. P. (2003). The status of the Sumatran orangutan *Pongo abelii*: An update. *Oryx*, *37*(1), 49–54. doi:10.1017/S0030605303000115

Wösten, J. H. M., Ismail, A. B., & van Wijk, A. L. M. (1997). Peat subsidence and its practical implications: A case study in Malaysia. *Geoderma*, *78*, 25–36. doi:10.1016/S0016-7061(97)00013-X

Wright, J. A., DiNicola, A., & Gaitan, E. (2000). Latin American forest plantations: Opportunities for carbon sequestration, economic development, and financial returns. *Journal of Forestry*, *98*(9), 20–23.

Wright, S. J. (2002). Plant diversity in tropical forests: A review of mechanisms of species coexistence. *Oecologia*, *130*, 1–14.

Wright, S. J., Muller-Landau, H. C., & Schipper, J. (2009). The future of tropical species on a warmer planet. *Conservation Biology*, *23*(6), 1418–1426. doi:10.1111/j.1523-1739.2009.01337.x

Xiubin, H., Tang, K., & Zhang, X. (2004). Soil erosion dynamics on the Chinese Loess Plateau in the last 10'000 years. *Mountain Research and Development*, *24*(4), 342–347. doi:10.1659/0276-4741(2004)024[0342:SEDOTC]2.0.CO;2

Yahner, R. H., & Mahan, C. G. (1997). Effects of logging roads on depredation of artificial ground nests in a forested landscape. *Wildlife Society Bulletin*, *25*(1), 158–162.

Yu, X., Gillespie, A., Burns, W., & Taplin, R. (Eds.). (2006). *Climate change in the South Pacific: Impacts and responses in Australia, New Zealand, and small island states*. Dordrecht, Netherlands: Springer.

Zachos, J., Pagani, M., Sloan, L., Thomas, E., & Billups, K. (2001). Trends, rhythms, and aberrations in global climate 65 Ma to present. *Science*, *292*(5517), 686–693. doi:10.1126/science.1059412

Zheng, N., & Yoon, J. (2009). Expansion of the world's deserts due to vegetation-albedo feedback under global warming. *Geophysical Research Letters*, *36*(L17401).

Zhou, C., Wei, X., Zhou, G., Yan, J., Wang, X., & Wang, C. (2008). Impacts of a large-scale reforestation program on carbon storage dynamics in Guangdong, China. *Forest Ecology and Management*, *255*(3-4), 847–854. doi:10.1016/j.foreco.2007.09.081

ZVS. (2010). *Zürcher Vogelatlas*. Retrieved August 9th, 2010, from http://www.birdlife-zuerich.ch/vogelfinder/atlas-nach-vogelarten.html

Chapter 3
Geo-Informatics for Land Use and Biodiversity Studies

P. K. Joshi
TERI University, India

Neena Priyanka
TERI University, India

ABSTRACT

The dynamics of land use/land cover (LU/LC) is a manifestation of the cyclic correlation among the kind and magnitude of causes, impacts, responses and resulting ecological processes of the ecosystem. Thus, the holistic understanding of the complex mechanisms that control LU/LC requires synergetic adoption of measurement approaches, addressing issues, and identifying drivers of change and state of art technologies for mitigation measures. As the spatio-temporal heterogeneity of the LU/LC increases, its impact on biodiversity becomes even more difficult to anticipate. Thus, in order to understand the spatio-temporal dynamics of change in landscape and its relationship to biodiversity, it is necessary to reliably identify and quantify the indicators of change. In addition, it is also important to have better understanding of the technologies and techniques that serve as complimentary tool for land mitigation and conservation planning. Against this background, the chapter aims to synthesize LU/LC studies worldwide and their impacts on biodiversity. This chapter explores identification and analysis of key natural, socio-economic and regulatory drivers for LU/LC. Finally, it attempts to collate some LU/LC studies involving usage of geospatial tools, such as satellite remote sensing, Geographic Information System (GIS), Global Positioning System (GPS), and integrative tools, besides conventional approaches that could assist decision makers, land managers, stakeholders and researchers in better management and formulation of conservation strategies based on scientific grounds.

DOI: 10.4018/978-1-60960-619-0.ch003

1. INTRODUCTION

Land use (LU) term entails the manner in which human beings employ the land and its resources (Ramachandra & Kumar, 2004; GLP, 2005; Castella *et al.*, 2007; Encyclopedia of Earth, 2007) whereas Land cover (LC) implies the physical or natural state of the Earth's surface (GLP, 2005; Castella *et al.*, 2007; Encyclopedia of Earth, 2007). The change in both LU and LC is intertwined with multiple socio-economic issues such as loss of biodiversity (Murthy *et al.*, 2003), sustainability of agriculture (Gordon *et al.*, 2008), provision of maintaining water and soil quality (NRCS, 2007), climate change and carbon cycle (Turner, 2004). Hence, in order to use land optimally, it is not only necessary to have the information on existing LU/LC but also the capability to monitor the dynamics of LU resulting from both changing demands of increasing population and forces of nature acting to shape the landscape. Conventional ground methods based on sampling techniques of LU/LC mapping are labor intensive, time consuming and are done relatively infrequently and thus become outdated rather soon with the passage of time, particularly in a rapid changing environment. In fact monitoring changes and time series analysis is quite difficult with traditional methods of surveying. In recent years, technologies such as satellite remote sensing, Global Positioning System (GPS), and integrative tools, such as Geographical Information System (GIS) and information systems, together form the basis for Geo-informatics, facilitate the synoptic analyses and monitoring of such dynamic land system function, pattern, and change at local, regional and global scales over time (Lee *et al.*, 1999; Sedano *et al.*, 2005; Navalgund *et al.*, 2007; Roy & Giriraj, 2008). The data assembled using such techniques also provide an important link between intensive, localized ecological research and regional, national and international conservation and management of biological diversity (Roy & Tomar, 2000, Sharma *et al.*, 2008). In case of inaccessible regions, these techniques are perhaps the best methods for obtaining the required data in a cost effective and efficient way.

Information on LU/LC at various scales is found in a widely scattered literature, some of it refereed and some in other grey literature and others unpublished as yet. Although information is incomplete globally, several products are now available that depict LU/LC scenarios worldwide (Global Land Cover Network, 2000; GLP, 2005; IGBP, 2007). A similar condition holds true for regional analysis whereby snapshots of many important regions with substantial LU/LC have been developed, for example, in Russia, South America and Africa, parts of East Asia and Southeast Asia, and the continental US and Canada for future sustainable planning and management of their land (Corves & Place, 1994; Cohen *et al.*, 2003; FAO, 2004; GEO, 2005). There are numerous instances of studies and resultant databases of rapid LU/LC and ecosystem disturbances at local scales in many parts of the world: deforestation and fragmentation in the pan-tropical forest belt, fire frequency in parts of South America, Southern Africa, and parts of Russia, influence of urbanization in selected cities worldwide, biodiversity assessment in parts of India etc (Roy & Tomar, 2000; Roy & Giriraj, 2008; Sharma *et al.*, 2008). In addition, there have been regions with concomitant rapid expansion of the availability of data and information but there has not yet been a systematic assessment of the status and trends in LU/LC of terrestrial, coastal or other ecosystem processes (Townshend *et al.*, 1991; Lambin *et al.*, 2003; Lillesand *et al.*, 2007). Henceforth, there is an apparent need to determine the inter-relationships between inventory data, geo-informatics and statistics and therefore synthesizing information about LU/LC augmented by indicators of condition, status, and trends of the change. In addition to the scientific needs for a systematic documentation of LU/LC from past to present for understanding the current state and potential future, there is a pressing need to understand these changes from the standpoint of their consequences for ecosystem sustenance and human welfare.

In purview of the above, for holistic understanding of the complex mechanisms that control LU/LC as well as their spatial and temporal dynamics, several initiatives have been launched, at both global and local scales for assessment, preservation, management and sustenance of LU/LC. At International level, efforts include the activity of the National Land Cover Data (NLCD), United Nations Environment Programme (UNEP), United Nations Conference on Environment and Development (UNCED), Global Land Cover Facility (GLCF), World Conservation Monitoring Centre (WCMC); Global Change and Terrestrial Ecosystem (GCTE), World Health Organization (WHO), World Research Institute (WRI), Food and Agricultural Organization (FAO) etc. At national levels, for example, in India, the Ministry of Environment and Forest (MoEF), National Remote Sensing Center (NRSC), Wildlife Institute of India (WII), Ashoka Trust for Research in Energy and Environment (ATREE), The Energy and Resources Institute (TERI), Botanical Survey of India (BSI), Forest Survey of India (FSI) as well as Universities and Non-Governmental Organizations (NGOs) are taking initiatives in launching several programs for ecosystem assessment and conservation of land at large. The investigations conducted have confirmed the importance of the contributions and interdisciplinary understanding of LU/LC patterns. Towards this, it is necessary that the land management apply a synoptic view on the indicators and the tools through which the processes could be understood. Against this backdrop, the present section attempts to synthesize studies based on general overviews of LU/LC, its implications to biodiversity, drivers of LU/LC change, current and potential state of art of Geo-informatics in LU/LC assessment, and role of spatial modeling and SDI for better understanding and synergetic adoption of measurement approaches for LU/LC. This will facilitate people's awareness of fundamental and core advantages and issues of application of geospatial tools for LUCC studies.

2. LAND USE/LAND COVER CHARACTERIZATION

Earth ecosystems are being dominated by humans with spectacular impacts on LU/LC patterns and biodiversity (Vitousek *et al.*, 1997). For example, LU/LC over the past 50–100 years contributed to significant changes in local-to-global climate conditions, soil and water degradation, habitat alteration and fragmentation, loss of biotic diversity, decline in ecosystem health and functioning and consequently decline of the environment (Matson *et al.*, 1997; Lambin *et al.*, 2001; Gordon *et al.*, 2008).

Geo-informatics has since the mid 1970's made an imperative contribution in characterization of LU/LC and illustrating the change in the ecosystems at global and local spatial scales (DeFries *et al.*, 2002; Achard *et al.*, 2002; Lambin *et al.*, 2003). The technologies and methods herein have evolved dramatically to include a suite of sensors operating at a wide range of imaging scales with potential interest and importance to planners and land managers. Coupled with the readily available GIS data, the reduction in data cost and increased resolution from satellite platforms, Geo-informatics is poised to make an even greater impact on planning agencies and land management initiatives involved in monitoring LU/LC at varied spatio-temporal scales. Beside this, it has played pivotal role in evaluating indices of change in ecosystem processes and functions such as biodiversity assessment, habitat analysis, monitoring soil and water use efficiency, assessing forest cover change, ecosystem disturbances, such as fire and disease outbreak, and assessing impacts of regional and global climate through the surface-energy budget.

The impact of LU/LC is henceforth unparalleled in its combination of spatial extent and intensity of influence. Moreover, local alteration of LU/LC can have global consequences that require local and regional solutions for better LU policy, projection of transportation and utility demand, identification of future development pressure points and areas, implementation of effective plans for regional development, as well as the cooperation of the world's legislators, planners, state and local governmental officials, land managers, and other stakeholders in land management at local, regional and global scales (Encyclopedia of Earth, 2007). A documentation of global patterns of LU change from 1700 to 2000 is presented in Goldewijk (2001). The author reported worldwide changes of other land uses to crops of 136, 412 and 658 million ha in the periods 1700-1799, 1800-1899 and 1900-1990, respectively depicting an acceleration of tropical deforestation during the 20th century. One of the prime prerequisites for better LU/LC characterization is to extract information on existing LU patterns and changes in LU through time. Knowledge of the present distribution and area of such agricultural, recreational, and urban lands, as well as information on their changing proportions, is needed by legislators, planners, and state and local governmental officials to determine better LU policy, to project transportation and utility demand, to identify future development pressure points and areas, and to implement effective plans for regional development. As Clawson & Stewart (1965) have stated "*in this dynamic situation, accurate, meaningful, current data on land use are essential. If public agencies and private organizations are to know what is happening, and are to make sound plans for their own future action, then reliable information is critical*".

2.1 Drivers of LU/LC

Changes in LU/LC have first occurred with the burning of areas to enhance the availability of wild game and accelerated dramatically with the origin of agriculture, resulting in extensive clearings (deforestation) and management of earth's terrestrial surface. More recently, industrialization, accompanied by the intensification of urbanization, has facilitated the ongoing process of change whose causes and their consequences are observable simultaneously throughout the world. Most LU/LC occur at the local scale and henceforth the driving forces appear to be basically revolving around social, economic and political systems as illustrated in Figure 1 (Baudry & Thenail, 2004).

Figure 1. Multi system driven LULCC

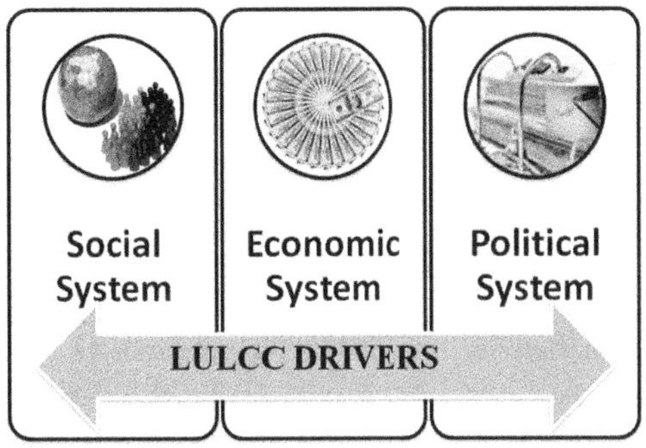

Amongst all, the LU/LC are driven mainly by multi-scale driving forces (Table 1 lists some of the LU/LC drivers) including local societal preferences and practices (food, farming, livelihood etc.), the global economy (demand for specific products, financial incentives), environmental conditions (soil quality, terrain, moisture availability), land policies, various development programs (agricultural programs, road building, zoning, construction), and feedbacks between these factors, including past human activity on the land (degradation, irrigation, forest fragmentation, and deforestation) (Veldkamp *et al.*, 2001; Lambin & Geist, 2006). Investigation of these drivers of LU/LC requires a full range of methods from the natural and social sciences, including climatology, soil science, ecology, environmental science, hydrology, geography, information systems, computer science, anthropology, sociology, and policy science.

Table 1. Proximate drivers of LU/LC

LU/LC Drivers	Illustrations
Population growth	As a region's population grows, the new residents need housing, as well as places to work and shop. In a region with declining accelerating population, there will be more new construction of homes and business centers to satisfy the demands.
Economic growth	A booming regional economy will result in construction of new commercial and industrial buildings to house that activity. As the economy grows, the new jobs created will attract workers, leading to population growth, leading to construction of new homes and places to shop. As incomes rise, household may choose to build new larger homes on larger lots, leaving smaller, older houses vacant.
Demographics	The average number of people living in a household has been decreasing over time. Therefore, more housing units are needed to house the same number of people. The number of retired households is increasing, and these households tend to have few members. Meanwhile, the proportion of non-white households is also increasing. These households tend to have more members on average than white households.
Agricultural and forest products prices	A change in the price of agricultural or forest products can affect landowners' decisions whether to keep the land in those uses. Policies aimed at supporting agricultural prices provide an incentive to keep land in farming.
Regional and local planning and policies	Regions can influence the rate at which land use and land cover change through a variety of means.

2.2 Summarization (Issues, Controversies and Problems)

Most data sources do not use on standard definitions of LU/LC, even though some definitions are more commonly accepted. For example, more than 50 different definitions of LU/LC are proposed and are in common use throughout the world, complicating the effort to measure and evaluate the change data. LU/LC datasets are not innumerable as data acquisition through satellite based remote sensing is usually very expensive and the classification process is very labour intensive. Moreover, most LU/LC data products are released several years after the satellite images were taken, and thus out of date to a certain extent. Subtle LU/LC are often disregarded, as the changes are quantitatively not significant so their impacts are often underestimated and neglected. Also, their detection from remotely sensed data is difficult and boundary line segmentation is cumbersome (limited spatial extent, hidden by another land cover, etc.). Few LU/LC studies are carried out at fine scales such as the local scale, whereas most of them are conducted at landscape or regional scales using pixels of various sizes as units of observations. Fine-scale analyses of LC change is required to refine the result in order to anticipate where changes are more likely to occur and proposing and implementing local sustainable environmental policies and efficient in-situ action.

3. LU/LC AND BIODIVERSITY

3.1 Biodiversity Studies

World's ecosystems are in a state of constant change at various spatio-temporal scales by a variety of socio-economic and environmental drivers. The changes today are more extensive and occur at more rapid rate on facets of landscape systems than ever before. The ramifications of these changes have become very apparent as these have altered ecosystem functioning and have resulted in LU/LC. The signs of these changes include biodiversity loss, degradation of water and soil resources, and destruction of habitats etc. Biodiversity and the land ecosystem are intertwined in each other i.e. change in one results in change of the other. Therefore, biodiversity characterization at landscape level is an important requirement for landscape conservation planning and vice versa. The recent developments in the field of remote sensing with its wide spectrum of sensor systems, provide an opportunity to gather information on biodiversity in spatio-temporal domain, varied spatial resolution and scales, quantization levels and spectral resolution enable precise and accurate measurements of change. The advancement also encompasses various application areas, information extraction techniques, multi-thematic information analysis and geospatial modeling for characterization of elements of biodiversity. The importance of biodiversity characteristics in a particular area should be realized before the implementation of conservation measures.

3.2 Regulators

Biodiversity generally refers to vast diversification of flora and fauna from all ecosystems viz. terrestrial, marine and other aquatics and their interactions in the ecological complexes of which they are a part. Biodiversity can also be defined as varieties of genes, species, ecosystems and habitats in a region that has evolved through millions of years of evolutionary history. In this context, inventorying and monitoring of biodiversity should be carried out at different organizational levels from genes to ecological systems (landscapes), and at different spatial scales from local to global for better characterization. The recognition and characterization of biodiversity depends critically on taxonomy, genetics and ecological systems (landscapes). Amongst these, ecological systems incorporate knowledge of the varied landscape systems together with taxonomic and genetic diversity. Landscape elements such as patch sizes, patch shapes, patch isolation, interspersion (adjacency of various LU/LC), juxtaposition (relative importance of adjacent patches), fragmentation, patchiness, etc. have direct bearing on the status of biodiversity and are found to be useful for generating scientific grounds and understanding of biodiversity characterization. Based on these parameters of ecological systems, prospecting of regions for conservation, planning, prioritization, resource mobilization and sustenance of the ecosystem can be carried out with the integration of landscape ecology and geospatial tools and techniques.

Anthropogenic activities coupled with the burgeoning human population, have led to the grim biodiversity scenario. It is, in order to bring about sustainable resource conservation and management, essential to adopt several different approaches for managing the ecosystem and biodiversity. To arrest the process of degradation and species-loss requires specialized solutions and an understanding of ecological processes besides information resources pertaining to forests, biodiversity – flora and fauna, causative factors for their degradation, and major threats. The available data are alarmingly inadequate to provide a lucid picture of the current status and ongoing losses/gains. Today, there is a shift from broad inventory surveys due to the high cost and impracticality of such an approach. Instead, there is much interest in techniques that can predict species occurrence, habitat type and genetic impacts with the help of spatial and temporal tools – Geographic Information System (GIS) and Remote Sensing (RS). This would ensure in establishment of associations groups of species with different landscape elements on the basis of field surveys. Temporal data (i.e. the data acquired at different time periods for the same location) help in assessing rate of transformations of habitat and the threats to different species as a result of ongoing landscape changes. However, such tools and techniques suffer the problem of the scale and resolution at which the study needs to be carried out and problems of availability and cost.

4. GEO-INFORMATICS

The complex mechanisms controlling LU/LC and biodiversity and their spatio-temporal dynamics, require synergetic adoption of measurement approaches, sampling designs and technologies.

The satellite remote sensing, GPS, and integrative tools (such as GIS and information systems) are important complimentary systems for such studies (Table 2). Together these technologies form the basis for geo-informatics. Satellite remote sensing technique is useful in characterization, mapping and monitoring the LU/LC in spatio-temporal domains in a cost effective and unbiased way (Lillesand & Kieffer, 2007). Furthermore, remote sensing technology in association with GIS and GPS is able to address many such issues pertaining to the management of natural resources and environment (Burrough, 2000). Ground-based measurements of various parameters are vital inputs for identification/cross-validation of geographical locations and positions of even smaller units of features on satellite imagery. However, the translation of the ground-measured data into a spatial domain, or the linking with any other spatial or non-spatial parameters to analyze the relationships and understand the trends, precise locations, and areas under consideration are of primary importance. The improvement in accuracy and precision of GPS data and availability

Table 2. Tools for LU/LC and biodiversity characterization

Scale	LU/LC Studies	Sensor	Spatial Resolution	Temporal revisit	References
Global	Land cover degradation; Land use mapping, Land cover Mapping, LU/LC; Change detection, LU/LC Modelling	MODIS	0.5km, 1.0km	~daily	Justice *et al.* 1998; Friedl *et al.*, 2002; Cohen *et al.*, 2003; Herold, 2000; Gonclaves *et al.*, 2006.
		AVHRR	10km 1.0km	~daily	Matthews, 1983; DeFries *et al.*, 1995; Foody *et al.*, 1996; GLCN, 2000; Lambin *et al.*, 2003; FAO, 2004; Lee & Olson, 2004; Loveland *et al.*,2000; Lillesand *et al.*, 2007
Regional/ National	Monitoring LU/LC in Urban - Rural Fringe; LU/LC Classification; Hazard study; Change detection; LU/LC Mapping; Spatial segregation of LU/LC types under shifting agriculture; Drought Monitoring; Irrigated Area Mapping; Forest Cover Change; LU/LC Modelling, Wetland Loss estimation, Modeling Forest change; Phenology assessment	MODIS	0.5km	~daily	Gopal *et al.*, 2002; Brown *et al.*, 2000; Zhan *et al.*, 2000; Knight *et al.*, 2006.
		AVHRR	1.0km	~dialy	Kerber, 1986; Roberts *et al.*, 2003; Gutam *et al.*, 2004; Fraser *et al.*, 2005.
		Landsat (MSS, TM, ETM+)	30m	18 days	Somporn., 1995; Mickelson *et al.*, 1998; Vogelmann *et al.*,2001; Yang *et al.*,2001;Dymond, *et al.*, 2002; Reese *et al.*, 2002; Homer *et al.*, 2004; Emch *et al.*, 2005; Yemefack, *et al.*, 2006; McRoberts & Tomppo, 2007; Reis, 2008; Tan *et al.*, 2009.
Local	Urban Expansion, City Planning; LU/LC change detection, Encroachment Analysis, LU/LC assessment based on Multi criteria Decision tree, Demographics profile and LU/LC assessment; LU/LC characterization and mapping in Riparian Zones; Impervious Surface Area Mapping	LISS III and LISS IV	23.5km	23 days	Chaurasia *et al.*, 1996; Kunwar *et al.*, 2001; Jayakumar & Arockiasmy, 2003; Auch *et al.*, 2004; Mundia & Aniya, 2005.
		ASTER	15m, 30m, 90m	16 days	Zsuzsanna *et al.*, 2005; Moller & Blaschke, 2005, Xian & Crane, 2005;
		SPOT	10m, 20m	26 days	Stefanov *et al.*, 2001; Pastor & Wolter, 2002
		IKONOS	1m, 4m	16 days	Thenkabail *et al.*, 2003; Thenkabail, 2004; Nichol & Wang, 2007; Jain, 2007.

of high resolution satellite data and GIS technologies, are certainly going to provide better results for mankind as they empower land managers, stakeholders, researchers and decision makers to expeditiously acquire, store, analyze, and display spatial data on LU/LC and biodiversity (Johnston *et al.*, 2007; Wadsworth & Treweek, 1999).

5. SPATIAL MAPPING – MODELING

5.1 LU/LC Analysis

Worldwide LU/LC is one of the most important ecological tools. The situation is particularly apparent in regions/countries based on rural economies and relying heavily upon the natural resources (e.g. for food, fodder, fuel wood, commodity exports). Degradation of these resources can result in rapid decline in the socio-economic profile of the community. There are certain drivers that bring about change in the facets of landscape such as demographic pressures, international economies, industrialization, mechanized agriculture and accelerating developmental activities. These along with ancillary forces are bringing changes to LU practices that are often inappropriate, causing degradation that are not irrecoverable and are unsustainable in the long term. In addition landscape change also demonstrates the influence of regional and global scales phenomenon such as deforestation, desertification, erosion, loss of biodiversity and very likely global climate change among other types of changes. Landscape phenomenon together with drivers of change, result in reduced resource bases leading to regional food shortages, political instability, and the humanitarian apprehension. Thus, analysis of changes in landscape and their consequences for the future availability of resources and other effects is important. In this context, of particular importance is predictions of magnitude and direction of changes in the landscape, which must be determined before any causal relationship can be postulated. Spatial modeling of LU/LC is therefore inevitable for such a study, with attention paid to incorporation of physical, biological, and social conditions (Lin *et al.*, 2009) to LU/LC models. Fine-scale analyses therefore have to be performed to better understand the LC change processes. At the same time, models of LC change have to be developed in order to anticipate where changes are more likely to occur next. Such predictive information is essential to propose and implement sustainable and efficient environmental policies. Future landscape studies can provide a framework to forecast how LU/LC changes are likely to react differently to subtle changes. Investigation of the future LU/LC studies are based on approaches that draw on coupled approaches to integrate various techniques and tools including landscape and LU/LC models, participatory analyses, and scenarios to achieve this goal. Decision makers can better adapt to uncertain conditions, if they have tools to explore alternative futures (Godet, 1986) thereby changing the world by changing the vision.

The immediate purpose of collating studies on projections of LU described in this section is to assess the techniques employed to understand future prospects of land resources at global, regional or national levels. This chapter broadly analyses the studies conducted at global, regional or national scale by organizations, the adopted classification scheme and impacts and consequences of studies on environmental and ecological processes at the landscape level. In addition the study also highlights various predictive models developed to understand LU/LC dynamics that enable characterization of spatio-temporal changes to understand landscape sustenance for future generations.

Most international research programs, e.g. International Geosphere Biosphere Programme (IGBP) (Lambin *et al.*, 1999), Millennium Ecosystem Assessment (MEA, 2003) and Global Land Cover (GLC) (GLCN, 2000) focus upon monitoring LU/LC through various levels of classification schemes (Table 3 illustrates a few glob-

Table 3. Classification scheme adopted by IGBP to study LU/LC

Global Land cover (Anderson *et al*., 1976)	IGBP Land cover (Belward, 1996)	Simple Biosphere (SiB) Model
Urban and built-up land	Evergreen needleleaf forest	Evergreen Broadleaf Trees
Dryland cropland and pasture	Evergreen broadleaf forest	Broadleaf Deciduous Trees
Irrigated cropland and pasture	Deciduous needleleaf forest	Deciduous and Evergreen Trees
Mixed dryland/irrigated cropland and pasture	Deciduous broadleaf forest	Evergreen Needleleaf Trees
Cropland/grassland mosaic	Mixed forest	Deciduous Needleleaf Trees
Cropland/woodland mosiac	Closed shrublands	Ground Cover with Trees and Shrubs
Grassland	Open shrublands	Groundcover Only
Shrubland	Woody savannas	Broadleaf Shrubs with Perennial Ground Cover
Mixed shrubland/grassland	Savannas	Broadleaf Shrubs with Bare Soil
Savanna	Grassland	Groundcover with Dwarf Trees and Shrubs
Deciduous broadleaf forest	Permanent wetlands	Bare Soil
Deciduous needleleaf forest	Croplands	Agriculture or C3 Grassland
Evergreen needleleaf forest	Urban-builtup	Persistent Wetland
Evergreen broadleaf forest	Cropland/natural vegetation mosaic	Ice Cap and Glacier
Mixed forest	Snow and ice	Water Bodies
Water bodies	Barren or sparsely vegetated	Missing Data
Herbaceous wetland	Water bodies	
Barren or sparsely vegetated	Interrupted areas (goodes homolosine projection)	
Herbaceous tundra	Missing data	
Wooded tundra		
Mixed tundra		
Bare ground tundra		
Snow or ice		
Interrupted area (goodes homolosine projection)		
Missing data		

ally recognized classification schemes adopted to study LU/LC) and data interpretation techniques. These systems satisfy three major attributes of the classification process as outlined by Grigg (1965): (1) give names to categories by simply using accepted terminology; (2) enable information to be transmitted; and (3) allow inductive generalizations to be made. The classification system is capable of further refinement on the basis of more extended and varied use. At the more generalized levels the classification system should meet the principal objective of providing a LU/LC used in planning and management activities. Attainment of the more fundamental and long-range objective of providing a standardized system of LU/LC classification for national and regional studies depends on the local specific needs and structures.

These Organizations and other regional/national institutions are working towards increasing

Table 4. Some examples of freely available LU/LC data provided by various Organizations

Organization	URL address	Study Level
FAO	http://faostat.fao.org/default.jsp?language=EN	Global
WRI	http://earthtrends.wri.org/	Global
GLCC	http://www-gvm.jrc.it/glc2000/	Global
GLC	http://edcdaac.usgs.gov/glcc/globdoc2_0.asp	Global
University of Boston	http://geography.bu.edu/landcover/index.html, http://edcdaac.usgs.gov/modis/dataproducts.asp	Global
University of Maryland	http://glcf.umiacs.umd.edu/data/	Global
PELCOM	http://www.geo-informatie.nl/projects/pelcom/public/index.htm	Continent
Africover	www.africover.org	Continent
FAO's FORIS	http://www.fao.org/forestry/site/fra/en	Continent
Corine Land Cover Database	http://www.eea.eu.int/	Country
CEReS	http://www.cr.chiba-u.jp/database.html	Country
Miombo	http://www.creaf.uab.es/miramon/mmr/examples/miombo/docs/index.htm	Country
LADA	http://lada.virtualcentre.org/pagedisplay/display.asp?section=ladahome	Country

availability of geographically referenced databases (as illustrated in Table 4). Availability of data on LU/LC data at global scale is generally much better as compared to regional and national scale. The FAO databases provide country level data on agricultural land and forests for most countries, and the GLC 2000, MOD12 (University of Boston/NASA), GLCF, GLCC, and CEReS databases also provide data on several other LU/LC categories. The World Resources Institute (WRI) provides some of these data in an analyzed form. In addition, there are several other databases such as Africover or Corine land cover databases that provide data for smaller regions. Data developed using various approaches are available at global and regional scale. One of the main sources for country level data on agricultural land and forests and other wooded land is FAO. Furthermore, the Pan- European Land Cover Monitoring (PELCOM) portal provides GLCC data in a readily available analyzed form. The Global Land Cover Facility (GLCF), which is housed at the University of Maryland, also provides earth science data and products with emphasis on determining where, how much and why LC changes around the world. In addition, LU/LC project, a program element of the International Geosphere-Biosphere Programme (IGBP) and the International Human Dimensions Programme (IHDP) on Global Environmental Change provide databases of global LC activities. The GLC 2000 database uses a flexible LCCS classification system that was developed by FAO and UNEP and adopted by the same organizations as the standard land cover classification system. LCCS is now proposed ISO standard. MOD12Q1 is a continuously updated database and provides data that is quite suitable for UNSD purposes. FAO data are, to a large extent, based on country reporting and therefore include local knowledge. All reporting countries also use, at least in principle, the same definitions. Databases such as MOD12 and GLC 2000 are, on the other hand, based on remote sensing. This greatly increases the objectivity and comparability of the data, but decreases the utilization of the knowledge of local conditions.

5.2 Predictive Modeling

These GIS databases coupled with growing interest in conducting interdisciplinary landscape-level analyses of ecological issues, have motivated the development of predictive models to project the rate and location of LU change on a pixel-by-pixel basis, by exploiting the additional information contained in spatially referenced LU data increasingly available in GIS. These models generally rely on LU data, constructed from satellite imagery or aerial photographs and combined with other spatially referenced data describing socioeconomic drivers and geographic and physical land variables. These data are used to derive models describing the transition potential or probability of a given LU/LC at a given location. Often, at least some explanatory variables are included to account for the spatial and socioeconomic factors hypothesized to affect LU, such as the distance to roads, markets, population pressures, and other developmental activities. Using the framework previously described an account of the LU models; the model type and its brief description are exemplified in Table 5 employing spatial modeling technique.

The illustrated LU/LC models below collectively reviewed in the above section represent the spatial covering extending from less than 1 ha to more than 1 million km^2 and the temporal scale ranging from less than a day to more than 100 years. Yet this range of extent and temporal variation is not covered by one model. Clearly, models seem to be associated with a particular spatio-temporal niche. Many models with separate ecological modules operate at fine time steps, e.g., a day or a month (except certain climate-focused models). This fine temporal resolution allows these models to more accurately represent rapid ecological changes with time in certain biophysical spheres, e.g., hydrology. Models with multiple time steps can span over both fine and coarse time steps and reflect the temporal complexity of different socioeconomic and biophysical sectors more effectively. More than half of the models provide spatial interaction and demonstrate the advantages of spatially explicit models that move beyond simple spatial representation. These models include the impact of variations across space and time of different bio-physical and socioeconomic factors on LU change.

Spatially explicit predictive models have now become an important tool for LU/LC modelers, planners, managers and decision makers. This has been possible due to a collaboration framework of various disciplines of the science. Development in the IT sector has boosted the computing power of the system and hence contributed immeasurably to expanding LU modeling efforts through desktop PCs that now has capability to run models that would have required a roomful of computers a decade ago. This facilitates models' ability to expand their extents and durations and, at the same time, make resolutions and time steps smaller. In addition, the modeling algorithms have advanced over a short time due to integration of socio-economic sciences. Community science has brought about further revolution in the improvement of the modeling tool and techniques and development of user-friendly interfaces which has enabled the building of more sophisticated models incorporating three major dimensions viz., spatial, temporal and human dimensions. Predictive models although have shown rapid progression in understanding LU/LC dynamics however endure various constraints. Above all, the text of availability of data for model validation imposes serious constraints in considering drivers for inclusion. Models using a significant amount of primary data are constrained in extent or duration, or both. Incorporation of human dimensions to LU/LC is still in its infant stage. Some model development approaches deliberately have restricted themselves to publicly available data for spatial replicability.

A general conclusion is drawn from reviewing predictive models that capturing spatio-temporal footprint of future scenarios is not sufficient for

Table 5. Some illustrations of Land-use models employing spatial modeling technique

Model	Details	Type	Reference
General Ecosystem Model (GEM)	Captures feedback among abiotic and biotic ecosystem components	Dynamics Systems Model (DSM)	Fitz *et al.*, 1996
Patuxent Landscape Model (PLM)	Predicts fundamental ecological processes and land use patterns at the watershed level	DSM	Volnov *et al.*, 1999
Conversion of Land Use and its Effects (CLUE) Model	Predicts land cover in the future	DFSM	Veldkamp and Fresco, 1995
Conversion of Land Use and its Effects – Costa Rica (CLUE-CR) Model	Simulated to-down and bottom-up effects of change in Costa Rica	DFSM	Veldkamp & Fresco, 1996
Area Base Model	Predicts land use proportions at country level	Area base model, using a modified multinomial logit model	Hardle & Parks, 1997
Univariate Spatial Models	Frequency of deforestation	Univariate spatial models	Mertens & Lambin, 1997
Econometric (multinomial logit) model	Predicts land use, aggregated in three classes: Natural vegetation, semi-subsistence agriculture and commercial farming	Econometric (multinomial logit) model	Chomitz & Gray, 1996
Spatial dynamic model	Predicts sites used for shifting cultivation in terms of topography and proximity to population centers	Spatial dynamics model	Gilruth *et al.*, 1995
Spatial Markov Model	Land Use change	Spatial Markov model	Wood *et al.*, 1997
California Urban Futures (CUF)	Explains land use in a metropolitan setting, in terms of demand (population growth) and supply of land (underdeveloped land available for redevelopment)	Spatial simulation model	Landis, 1995
Land Use Change Analysis System (LUCAS)	Transition probability matrix (TMP) (of change in land cover), simulates the landscape change and assessing the impact of species habitat	Spatial stochastic model	Berry *et al.*, 1996
Simple log weights	Predicts are of timberland adjusted for population density	Simple log weights	Wear *et al.*, 1998
Logit Model	Predicts the probability of land being classified as potential timberland	Logit Model	Wear *et al.*, 1999
Dynamic model	Simulates an optimal harvest sequence	Dynamic Model	Swallow *et al.*, 1997
NELUP-Natural Environment Research Council (NERC)–Economic and Social Research Council (ESRC): NERC/ESRC Land Use program (NELUP)	Explains patterns of agricultural and forestry land use under different scenarios	General systems framework Economic component uses a recursive linear planning	O'Callaghan, 1995
NELUP - Extension	Maximizes income, profit is the dependent variable	Linear planning model	Oglethorpe & O'Callaghan, 1995
Forest and Agriculture Sector Optimization Model (FASOM)	Allocation of land in the forest and agriculture sectors, Objective function maximizes the discounted economic welfare of producers and consumers in the US agriculture and forest sectors over a nine decade time horizon	Dynamic, nonlinear, price endogenous, mathematics programming model	Adams *et al.*, 1996
California Urban and Biodiversity Analysis Model (CURBA)	The interaction among the probabilities of urbanization, its interaction with habitat type and extent, and impacts of policy changes on the two	Overlay of GIS layers with statistical urban growth projections	Landis *et al.*, 1998
Cellular automata model	Changes in urban area over time	Cellular automata model	Clarke *et al.*, 1998; Kirtland *et al.*, 1994

future prediction. Models focusing on socio-economic and political drivers of change must be accounted for true empirical modeling of future as it relates to actual LU/LC. Building on the National Research Council's report (NRC, 2001) on "Global Environmental Change: Understanding Human Dimensions", articulation of core social, economic and political science areas need to be studied to understand variations in LU/LC patterns. To further illustrate this need, human decision making does not occur in a vacuum rather it takes place in a particular spatial and temporal context, and, since decision making about LU/LC usually concerns some biophysical processes therefore these must be included in spatially explicit predictive models. Increasingly, the policy community is interested in predictive models that are relevant to sustainable planning and management and hence LU/LC modelers will have to translate those needs by incorporating implicit and explicit temporal, spatial, and human dimensions of scale and complexity.

6. BIODIVERSITY CHARACTERIZATION

6.1 Mapping

Using judicious combination of satellite datasets along with field-survey based studies, makes it possible to carry out detailed mapping and monitoring of biodiversity. Most extensive application of satellite remote sensing technique has been reported using coarse and medium resolution datasets from sensors like, NOAA-AVHRR, SPOT-VEGETATIO, MODIS, ERS and IRS-WiFS. These satellites not only provide multispectral data but also have very high temporal resolution, allowing reconstruction of phenological trends and use it for discriminating major communities of the forest. These applications are most suitable for global, continental and regional estimations. They have been used for large-scale deforestation in Amazonia, Thailand, Indonesia and North-east India. The national forest cover assessment, however, requires medium resolution sensors (in order of ~50m), depending on the areas of investigation. Such a study was carried out in India by National Remote Sensing Agency (1983), the first of its kind at national level to assess forest cover during the periods of 1972-75 and 1980-82. The multispectral digital data from satellite viz., IRS LISS or SPOT or Landsat MSS/TM have also been used comprehensively to stratify forest types on the basis of community formations. Advanced digital image processing techniques like, artificial intelligence and neural network, further improve the accuracy of the derived thematic layers from satellite image. In such an approach, it is also possible to include altitude, climate and contextual details for accurate classification. Digital change detection has also been used to study the human dimensions (encroachment, deforestation, and shifting cultivation) in the forest landscape. This has provided new dimensions to the understanding of ecosystem dynamics and bio-physical parameters in the forested landscape. Finally, these satellite derived forest/ vegetation cover maps provide authentic basis for designing sampling and sampled distribution for detailed ground inventories. This study envisages the use of satellite remote sensing and its kindred technologies like GIS and GPS supplemented by ground-based limited field survey for characterizing forest vegetation cover.

Various studies have been carried out to analyze explicit relationship of landscape elements particularly LU/LC and biodiversity. One such investigation was carried out in eco-sensitive sites of Doon valley in Indian Himalaya with one of the objective to study the biodiversity characterization with reference to phytodiversity (Gupta and Sas, 1997). The study revealed various factors influencing the biodiversity of the area such as agricultural field, settlements, invasion of weeds, grazing, road networks, heavy erosion and pilgrimage which are directly or indirectly linked to LU/LC. In another study, remote sensing data was used primarily to stratify habitats, vegetation types, LU and their association for mapping and studying of tropical deciduous vegetation types and other land uses of the Warangal district as these threatened forests ecosystems are of immense potential value for timber, fuel wood, food and medicine. GIS has been used to develop spatially explicit model to characterize disturbance regimes and to integrate the ground based non-spatial data with the spatial characteristics of the landscape. The approach of disturbance gradient analysis using geospatial techniques provides insight into the disturbance status of forests of Warangal district, which could be useful for forest management and biodiversity conservation (Reddy *et al.*, 2008). A biodiversity study at Gori-Ganga valley at landscape level was done by analyzing several landscape characterization parameters derived using LU/LC (Mathur *et al.*, 2003). Using a combination of image processing and GIS techniques, 17 LC classes were delineated which led to the characterization of the forest and avifaunal biodiversity of an area with high conservation significance in the Uttaranchal State (Rawal & Dhar, 2001). The use of LU/LC in biodiversity characterization is a significant step in the methodological advancement for understanding the patterns, processes and correlation that can help in determining the conservation significance of an area. Conservation of biodiversity should be given the highest priority for the sustainable use of natural resources and safeguarding the future. Identification and prioritization of the conservation employing landscape matrices and geospatial tool are important for the landscape and biodiversity assessment. In the study carried out in part of Hoshiarpur district of Punjab, species diversity was characterized and results indicated that the land under participatory forest programme have the higher species diversity. The impacts of human activities such as forestland conversion have created a negative impact on biodiversity in the Hoshiarpur forest. The study carried out on widely fragmented land cover of Sonitpur (Srivastava *et al.*, 2002) revealed that the fragmentation has caused loss of connectivity, ecotones, corridors and the meta population structure along with high degree of propagation of the disturbance. This is important for understanding biodiversity and landscape patterns in the long-term success of conservation policies. This study has created an information base, which will help design conservation schemes for long-term maintenance of biodiversity.

In conclusion the geospatial tool together with LU/LC data help in extracting maximum amount of information and gives an input for biodiversity characterization that describes ecosystem diversity, i.e. extent, structure, composition, biomass, condition and maps of vegetation, species distribution, habitat status etc. which is essential for planning and management.

6.2 Modeling

Landscape ecological principles provide insight to the natural and anthropogenic factors that influence the biodiversity characterization. In this regard, the application of GIS proved successful in integrating spatial data like LU, LC, disturbance regimes and biological richness maps with non-spatial data like taxonomic and genetic information and creating landscape level information linked with a comprehensive database, which can be further integrated for providing modeling base solutions.

In this context, various modeling tools and techniques have been developed to understand the elements of a biodiversity and landscape system using varied approaches. One of such initiative was undertaken by the Department of Space (DOS) and the Department of Biotechnology (DBT) of the Government of India (GOI) for implementation of Genes to Ecosystem concept in biodiversity conservation and prospecting (Roy & Tomar, 2000). It presents the geospatial database on vegetation cover types, biological richness and disturbance regimes at landscape level for North-East India, Western Himalaya, Western Ghats, Andaman and Nicobar Islands, Central India, Eastern Coasts and other parts of the country. The selected region is often referred to as a bowl of biodiversity due to its broad range of ecological habitats, floristic richness and high levels of endemism. The study has been undertaken after reviewing similar endeavors and adopting improved strategies considering the limitations of the conventional approaches. Preserving a regions biodiversity (its genetic and evolutionary capital) is paramount to the sustainability of both human and wild populations. The first step in preserving biodiversity is being able to speak with authority on where the biodiversity resides in the biosphere. Like other ecological phenomena such as species and habitats, biodiversity is not distributed uniformly in either space or time but rather, clumped and organized into "hotspots" and these hotspots can operate at different scales of organization from the global (such as equatorial rainforests) to the local (such as habitat edges and ecotones). While biodiversity preservation is arguably among the most important tasks faced by ecologists, a great irony is that identifying the location of biodiversity hotspots is among the most expensive (both monetarily and intellectually) activities that ecologists can undertake. Great expense can be incurred during intensive field surveys and the taxonomic expertise required across diverse assemblages of species can be daunting. For this reason, any method that can help in identifying biodiversity hotspots faster, better and cheaper, would be welcome. Diversidad is a software tool developed for ecologists and land managers that enable automatic identification of candidate biodiversity hotspots by filtering digital earth images and automatically identifying those sub-regions with the greatest pixel-class richness. In both applications of the model, heterogeneity is favored over homogeneity. Such tools hypothesize that information rich regions of the image will prove to be field sites with high biological richness. This leap of faith is based on the underlying assumption that pixel heterogeneity is a reasonable surrogate for ecological diversity. Field validation projects have supported this assumption.

In another modeling approach, BioCAP, a customized software developed at Indian Institute of Remote Sensing (IIRS) for biodiversity characterization, was used to carry out multi-criteria spatial analysis that facilitate the rapid assessment of biodiversity and its monitoring (loss and/or gain). It was upgraded to Spatial Landscape Analysis Model (SPLAM) with enhance functionalities and robustness in multi-criteria spatial analysis for biodiversity conservation and planning. This was under the aegis of DOS-DBT program (as stated above) to characterize the biodiversity at landscape level. Under three different phases it has developed the most comprehensive and unparallel database on the LU/LC, forest fragmentation, disturbance regimes, biodiversity richness and detailed insitu phytosociological information. Assessment of the nature of habitats and the disturbance regimes therein, evolving, species–habitat relationships, mapping biological richness and gap analysis, prioritizing conservation and bio-prospecting and redefining ecological zones required for biodiversity conservation were also carried out.

6.3 Databases/WebGIS

Spatial ecological databases and Geomatics technology, together called Eco-Informatics, have been exploited by the conservationists, managers, decision makers to great extent and the potential of technology has been exemplified in some aspects of biodiversity characterization. Conservation studies are discussed below:

1. Forest mapping and monitoring changes: Nationwide mapping of the forest was undertaken first by the National Remote Sensing Agency and subsequently taken up as scheduled task by the Forest Survey of India. 'State of Forests' is valuable publication by Forest Survey of India, which provides

authentic information on status of forest cover in India. Forest cover map is readily available information for defining protected area boundaries, planning ecological corridors, performing environmental impact assessment for development projects. The forest cover type maps are also developed at various scales using SPOT Vegetation cover type map (Agarwal *et al.*, 2003), IRS 1C WiFS (Joshi *et al.*, 2006), IRS LISS III (under DOS-DBT project and also at FSI). Decision support systems can be built around temporal database on forest cover so as to highlight areas under drastic changes.

2. Biomass and productivity estimation: Biomass and productivity models have been developed and tested by the researcher at Indian Institute of Remote Sensing and National Remote Sensing Agency. The well-tested models can become part of Decision Support System, thus, offering rapid and customized methods for biomass and productivity estimation for providing inputs to decision makers to deal with global issues such as global warming and understanding carbon flux.
3. Biodiversity characterization: The concept based on Geo-informatics differs from traditional methods of inventory of the flora and fauna in certain locations. The biodiversity characterization using Geo-informatics account much wider aspects by analyzing threats to the biodiversity in long-term, thus decides places where biodiversity will sustain for longer period. The model works on the 'Principles of Landscape Ecology', integrated with field based inventory of flora/fauna. Thus resultant maps show spatial distribution of biodiversity richness. Biodiversity Information System (BIS) is one of the classic examples from India to develop and portrait such type of database and information system.
4. Wetland conservation planning: Although wetland mapping has been carried out at 1:250,000 scale, many more small wetlands have not been mapped and nation-wide prioritization of wetlands has concluded that as many as 700 wetlands do not have any data of use for prioritization. Development of GIS database on network of wetlands make lots of sense to prioritize inland wetlands for a network of protected areas. The initiative taken up by Salim Ali Centre for Ornithology and Natural History towards providing basic information on wetland is helpful to build GIS based Decision Support System for wetland conservation.
5. Forest fire modeling and mitigation planning: The use of Geomatics based Decision Support System have two aspects towards management of forest fires. The *first* focuses on modeling the spatial data to identify fire prone places, whereas as the latter focuses on providing near real time information on forest fire spread. First one also provides inputs for preparedness, while latter provides information for controlling fire. Both aspects are important and provide valuable inputs for decision making in order to save forest damage due to fires. Space Application Centre has provided near -real time forest fire monitoring to the forest department in Gir Forests of Gujrat.
6. Protected area networking: Protected areas maintain biodiversity by maintaining the habitat and ecosystem processes that species require for their existence. However, the habitat requirements of most species are not known. For this reason an individual-species approach to habitat conservation is unworkable. To ensure ecological integrity, connectivity among protected areas must be maintained in order to maintain biodiversity within the system of reserves. Managing protected area network of a large country like India calls for GIS based Information

Figure 2. A prototype of Spatial Decision Support System

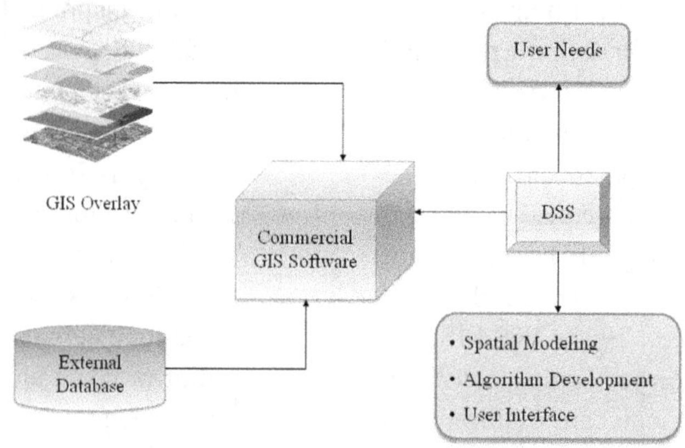

System to study distribution of protected area in a given landscape. The initiative taken by Wildlife Institute of India for development of Protected Area Information System (PAIS) is remarkable.

While appreciating the potential of technology, it may be noted that technology has not been received in totality by the decision makers and implementers, who are responsible for executing biodiversity conservation plans in the field. Spatial Decision Support System (Figure 2) offers the system, which captures knowledge of scientists/ conservationists and requirements of decision makers. In order to translate efforts of scientists/ conservationists/ technologists in reality, the role of Spatial Decision Support System is enormous.

7. FUTURE RESEARCH DIRECTIONS

LU/LC change is a pervasive, accelerating and irreversible process, which is driven by a multitude of natural and human induced processes. Analyzing these changes is therefore inevitable for formulating effective environmental policies and management strategies. The chapter explained the role of geospatial technology in facilitating and enabling in understanding the spatial and temporal patterns of LU/LC and drivers of change as a precursor for landscape assessment. Additionally, future research work has been proposed in terms of requirements to assess the LU/LC.

1. A quantitative accuracy assessment of the coarse-scale data should be performed with finer resolution satellite imagery of a subset of locations integrated with ground-truth data on actual land-use conversions for better estimation.
2. A hierarchical standardized LU/LC classification system to be adopted and validated at a fine spatial resolution and to time series of data integrated at the appropriate scale.
3. Operational monitoring of LC should be extended to regions that are not known as "hot spots" but where rapid changes are occurring or have potential to take place.
4. New empirical work is required based on urgent need for systematic observations and data availability of LU/LC at local measured scale for involvement and planning of landscape for better sustenance.
5. A modular open source approach for assessing complex LU/LC issues is proposed because it urgently needs the collective resource. The published and authentic reports on social, economic and political patterns and

processes of LU/LC and biodiversity as a practical guide should be used as subordinate information for formulation of landscape planning and management.

6. Open-source modeling offers additional hope for future LU/LC modeling and analysis. There have been several very successful, complex programming endeavors using the open-source concept. These methods have potential to spur the development of modeling the LU/LC coupled with biodiversity and spatial decision support system as well.
7. An integrative approach incorporating landscape assessment and policy formulation should be adopted by stakeholders and decision makers for better management actions. For example land-use modelers will need to consider the relative significance of different drivers on land-use change within the context of policy makers' needs. Even the land-use modelers will have to translate the needs with particular attention to implicit and explicit temporal, spatial, and human decision making scale and complexity and the interactions between scale and complexity.

8. CONCLUSION

The chapter highlights the available information on LU/LC changes and biodiversity characterization studies, carried out from local to global scale. It was based exclusively on literature review. As for any global map, one should look at the broad scale patterns. Local scale scrutiny of the maps is likely to reveal anomalies caused by heterogeneous data sources. Finer resolution data show more change than coarse resolution datasets. Despite limitations in the data, the synthesized report helps to focus attention on the rapid land-cover changes and theirs link to biodiversity. The products also reveal the global geographic patterns of land-cover change. Most notably, this chapter revealed that: LU/LC maps for many parts of the world are not adequately represented or are unavailable despite huge databases. It is also possible that ecological impacts of change are large even though observable LC changes maps were not able to capture such subtle changes due to current limits in availability of data. Rapid LU/LC is not randomly or uniformly distributed but is clustered in some locations. There are different trajectories of LC change in different parts of the world (e.g. decrease in cropland in temperate and increase in tropics), as well as in its drivers. Thus uniform drivers cannot be applied to all studies. Asia currently has the greatest concentration of areas of rapid land-cover changes but data is not available to support the claim. Much of our information on tropical land-cover change comes from remotely-sensed LU/LC data, while information on change in the non-tropical regions comes predominantly from census data. Systematic analysis to identify land-cover change is possibly missing due to the lower availability and reliability of census data in the tropics. There are other forms of rapid land-cover change that are thought to be widespread, but they are still poorly documented at the global scale. Local-to national-scale, however, demonstrate their importance and ecological significance. But a quantitative accuracy assessment of the data has not been performed. Data producers should use a hierarchical standardized LU/LC classification system for validating land cover data at a fine spatial resolution and to time series data for integrating at the appropriate scale. As an alternative or a complement to categorical land cover representations, a continuous description of the LU/LC should be more widely adopted whenever possible as it offers greater ease for comparison of different databases (DeFries *et al.*, 2002; Ramankutty & Foley, 1999). New empirical work is required based on advances in geospatial tools & techniques and Spatial Decision Support System (SDSS). There is an urgent need for systematic observations on the still poorly measured processes of LU/LC and Biodiversity characterization.

REFERENCES

Achard, F., Eva, H. D., Stibig, H. J., Mayaux, P., Gallego, J., Richards, T., & Malingreau, J. P. (2002). Determination of deforestation rates of the world's humid tropical forests. *Science, 297*, 999–1002. doi:10.1126/science.1070656

Adams, D. M., Alig, R. J., & McCarl, B. A. (1996). An analysis of the impacts of public timber harvest policies on private forest management in the U.S. *Forest Science*, *42*(3), 343–358.

Agarwal, S., Joshi, P. K., Shukla, Y., & Roy, P. S. (2003). SPOT VEGETATION multitemporal data for classifying vegetation in south central Asia. *Current Science*, *84*(11), 1440–1448.

Anderson, J. R., Hardy, E. E., Roach, J. T., & Witmer, R. E. (1976). *A land use and land cover classification system for use with remote sensor data, (US Geological Survey Professional Paper, No. 964)*. Washington, DC: USGS.

Auch, R., Taylor, J., & Acevado, W. (2004). *Urban growth in American cities: Glimpses of U.S. urbanization*. U.S. Geological Survey Circular 1252. U.S. Geological Survey, Sioux Falls, SD.

Bartholomé, E., Belward, A. S., Achard, F., Bartalev, S., Carmona-Moreno, C., & Eva, H. (2002). *Global land cover mapping for the year 2000—project status November 2002. Publications of the European Communities, EUR 20524 EN*. Luxembourg: European Commission.

Baudry, J., & Thenail, C. (2004). Interaction between farming systems, riparian zones, and landscape patterns: A case study in western France. *Landscape and Urban Planning*, *67*, 121–129. doi:10.1016/S0169-2046(03)00033-1

Berry, M., Flamm, R., Hazen, B., & Macintyre, R. (1996). LUCAS: A system for modeling land use change. *IEEE Computational Science & Engineering*, *3*, 24–35. doi:10.1109/99.486758

Brown, D. G., Pijanowski, B. C., & Duh, J. D. (2000). Modeling the relationships between land-use and land-cover on private lands in the Upper Midwest, USA. *Journal of Environmental Management*, *59*, 247–263. doi:10.1006/jema.2000.0369

Burrough, P. A. (2000). *Principles of Geographical Information systems*. London, UK: Oxford University Press.

Castella, J. C., Kam, S. P., Quang, D. D., Verburg, P., & Hoanh, C. T. (2007). Combining top-down and bottom-up modelling approaches of land use/cover change to support public policies: application to sustainable management of natural resources in northern Vietnam. *Land Use Policy*, *24*, 531–545. doi:10.1016/j.landusepol.2005.09.009

Chaurasia, R., Loshali, D. C., Dhaliwal, S. S., Shrma, M., Kudrat, P. K., & Tiwari, A. K. (1996). Land use change analysis for agriculture management – a case study of tehsil Talwandi Sabo, Punjab. Photonirvachk. *Journal of the Indian Society of Remote Sensing*, *24*(2), 89–98. doi:10.1007/BF03016124

Chomitz, K. M., & Gray, D. A. (1996). Roads, land use, and deforestation: A spatial model applied to Belize. *The World Bank Economic Review*, *10*(3), 487–512.

Clarke, K. C., Hoppen, S., & Gaydos, L. J. (1998). Loose-coupling a cellular automata model and GIS: Long term urban growth prediction for San Francisco and Washington, Baltimore. *International Journal of Geographical Information Science*, *12*(7), 699–714. doi:10.1080/136588198241617

Clawson, M., & Stewart, C. L. (1965). *Land use information. A critical survey of U.S. statistics including possibilities for greater uniformity*. Baltimore, MD: The Johns Hopkins Press for Resources for the Future, Inc.

Cohen, W. B., Maiersperger, T. K., Yang, Z. Q., Gower, S. T., Turner, D. P., & Ritts, W. D. (2003). Comparisons of land cover and LAI estimates derived from ETM plus and MODIS for four sites in North America: A quality assessment of 2000/2001 provisional MODIS products. *Remote Sensing of Environment*, *88*(3), 233–255. doi:10.1016/j.rse.2003.06.006

Corves, C., & Place, C. J. (1994). Mapping the reliability of satellite-derived land cover maps - an example from central Brazilian Amazon Basin. *International Journal of Remote Sensing*, *15*, 1283–1294. doi:10.1080/01431169408954161

DeFries, R., Bounoua, L., & Collatz, G. J. (2002). Human modification of the landscape and surface climate in the next fifty years. *Global Change Biology*, *8*, 438–458. doi:10.1046/j.1365-2486.2002.00483.x

DeFries, R., Hansen, M., & Townshend, J. G. R. (1995). Global discrimination of land cover types from metrics derived from AVHRR Pathfinder data. *International Journal of Remote Sensing*, *19*, 3141–3168.

Dymond, C. C., Mladenoff, D. J., & Radeloff, V. C. (2002). Phenological differences in Tasseled Cap indices improve deciduous forest classification. *Remote Sensing of Environment*, *80*, 460–472. doi:10.1016/S0034-4257(01)00324-8

Emch, M., James, W., Quinn, M. P., & Alexander, M. (2005). Forest cover change in the Toledo district, Belize from 1975 to 1999: A remote sensing approach. *The Professional Geographer*, *57*(2), 256–267. doi:10.1111/j.0033-0124.2005.476_1.x

Encyclopedia of Earth. (2007). *Land-use and land-cover change. Washington DC*. Retrieved March 20, 2010, from http://www.eoearth.org/article/Land-use_and_land-cover_change

FAO. (2004). *FAO—The Africover Initiative*. Food and Agricultural Organization of the United Nations. Retrieved March 25, 2010, from http://www.africover.org/africover_initiative.htm

Fitz, H. C., DeBellevue, E. B., Costanza, R., Boumans, R., Maxwell, T., Wainger, L., & Sklar, F. H. (1996). Development of a general ecosystem model (GEM) for a range of scales and ecosystems. *Ecological Modelling*, *88*, 263–297. doi:10.1016/0304-3800(95)00112-3

Foody, G. M., Boyd, D. S., & Curran, P. J. (1996). Relations between tropical forest biophysical properties and data acquired in AVHRR channels 1± 5. *International Journal of Remote Sensing*, *17*, 1341–1355. doi:10.1080/01431169608948707

Fraser, R. H., Abuelgasim, A., & Latifovic, R. (2005). A method for detecting large-scale forest cover change using coarse spatial resolution imagery. *Remote Sensing of Environment*, *95*, 414–427. doi:10.1016/j.rse.2004.12.014

Friedl, M. A., Mclver, D. K., Hodges, J. C. F., Zhang, X. Y., Muchoney, D., & Strahler, A. H. (2002). Global land cover mapping from MODIS: Algorithms and early results. *Remote Sensing of Environment*, *83*(1-2), 287–302. doi:10.1016/S0034-4257(02)00078-0

GEO. (2005). *The g=Global Earth Observation System of Systems (GEOSS)—10-year implementation plan and reference document*. Retrieved on May 12, 2010, from http://earthobservations.org

Gilruth, P. T., Marsh, S. E., & Itami, R. (1995). A dynamic spatial model of shifting cultivation in the highlands of Guinea, West Africa. *Ecological Modelling*, *79*(1-3), 179–197. doi:10.1016/0304-3800(93)E0145-S

Global Land Cover Network (GLCN). (2000). *Global land cover 2000*. Retrieved March 15, 2010, from www.glcn.org

GLP. (2005). *Global land project science plan and implementation strategy*. (IGBP Report No. 53/IHDP Report No. 19). IGBP Secretariat, Stockholm. Retrieved March 27, 2010, from http://www.globallandproject.org/Documents/report_53.pdf

Goldewijk, K. (2001). Estimating global land use change over the past 300 years: The HYDE database. *Global Biogeochemical Cycles, 15*, 417–434. doi:10.1029/1999GB001232

Gonçalves, P., Carrão, H., Pinheiro, A., & Caetano, M. (2006). Land cover classification with Support Vector Machine Applied to MODIS imagery. In Marçal, A. (Ed.), *Global developments in environmental earth observation from space* (pp. 517–526). Portugal: MillPress.

Gopal, S., Friedl, M. A., McIver, D. K., Hodges, J. C. F., Zhang, X. Y., & Muchnoy, D. (2002). Global land cover mapping from MODIS: Algorithms and early results. *Remote Sensing of Environment, 83*, 287–302. doi:10.1016/S0034-4257(02)00078-0

Gordon, L. J., Peterson, G. D., & Bennett, E. M. (2008). Agricultural modifications of hydrological flows create ecological surprises. *Trends in Ecology & Evolution, 23*, 211–219..doi:10.1016/j.tree.2007.11.011

Grigg, D. (1965). The logic of regional systems. *Annals of the Association of American Geographers. Association of American Geographers, 55*(3), 465–991. doi:10.1111/j.1467-8306.1965.tb00529.x

Gupta, B.N. and Biswas Sas (1997). Biodiversity characterization at landscape level using satellite remote sensing. Paper presented in a workshop, " Biodiversity characterization using remote sensing " project of National Remote sensing agency, Hyderabad, sponsored by department of space and biotechnology, Govt. of India organized by NRSA at Hyderabad, India.

Gutman, G., Janetos, A., Justice, C., Moran, E., Mustard, J., & Rindfuss, R. ... Turner, II B. J. (2004). *Land change science: Observing, monitoring, and understanding trajectories of change on the Earth's surface.* New York, NY: Kluwer Academic Publishers.

Hardie, I. W., & Parks, P. J. (1997). Land use with heterogeneous land quality: An application of an area-base model. *American Journal of Agricultural Economics, 79*(2), 299–310. doi:10.2307/1244131

Herold, M. (2000). *UN Global land cover network: An international framework for standardized development of land cover data.* Global Terrestrial Observing System (GTOS) of the United Nations.

Homer, C. G., Huang, C., Yang, L., Wylie, B., & Coan, M. (2004). Development of a 2001 national land-cover database for the United States. *Photogrammetric Engineering and Remote Sensing, 70*(7), 829–840.

International Geosphere Biosphere Programme. (2010). *IGBP projects*. Retrieved March 22, 2010, from http://www.igbp.net

Jain, S. (2007). Technical note: Use of IKONOS satellite data to identify informal settlements in Dehradun, India. *International Journal of Remote Sensing, 28*, 3227–2333. doi:10.1080/01431160600705122

Jayakumar, S., & Arockiasmy, D. I. (2003). Landuse/landcover mapping and change detection in part of Eastern Ghats of Tamil Nadu using remote sensing and GIS. *Journal of Indian Society of Remote Sensing, 31*, 251–260. doi:10.1007/BF03007345

Johnston, C. A., Watson, T., & Wolter, P. T. (2007). Sixty-three years of land alteration in Erie Township. *Journal of Great Lakes Research, 33*, 253–268. doi:10.3394/0380-1330(2007)33[253:SYOLAI]2.0.CO;2

Joshi, P. K., Roy, P. S., Singh, S., Agrawal, S., & Yadav, D. (2006). Vegetation cover mapping in India using multi-temporal IRS Wide Field Sensor (WiFS) data. *Remote Sensing of Environment, 103*(2), 190–202. doi:10.1016/j.rse.2006.04.010

Justice, C. O., Vermote, E., Townshend, J. R. G., Defries, R., Roy, D. P., & Hall, D. K. (1998). The Moderate Resolution Spectroradiometer (MODIS): Land remote sensing for global change research. *IEEE Transactions on Geoscience and Remote Sensing, 36*, 1228–1249. doi:10.1109/36.701075

Kerber, A. G., & Schutt, J. B. (1986). Utility of AVHRR channel 3 and 4 in land-cover mapping. *Photogrammetric Engineering and Remote Sensing, 52*, 1877–1883.

Kirtland, D., Gaydos, L., Clarke, K., DeCola, L., Acevedo, W., & Bell, C. (1994). An analysis of human-induced land transformations in the San Francisco Bay/Sacramento area. *World Resource Review, 6*(2), 206–217.

Knight, J. F., Lunetta, R. L., Ediriwickrema, J., & Khorram, S. (2006). Regional scale land-cover characterization using MODIS-NDVI 250 m multi-temporal imagery: A phenology based approach. *GIScience and Remote Sensing, 43*, 1–23. doi:10.2747/1548-1603.43.1.1

Kunwar, P., Kachchwaha, T. S., Kumar, A., Aggarwal, A. K., Singh, A. N., & Mendiratta, N. (2010)... *Current Science, 98*(2).

Kunwar, P., & Kachhwaha, T. S. (2001). Transformation of landuse-landcover for optimal utilization of natural resources on sustainable basis in Mandawara Block, Lalitpur District, Uttar Pradesh using remote sensing and GIS techniques. In Singh, S. B., & Singh Rana, P. B. (Eds.), *National Geographical Society of India, BHU* (pp. 151–156). Varanasi: Environment and Development.

Lambin, E. F., Baulies, X., Bockstael, N., Fischer, G., Krug, T., & Leemans, R. ... Vogel, C. (1999). *Land-use and cover change: Implementation strategy*. (IGBP Report No. 48/IHDP Report No. 10). IGBP, Stockholm, 125 pp.

Lambin, E. F., & Geist, H. (Eds.). (2006). *Land-use and land-cover change: Local processes and global impacts*. New York, NY: Springer.

Lambin, E. F., Geist, H., & Lepers, E. (2003). Dynamics of land use and cover change in tropical regions. *Annual Review of Environment and Resources, 28*, 205–241. doi:10.1146/annurev.energy.28.050302.105459

Lambin, E. F., Turner, B. L. II, Geist, H., Agbola, S., Angelsen, A., & Bruce, J. W. (2001). The causes of land use and land cover change: Moving beyond the myths. *Global Environmental Change, 11*(4), 261–269. doi:10.1016/S0959-3780(01)00007-3

Landis, J. D. (1996). Imagining land use futures: Applying the California urban futures model. *Journal of the American Planning Association. American Planning Association, 61*(4), 438. doi:10.1080/01944369508975656

Landis, J. D., Monzon, J. P., Reilly, M., & Cogan, C. (1998). *Development and pilot application of the California Urban and Biodiversity Analysis (CURBA) Model*. Paper presented at the 1998 ESRI International User Conference, October 7–9. Retrieved from http://www.esri.com/library/userco...oc98/PROCEED/TO600/PAP571/P571.HTM

Lee, J. T., Elton, M. J., & Thompson, S. (1999). The role of GIS in landscape assessment: using land use based criteria for an area of the Chiltern Hills Area of outstanding natural beauty. *Land Use Policy, 16*, 23–32. doi:10.1016/S0264-8377(98)00033-7

Lee, P., & Olson, C. E. (2004). *Landscape changes projection using remote sensing and geographic information system.* Retrieved January 27, 2009, from http://wagner.zo.ntu.edu.tw/Download/Landscape%20changes%20projection%20by%20remote%20sensing.pdf

Lillesand, T. M., Kiefer, R. W., & Chipman, J. W. (2007). *Remote sensing and image interpretation* (5th ed.). New Delhi, India: Wiley India Pvt. Ltd.

Lin, W., Zhang, L., Du, D., Yang, L., Lin, H., Zhang, Y., & Li, J. (2009). Quantification of land use/land cover changes in Pearl River Delta and its impact on regional climate in summer using numerical modeling. *Regional Environmental Change, 9*(2), 75–82. doi:10.1007/s10113-008-0057-5

Loveland, T. R., Reed, B. C., Brown, J. F., Ohlen, D. O., Zhu, Z., Yang, L., & Merchant, J. W. (2000). Development of a global land cover characteristics database and IGBP DISCover from 1-km AVHRR data. *International Journal of Remote Sensing, 21*(6-7), 1303–1330. doi:10.1080/014311600210191

Mathur, V. B., Raza, R., Lal, P., Agarwal, M. K., & Singsit, S. (2003). Biodiversity characterization in Middle Goriganga Valley, Askot Wildlife Sanctuary, Uttarakhand State. *Technical report submitted to Indian Institute of Remote Sensing, Dehradun.* Retrieved July 7, 2010 from http://www.wii.gov.in/publications/researchreports/2005/diversity&rarityinfloralwh.pdf

Matthews, E. (1983). Global vegetation and land use: New high resolution data bases for limited studies. *Journal of Climate and Applied Meteorology, 22*, 474–487. doi:10.1175/1520-0450(1983)022<0474:GVALUN>2.0.CO;2

McRoberts, E., & Tomppo, E. (2007). Remote sensing support for national forest inventories. *Remote Sensing of Environment, 110*(4), 412–419. doi:10.1016/j.rse.2006.09.034

Millennium Ecosystem Assessment (MEA). (2002). *Ecosystems and human well-being.* Retrieved from http://www.millenniumassessment.org

Mertens, B., & Lambin, E. (1997). Spatial modelling of deforestation in southern Cameroon: Spatial disaggregation of diverse deforestation processes. *Applied Geography (Sevenoaks, England), 17*(2), 143–162. doi:10.1016/S0143-6228(97)00032-5

Mickelson, J. G., Civco, D. L., & Silander, J. A. Jr. (1998). Delineating forest canopy species in the Northeastern United States using multi-temporal TM imagery. *Photogrammetric Engineering and Remote Sensing, 64*(9), 891–904.

Moller, M. S., & Blaschke, T. (2005). *Monitoring LU/LC dynamics in the urban-rural fringe.* Paper presented at the Anais XII Simpósio Brasileiro de Sensoriamento Remoto, Goiânia, Brasil, 16-21 abril 2005, INPE, (pp. 3821-3828).

Mundia, C. N., & Aniya, M. (2005). Analysis of land use/cover changes and urban expansion of Nairobi city using remote sensing and GIS. *International Journal of Remote Sensing, 26*(13), 2831–2849. doi:10.1080/01431160500117865

Murthy, M. S. R., Giriraj, A., & Dutt, C. B. S. (2003). Geoinformatics for biodiversity assessment. *Biology Letters, 40*(2), 75–100.

National Research Council (NRC). (2001). *Compensating for wetland losses under the Clean Water Act.* Washington, DC: National Academy Press.

National Resources Inventory (NRCS). (2001). *Percent change in cropland area* (pp. 1982-1997), USDA-NRCS-RID. Retrieved January 15, 2010, from http://www.nrcs.usda.gov/technical/land/index.html

National Resources Inventory (NRCS). (2007). *Change in average annual soil erosion by wind and water on cropland and CPR land,* (pp. 1982-1992). USDA-NRCS-RID. Retrieved January 15, 2010, from http://www.nrcs.usda.gov/technical/land/index.html

Navalgund, R. R., Jayaraman, V., & Roy, P. S. (2007). Remote sensing applications: An overview. *Current Science*, *93*(12), 1747–1766.

Nichol, J., & Wang, M. S. (2007). Remote sensing of urban vegetation life from by spectral mixture analysis of high resolution IKONOS satellite images. *International Journal of Remote Sensing*, *28*, 985–1000. doi:10.1080/01431160600784176

O'Callaghan, J. R. (1995). NELUP: An introduction. *Journal of Environmental Planning and Management*, *38*(1), 5–20. doi:10.1080/09640569513084

Oglethorpe, D. R., & O'Callaghan, J. R. (1995). Farmlevel economic modelling within a river catchment decision support system. *Journal of Environmental Planning and Management*, *38*(1), 93–106. doi:10.1080/09640569513138

Pastor, J., & Wolter, P. T. (2002). Mapping and modeling forest change in a Boreal Landscape. In *Proceeding NASA LCLUC*, Science Team Meeting Washington D.C. 19–21. November.

Ramachandra, T. V., & Kumar, U. (2004). Geographic resources decision support system for land use, land cover dynamics analysis. In *Proceedings of the FOSS/GRASS Users Conference*, Bangkok, Thailand

Ramankutty, N., & Foley, J. A. (1999). Estimating historical changes in global land cover: Croplands from 1700 to 1992. *Global Biogeochemical Cycles*, *13*(4), 997–1028. doi:10.1029/1999GB900046

Reddy, C. S., Babar, S., Sudha, K., & Raju, V. S. (2008). Vegetation cover mapping and landscape level disturbance gradient analysis in warangal district, Andhra Pradesh, India using satellite remote sensing and GIS. *Space Research Journal*, *1*, 29–38. doi:10.3923/srj.2008.29.38

Reese, H. M., Lillesand, T. M., Nagel, D. E., Stewart, J. S., Goldmann, R. A., & Simmons, T. E. (2002). Statewide land cover derived from multiseasonal Landsat TM data—a retrospective of the WISCLAND project. *Remote Sensing of Environment*, *82*(2), 3224–3237. doi:10.1016/S0034-4257(02)00039-1

Reis, S. (2008). Analyzing land use/land cover changes using remote sensing and GIS in Rize. *North-East Turkey Sensors*, *8*(10), 6188–6202.

Roberts, D. A., Keller, M., & Soares, J. V. (2003). Studies of land-cover, land-use and biophysical of vegetation in the large scale biosphere atmosphere experiment in Amazonia. *Remote Sensing of Environment*, *87*, 377–388. doi:10.1016/j.rse.2003.08.012

Roy, P. S., & Giriraj, A. (2008). Land use and land cover analysis in Indian context. *Journal of Applied Sciences*, *8*(8), 1346–1353. doi:10.3923/jas.2008.1346.1353

Roy, P. S., & Tomar, S. (2000). Biodiversity characterization at landscape level using geospatial modelling technique. *Biological Conservation*, *95*, 95–109. doi:10.1016/S0006-3207(99)00151-2

Sedano, F., Gong, P., & Ferrão, M. (2005). Land cover assessment with MODIS imagery in southern African Miombo ecosystems. *Remote Sensing of Environment*, *98*, 429–441. doi:10.1016/j.rse.2005.08.009

Sharma, P. K., Lahkar, B. P., Ghosh, S., Rabha, A., Das, J. P., & Nath, N. K. (2008). Land use and land cover change and future implication analysis in Manas National Park, India using multitemporal satellite data. *Current Science*, *95*(2), 223–227.

Somporn, S. (1995). *Land use/land cover change detection in the Chiang Mai Area using Landsat TM.* Department Of Geography and Earth Sciences- A Report.

Srivastava, S. K., & Gupta, R. D. (2003). *Monitoring of changes in land use/ land cover using multi- sensor satellite data.* Map India Conference 2003.

Srivastavata, S., Singh, T. P., Singh, H., Kushwaha, S. P. S., & Roy, P. S. (2002). Large scale deforestation in Sonitpur district of Assam. *Current Science, 82*(12), 1479–1484.

Stefanov, W. L., Ramsey, M. S., & Christensen, P. R. (2001). Monitoring urban land cover change: An expert system approach to land cover classification of semiarid to arid urban centers. *Remote Sensing of Environment, 77*, 173–185. doi:10.1016/S0034-4257(01)00204-8

Swallow, S. K., Talukdar, P., & Wear, D. N. (1997). Spatial and temporal specialization in forest ecosystem management under sole ownership. *American Journal of Agricultural Economics, 79*, 311–326. doi:10.2307/1244132

Tan, K. C., Lim, H. S., & Matjafri, M. Z. (2009). Landsat data to evaluate urban expansion and determine land use/land cover changes in Penang Island, Malaysia. *Environmental Earth Sciences*. doi. 10.1007/s12665-009-0286-z

Thenkabail, P. S. (2004). Inter-sensor relationships between IKONOS and Landsat-7 ETM+ NDVI data in three ecoregions of Africa. *International Journal of Remote Sensing, 25*(2), 389–408. doi:10.1080/0143116031000114842

Thenkabail, P. S., Hall, J., Lin, T., Ashton, M. S., Harris, D., & Enclona, E. A. (2003). Detecting floristic structure and pattern across topographic and moisture gradients in a mixed species Central African forest using IKONOS and Landsat-7 ETM+ images. *International Journal of Applied Earth Observation and Geoinformation, 4*, 255–270. doi:10.1016/S0303-2434(03)00006-0

Townshend, J., Justice, C., Li, W., Gurney, C., & McManus, J. (1991). Global land cover classification by remote sensing—present capabilities and future possibilities. *Remote Sensing of Environment, 35*(2-3), 243–255. doi:10.1016/0034-4257(91)90016-Y

Turner, M. G. (2004). Landscape ecology: The effect of pattern of process. *Annual Review of Ecology and Systematics, 20*, 171–197. doi:10.1146/annurev.es.20.110189.001131

Veldkamp, A., & Fresco, L. O. (1996a). CLUE: A conceptual model to study the conversion of land use and its effects. *Ecological Modelling, 85*, 253–270. doi:10.1016/0304-3800(94)00151-0

Veldkamp, A., & Fresco, L. O. (1996b). CLUE-CR: An integrated multi-scale model to simulate land use change scenarios in Costa Rica. *Ecological Modelling, 91*, 231–248. doi:10.1016/0304-3800(95)00158-1

Veldkamp, A., Kok, K., de Koning, G. H. J., Verburg, P. H., Priess, J., & Bergsma, A. R. (2001). The need for multi-scale approaches in spatial specific land use change modelling. *Environmental Modeling and Assessment, 6*, 111–121. doi:10.1023/A:1011572301150

Vitousek, P. M., Mooney, H. A., Lubchenco, J., & Melillo, J. M. (1997). Human domination of Earth's ecosystems. *Science, 277*, 494–499. doi:10.1126/science.277.5325.494

Vogelmann, J. E., Howard, S. M., Yang, L., Larson, C. R., Wylie, C. R., & Van Driel, N. (2001). Completion of the 1990s national land cover data set for the Conterminous United States from Landsat Thematic map-per data and ancillary data sources. *Photogrammetric Engineering and Remote Sensing, 67*, 6650–6662.

Wadsworth, R. A., & Treweek, J. (1999). *Geographical information systems for ecology: An introduction*. Harlow: Addison Wesley Longman.

Wear, D. N., Apt, R., & Mangold, R. (1998). People, space, time: Factors that will govern forest sustainability. *Proceedings of the 63rd North American Wildlife and Natural Resources Conference* (pp. 348–361). Wildlife Management Institute, Washington, D.C.

Wear, D. N., Liu, R., Foreman, J. M., & Sheffield, R. M. (1999). The effects of population growth on timber management and inventories in Virginia. *Forest Ecology and Management, 118*, 107–115. doi:10.1016/S0378-1127(98)00491-5

Wolter, P. T., & White, M. A. (2002). Recent forest cover type transitions and landscape structural changes in northeast Minnesota. *Landscape Ecology, 17*, 133–155. doi:10.1023/A:1016522509857

Wood, E. C., Lewis, J. E., Gray Tappan, G., & Lietzow, R. W. (1997). *The development of a land cover change model for southern Senegal.* Paper presented at the Land Use Modeling Workshop, June 1997, Sponsored by the USGS and NCGIA. Retrieved June 5, 2010, from http://ncgia.ncgia.ucsb.edu/conf/landuse97/papers/wood eric/pecdoc.html

Xian, G., & Crane, M. (2005). Assessments of urban growth in the Tampa Bay watershed using remote sensing data. *Remote Sensing of Environment, 97*, 203–215. doi:10.1016/j.rse.2005.04.017

Yang, L., Homer, C. G., Hegge, K., Huang, C., Wylie, B., & Reed, B. (2001). A Landsat 7 scene selection strategy for a National Land Cover Database. In *Proceedings IEEE 2001 International Geoscience and Remote Sensing Symposium* Sydney, Australia.

Yemefack, M., Bijker, W., & Jong, S. M. (2006). Investigating relationships between Landsat-7 ETM+ data and spatial segregation of LU/LC types under shifting agriculture in southern Cameroon. *International Journal of Applied Earth Observation and Geoinformation, 8*(2), 2234–2256. doi:10.1016/j.jag.2005.08.003

Zhan, X., DeFries, R. S., Townshend, J. R. G., Dimiceli, C., Hansen, M., Huang, C., & Sohlberg, R. (2000). The 250m global land cover change product from the moderate resolution imaging spectroradiometer of NASA's Earth Observing System. *International Journal of Remote Sensing, 21*(6-7), 1433–1460. doi:10.1080/014311600210254

Zsuzsanna, D., Bartholy, J., Pongracz, R., & Barcza, Z. (2005). Analysis of land-use/land-cover change in the Carpathian region based on remote sensing techniques. *Physics and Chemistry of the Earth, 30*, 109–115.

Chapter 4
Monitoring Biodiversity Using Remote Sensing and Field Surveys

C. A. Mücher
Wageningen University and Research Centre, The Netherlands

ABSTRACT

The world population has grown rapidly in conjunction with technological developments, especially in the last two centuries, which has led to a significant expansion of industrialisation, urbanisation, and agricultural intensification. As a result, land use and associated land cover have changed at an increasing rate, intensifying the pressures on habitats and landscapes, and biodiversity in general. The steady decline of habitats and landscapes demonstrates the need for protection. Monitoring the extent and quality is also required in a more comprehensive fashion across the countryside, ranging from regional to global scales. The Rio Declaration in 1992 confirmed the need to work towards international agreements to protect the integrity of the global environment. The associated Convention on Biological Diversity (CBD) draws attention to the need to identify and monitor ecosystems, habitats, species, communities, genomes, and genes. All CBD parties have committed themselves in achieving the 2010 Biodiversity Target: to protect and restore habitats and natural systems and halt the loss of biodiversity by 2010. All these policies require quantitative figures on the extent of habitats and their degree of fragmentation. Unfortunately 'hard' figures on the extent of landscapes and associated habitats (inside and outside protected areas) are currently not available. Therefore, the main objective is to develop quantitative methodologies for the spatial identification and monitoring of European landscapes and their habitats.

DOI: 10.4018/978-1-60960-619-0.ch004

This chapter concludes that, in combination with additional environmental data sets, it is now possible to model quantitatively the spatial extent of widespread habitats and landscapes on the basis of land cover information derived from satellite imagery. Although it is now possible to model the spatial extent of widespread European habitats, these patterns cannot be directly translated into area estimates. The retrieval of accurate land cover information is not only crucial for the spatial modelling of European landscapes and habitats, but also for their monitoring. Operational remote sensing enables land cover characterization at various scales but the classification accuracies are still insufficient at continental and global scales for monitoring purposes. Instead, the use of continuous thematic fraction layers, as derived from linear unmixing, provides a good basis for monitoring land cover changes of Europe's complex landscapes. However, gradual and small changes in habitats and their quality are not easily detected from space by satellite imagery, and therefore, additional information from field surveys is needed. Protocols for rapid field surveying of habitats have been developed that can provide a European baseline based on a sampling design across European landscapes. The information from the field samples (e.g. square kilometres) can be used for the validation and calibration of the obtained distribution maps of European habitats. The field surveying method is amongst others based on the estimation of the main plant life forms, which are highly correlated with vegetation structure. The latter has been shown to have a good relationship with satellite imagery. Field surveys are always limited to relatively small areas in Europe, and therefore, the spatial modelling of habitats and landscapes with the help of remotely sensed information remains important for providing a synoptic overview.

1. INTRODUCTION

During the last two centuries in particular, the world population grew rapidly, in conjunction with technological developments, which led to a significant expansion of industrialisation, urbanisation and agricultural activity (Stanners & Bordeaux, 1995; Moran *et al.*, 2004; EEA, 2005). As a result, land use and associated land cover changed at an increasing rate, intensifying the pressures on landscapes, habitats and biodiversity in general. A global analysis by Klein Goldewijk & Ramankutty (2004) showed that between 1700 and 1990 the area of arable land increased by approximately 500%, from 3 million km^2 to 15 million km^2, and that of grassland by approximately 600%, from 5 million km^2 to 31 million km^2, both at the expense of semi-natural vegetation and forests. Over the same period, forest area decreased by approximately 17%, from 53 million km^2 to 44 million km^2. Types and rates of land cover change vary over time and space. Europe, for example, has experienced an opposite trend over the last 40 years, which included a net forest increase of approximately 10%, a net loss of arable land of about 11% and a net loss of permanent grassland of about 11% (source: FAO land use statistics). The EU project BIOPRESS showed, by analysis of historical aerial photographs over the period 1950-1990-2000, that of these land cover changes urbanisation was predominant. Alarmingly, the project showed that in the 59 transects across Europe the rate of land cover change remained almost constant; respectively, 15% and 14% per decade over the periods 1950-1990 and 1990-2000 (Köhler *et al.*, 2006; Gerard *et al.*, 2010). In The Netherlands, between 1950-1990, in parallel with a net loss of agricultural land and a net increase of forest and urbanisation, there was a dramatic 44% decline of natural areas (Van Duuren *et al.*, 2003). The amount of heathland was reduced by 68%, of salt marshes by 60%, of raised bogs (moors and peat-land) by 81% and of inland sand dunes by 52%. Only wetlands increased, by 9% (http://

www.pbl.nl/nl/publicaties/mnp/2003/Natuurcompendium_2003.html) due to land reclamation from the sea resulting in the creation of new wetlands (e.g., Oostvaardersplassen).

Global biodiversity is declining, and habitat destruction and degradation are caused mainly by changes in land use which, next to climate change, remains the most important driver of biodiversity loss (Hansen *et al.*, 2004). Changes in land use that are related to intensification and marginalization in agriculture are seen as major threats to European landscapes and their biodiversity (Jongman, 1996).

Therefore, there is an increasing need for reliable, up-to-date, Europe-wide data on land use and land cover to inform current environmental policies and nature conservation planning (Stanners & Bourdeaux, 1995). The impact of land use change is widely recognised and has forced national and international agencies to take policy measures to afford a higher degree of protection to our landscapes and habitats, in association with an increasing demand for monitoring and identification of potential sites for nature conservation. In Europe, the Convention on the Conservation of European Wildlife and Natural Habitats (the Bern Convention) that was adopted in Bern, Switzerland, in 1979 was a step forwards. The principal aim of the Convention is to ensure conservation and protection of wild plant and animal species and their natural habitats. To implement the Bern Convention in Europe, the European Community adopted Council Directive 79/409/EEC on the Conservation of Wild Birds (the EC Birds Directive), in 1979, and Council Directive 92/43/EEC on the Conservation of Natural Habitats and of Wild Fauna and Flora (the EC Habitats Directive), in 1992. The Directives facilitate, among other things, the establishment of a European network of protected areas (Natura, 2000), to tackle the continuing losses of European biodiversity due to human activities.

The loss of biodiversity has a clear global dimension. The United Nations Conference on Environment and Development (UNCED) in Rio de Janeiro, in 1992, led to the Rio Declaration, confirming the need to work towards international agreements to protect the integrity of the global environment. Countries acknowledged the responsibility that they bear in the international pursuit of sustainable development, in view of the pressures their societies place on the global environment and of the technologies and financial resources they command. In addition to the Rio Declaration, the 1992 Rio Earth Summit resulted in other important documents, such as the Agenda 21 and the Convention on Biological Diversity (CBD, 1992). The objectives and activities in Chapter 15 of Agenda 21 are intended to improve the conservation of biological diversity and the sustainable use of biological resources, and also to support the CBD (http://www.un.org/esa/sustdev/documents/agenda21/english/agenda21toc.htm). The CBD draws attention to the need to identify and monitor ecosystems, habitats, species, communities, genomes and genes (Spellenberg, 2005). Article 7 of the CBD (*Identification and Monitoring*) pursues monitoring the components of biological diversity through sampling and other techniques. Biological diversity – or biodiversity – is defined here as the variety of life on Earth and the natural patterns it forms. In 1995, at the 3rd Conference of Ministers *An Environment for Europe* in Sofia, a Pan-European response to the CBD was approved through the endorsement of the Pan-European Biological and Landscape Diversity Strategy (PELBDS) by 55 states present at the conference (Council of Europe, 1996). The PEBLDS strategy provided the only platform for Pan-European cooperation on tackling biodiversity loss (EEA, 2007). The PEBLDS Strategy aims to ensure the conservation of habitats and species, maintain genetic diversity and preserve important European landscapes. The Action Plan for European Landscapes (Theme 4) included the objective to establish of a Pan-European Landscape Map, next to the development of landscape assessment criteria, and a Strengths-Weaknesses-

Opportunities-Threats (SWOT) analysis of European landscapes (Council of Europe, 1996). The PEBLDS Strategy was reconfirmed by the leaders of the European Union at the Gothenburg Summit in 2001 and was adopted in 2003 in the Kyiv Resolution on Biodiversity at the fifth Ministerial Conference *An Environment for Europe*.

Conventions become especially focused when specific targets are set, such as the *2010 Biodiversity Target*, adopted in 2002 by CBD (CBD, 2002; Secretariat of the Convention on Biological Diversity, 2006). All CBD parties have committed themselves to achieving the 2010 Biodiversity Target: to protect and restore habitats and natural systems and halt the loss of biodiversity by 2010. To fulfil these targets, a Pan-European initiative; Streamlining European Biodiversity Indicators 2010 (SEBI 2010); was launched in 2004. This initiative is co-ordinated by the European Environment Agency (EEA) in collaboration with Directorate General (DG) Environment of the European Commission (EC), the European Centre for Nature Conservation (ECNC), United Nations Environment Programme – World Conservation Monitoring Centre (UNEP-WCMC) and the UNEP/PELBDS secretariat. An important objective of SEBI 2010 is the development of indicators to monitor and promote progress towards the achievement of the 2010 target. The SEBI process (EEA, 2007) proposed 26 indicators, with amongst others two important headline indicators: i) trends in extent of selected biomes, ecosystems and habitats, and ii) fragmentation of these selected classes.

All these policies show that the provision of quantitative figures on fragmentation and extent of habitats and their trends is fundamental for general policy formulation in relation to the maintenance and enhancement of biodiversity across Europe (Bunce *et al.*, 2008). The development of the series of Natura 2000 sites based on the above mentioned Directives is the major EU initiative for the protection of primary nature conservation areas (EU Council Directive, 1992; Ostermann, 1998). However, at the same time, these sites do not guarantee the maintenance of biodiversity in the wider countryside, because inevitably many habitats and species are outside protected areas (Bunce *et al.*, 2008). Therefore, there is a need to develop additional policy instruments for nature conservation outside protected areas that are equally appropriate to those applied within. The development of the Pan-European Ecological Network (PEEN) is the most significant tool in the implementation of PEBLDS (ECNC, 2004). The PEEN concept (Jones-Walters, 2007) is designed to strengthen the ecological coherence of Europe as a whole, with a common set of criteria consisting of core areas, corridors, buffer zones and nature development areas. One of the major goals of PEEN is to develop an indicative map of the Pan-European Ecological Network for the whole of Europe (van Opstal, 1999). The design of such an indicative PEEN map requires information about the spatial distribution of habitats and species in Europe, both inside and outside protected areas (Mücher *et al.*, 2005). This spatial information is also necessary to determine the spatial cohesion of habitat networks for viable populations in the landscape (Opdam *et al.*, 2003). Information about the spatial distribution of species is already being collected by many international organisations (e.g., Birdlife International), but methodologies for spatial modelling of European habitats and landscapes need to be developed, because there are currently no quantitative figures available for these.

In this chapter methodologies are proposed to identify the spatial distribution and extent of habitats and landscapes at a Pan-European scale, but there is also an urgent need for monitoring. Remote sensing provides excellent methods towards this objective, especially with regard to large areas such as Pan-Europe. These methods have merits, but also limitations, especially when considering small and fragmented habitats and gradual changes within them. Therefore it is additionally necessary to monitor the components of European landscapes, by the use of standardised procedures for the surveillance of habitats (points, lines and patches), in order to enable habitat changes to be assessed. The proposed field surveying method can facilitate the integration with remote sensing for baseline monitoring of habitats with a regional to global extent (Bunce et al., 2008; Mücher, 2009).

2. GEO-SPATIAL MODELLING OF EUROPEAN LANDSCAPES AND HABITATS

For the spatial modelling of European landscapes and habitats, use has been made of Geographic Information Science defined as Geographic Information Systems (GIS) combined with remote sensing methods and exploiting digitally available environmental data sets to indentify the spatial patterns or spatial distribution of landscapes and habitats. Burrough & McDonnell (1998) define GIS as a powerful set of tools for collecting, storing, retrieving at will, transforming and displaying spatial data from the real world for a particular set of purposes. Remote sensing is strongly related to GIS, since it is the science of obtaining information about an object, an area, or phenomenon through the analysis of data acquired by a device that is not in contact with the object, area or phenomenon under investigation (Lillesand *et al.*, 2008). Landscape ecology makes use of these methods and techniques to study and describe spatial configurations (Groom *et al.*, 2006). The spatial configurations are scale dependent. For example in landscape ecology, landscapes are conceived as a mosaic of land cover or habitat patches whose spatial pattern was significant in some profound sense (Potschin & Haines-Young, 2006). The definition of our objects of interest, namely landscapes and habitats is not that straightforward, since the interpretation of these concepts is very divergent, and differs according to the context and type of application. In this chapter landscapes are defined as recognizable, although often heterogeneous, parts of the earth's surface, which show a characteristic ordering of elements (Vos & Stortelder, 1992). Landscapes result from long-term interactions of natural abiotic, biotic and anthropogenic processes and are complex systems in which many components are interdependent (Mücher *et al.*, 2010). Habitats are defined on the European Nature Information System (EUNIS) website (http://eunis.eea.europa.eu) as follows: plant and animal communities as the characterising elements of the biotic environment, together with abiotic factors (soil, climate, water availability and quality, and others), operating together at a particular scale. More strictly habitats can be defined as ecotopes, defined by Runhaar & De Haes (1994) as spatial units that are homogenous in vegetation structure, succession stage and site factors that determine the species composition of the vegetation.

Ecological systems are characterized by diversity, heterogeneity and complexity (Wu & David, 2002) and need a multi-scale or hierarchical approach to their analysis, monitoring, modelling and management (Hay *et al.*, 2002). Wu & David (2002) advocate the Hierarchical Patch Dynamics Model (HPDM) which provides a powerful framework for breaking down complexity and integrating pattern with process (Wu & Marceau, 2002). HPDM uses a spatially nested patch hierarchy which consists of local ecosystems, local landscapes and regional landscapes. Jongman & Bunce (2000) propose a more comprehensive hierarchy, which is adapted here into the following hierarchical levels: (1) *biosphere* as the global sum of all ecosystems including its interactions with the lithosphere, hydrosphere and atmosphere; (2) *biogeographic regions* or environmental zones such as the Atlantic region which is dominated by a specific climate regime; (3) *landscape,* e.g., Atlantic lowlands dominated by clayey sediments and arable land such as the Dutch polders, characterized by a dominant biome and land use pattern at the regional scale. This is similar to the regional landscape of Wu & David (2002). (4)

Ecosystem or *habitat* such as a fresh water habitat. In principle these ecosystems or habitats consist of relatively homogenous vegetation-soil complexes and resemble the local ecosystem in HPDM; (5) *species* and *ecotypes*. Within a species, an ecotype is a genetically unique population that is adapted to its local environment. In this chapter, we adopt the above mentioned modification of HPDM and use its terminology as discussed above.

There have been many modelling studies on components of the European environment at the landscape level. Examples of these components are: composition, pattern and complexity (Perry & Enright, 2002; Papadimitriou, 2009), soil genesis (Sommer *et al.*, 2008), landscape change (De Aranzabal *et al.*, 2008), potential change (Brown, 2006) and nitrogen fluxes (Theobald *et al.*, 2004). Moreover, most of these studies concern a study area in one particular landscape type. Strikingly, there are no spatial modelling studies of the landscapes themselves at the European scale. Only the Burnett & Blaschke (2003) and Blaschke (2006) methodology for analysis of multi-scale segmentation/object relationship provides linkages for small-scale and large-scale landscape modelling. However, it is limited to the use of very high resolution satellite imagery. There are a number of regional and national landscape classifications, but they differ widely in methodological approaches, data sources and nomenclatures (Groom, 2005), and as a consequence they can not be integrated for Europe as a whole. Landscape classifications that are available for the whole of Europe, such as the ones from Meeus (1995) and Milanova & Kushlin (1993), are based on environmental data sets with coarse spatial resolution, and do not incorporate satellite imagery combined with modern GIS and remote sensing methods.

There are many more studies existing at the habitat level. Guisan & Zimmermann (2000) give an extensive review of predictive, niche, and species distribution modelling (see also Guisan & Thuiller, 2005). Niche-based species distribution models (Guisan & Zimmerman, 2000; Guisan & Thuiller, 2005; Dullinger *et al.*, 2009) have become an important tool for assessing the potential range of species under current as well as predicted future environmental conditions. The quantification of such species/environment relationships represents the core of predictive geographical modelling in ecology (Guisan & Zimmermann, 2000). Conservation biologists increasingly rely on spatial predictive models of biodiversity to support decision making (Steinmann *et al.*, 2009). Guisan & Zimmermann (2000) give an overview of the wide range of statistical methods that is in use to simulate the spatial distribution of terrestrial plant and animal species, biomes and other global vegetation groups, and plant functional types. In the majority of cases, the purpose of the statistical modelling is to predict species distribution (Austin, 2002). Studies that concentrate on the spatial modelling of European plant communities or vegetation types are less common. The paper by Zimmermann & Kienast (1999) concerns the predictive mapping of alpine grasslands using a species versus community approach, but is limited to the Swiss Alps. The two types of models presented in that paper yield patterns that are significantly correlated with real patterns observed in the field. Most of the statistical models in niche modelling rely to a large degree on bioclimatic and topographic data, and to some extent of soil properties. Almost no information is used on land use and land cover which determine to a large extent the actual distribution of species and habitats. Zimmermann & Kienast (1999) conclude that major problems arose from the lack of spatially explicit information of land use/history and the associated influence of soil development and secondary succession. Aready several studies included remotely sensed information for predictive habitat distribution modelling. Thuiller *et al.* (2004) investigated the extent to which the remotely sensed land cover classification PELCOM (Mücher *et al.*, 2000; 2001) improved the predictive power when added to bioclimatic predictors in models for a range of taxonomic groups. Although they found that remotely sensed predictors clearly improve the fit of individual species models, it did not improve the cross-validated accuracy of the models. Zimmermann *et al.* (2007) interpret this as an indication that land cover patterns are highly correlated with bioclimatic gradients. In addition, Pearson *et al.* (2004) state that remotely sensed habitat information helps to discriminate between suitable and unsuitable sites which cannot be distinguished from bioclimatic layers alone. Pearson *et al.* (2004) show that that there is good potential for integrating land cover into the existing bioclimatic modelling frameworks. Land cover determines habitat availability and its interaction with climate plays an important role in determining the biogeography of species.

Nevertheless, most of these studies concentrate on particular species, have a limited extent, or use coarse resolution spatial maps for large areas and they do not include high resolution land cover data. Since up-to-date quantitative figures on European habitats were missing, a methodology was developed to predict the actual distribution of habitats (and not individual species), as defined in the Annex I of the Habitats Directive, at a European scale, using environmental data sets with a high spatial resolution in rule-based classifications. Guisan & Zimmermann (2000) state in relation to this aspect that higher accuracy and resolution of biophysical input maps, e.g. land use and soil units that can act as powerful 'filters', are still considered as primary requirements for improving model predictions. Finally, they state that progress in GIS-modelling and in remote sensing could pave the way for obtaining more accurate information.

3. MONITORING EUROPEAN HABITATS USING REMOTE SENSING

The increasing deterioration of many landscapes, habitats and landscape elements demonstrates that they need to be protected and monitored in a more comprehensive fashion, ranging from regional to global scales. Monitoring is defined here as a procedure that involves the systematic measurement of a targeted object in time (at least two times) to be able to assess changes and trends in quantity and/or quality of the targeted object. And finally to understand the processes that are behind these changes. The use of remote sensing is an obvious means of providing the necessary information (Nagendra, 2001; Battrick, 2005; Battrick, 2006; Groom *et al.*, 2006) because, compared to other survey techniques, it is unique in its potential for providing census data; i.e. complete coverage of large areas which is able to complement sample data (Inghe, 2001). Amongst other things, the synoptic overview represents more for landscape ecology than the mere possibility of capturing a large area at one moment (Groom *et al.*, 2006). More fundamentally, it represents the possibility of identifying spatial-temporal patterns that are only discernible when a larger part of the landscape is repeatedly in view. Given that each nation state has its own history in surveying and mapping; the relevance of remote sensing for the coordination of Europe-wide landscape and habitat monitoring is significant, since satellite imagery operates irrespective of borders. Field surveys provide higher levels of accuracy than remote sensing, but its use makes it possible to increase the speed and frequency with which one can analyse a landscape (Strand *et al.*, 2007). Groom *et al.* (2006) state that the relationship between remote sensing and landscape ecology is an evolving relationship, because new possibilities for exploration are emerging through technological advancements, including those represented by newly launched satellite sensors and novel image interpretation methods. The wide array of satellite sensors differ in their spatial, temporal, spectral, and radiometric resolution. Developments in multi-angle viewing (Chen et al., 2003; Su *et al.*, 2007), radar (Bugden *et al.*, 2004), imaging spectroscopy (Foody *et al.*, 2004) and Lidar (Hall *et al.*, 2009) all have considerable potential relevance for monitoring.

However, consistent measurements are vital for long term monitoring of the environment. Therefore, it is important that consistent products are used throughout a project.

Noss (1990) describes a hierarchy concept for monitoring biodiversity. The different levels of information that can be considered for biodiversity and ecosystems studies are the *compositional*, *structural* and *functional* aspects of the landscape at multiple levels of ecological complexity. The compositional aspects discussed in this chapter are landscape and habitat types including structural aspects like habitat structure and physiognomy. Functional aspects are landscape and habitat processes, which can be monitored by habitat field surveying techniques, and the study of land cover changes. The conceptual framework of Noss (1990) may facilitate the selection of indicators to represent the different dimensions of biodiversity that provide a basis for monitoring. An indicator can be defined as a measure used to determine the performance of functions, processes, and outcomes over time (Strand *et al.*, 2007). Important 2010 biodiversity indicators selected by the Secretariat of the Convention on Biological Diversity and SEBI 2010 (EEA, 2007) to which this chapter can contribute include: (1) trends in the extent of selected biomes, ecosystems and habitats, (2) their fragmentation and (3) threats to biodiversity, such as land use and land cover changes. There are already a number of successful remote sensing studies which concentrate on a specific habitat, vegetation, or plant functional type using very high resolution satellite data (Küchler *et al.*, 2004; Mander *et al.*, 2005; Keramitsoglou *et al.*, 2005, Kobler *et al.*, 2006; Förster *et al.*, 2008.; Schaepman-Strub *et al.*, 2009), but they are limited in their spatial extent. Even for the majority of habitat types that could be mapped with high resolution image data, the lack of a simple relationship to a single biophysical parameter restricts the possibilities for many forms of automated image classification (Groom *et al.*, 2006). The possibilities for direct mapping from satellite imagery for general sets of habitats, therefore have limitations. Instead, it is possible to identify components of the habitat complexity that satellite imagery can more directly map and develop actual habitat mapping procedures accordingly. One such component is land cover, which has the capability of acting as a surrogate parameter between several major sets of habitat types. Examples are those that are primarily associated with certain parts of the landscape, such as forest, arable land, grassland and wetlands (Groom *et al.*, 2006; Duro *et al.*, 2007). A spatial modelling approach starting with remotely derived land cover is appropriate to identify the likely locations of specific habitats.

Land cover provides essential information for the spatial identification of landscapes and habitats and is the most dynamic part capable of being monitored using remote sensing. Duro *et al.* (2007) give a good overview with referring to studies in which indicators of biodiversity have been modelled or mapped from Earth Observation (EO), and show that land cover is a key component. As mentioned before, land use and climate change are the most important drivers of biodiversity loss. Habitat destruction and degradation are caused mainly by changes in land use. At the same time, land use and associated land cover have been changing at an increasing rate over recent centuries and decades, causing increasing pressures on landscapes, habitats, and biodiversity in general. Therefore, land cover monitoring is a central issue in biodiversity monitoring. Land cover is not the same as land use. In the simplest case, land cover is an expression of a specific land use intervention – including no intervention at all – on a specific type of land at a specific point of time (Stomph *et al.*, 1997). As stated by Stomph *et al.* (1997), the problem with the term land use is that land use refers both to the way land is used i.e. manipulated (the interventions by man) and to the use or economic function that land has to man (the purpose of these interventions). Land cover can be defined as 'the attributes occupying a part of the earth's surface, such as vegetation, artificial constructions, rocks and water which can be distinguished from a distance' (Anderson *et al.*, 1976). In principle everything that is seen by a satellite sensor is land cover. However, in many cases the land use can be inferred from the land cover by its spatial configuration and context. Sports fields, as an example, can be distinguished from grassland by their specific size and shape and the fact that they are often located within urban areas. Urban area is also a land use, as inferred from the built-up area seen from a distance. Land use and land cover have a many-to-many relationship and as such should be used as separate terms.

Important past and current activities in the derivation of Pan-European land cover information from remotely sensed data include: (1) the on-going CORINE (Coordination of Information on the Environment) land cover project (CEC, 1994) under the co-ordination of the European Environment Agency (EEA) that was initiated in 1985, (2) the 1 km global land cover product DISCover (Loveland *et al.*, 2000) established under the coordination of the International Geosphere and Biosphere Programme's Data and Information System (IGBP-DIS), (3) the 1 km Pan-European land cover database PELCOM established under the coordination of Alterra (Mücher *et al.*, 2000), (4) the 1 km GLC2000 global land cover data for the year 2000 established under the coordination of the Joint Research Centre (JRC) of the European Commission (Bartholomé & Belward, 2005), and (5) the recently finished 300 m global GLOBCOVER database (Arino *et al.*, 2008). Accuracy assessments are of utmost importance for the use of these land cover data sets.

Validation of the CLC2000 (CORINE land cover database for the year 2000) with LUCAS field samples from Eurostat indicated an average accuracy of 74.8% (Büttner & Maucha, 2006). Validation of the IGBP DISCover global land cover set indicated an area-weighted global accuracy of 66.9% (Scepan *et al.*, 1999). Validation of the PELCOM land cover database showed an overall accuracy of 69.2% (Mücher et al., 2001). Validation of the GLC2000 global land cover set indicated an area-weighted global accuracy of 68.6% (Mayaux *et al.*, 2006; Herold *et al.*, 2008). Validation of the 300 m GLOBCOVER indicated an area-weighted global accuracy of 73% (Defouney *et al.*, 2009). As stated already by Mücher *et al.*, (2000) and reconfirmed by Herold *et al.* (2008) the overall accuracy of continental or global land cover databases with low resolution satellite imagery barely exceeds 70% and medium resolution only achieves 73%. Such levels make it impossible to detect changes by comparing different land cover maps, while for biodiversity and environmental monitoring it is a prerequisite that the land cover databases can be easily updated. This means that additional techniques have to be developed to detect changes for Europe as a whole. Remote sensing definitely has limitations, especially with regard to habitats, and therefore needs to be complemented by field surveys. Sampling strategies or designs as proposed in Appendix 1 are crucial for the monitoring of habitats. Consistent biodiversity measurements in time and space are rare in Europe, with almost no consistent quantitative figures apart from butterflies and birds. Therefore a standardized procedure for the surveillance and monitoring of European habitats has been proposed (Bunce *et al.*, 2008).

4. QUESTIONS IN RELATION TO SPATIAL IDENTIFICATION AND MONITORING

The main objective of this chapter is to identify quantitative methodologies for the spatial identification and monitoring of European landscapes and habitats. In a broader context, it concerns biodiversity monitoring using Earth Observation data and methods as well as geo-information tools integrated with available European environmental data sets and field surveying techniques, with emphasis on habitats across European landscapes.

The increasing deterioration of many European landscapes, habitats and landscape elements has created the awareness that they need to be protected and monitored in more comprehensive ways. However, there are currently no quantitative figures about the extent and trends of European habitats and landscapes. To achieve this objective, the following specific research questions have been formulated:

A. What is the added value of remote sensing for landscape ecology in Europe, with special emphasis on mapping and monitoring of habitats and landscapes? And more specific: do uses of remote sensing provide principles for classification within European landscape ecology?
B. Is it possible to model the spatial distribution of European landscapes using remote sensing and additional spatial information?
C. Is it possible to model the spatial distribution of European habitats using remote sensing and additional spatial information?
D. Since land cover information plays a crucial role in the spatial modelling of European landscapes and habitats, can we monitor Europe's land cover?
E. If it is possible to monitor European habitats using standardized procedures for field surveillance, can this be integrated with remote sensing to mitigate the latter's limitations?

In relation to the main objective of this chapter, which was to identify quantitative methodologies for the spatial identification and monitoring of European landscapes and habitats, it can be concluded that; in combination with other environmental data sets; it is possible to model quantitatively the spatial extent of widespread habitats and landscapes on the basis of remotely sensed land cover information derived from satellite imagery (Mücher *et al.*, 2009, 2010). The lack of consistent cultural-historical digital data sets for Europe still is a major limitation in relation to the spatial modelling and characterization of European landscapes, and this might lead to the underestimation of regional identity (Mücher *et al.*, 2010). Although it is possible now to model the spatial extent of widespread European habitats, these patterns cannot be directly translated to area estimates of those habitats (Mücher *et al.*, 2009). This purpose requires validation and calibration with ground-truth sample sites across the European countryside as obtained from the field surveying methodology (Bunce *at al.*, 2008). The retrieval of accurate land cover information is not only crucial for the spatial modelling of European landscapes and habitats, but also for their monitoring, since their destruction and degradation are mainly caused by changes in land management, which remains the most important driver of biodiversity loss. Operational remote sensing enables land cover characterization at various scales but the classification accuracies are still insufficient at continental and global scales for monitoring purposes (Mücher *et al.*, 2000; Herold *et al.*, 2008). The use of continuous thematic fraction layers, as derived from linear unmixing, provides a good basis for monitoring land cover changes of Europe's complex landscapes (Mücher *et al.*, 2000). However, gradual or small changes in habitats and their quality are not easily detected by such images and therefore additional information from field surveying is needed. The field procedures developed for mapping patches as well as for linear and point habitats are sufficiently robust to provide a consistent baseline (Bunce *et al.*, 2008). They also provide perspectives for further integration with remotely sensed information. However, cost dictates that field surveys always need to be implemented using a sampling framework in which the samples are limited to small areas, e.g., one square kilometre. Spatial modelling of habitats is therefore required to provide a synoptic overview of their spatial distribution (Mücher *et al.*, 2009).

5. DISCUSSION

5.1 Spatial Modelling of European Landscapes and Habitats

Until recently there were few quantitative approaches to European landscape classification. Those that were available for Europe as a whole (e.g., Meeus, 1995), were coarse in spatial resolution and were not based on modern data acquisition and analysis. The newly established European landscape classification LANMAP was a major breakthrough, because a consistent methodology was used to integrate various thematic layers. It therefore provides a consistent view across Europe as well as a common language and classification system (Mücher *et al.*, 2010). However, there is still enough room for improvement. Firstly, LANMAP includes no information on socio-economic and cultural-historical aspects and, particularly with regard to spatial information, it is not expected that much of these aspects will become available consistently across Europe with sufficient regional detail. Nevertheless, it has been shown that information on landscape patterns can be derived in a consistent way from satellite imagery by segmentation techniques (Mücher *et al.*, 2007). Burnett & Blaschke (2003) have already shown the possibilities of multi-scale segmentation for landscape analysis. In the Austrian research project SINUS, Austrian cultural landscape types have been identified on the basis of segmentation of Landsat TM images (Peterseil *et al.*, 2004). Landscape structure provides a good basis for many indicators that can link patterns to processes within landscapes (Wrbka *et al*, 2004; Renetzeder *et al.*, 2010). Obtaining consistent landscape structure information for the whole of Europe can become a reality, but needs a higher resolution than is provided by Landsat, e.g. by the use of current SPOT satellite imagery. It would be interesting to investigate the added value of landscape-based metrics such as landscape heterogeneity, as expressed by the information-entropy of the Shannon index as extra parameters, to identify and describe European landscapes as has been done by Van Eetvelde & Antrop (2009a, 2009b) for Belgium. Integration of LANMAP with socio-economic data also took place in the SENSOR (EU FP-6 project Sustainability Impact Assessment: Tools for Environmental, Social and Economic Effects of Multifunctional Land Use in European Regions) project (Renetzeder *et al.*, 2008), but the selection of the appropriate parameters and their disaggregation to regional scales needs more research. Improvements are also needed in cases of specific landscape types (e.g., coastal dunes), by exploiting detailed digital elevation data within the coastal regions. Recently the landscape types in LANMAP have been described more extensively which was urgently needed (Van der Heijden, 2007). In the end, it will be important for national concepts to be nested within a hierarchy of scales that build upon each other. Regional, national and European units should therefore be part of the same methodological system and LANMAP should be able to provide such a framework at the highest level.

Until recently, spatial distribution maps of European habitats were not available. However, recent improved quantitative methologies have made it possible to model the spatial extent of widespread examples with unprecedented accuracy (Mücher *et al.*, 2009). Evans (2006) indicated that for the implementation of the Habitats Directive much information is still missing on habitat distribution. In this perspective, Evans indicated in October 2008 (pers. comm.) that a significant part of the habitat reports under Article 17 of the Habitats Directive provided limited or no information on a habitat's area and its trends. Therefore, the developed methodology and resulting habitat distribution maps are not only crucial for the design of ecological networks in Europe, but could also support individual countries in the production of distribution maps and area estimates. However, it is only possible to estimate the likely occurrence of the habitats if all spatial information layers are available. In cases where crucial information is lacking, e.g., on water quality, the inclusion of geo-referenced vegetation relevés as an additional information source is a possible methodological improvement, which would also be useful in cases of local and dispersed habitats. Nevertheless, the distribution maps cannot be directly translated into area estimates (number of hectares) of the specific habitat. For this, interpolation is needed between the remotely sensed data and in-situ information across Europe, which is currently investigated in the European projects ECOCHANGE (Challenges in assessing and forecasting biodiversity and ecosystem changes in Europe- EU FP6 project) and EBONE (European Biodiversity Observation Network: Design of a plan for an integrated biodiversity observing – EU FP7 project) in collaboration with SynBioSys Europe (Schaminée *et al.*, 2007). Precisely located geo-referenced vegetation relevés (point location) will provide suitable information for the further improvement of the knowledge rules with regard to site conditions. Due to the very limited surface of most vegetation relevés (much smaller than the spatial resolution of most sensors and more likely to represent a point than an area), they cannot be used easily to produce the confusion matrices that are needed to produce robust area estimates. Moreover, these vegetation relevés will miss most of the information on the presence of various landscape elements, like, hedgerows and small streams. The methodology for the field surveillance of habitats provides a basis for robust ground-truth measurements. It gives useful information for the validation and calibration (correspondence analysis) of the habitat distribution maps, as obtained from the spatial modelling methodology. This results in better area (stock) estimates of habitats than using land cover information alone. In the Flemish-Dutch project HABISTAT (A Classification Framework for Habitat Status Reporting with Remote Sensing Methods – STEREO II project) the proposed habitat recording methodology is currently being used for the training and validation of hyperspectral imagery (Haest *et al.*, 2009). With regard to the input data for spatial models there remains a serious shortage of validated European data sets on e.g. groundwater tables and water quality. The Atlas Florae Europaeae (AFE) should be expanded to include all European species.

In general, much work still remains to be done on the spatial and thematic improvements of the spatial input data sets and their accuracy assessments. For satellite sensors and derived products the CEOS (Committee on Earth Observation Satellites) working group on calibration & validation (WGCV) has an important role (Belward, 1999). However, environmental data sets that have not been derived from EO data need also standardized and robust accuracy assessements, which is unfortunately in many cases absent. Testing the range of uncertainties in the input data would be very valuable in relation to error propagation.

Higher spatial resolutions, especially of land cover information, elevation and soil data, would improve the modelling results to a large extent, because most European habitats are fragmented. The SRTM global elevation data set (Chen, 2005) already has a much higher spatial resolution (~ 90 m) than the GTOPO30 data set (~ 1 km), but has too many internal distortions caused by its acquisition procedure and processing chain. Further, the development of the expert system approach, by combining local ecological knowledge with available spatial information, would improve the identification of European habitats (http://www.synbiosys.alterra.nl/ecochange/single classes.aspx).

To achieve public appreciation and acceptance, the landscape and habitat maps resulting from the spatial modelling require high quality cartographic presentation. This process needs further development on generalization of e.g. gridcell derived polygons and lines (Chen & Chen, 2005).

As has been demonstrated, when using remote sensing based methods for habitat classification in Europe, current satellites (or combinations of different satellites) do not provide measurements of the Earth surface at the typical length scale of the existing habitats and their fragmentation levels. It might be suggested that forthcoming satellite initiatives could be based on summarizing the typical temporal, spectral and geometric resolutions needed for European habitat inventories. In this case user driven requirements, e.g., adequate instruments and platforms, could be used for a Pan-European habitat mapping at unprecedented accuracy. Currently, as has been shown throughout this work, mapping is limited by the nature of existing instruments, which were primarily designed for different purposes, a deficiency that significantly influences the accuracy of this work.

Since there are many possibilities for improving spatial identification of European habitats and estimates of their area, a priority ranking should be given in the following order:

- completing a baseline field survey of European habitats to enable validation and calibration of the habitat distribution maps and associated area estimates;
- finishing the Atlas Florae Europaeae for all European plant species;
- collecting, harmonizing and making available existing geo-referenced European vegetation relevés with a high spatial precision (geo-referenced to a point and not to a grid);
- obtaining more detailed land cover and digital elevation models;
- making use of forthcoming satellite initiatives that might fulfil typical temporal, spectral and geometric needs for European habitat inventories;
- collecting additional validated environmental data sets, e.g., on water quality and groundwater tables;
- improved methods for the generalization of gridcell polygons and lines to provide better cartographic products.

5.2 Monitoring of European Habitats

Accurate land cover information is crucial for monitoring as well as for spatial modelling of landscapes and habitats, whose destruction and modification are to a large extent caused by changes in land management. Monitoring is therefore essential for determining changes and trends in the extent and quality of a habitat. Land cover is the visual reflection of the land use at a certain moment in time and can be monitored very well by remote sensing. However, the use of remote sensing for monitoring is restricted by classification accuracies of only 70% maximum at continental and global scales. This limitation has two origins: First the complexity of the legend of land cover that does not reflect a physical measurement (satellites measure radiance and not categorical classes such as land cover, so you must always translate) and second, the perfect spectral, temporal and spatial satellite configuration is not yet available for this task. Due to limited land cover classification accuracies, land cover monitoring requires specific approaches towards change detection, such as in the CORINE land cover project (Perdigão & Annoni, 1997; Büttner *et al.*, 2002; Feranec *et al.*, 2007) or by thematic fraction techniques. However, severe limitations remain. The CORINE land cover database still has a limited spatial resolution (scale 1:100,000 and minimum of 5 ha for change detection) and the use of fraction images limits the number of thematic classes. More recent trends show that the construction of land cover databases can be based on the automatic classification of high resolution satellite imagery, e.g., Landsat imagery with a 25m spatial resolution, for very large areas such as Europe (Pekkarinen *et al.*, 2009). In this perspective, also a change detection method based on change vector analysis – decision tree classification (CVA-DTC method) of Xian *et al.* (2009) seems to be very promising. However, land cover change assessments in large areas still face many challenges, e.g., cost effectiveness, timely acquisition of data, minimizing inter- and intra-annual vegetation phenology variance, removal of image noises caused by atmospheric effects and the availability of appropriate analytical techniques (Coppin *et al.*, 2004; Xian *et al.*, 2009) A sampling approach, using statistically sound sample designs, would be a solution that provides a methodology for land cover monitoring at such scales. A thorough knowledge of existing land cover conditions is also needed to be integrated with the remotely sensed change detection. A sampling approach also provides opportunities for using newer sensors which have high spatial and spectral resolutions, e.g., imaging spectrometers. At the same time it must be noted, when using much higher spatial resolution satellite data, the complexity of signal interpretation usually increases. This is due to the fact that shaded components increase in area fraction when striving for higher spatial resolution. Shaded parts of canopies can extend to more than 50% cover within a pixel rendering habitat classification approaches significantly worse that 50% accuracy. In general, a sampling approach can bridge scaling gaps, allowing spatial-temporal continuous sampling with limited discontinuities, using a multitude of sensors with varying spatial-temporal characteristics, in combination with solid and continuous ground observations (Schaepman *et al.*, 2007). This requirement is also in line with the recently postulated complete observing system within the Global Earth Observation System of Systems (GEOSS). Sampling units for remotely sensed change detection can still be much larger than those used in most field surveys. In addition, once the objects are identified within the samples, remote sensing can provide excellent methods for the monitoring of specific bio-physical and bio-chemical parameters of objects, e.g., albedo, leaf area index, fractional cover, vegetation height, plant pigment and non-pigment retrieval at leaf or canopy level (Turner *et al.*, 1999; Cohen *et al.*, 2003; Schaepman-Strub *et al.*, 2006; Zimmerman *et al*, 2007; Joshi *et al.*, 2008; Ustin *et al.*, 2009). Time-series analysis of satellite imagery as a special case of change detection is especially suited for the identification of trends in phenology (e.g. length of the growing season), as White & Nemani (2006) have shown for real-time monitoring of land surface phenology; White *at al.* (2009) for the long-term changes in phenology in North-America and De Wit & Mücher (2009) for phenological trends in Europe. There are also improvements possible in thematic land cover, e.g., separation of evergreen from deciduous forests as different land cover types or plant functional groups (Vancutsem *et al.*, 2009). However, remote sensing can not solve the whole information chain. Remote sensing will always require ground truth information, not only for training and calibration of the used methodology but also for validation, since, although it addresses spatial and temporal scales inaccessible to traditional field surveys, it cannot match the accuracy and detail of in-situ measurements (Gross *et al.*, 2009). For field surveys involving estimates of the percentage cover of each life form and associated percentage of dominant species (both in vertical projection), efficient protocols in field recording are important for integration with remotely sensed information. Spatial accuracy and scale of the field measurements remain crucial for the integration with remote sensing (Zimmermann *et al.*, 2007) and are important characteristics of the proposed field methodology (Bunce *at al.*, 2008). Field surveys are indispensable because many changes in habitats are gradual shifts in habitat quality, such as changes in species abundance. Changes in land management such as adaptation to organic farming are also difficult to detect directly by remote sensing. While detailed vegetation records are not required for monitoring the habitat extent, such data are essential in determining habitat quality and condition, i.e. conservation status (Bunce *et al.*, 2008). Nevertheless, measuring step-wise changes in habitat quality remains as important as measuring changes in habitat quantity. A good principle is the concept that is provided by the

Natural Capital Index (Ten Brink & Tekelenburg, 2002).

As long as an appropriate sampling scheme is used, the methodology for field surveys provides a robust baseline for monitoring changes in habitats, and although its cost may seem high, it is relatively low in comparison with the estimation of Lengyel *et al.* (2008) that 80 million Euro are spent annually on 123 national habitat monitoring schemes. A stratified random sampling of 1 km^2 sample units is proposed for Europe (see also Appendix 1). Much can be said about the sample size, but smaller sample sites are not suitable for the integration with satellite sensors having a range of spatial resolutions (from 0.5 m to 1000 m) and are not cost-efficient since travel time may become expensive. Larger sample sizes could be suitable, but then it is recommended to use more sample sites instead of larger samples to reduce the standard deviation of error, as discussed in Jongman *et al.* (2006).

Although design-based sampling is less flexible than model-based sampling (Gruijter *et al.*, 2006), the former is preferred since assumptions can be limited and therefore more robust. Such a survey of habitats is essential in Europe as a baseline to compare the widely different national activities on habitat monitoring. Moreover, existing long-term (national) integrated monitoring programmes are difficult to harmonize and have been basically designed for national priorities. Failure to achieve an appropriate statistical structure for a monitoring programme will jeopardize the credibility of the results and support for the programme itself (Parr *et al.*, 2002).

The methodology for field surveillance can provide suitable in-situ sites for the validation and calibration of the habitat distribution maps described above, but it can also be used to calibrate land cover changes, as detected by remotely sensed information, with the changes in habitats obtained from the field survey to produce, not only trends in habitats for Europe, but also to anticipate the implications of actual and future land cover changes.

Although the frequency of remote sensing measurements is usually higher than for field measurements, decisions have to be made about the frequency of recording from space and in the field. Landscapes and habitats differ widely in their dynamics and may therefore require different frequencies of recordings. However, in the case of sample sites across Europe, a fixed frequency is suggested, to avoid misleading conclusions. A six-year cycle, as required for reporting under Article 17 of the Habitat Directive, seems to be optimal. However, in many European regions, within a given year, three high-resolution satellite images may be required to interpret the highly seasonal vegetation cover. In terms of the habitat types and life forms as the basis for the GHCs it is recommended to investigate more the possibilities of Lidar data in combination with ESA's Sentinel satellite family of optical and radar sensors (see also www.esa.int) to discriminate these classes.

Identified changes in land cover and associated habitats need to be analysed, summarised and reported at the different scales, e.g., by using the different levels of the European landscape classification (LANMAP) combined with possible driving forces that can be derived from e.g. socio-economic data and scenario studies (Mücher *et al.*, 2008). Nowcasting (actual monitoring), as well as hindcasting (historical monitoring), e.g EU project BIOPRESS (Linking pan-European land cover changes to pressures on biodiversity; Gerard *et al.*, 2010) and forecasting (scenario building; Kok *et al.*, 2007) are equally important. Knowledge of trends in land cover changes (land cover flows), not only *how much* but also *where* and *when* changes have occurred, can help land managers to identify key resource and ecosystem stressors, as well as prioritize management efforts (Wang *et al.*, 2009). Unfortunately, within European programmes currently more effort and resources are invested in scenario building than in actual monitoring of land cover and habitats.

Further research is therefore necessary in the future in order to understand the interaction of changes at diffent scales in our landscapes, and to assess the uncertainties in measurements and their propagation in time and space. Robust biodiversity observation networks that exploit both remote sensing and field surveys, in combination with appropriate data infrastructures, are essential to facilitate operational monitoring, not only at the European level, but also at global scales. This is also anticipated by USA National Ecological Observatory Network (NEON), in which the observatory design (NRC, 2003) has the overarching goal to enable understanding and forecasting of the impacts of climate change, land use change and invasive species on continental-scale ecology by providing infrastructure, and incorporating long term observation sites to support research in these areas. The NEON observation sites unfortunately, do not follow the principles of a proper sampling design, which is the same problem for the European Long Term Ecological Research (LTER) sites. LTER-Europe is Europe's long-term ecosystem research and monitoring (LTER) network. It was formally launched in June 2007, as a result of ALTER-Net work to develop the network (http://www.alter-net.info). Therefore, next to these LTER sites, a baseline monitoring system of our habitats remains an urgent requirement next to LTER sites and national monitoring programmes. The approach requires organizational skills that can be facilitated by incorporation into international programmes such as GMES and GEO.

REFERENCES

Anderson, J. R., Hardy, E. E., Roach, J. T., & Witmer, R. E. (1976). *Land use and land cover classification system for use with remote sensor data*. U.S. Geological Survey professional paper (964).

Arino, O., Bicheron, P., Achard, F., Latham, J., Witt, R., & Weber, J. L. (2008). GlobCover: The most detailed portrait of Earth. *European Space Agency Bulletin, 136*, 24–31.

Austin, M. P. (2002). Spatial prediction of species distribution: An interface between ecological theory and statistical modelling. *Ecological Modelling, 157*(2-3), 101–118. doi:10.1016/S0304-3800(02)00205-3

Bartholomé, E., & Belward, A. S. (2005). GLC2000: A new approach to global land cover mapping from earth observation data. *International Journal of Remote Sensing, 26*(9), 1959–1977. doi:10.1080/01431160412331291297

Battrick, B. (Ed.). (2005). *Global Earth Observation System of Systems GEOSS. 10-Year implementation plan reference document. Group on Earth Observations. GEO 1000R / ESA SP – 1284, ESA Publication Division*. Noordwijk, the Netherlands: ESTEC.

Battrick, B. (Ed.). (2006). *The changing Earth. New scientific challenges for ESA's living planet programme. SP-1304, ESA Publication Division*. Noordwijk, the Netherlands: ESTEC.

Belward, A. S. (1999). International co-operation in satellite sensor calibration: The role of the CEOS Working Group on calibration and validation. *Advances in Space Research, 23*(8), 1443–1448. doi:10.1016/S0273-1177(99)00296-3

Blaschke, T. (2006). The role of the spatial dimension within the framework of sustainable landscapes and natural capital. *Landscape and Urban Planning, 75*(3-4), 198–226. doi:10.1016/j.landurbplan.2005.02.013

Brown, I. (2006). Modelling future landscape change on coastal floodplains using a rule-based GIS. *Environmental Modelling & Software, 21*(10), 1479–1490. doi:10.1016/j.envsoft.2005.07.011

Bugden, J. L., Andrey, J., & Howarth, P. J. (2004). A SAR process model for land-cover mapping. *Canadian Journal of Remote Sensing, 30*(2), 195–204.

Bunce, R. G. H., Metzger, M. J., Jongman, R. H. G., Brandt, J., De Blust, G., & Elena-Rossello, R. (2008). A standardized procedure for surveillance and monitoring European habitats and provision of spatial data. *Landscape Ecology, 23*(1), 11–25. doi:10.1007/s10980-007-9173-8

Burnett, C., & Blaschke, T. (2003). A multi-scale segmentation/object relationship modelling methodology for landscape analysis. *Ecological Modelling, 168*(3), 233–249. doi:10.1016/S0304-3800(03)00139-X

Burrough, P. A., & McDonnell, R. A. (1998). *Principles of Geographical Information Systems*. Oxford, UK: Oxford University Press.

Büttner, G., Feranec, J., & Jaffrain, G. (2002). *Corine land cover update 2000: Technical guidelines*. (Technical report 89). EEA, Copenhagen.

Büttner, G., & Maucha, G. (2006). *The thematic accuracy of CORINE Land Cover 2000*. Assessment using LUCAS (land use/cover area frame statistical survey). (EEA Technical report no.7/2006). Copenhagen, Denmark.

CBD. (1992). *The United Nations Convention on Biological Diversity*. Retrieved December 29, 2009, from http://www.cbd.int/

CBD. (2002). *The United Nations Convention on Biological Diversity*. 2010 Biodiversity Target. Retrieved December 28, 2009, from http://www.cbd.int/2010-target/

CEC. (1994). *CORINE land cover technical guide*. European Union, Directorate-General Environment, Nuclear Safety and Civil Protection, Office for Official Publications of the European Communities, Luxembourg. EUR, ISSN 1018-5593, ISBN 92-826-2578-8.

Chen, C. F., & Chen, M. H. (2005). Generalization of GIS polygon data using curvature-based approach. [IGARSS]. *International Geoscience and Remote Sensing, 5*, 3498–3501.

Chen, J. (2005). Quality evaluation of topographic data from SRTM3 and GTOPO30. *Geomatics and Information Sciences, 30*(11), 941–944.

Chen, J. M., Liu, J., Leblanc, S. G., Lacaze, R., & Roujean, J. L. (2003). Multi-angular optical remote sensing for assessing vegetation structure and carbon absorption. *Remote Sensing of Environment, 84*(4), 516–525. doi:10.1016/S0034-4257(02)00150-5

Cohen, W. B., Maiersperger, T. K., Gower, S. T., & Turner, D. P. (2003). An improved strategy for regression of biophysical variables and Landsat ETM+ data. *Remote Sensing of Environment, 84*(4), 561–571. doi:10.1016/S0034-4257(02)00173-6

Coppin, P., Jonckheere, I., Nackaerts, K., Muys, B., & Lambin, E. (2004). Digital change detection methods in ecosystem monitoring: A review. *International Journal of Remote Sensing, 25*(9), 1565–1596. doi:10.1080/0143116031000101675

Council of Europe. UNEP, ECNC. (1996). *The Pan-European biological and landscape diversity strategy: A vision for Europe's natural heritage*. Strasbourg, Austria: Council of Europe Publishing.

De Aranzabal, I., Schmitz, M. F., Aguilera, P., & Pineda, F. D. (2008). Modelling of landscape changes derived from the dynamics of socio-ecological systems. A case of study in a semiarid Mediterranean landscape. *Ecological Indicators, 8*(5), 672–685. doi:10.1016/j.ecolind.2007.11.003

De Wit, A., & Mücher, C. A. (2009). *Satellite-derived trends in phenology over Europe: Real trends or algorithmic effects*. Poster presented at the General Assembly 2009 of the European Geosciences Union (EGU), from 19 – 24 April 2009, Vienna, Austria.

Defourny, P., Schouten, L., Bartalev, S., Bontemps, S., Cacetta, P., de Wit, A. J. W., et al. Arino, O. (2009). *Accuracy assessment of a 300 m global land cover map: The GlobCover experience*. 33rd International Symposium on Remote Sensing of Environment (ISRSE), May 4-8.2009, Stresa, Italy

Dullinger, S., Kleinbauer, I., Peterseil, J., Smolik, M., & Essl, F. (2009). Niche based distribution modelling of an invasive alien plant: Effects of population status, propagule pressure and invasion history. *Biological Invasions*, *11*(10), 2401–2414. doi:10.1007/s10530-009-9424-5

Duro, D. C., Coops, N. C., Wulder, M. A., & Han, T. (2007). Development of a large area biodiversity monitoring system driven by remote sensing. *Progress in Physical Geography*, *31*(3), 235–260. doi:10.1177/0309133307079054

ECNC, IUCN and Council of Europe. (2004). *Pan-European biological and landscape diversity strategy*. Council for the European Biological and Landscape Diversity Strategy. Third Intergovernmental Conference 'Biodiversity in Europe'. Follow up of the Kyiv Biodiversity Resolution: Pan-European Ecological Network Action Plan Proposal. STRA-CO (2004) 3c rev. Retrieved October 25, 2009, from http://www.peblds.org

EEA. (2005). *The European environment - state and outlook 2005*. Copenhagen.

EEA. (2007). *Halting the loss of biodiversity by 2010: Proposal for a first set of indicators to monitor progress in Europe*. (EEA Technical report No 11/2007). ISSN 1725–2237, EEA, Copenhagen.

EU Council Directive. (1992). *Natura 2000: Network on the conservation of natural habitats and of wild fauna. COUNCIL DIRECTIVE 92/43/EEC (1) of 21 May 1992*. The Council of the European Communities.

Evans, D. (2006). The habitats of the European union habitats directive. *Proceedings of the Royal Irish Academy- Section B, Biological Environments, 106*(3), 167-173.

Feranec, J., Hazeu, G., Christensen, S., & Jaffrain, G. (2007). Corine land cover change detection in Europe (case studies of The Netherlands and Slovakia). *Land Use Policy*, *24*(1), 234–247. doi:10.1016/j.landusepol.2006.02.002

Foody, G. M., Sargent, I. M. J., Atkinson, P. M., & Williams, J. W. (2004). Thematic labelling from hyperspectral remotely sensed imagery: Trade-offs in image properties. *International Journal of Remote Sensing*, *25*(12), 2337–2363. doi:10.1080/01431160310001654969

Förster, M., Frick, A., Walentowski, H., & Kleinschmit, B. (2008). Approaches to utilising QuickBird data for the monitoring of NATURA 2000 habitats. *Community Ecology*, *9*(2), 155–168. doi:10.1556/ComEc.9.2008.2.4

Gerard, F., Petit, S., Smith, G., Thomson, A., Brown, N., & Manchester, S. (2010). Huitu, Tuominen, S., Köhler, R., Olschofsky, K., Ziese, H., Kolar, J., Sustera, J., Luque, S., Pino, J., Pons, X., Roda, F., Roscher, M., Feranec, J., 2010. Land cover change in Europe between 1950 to 2000 determined employing aerial photography. *Progress in Physical Geography*, *34*(2), 183–205. doi:10.1177/0309133309360141

Groom, G. (2005). Methodological review of existing classifications. In D. M. Wascher (Ed.), *European landscape character areas–typology, cartography and indicators for the assessment of sustainable landscapes* (pp. 32-45). Final ELCAI project report, Landscape Europe.

Groom, G., Mücher, C. A., Ihse, M., & Wrbka, T. (2006). Remote sensing in landscape ecology: Experiences and perspectives in a European context. *Landscape Ecology*, *21*, 391–408. doi:10.1007/s10980-004-4212-1

Gross, J. E., Goetz, S. J., & Cihlar, J. (2009). Application of remote sensing to parks and protected area monitoring: Introduction to the special issue. *Remote Sensing of Environment*, *113*, 1343–1345. doi:10.1016/j.rse.2008.12.013

Gruijter, J. J., Bierkens, M. F. P., Brus, D. J., & Knotters, M. (2006). *Sampling for natural resource monitoring*. Berlin/ Heidelberg, Germany: Springer-Verlag.

Guisan, A., & Thuiller, W. (2005). Predicting species distribution: Offering more than simple habitat models. *Ecology Letters, 8*(9), 993–1009. doi:10.1111/j.1461-0248.2005.00792.x

Guisan, A., & Zimmermann, N. E. (2000). Predictive habitat distribution models in ecology. *Ecological Modelling, 135*(2-3), 147–186. doi:10.1016/S0304-3800(00)00354-9

Haest, B., Kempeneers, P., Cheung-Wai Chan, J., Vanden Borre, J., Kooistra, L., Mücher, C. A., & Deronde, B. (2009). *Spectral comparison of Chris/Proba and AHS hyperspectral imagery in the framework of NATURA 2000 habitat monitoring*. 6th EARSeL SIG IS workshop "Imaging Spectroscopy: Innovative tool for scientific and commercial environmental applications" Tel-Aviv, Israel, 16-19 March 2009.

Hall, R. K., Watkins, R. L., Heggem, D. T., Jones, K. B., Kaufmann, P. R., Moore, S. B., & Gregory, S. J. (2009). Quantifying structural physical habitat attributes using LIDAR and hyperspectral imagery. *Environmental Monitoring and Assessment, 159*(1-4), 63–83. doi:10.1007/s10661-008-0613-y

Hansen, A. J., deFries, R. S., & Turner, W. (2004). Land use change and biodiversity: A synchapter of rates consequences during the period of satellite imagery. In G. Gutman, A. C. Janetos, C. O. Justice, E. F. Moran, J. F. Mustard, R. R. Rindfuss,... M. A. Cochrane (Eds.), *Land change science: Observing, monitoring and understanding trajectories of change on the Earth's surface* (vol. 6) (pp. 277-300). Dordrecht, The Netherlands: Kluwer Academic Publishers.

Hay, G. J., Dubé, P., Bouchard, A., & Marceau, D. J. (2002). A scale-space primer for exploring and quantifying complex landscapes. *Ecological Modelling, 153*(1-2), 27–49. doi:10.1016/S0304-3800(01)00500-2

Herold, M., Woodcock, C. E., Loveland, T. R., Townshend, J., Brady, M., Steenmans, C., & Schmullius, C. C. (2008). Land-cover observations as part of a global earth observation system of systems (GEOSS): Progress, activities, and prospects. *IEEE Systems Journal, 2*(3), 414–423. doi:10.1109/JSYST.2008.925983

Inghe, O. (2001). The Swedish landscape monitoring programme: Current status and prospects for the near future. In Groom, G., & Reed, T. (Eds.), *Strategic landscape monitoring for the Nordic countries (TemaNord 2001: 523)* (pp. 61–67). Copenhagen: Nordic Council of Ministers.

Jones-Walters, L. (2007). Pan-European ecological networks. *Journal for Nature Conservation, 15*(4), 262–264. doi:10.1016/j.jnc.2007.10.001

Jongman, R. H. G. (Ed.). (1996). Ecological and landscape consequences of land use change in Europe. *Proceedings of the First ECNC Seminar on L and Use Change and its Ecological Consequences,* Tilburg, the Netherlands, 16–18 February 1995. ECNC Publication Series on Man and Nature, Volume 2, November 1996.

Jongman, R. H. G., & Bunce, R. G. H. (2000). Landscape classification, scales and biodiversity in Europe. In Mander, U., & Jongman, R. H. G. (Eds.), *Consequences of land use changes: Advances in ecological sciences* (pp. 11–38). Southampton/ Boston, MA: Wessex Institute of Technology Press.

Jongman, R. H. G., Bunce, R. G. H., Metzger, M. J., Mücher, C. A., Howard, D. C., & Mateus, V. L. (2006). Objectives and applications of a statistical environmental stratification of Europe. *Landscape Ecology*, *21*(3), 409–419. doi:10.1007/s10980-005-6428-0

Joshi, P. K., Gupta, B., & Roy, P. S. (2008). Spectral evaluation of vegetation features using multi-satellite sensor system (Terra ASTER, Landsat ETM+ and IRS 1D LISS III) in man-made and natural landscape. *Sensor Review*, *28*(1), 52–61. doi:10.1108/02602280810850035

Keramitsoglou, I., Kontoes, C., Sifakis, N., Mitchley, J., & Xofis, P. (2005). Kernel based re-classification of Earth observation data for fine scale habitat mapping. *Journal for Nature Conservation*, *13*(2-3), 91–99. doi:10.1016/j.jnc.2005.02.004

Klein Goldewijk, K., & Ramankutty, N. (2004). Land cover change over the last three centuries due to human activities: The availability of new global data sets. *GeoJournal*, *61*(4), 335–344. doi:10.1007/s10708-004-5050-z

Kobler, A., Džeroski, S., & Keramitsoglou, I. (2006). Habitat mapping using machine learning-extended kernel-based reclassification of an Ikonos satellite image. *Ecological Modelling*, *191*(1), 83–95. doi:10.1016/j.ecolmodel.2005.08.002

Köhler, R., Olschofsky, K., & Gerard, F. (Eds.). (2006). *Land cover change in Europe from the 1950s to 2000. Aerial photo interpretation and derived statistics from 59 samples distributed across Europe. Hamburg, Germany: University of Hamburg*. Germany: World Forestry Institute.

Kok, K., Verburg, P. H., & Veldkamp, T. (2007). Integrated assessment of the land system: The future of land use. *Land Use Policy*, *24*(3), 517–520. doi:10.1016/j.landusepol.2006.04.007

Küchler, M., Ecker, K., Feldmeyer-Christe, E., Graf, U., Küchler, H., & Waser, L. T. (2004). Combining remotely sensed spectral data and digital surface models for fine-scale modelling of mire ecosystems. *Community Ecology*, *5*(1), 55–68. doi:10.1556/ComEc.5.2004.1.6

Lengyel, S., Déri, E., Varga, Z., Horváh, R., Tóthmérész, B., & Henry, P. Y. (2008). Habitat monitoring in Europe: A description of current practices. *Biodiversity and Conservation*, *17*(14), 3327–3339. doi:10.1007/s10531-008-9395-3

Lillesand, T. M., Kiefer, R. W., & Chipman, J. W. (2008). *Remote sensing and image interpretation* (6th ed.). New York, NY: Wiley.

Loveland, T. R., Reed, B. C., Brown, J. F., Ohlen, D. O., Zhu, Z., Yang, L., & Merchant, J. W. (2000). Development of a global land cover characteristics database and IGBP DISCover from 1 km AVHRR data. *International Journal of Remote Sensing*, *21*(6-7), 1303–1330. doi:10.1080/014311600210191

Mander, Ü., Mitchley, J., Xofis, P., Keramitsoglou, I., & Bock, M. (2005). Earth observation methods for habitat mapping and spatial indicators for nature conservation in Europe. *Journal for Nature Conservation*, *13*(2-3), 69–73. doi:10.1016/j.jnc.2005.04.001

Mayaux, P., Eva, H., Gallego, J., Strahler, A. H., Herold, M., & Agrawal, S. (2006). Validation of the global land cover 2000 map. *IEEE Transactions in Geoscience Remote*, *44*(7), 1728–1737. doi:10.1109/TGRS.2006.864370

Meeus, J. H. A. (1995). Pan-European landscapes. *Landscape and Urban Planning*, *31*(1-3), 57–79. doi:10.1016/0169-2046(94)01036-8

Milanova, E. V., & Kushlin, A. V. (Eds.). (1993). *World map of present day landscapes. An explanatory note.* Department of World Physical Geography and Geoecology, Moscow State University, in collaboration with UNEP.

Moran, E. F., Skole, D. L., & Turner, B. L. II. (2004). The development of the international land use and land cover change (LCLUC) initiative. In Gutman, G., Janetos, A. C., Justice, C. O., & Moran, E. F. (Eds.), *Land change science: Observing, monitoring and understanding trajectories of change on the Earth's surface* (pp. 1–16). Dordrecht, The Netherlands: Kluwer Academic Publishers.

Mücher, C. A. (2009). *Geo-spatial modelling and monitoring of European landscapes and habitats using remote sensing and field surveys*. PhD thesis, Wageningen University, Wageningen, The Netherlands. ISBN 978-90-8585-453-1, 269 pp.

Mücher, C. A., Champeaux, J. L., Steinnocher, K. T., Griguolo, S., Wester, K., Heunks, C., et al. Nieuwenhuis, G. J. A. (2001). *Development of a consistent methodology to derive land cover information on a European scale from remote sensing for environmental monitoring: The PELCOM report*. Alterra-rapport 178, Wageningen, the Netherlands.

Mücher, C. A., Hazeu, G. W., Swetnam, R., Pino, J., Halada, L., & Gerard, F. (2008). Historic land cover changes at Natura 2000 sites and their associated landscapes across Europe. *Proceedings 28th EARSeL Symposium: Remote Sensing for a Changing Europe* (pp. 226-231). Istanbul, Turkey.

Mücher, C. A., Hennekens, S. M., Bunce, R. G. H., & Schaminée, J. H. J. (2005). Spatial identification of European habitats to support the design and implementation of a Pan-European Ecological Network. In D. McCollin & J. I. Jackson (Eds.), *Planning, people and practice. The landscape ecology of sustainable landscapes* (pp. 217-225). 13th Annual IALE(UK) Conference held at the University of Northampton.

Mücher, C. A., Hennekens, S. M., Bunce, R. G. H., Schaminée, J. H. J., & Schaepman, M. E. (2009). Modelling the spatial distribution of Natura 2000 habitats across Europe. *Landscape and Urban Planning*, *92*(2), 148–159. doi:10.1016/j.landurbplan.2009.04.003

Mücher, C. A., Klijn, J. A., Wascher, D. M., & Schaminée, J. H. J. (2010). A new European Landscape Classification (LANMAP): A transparent, flexible and user-oriented methodology to distinguish landscapes. *Ecological Indicators*, *10*, 87–103. doi:10.1016/j.ecolind.2009.03.018

Mücher, C. A., Steinnocher, K. T., Kressler, F. P., & Heunks, C. (2000). Land cover characterization and change detection for environmental monitoring of pan-Europe. *International Journal of Remote Sensing*, *21*(6-7), 1159–1181. doi:10.1080/014311600210128

Mücher, C. A., Vos, C. C., Renetzeder, C., Wrbka, T., Kiers, M. A., van Eupen, M., & Bugter, R. J. F. (2007). *The application of satellite imagery to identify landscape structure. Proceedings of the IALE 2007 World Congress* (pp. 590-591). Wageningen, the Netherlands.

Mücher, C. A., & Wascher, D. M. (2007). European landscape characteracterization. In Pedroli, B., van Doorn, A., de Blust, G., Paracchini, M. L., Wascher, D., & Bunce, F. (Eds.), *Europe's living landscapes* (pp. 37–43). KNNV Publishing.

Nagendra, H. (2001). Review article. Using remote sensing to assess biodiversity. *International Journal of Remote Sensing*, *22*(12), 2377–2400. doi:10.1080/01431160117096

Noss, R. F. (1990). Indicators for monitoring biodiversity: A hierarchical approach. *Conservation Biology*, *4*(4), 355–364. doi:10.1111/j.1523-1739.1990.tb00309.x

NRC (National Research Council). (2003). *NEON: Addressing the nation's environmental challenges*. Washington, DC: National Academies Press.

Opdam, P., Verboom, J., & Pouwels, R. (2003). Landscape cohesion: An index for the conservation potential of landscapes for biodiversity. *Landscape Ecology*, *18*, 113–126. doi:10.1023/A:1024429715253

Ostermann, O. P. (1998). The need for management of nature conservation sites designated under Natura 2000. *Journal of Applied Ecology*, *35*(6), 968–973. doi:10.1111/j.1365-2664.1998.tb00016.x

Papadimitriou, F. (2009). Modelling spatial landscape complexity using the Levenshtein algorithm. *Ecological Informatics*, *4*(1), 48–55. doi:10.1016/j.ecoinf.2009.01.001

Parr, T. W., Ferretti, M., Simpson, I. C., Forsius, M., & Kovács-Láng, E. (2002). Towards a long-term integrated monitoring programme in Europe: Network design in theory and practice. *Environmental Monitoring and Assessment*, *78*(3), 253–290. doi:10.1023/A:1019934919140

Pearson, R. G., Dawson, T. P., & Liu, C. (2004). Modelling species distributions in Britain: A hierarchical integration of climate and land-cover data. *Ecography*, *27*(3), 285–298. doi:10.1111/j.0906-7590.2004.03740.x

Pekkarinen, A., Reithmaier, L., & Strobl, P. (2009). Pan-European forest/non-forest mapping with Landsat ETM+ and CORINE Land Cover 2000 data. *ISPRS Journal of Photogrammetry and Remote Sensing*, *64*(2), 171–183. doi:10.1016/j.isprsjprs.2008.09.004

Perdigão, V., & Annoni, A. (1997). *Technical and methodological guide for updating CORINE land cover database*. Joint Research Centre. Ispra.

Perry, G. L. W., & Enright, N. J. (2002). Spatial modelling of landscape composition and pattern in a maquis-forest complex, Mont Do, New Caledonia. *Ecological Modelling*, *152*(2-3), 279–302. doi:10.1016/S0304-3800(02)00004-2

Peterseil, J., Wrbka, T., Plutzar, C., Schmitzberger, I., Kiss, A., & Szerencsits, E. (2004). Evaluating the ecological sustainability of Austrian agricultural landscapes-the SINUS approach. *Land Use Policy*, *21*(3), 307–320. doi:10.1016/j.landusepol.2003.10.011

Potschin, M. B., & Haines-Young, R. H. (2006). Landscapes and sustainability. *Landscape and Urban Planning*, *75*(3-4), 155–161. doi:10.1016/j.landurbplan.2005.03.006

Renetzeder, C., Schindler, S., Peterseil, J., Prinz, M. A., Mücher, C. A., & Wrbka, T. (2010). Can we measure ecological sustainability? Landscape pattern as an indicator for naturalness and land use intensity at regional, national and European level. *Ecological Indicators*, *10*, 39–48. doi:10.1016/j.ecolind.2009.03.017

Renetzeder, C., van Eupen, M., Mücher, C. A., & Wrbka, T. (2008). A spatial regional reference framework for sustainability assessment in Europe. In Helming, K., Perez-Soba, M., & Tabbush, P. (Eds.), *Sustainability impact assessment of land use changes* (pp. 249–268). Berlin/Heidelberg, Germany & New York, NY: Springer. doi:10.1007/978-3-540-78648-1_13

Runhaar, J., & De Haes, H. A. (1994). The use of site factors as ecosystem classification characteristics. In Klijn, F. (Ed.), *Ecosystem classification for environmental management* (pp. 139–172). Dordrecht, The Netherlands: Kluwer.

Scepan, J., Menz, G., & Hansen, M. C. (1999). The DISCover validation image interpretation process. *Photogrammetry Engineering and Remote Sing*, *65*(9), 1075–1081.

Schaepman, M. E., Malenovsky, Z., Mücher, C. A., Kooistra, L., & Thullier, W. (2007). Bridging scaling gaps for the assessment of biodiversity from space. In G. E. O. Secretariat (Ed.), *The full picture. A publication for the GEO Ministerial Summit, Earth Observation for Sustainable Growth and Development* (pp. 258-161). Cape Town, Geneva (CH) and Tudor Rose.

Schaepman-Strub, G., Limpens, J., Menken, M., Bartholomeus, H. M., & Schaepman, M. E. (2009). Towards spatial assessment of carbon sequestration in peatlands: Spectroscopy based estimation of fractional cover of three plant functional types. *Biogeosciences*, *6*(2), 275–284. doi:10.5194/bg-6-275-2009

Schaepman-Strub, G., Schaepman, M. E., Martonchik, J., & Schaaf, C. (2006). What's in a satellite albedo product? *International Geoscience and Remote Sensing* (IGARSS), 2848-2851.

Schaminée, J. H. J., Hennekens, S. M., & Ozinga, W. A. (2007). Use of the ecological information system SynBioSys for the analysis of large datasets. *Journal of Vegetation Science*, *18*, 463–470. doi:10.1111/j.1654-1103.2007.tb02560.x

Secretariat of the Convention on Biological Diversity. (2006). *Global biodiversity outlook 2*. Montreal.

Shi, W. Z., Ehlers, M., & Molenaar, M. (2005). Uncertainties in integrated remote sensing and GIS. *International Journal of Remote Sensing*, *26*(14), 2911–2915. doi:10.1080/01431160500197537

Sommer, M., Gerke, H. H., & Deumlich, D. (2008). Modelling soil landscape genesis - a time split approach for hummocky agricultural landscapes. *Geoderma*, *145*(3-4), 480–493. doi:10.1016/j.geoderma.2008.01.012

Spellenberg, I. F. (2005). *Monitoring ecological change* (2nd ed.). Cambridge, UK: Cambridge University Press. doi:10.1017/CBO9780511614699

Stanners, D., & Bordeaux, P. (Eds.). (1995). *Europe's environment; The Dobříš assessment*. Luxembourg: European Environmental Agency, Copenhagen. Office for Official Publications of the European Communities.

Steinmann, K., Linder, H. P., & Zimmermann, N. E. (2009). Modelling plant species richness using functional groups. *Ecological Modelling*, *220*(7), 962–967. doi:10.1016/j.ecolmodel.2009.01.006

Stomph, T. J., Mücher, C. A., & Fresco, L. O. (1997). Environmental impact of land use: A new basis for analysis. *The Land*, *1*(2), 129–142.

Strand, H., Höft, R., Strittholt, J., Miles, L., Horning, N., Fosnight, E., & Turner, W. (Eds.). (2007). *Sourcebook on remote sensing and biodiversity indicators*. Secretariat of the Convention on Biological Diversity, Montreal, Technical Series no. 32.

Su, L., Chopping, M. J., Rango, A., Martonchik, J. V., & Peters, D. P. C. (2007). Differentiation of semi-arid vegetation types based on multi-angular observations from MISR and MODIS. *International Journal of Remote Sensing*, *28*(6), 1419–1424. doi:10.1080/01431160601085995

Ten Brink, B. J. E., & Tekelenburg, T. (2002). *Biodiversity: How much is left*? The Natural Capital Index framework (NCI). (RIVM report 402001014). Bilthoven.

Theobald, M. R., Dragosits, U., Place, C. J., Smith, J. U., Sozanska, M., & Brown, L. (2004). Modelling nitrogen fluxes at the landscape scale. *Water Air and Soil Pollution Focus*, *4*(6), 135–142. doi:10.1007/s11267-004-3023-3

Thuiller, W., Araujo, M. B., & Lavorel, S. (2004). Do we need land-cover data to model species distributions in Europe? *Journal of Biogeography*, *31*, 353–361. doi:10.1046/j.0305-0270.2003.00991.x

Turner, D. P., Cohen, W. B., Kennedy, R. E., Fassnacht, K. S., & Briggs, J. M. (1999). Relationships between leaf area index and Landsat TM spectral vegetation indices across three temperate zone sites. *Remote Sensing of Environment*, *70*(1), 52–68. doi:10.1016/S0034-4257(99)00057-7

Ustin, S., Gitelson, A. A., Jacquemoud, S., Schaepman, M. E., Asner, G., Gamon, J. A., & Zarco-Tejada, P. (2009). Retrieval of foliar information about plant pigment systems from high resolution spectroscopy. *Remote Sensing of Environment, 113*(1), S67–S77. doi:10.1016/j.rse.2008.10.019

Van der Heijden, R. B. J. (2007). *Characterization of European landscapes and analysis of their dynamics. Internal CGI Report*. Wageningen, the Netherlands: Alterra.

Van Duuren, L., Eggink, G. J., Kalkhovan, J., Notenboom, J., van Strien, A. J., & Wortelboer, R. (Eds.). (2003). *Natuurcompendium 2003*. Natuur in cijfers. CBS (Voorburg), MNP Bilthoven en Wageningen. Retrieved from November 20, 2009, from http://www.pbl.nl/nl/publicaties/mnp/2003/Natuurcompendium_2003.html, ISBN 906960101X

Van Eetvelde, V., & Antrop, M. (2009a). Indicators for assessing changing landscape character of cultural landscapes in Flanders (Belgium). *Land Use Policy, 26*(4), 901–910. doi:10.1016/j.landusepol.2008.11.001

Van Eetvelde, V., & Antrop, M. (2009b). A stepwise multi-scaled landscape typology and characterization for trans-regional integration, applied on the federal state of Belgium. *Landscape and Urban Planning, 91*(3), 160–170. doi:10.1016/j.landurbplan.2008.12.008

Van Opstal, A. (1999). *The architecture of the Pan-European Ecological Network: Suggestions for concept and criteria*. Discussion report on behalf of the committee of experts of the Pan European Ecological Network. Report IKC Natuurbeheer Nr.37, Wageningen, The Netherlands.

Vancutsem, C., Pekel, J. F., Evrard, C., Malaisse, F., & Defourny, P. (2009). Mapping and characterizing the vegetation types of the Democratic Republic of Congo using SPOT vegetation time series. *International Journal of Applied Earth Observation and Geoinformation, 11*(1), 62–76. doi:10.1016/j.jag.2008.08.001

Vos, W., & Stortelder, A. H. F. (1992). *Vanishing Tuscan landscapes: Landscape ecology of a Submediterranean-Montane area (Solano Basin, Tuscany, Italy)*. Wageningen, The Netherlands: Pudoc Scientific Publishers.

Wang, Y., Mitchell, B. R., Nugranad-Marzilli, J., Bonynge, G., Zhou, Y., & Shriver, G. (2009). Remote sensing of land-cover change and landscape context of the National Parks: A case study of the Northeast Temperate Network. *Remote Sensing of Environment*..doi:10.1016/j.rse.2008.09.017

White, M. A., de Beurs, K. M., Didan, K., Inouye, D. W., Richardson, A. D., & Jensen, O. P. (2009). Intercomparison, interpretation, and assessment of spring phenology in North America estimated from remote sensing for 1982-2006. *Global Change Biology, 15*(10), 2335–2359. doi:10.1111/j.1365-2486.2009.01910.x

White, M. A., & Nemani, R. R. (2006). Real-time monitoring and short-term forecasting of land surface phenology. *Remote Sensing of Environment, 104*(1), 43–49. doi:10.1016/j.rse.2006.04.014

Wrbka, T., Erb, K. H., Schulz, N. B., Peterseil, J., Hahn, C., & Haberl, H. (2004). Linking pattern and process in cultural landscapes. An empirical study based on spatially explicit indicators. *Land Use Policy, 21*(3), 289–306. doi:10.1016/j.landusepol.2003.10.012

Wu, J., & David, J. L. (2002). A spatially explicit hierarchical approach to modeling complex ecological systems: Theory and applications. *Ecological Modelling, 153*(1-2), 7–26. doi:10.1016/S0304-3800(01)00499-9

Wu, J., & Marceau, D. (2002). Modeling complex ecological systems: An introduction. *Ecological Modelling, 153*(1-2), 1–6. doi:10.1016/S0304-3800(01)00498-7

Xian, G., Homer, C., & Fry, J. (2009). Updating the 2001 National Land Cover Database land cover classification to 2006 by using Landsat imagery change detection methods. *Remote Sensing of Environment, 113*(6), 1133–1147. doi:10.1016/j.rse.2009.02.004

Zimmermann, N. E., Edwards, T. C. Jr, Moisen, G. G., Frescino, T. S., & Blackard, J. A. (2007). Remote sensing-based predictors improve distribution models of rare, early successional and broadleaf tree species in Utah. *Journal of Applied Ecology, 44*(5), 1057–1067. doi:10.1111/j.1365-2664.2007.01348.x

Zimmermann, N. E., & Kienast, F. (1999). Predictive mapping of alpine grasslands in Switzerland: Species versus community approach. *Journal of Vegetation Science, 10*(4), 469–482. doi:10.2307/3237182

Zimmermann, N. E., Washington-Allen, R. A., Ramsey, R. D., Schaepman, M. E., Mathys, L., Kötz, B., et al. Edwards, T. C. (2007). Modern remote sensing for environmental monitoring of landscape states and trajectories. In F. Kienast, O. Wildi & S. Ghosh (Eds.), *A changing world: Challenges for landscape research* (pp.65 – 91). Dordrecht, The Netherlands: Springer.

Section 3
Methods:
Land Use and Biodiversity Modeling

Chapter 5
Integrated Modeling of Global Environmental Change (IMAGE)

T. Kram
PBL Netherlands Environmental Assessment Agency, The Netherlands

E. Stehfest
PBL Netherlands Environmental Assessment Agency, The Netherlands

ABSTRACT

Continued population growth, rising per capita income, industrialization and ever-increasing flows of materials, have created growing concern over how to ensure a more sustainable form of global human development. It is widely accepted that human development in currently less developed countries, following a similar path of many industrialized countries in coming decades, will lead to an unsustainable future. In particular, problems associated with climate change, loss of biodiversity, water scarcity, and the accelerated nitrogen cycle will be encountered at global, continental, and regional scales. Solving them will demand a comprehensive understanding of the Earth system. Integrated assessment models such as the Integrated Model to Assess the Global Environment (IMAGE) is a helpful tool for investigating these changes, their causes, and interlinkages in a comprehensive framework. This includes the major feedback mechanisms in the biophysical system. This chapter describes briefly the history of IMAGE, data and sub-models, and how they are linked together It is adapted from Kram & Stehfest (2006). IMAGE starts from basic driving forces like demographics and economic development, energy consumption and production, and agricultural demand, trade, and production. Important elements in the bio-physical modeling are addressed, such as land cover and land use processes, the global current and historical carbon cycle, the global nitrogen cycle, management of nutrients in agricultural systems, and climate variability including interaction with land use. A short discussion on uncertainty and sensitivity is presented, and finally, an overview of major applications is given.

DOI: 10.4018/978-1-60960-619-0.ch005

1. INTRODUCTION

The current version of the Integrated Model to Assess the Global Environment (IMAGE 2.4), described in this chapter, represents the latest incarnation of a development that goes back as far as the late 1980s. Then a team at the National Institute for Public Health and the Environment (RIVM) in Bilthoven, the Netherlands, embarked on developing a global model to explore relevant aspects of climate change, emerging in those years as an important case for internationally concerted policy deliberations. The first version (1.0), formerly known as the Integrated Model to Assess the Greenhouse Effect (IMAGE), was a global, single-region model describing global trends in driving forces and the ensuing consequences for climatic change and impacts on key sectors, through a coupled set of modules representing the main processes involved (Rotmans, 1990). At the time, IMAGE 1.0 was among the first pioneering examples of Integrated Assessment Models addressing climate change.

Since then, IMAGE has evolved through a series of new versions, each introducing major revisions, enhancements and extensions up to the current version (2.4) briefly described here. This version marks an important milestone on the development path towards a next generation model, referred to as IMAGE 3, aimed at capturing – to a larger extent – the different aspects and domains of sustainability, with emphasis on the ecological domain but also related to the economic and social domains.

Specific features of the IMAGE model include comprehensive coverage of direct and indirect pressures on human and natural systems, closely related to human activities in industry, housing, transport, agriculture and forestry. The socio-economic activities and drivers of change are elaborated at the 24 region level (Figure 1), while the climate, land-cover and land-use change-related processes are represented in a geographically explicit manner on the 0.5 by 0.5 degree grid scale. It is this latter characteristic, relatively rare in integrated assessment models, that makes IMAGE particularly suited to exploring interactions between human and natural systems.

Key elements of sustainable development include provision of affordable energy while keeping air pollution and climate change under control; management of water systems in support of agriculture, industry and human settlements; increasing agricultural production while protecting soil, groundwater and surface water quality, and slowing down and eventually halting further loss of biodiversity. More generally, these issues can be described as the challenge to strike the balance between the increased use of natural systems for human development and the goods and services provided to humans by natural ecosystems, which are put at risk by human activities (Millennium Ecosystem Assessment, 2005a). An integrated assessment model like IMAGE 2.4 is a helpful tool for investigating these interactions in a comprehensive framework and understanding the major feedback mechanisms within the biophysical systems.

As stated earlier, the current version of IMAGE is the result of many years of development at the National Institute for Public Health and the Environment (RIVM), and –following a recent re-organization – the now separate Netherlands Environmental Assessment Agency (PBL). The development stages can be followed in a series of three books (Rotmans, 1990; Alcamo, 1994; Alcamo *et al.*, 1998). Substantive further development work was undertaken between 1998 and 2001, resulting in the version 2.2 model used to elaborate the IPCC-SRES scenarios (Nakicenovic *et al.*, 2000). The documentation on version 2.2 covering the implementation of the SRES scenarios is included on two CD-ROMs (IMAGE-team, 2001a; IMAGE-team, 2001b).

This chapter summarizes version 2.4 of the IMAGE model and is largely based on Kram & Stehfest (2006). After highlighting some key features of version 2.3 in the framework of a

Figure 1. Breakdown in regions in IMAGE 2.4

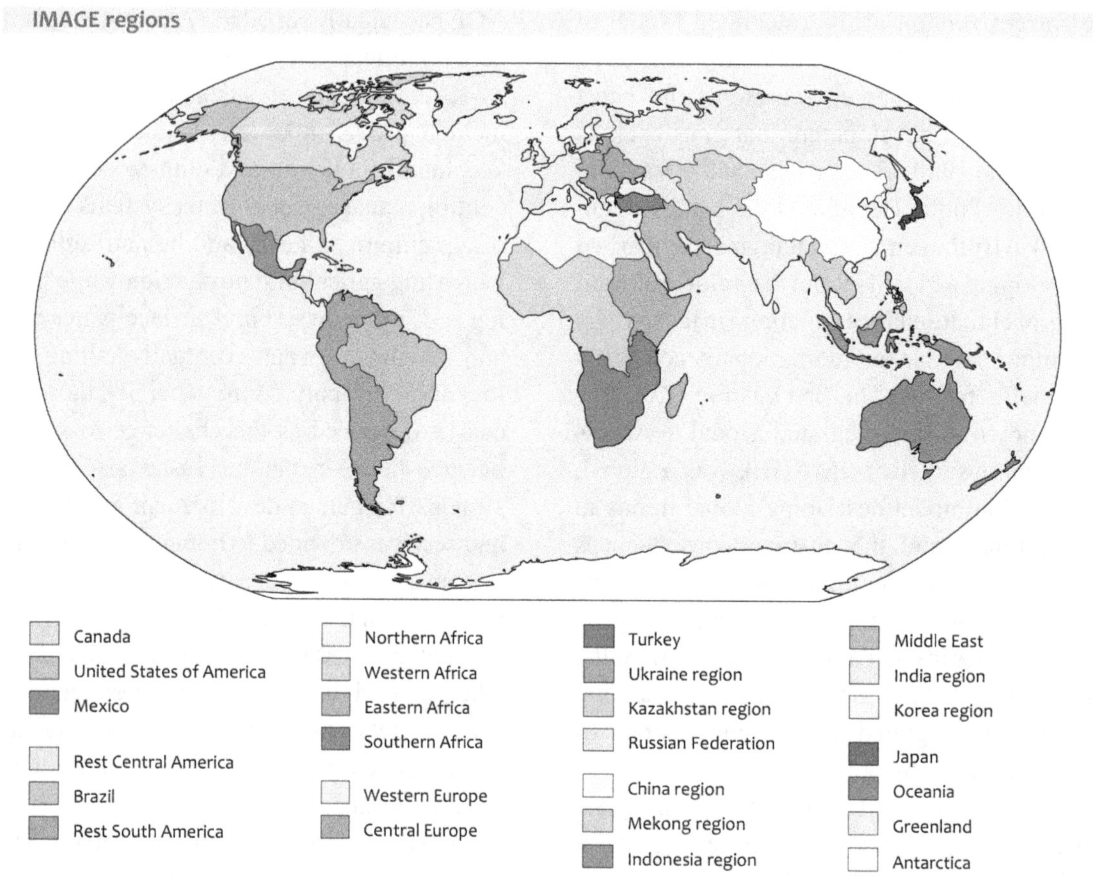

historical overview, the first section summarizes the development process of the model until 2010. It goes on to compile uncertainty and sensitivity analyses of the main model components, which is followed by an overview of recent and current IMAGE applications.

2. HISTORY OF THE IMAGE MODEL

The IMAGE model had its beginnings back in the mid 1980s, when RIVM decided to build a simple prototype model to capture the relationships between human activities and climate change. The experience gained from the prototype was subsequently used to build the Integrated Model to Assess the Greenhouse Effect (IMAGE 1.0, Rotmans (1990). IMAGE 1.0 was a global (single-region) model to capture major cause–effect relationships for the complex greenhouse problem. It constituted a global-averaged integrated structure, combining energy and agriculture models for greenhouse gas emissions, a global carbon cycle model, parameterized global radiative forcing, temperature change and sea level rise. IMAGE 1.0 was used to explore global scenarios for further developing the first set of scenarios for IPCC. With regard to effects of climate change and possible feedbacks, a regional set of modules was implemented to drive grid-based impact calculations as part of the ESCAPE framework (European Commission, 1992).

Building further on innovative approaches taken in the ESCAPE framework to estimate emissions resulting from energy and land use for world regions, IMAGE's focus was shifted to a regional base, known as version 2.0 (Alcamo, 1994). In this version land-cover and land-use modelling was done on a resolution of 0.5 by 0.5 degrees, drawing on experience with geographically explicit global models. At the time (1994), IMAGE 2.0 was the first published global integrated model having geographic resolution. All subsequent versions of IMAGE 2 have retained this two-strand approach of regional drivers and grid-based biophysical modelling.

In essence, IMAGE 2.0 consisted of three linked clusters of modules: the Energy-Industry System (EIS), the Terrestrial Environment System (TES) and the Atmosphere-Ocean System (AOS). The EIS generates industrial and energy emissions for 13 regions using simplified energy-economy relationships. TES, which has hardly changed in subsequent versions, establishes global land-cover change on the grid scale, taking agro-economic and climate factors into account. Changing land cover and other factors are used to compute the (net) flux of carbon dioxide (CO_2) and other greenhouse gases to the atmosphere. The BIOME model (Prentice *et al.*, 1992), the terrestrial carbon model and an FAO-based crop-growth model are important determinants of the changes in land-cover and associated emissions. The collective emissions from EIS and TES are then fed into AOS, which subsequently computes the build-up of greenhouse gases in the atmosphere. The zonal average temperature and precipitation patterns are calculated from the atmospheric composition changes.

Guided by recommendations from international review meetings, further refinements and extensions were implemented in IMAGE 2.1 (Alcamo *et al.*, 1998). Here, the aim was to enhance the model's performance and broaden its applicability. Major steps included improved computation of future regional energy use in EIS. Since the development of IMAGE 2.1 future fuel prices have influenced the selection of fuels in the model, depending on resource depletion on the supply side and price-dependent energy conservation on the demand side. The initial land-cover map, from which the global simulations start, was updated on the basis of DISCover version 1 (Belward & Loveland, 1995), together with improved allocation of agricultural land, computation of vegetation responses to climate change. The map also included demand of land for timber production.

The third session of the IMAGE Advisory Board in 1999 resulted in a list of recommendations and suggestions for further development work on IMAGE (Tinker, 2000). The board recommended making Global Change the target area, extending it beyond climate change, and building on integration of socio-economic and natural systems. While development up to IMAGE 2.1 had up to this point been largely an in-house effort by RIVM staff, collaboration with other domestic and international research groups was now suggested for further steps. Scientific recommendations included the development of cost curves for land-use emission reductions, meta-models and scanner models to address policy discussions and a revision of agro-economic modelling to be more in concert with approaches in other sectors. Furthermore, it was recommended to include interannual climate variability in relation to vegetation and water, and its effects on climate impacts, and to replace the zonal-mean climate-ocean model with a two-track approach. Here, a fast track would employ a simple climate model and a second track use a climate model of intermediate complexity. In addition, a list of more detailed recommendations and suggestions were proposed to take concrete steps for development of IMAGE 2. As previously described, the recommendations and suggestions from the Advisory Board formed a welcome guide for the IMAGE team, with the majority of recommendations being incorporated in the subsequent work.

One of the major changes in IMAGE 2.2 was the recommended two-track strategy for the climate model. The earlier zonal-mean climate-ocean model in IMAGE was replaced by a combination of the simple MAGICC climate model and the Bern ocean model. In the new approach, the resulting global average temperature and precipitation changes were scaled using temperature and precipitation patterns generated by complex coupled Global Circulation Models (GCMs). The widely accepted method of Schlesinger *et al.* (2000) for scaling patterns of aerosol-induced climate change was also adopted. This new approach is now the standard method for the first, simple and fast tracks to deal with climate change in the IMAGE model. A specific advantage is that patterns from different GCMs can be used to explore the uncertainties in the behaviour of the global climate system (IMAGE-team, 2001b). Parallel to this development, a second track – aiming to couple a climate model of intermediate complexity – was explored in co-operation with the Netherlands Meteorological Institute (KNMI). To date, this track currently operates in a parallel mode through the SPEEDY model.

On the economy–energy side, the linkages between the TIMER energy model, which had replaced the EIS, and the macro-economic model Worldscan were improved; this included downscaling from the 12 regions in Worldscan to the by then 17 active regions in TIMER and the rest of the IMAGE framework.

This IMAGE 2.2 version (IMAGE-team, 2001a) was used for preparing the Special Report on Emissions Scenarios of the Intergovernmental Panel on Climate Change (Nakicenovic *et al.*, 2000), in particular the B1 scenario (De Vries *et al.*, 2000). Besides changes in the model structure, much effort was devoted to generate model input parameters in line with the overall story lines that required harmonization of key input data of the SRES exercise. Special efforts were made to attune emission factors to available data in the start-year and their scenario-specific development over time. In this sense, the participation in the IPCC-SRES process has greatly enhanced the capacity of both the model and the IMAGE team to explore scenarios and obtain results geared to the requirements of various international assessment processes.

After completion of version 2.2 and the SRES scenarios, the model and model results were published in the form of a CD-ROM, facilitated by the development of the User Support System (USS) and allowing for interrogation of the model structure, input data and the vast number of results through a user-friendly interface.

The experiences gained from the SRES process had, however, reinforced the desire to seriously reconsider the future of the IMAGE model. It had become abundantly clear that further major steps in model development would be beyond the capacity of the IMAGE team, both in terms of expertise to support in-house development and in resources to simultaneously pursue further applications of the model. This tension had already been flagged by the Advisory Board in 1999 and required firm decisions on priorities and operational structures to pursue the overall goals and ambitions. As part of the process, the Dutch Ministry of Environment was involved in setting out strategic directions for IMAGE.

One of the main conclusions is that the IMAGE framework had received adequate international recognition to warrant or justify further investment in parallel with policy-relevant applications. Furthermore, broadening the scope to serve the emerging demand for analyses of global sustainability debates was adopted as the main challenge and a more active stand for setting up co-operative arrangements with other research groups was seen as indispensable. A set of model enhancements was identified and later initiated; these enhancements taken jointly will constitute the next generation model IMAGE 3. However, parallel to this, a small set of model changes, internally referred to as version 2.3, mainly on the integration of energy crops and carbon plantations was implemented for the analysis of mitigation options (Van Vuuren *et al.*, 2007).

The main milestone on the road to realizing IMAGE 3, however, is the IMAGE 2.4 version described in this book, which already addresses much of the overall development strategy and new challenges.

3. IMAGE 2.4

In the last decade, a series of improvements, enhancements and extensions of the IMAGE model have been initiated and framed in an overall model strategy towards broader coverage of sustainable development issues. The development activities increasingly take place in close collaboration with national and international partner institutes, with the aim of jointly benefiting from shared expertise and models. A scheme of the current model structure is given in Figure 2. Recent developments of IMAGE 2.4 are described in detail in Bouwman *et al.* (2006a).

Looking at the top of the scheme in Figure 2, we see a description of the basic driving forces, including demographics, energy supply and de-

Figure 2. Schematic diagram of IMAGE 2.4

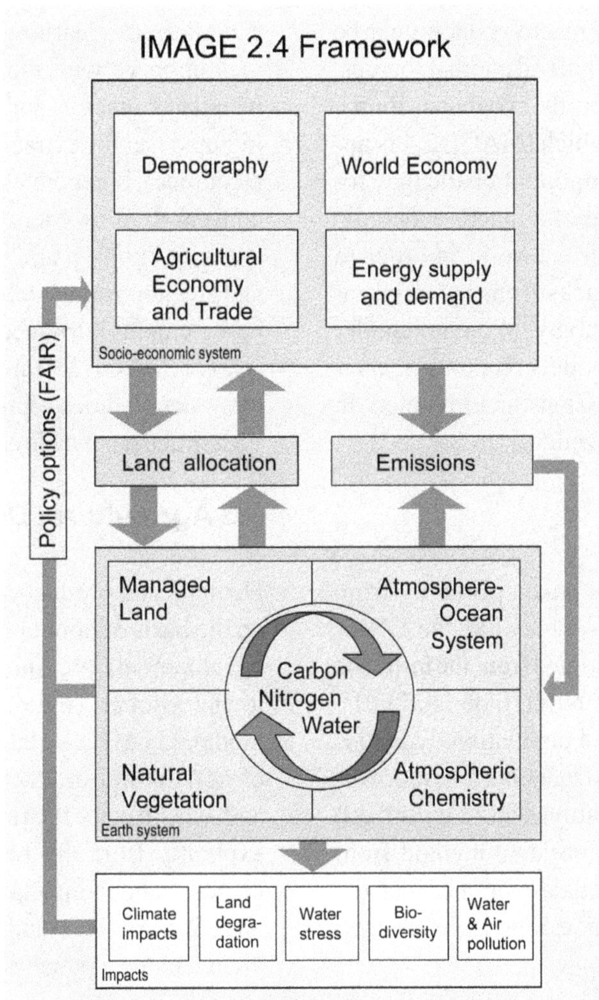

mand, and agricultural demand, trade and production. All of these interact through land use and emissions with the Earth systems. Subsequently, important elements in the biophysical modelling of land-cover and land-use processes are also addressed, i.e. land-cover and land use, contemporaneous and historical land cover, the carbon cycle and nutrients, followed by climate and climate variability, including its interaction with land cover. Finally, the use of data and information from IMAGE as input for broader policy-exploring tools is discussed for both global biodiversity and comprehensive climate mitigation strategies and regimes.

The scheme of the IMAGE 2.4 model framework in Figure 2 shows many of the basic structural components of its predecessors. The key drivers of change, population and the macro-economy can be derived from various external and internal sources. For macro-economic drivers the exogenous source depends on the study in which IMAGE 2.4 is applied. One of the most important challenges for IMAGE is the integration of a macro-economic model in the modelling framework, in order to be able to address feedbacks from the environmental system to the economy. In the remainder of this section the main model components, their improvements and extensions incorporated in IMAGE 2.4 are summarized.

3.1 Demographics

Population projections are taken primarily from authoritative exogenous sources like the UN or IIASA, but may also be adopted from the in-house demographic model PHOENIX (Hilderink, 2001). In IMAGE 2.4, grid-based population dynamics have been improved by introducing a new downscaling algorithm. Population within a grid cell is calculated using a proportional method from available country-specific data combined with the trends on the level of world regions, as determined by PHOENIX. This approach allows for simulating shifts in population within IMAGE regions.

3.2 Energy Supply and Demand

In the TIMER model, aggregated economic indicators like GDP, household consumption and value added in industry, services and agriculture are used to estimate the demand for energy services. Energy supply chains with substantial technological detail are then selected on the basis of relative costs to meet the final energy demand after autonomous and price-induced energy savings. Market shares for energy resources and technologies are calculated via a multinomial logit distribution function (De Vries *et al.*, 2001). TIMER includes explicit treatment of traditional biofuels, vintages of capital stock, learning-by-doing (i.e. technologies improve with their cumulative build-up of installed capacity) and resource depletion (driving up costs for extraction of exhaustible energy resources). It generates primary and final energy consumption by energy type, sector and region; capacity build-up and utilization; cost indicators and greenhouse gas and other emissions. Important new elements introduced in the TIMER 2.0 model (part of IMAGE 2.4) are hydrogen production and more detailed descriptions of the electric power system and renewable energy, including bioenergy.

3.3 Agricultural Demand and Trade

Demand and production of agricultural products on the basis of population changes and economic developments are simulated through a linkage to the Global Trade Analysis Project (GTAP) model. GTAP calculates consumption and trade of agricultural products by accounting regional and world market prices, which are calculated explicitly from production functions including capital, labour and land prices. In return, IMAGE 2.4 provides land-supply curves, yields and yield changes, which result from climate change and expansion of agriculture to less productive areas, and simulates the geographically explicit environmental impacts. This iterative coupling between GTAP and IMAGE allows assessment of the economic and environmental consequences of specific trade policies.

3.4 Land Use and Land Cover

One of the most striking parts of IMAGE 2.4 is the geographically explicit land-use modelling, considering both cropping and livestock systems on the basis of demand of agricultural crops and energy crops. The rule-based allocation accounts for crop productivity (Agro-Ecological Zones approach; FAO, 1978-1981), and other suitability factors like proximity to existing agricultural land and water bodies. The land-cover type 'energy crops' is now included in IMAGE 2.4. A more detailed description of animal production systems has been introduced in IMAGE 2.4 to portray the spatial variability in grazing systems and to address the rapid development of intensive ruminant production on managed grassland and rapidly increasing use of various feedstuffs. Moreover, a new initial land-use map for 1970 is incorporated on the basis of satellite observations combined with statistical information. Historical land cover (HYDE 3) for the period during 1700-1970 is based on census data, land's suitability for agricultural production and historical population density distributions. Changes in natural vegetation cover on undisturbed or abandoned land are simulated in IMAGE 2.4 on the basis of a static natural vegetation model (Prentice *et al.*, 1992).

Recently a new module for estimating forestry is added.

In conjunction with the dynamic climate model SPEEDY (see above), the current BIOME vegetation model of IMAGE will be replaced by the dynamic vegetation model (LPJ) of the Potsdam Institute for Climate Impact Research. The linkage to LPJ will allow a better representation of biogeochemical cycles and analysis of the compounded effect of changes in these cycles and biogeophysical changes associated with land use and hydrology.

3.5 Carbon Cycle

The consequences of these land-use and land-cover changes for the carbon cycle are simulated by a geographically explicit terrestrial carbon cycle model. If agricultural land is abandoned, it is assumed to revert gradually to its more natural state, with implications for the carbon stock. The carbon cycle model implemented in the IMAGE framework, since version 2.0 has been subjected to a thorough evaluation, which showed that the model is suitable for simulating global and regional carbon pools and fluxes. The model accounts for important feedback mechanisms related to changing climate, CO_2 concentrations and land use. In addition, it allows for evaluating the potential for carbon sequestration in natural vegetation and carbon plantations.

3.6 Nitrogen Cycle

IMAGE 2.4 also includes a new module to assess the consequences of the changing population, economy, land use and technological development for surface-nutrient balances and reactive nitrogen emissions from point sources and non-point sources. These surface balances are the basis for describing the major fluxes in the global and regional nitrogen cycle, as well as the effects on water and air quality (Figure 2). Processes that are accounted for in this module are human emissions, wastewater treatment, surface nitrogen and phosphorous balances for terrestrial systems, ammonia emissions, denitrification and N_2O and NO emissions from soils, nitrate leaching, and transport and retention of nitrogen in groundwater and surface water. In order to derive spatially explicit scenarios, tools were developed to translate regional or country-specific information to grid-specific input parameters.

3.7 Atmosphere – Ocean System

Emissions from the energy system and emissions due to land-use changes determine the composition of the atmosphere. IMAGE 2.4 uses the Atmosphere–Ocean System model developed for IMAGE 2.2 (Eickhout *et al.*, 2004). However, important non-linear interactions between the land, the atmosphere and the ocean cannot be studied with IMAGE 2.4 due to limitations of the current climate model and the natural vegetation module. Therefore, a series of studies was carried out to explore a possible pathway to include a more detailed climate model in IMAGE. As an outcome of this exploration, the detailed climate circulation model (SPEEDY) coupled to the Dynamic Global Vegetation Model LPJ, including the global water cycle, will be part of future IMAGE versions.

3.8 Biodiversity

In addition to these environmental impacts of global change calculated within the core biophysical modules, results are also used as input to drive impact models in the broader IMAGE 2.4 framework, such as the biodiversity model GLOBIO 3. GLOBIO (Alkemade *et al.*, 2009; Chapter 8) can be used to assess the impacts of climate and land-use change, infrastructure, fragmentation and nitrogen deposition on biodiversity and ecosystems. Likely effects of scenario assumptions or political interventions are estimated by calculating trends in mean species abundance. A parallel linkage to the module for aquatic biodiversity is now available.

3.9 Climate Policy Options

IMAGE results are also used for the evaluation of climate policies in conjunction with the policy decision-support model FAIR. FAIR is widely used to assess the environmental and abatement cost implications of international regimes for the differentiation of future emission reductions of greenhouse gases. The model links long-term climate targets and global reduction objectives with regional emission allowances and abatement costs, accounting for the Kyoto Mechanisms.

4. UNCERTAINTY AND SENSITIVITY

Obviously, numerous sources of uncertainty in the various components of IMAGE 2.4 influence its analytical results, ranging from data imprecision and model uncertainties to scenario assumptions in applications. To date, no comprehensive and systematic exploration has been performed of key uncertainties and how they are propagated throughout the entire IMAGE model to influence the final results. Currently, plans are being developed to undertake this demanding exploratory task. What has been done in many instances is to look at uncertainties of underlying data and model formulations in subsystems of the overall framework, thus providing partial sensitivity analyses for IMAGE 2.4 framework. An overview of the available sensitivity studies for the main modules is given below.

An earlier version of the TIMER energy and emissions model was systematically examined to establish the most important parameter settings and model assumptions influencing model results. This exploration uses the Numeral Spread Assessment Pedigree (NUSAP) system (Van der Sluijs *et al.*, 2005). Input variables and model components most sensitive to projected CO_2 emissions were population and economic growth; shifts in economic structure; technology improvement factors; fossil and renewable resource cost/supply curves, and autonomous and price-induced efficiency gains. Combined with the outcome of expert appraisal of the parameter 'pedigree', estimates of the 'strength' of the parameters were added to their sensitivity.

Obviously, any projection of future environmental conditions rests critically on the underlying emission factors and their relationship with relevant human activities or drivers. The IMAGE model has incorporated the most recent and authoritative sources. Despite ongoing efforts to collect data and enhance statistical procedures and modelling, many emission sources of greenhouse gases and other anthropogenic trace gases remain uncertain. Van Aardenne *et al.* (2001) have overviewed the qualitative analysis of activity data, emission factors and grid maps as in IMAGE. As a rule, emissions from large point sources like power plants tend to be of acceptable quality, while smaller and dispersed sources are typically poor to very poor. Whereas global or large-scale regional aggregate budgets are generally reasonably well known, the contribution of sectors and activities by geographic location is for the most part much more uncertain. Emission factors that depend on fuel properties, like CO_2 and sulphur dioxide (SO_2), can be estimated within narrow ranges, but others are very sensitive to technological details, local conditions like soil properties and management practices. This induces not only uncertainties in the initial inventories, but also in future emission projections.

In the coupled application of the agro-economic GTAP model and IMAGE, land-supply curves play a crucial role in establishing agricultural demands, production and trade flows. Derived from biophysical properties in IMAGE, land-supply curves are used in GTAP to find solutions of equilibrium for agricultural land volumes and the associated land rental rate. To test the sensitivity, simulation experiments were run with the asymptote 2.5% lower and 2.5% higher than the central estimate and the impacts on model results for land supply, the real land rental rate and production changes were investigated (Tabeau *et al.*, 2006). Analyses show that changing the asymptote of the land supply function leads to significant changes of land supply for countries where agricultural land is relatively scarce. However, the induced production changes are rather low. The aggregated agricultural production elasticity with respect to the asymptote change varies from 0.1 for countries where agricultural land is abundant to 0.5 for countries where the agricultural land is scarce. This means that the simulation results for production development are rather robust with regard to the estimated land supply-curve parameters.

The sensitivity of ammonia (NH_3) volatilization in agricultural production systems to variation in input parameters was investigated by Bouwman *et al.* (2006b). Various parameters were selected, including nitrogen excretion per head, animal stocks, distribution of production over pastoral, mixed and landless systems, fertilizer inputs and the NH_3 emission factors for animal housing, etc. The results suggest that on the global scale, excretion of nitrogen per head and animal stocks are the most important parameters in the model. However, the importance of the various parameters varies among world regions and countries. For example, in China fertilizer use is a far more important determinant for total ammonia loss than in other regions of the world, while in India the use of manure as fuel is a very important process. The overall conclusion was that nitrogen excretion per head merits our attention in future research. Research will focus on the difference between N excretions in extensive versus intensive system, and modelling excretion as a function of production characteristics such as milk production per head and nutrient intake by feed category. This study also made clear that the spatial modelling of nutrient use allows for analysis of various policy alternatives and consequences for the nitrogen cascade.

A series of experiments examined the role of the terrestrial carbon cycle in overall climate change scenarios implemented in IMAGE 2.2 (Leemans *et al.*, 2002). The experiments yielded a broad spectrum of atmospheric CO_2 concentrations, ranging, for example, in the IPCC-SRES A1B scenario from 714 to 1009 ppmv in 2100. The spread of this range is comparable to the full range arising from the different SRES scenarios with respect to the IMAGE 2.2 default settings for the carbon cycle:

515-895 ppmv. The most important negative and positive feedback processes are CO_2 fertilization and soil respiration, respectively. In recognition of the importance of a proper parameterization of the major feedbacks on the carbon cycle and land use and thus in determining the future state of the climate system, the issue has been further pursued in more recent years.

With regard to the response of the climate system to changes in atmospheric composition and associated radiative properties, two core aspects were tested. The first parameter addressed was climate sensitivity, which describes by how many degrees the equilibrium global mean temperature will rise if the CO_2-equivalent concentration of greenhouse gases in the atmosphere doubles compared to the pre-industrial level. The simple climate model in the Atmosphere-Ocean System (AOS) of IMAGE 2, attuned to represent the generally accepted central estimate for the climate sensitivity of 2.5 degrees, was adjusted to explore the range from 1.5 to 4.5 degrees. As expected, this amplified or reduced all climate-related impacts very tightly for any given emission projection. For climate change impacts, however, global mean effects are of limited significance. Therefore, a second sensitivity analysis addressed the spatial patterns of temperature and precipitation projections. IMAGE employs exogenous patterns from complex climate models (GCMs) to scale the impacts of the endogenously derived global mean temperature change. The robustness of regional impacts to different GCM patterns was tested by UNEP/RIVM (2004). Results indicate that while GCM outcomes for some regions are fairly consistent, in other regions the temperature effect is very different. With regard to annual precipitation the disagreement between models is even stronger. In some regions, e.g. in South America, they do not even agree on the direction of change.

Estimates of the costs of emission reductions, even within a well-defined scenario context are subject to considerable uncertainties, as the potential contribution and cost of abatement options are spread across wide ranges. A sensitivity analysis was performed for a scenario that stabilizes emission at 550 ppm CO_2-equivalent (Van Vuuren *et al.*, 2007) in order to identify for which abatement options the alternative assumptions had a significant impact on overall abatement costs. Selected options were tested one by one as well as in combination and simultaneously. Most individual options did not affect the total abatement costs by more than 10% (up or downwards) until 2050, with the exception of energy crops. Accepting the high end of the literature estimates on the supply potential and introducing the option to capture and store CO_2 from bioenergy, costs dropped by up to 40%. The compounded effect of all options taken together, however, results in 40% lower to almost 100% higher costs in 2050. Beyond 2050, the impact of uncertainties in options increases further. This applied particularly to options that are expected to become viable on a large scale in the longer term, such as hydrogen (± 20% in 2100). The compounded effect of all options considered collectively falls into the range of -40% to +250% by 2100.

5. APPLICATIONS

In parallel with the development steps outlined in the previous section, the IMAGE model has been applied to a variety of global studies. The specific issues and questions addressed in these studies have inspired the introduction of new model features and capabilities, and in turn, the model enhancements and extensions have broadened the range of applications that IMAGE can address. Since the publication of IMAGE 2.1 (Alcamo *et al.*, 1998), subsequent versions and intermediate releases have been used in most of the major global assessment studies and other international analyses, as listed below:

- **IPCC-Special Report on Emissions Scenarios (SRES)**: implementation of the

B1 marker scenario and calculation of the other harmonized set of comprehensive emissions scenarios up to 2100 (De Vries *et al.*, 2000; Nakicenovic *et al.*, 2000; IMAGE-team, 2001a);

- **UNEP Third Global Environment Outlook (GEO-3)**: assessment of environmental impacts from four global scenarios to 2030 (UNEP, 2002; UNEP/RIVM, 2004);
- **Millennium Ecosystem Assessment (MA)**: development of four global scenarios for the development of ecosystem services up to 2050 (Millennium Ecosystem Assessment, 2005b);
- **EuRuralis-1**: assessment of future prospects for agriculture and the rural areas in the EU-25 countries (Eickhout *et al.*, 2007);
- **Fourth Assessment Report of the IPCC (AR4)**: comprehensive global mitigation scenarios explored using IMAGE/TIMER/FAIR (Van Vuuren *et al.*, 2007). Besides participating in the mitigation scenarios study, several MNP experts serve as contributing/lead authors to the Working Group III report. A sensitivity study on the terrestrial carbon cycle was also done with IMAGE to obtain an adequate baseline against which to evaluate the potential for carbon sequestration options.
- **Greenhouse Gas Reduction Policy (GRP)** study: exploration of alternative climate change abatement goals and regimes in support of EU policy making using IMAGE/TIMER/FAIR (European Commission, 2005);
- **Second Global Biodiversity Outlook (GBO-2):** background report for the UN Convention on Biodiversity: evaluation of baseline trends in biodiversity loss and effects of policy actions in different fields with IMAGE/GLOBIO up to 2050 (Alkemade *et al.*, 2009);(sCBD and MNP, 2007)
- **MNP Sustainability Outlook (DV)**: assessment of sustainability issues in land use and energy resulting from different scenarios reflecting various perspectives on future directions for Dutch society (MNP, 2004);
- **Global Nutrients from Watersheds (NEWS)**: preparation of data on global nutrient surface balances for the UNESCO-Intergovernmental Oceanographic Committee NEWS project (Seitzinger *et al.*, 2005).
- **Fourth Global Environment Outlook of UNEP (GEO-4)** focuses on 'Environment for Human Well-being' linking environment and development. IMAGE evaluated the four updated GEO scenarios (UNEP, 2007).
- **International Assessment of Agricultural Science and Technology for Development (IAASTD)** The IMAGE team along with the International Food Policy Research Institute (IFPRI) played a pivotal role in the quantification of agricultural markets and environmental consequences. The scenarios of the Millennium Ecosystem Assessment (Millennium Ecosystem Assessment, 2005b) are as a basis (IAASTD, 2009).
- **Second Environmental Outlook of OECD** IMAGE's task was to develop the environmental baseline, according to the economic projections of OECD's economic model i.e. ENV-Linkages and analyses the impacts of several policy packages (OECD, 2008)
- **Global Biodiversity Outlook 3**. Some results from the IMAGE model are presented and related to the provision of Ecosystem Goods and Services (Leadley *et al.*, 2010)

In addition to these global assessments, IMAGE is also widely used in other projects and studies at sub-global scale, mostly European. IMAGE proved to be an invaluable tool to analyse the impacts of scenarios and policy options in an integrative way, including economic development, environmental changes and changes in human well being. IMAGE is a key model to evaluate consequences of development and policies on biodiversity on regional and global levels.

REFERENCES

Alcamo, J. (Ed.). (1994). *IMAGE 2.0: Integrated modeling of global change*. Dordrecht, The Netherlands: Kluwer Academic Publishers.

Alcamo, J., Leemans, R., & Kreileman, E. (Eds.). (1998). *Global change scenarios of the 21st century. Results from the IMAGE 2.1 model*. Oxford, UK: Elsevier Science.

Alkemade, R., van Oorschot, M., Miles, L., Nellemann, C., Bakkenes, M., & ten Brink, B. (2009). GLOBIO3: A framework to investigate options for reducing global terrestrial biodiversity loss. *Ecosystems, 12*, 374–390. doi:10.1007/s10021-009-9229-5

Belward, A., & Loveland, T. (1995). The IGBP-DIS 1 km land cover project. In Curran, P. J., & Robertson, C. (Eds.), *Remote sensing in action* (pp. 1099–1106). Southampton, UK: University of Southampton.

Bouwman, A. F., Kram, T., & Klein Goldewijk, K. (Eds.). (2006a). *Integrated modelling of global environmental change. An overview of IMAGE 2.4*. Bilthoven, the Netherlands: Netherlands Environmental Assessment Agency (MNP).

Bouwman, A. F., Van Der Hoek, K. W., Van Drecht, G., & Eickhout, B. (2006b). World livestock and crop production systems, land use and environment between 1970 and 2030. In Brouwer, F., & McCarl, B. (Eds.), *Rural lands, agriculture and climate beyond 2015: A new perspective on future land use patterns* (pp. 75–89). Dordrecht, The Netherlands: Springer.

Carpenter, S. R., Pingali, P. L., Bennett, E. M., & Zurek, M. B. (Eds.). (2005b). *Ecosystems and human well-being: Scenarios* (*Vol. 2*). Washington, DC: Island Press.

De Vries, B., Bollen, J., Bouwman, L., Den Elzen, M., Janssen, M., & Kreileman, E. (2000). Greenhouse gas emissions in an equity-environment- and service-oriented world: An IMAGE-based scenario for the 21st Century. *Technological Forecasting and Social Change, 63*, 137–174. doi:10.1016/S0040-1625(99)00109-2

Eickhout, B., Den Elzen, M., & Kreileman, E. (2004). *The atmosphere-ocean system in IMAGE 2.2. Report 481508017*. Bilthoven, The Netherlands: National Institute for Public Health and the Environment.

Eickhout, B., Van Meijl, H., Tabeau, A., & Van Rheenen, R. (2007). Economic and ecological consequences of four European land use scenarios. *Land Use Policy, 24*(3), 562–575. doi:10.1016/j.landusepol.2006.01.004

European Commission. (1992). *Development of a framework for the evaluation of policy options to deal with the greenhouse effect*. Brussels: Commission of the European Community, Directorate General for Environment, Nuclear Safety and Civil Protection.

Hilderink, H. B. M. (2001). *World population in transition: An integrated regional modeling framework*. Unpublished Thela thesis, Rozenberg, Amsterdam.

IMAGE-team. (2001a). The IMAGE 2.2 implementation of the SRES scenarios. *A comprehensive analysis of emissions, climate change and impacts in the 21st century*. (CD-ROM publication 481508018). National Institute for Public Health and the Environment, Bilthoven, The Netherlands; reprinted as CD-ROM publication 500110001, Netherlands Environmental Assessment Agency (MNP), Bilthoven, The Netherlands.

IMAGE-team. (2001b). The IMAGE 2.2 implementation of the SRES scenarios. *Climate change scenarios resulting from runs with several GCMs*. (CDROM publication 481508019). National Institute for Public Health and the Environment, Bilthoven, The Netherlands.

Kram, T., & Stehfest, E. (2006) The IMAGE model: History, current status and prospects. In A. F. Bouwman, T. Kram & K. Klein Goldewijk (Eds.), *Integrated modelling of global environmental change. An overview of IMAGE 2.4* (pp. 7-24). Netherlands Environmental Assessment Agency (MNP), Bilthoven, the Netherlands.

Leadley, P., Pereira, H. M., Alkemade, R., Fernandez-Manjarrés, J. F., Proenca, V., Scharlemann, J. P. W., & Walpole, M. (2010). *Biodiversity scenarios: Projections of 21st century change of biodiversity and associated ecosystem services*. Secretariat of the Convention on biological Diversity, Montreal.

Leemans, R., Eickhout, B. J., Strengers, B., Bouwman, A. F., & Schaeffer, M. (2002). The consequences for the terrestrial carbon cycle of uncertainties inland use, climate and vegetation responses in the IPCC SRES scenarios. *Science in China*, *43*, 1–15.

Millennium Ecosystem Assessment. (2005). *Ecosystems and human well-being. Synthesis report*. Washington, DC: Island Press.

MNP. (2004). *Quality and the future. Sustainability outlook (summary)*. Bilthoven, The Netherlands: Netherlands Environmental Assessment Agency.

Nakicenovic, N., Alcamo, J., Davis, G., De Vries, B., Fenhann, J., & Gaffin, S. ... Dadi, Z. (2000). *Special report on emissions scenarios*. IPCC Special Reports. Cambridge, UK: Cambridge University Press.

OECD. (2008). *Environmental outlook to 2030*. Paris, France: Organisation for Economic Cooperation and Development.

Potting, J., & Bakkes, J. (Eds.). (2004). *The GEO-3 scenarios 2002-2032: Quantification and analysis of environmental impacts*. Report UNEP/DEWA/RS.03-4 and RIVM 402001022, Division of Early Warning and Assessment (DEWA), United Nations Environment Programme (UNEP) / National Institute for Public Health and the Environment, Nairobi / Bilthoven.

Prentice, I. C., Cramer, W., Harrison, S., Leemans, R., Monserud, R. A., & Solomon, A. M. (1992). A global biome model based on plant physiology and dominance, soil properties and climate. *Journal of Biogeography*, *19*, 117–134. doi:10.2307/2845499

Rotmans, J. (1990). *IMAGE. An integrated model to assess the greenhouse effect*. Dordrecht, The Netherlands: Kluwer Academic Publishers.

sCBD & MNP. (2007). *Cross-roads of life on earth—exploring means to meet the 2010 Biodiversity target*. Solution-oriented scenarios for global biodiversity outlook 2. Technical Series no 31. Secretariat of the Convention on Biological Diversity, Montreal.

Schlesinger, M. E., Malyshev, S., Rozanov, E. V., Yang, F., Andronova, N. G., & De Vries, B. (2000). Geographical distributions of temperature change for scenarios of greenhouse gas and sulphur dioxide emissions. *Technological Forecasting and Social Change*, *65*, 167–193. doi:10.1016/S0040-1625(99)00114-6

Seitzinger, S. P., Harrison, J. A., Dumont, E., Beusen, A. H. W., & Bouwman, A. F. (2005). Sources and delivery of carbon, nitrogen, and phosphorus to the coastal zone: An overview of Global NEWS models and their application. *Global Biogeochemical Cycles, 19*, GB4S. doi:10.1029/ 2004GB002453

Tabeau, A., Eickhout, B., & van Meijl, H. (2006). *Endogenous agricultural land supply: Estimation and implementation in the GTAP model.* Ninth Annual Conference on Global Economic Analysis, June 2006, Addis Ababa, Ethiopia.

Tinker, B. (2000). *Report of the third session of the IMAGE advisory board. Report 481508014.* Bilthoven, The Netherlands: National Institute of Public Health and the Environment.

UNEP. (2002). *GEO-3. Past, present and future perspectives*. London, UK: Earthscan.

Van Aardenne, J. A., Dentener, F. J., Olivier, J. G. J., Klein Goldewijk, C. G. M., & Lelieveld, J. (2001). A 1 x 1 degree resolution dataset of historical anthropogenic tracé gas emissions for the period 1890-1990. *Global Biogeochemical Cycles, 15*, 909–928. doi:10.1029/2000GB001265

Van der Sluijs, J. P., Craye, M., Funtowicz, S., Kloprogge, P., Ravetz, J., & Risbey, J. (2005). Combining quantitative and qualitative measures of uncertainty in model based environmental assessment: The NUSAP system. *Risk Analysis, 25*(2), 481–492. doi:10.1111/j.1539-6924.2005.00604.x

Van Vuuren, D. P., Den Elzen, M., Lucas, P., Eickhout, B., Strengers, B., & Van Ruijven, B. (2007). Stabilizing greenhouse gas concentrations at low levels: An assessment of reduction strategies and costs. *Climatic Change, 81*, 119–159. doi:10.1007/s10584-006-9172-9

Chapter 6
Simulating Land Use Policies Targeted to Protect Biodiversity with the CLUE-Scanner Model

Peter H. Verburg
VU University Amsterdam, The Netherlands

Jan Peter Lesschen
Alterra Wageningen UR, The Netherlands

Eric Koomen
VU University Amsterdam, The Netherlands

Marta Pérez-Soba
Alterra Wageningen UR, The Netherlands

ABSTRACT

This chapter presents an integrated modelling approach for assessing land use changes and its effects on biodiversity. A modelling framework consisting of a macro-economic model, a land use change model, and biodiversity indicator models is described and illustrated with a scenario study for the European Union. A reference scenario is compared to a scenario in which a number of possible policies for conservation and protection of biodiversity are assumed to have been implemented. The results are evaluated by an indicator of the habitat quality for biodiversity and an indicator of landscape connectivity. The results illustrate that land use change has spatially diverse impacts on biodiversity. The effectiveness of the assumed policies is region and context dependent. The modelling framework can thus provide ex-ante assessments of policies and identify critical regions for biodiversity conservation and assist in targeting policies and incentives to protect biodiversity in vulnerable areas.

DOI: 10.4018/978-1-60960-619-0.ch006

1. INTRODUCTION

Land use change is an important determinant of biodiversity loss and changes in the availability of natural resources. Many studies have indicated the importance of land use change research to assess the impacts on biodiversity and other environmental and social consequences of land use change (Reidsma *et al.*, 2006; Verboom *et al.*, 2007; Trisurat *et al.*, 2010; DeFries *et al.*, 2004). These studies have indicated that not only the total area of the ecosystem converted, but also the spatial pattern and the location of change, are important factors determining the impact of land use changes on biodiversity. Fragmentation of habitats, the conversion of critical locations for threatened species and the blocking of migration routes are important processes that make the effects of land use change on biodiversity more important than the conversion of the habitat by itself. Therefore, the relation between land use change and biodiversity can only be adequately assessed when a spatial perspective is taken. Measures taken to avoid or reduce biodiversity loss are, in many cases, also related to specific locations, e.g. the establishment of natural parks, ecological corridors, buffer zones etc. Planning of such measures needs to consider not only the types of pressure that the anticipated land use changes will cause, but also where they will happen, as well as the spatial determinants of biodiversity.

Land use modelling in scenario studies has become an important tool in ex-ante evaluation of policy and spatial planning (Koomen *et al.*, 2008b). Land use modelling facilitates the identification of the possible consequences of different types of development and helps to evaluate the effectiveness of policies. In addition, integrated land use modelling can identify the trade-offs of policies in other sectors on biodiversity. Examples of such ex-ante assessments of land use change are available for different regions in the world (Verburg *et al.*, 2006; Hellmann & Verburg, 2010; Voinov *et al.*, 1999). In such studies a range of different land use models are used, often adapted to the local situation and its specific conditions. Also, the scale of analysis is an important determinant of the type of modelling chosen. A major challenge in such studies is to consistently link the methods of land use analysis with methods to derive the scenarios and assessments of the impacts on biodiversity indicators. Often the amount of information on ecosystem changes provided by land use models is limited by data availability and the abstractions made during the modelling process. The biodiversity assessment methods need to make best use of the limited information available on the specific ecosystem conditions, while the land use modelling needs to be tailored towards an output that makes an evaluation of the possible consequences of policies related to biodiversity conservation possible.

The objective of this chapter is to present a consistent method for evaluating the effects of policy scenarios affecting land use for the complete European Union (EU27), using as illustration a scenario aimed at conserving and protecting biodiversity.

2. METHODS

2.1 The CLUE-Scanner Model

The overall methodology for assessment is based on a multi-scale, multi-model approach that integrates the economic, demographic and environmental drivers of land change in a consistent modelling framework as described by Verburg *et al.* (2008). Figure 1 provides an overview of the modelling methodology. Global scale drivers of land use change originating from changes in demography, consumption patterns, economic development, trade and climate change are analyzed with the combined application of the global economy model LEITAP and the global integrated assessment model IMAGE. A detailed description of the interaction between these two

Figure 1. Overview of the integrated modelling framework

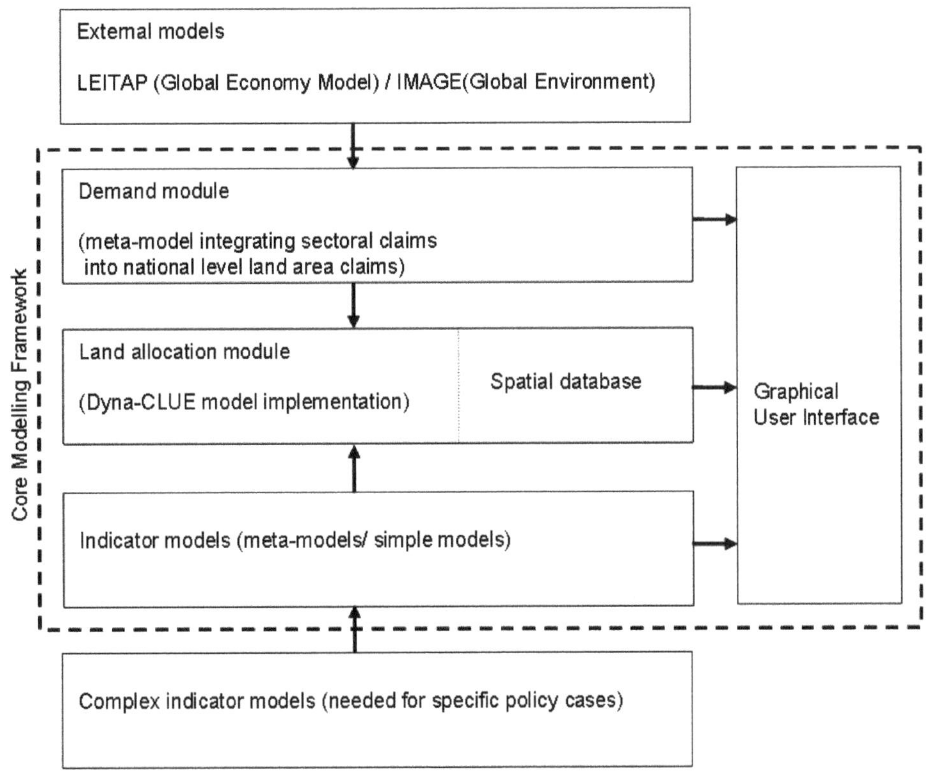

models is provided by van Meijl *et al.* (2006) and Eickhout *et al.* (2007). These global scale models provide output in terms of changes in agricultural area at the level of individual countries within the European Union. In a demand module these changes in agricultural area are integrated with claims from the urban/industry and forestry sector. Land cover areas at a national scale are then input to the land allocation model. The land allocation model translates the national scale land areas to a 1 km^2 grid. Based on the thus derived land cover maps a number of indicators for the impacts of the land use changes can be calculated, either by simple indicator models or more complex models linked for specific applications to the modelling framework (Hurkmans *et al.*, 2009). The core of the modelling framework (indicated in Figure 1) is integrated into a consistent modelling interface called the CLUE-Scanner based on the land allocation methodology of the Dyna-CLUE model (Verburg and Overmars, 2009) and the numerical algorithms of the Land Use Scanner model (Koomen *et al.*, 2008a).

The translation of the national level changes in agricultural area from the LEITAP model to input of the Dyna-CLUE model requires a number of corrections to ensure consistency between the models. While LEITAP is based on agricultural statistics the Dyna-CLUE simulations are based on land cover data derived from CLC2000. Large differences in agricultural areas between the two data sources are the result of differences in definition, observation technique, data inventory bias etc. (Verburg *et al.*, 2009). To some extent these differences are structural and can be corrected. Absolute changes in agricultural area in LEITAP are corrected for some of these differences and then serve as input to the Dyna-CLUE model. Changes in urban area are calculated based on trends in demography and economic growth

Figure 2. Land use allocation procedure in Dyna-CLUE (based on: Verburg & Veldkamp (2004))

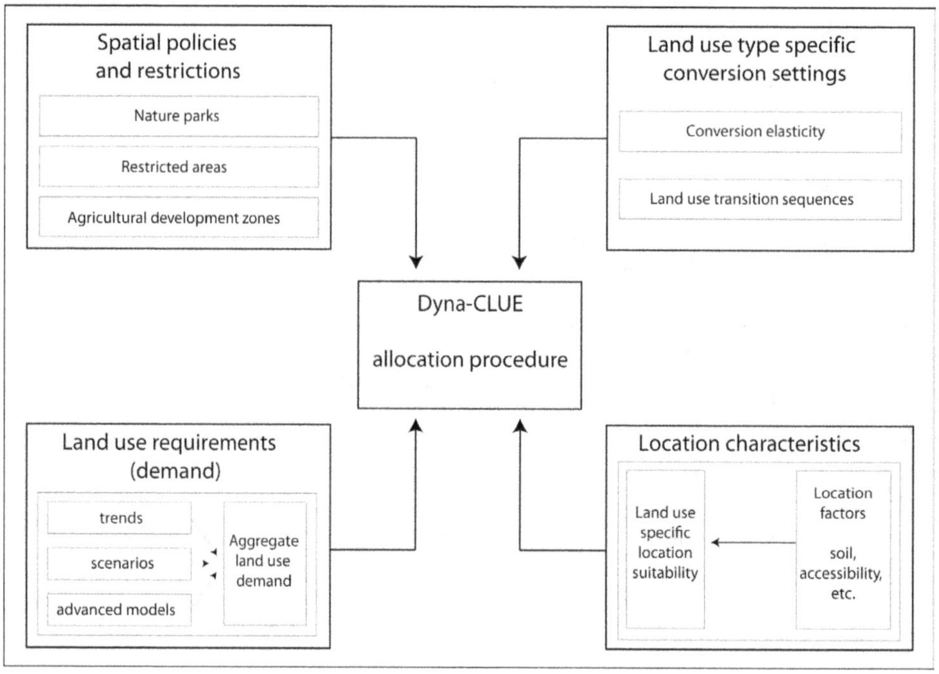

projections. The net change in agricultural and urban area will determine the overall area left for semi-natural land use types and forestry. From the IMAGE model climate change data are used as one of the location factors considered in the Dyna-CLUE model. The simulated changes in climate at coarse spatial resolution (50x50 km) are downscaled to 1x1 km and superimposed on the more detailed Worldclim data for use in the simulations.

For the land use allocation module, use is made of the Dyna-CLUE model which is a recent version of the CLUE model (Verburg *et al.*, 1999; Verburg *et al.*, 2002). CLUE is one of the most used land allocation models globally and is highly applicable for scenario analysis (Pontius *et al.,* 2008). The use of the model in many case studies at local and continental scale by different institutions worldwide (e.g., Castella *et al.*, 2007; Wassenaar *et al.*, 2007) has proven its capacity to model a wide range of scenarios and provide adequate information for indicator models. Figure 2 shows the land use change allocation procedure.

There are 'four boxes' that provide the information to run the model:

- Spatial policies and restrictions (e.g. nature reserves);
- Land use demand (i.e. agriculture, urban and forest);
- Location characteristics, maps that define the suitable location for each land use type based on empirical analysis; for example, the European soil map is translated into functional properties such as soil fertility, water retention capacity. In addition to the soil map there is a set of 100 factors that range from accessibility to bio-physical properties; the factors can be dynamic in time. A full list of factors considered can be found in Verburg *et al.*(2006);
- Set of rules for possible conversions (conversion elasticity, land use transition sequences).

Table 1. Reference scenario socio-economic assumptions and key characteristics for the EU (based on:Westhoek et al., 2006and www.eururalis.eu)

Aspect	Scenario assumptions
Population EU-27 in 2030	500 million
Population change since 2000	4%
EU-15 GDP yearly growth	1.3%
EU-12 GDP yearly growth	3.4%
EU enlargement	Turkey enters EU
Trade of agricultural products	Export subsidies and import tariffs phased out. Slight increase in non-tariff barriers
Product quota	Phased out; abolished by 2020
Farm payments	Fully decoupled and gradually reduced (by 50% in 2030)
Intervention prices	Phased out; abolished by 2030
Compulsory set-aside of arable land (excl. organic farms)	Set-aside target remains at 10% level

A detailed description of the functioning of the Dyna-CLUE land allocation procedure is provided by Verburg & Overmars (2009).

Finally, a series of indicator models corresponding to the demands of the policy scenarios are implemented. The modelling framework contains a balanced set of indicators focussing on the land use and environmental domains that are calculated based on the results of the economic and land use modelling. In section 2.3 the two indicator models related to biodiversity are described.

2.2 Scenarios

2.2.1 Description of the Scenario Assumptions

A reference scenario of foreseen future developments is constructed accounting for exogenous global drivers like:

- increasing food and feed demand in emerging countries, i.e. the BRIC countries (Brazil, Russia, India and China);
- changing trade regimes because of increasing competitiveness of Asian and Latin-American regions;
- changing environmental constraints because of resource scarcity and climate change;
- demographic changes.

For the development of the reference scenario use is made of the well-known B1 scenario of IPCC-SRES (IPCC, 2000) and elaborated for the European conditions by Westhoek *et al.* (2006). The B1 scenario (global co-operation) includes many policy developments that correspond to ongoing changes in policy context and discussions and includes a modest economic growth which seems realistic after the economic crisis. As such it may be interpreted as a business-as-usual type of scenario. The B1 scenario combines a global orientation with a preference for social, environmental and more broadly defined economic values. Economic profit is not the only objective. Governments are actively regulating, ambitiously pursuing goals related to, for example, equity, environmental sustainability and biodiversity.

An overview of the most important socio-economic assumptions and key characteristics for the EU is provided in Table 1.

The B1 reference scenario is useful as reference point for the assessment of the specific potential impacts of future spatial EU-polices, as it already contains many current spatially explicit EU policies. This refers especially to the *Less Favoured Areas* (LFA) support (compensation to farmers in regions with constraints for agricultural use), which is maintained, and current protected nature areas (including the EU defined Natura 2000 areas, forests and other natural areas), that remain protected from development. In this way the reference scenario offers business-as-usual baseline conditions that allow a proper assessment of the impacts of new policy alternatives.

In addition to the reference scenario an alternative scenario is defined that introduces a number of ambitious policies to increase the protection of specific ecological and landscape related values. It builds on existing policy options that are currently being discussed (Table 2) within the European Union. The scenario options were interactively discussed between the modellers and policy makers at the European Union. This process ensured a good correspondence between the scenario assumptions and the ongoing policy discussions (Pérez-Soba *et al.,* 2010).

Urban growth is a threat to biodiversity and controlling this growth is an important policy issue in many Member States. Some urban growth control measures are included in this policy alternative to demonstrate their potential impact, i.e. what could be the consequences of more active policies controlling urban growth. Another concern is the fragmentation of natural habitats. This issue has become even more pressing in view of climate change which is likely to cause many plant and animal species to migrate, in general from south-west to north-east Europe. To allow this migration to actually take place and help to create robust habitats, strategies for establishing natural corridors have been suggested. For the biodiversity protection alternative the following alternatives were considered: enlarging current nature areas and creating networks of intercon-

Table 2. Overview of the current spatial policy ambition level incorporated in the reference scenario and the more ambitious policies in the biodiversity protection alternative

Policy theme	Reference scenario	Biodiversity protection alternative
Controlling urban growth	No European-wide policy	Spatial planning to promote more compact forms of urbanisation; prevention of urbanisation in semi-natural and forest areas
Fragmentation control and promotion of clustering of nature	Current fragmentation control following EIA legislation, no active promotion of clustering	Policy targeted at clustering natural land use types towards large robust natural areas
Natural corridors	No European-wide policy (except what is done in Natura 2000)	Create a coherent European-wide approach to give space to ecosystems; as an example we use the main Pan–European Ecological Network (PEEN) corridors (incentives to convert land in specified corridor areas to nature)
Natura 2000	Some incentives to continue extensive land use in Natura 2000 areas (2nd pillar funds)	More funds through 2nd pillar payments to continue extensive land use in Natura 2000 areas (incentive approx. three times as strong)
High Nature Value (HNV) protection	No specific protection	Compensation of extensive farming (especially permanent pastures) in HNV areas to prevent abandonment or intensification (compensation for pasture similar to current LFA support, for arable land 50% of current LFA support)
Less Favoured Areas (LFA)	Current LFA support	Targeted LFA support to HNV within LFA, increased level of 2nd pillar payments
Protection peat land	No policies	Land conversion in peat areas are not allowed

nected nature areas by the protection and development of ecological corridors. Although there is already an established European wide strategy for protected areas (Natura 2000) our policy alternative considers the increase of current funding to promote the sustainable land use in these protected areas and possibly establish buffer zones around these areas.

Besides the value of natural ecosystems for biodiversity high values are assigned to biodiversity related to extensive farmlands and mosaic landscapes. Therefore, policies are assumed that stimulate continuation of extensive farming with associated high nature values in specified areas. Finally, peatlands that contain specific biodiversity values are protected from conversion to agricultural or urban use in this alternative. This also limits the emission of greenhouses gasses that is associated with such conversions.

2.2.2 Implementation of Scenarios in the Modelling Framework

The reference scenario and the biodiversity protection alternative described in the preceding section were translated into model input in a policy-science iterative process, which involved the model operators and policy developers. Initial implementation suggestions were offered by the modellers and adjusted after consultation with the relevant experts. The policy alternatives are implemented in the CLUE-Scanner model through changing several input parameters. More specifically these relate to:

1. specification of location-specific preference additions, indicating where the suitability of a location is enhanced (e.g. through a subsidy) or restricted;
2. conversion matrices that specify which land use transitions are allowed at specified locations;
3. conversion elasticities that regulate the ease of land use transitions;
4. neighbourhood settings specifying the importance of the surrounding land use for simulation;

2.3 Biodiversity Indicators

Two indicators were selected to evaluate the effects of land use changes on biodiversity. The two indicators were designed to capture the biodiversity effects given the spatial resolution and thematic content of the results of the land use modelling The first indicator is a measure of the completeness of the habitat for maintaining biodiversity while the second indicator aims to provide a measure of the connectivity of the habitats.

The Mean Species Abundance (MSA) index is derived from land use, land use intensity (agriculture and forestry), nitrogen deposition, spatial fragmentation, infrastructure developments and policy assumptions on high nature value (HNV) farmland protection and organic agriculture. The methodology used is the GLOBIO3 approach initially developed for biodiversity assessments at a global scale (Alkemade *et al.*, 2009), but also applied to the level of Europe (Verboom *et al.*, 2007). The indicator provides an approximation of the land use related changes in biodiversity. As it is not able to discern actual habitats, it applies a 1x1 km resolution that is too coarse to capture detailed ecological processes and only uses a limited range of factors that influence biodiversity. The index ranges from 0 to 100, and represents the species abundance compared to species abundance in the natural system without human disturbances. The results do not provide a precise, local account of biodiversity. It does, however, allow for the comparison between the current and different future situations.

The second indicator measures the connectivity of individual patches of natural area. This newly developed indicator assesses the difficulty to reach the nearest larger sized habitat from smaller habitats based on the land use allocation results. It offers an approximation of the connectivity of the landscape for species and the viability of smaller habitats within the landscape matrix. The difficulty to reach other habitats is differentiated between land use types, assuming a high resistance for urban and arable areas to allow migration of species, a medium to low resistance of permanent grassland areas and a low resistance of small patches of (semi-) natural area. The connectivity of a larger area is assessed by calculating the average resistance (or travel time) to reach the larger patches of natural vegetation from the smaller patches within a neighbourhood or administrative region. As the indicator is not including information on the quality of different land use types, it only offers an indication of the potential coherence of possibly valuable natural areas.

The indicator has been defined in such a way to as much as possible be independent of the area of natural land use types in the region and solely capture the spatial arrangement. Therefore, also areas with limited natural area may still have, in theory, a good connectivity potential. This way the indicator has added value to the biodiversity indicator. Alternative indicators for landscape connectivity such as the frequently used proximity indicator (Gustafson and Parker, 1994) are not sufficiently sensitive to the data used at the spatial and thematic resolution of analysis.

3. RESULTS

Figure 3 shows the main land use change processes for the reference scenario and the biodiversity policy alternative. Because it is assumed that the policies included in the biodiversity alternative do not lead to different locations of agriculture or urbanization the overall picture across Europe is similar for both scenarios. During the thirty years of the simulation we see that urbanization is concentrated around the main urban centres with a strong focus on the economically strong regions. Land abandonment is, in both scenarios, concentrated at the marginal mountain landscapes

Figure 3. Main land use change processes for the reference scenario (left) and biodiversity alternative (right). For visualization purposes the areas of the land use change processes are somewhat exaggerated. Colour version of the figure at http://www.ivm.vu.nl/Picturesbiodiv

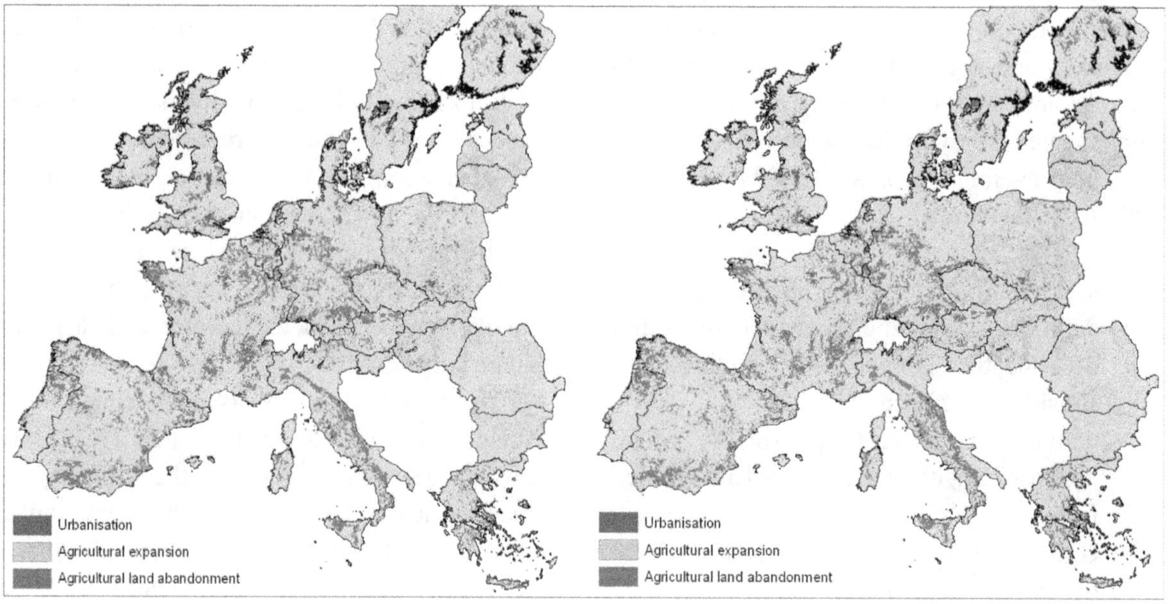

of the Carpathians, Alps and Pyrenees and some smaller mountain regions across Europe. Expansion of agriculture is seen in Eastern Europe. This is basically resulting from the global economic model that foresees a competitive position of agriculture in this region.

To see the effects of the spatial policies for the Biodiversity alternative, one has to zoom in to a higher level of detail, where clear impacts of the spatial policies can be observed. Figure 4 shows an area in Brittany (North western France). Here several ecological corridors are located, where incentives are provided to convert arable land to nature. In the biodiversity scenario, conversion of arable land to permanent grassland or nature is indeed occurring mainly within the ecological corridors. Most of this 'new' nature originates from abandonment of current arable lands. Since the total area of arable land at the national level is the same as in the reference scenario concentration of abandonment within the ecological corridors leads to less abandonment outside the designated corridors.

An important indicator to assess the impact of the different scenarios on biodiversity is the Mean Species Abundance (MSA) index. In Figure 5 the MSA index is given per country for the different scenarios. For countries with a lot of forest, e.g. Sweden and Finland, the index is highest. In these countries the (semi-)natural areas are less disturbed, whereas highly populated countries, e.g., Belgium have the lowest index. In most countries the changes in land use between 2000 and 2030 have a positive effect on biodiversity. The decrease in agricultural activities is reflected in an increase of of agricultural land abandonment, especially within areas at the fringe of nature reserves and in mosaic landscapes. This results in larger and less fragmented natural areas that favour biodiversity, as measured by the MSA index. At the same time, it is mostly extensively managed lands that are abandoned that may have high values of agro-biodiversity which is not accounted for in the MSA index (Falcucci *et al.*, 2007; Burel & Baudry, 1995). In a number of eastern European countries the effects of the expansion of agricul-

Figure 4. Land use patterns in 2030 for an area in Brittany (North western France) for the reference scenario (left) and the biodiversity alternative (right). The marked areas indicate the ecological corridors. Colour version of the figure at http://www.ivm.vu.nl/Picturesbiodiv

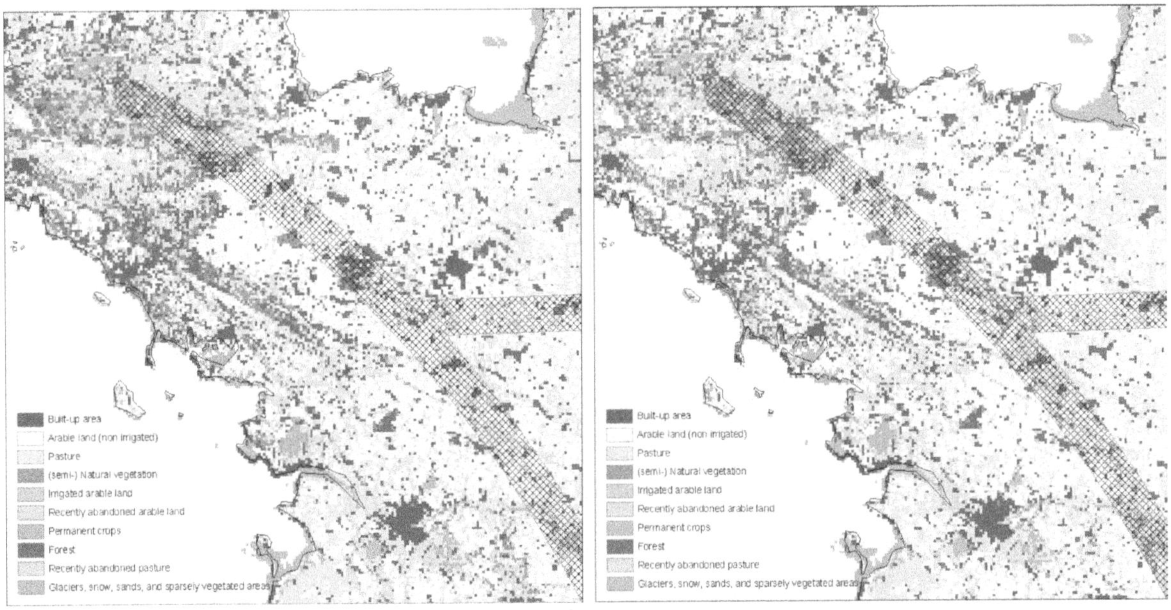

ture can be seen in the MSA values as a small decrease. The graph also shows that the differences between the scenarios are small for the different scenarios, since the MSA index is mainly determined by the total areas of the different land uses, and to lesser extent influenced by their spatial distribution. But, even while the total areas of the different land uses at the national scale are similar between the two scenarios it is clear that the spatial configuration influenced by the assumed policies has, in all countries, a positive effect on this measure of the biodiversity. The decrease in MSA in the reference scenario observed in some countries is offset by the spatial policies aimed at conserving biodiversity.

Figure 6 shows the change in connectivity index between 2000 and 2030 for both the reference scenario and the biodiversity alternative. The results are aggregated for NUTS_2 units that correspond to administrative regions in the different countries. The maps show that an increase in MSA index does not always mean that the connectivity of the landscape improves. Urbanization and intensification of agriculture associated with the removal of landscape elements and remnant patches of semi-natural vegetation cause a decrease in landscape connectivity in a number of regions. In other NUTS regions the connectivity increases as a result of land abandonment. Especially in the eastern European countries the effect of intensification and removal of natural vegetation within the main agricultural areas has a negative effect on the landscape connectivity. It is especially in these regions that the measures taken in the biodiversity scenario have a positive effect. In the biodiversity alternative the expansion of agricultural area is the same as in the reference scenario, however, the spatial measures offset in some regions the negative effects on landscape connectivity. At the same time, regions where no spatial policies are implemented have less im-

Figure 5. Mean Species Abundance index (MSA) per country for the reference scenario and biodiversity alternative. Colour version of the figure at http://www.ivm.vu.nl/Picturesbiodiv

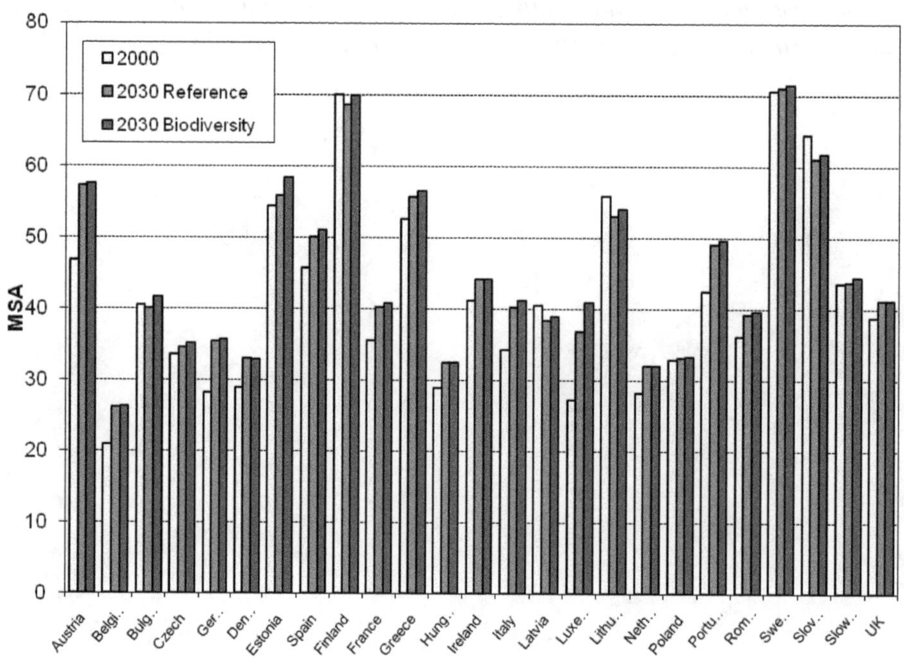

provement or even more loss of landscape connectivity in the biodiversity scenario.

4. DISCUSSION AND CONCLUSION

This chapter presents a state-of-the-art modelling framework for analyzing land use changes at the European scale for different policy scenarios. Specific indicators for the possible effects on biodiversity that account for the spatial resolution and thematic content of the addressed land use classes are included. Both indicators address different aspects of biodiversity: the first indicator focuses on the habitat characteristics while the second indicator focuses on the connectivity of habitats. The two indicators thus show a different response to the scenario conditions. We have also chosen these two specific indicators because alternative indicators for biodiversity are difficult to link to the land use modelling results as these indicators may require a different type of information than available in the land use modelling. The main advantage of the modelling framework is its flexibility in analyzing different scenarios. The scenario described in this chapter contained different types of policy that are implemented in different ways in the modelling framework. Alternative scenarios can include variations in economic development of global trade affecting the demand for land resources within the study area. In the presented study only variations in a number of spatial policies including land use planning and agri-environmental subsidies were analyzed while assuming that such measures would not affect the overall areas of the different land use types. The framework allows for the evaluation of individual policies as well as packages of interacting policies such as addressed in this study. Scenario descriptions should be translated in settings of the model that fit in one of the four types of model input as shown in Figure 2. The translation of scenario descriptions to model settings should be done with great care to correctly represent the scenario description in the model.

Especially important is the comparison of different scenarios in the context of the developments in the reference scenario. Such a comparison

Figure 6. Change in habitat connectivity between 2000 and 2030 for administrative regions for the reference scenario (left) and the biodiversity alternative (right). Colour version of the figure at http://www.ivm.vu.nl/Picturesbiodiv

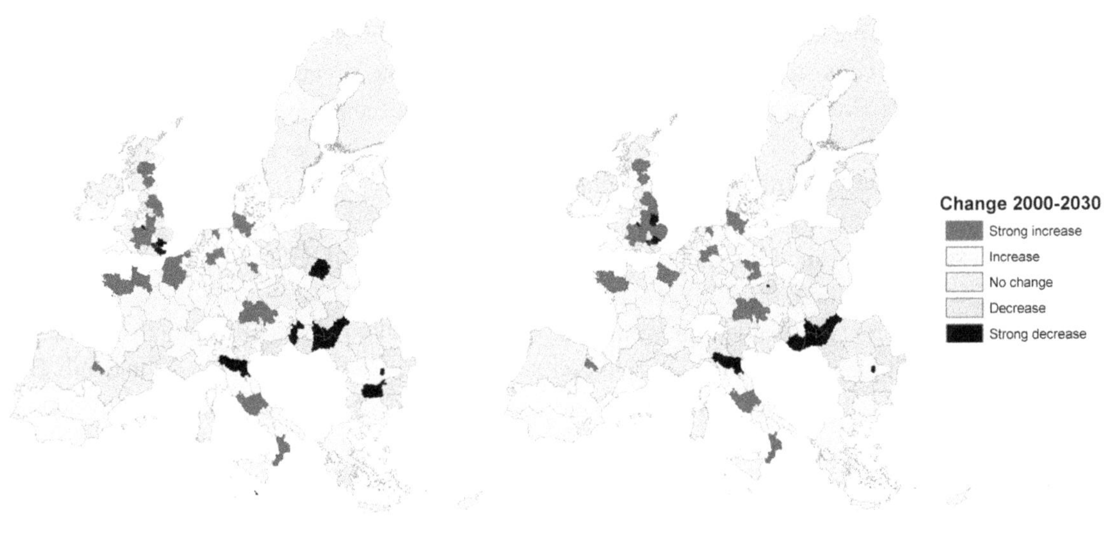

makes the evaluation of specific measures possible in the context of ongoing processes. The results of this study make clear that in the context of increases in agricultural area all kinds of (voluntary) incentives to take land out of agricultural production in order to enhance biodiversity are likely to be much less successful as compared to conditions in which marginalisation of agriculture is happening. The difference in dynamics of agricultural area between eastern and western Europe under the reference scenario has a clear effect to the success of the spatial policies to conserve biodiversity. In Western Europe the land use change conditions are more favourable for implementation of these policies as compared to Eastern Europe. Therefore, the effects on the indicators are more favourable in Western Europe. At the same time, in the context of intensification and expansion of agriculture in Eastern Europe the measures assumed in the biodiversity scenario are capable to off-set some of the negative consequences for biodiversity.

The results also illustrate that the effects of policies aimed at conserving and developing biodiversity values are strongly context specific and thus require spatially explicit modelling techniques.

The validity of the land use projections is difficult to check. A common way to test and validate the results of land use models is to use a historic period with observed data at the start and end of the period to compare the model performance. Available land use data for Europe are not sufficiently consistent to make a reliable validation possible (Verburg *et al.*, 2009). The allocation algorithm of the Dyna-CLUE model underlying the modelling framework presented in this chapter has been validated in a number of different case studies around the globe (Castella & Verburg, 2007; Pontius *et al.*, 2008). Although the performance largely depended on input data quality and the complexity of the land use change processes simulated, the model proved to be capable of capturing the important patterns of land use change.

The use of integrated modelling frameworks may assist the further design and optimization of policies on biodiversity. Regions of limited success of the policies may be identified and the designed measures may be adapted. Furthermore, hotspots of biodiversity loss may be identified and new measures may be targeted at these regions. Moreover, the modelling framework allows the evaluation of policy proposals in other sectors, e.g. agriculture or transport, on its effects on biodiversity and may thus help to evaluate cross-sectoral tradeoffs. Such analysis can benefit the design of improved cross-sectoral policies.

ACKNOWLEDGMENT

The authors would like to thank everyone that has contributed to the design and implementation of the modelling framework and scenarios presented in this chapter. The macro-economic modelling results of the LEITAP model have been prepared by Martin Banse and Geert Woltjer while the IMAGE results were provided by Anne-Gerdien Prins. Igor Startitsky assisted in preparing the CLUE-Scanner simulations while Maarten Hilferink and Martin van Beek are responsible for the programming of the software. The work presented in this chapter is based on research within the 'Land Use Modelling – Implementation' project commissioned by DG Environment of the European Commission. We thank Viviane André for her contribution in guiding the project and the project steering committee for the specification of the policy scenario.

REFERENCES

Alkemade, R., van Oorschot, M., Miles, L., Nellemann, C., Bakkenes, M., & ten Brink, B. (2009). GLOBIO3: A framework to investigate options for reducing global terrestrial biodiversity loss. *Ecosystems*, *12*, 374–390. doi:10.1007/s10021-009-9229-5

Burel, F., & Baudry, J. (1995). Species biodiversity in changing agricultural landscapes: A case study in the Pays d'Auge, France. *Agriculture Ecosystems & Environment, 55*, 193–200. doi:10.1016/0167-8809(95)00614-X

Castella, J. C., Pheng Kam, S., Dinh Quang, D., Verburg, P. H., & Thai Hoanh, C. (2007). Combining top-down and bottom-up modelling approaches of land use/cover change to support public policies: Application to sustainable management of natural resources in northern Vietnam. *Land Use Policy, 24*, 531–545. doi:10.1016/j.landusepol.2005.09.009

Castella, J. C., & Verburg, P. H. (2007). Combination of process-oriented and pattern-oriented models of land-use change in a mountain area of Vietnam. *Ecological Modelling, 202*, 410-420.

DeFries, R. S., Foley, J. A., & Asner, G. P. (2004). Land-use choices: Balancing human needs and ecosystem function. *Frontiers in Ecology and the Environment, 2*, 249–257. doi:10.1890/1540-9295(2004)002[0249:LCBHNA]2.0.CO;2

Eickhout, B., Van Meijl, H., Tabeau, A., & van Rheenen, T. (2007). Economic and ecological consequences of four European land use scenarios. *Land Use Policy, 24*, 562–575. doi:10.1016/j.landusepol.2006.01.004

Falcucci, A., Maiorano, L., & Boitani, L. (2007). Changes in land-use/land-cover patterns in Italy and their implications for biodiversity conservation. *Landscape Ecology, 22*, 617–631. doi:10.1007/s10980-006-9056-4

Gustafson, E. J., & Parker, G. R. (1994). Using an index of habitat patch proximity for landscape design. *Landscape and Urban Planning, 29*, 117–130. doi:10.1016/0169-2046(94)90022-1

Hellmann, F., & Verburg, P. H. (2010). Impact assessment of the European biofuel directive on land use and biodiversity. *Journal of Environmental Management, 91*, 1389–1396. doi:10.1016/j.jenvman.2010.02.022

Hurkmans, R. T. W. L., Terink, W., Uijlenhoet, R., Moors, E. J., Troch, P. A., & Verburg, P. H. (2009). Effects of land use changes on streamflow generation in the Rhine basin. *Water Resources Research, 45*, W06405..doi:10.1029/2008WR007574

IPCC. (2000). *Special report on emissions scenarios - a special report of working group III of the Intergovernmental Panel on Climate Change*. Cambridge, UK: Cambridge University Press.

Koomen, E., Loonen, W., & Hilferink, M. (2008a). Climate-change adaptations in land-use planning: A scenario-based approach. In Bernard, L., Friis-Christensen, A., & Pundt, H. (Eds.), *The European information society: Taking geoinformation science one step further* (pp. 261–282). Berlin, Germany: Springer.

Koomen, E., Rietveld, P., & De Nijs, T. (2008b). Modelling land-use change for spatial planning support [Editorial]. *The Annals of Regional Science, 42*, 1–10. doi:10.1007/s00168-007-0155-1

Meijl, Hv., van Rheenen, T., Tabeau, A., & Eickhout, B. (2006). The impact of different policy environments on agricultural land use in Europe. *Agriculture Ecosystems & Environment, 114*, 21–38. doi:10.1016/j.agee.2005.11.006

Pérez-Soba, M., Verburg, P. H., & Koomen, E. (2010). *Land use modelling-implementation: Preserving and enhancing the environmental benefits of land-use services. Final report to the European Commission, DG Environment*. Wageningen: Alterra Wageningen UR/ Geodan Next/ Object Vision/ BIOS/ LEI and PBL.

Pontius, R. G., Boersma, W., Castella, J.-C., Clarke, K., de Nijs, T., Dietzel, C., & Verburg, P. H. (2008). Comparing the input, output, and validation maps for several models of land change. *The Annals of Regional Science, 42*, 11–37. doi:10.1007/s00168-007-0138-2

Reidsma, P., Tekelenburg, T., van den Berg, M., & Alkemade, R. (2006). Impacts of land use change on biodiversity: An assessment of agricultural biodiversity in the European Union. *Agriculture Ecosystems & Environment, 114*, 86–102. doi:10.1016/j.agee.2005.11.026

Trisurat, Y., Alkemade, R., & Verburg, P. (2010). Projecting land-use change and its consequences for biodiversity in Northern Thailand. *Environmental Management, 45*, 626–639. doi:10.1007/s00267-010-9438-x

Verboom, J., Alkemade, R., Klijn, J., Metzger, M. J., & Reijnen, R. (2007). Combining biodiversity modelling with political and economic development scenarios for 25 EU countries. *Ecological Economics, 62*, 267–276. doi:10.1016/j.ecolecon.2006.04.009

Verburg, P., Eickhout, B., & van Meijl, H. (2008). A multi-scale, multi-model approach for analyzing the future dynamics of European land use. *The Annals of Regional Science, 42*, 57–77. doi:10.1007/s00168-007-0136-4

Verburg, P., & Overmars, K. (2009). Combining top-down and bottom-up dynamics in land use modelling: Exploring the future of abandoned farmlands in Europe with the Dyna-CLUE model. *Landscape Ecology, 24*, 1167–1181. doi:10.1007/s10980-009-9355-7

Verburg, P. H., Overmars, K. P., Huigen, M. G. A., de Groot, W. T., & Veldkamp, A. (2006). Analysis of the effects of land use change on protected areas in the Philippines. *Applied Geography (Sevenoaks, England), 26*, 153–173. doi:10.1016/j.apgeog.2005.11.005

Verburg, P. H., Soepboer, W., Limpiada, R., Espaldon, M. V. O., Sharifa, M., & Veldkamp, A. (2002). Land use change modelling at the regional scale: The CLUE-S model. *Environmental Management, 30*, 391–405. doi:10.1007/s00267-002-2630-x

Verburg, P. H., van de Steeg, J., Veldkamp, A., & Willemen, L. (2009). From land cover change to land function dynamics: A major challenge to improve land characterization. *Journal of Environmental Management, 90*, 1327–1335. doi:10.1016/j.jenvman.2008.08.005

Verburg, P. H., & Veldkamp, A. (2004). Projecting land use transitions at forest fringes in the Philippines at two spatial scales. *Landscape Ecology, 19*, 77–98. doi:10.1023/B:LAND.0000018370.57457.58

Verburg, P. H., Veldkamp, A., de Koning, G. H. J., Kok, K., & Bouma, J. (1999). A spatial explicit allocation procedure for modelling the pattern of land use change based upon actual land use. *Ecological Modelling, 116*, 45–61. doi:10.1016/S0304-3800(98)00156-2

Voinov, A., Costanza, R., Wainger, L., Boumans, R., Villa, F., Maxwell, T., & Voinov, H. (1999). Patuxent landscape model: Integrated ecological economic modelling of a watershed. *Environmental Modelling & Software, 14*, 473–491. doi:10.1016/S1364-8152(98)00092-9

Wassenaar, T., Gerber, P., Verburg, P. H., Rosales, M., Ibrahim, M., & Steinfeld, H. (2007). Projecting land use changes in the Neotropics: The geography of pasture expansion into forest. *Global Environmental Change, 17*, 86–104. doi:10.1016/j.gloenvcha.2006.03.007

Westhoek, H. J., van den Berg, M., & Bakkes, J. A. (2006). Scenario development to explore the future of Europe's rural areas. *Agriculture Ecosystems & Environment, 114*, 7–20. doi:10.1016/j.agee.2005.11.005

Chapter 7
Landscape Biodiversity Characterization in Ecoregion 29 Using MODIS

Nitin Kumar Tripathi
Asian Institute of Technology, Thailand

Aung Phey Khant
Asian Institute of Technology, Thailand

ABSTRACT

Biodiversity conservation is a challenging task due to ever growing impact of global warming and climate change. The chapter discusses various aspects of biodiversity parameters that can be estimated using remote sensing data. Moderate resolution satellite (MODIS) data was used to demonstrate the biodiversity characterization of Ecoregion 29. Forest type map linked to density of the study area was also developed by MODIS data. The outcome states that remote sensing and geographic information systems can be used in combination to derive various parameters related to biodiversity surveillance at a regional scale.

1. INTRODUCTION

A natural environment is self-renewing, self-perpetuating and stable one, in which every organism contributes in some way, however, small to the overall stability. In natural ecosystems, the plants and animals have evolved at their own pace and in their own way under the influence of natural selection to fit in the constellation of certain environmental factors or niches. In the process, they help to sustain others, each species controlling its own population growth and at the same time limiting of other species, so that a reasonable ecological balance may be achieved and maintained for hundreds of years The satellite remote sensing can identify the important parameters for biodiversity characterization like size, fragmentation, porosity, patchiness, interspersion and juxtaposition at the landscape level. The role of remote sensing is emphasized in quick appraisal of regional biodiversity surveillance. This becomes of high importance in present context of biodiversity loss due to climate change.

DOI: 10.4018/978-1-60960-619-0.ch007

Table 1.

Ecoregion Name	:	Kayah-Karen Montane Rain Forests
Bioregion	:	Indochina
Major Habitat Type	:	Tropical and subtropical moist broadleaf forest
Ecoregion Number	:	51
Political Unit(s)	:	Myanmar, Thailand
Ecoregion Size	:	119,200 km^2
Biological Distinctiveness	:	Globally outstanding
Conservation Status	:	Relatively intact
Conservation Assessment	:	III

Table 2.

Ecoregion Name	:	Tenasserim-South Thailand Semi-evergreen Rain Forests
Bioregion	:	Indochina
Major Habitat Type	:	Tropical and subtropical moist broadleaf forest
Ecoregion Number	:	53
Political Unit(s)	:	Myanmar, Thailand, Malaysia
Ecoregion Size	:	96,900 km^2
Biological Distinctiveness	:	Globally outstanding
Conservation Status	:	Relatively intact
Conservation Assessment	:	III

There is an urgent need to inventory and monitor indicators of biological diversity such as species richness and habitats. Remotely sensed data provide a means to accomplish part of this task, but there has been no comprehensive scientific framework to guide its effective application (Stoms & Estes, 1993). Most of the discussion concerning potential roles for remote sensing in biodiversity assessment has come from conservation biologist and ecologists (Soule & Kohm, 1989; Noss, 1990; Lubchenco *et al.*, 1991). The remote sensing community has had little involvement to date in supporting biodiversity research, largely concentrating instead in the global change domain (Stoms & Estes, 1992). Very little quantitative analysis has been accomplished to determine the actual value of remote sensing and geographic information systems in biological research.

Remote sensing provides spatial data, which are less used but they are the powerful source to acquire accurate, up-to-date information essential for conserving biodiversity and wildlife habitat mapping. Although technically complex, the remote sensing techniques have revolutionized the process of data gathering and map making. The combinations of Remote Sensing (RS) and Geographical Information System (GIS) have proven to be very effective tools to analyze the landscape patterns for biodiversity characterization at various levels.

2. STUDY AREA

Kayah-Kayin and Tenisserim ecoregion (Ecoregion 29) are the richest in species in mainland Southeast Asia, for this area is the cross road to exchange species among different geographic regions of Holartic, Oriental and Greater Sundas Island. On the other hand, this ecoregion forming a juncture of the Indo-Chinese, Indo-Burmese, and Malaysian floral and faunal elements. Formerly, it is divided into the Kayah-Karen Montane Rain Forests (ecoregion 51: Table 1) and the Tenasserim-South Thailand Semi-evergreen Rain Forests (ecoregion 53: Table 2). In the year 2000, World Wide Fund For Nature (WWF) scientists team combined these two regions and defined as ecoregion (29). The flora and fauna in this region is distinct and includes several endemic species. Among the ecoregions of Indochina, this ecoregion contains some of the highest diversity of both bird and mammal species found in the Indo-Pacific region

This ecoregion encompasses the mountainous, semi-evergreen rain forests of Thailand, Myanmar and Malaysia, and includes the extensive lowland

Figure 1. Map of Ecoregion 29

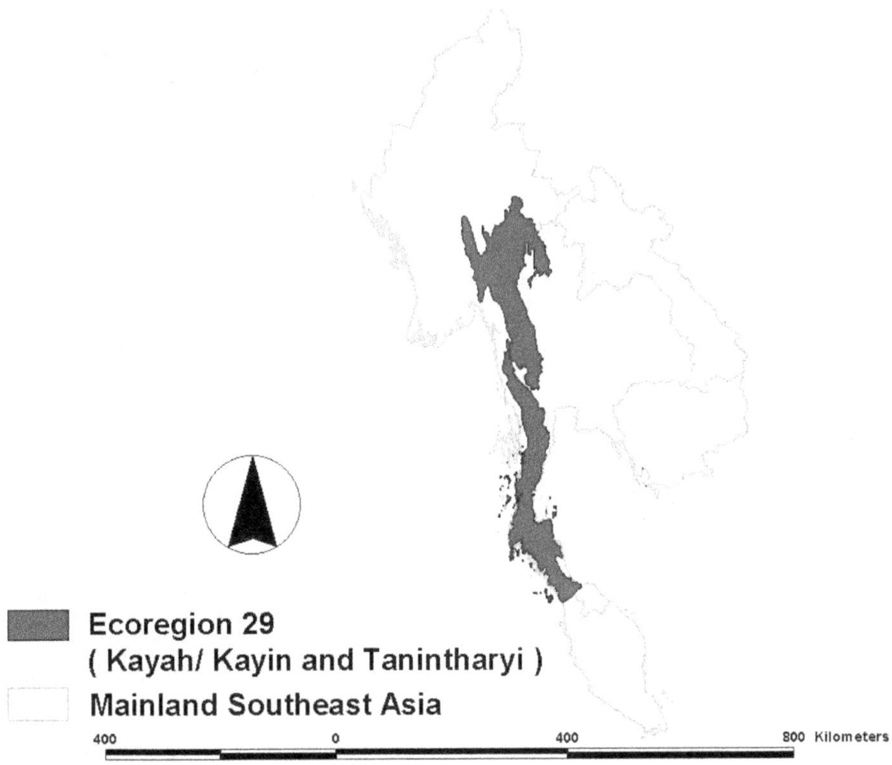

plain that lie between the peninsular mountains and which until recent decades supported extensive lowland forest. The southern margin of this ecoregion is defined by the Kangar-Pattani floristic boundary (Whitmore & Sayer, 1992). Figure 1 shows the geographic extent of the study area.

2.1. Vegetation

The vegetation of this ecoregion includes both tropical and subtropical moist broadleaf forest, montane forests, lowland rainforest; and a higher proportion of evergreen broad-leaved species. This ecoregion represents the semi-evergreen forests of the Kayah-Karen Mountains in the broad transition zone between the subtropical broadleaf evergreen forests in the north and the southern tropical and dry deciduous forests in Tenesserim and southern Thailand (Figure 2). Tropical hardwood trees in the family *Dipterocarpaceae* dominate forests throughout the ecoregion. Forests to the east are dominated, especially at the lower elevations, trees that have a drought-deciduous phenology, while the west-facing slopes are a mixture of deciduous and evergreen species. At low elevation (<= 1000 m) on the east side of the Tenasserim Hills, the potential vegetation consists of drought deciduous forest or savanna woodland. Higher elevations support much richer broad-leaved forest communities with a mixture of evergreen and deciduous species. Forests of teak *Tectona grandis* represent the climax vegetation at low elevation in the absence of fire, but today the teak forests are nearly extirpated in Thailand and also declining rapidly in Myanmar. At 800-1200m, a well-developed undestroyed shrub grows beneath a tall, closed forest canopy that includes some very large, buttressed trees that share an affinity with tropical Asia together with temperate tree taxa in the families *Magnoliaceae* and *Lauraceae* (IUCN, 1991). Although fire is frequent today, there is little consensus as to the historical frequency of fire or its importance in this ecosystem. An important unresolved question is whether fire (mostly anthropogenic) or pre-monsoon drought stress (non-anthropogenic) is primarily responsible for limiting species diversity in these places. This ecoregion remains relatively unexplored scientifically, especially those parts that lie in Myanmar, it will very likely yield more biological surprises. However, after Thailand banned timber exploitation in its forests in 1988, Myanmar granted large logging concessions to Thai companies, and illegal timber extraction in Myanmar by Thai loggers has become common in recent years (WWF & IUCN 1995).

2.2. Fauna

This ecoregion contains one of the most intact vertebrate faunas of Indochina, including one of the richest assemblages of mammals in Asia. The fauna is also distinctive, with characteristics of the islands of the Malay Archipelago as well as the mountains of China and India. The relatively intact and contiguous hill and montane habitat has potential for conserving large landscapes that will provide adequate habitat to maintain a viable population of Asia's largest carnivore, the tiger (*Panthera tigris*), and Asian elephant (*Elephas maximus*). This ecoregion lies within a high priority, Level (I) Tiger Conservation Unit (Dinerstein et al. 1997). This range of forests in conjunction with the Kayah-Karen Mountains represents some of the best landscapes for Asian elephant conservation in Indochina. Numerous other mammals are of conservation significance, primarily the elusive and endemic Fea's muntjac (*Muntiacus feae*). The population of the Malayan tapir (*Tapirus indicus*), the only Old World tapir representative, has been drastically reduced. It survives in the hill and montane protected areas of this ecoregion and scattered pockets throughout peninsular Malaysia and Sumatra. More than 25

Figure 2. Land use map of study area

pairs have been found in some of the last remaining forest, and that forest is now contained within Bang Kram wildlife sanctuary (Stewart-Cox, 1995). Several primate species are found in these forests, and include the threatened banded langur (*Trachypithecus melalophus*) and slow loris (*Loris nycticebus*), a small, nocturnal prosimian. Other species of conservation concern include the Dyak fruit bat (*Dyacopterus spadiceus*), the endangered clouded leopard (*Pardofelis nebulosa*), common leopard (*Panthera pardus*), sun bare, binturong (*Arctictis binturong*), gaur (*Bos gaurus*), and banteng (*Bos javanicus*) (Stewart-Cox, 1995).

The wide diversity of habitats within this ecoregion, from deciduous forests in the north to seasonal evergreen forests in the south, habitats lowland to montane, make it one of the richest in bird species for the entire Indo-Pacific. A total of 560 bird species have been recorded there. The ecoregion is the fourth richest in the Indo-Pacific region for mammals with 168 known species. These include one ecoregional endemic species, the tiny Kitti's hog-nosed bat (*Craseonycteris thonglongyai*). Some of the other mammals of conservation importance include several threatened species such as the tiger (*Panthera tigris*), Asian elephant (*Elephas maximus*), gaur (*Bos gaurus*), banteng (*Bos javanicus*), wild water buffalo (*Bubalus arnee*), southern serow (*Naemorhedus sumatraensis*), clouded leopard (*Pardofelis nebulosa*), Malayan tapir (*Tapirus indicus*), wild dog (*Cuon alpinus*), Asiatic black bear (*Ursus thibetanus*), Assamese macaque (*Macaca assamensis*), stump-tailed macaque (*Macaca arctoides*), smooth-coated otter (*Lutrogale perspicillata*), great Indian civet (*Viverra zibetha*), and Particoloured flying squirrel (*Hylopetes alboniger*). Sumatran rhinoceros is believed to have inhabited remote regions of the Tenasserim Hills in recent years, but this critically endangered species is now thought to have been extirpated from this ecoregion. Hunting has decimated most of the large mammal populations, such as elephant, banteng, gaur and tiger (IUCN, 1991). On the other hands, many species are declining to unsafe population levels, important habitats are destroyed, fragmented and degraded, and the ecosystems are destabilized through climate change, population, invasive species, and human impact.

The relatively intact, contiguous habitat has potential to conserve large landscape that will provide adequate habitat to maintain a viable population of Asia's largest carnivore, the tiger, as well as other species of critical conservation significance. Several of Myanmar and Thailand's largest and most intact wildlife reserves lie within this ecoregion, including Myinmo-let-khat National Park (proposed) and Huai Kha Khaeng Wildlife Sanctuary (2,575 km^2) and several other protected areas with which it forms a contiguous network. Huai Kha Khaeng is prized for the high diversity of cat species it supports, and it's relatively intact vertebrate communities and intact lowland dipterocarp forests. Moister habitats on the Myanmar side of the Tenasserim ranges also include significant amounts of intact habitat, probably still in better condition overall than the forest on the eastern (Thailand) side of the range. However it is difficult to assess ecological conditions in the forests of eastern Myanmar at this time.

The existing (50) protected areas that cover 32% (35,030 km^2) of the ecoregion and most of these protected areas are located in Thailand. Large blocks of intact seasonal evergreen forest habitats are still remaining in Myanmar, but these are mostly not protected. Some protected areas have been designated in the portion of this ecoregion that lies within Myanmar, but their effectiveness is difficult to assess at this time due to the political instability of the region.

2.3. Agriculture

Shifting cultivation is the main cause of deforestation throughout the region. But in areas like northern Thailand, where previously nomadic tribal peoples have been settled, pressure exists to convert forest into more intensive agricultural land devoted to cash crops like cabbage, coffee, and lychee. Opium replacement efforts, in Thailand, have compelled local people to grow alternative crops that require more cultivated land area and higher pesticide inputs. The land requirements of an increasing population have forced itinerant farmers to reduce the cycle of cultivation-fallow periods, and have pushed them deeper into the forest and into more marginal areas.

Lowland area in Peninsular Thailand (Bang Kram) still supports a significant amount of late successional forest. Other areas support vast tracts of rubber plantation, monocultures of a fast-growing, short-lived tree species native to South America, and plantations of oil palm. Pineapple is grown there as a rotation crop following the removal of senescent rubber trees. Paddy is also grown in some lowland areas. Unfortunately, none of these crops with the possible exception of paddy provide any significant support for natural biological diversity. Hill slopes support more native forest than the lowland areas, and the hill forests of Southern Thailand are relatively intact, although swidden (slash and burn) agriculture is still practiced in hill areas in the northern part of the ecoregion. Mature forest cut for swidden agriculture is generally succeeded in this ecoregion by a grassy sub-climax that supports far fewer species than the mature forest (IUCN, 1991). General land-use map of the study area is shown in Figure 2.

2.4. Geology

Ecoregion (29) consists of hills of Paleozoic limestone that have been much dissected by chemical weathering. The overhanging cliffs, sinkholes, and caverns characteristic of tropical karst landscapes are all-present in this ecoregion.

2.5. Landscape Structure

Terrain throughout much of this ecoregion is rugged and intricately folded. Hillsides tend to be steep and ridges exceed 2,500 m elevation. Valley bottoms are narrow but fertile and tend to lie at about 300 m elevation. The western slopes drain into the Salween River that flows through Myanmar and entered Gulf of Martaban in the Indian Ocean. The eastern slopes drain into the Chao Phraya River that drains into the gulf of Thailand.

2.6. Climate

Annual precipitation increases southward as the length of the dry season and the magnitude of pre-monsoon drought stress declines. The southern mountain ranges receive rain from both the northeast and southwest monsoons so that, unlike mountain ranges further north, there is no significant rainshadow. The entire region has a monsoon climate with warm, moist summers and mild winters that tend to be dry. Overall annual rainfall average is 1,500 mm to 2,000 mm. Although this ecoregion lies within the Tropic of Cancer, winter temperatures can be quite cool, especially at the higher elevations where frost has been recorded from the northern part of the ecoregion. West-facing slopes (Myanmar side) face the Bay of Bengal and receive more precipitation; and East-facing slopes (Thailand side) lie within a partial rainshadow and tend to be drier. This climatic difference is clearly reflected in the vegetation. The Köppen climate system placed this ecoregion in the Tropical Wet Climate Zone (Townsend & Cohoon, 1999).

3. DATA USED

Fortunately, enough data is available with institutions and agree to share the information. These are Biogeographical zone map, protected areas networks and species list from WWF-Thailand, topographic/ vegetation/ LU & LC/ geology/ soil maps from Department of Forestry (Myanmar and Thailand); and annual rainfall and temperature for ten years from Department of Geography. Lists of data used for this study are as follows:

- MODIS satellite data for study area with the spatial resolution of 250 m.
- Digital map of entire ecoregion with the scale of 1: 250000
- Topographic/ Vegetation/ LU&LC/ Geology/ Soil map
- Annual rainfall and temperature for 10 years
- Other ancillary data

MODIS image in Figure 3 shows the overview of the ecoregion 29 and surrounding. This image was utilized to extract the biodiversity characteristic of the ecoregion.

Figure 3. Kayah-Kayin and Tenisserim ecoregion (Ecoregion 29) in MODIS image

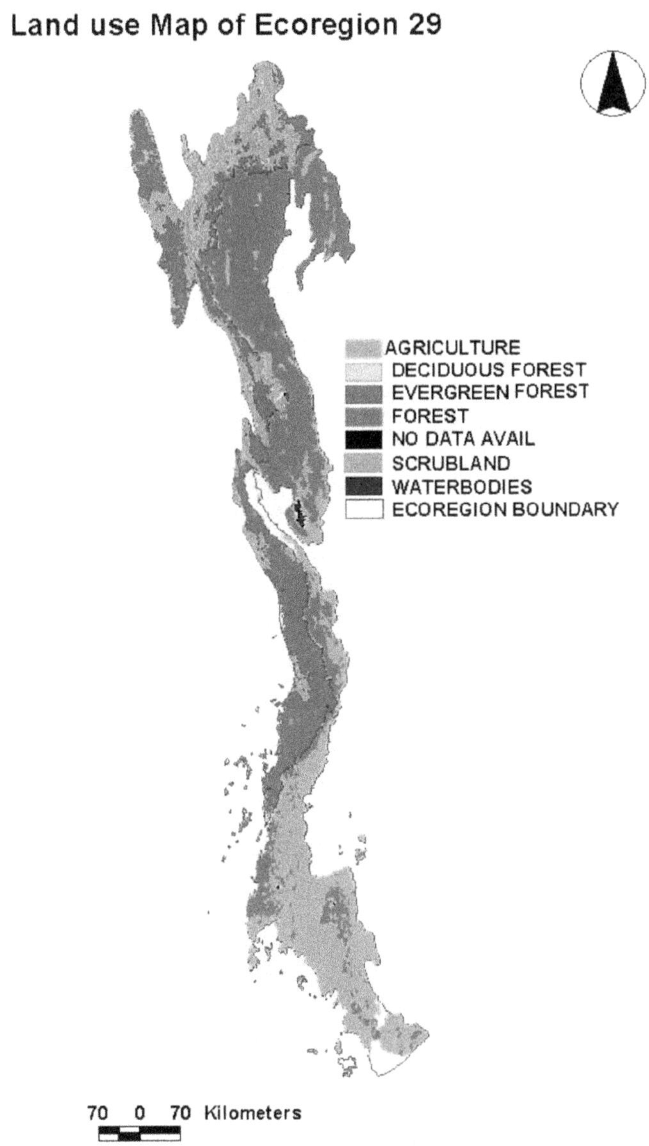

4. BIODIVERSITY RICHNESS ESTIMATION MODEL

Growing concerns over the loss of biodiversity has spurred land managers to seek better ways of managing landscapes at a variety of spatial and temporal scales. A number of developments have made possible the ability to analyze and manage entire landscapes to meet multi-resource objectives. The developing field of landscape ecology has provided a strong conceptual and theoretical basis for understanding landscape structure, function, and change (Forman & Godron 1986; Urban *et al.*, 1987; Turner, 1989). Growing evidence that habitat fragmentation is detrimental to many species and may contribute substantially to the loss of regional and global biodiversity (Saunders *et al.*, 1991; Harris, 1984) has provided empirical justification for the need to manage entire landscapes, not just the components.

As growing numbers of researchers and resource managers rely on digital geographic data, and look to remotely sensed imagery as a source of data for their GIS, issues of geographic data models and remote sensing scene models (Goodchild, 1992, 1994; Strahler *et al.*, 1986), as well as image processing algorithm development, are central to research on the integration of remote sensing and GIS.

A rotational approach to the management of biological resources requires advanced technology in many areas of information processing. The conceptual framework is quite complex, with many interactions between species and environments. Advanced computation and massive data archives would produce quantitative improvement in analysis of biodiversity through more accurate and efficient solution. Visualization and interactivity would produce greater understanding and clarity on the status of the environment and effort to maintain it.

A landscape is composed typically of several types of landscape elements. Of these, the matrix is the most extensive and most connected landscape element type, and therefore plays the dominant role in the functioning of the landscape. As landscape patterns are complex and heterogeneous, variety of matrices are needed to measure different aspect of patterns to characterize biodiversity. Digital satellite data provides a consistent and complete spatial data for a very large area and may be considered very useful tool for measuring landscape patterns (patch, fragmentation, terrain complexity, porosity, interspersion and juxtaposition) to estimate biodiversity richness together with species diversity index (Shannon and Simpson). And also, the landscape spatial configurations influence ecological processes such as biodiversity, habitat or animal population dispersal and abundance.

Landscapes are distinguished by spatial relationships among component parts. A landscape can be characterized by both its composition and configuration, and these two aspects of a landscape can independently or in combination affect the ecological processes and organisms. The difference between landscape composition and configuration is analogous to the difference between floristic and vegetation structure that are commonly considered in biodiversity studies at the within-patch scale.

Spatial data layers are generated using digital satellite data through digital image processing and classification. In addition to that spatial and non-spatial data from other ancillaries data sources are collected. Database is generated using GIS software and analyzed. Then biodiversity characterization was done analyzing the landscape parameters of patch, fragmentation, porosity, interspersion and juxtaposition, and using ground base data on species richness, terrain complexity and diversity index.

First, and more importantly, subsets of particular landscape have to be analyzed in order to evaluate the distribution of spatial characteristics (to detect the high diversity or where the habitat is more endangered because of the loss of connectivity).

Therefore, it is necessary to know how the extent of analyzed pattern is going to influence the results of estimation and whether the obtained values are reliable to the entire region under analysis.

4.1 Indices for Biodiversity Characterization

Species Diversity Index

Species diversity can be described as the number of species in a sample or habitat per unit area. The indices are as follow:

Shannon's diversity index (log base 10 and natural log)

$$H = -\Sigma(P_i \log[P_i])$$

$$H = -\Sigma(P_i \ln[P_i])$$

Simpson's diversity index (*D*)

$$D = \Sigma(P_i^2)$$

$$D = 1-\Sigma(P_i^2)$$

Where, Pi = the total number of species
H = Shannon's index value
D = Simpson's index value

Fragmentation

Forest fragmentation occurs when large, continuous forests are divided into smaller blocks, either by roads, clearing for agriculture, urbanization, or other human development.

$$\mathbf{F} = \sum_{i=1}^{N} \mathbf{D}_i$$

Where, **F** = Index value of Fragmentation
$\mathbf{D}_i$ = dissimilarity value for the ith boundary between adjacent cells

N = number of boundaries between the adjacent cell

Generally, forest type map is classified into two classes such as forest and non-forest. A grid cell (n * n) is convolved with the spatial data layer with a deriving number of forest patches within the grid cell. This will be repeated by moving the grid cell through the entire spatial layer. The output layer with patches was derived.

Patchiness

Patchiness is a measure of the density of patches of all types or number of clusters in given mask. It is a measurement of number of polygons over a particular area.

$$Pj = ((\sum Di) / Nj) * 100$$

Where, N = Nj is the number of boundaries between adjacent intervals along Transect j
Di = Dissimilarity value for the i th boundary between adjacent Cells

In landscape ecology, patches are spatial units at the landscape scale. Patches are areas surrounded by matrix, and may be connected by corridors. The geomorphology of the land interacting with climate factors, along with the other factors such as the establishment of flora and fauna, soil development, natural disturbances, and human influences work to determine patch size, shape, location, and orientation (Forman & Godron, 1986).

Five major types of patches:

(1) spot disturbance patch
(2) remnant patch
(3) environmental resource patch
(4) introduced patch
(5) ephemeral patch

Porosity

Porosity is the measure of number of patches or density of patches within a landscape, considers only number of patches, not size.

$$\mathbf{PO} = \sum_{I=1}^{n} \mathbf{Cpi}$$

Where, Cpi = number of closed patches of ith cover class

Interspersion

Interspersion is count of dissimilar neighbors with respect to central pixel or measurement of the spatial intermixing of the vegetation types. It can also be used to represent the *landscape diversity.*

$$\mathbf{I} = \sum_{I=1}^{n} (\sum \mathbf{Fi} / \mathbf{n})$$

where Fi is shape of factor

For the determination of interspersion a 'Convolution' will be used with the forest type map to compute the number of dissimilar pixels in the nearest neighborhood. The computation is performed in an interactive mode through the entire spatial layer to derive an output interspersion layer. Calculation of interspersion gives the magnified view of resistance; the central pixel or class has with respect to its surroundings.

Juxtaposition

Juxtaposition is defined as measure of proximity of the vegetation of vegetation. It's measurement includes relative weight assigned by the importance of the adjacency of two cover types for the species in question. It is species-specific measurement i.e. edge between cover and quality of edge.

$$J = \sum Di\,(Ji) / Jmax$$

I=1

where, Di = the shape desirability weight for each cover type combination

Ji = the length of edge between combination of cover types on either side of an edge

Jmax = the average total weighted edge per habitat unit of good habitat

A grid cell of size 5 (e.g. 5*5) is convolved with the derived layer in an interactive manner by assigning higher weight to natural vegetation and lower weight to unnatural vegetation. The juxtaposition helps in characterizing the parameter porosity with respect to natural or unnatural (manmade) vegetation type. Since porosity can be understood as one of the important factors influencing the disturbance index, the added weightage through juxtaposition gives the right perspective in ultimately deriving disturbance index.

4.2 Landscape Configuration Metrics for Biodiversity Characterization

Landscape Similarity Index (LSIM)

$$\text{LSIM} = P_i = \frac{\sum_{j=1}^{n} a_{ij}}{A}(100)$$

LSIM equals total class area (m^2) divided by total landscape area (m^2), multiplied by 100 (to convert to a percentage); in other words, LSIM equals the percentage of the landscape comprised of the corresponding patch type.

Patch Density (PD)

$$\text{PD} = \frac{n_i}{A}(10,000)(100)$$

PD equals the number of patches of the corresponding patch type (NP) divided by total landscape area, multiplied by 10,000 and 100 (to convert to 100 hectares).

Landscape Shape Index (LSI)

$$LSI = \frac{\sum_{k-1}^{m} e_{ik}}{2\sqrt{\grave{A}\underline{\ }A}}$$

When LSI = 1, the landscape consists of a single patch of the corresponding type and is circular (vector) or square (raster); LSI increases without limit as landscape shape becomes more irregular and/or as the length of edge within the landscape of the corresponding patch type increases.

There are a number of landscape configuration metrics to characterize the biodiversity spatially. The spatial statistics output table created from a landscape pattern analysis contains all the results of the analysis (Class area, Total landscape area, number of patches, Total edges, Edge density, Mean shape Index, Fractal Dimension, Mean Nearest Neighbor Distance, Mean Proximity Index, Interspersion and Juxtaposition Index, Shannon's Diversity Index, Shannon's Evenness Index).

Mean Nearest Neighbor (MNN), Measure of patch isolation, means the nearest neighbor distance of an individual patch is the shortest distance to a similar patch (edge to edge). The mean nearest neighbor distance is the average of these distances (meter) for individual classes at the class level and the mean of the class nearest neighbor distances at the landscape level.

5. RESULTS AND DISCUSSIONS

The improvement in spectral, spatial and temporal resolutions and better digital image processing of remotely sensed data has kept pace with the information needs of the characterizing biodiversity. The geographic information system (GIS) tool has developed fast with greater facility of large area, multiple spatial and non-spatial data integration and analysis.

5.1 Results Discussion on Biodiversity Richness Estimation Model

In the biodiversity richness estimation model, the most significant landscape patterns, which can be extracted from the digital satellite data were used for biodiversity characterization.

For meaningful estimation and characterization of biodiversity, the degree of variation of each index is related to landscape patterns that are used for biodiversity richness estimation model and not to artifacts derived from the methodological problems involved in the measurement. This uncertainty associated with the index estimated is arguably one of the major limitations of this kind of quantitative analysis for biodiversity characterization. And it is also greatly related to the scale such as spatial resolution of satellite data and the spatial extent (i.e. total area).

The biodiversity richness estimation model can provide variety of landscape value and allowing obtaining wide range of patterns with intermediate level of spatial differences, in which patchiness and fragmentation is generated.

The use of these indices and landscape configuration matrices instead of real world landscape data is preferable in this study because this approach is possible to generate and isolate different factors affecting the landscape to characterize the biodiversity.

The influences of fragmentation, patchiness, porosity, interspersion and juxtaposition can be adequately separated, and thus the biodiversity richness estimation using landscape patterns and configuration metrics can be specifically analyzed avoiding confusion with other land cover data (Table 3).

Patch shape and orientation also play an important ecological role. An ecologically optimum patch shape usually has a large core with some

Table 3. Indices for Biodiversity characterization of study area

No.	Index name	Result value
1	Patchiness	60.98
2	Fragmentation	2.009
3	Porosity	2,962
4	Interspersion/ Juxtaposition	49.015
5	Landscape similarity	48.07
6	Patch fractal dimension	33

curvilinear boundaries and narrow lobes (Forman, 1995). This shape may allow both interior species and edge species to flourish. Patch shapes also determine the edge length.

Porosity is very important parameter for conservation aspect at landscape and regional level. The porosity (1) provides an over all clue to the change of species isolation present and to the potential genetic variability present within population of animals and plants in regional level, (2) sometime indicates the presence of high conservation value and lower interaction among landscape elements, homogeneity and low fragmented habitats, (3) higher porosity value indicates the higher interaction among landscape elements, heterogeneous and high fragmented habitats, and (4) disturbance in natural landscape increases patch density and decrease matrix connectivity.

Interspersion Juxtaposition Index, Measure of patch adjacency, the result will be zero when the distribution of unique patch adjacencies becomes uneven and 100 when all patch types are equally adjacent. At the landscape level it is a measure of the interspersion of the each patch in the landscape.

The dispersal ability of species depends on spatial organization of landscape. The fragile or rare species occur only in highly connected landscape mosaic. Higher value of interspersion means dispersal ability of the central class will be low or reduce.

Landscape Similarity Index Matrices approaches 0 when the corresponding patch type (class) becomes increasingly rare in the landscape. LSIM equal to 100 when the entire landscape consists of the corresponding patch type; that is, when the entire image is comprised of a single patch.

Fractal dimension is a measure of shape complexity that can be computed for each patch and then for landscape, or it can be computed from the landscape as a whole.

These diversity measures are influenced by two components: (1) richness and (2) evenness. Richness refers to the number of patch types present and evenness refers to the distribution of area among different types. Richness and evenness are generally referred to as the compositional and structural components of diversity, respectively. Simpson's diversity index is more sensitive to richness than evenness. Thus, rare types have a disproportionately large influence on the magnitude of the index. This diversity index has been applied to measure one aspect of landscape structure and landscape composition to characterize biodiversity. Thus, the higher the value the greater the likelihood that any randomly drawn patches would be different patches that is greater in diversity.

In many biodiversity applications, the primary interest is in the amount and distribution of a particular patch. Forest fragmentation is a landscape-level process in which forest tracts are progressively sub-divided into smaller, geometrically more complex, and more isolated forest fragments as a result of both natural processes and human land use activities. This process involves changes in landscape composition, structure, and function and occurs on a backdrop of a natural patch mosaic created by changing landforms and natural disturbances. Forest fragmentation is the prevalent trajectory of landscape change in several human dominated forest regions, and is increasingly becoming recognized as a major cause of declining biodiversity.

The use of diversity measures in community ecology has been heavily criticized because diversity conveys no information on the actual species composition of a community. Species diversity is a community summary measure that does not take into account the uniqueness or potential ecological, social, or economical importance of individual species. A community may have high species diversity yet be comprised largely of common or undesirable species. Conversely, a community may have low species diversity yet be comprised of especially unique, rare, or highly desired species. Although these criticisms have not been discussed explicitly with regards to the landscape ecological application of diversity measures, these criticisms are equally valid when diversity measures are applied to patch types instead of species. In addition, these diversity indices combine richness and evenness components into a single measure, even though it is usually more informative to evaluate richness and evenness independently (Tables 4 and 5, Figures 4, 5 and 6).

Therefore, a landscape with many habitats will be richer than a less heterogeneous one. However, if habitat patches become too fragmented and disjunctive, as typically results from human-induced land use conversion, regional richness declines.

5.2 Forest Type Distribution

The potential of remote sensing and Geographic Information Systems is displayed in providing accurate and timely information on forest density map. The forest types are classified base on the dominant species composition. As a result, in the study area various forest types were classified and mapped with reasonable amount of accuracy. The vegetation type map of ecoregion 29 is classified into nine types: (1) Deciduous Broadleaf Forest, (2) Semi-deciduous Broadleaf Forest, (3) Evergreen Needleleaf Forest, (4) Lowland Evergreen Broadleaf Rain Forest, (5) Semi-evergreen Moist Broadleaf Forest, (6) Upper Mountain Forest, (7) Lower Mountain Forest, and (8) Mangrove Forest (Table 6).

Table 4. The results of Shannon's and Simpson's species richness index

Index	Plot 1	Plot 2	Plot 3	Plot 4	Plot 5	Plot 6	Plot 7	Plot 8	Plot 9	Plot 10	Plot 11	Plot 12	Summarized Index
Shannon H' Log 10.	1.101	1.229	1.287	1.232	1.19	1.27	1.228	1.192	1.052	1.107	1.196	1.058	1.179
Shannon Hmax Log10.	1.431	1.556	1.519	1.58	1.415	1.431	1.519	1.491	1.556	1.505	1.519	1.505	1.502
Shannon J'	0.77	0.789	0.848	0.78	0.841	0.887	0.808	0.799	0.676	0.735	0.788	0.703	0.785
Simpsons Diversity (D)	0.136	0.104	0.071	0.1	0.096	0.062	0.086	0.095	0.129	0.147	0.096	0.195	0.110
Simpsons Diversity (1/D)	7.343	9.577	14.046	10.008	10.383	16.228	11.686	10.575	7.727	6.826	10.45	5.12	9.997

Table 5. Overall species richness estimation

No. of plots	Estimation
1	31.6
2	45.4
3	55.4
4	59.4
5	61.6
6	64.4
7	66.2
8	67.2
9	67.8
10	68.4
11	68.6
12	69

The degradation activities such as extensive shifting cultivation, excessive logging, increasing in human population, development of infrastructure, and conversion of forest habitats to agricultural land have altered the natural landscape to great extent in ecoregion 29. Because of these increased anthropogenic activities, the natural landscape have become fragmented. In addition to forest cover lost, fragmentation has a great impact on biodiversity. Bio-habitat degradation due to fragmentation is also considered to be one of the major threats to biodiversity and often reduced genetic diversity, consequently.

The fragmentation of forest communities leads to high patch density and also affected to

Figure 4. Graph of Shannon's species richness index

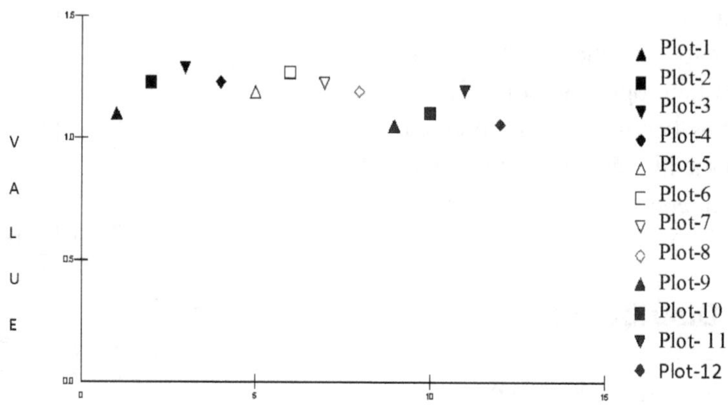

Figure 5. Graph of Simpson's species richness index

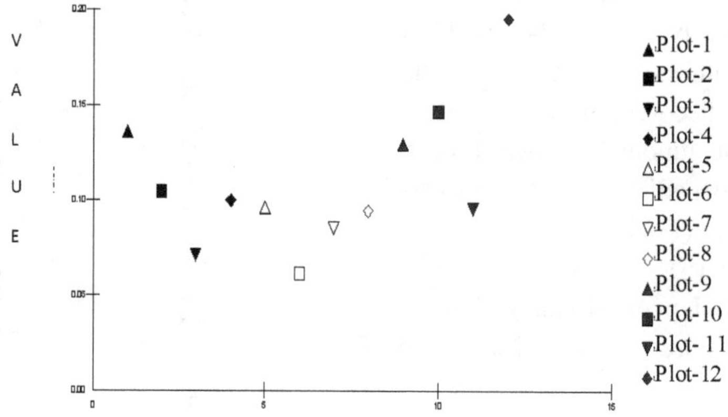

Figure 6. Graph of overall species richness estimation

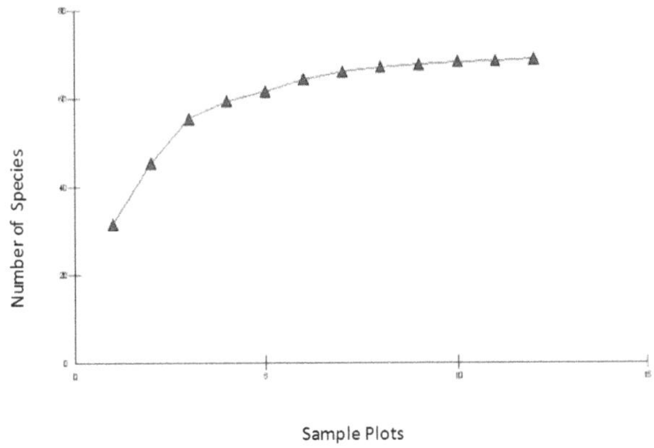

Table 6. Forest density distribution

No.	Forest Type	Area (km²)	Percentage (%)
1	Deciduous/semi-deciduous broadleaf forest	26,886.55	12.45
2	Disturbed natural forest	20,438.47	9.46
3	Evergreen forest	4,597.43	2.13
4	Semi-evergreen moist broadleaf forest	55,246.34	25.58
5	Lower montane forest	22,093.60	10.29
6	Upper montane forest	1,327.88	0.67
7	Mangrove	839.82	0.38
8	Others	84,521.71	39.04
	Total	215,951.79	100.00

various ecological processes such as species distribution and degradation of habitats of keystone and indicator species. For core habitat zonation, as well as mapping of habitat for endangered species deals with these transformations. These land transformations have resulted in the alteration of natural habitats and have brought in lost of biodiversity. Figures 7 and 8 show the status of forest in ecoregion 29.

6. CONCLUSION

Study demonstrates that MODIS data can be used to characterize biodiversity and help the understanding of landscape diversity and terrestrial ecosystem functions at the regional or landscape level.

The assessment to calculate the biodiversity richness using important indices (Fragmentation, Patch, Terrain complexity, Porosity, Interspersion and Juxtaposition, Species richness, etc.) including landscape configuration metrics generated important quantitative information about the status of different resources and landscape patterns of Ecoregion 29. The use of these index indicate the status of entire landscape and, therefore, can be assumed as essential tools for biodiversity application towards conservation strategies to prevent biodiversity loss.

As known, the biodiversity is a reflection of several biotic, abiotic and climatic factors, there-

Figure 7. General status of Ecoregion 29

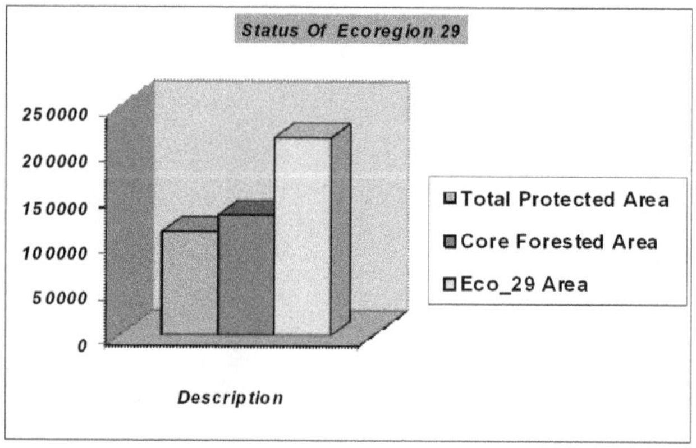

Figure 8. Forest density distribution

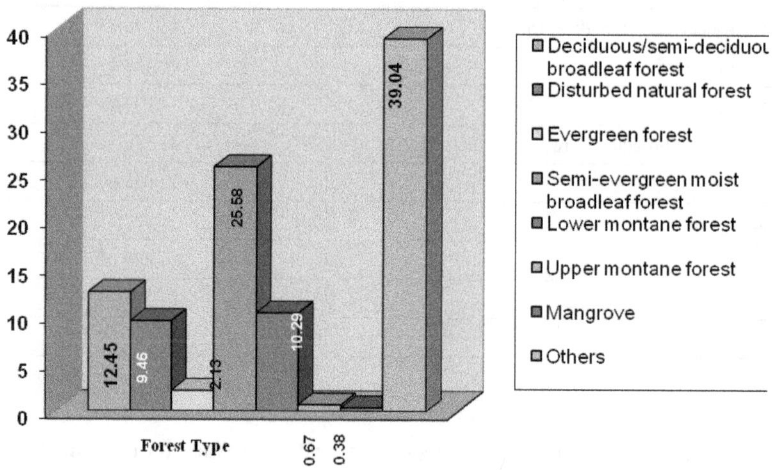

fore the ecological relationships was considered as the most important underlying factor for their *in situ* conservation.

Earlier study lack using moderate resolution satellite data such as MODIS towards biodiversity mapping. This data is regularly available and is well suited for regional level biodiversity surveillance (inventory and monitoring) to extract the data and information of landscape patterns, land cover, and vegetation. Conventional methods for biodiversity mapping are very costly and time consuming. Since biodiversity loss and ecological imbalance are the major concern of climate change impact on biological richness of ecoregion 29, the surveillance based on MODIS satellite data and limited ground observations will facilitate the prioritization for conservation.

The results from biodiversity richness estimation model indicate that the overall biodiversity richness is still high except the southern part of Thailand.

REFERENCES

David, M. S., Davis, F. W., & Hollander, A. D. (1994). Hierarchical representation of species distribution for biological survey and monitoring. In Goodchild, M., Parks, B. O., & Steyaert, L. (Eds.), *Environmental modeling: Progress and research issues* (pp. 445–449). Fort Collins, CO: GIS World Books.

Dinerstein, E. (1997). *A framework for identifying high priority areas and actions for the conservation of tigers in the wild*. World Wildlife Fund-US.

Foreman, R. T. T. (1995). Some general principles of landscape and regional ecology. *Landscape Ecology, 10*(3), 133–142. doi:10.1007/BF00133027

Foreman, R. T. T., & Gordon, M. (1986). *Landscape ecology*. New York, NY: John Wiley & Sons.

Goodchild, M. F. (1994). Integrating GIS and remote sensing for vegetation analysis and modeling: Methodological issues. *Journal of Vegetation Science, 5*(5), 615–626. doi:10.2307/3235878

Harris, L. D. (1984). *The fragmented forest: Island biogeography theory and the preservation of biotic diversity*. Chicago, IL: University of Chicago Press.

IUCN. (1991). *Caring for the Earth*. The World Conservation Union, Gland.

Lubchenco, J., Olson, A. M., Brubaker, L. B., Carpenter, S. R., Holland, M. M., & Hubbell, S. P. (1991). The sustainable biosphere initiative: An ecological research agenda: A report from the Ecological Society of America. *Journal of Ecology, 72*(2), 371–412. doi:10.2307/2937183

Noss, R. F. (1990). Indicators for monitoring biodiversity: A hierarchical approach. *Conservation Biology, 4*(4), 355–364. doi:10.1111/j.1523-1739.1990.tb00309.x

Saunders, D. A., Hobbs, R. J., & Margules, C. R. (1991). Biological consequences of ecosystem fragmentation: A review. *Conservation Biology, 5*(1), 18–32. doi:10.1111/j.1523-1739.1991.tb00384.x

Soule, M. E., & Kohm, K. A. (Eds.). (1989). *Research priorities for conservation biology*. Washington, DC: Island Press.

Stewart-Cox, B. (1995). *Wild Thailand*. Cambridge, MA: MIT Press.

Stoms, D. M., & Estes, J. E. (1993). A remote sensing research agenda for mapping and monitoring biodiversity. *International Journal of Remote Sensing, 14*, 1839–1860. doi:10.1080/01431169308954007

Strahler, A. H., Woodcok, C. E., & Smith, J. A. (1986). On the nature of models in remote sensing. *Remote Sensing of Environment, 30*(2), 121–139. doi:10.1016/0034-4257(86)90018-0

Townsend, P. A., & Cohoon, K. P. (1999). Sensitivity of distributional prediction algorithms to geographic data completeness. *Ecological Modelling, 117*(1), 159–164. doi:10.1016/S0304-3800(99)00023-X

Turner, M. G. (1989). Landscape ecology: The effect of pattern on process. *Annual Review of Ecology and Systematics, 20*(1), 171–197. doi:10.1146/annurev.es.20.110189.001131

Urban, D. L., O'Neill, R. V., & Shugart, J. R. (1987). Landscape ecology. *Bioscience, 37*, 119–127. doi:10.2307/1310366

Whitemore, T. C., & Sayer, J. A. (1992). *Tropical deforestation and species extinction*. IUCN, Gland.

Chapter 8
Applying GLOBIO at Different Geographical Levels

Rob Alkemade
PBL Netherlands Environmental Assessment Agency, The Netherlands

Jan Janse
PBL Netherlands Environmental Assessment Agency, The Netherlands

Wilbert van Rooij
AIDEnvironment, The Netherlands

Yongyut Trisurat
Kasetsart University, Thailand

ABSTRACT

Biodiversity is decreasing at high rates due to a number of human impacts. The GLOBIO3 model has been developed to assess human-induced changes in terrestrial biodiversity at national, regional, and global levels. Recently, GLOBIO-aquatic has been developed for inland aquatic ecosystems. These models are built on simple cause–effect relationships between environmental drivers and biodiversity, based on meta-analyses of literature data. The mean abundance of original species relative to their abundance in undisturbed ecosystems (MSA) is used as the indicator for biodiversity. Changes in drivers are derived from the IMAGE 2.4 model. Drivers considered are land-cover change, land-use intensity, fragmentation, climate change, atmospheric nitrogen deposition, excess of nutrients, infrastructure development, and river flow deviation. GLOBIO addresses (1) the impacts of environmental drivers on MSA and their relative importance; (2) expected trends under various future scenarios; and (3) the likely effects of various policy-response options. The changes in biodiversity can be assessed by the GLOBIO model at different geographical levels. The application depends largely on the availability of future projections of drivers. From the different analyses at the different geographical levels, it can be seen that biodiversity loss, in terms of MSA, will continue, and current policies may only reduce the rate of loss.

DOI: 10.4018/978-1-60960-619-0.ch008

1. INTRODUCTION

Biodiversity is decreasing at high rates due to a number of human impacts. The changes in biodiversity include the shifts of entire biomes due to climate change, the appearance of new 'alien' species, that may become invasive and the decrease in abundance of species, eventually leading to local and global extinction of some of them. The recorded losses of species and habitats urged policy makers to take actions at national regional and global levels. The Convention on Biological Diversity (CBD) was formed in 1992 and in 2002 the 2010 target of significantly reducing the rate of biodiversity loss was formulated. A series of indicators was proposed in order to measure the changes of biodiversity and to be able to evaluate the biodiversity targets. The 198 parties to the Convention adopted the target. The EU decided to sharpen the target to halt the loss by 2010. By the year 2010, proclaimed as the International Year of Biodiversity by the UN, the COP – CBD admitted that the target was not met (sCBD, 2010). Several reports concluded that biodiversity loss continues and will continue in the coming decades, if major actions fail to materialize (Leadley *et al.*, 2010; Pereira *et al.*, 2010).

In 2010 the CBD and other bodies are formulating new and achievable targets on biodiversity protection. An initiative to prepare appropriate indicators was launched in 2007 (BIP; www.twentyten.net), an initiative was started for designing and coupling global monitoring systems (GEO BON; www.earthobservations.org/geobon) and an Intergovernmental Science-Policy Platform on Biodiversity and Ecosystem services (IPBES) was launched.

Models describing impacts of human induced environmental changes (drivers) on biodiversity are essential tools for analyzing the relative importance of drivers, to describe expected trends under future scenarios and to evaluate the likely effects of various responses or policy options. Models can thus be used to predict whether biodiversity targets can be met or not if necessary policy actions are taken.

For this purpose an international consortium, made up by UNEP World conservation and monitoring centre (WCMC), UNEP GRID – Arendal and the PBL-Netherlands Environmental assessment Agency, developed the GLOBIO3 model (Alkemade *et al.*, 2009). GLOBIO3 is a simple method linking multiple drivers to a metric of biodiversity: the remaining mean species abundance (MSA) of original species, relative to their abundance in pristine or primary ecosystems. (see Table 1). MSA is comparable to the Biodiversity Integrity Index (Majer & Beeston, 1996) and the Biodiversity Intactness Index (Scholes & Biggs, 2005) and can be considered as a proxy for the CBD indicator on trends in species abundance (UNEP, 2004). The main difference between MSA and BII is that every hectare is given equal weight in MSA, whereas BII gives more weight to species rich areas. MSA also bears some analogy to the Living Planet Index (LPI; Loh, *et al.*, 2005), which relates changes in selected populations to a 1970 baseline, rather than to the pristine situation. MSA represents the average response of the total set of native species belonging to an ecosystem. It should be emphasized that MSA does not completely cover the complex biodiversity concept, and a combination of complementary indicators should be used in biodiversity assessments (Faith *et al.*, 2008).

The intensity of drivers is linked to changes of the abundance and occurrence of species, calculated as MSA, in simple cause-effect relationships. Observational data and data derived from experiments were used to generate these relationships. A major advantage of this approach is its generality. The relationships can be applied in combination with spatial maps, tabulated summaries and environmental model outcomes. It can also be applied at different geographical levels, ranging from sub-national to global level. It can be used in different types of studies from global

Table 1. Relative Mean Species Abundance (MSA)

The relative Mean Species Abundance of originally occurring species (MSA) is an indicator describing biodiversity changes with reference to the original state of ecosystems. It is defined as the average abundances of originally occurring species relative to their abundance in the original, pristine or mature state as the basis. As such it is an indicator of biodiversity intactness, and also describes the process of homogenization.

When people intervene in ecosystems, some species decrease in abundance and distribution. At the same time a few other, opportunistic, species increase in abundance, replacing the original ones. If the intervention or disturbance by humans increases, many more species will decrease in abundance with extinction as the final step for some of them. Other species will increase and some new species replacing the original ones. If these new species are the same species as everywhere else, homogenization of ecosystems occurs. MSA describes this process by tracking the abundances of the original species.

MSA is quantified by using datasets derived from peer reviewed publications comparing disturbed situations with the original ones. The observed abundances of species in the disturbed situations are divided by the abundances found in the original system described in the same publication. These relative values are capped at 1, as to avoid compensation by increasing species beyond their 'original' abundance over decreasing species.

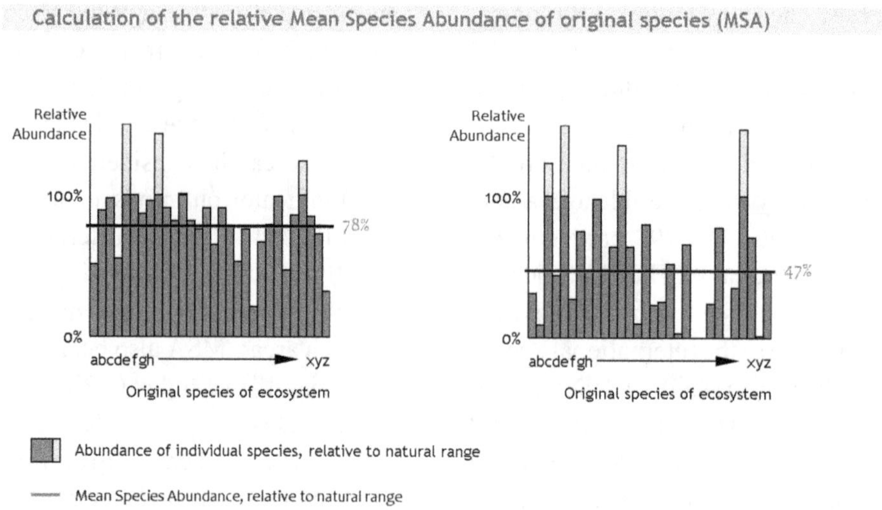

integrated environmental assessments to conservation planning and Life Cycle Analysis.

The GLOBIO model has originally been developed for terrestrial ecosystems (GLOBIO3). The current version of the model excludes the ice biome (Antarctica and Greenland). The drivers for terrestrial systems include: land use; nitrogen deposition fragmentation, infrastructure and climate change. At the global (and regional) level land-use change and harvesting (mainly forestry), atmospheric nitrogen deposition, fragmentation, and climate change are sourced from the Integrated Model to Assess the Global Environment (IMAGE; MNP, 2006). For infrastructure development the module developed in an earlier version (UNEP, 2001) was updated. At the (sub-)national level local data are used with more detail. Data on climate change and nitrogen deposition are often lacking on national scale these are being derived from the global IMAGE model.

Recently a separate GLOBIO aquatic model for inland waters (rivers, lakes and wetlands) was developed, based on a similar approach. Aquatic ecosystems contain a huge and often unique biodiversity, and deliver important ecosystem services. World-wide freshwater biodiversity is declining due to many interacting drivers, such as constructions of dams and other structures, wetland conversion, pollution, overexploitation and invasive species (MA, 2005; Revenga *et al.*,

2005). GLOBIO aquatic currently describes the impacts of land use, pollution by nutrients, the impacts of the flow regime due to dams,, canalization and water abstraction, and climate change effects on that.

Since the development of GLOBIO3 in 2005, it has been applied for several global assessments and scenario studies. An extensive scenario analysis using the GLOBIO3 model is reported in GEO4 (UNEP, 2007). Four scenarios are described and their consequences evaluated by using the IMAGE model and GLOBIO3. For the second Global Biodiversity Outlook a series of policy options to reduce the rate of biodiversity loss were evaluated (sCBD & MNP, 2007; Alkemade *et al*., 2009). Other assessments include a European scenario analysis and illustrations within broader assessments (e.g. Verboom *et al*., 2007; UNEP, 2006). At national level several applications were carried out to explore the possibilities of the method (e.g. Trisurat *et al*., 2010). The model has been applied in combination with the CLUE-s model (Verburg *et al*., 2002; Chapter 6) enabling the use of national future land use maps. An example of a national application of GLOBIO3 and integration with the CLUE-s model is shown in Chapter 19 of this book. More information on assessment and applications of GLOBIO3 can be found on www.globio.info.

This chapter briefly describes the GLOBIO3 methodology and how it can be applied at different geographical levels and for different objectives. At first, we described the GLOBIO3 model, cause-effect relationships and the biodiversity metric used, which is mainly derived from Alkemade *et al*., (2009). Secondly we describe the process of applying the GLOBIO3 model by first preparing the maps of environmental drivers, both in the current state and future projections and finally we describe the aggregation protocol. The use of GLOBIO3 is illustrated by an example at the global level. We finally discuss the major advantages, drawbacks and suggestions for improvements.

2. MODEL DESCRIPTION

GLOBIO3 and GLOBIO aquatic are built on a set of equations linking environmental drivers and biodiversity impact, so called cause–effect relationships. Maps of environmental drivers are based on the outcome of integrated assessment models (IAM's; for example IMAGE; Chapter 5), of land use allocation models (for example CLUE-s; Chapter 6; Verburg *et al*., 2002) and on additional information on, for example, the intensity of land use and water quality. The different drivers are assumed to be independent, hence are combined by multiplication. The resulting terrestrial MSA and aquatic MSA are not further combined but reported separately.

2.1 Cause-Effect Relationships

2.1.1 Terrestrial

Cause–effect relationships for each driver were derived from published data in peer-reviewed literature by using meta-analyses (see also Alkemade *et al*., 2009). Meta-analysis is the quantitative synthesis, analysis, and summary of a collection of studies and requires that the results be summarized in a common estimate of the 'effect size' (Osenberg *et al*., 1999). MSA is considered to be the effect size in our analyses. Meta-analyses were performed by first scanning the peer-reviewed literature using a relevant search profile Secondly, we selected papers that present data on species composition in disturbed and undisturbed situations. Thirdly, these data were extracted from the paper and MSA values and their variances were calculated (see Table 1).

Land Use

For finding relevant papers for land use, land-use intensity, and harvesting (including forestry), SCI—Web of Science was queried using key words like species diversity, biodiversity, richness, or abundance; land use, or habitat conversion; and

Figure 1. Relative mean species abundance values for each land-use category, derived from Alkemade et al., (2009)

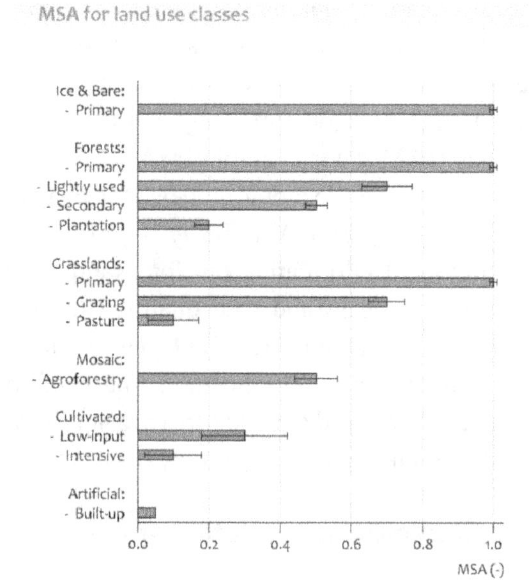

pristine, primary, undisturbed, or original (see Alkemade *et al.*, 2009). The land-use types were categorized into 10 classes: primary vegetation, lightly used forests, secondary forests, forest plantations, livestock grazing, manmade pastures, agroforestry, low-input agriculture, intensive agriculture, and built-up areas (Figure 1).

Nitrogen Deposition

The analysis for N deposition in excess of critical loads (N exceedance) was based on data from empirical N critical-load studies (Bobbink *et* al., 2003). Additional data were obtained from SCI—Web of Science queries in 2007. Data were analyzed for separate biomes using linear or loglinear regression (Figure 2, derived from Bobbink *et al.*, 2010).

Infrastructure

In addition to papers used for the previous model version (UNEP, 2001), literature databases were queried using the key words: road impact, infrastructure development, road effect, road disturbance, and road avoidance. For each impact zone derived from UNEP/RIVM (2004) we estimated MSA using generalized linear mixed models. The impact zones include effects of disturbance on wildlife, increased hunting activities, and small-scale land-use change along roads (Benitez-Lopez *et al.*, 2010) (Figure 3).

Fragmentation

The relationship between MSA and patch size was built upon data on the minimum area requirement of animal species defined as the area needed to support at least a minimum viable population

Figure 2. Relative mean species abundance values and regression lines for nitrogen exceedance. Each point represents a data point from a single study

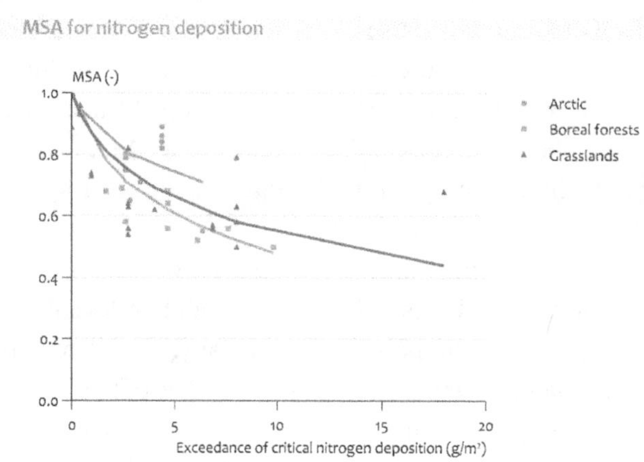

Figure 3. (a) Relative mean species abundance values for the impact zones along roads derived from Alkemade et al. (2009); (b) width of Impact zones for different ecosystems, derived from UNEP/RIVM (2004)

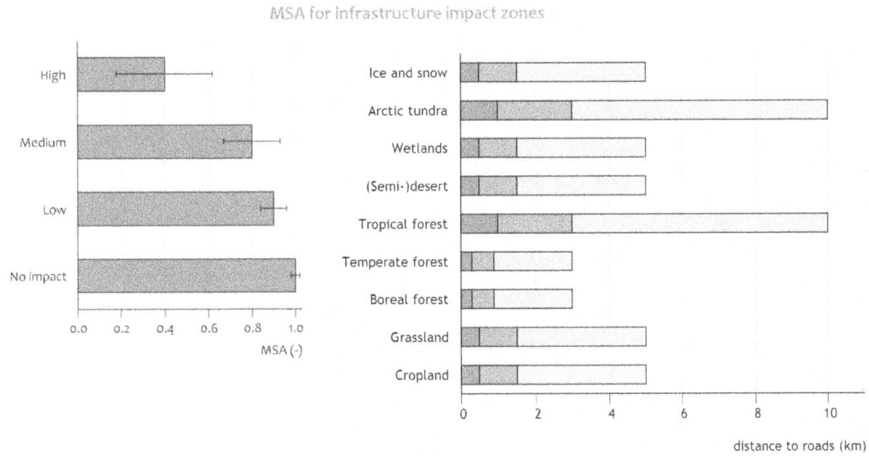

(a) (b)

(Verboom *et al.*, 2007). The proportion of species for which a certain area is sufficient for their MVP is calculated and considered as a proxy for MSA (Figure 4).

Climate Change

The cause–effect relationships for climate change are based on model studies. Species Distribution Models for plant species were used to estimate species distributions for the situation in 1995 and the forecasted situation in 2050 for three different climate scenarios (Bakkenes *et al.*, 2002). For each grid cell the proportion of remaining species were calculated by comparing the species distribution maps for 1995 and for 2050 (Alkemade *et al.*, 2011). For each biome, a linear regression equation was estimated between the proportion of remaining species and the Global Mean Temperature Increase (GMTI, relative to pre-industrial), corresponding to the different climate scenarios. Additionally, the expected stable area for each biome calculated for different GMTIs was derived from Leemans & Eickhout (2004). They presented percentages of stable area of biomes at 1, 2, 3, and 4_C GMTI. Linear regression analysis was used to relate the percentages and GMTI. Stable areas for each biome (IMAGE), or group of plant species occurring within a biome (EUROMOVE) are considered proxies for MSA. The different relationships for each biomes include the differences in climate change projected for each biome (Figure 5).

Figure 4. Relative mean species abundance values for different patch size categories derived from Alkemade et al., (2009)

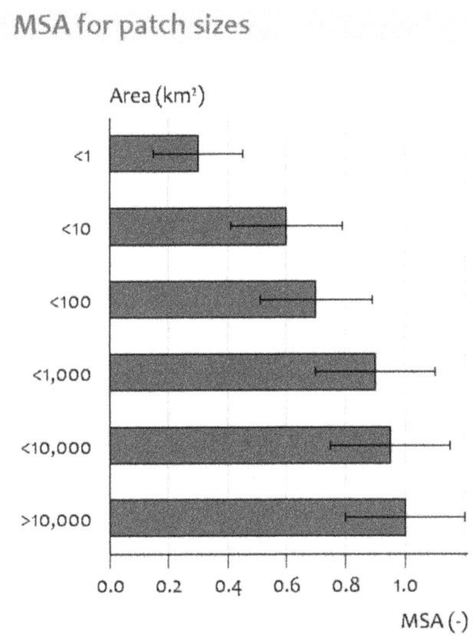

Figure 5. Relative mean species abundance values for temperature increases derived from Alkemade et al., (2009)

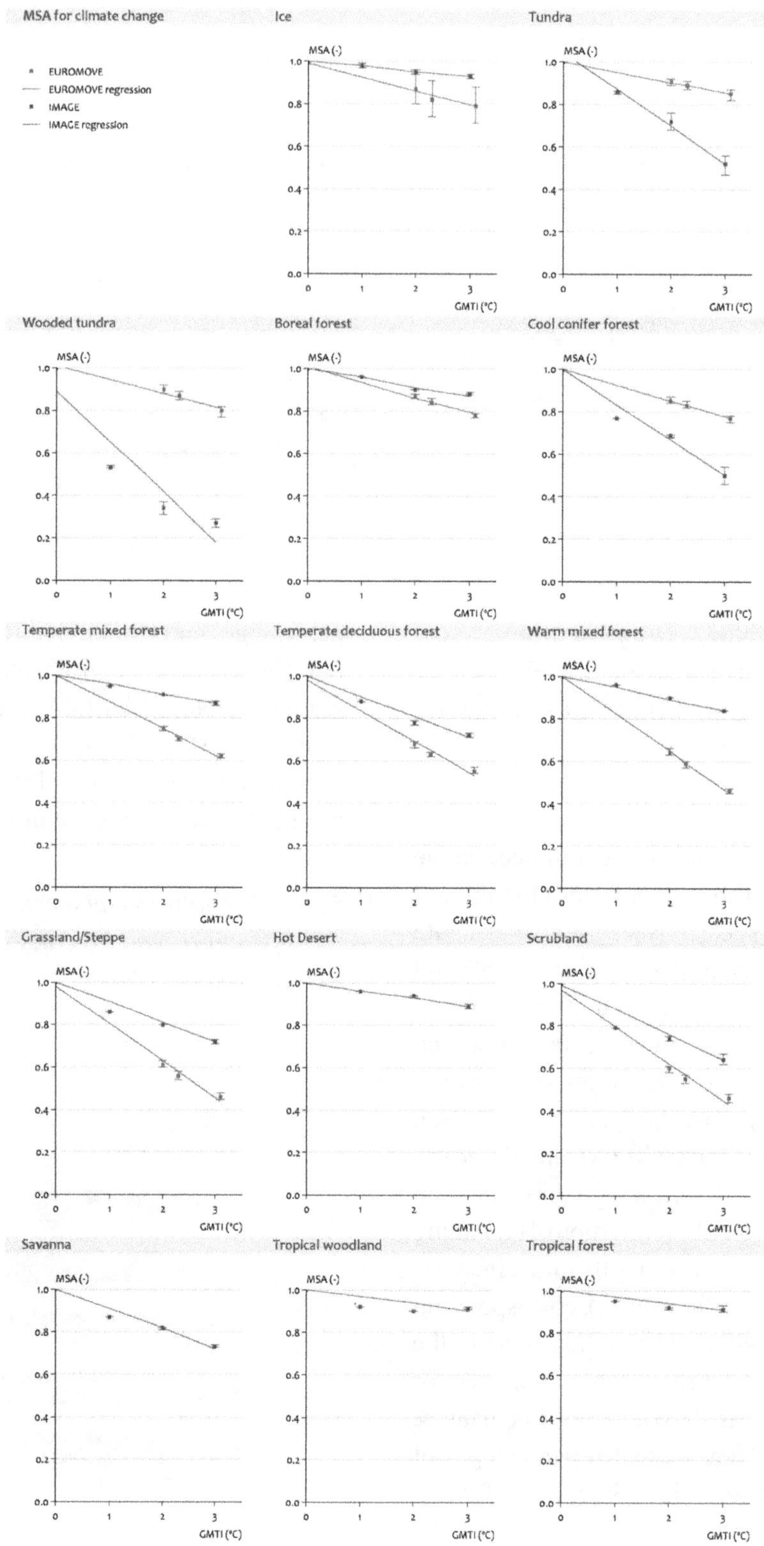

2.1.2 Aquatic (Inland Waters)

The aquatic module of GLOBIO describes the relation between environmental drivers and biodiversity in rivers, lakes and wetlands, based on meta-analyses of literature data. The model is based on the catchment approach, which implies that spatial relations based on flow direction are included. Drivers currently included are: catchment land use changes and eutrophication, physical alteration by river damming and water withdrawal (all leading to habitat losses) and climate change effects on hydrology. Water temperature, other pollutants, overexploitation and invasive species are not yet included. The effects of drivers are described separately for lakes, rivers and wetlands.

Catchment Land Use Change and Eutrophication

Studies on biodiversity in rivers and streams in (sub) catchments with different forms of land use (forest, agricultural, urban, etc.) were combined and the results were expressed as MSA (Weijters et al., 2009). The data were fitted by linear regression (Figure 6).

A comparable analysis is being performed for wetlands. For lakes, the analysis was based on phosphorus and nitrogen loadings rather than land use, as the land use effects in lakes often occur via eutrophication, and also to better quantify land use intensity (e.g. Johnes, 1996). Nutrient loading to surface waters in general highly correlates with type and intensity of (agricultural) land use. Literature data on biodiversity related to P and N were combined and fitted by logistic regression, for deep and shallow lakes separately (Figure 7).

River Damming and Flow Changes

Seasonal river flow patterns, both in pristine and in actual or future situations (affected by river dams or water abstraction), are derived from the hydrological module of LPJ (Biemans *et al.*, 2011), and the deviation between the affected and natural pattern is calculated. Literature data on biodiversity in rivers at different degrees of regulation (e.g. by dams) were combined and expressed

Figure 6. Linear regression between the percentage of non-natural land-use in the catchment and relative taxon diversity of EPT, Macroinvertebrate and Fish in rivers and streams., derived from Weijters et al. 2009)

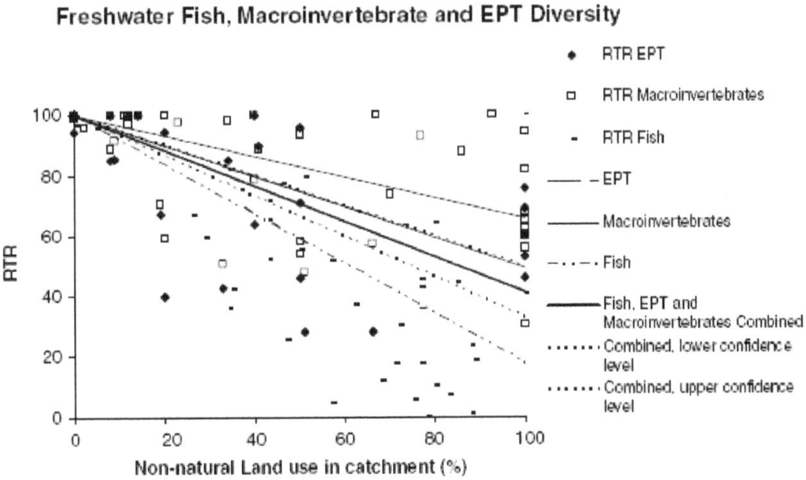

Figure 7. Logistic regression between MSA in lakes and total phosphorus (TP) concentration, for deep and shallow lakes

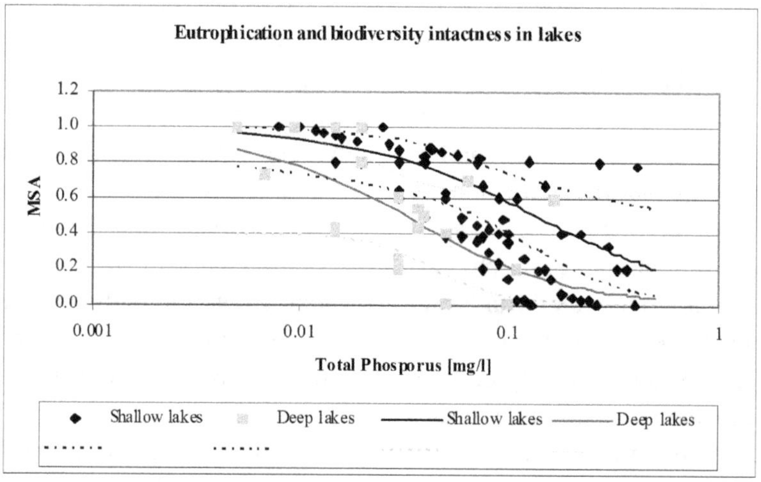

Figure 8. Logistic regression between MSA and river regulation (dams and other infrastructure)

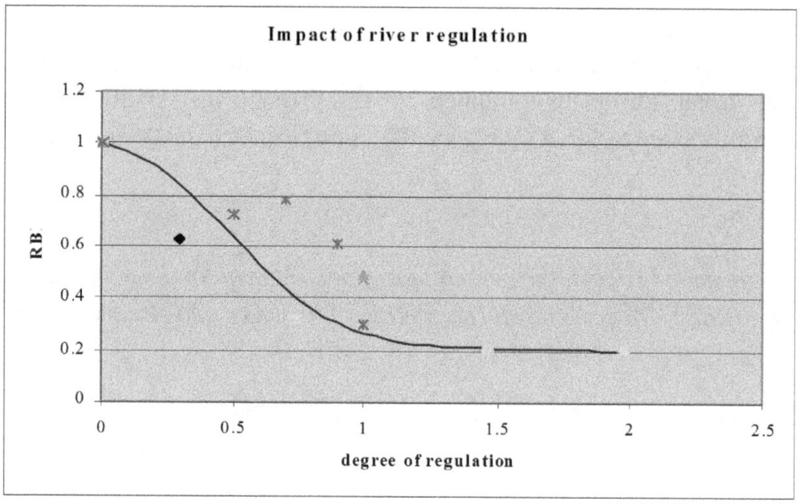

as MSA (Figure 8). A comparable study is being performed on the effects in wetlands.

Climate Change

Climate change is affecting aquatic ecosystems in two ways: by rising water temperature and by changing hydrological patterns, such as the amount and timing of rainfall and evaporation. The latter aspect is covered by the flow module described above. The effects of rising temperatures will be included later.

2.2 Combining MSA and Aggregation

After having all the driver maps ready the GLOBIO cause relationships are applied to the these maps to obtain MSA maps for each separate factor (see Figures 9 and 10). A combined MSA map can be

Figure 9. General flowchart of a GLOBIO3 analysis

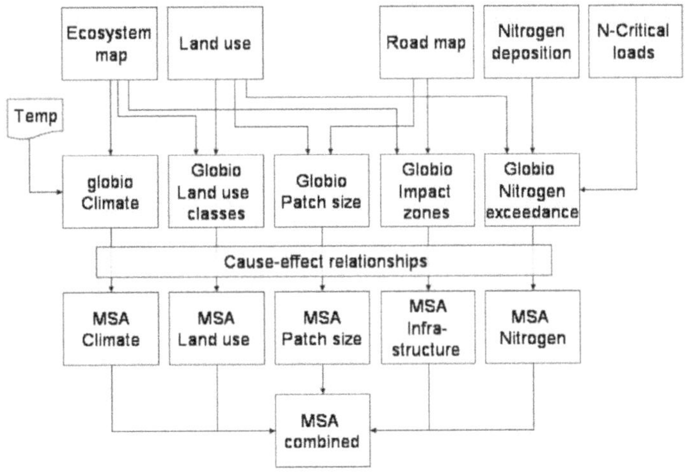

Figure 10. Overview of aquatic module

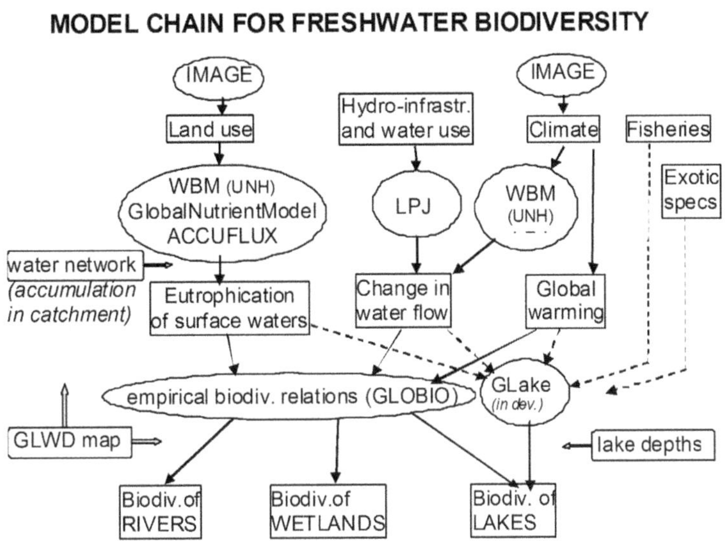

obtained by multiplying the separate MSA maps to one single MSA-total map (Alkemade et al., 2009).

The aggregated value for each region or a global average is obtained by calculating the area weighted mean of MSAi values of all grid cells within a region (MSA_r).

$$MSA_r = \sum_i MSA_i * A_i / \sum_i A_i$$

where Ai is the land area of grid cell i. The relative contribution of each driver to the loss in MSA may be calculated by using this formula for each driver separately and calculate the proportion of each driver to the total MSA loss.

We assumed that N deposition does not affect MSA in croplands, because the addition of N in agricultural systems was expected to be much higher than the atmospheric N deposition, and should have already been accounted for in the estimation of agricultural impacts. Furthermore, climate change and infrastructure are assumed to affect only natural and semi-natural areas, and effects of infrastructure are reduced in protected areas.

For GLOBIO aquatic a similar procedure is followed.

3. HOW TO APPLY GLOBIO

The application of the GLOBIO model requires maps for all relevant drivers and an aggregation protocol. GLOBIO can be applied for a contemporary estimate of biodiversity status in terms of MSA and for future projections of biodiversity status based on changing drivers.

For the current status basically three steps are followed

1. Translate maps, tables and output from models, to GLOBIO environmental driver maps
2. Apply the cause effect relationship for each factor
3. Multiply these maps and aggregate to estimate average MSA values for specific parts and regions and for the complete area under study. The contributions of each factor to MSA-loss can also be extracted.

Figure 9 shows a general scheme of a GLOBIO3 analysis for terrestrial systems.

To construct a GLOBIO MSA map we need a land use map derived from remotely sensed data, a natural ecosystem map and some additional information. Land use maps are mostly available from official institutions. At global level the Global Land Cover database (Bartholome, 2002) is frequently used, as is the newer GLOBCOVER map (GLOBCOVER, 2008). At national levels land cover maps are available, mostly on various levels of precision and resolution. The majority of these maps are based on remote sensing, using satellite imagery (e.g., Landsat results). Available land use maps may differ in their purposes: land use maps produced for facilitate forestry (control) are different from maps produced for agricultural purposes (see for example the Viet Nam case in Chapter 19 and the Central America case in Chapter 17 of this Volume). These maps do not always use the same legends and distinctions between land cover types.

In general available land use maps at global and regional level need to be translated to the GLOBIO categories, as these maps mostly focus on broad distinctions between different land use types and do not distinguish between the intensity of use within a land use type. It is often a challenge to translate available maps into a land use map with classes similar to those used in GLOBIO. A natural vegetation or ecosystem map, and some additional information from experts or other sources are needed to do so. At global level the distinction of land use intensity classes is based on FAO reports for forestry and agriculture (Brown, 2000; Dixon *et al.*, 2001; FAO, 2006). In general the higher the land use intensity is the higher the impact. A biome map is used to describe the natural vegetation (Prentice *et al.*, 1992). In case a detailed land use map is available at national level, more land use classes can be included to preserve information embedded in the local classes themselves. For example in the case of Vietnam (see Chapter 19) the used land use map had more than 17 forest classes. An aggregation of these classes into 5 global forestry categories may result in information loss. This detailed land use information can be used if experts can fill in the MSA values for the extra classes and these additional classes can be dealt with for future projections.

The GLOBIO map for infrastructure is basically formed by a map that includes impact zones along roads. The Digital Chart of the World infrastructure map is used at global level (DMA, 1992). At national level other road maps may be available. Unfortunately recent digitized road maps are difficult to acquire, as they are mostly owned by companies for navigation purposes.

The size of the impact zones depends on the type of road (minor roads are often omitted as it can be assumed that they do not have large impact), the human population density near the road and on the biome or ecosystem that is crossed by the road (Figure 3; UNEP/RIVM, 2004). For analyses at global levels the impact zones are first calculated and subsequently stored as the area per impact zone in a grid cell, to facilitate the calculation of the overall MSA values.

The map of GLOBIO patch sizes consists of a map that attributes the size of the patch to which the individual grid cells belong. The patch size is calculated for the grid cells containing natural vegetation only. The patch size map is derived from the GLOBIO land use map, which is first converted into a map of two classes: natural and non-natural (agriculture and urban areas). This map is subsequently overlaid with the road map, where only the main roads are selected. The size of the patches are calculated using standard options in GIS packages and attributed to each grid-cell.

The GLOBIO map for Nitrogen exceedance is derived from a model describing the spatial distribution of nitrogen deposition and a map denoting critical load values for each ecosystem. Nitrogen deposition is difficult and expensive to measure. Direct measurements are therefore restricted to specific research projects and to regions with large nitrogen problems (e.g. Europe). Maps of nitrogen deposition are therefore derived from interpolations of these data and models (MNP, 2006). At global level, Critical load maps are based on long term field studies and experiments of Nitrogen additions of different levels in plots of natural vegetation (Bobbink *et al.*, 2003) The overall Critical load map is derived from these studies, combined with an ecosystem and a soil map (Bouwman *et al.*, 2002). For national analyses Nitrogen deposition maps and Critical load maps can be used if available. In many tropical countries, however, the problem of atmospheric nitrogen deposition is not a big problem yet to be a relevant factor. If needed in some analyses the global maps of nitrogen exceedance can be downscaled and used.

For constructing the climate change map for GLOBIO we only need the map depicting change in global mean temperature and an ecosystem map. The ecosystem map needs to have a similar legend as the biome map used for deriving the GLOBIO relationships. So for both global and regional, and national analyses the same source of information can be used. The distinction between the biomes can however be much more precise on national levels than on global level. The differences of impact on climate change in the different biomes are included in the dose response relationships.

In contrast to the GLOBIO3 terrestrial model, the aquatic module is based on the catchment approach: the drivers, such as land use, water flow and nutrient loadings are spatially accumulated according to the river catchments (ACCUFLUX module). These are described by an LDD map ('local drain direction') based on altitudes. The location and types of inland water bodies in these catchments are based on the GLWD (Global Lakes and Wetlands Database'; Lehner & Döll, 2004). The MSA values for lakes, rivers and wetlands are combined (by weighted averaging) to an overall MSA-aquatic, which may be reported aside the terrestrial MSA. Figure 10 gives an overview of the aquatic module.

4. FUTURE PROJECTIONS

4.1 Terrestrial

For policy makers projections of possible futures can be useful information for selecting strategies to reduce rates of biodiversity loss. Information on developments in demography, economy, and its consequences on the environment may help policy makers to formulate measures for adaptation, mitigation or biodiversity protection.

Future projections can be based on trend analysis of the near past in combination with scenarios. Scenarios consist of a narrative, which describe a possible future for the most important socio-economic sectors connected to a world vision, and estimates of its consequences for the environment, see for example the IPCC scenarios (Nakicenovic *et al.*, 2000) and the scenarios described in the Global Environmental Outlook (UNEP, 2007). Policy options are formulated in the context of one or more scenarios, and can be considered as the possibilities policy makers have to contain, avoid or reduce undesired consequences of the autonomous change described in a scenario (see for example sCBD & MNP, 2007).

Scenarios and policy options can be considered at local, national, regional and global level. Besides their impacts on biodiversity, policy options need to be evaluated for their (positive or negative) impacts on other societal sectors, as well as on their achievability in the political context of each country.

Land use projections are derived from future demands of food, feed and fiber. An economic model that accounts for trade between countries or regions, distributes the production of agricultural products and allow for assumptions on the ability to enhance productivity of agricultural land, together with the foreseen total production that meets the demand determines the demand for agricultural land in each country or region. For national scale scenario information can be extracted from statistical census data, forestry and agricultural development plans, national visions, socio-economic development plans and land use maps.

A land use allocation model can then be used to estimate the likely pattern of land use changes at the desired level. The IMAGE model uses a simple allocation algorithm for large 0.5 degree cells (see Chapter 5). At the country level models like the CLUE-s model are useful (see Chapter 6). Allocation follows some rules: 1) Existing land policies may imply that a certain area can not change from year to year (for example, if one assumes that protected areas are effectively implemented). 2) Future use is consistent with prior use (for example, it is unlikely to have agriculture in sectors previously used for mining). Rules define valid paths for land use change 3) The inertia or elasticity rules for land use (for example it is very likely that urban areas remain urban under almost all circumstances, whereas unprotected grasslands or forests are more easy to convert). 4) Probability rules for each type of land use conversion based on suitability maps constructed from topographic factors and neighborhood relationships.

Projection of roads is difficult to obtain directly. The exact routing of new roads is hard to forecast. Therefore future projection of the impact zones is used in the GLOBIO model. The increase of the impact is simply modeled by assuming that the impact zones along roads are broadened, based on expected economic and demographic growth (see Nellemann *et al.*, 2003). This is to mimic small scale road construction perpendicular on existing roads.

A new map on patch sizes can be constructed simply using the land use projection and the original infrastructure map.

The projection of nitrogen deposition is based on the new pattern of agricultural land combined with assumption of agricultural intensity and possible policy measures to reduce Nitrogen pollution (Bouwman *et al.*,2002).

The future temperature is simply derived from the projections of Global Circulation Models, for this. We used GCM's as applied in the IMAGE model (MNP, 2006).

An example of a global application of the terrestrial part is shown in Table 2 and is based on an analysis for the Global Biodiversity Outlook Examples (sCBD & MNP,2007). Examples of

Table 2. A global 'business as usual' scenario

A global 'business as usual scenario' includes autonomous developments in demography, the economy and technology, and current policies agreed upon in international treaties. The scenario presented here is based on moderate assumptions on population growth and economic development. The global population grows from 6.1 billion in 2000 to 9 billion in 2050, but at a declining growth rate. Over the same period, the global average income increases from $5,300 to $ 16,000 per capita. The compounded effect of population and economic growth represents more than a fourfold increase in global GDP in the next half century. Due to structural shifts of economies to less energy-intensive sectors and technological improvements leading to energy savings, total primary energy consumption increases by just over a factor of 2: from 400 to 900 EJ in 2050. In the baseline, energy supply continues to rely on fossil resources (coal, oil and gas) and thus emissions of greenhouse gases from combustion also keep rising. Together with emissions from land use and other sources, this leads to an ongoing rise in global temperature to 1.8 K compared to pre-industrial levels in 2050. This means that the rise in the next half century will exceed the observed increase in the last 130 years. After implementation of the Kyoto Protocol for 2008-2012, no further climate mitigation measures are be taken at the baseline. Consumption of agricultural products lags behind overall economic growth. However, the combined effect of more people taking in more calories, especially in currently undernourished regions, and the shift towards more animal products in the diet at higher income levels, imply a sharp increase in agricultural output. If we follow and extend the assumptions on agricultural productivity according to the FAO projection towards 2030, the total area required for food crops, grass and fodder remains fairly stable over the entire period. This illustrates that productivity assumptions here are relatively optimistic compared to other recent studies. For example, in the scenarios of the Millennium Ecosystem Assessment (MEA) the total crop area increases by 8% to 23% over the same period. As far as nature conservation policies are concerned, the current protected areas are not expanding in the baseline scenario, being close to 10%. Rising timber demand is met by production from use of (semi-) natural forests. Forest conversion for agricultural expansion (slash and burn) does not contribute to the wood demand, and there is no wood production from (current or future) plantations. GLOBIO3 estimates the current MSA at about 70% globally resulting from centuries of increased and intensified land use, infrastructural development, human settlement, pollution and fragmentation. A further reduction of MSA from about 70% to 63% is projected up to 2050. The most affected regions are drylands—grasslands and savannah—which show the largest deterioration, followed by tropical forests and tundra. Infrastructure (plus related settlement) and climate change are the dominant causes of the further loss in the baseline development. The share in biodiversity loss due to agricultural land use remains constant, as agricultural productivity is projected to show a considerable increase. Increasingly, agricultural products are traded between world regions, implying that food consumption in densely populated regions will affect land-use change in production regions more than in one's own region (see also van Vuuren en Bouwman, 2005).

Based on the Global Biodiversity Outlook (sCBD & MNP, 2007).

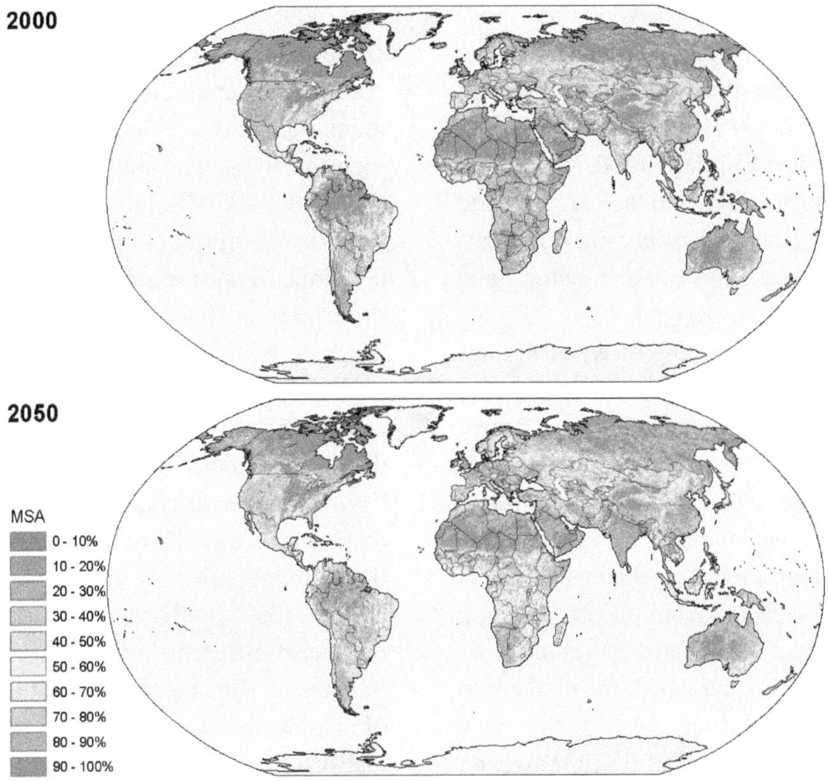

national analyses can be found in Chapters 9, 12, and 19 of this volume and in Trisurat *et al.* (2010).

4.2 Aquatic

The GLOBIO aquatic model uses output from several other models. The IMAGE model of land use and climate change (MNP, 2006); the WBM water network and discharge model (Vorosmarty *et al.*, 2000); the LPJ water flow module (Biemans *et al.*, 2011); the Global Nutrient Model (including ACCUFLUX) for diffuse and point sources of N and P (Bouwman *et al.*, 2009; Van Drecht *et al.*, 2009; Seitzinger *et al.*, 2010). These models translate future population size and land-use patterns into nutrient loadings to aquatic systems. Nutrient runoff from the land to the surface waters, both nitrogen and phosphorus, is modelled based on the agricultural area, the application of fertilizer and manure, precipitation and spatial characteristics like slope, soil texture, amount of groundwater, and others. Urban nutrient emissions are modelled as well, based on population, affluence, sanitation and the use of detergents (Van Drecht et al., 2009). The model also calculates the river discharge, based on precipitation and evaporation patterns, water abstraction, and the presence and management of dams and reservoirs (Biemans et al., 2011). The deviation between natural and impacted flow pattern is derived.

For the location and typing of water bodies, the Global Lakes and Wetlands Database map (Lehner & Doll, 2004) is available at different geographical levels. This map discerns the main inland water types: rivers, lakes, reservoirs and several types of wetlands including riverine marshes and swamps, fens, brackish and coastal wetlands. Drivers are modeled (at present) in a spatial resolution of 0.5° (lat/long) (approx. 50 km), and fluxes are accumulated downstream to calculate the values for rivers, lakes and riverine wetlands. Some of the wetland types, however, were assumed to be more isolated and hence to be influenced by the land use or nutrient emissions in the current grid cell only.

The GLOBIO aquatic model is, until now, only applied at global level as shown in figure 11 (PBL, 2010). It can, however be applied at a more detailed level if the information mentioned is available at a finer resolution. An application on the catchment of Lake Cocibolca (Nicaragua) is underway.

5. DISCUSSION AND CONCLUSION

The GLOBIO3 framework has now been applied in several global, regional and national assessments and proved to be a flexible and quick tool to evaluate scenarios and policy options. At the global level five major environmental factors are included for the terrestrial part and three for the aquatic part. The selection of these drivers for the terrestrial part is described in Alkemade *et al.* (2009).

Although conclusions derived from GLOBIO3 confirm earlier studies and recent global assessment, such as the Millennium Ecosystem Assessment and the second Global Biodiversity Outlook (MA, 2005; sCBD, 2006), we need to consider a series of uncertainties inherent to GLOBIO3. Uncertainties relate to the cause–effect relationships, the drivers and the models estimating the drivers, the underlying data, and the indicators used.

Being a compilation of existing knowledge the cause–effect relationships are based on limited data. The selected published datasets, although having a wide variety, do not cover all biomes or represent all important species groups. For land use our estimates are close to those found by Scholes & Biggs (2005) and Nichols *et al.* (2007), although they used different indicators. The BII used by Scholes & Biggs (2005) estimates the fractions of original species populations under a range of land-use types in southern Africa, based on expert knowledge. Nichols *et al.* (2007) used the Morisita Horn index of community similarity. Studies

Figure 11. Results of GLOBIO-aquatic for the year 2000 and for 2050, OECD baseline scenario, derived from PBL (2010)

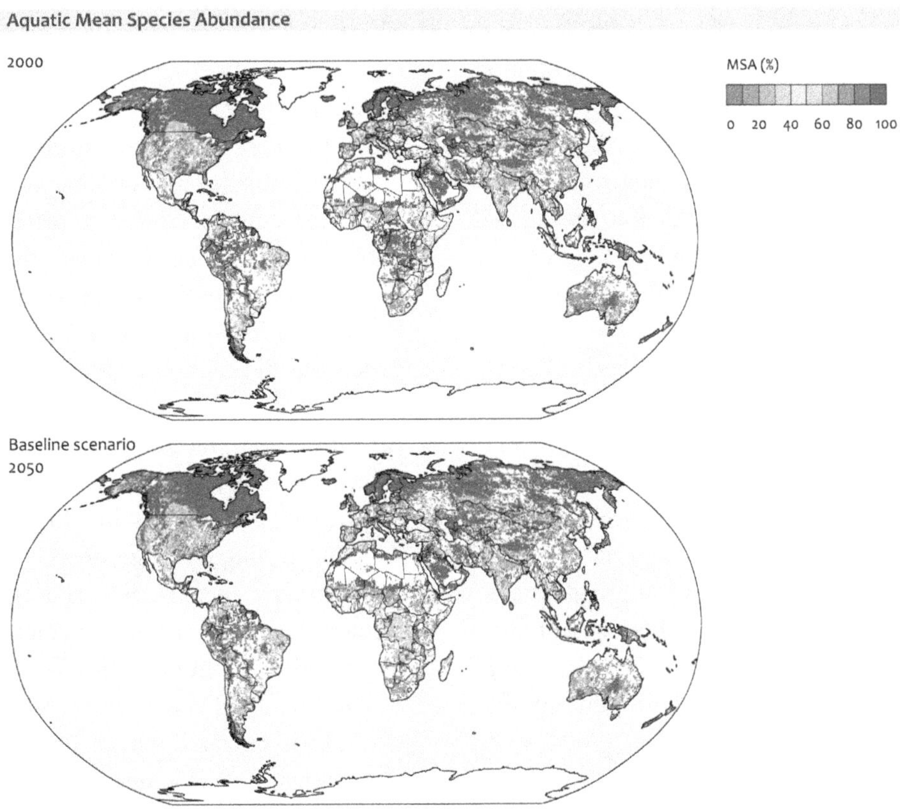

from currently heavily converted regions, such as Europe and East Asia, are under-represented.

For infrastructure data on birds and on mammals are used, and thus effects on, for example, plants and insects, are under-represented (Benitez-Lopez *et al.*, 2010). Effects of Nitrogen deposition are mainly based on studies of plant species composition from temperate regions (Bobbink *et al.*, 2010). The effects of fragmentation are restricted to the effects on populations of reduced patch-size. We chose to use data on the minimum area requirement of species, which is qualitatively similar to a direct relationships of species abundances and patch, but may well differ quantitatively (Bender *et al.*, 1998). Generalisations from model studies into a cause effect relationship for climate change are useful for quick assessments of the effects on biodiversity (Alkemade *et al.*, 2011). The one used in GLOBIO3 is based on plant species in Europe (Bakkenes *et al.*, 2002) and biomes (Leemans & Eickhout, 2004). More studies are available on shifts of species using climate envelopes and forecasted climate change and can potentially be used to improve the cause-effect relationships and research is going on to include these into the model. (e.g., Peterson *et al.*, 2002; Araujo *et al.*, 2006; Thuiller *et al.*, 2006).

Some factors indicated to have a possible major impact on biodiversity have not yet been included in the model. Factors not addressed include the impact of biotic exchange and the direct impact of increased CO_2 concentration in the atmosphere, fire incidence, extreme events and pollution (except atmospheric N deposition), see for example Sala *et al.* (2000). For these factors, cause–effect relationships have not yet been established in GLOBIO3, due to a lack of data. However, by adding more factors a problem may be the lack of interactions captured by the method used. For instance fire may be a management factor within the agricultural practice. Including fire as a separate independent factor will thus overestimate the effect of fire. The same holds for hunting and exploitation of forest which are partly included in the infrastructure part and in the lightly forest use category as land use type.

The GLOBIO3 model results depend largely on the quality and detail of the data input. The area and spatial distribution of the different land-use classes is of particular importance. Different methods are used to estimate the areas of cropland, grazed land, forests and other natural areas resulting in different estimates and allocations. Uncertainty remains about the total area of agricultural land as shown by the statistics available from the FAO (FAO, 2006) and different satellite imagery sources (Bartholome *et al.*, 2004; Fritz & See, 2008).

Similar uncertainties exist for the other drivers. Uncertainties in measurements and model forecasts for climate and N deposition are extensively documented in IPCC reports (IPCC, 2007). The DCW infrastructure map is quite out-dated and far from complete. This incompleteness of the map makes it difficult to adequately distinguish between important roads and small roads. Currently, however, the DCW map is the only global available map on infrastructure and several other studies used the map to assess effects on biodiversity (Sanderson *et al.*, 2002; Wackernagel *et al.*, 2002). At national level road maps are often not available or based on the same DCW map.

Ideally biodiversity assessments need to be based on a set of indicators, for example as proposed by the CBD. Other aspects of biodiversity loss may then be included (UNEP, 2004). A Red List index or indicator that is sensitive to uniqueness, will probably show different patterns than the MSA. For example by setting up a well-chosen network of protected areas, specific ecosystems may be conserved, containing the majority of the species, including rare and threatened species. This will obviously improve the status of numerous species, reflecting in a red list index, but may have a minor effect on MSA, because land use move to other sites, compensating the positive effects of protected areas. Moreover MSA weights every km^2 equally, so that an increase in species poor regions may be compensated by a decrease in species rich regions. Methods suggested by Faith *et al.* (2008) and Ferrier *et al.* (2007) may be developed and applied at global level to overcome this inequality between regions.

Furthermore recovery and restoration do not always develop towards an historic original situation, but to another possibly diverse ecosystem. Also due to climate change or other circumstances different, but species rich, ecosystems may be formed. The aspect of pure species richness may be a valuable addition to the GLOBIO model as an additional indicator next to MSA.

GLOBIO in itself is not a dynamic model. The changes in the environment are derived from other models (like IMAGE) and the MSA values are applied directly. This means that if any change occur in, for example, land use, there biodiversity will immediately change. In reality this will never be the case. Time is needed to go from one situation to the other, especially when slow processes are involved like nitrogen loading, climate change or forest recovery. A dynamic biodiversity model will be needed to address this feature, which may be linked to a global dynamic vegetation model (e.g. Sitch *et al.*, 2008).

The recent development of GLOBIO aquatic enables the evaluation of biodiversity in aquatic systems. The model still needs improvement. Land use changes and eutrophication in catchments have already resulted in considerable loss of original biodiversity in aquatic ecosystems of all types, and these will aggravate according to future scenarios. The results are often compatible with the terrestrial model. In regions with high human land use, downstream waters are most affected.

Damming and water extraction (irrigation) add to the biodiversity loss in rivers, also in regions with lower human land use. Future developments of the model comprise: refinement of cause-effect relationships; inclusion of fisheries; development of an integrated (functional) module for lakes and the inclusion of wetland conversion (e.g. by the use of historical maps).

REFERENCES

Alkemade, R., Bakkenes, M., & Eickhout, B. (2011). Towards a general relationship between climate change and biodiversity: an example for plant species in Europe. *Regional Environmental Change, 11*(1), S143–S150. doi:10.1007/s10113-010-0161-1

Alkemade, R., Van Oorschot, M., Miles, L., Nellemann, C., Bakkens, M., & Ten Brink, B. (2009). GLOBIO3: A framework to investigate options for reducing global terrestrial biodiversity loss. *Ecosystems (New York, N.Y.), 12*, 374–390. doi:10.1007/s10021-009-9229-5

Araújo, M. B., & New, M. (2006). Ensemble forecasting of species distributions. *Trends in Ecology & Evolution, 22*(1), 42–47. doi:10.1016/j.tree.2006.09.010

Bakkenes, M., Alkemade, J. R. M., Ihle, F., Leemans, R., & Latour, J. B. (2002). Assessing effects of forecasted climate change on the diversity and distribution of European higher plants for 2050. *Global Change Biology, 8*, 390–407. doi:10.1046/j.1354-1013.2001.00467.x

Bartholome, E., Belward, A., Beuchle, R., Eva, H., Fritz, S., & Hartley, A. (2004). *Global land cover for the year 2000*. European Commission.

Bender, D. J., Contreras, T. A., & Fahrig, L. (1998). Habitat loss and population decline: A meta-analysis of the patch size effect. *Ecology, 79*, 517–533. doi:10.1890/0012-9658(1998)079[0517:HLAPDA]2.0.CO;2

Benitéz-Lopéz, A., Alkemade, R., & Verweij, P. A. (2010). The impacts of roads and other infrastructure on mammal and bird populations: A meta-analysis. *Biological Conservation, 143*, 1307–1316. doi:10.1016/j.biocon.2010.02.009

Biemans, H., Haddeland, I., Kabat, P., Ludwig, F., Hutjes, R. W. A., & Heinke, J. (2011). (in press). Impact of reservoirs on river discharge and irrigation water supply during the 20th century. *Water Resources Research*. doi:10.1029/2009WR008929

Bobbink, R., Ashmore, M., Braun, S., Fluckiger, W., & Van den Wyngaert, I. J. J. (2003). *Empirical nitrogen critical loads for natural and semi-natural ecosystems: 2002 update.*

Bobbink, R., Hicks, K., Galloway, J., Spranger, T., Alkemade, R., & Ashmore, M. (2010). Global assessment of nitrogen deposition effects on terrestrial plant diversity: A synthesis. *Ecological Applications, 20*, 30–59. doi:10.1890/08-1140.1

Bouwman, A. F., & Beusen, A. H. W. (2009). Human alteration of the global nitrogen and phosphorus soil balances for the period 1970-2050. *Global Biogeochemical Cycles, 23*(4), 26–42.

Bouwman, A. F., Kram, T., & Klein Goldewijk, K. (Eds.). (2006). *Integrated modelling of global environmental change: An overview of IMAGE 2.4. Netherlands Environmental Assessment Agency*. Bilthoven, The Netherlands: MNP.

Bouwman, A. F., Van Drecht, G., Knoop, J. M., Beusen, A. H. W., & Meinardi, C. R. (2005). Exploring changes in river nitrogen export to the world's oceans. *Global Biogeochemical Cycles, 19*, 1–14. doi:10.1029/2004GB002314

Bouwman, A. F., Van Vuuren, D. P., Derwent, R. G., & Posch, M. (2002). A global analysis of acidification and eutrophication of terrestrial ecosystems. *Water, Air, and Soil Pollution, 141*, 349–382. doi:10.1023/A:1021398008726

Brown, C. (2000). *The global outlook for future wood supply from forest plantations*. Working Paper No GFPOS/WP/03 (p 141). FAO, Rome.

Dixon, J., Gulliver, A., & Gibbon, D. (2001). *Farming systems and poverty*. Rome and Washington DC: FAO and World Bank.

DMA. (1992). Digital chart of the world. Defence Mapping Agency. In Fairfax, V., & Duellman, W. E. (Eds.), *Patterns of distribution of amphibians: A global perspective*. Baltimore, MD: John Hopkins University Press.

Faith, D. P., Ferrier, S., & Williams, K. J. (2008). Getting biodiversity intactness indices right: Ensuring that biodiversity reflects diversity. *Global Change Biology, 14*, 207–221. doi:10.1111/j.1365-2486.2007.01500.x

FAO. (2006). *Global forest resources assessment 2005. Progress towards sustainable forest management. FAO Forestry Paper* (p. 320). Rome: FAO.

Ferrier, S., Manion, G., Eilth, J., & Richardson, K. (2007). Using generalized dissimilarity modelleing to analyse and predict patterns of beta diversity in regional biodiversity assessment. *Diversity & Distributions, 13*, 252–264. doi:10.1111/j.1472-4642.2007.00341.x

Fritz, S., & See, L. (2008). Identifying and quantifying uncertainty and spatial disagreement in the comparison of Global Land Cover for different applications. *Global Change Biology, 14*, 1057–1075. doi: 10.1111/j.1365-2486.2007.01519.x

GLOBCOVER. (2008). *Globcover products description validation report I2.1*. European Space Agency. Retrieved from http://www.esa.int/esaEO/ SEMGSY2IU7E_index_0.html.

IPCC. (2007). *Climate change 2007 – The physical science basis: Contribution of working group 1 to the fourth assessment report of the IPCC*. Cambridge, UK: Cambridge University Press.

IUCN. 2004. *The Durban action plan. March 2004*. Retrieved April 8, 2008 from http://www.iucn.org/themes/wcpa/ wpc2003/english/outputs/durban/daplan.html

Johnes, P. J. (1996). Evaluation and management of the impact of land use change on the nitrogen and phosphorus load delivered to surface waters: the export coefficient modelling approach. *Journal of Hydrology (Amsterdam), 183*, 323–349. doi:10.1016/0022-1694(95)02951-6

Leadley, P., Pereira, H. M., Alkemade, R., Fernandez-Manjarrés, J. F., Proenca, V., Scharlemann, J. P. W., & Walpole, M. (2010). *Biodiversity scenarios: Projections of 21st century change of biodiversity and associated ecosystem services*. Secretariat of the Convention on biological Diversity, Montreal.

Leemans, R., & Eickhout, B. (2004). Another reason for concern: Regional and global impact of ecosystems for different levels of climate change. *Global Environmental Change Part A, 14*, 219–228. doi:10.1016/j.gloenvcha.2004.04.009

Lehner, P., & Doll, P. (2004). *Global lakes and wetlands database*.

Loh, J., Green, R. E., Ricketts, T., Lamoreux, J., Jenkins, M., Kapos, V., & Randers, J. (2005). The living plant Index: Using species population time series to track trends in biodiversity. *Philosophical Transactions of the Royal Society of London. Series B, Biological Sciences, 360*, 286-295.

MA. (2005). *Millennium ecosystem assessment. Ecosystems and human well-being: Scenarios* (*Vol. 2*). Washington, DC: Island Press.

Majer, J. D., & Beeston, G. (1996). The biodiversity integrity index: An illustration using ants in Western Australia. *Conservation Biology, 10*, 65–73. doi:10.1046/j.1523-1739.1996.10010065.x

Nakicenovic, N., Alcamo, J., & Davis, G. De. Vries, B., Fenhann, J., Gaffin, S.,... Dadi, Z. (2000). *Special report on emissions scenarios*. Cambridge, UK: Cambridge University Press.

Nellemann, C., Vistness, I., Ahlenius, H., Rekacewicz, P., Kaltenborn, B. P., & Magomedova, M. … Furuhovde, T. (2003). Environment and security, 2050 scenarios. In R. O. Rasmussen & N. E. Koroleva (Eds.), *Social and environmental impacts in the North* (pp. 129-148). Dordrecht, The Netherlands: Kluwer Academic Publishers.

Nichols, E., Larsen, T., Spector, S., Davis, A. L., Escobar, F., Favila, M., & Vulinec, K. (2007). Global dung beetle response to tropical forest modification and fragmentation: A quantitative literature review and meta-analysis. *Biological Conservation, 137*, 1–19. doi:10.1016/j.biocon.2007.01.023

Osenberg, C. W., Sarnelle, O., Cooper, S. D., & Holt, R. D. (1999). Resolving ecological questions through meta-analysis: Goals, metrics and models. *Ecology, 80*, 1105–1117. doi:10.1890/0012-9658(1999)080[1105:REQTMA]2.0.CO;2

PBL. (2010). *Rethinking Global Biodiversity Strategies, Exploring structural changes in production and consumption to reduce biodiversity loss*. PBL-Netherlands Environmental Assessment agency, Bilthoven.

Pereira, H. M., Leadley, P. W., Proença, V., Alkemade, R., Scharlemann, J. P. W., & Fernandez-Manjarrés, J. F. (2010). Scenarios for Global Biodiversity in the 21st Century. *Science, 330*, 1496–1501. doi:10.1126/science.1196624

Peterson, A. T., Ortega-Huerrta, M. A., Bartley, J., Sanchez-Cordero, V., & Soberon, J., V., Buddemeier, R. H., & Stockwell, D. R. B. (2002). Future projections for Mexican faunas under global climate change scenarios. *Nature, 416*, 626–629. doi:10.1038/416626a

Prentice, C., Cramer, W., Harrison, S. P., Leemans, R., Monserud, R. A., & Solomon, A. M. (1992). A global biome model based on plant physiology and dominance, soil properties and climate. *Journal of Biogeography, 19*, 117–134. doi:10.2307/2845499

Revenga, C., Campbell, I., Abell, R., de Villiers, P. & Bryer, M. (2005). Prospects for monitoring freshwater ecosystems towards the 2010 targets. *Philosophical Transactions of the Royal Society B Biological Sciences, 360*(1454), 397-413.

Sala, O. E., Chapin, F. S. III, Armesto, J. J., Berlow, E., Bloomfield, J., & Irzo, R. (2000). Global biodiversity scenarios for the year 2100. *Science, 287*, 1770–1774. doi:10.1126/science.287.5459.1770

Sanderson, E. W., Jaiteh, M., Levy, M. A., Redford, K. H., Wannebo, A. V., & Woolmer, G. (2002). The human footprint and the last of the wild. *Bioscience, 52*, 891–904. doi:10.1641/0006-3568(2002)052[0891:THFATL]2.0.CO;2

sCBD., & MNP. (2007). *Cross-roads of life on Earth - exploring means to meet the 2010 biodiversity target. Solution-oriented scenarios for Global Biodiversity Outlook 2*. In Technical Series no 31. Secretariat of the Convention on Biological Diversity, Montreal.

sCBD. (2006). *Global biodiversity outlook 2*. Secretariat of the Convention on Biological Diversity, Montreal.

sCBD (2010) *Global Biodiversity Outlook 3*. Secretariat of the Convention on Biological Diversity Montréal, 94 pages.

Scholes, R. J., & Biggs, R. (2005). A biodiversity intactness index. *Nature, 434*, 45–49. doi:10.1038/nature03289

Seitzinger, S., Mayorga, E., Bouwman, A. F., Kroeze, C., Beusen, A. H. W., & Billen, G. (2010). Global river nutrient export: A scenario analysis of past and future trends. *Global Biogeochemical Cycles, 24*, GB0A08..doi:10.1029/2009GB003587

Sitch, S., Huntingford, C., Gedney, N., Levy, P. E., Lomas, M., & Piao, S. L. (2008). Evaluation of the terrestrial carbon cycle, future plant geography and climate-carbon cycle feedbacks using fice dynamic Global Vegetation Models (DGVMs). *Global Change Biology*, *95*, 2015–2039. doi:10.1111/j.1365-2486.2008.01626.x

Thuiller, W., Broennimann, O., Hughes, G., Alkemade, J. R. M., Midgley, G. F., & Corsi, F. (2006). Vulnerability of African mammals to anthropogenic climate change under conservative land transformation assumptions. *Global Change Biology*, *12*, 424–440. doi:10.1111/j.1365-2486.2006.01115.x

Trisurat, Y., Alkemade, R., & Verburg, P. H. (2010). Projecting land-use change and its consequences for biodiversity in Northern Thailand. *Environmental Management*, *45*, 626–639. doi:10.1007/s00267-010-9438-x

UNEP. (2001). *GLOBIO. Global methodology for mapping human impacts on the biosphere*. In: Report UNEP/DEWA/TR 25. United Nations Environmental Programme, Nairobi.

UNEP. (2004). *Decisions adopted by the Conference of the Parties to the Convention on Biological Diversity at its seventh meeting* (UNEP/CBD/COP/7/21/Part 2), Decision VII/30 (CBD 2004). Retrieved from http://www.biodiv.org/decisions/

UNEP. (2006). *Global deserts outlook. Devision of early Warning and Assessment*. Nairobi, Kenya: United Nations Environmental Programme.

UNEP. (2007). *Global environmental outlook 4*. Nairobi, Kenya: Environment for Development. United Nations Environmental Programme.

UNEP/RIVM. (2004). *The GEO-3 Scenarios 2002-2032: Quantification and analysis of environmental impacts*. In: Report UNEP/DEWA/RS03-04; RIVM 402001022. Division of Early Warning and Assessment. DEWA-UNEP. / National Institute for Public Health and the Environment. RIVM., Nairobi, Kenya; Bilthoven, The Netherlands.

Van Drecht, G., & Bouwman, A. F. (2009). Global nitrogen and phosphate in urban wastewater for the period 1970 to 2050. *Global Biogeochemical Cycles*, *23*(3).

Verboom, J., Alkemade, R., Klijn, J. A., Metzger, M. J., & Reijnen, R. (2007). Combining biodiversity modeling with political and economical development scenarios for 25 countries. *Ecological Economics*, *62*, 267–276. doi:10.1016/j.ecolecon.2006.04.009

Verburg, P. H., Soepboer, W., Limpiada, R., Espaldon, M. V. O., Sharifa, M., & Veldkamp, A. (2002). Land use change modelling at the regional scale: The CLUE-S model. *Environmental Management*, *30*, 391–405. doi:10.1007/s00267-002-2630-x

Vorosmarty, C. J., Green, P., Salisbury, J., & Lammers, R. B. (2000). Global water resources: Vulnerability from climate change and population growth. *Science*, *289*(5477), 284–288. doi:10.1126/science.289.5477.284

Wackernagel, M., Schulz, N. B., Deumling, D., Callejas Linares, A., Jenkins, M., Kapos, V., et al. Randers, J. (2002). Tracking the ecological overshoot of the human economy. *Proceedings of the National Academy of Sciences*, *99*, 9266-9271.

Weijters, M. J., Janse, J. H., Alkemade, R., & Verhoeven, J. T. A. (2009). Quantifying the effect of catchment land-use and water nutrient concentrations on freshwater river and stream biodiversity. *Aquatic Conservation: Marine and Freshwater Ecosystems*, *19*, 104–112. doi:10.1002/aqc.989

Chapter 9
Modeling Species Distribution

Yongyut Trisurat
Kasetsart University, Thailand

Albertus G. Toxopeus
University of Twente, The Netherlands

ABSTRACT

There are many methods for modeling species distribution in the landscape. In this chapter, the authors elaborate on the concepts of species modeling and present three popular techniques to generate species distribution: cartographic overlay, logistic multiple regression and maximum entropy (MAXENT). The cartographic overlay method is relevant to generate a habitat suitability index. Logistic multiple regression generates the probability of distribution based on presence and absence of data in relation to habitat factors. The authors use pseudo absence data selected randomly from low suitability classes, because real absence data were not available. The third technique, maximum entropy method (MAXENT), uses presence-only data. The Asian elephant (Elephas maximus) was selected as a proxy species for this study. The study was conducted in Bun Tharik-Yod Mon, a proposed wildlife sanctuary in northeast Thailand.

The results show that among the three approaches, the potentially suitable habitats derived from cartographic overlay cover the largest area and are likely to overestimate existing occurrence areas. The logistic regression model predicts approximately 56% as suitable area, while maximum entropy results covers approximately 9% of the sanctuary. Although the results show large differences in the suitable areas, it should not be concluded that any one method always proves better than the others. Utilization of any method is dependent on the situation and available information. If species observations are limited, the cartographic overlay or habitat suitability is recommended. The logistic regression method is recommended when adequate presence and absence data are available. If presence-only data is available, a niche-based model or the maximum entropy method (MAXENT) is highly recommended.

DOI: 10.4018/978-1-60960-619-0.ch009

1. INTRODUCTION

Science currently recognizes around 1.8 million species on Earth (out of an estimated total of five million to 30 million). The IUCN Red List of Threatened Species 2009 revealed that 17,291 species out of 47,677 assessed species, or 36 percent, are threatened with extinction (IUCN, 2009). Currently therefore, a lot of attention is focused on the conservation of nature. For example, the Convention on Biological Diversity, established after the Earth Summit in Brazil in 1992, has been ratified by nearly 200 countries. Beside the three broad objectives of the Convention, the Conference of the Parties (COP) adopted the Convention's Strategic Plan and committed themselves to a more effective and coherent implementation of the Convention objectives, to achieve a significant reduction of the current rate of biodiversity loss at the global, regional and national level by 2010 and thus to contribute to poverty alleviation and to the benefit of all life on Earth. The broad scope of biodiversity conservation is well represented in the Global Biodiversity Strategy (WRI *et al*., 1992).

In addition, a number of publications have focused upon the most practical and effective strategies for the conservation of nature in addition to the policy and strategy issues related to the conservation of biodiversity. For example, Sayer *et al*. (2000) proposed a Rapid Ecological Assessment (REA) method, which was developed by The Nature Conservancy, to provide comprehensive and reliable information about biodiversity resources in situations where time and financial resources are limited. REAs utilize a combination of remote-sensed imagery, reconnaissance overflights, field data collection, and visualization of spatial information to generate useful information for conservation planning. It provides researchers with the essential tools and techniques they need to conduct an REA, and offers valuable advice about the planning and implementation aspects. Moreover, Miller (1994) published a book on Mapping the Diversity of Nature. This book presents approaches used by some of the foremost active conservationists to map the patterns of species and habitat at local, regional and global scales during 1990s-2000s. Another book (Haines-Young *et al*., 1993) reviews the application of GIS to landscape ecology. The approaches presented in this book show applications of GIS developed with the intention of influencing the design, planning and implementation of programs to protect species and their habitats. In recent years, many species modeling techniques have been developed by scientists to predict species distribution in the natural landscape.

The objectives of this chapter are to present various methods for species modeling in a forested landscape and to describe briefly the elements of each approach with an example of biodiversity conservation in Bun Tharik-Yod Mon, a proposed wildlife sanctuary in Thailand. In addition, the authors also discuss some strengths and weaknesses, as well as when and how to use each method and draw perspectives of species modeling in the future.

2. CONCEPTS

2.1 Sources of Species Distribution Data

Species distribution data can be obtained either from primary data or secondary data. Sources of primary data include inventory data and field observations, while secondary data include herbarium collections, taxonomic literature and ecological communities.

2.1.1 Herbarium Specimen and Museum Collections

A *herbarium* is a collection of preserved plant specimens. These specimens may be whole plants or parts of plants. These will usually be in a dried condition, mounted on a sheet, but depending upon the material may also be kept in alcohol or other preservatives. Information from a herbarium is valuable because the specimens are permanently preserved and can be physically examined.

Notes provided by collectors provide insight into site conditions, associated species, species prominence, stature and uses. Santisuk *et al.* (1991) indicate that there are only 157,000 specimens held in Thailand; immediately neighboring countries have fewer. However, to locate and examine all herbarium material of a widely taxon is too time-consuming for the process to be feasible. Numerous herbaria are registered with the International Bureau for Plant Taxonomy and Nomenclature. This scientific network facilitates scientists to access species distribution in a short time.

It is stated that there is a lack at consistency in the operations of many herbaria with regard to resources, implementation of data collection, availability of specialist taxonomic experts and their designated services. Thus despite fairly rapid progress the Royal Thai Forest Department will not be able to collect all plant species in the country for many years to come (estimates ranging from about 100 years (Santisuk *et al.*,1991) to, a perhaps overoptimistic estimate of, 30 years (Parnell, 2000). In addition, relevant taxonomic literature is often not aware of name changes and incorporates miss-identified specimens. Even in major international herbaria, specimens in critical groups may remain wrongly assigned pending active flora work in the country of origin (Hall, 1984).

2.1.2 Taxonomic Literature

Publications and citing of specimens and geographical locations are major sources of data for mapping species distribution. This type of publication is available for much of the world and covers the majority of tree species. For instance, the Flora of Thailand documents forest tree species that have been documented by a major long-term project which was initiated in 1957–58 (Larsen & Warncke, 1966; Smitinand & Larsen, 1966; Larsen, 1979, 1988). To date, ten volumes of Flora of Thailand have been published and have documented approximately, 30–40% of the angiosperm flora in Thailand, with estimates of the size of the flora varying from *ca* 10,000 to as many as 12,500 species (Santisuk *et al.*, 1991). This document includes not only species name, its characteristic and key, but also the distribution of the species in Thailand and other parts of the world. Nevertheless, geographical locations are lacking in earlier volumes, but coarse distribution maps can be produced from this literature.

Besides lack of resources, there are many gaps in collecting activities which make a straightforward interpretation of bio-geographical patterns very difficult. According to Parnel *et al.* (2003) the spread of collecting activities in Thailand are markedly uneven; 20% of the collections come from a single province and 53% of provinces have fifty or fewer specimens. The distribution of collections by province and by quarter degree square is erratic with most squares and provinces having few collections, both in proportionate and absolute terms. Some of the most densely forested provinces and squares appear to be under-collected (Parnell *et al.,* 2003).

2.1.3 Ecological Literature

Ecological communities are useful sources to supplement information gathered from herbaria or taxonomic literature (Hall, 1984). Ecological literature represents a more extensive mass of data than inventories and covers all major vegetation formations. Most ecological literature focuses on plant communities rather than individual species. In addition, ecological literature can be classified into two kinds of representation namely, vegetation maps and phyto-sociological features. Vegetation maps can be obtained from existing forest type maps or land use/land cover maps or may be interpreted from remotely sensed data. The level of details concerning ecological features depends on the map scale and the spatial resolution.

Because ecological literature focuses on plant communities rather than on individual species, scientists can predict likely species inhabiting each community type according to habitat preferences. This is due to the fact that vegetation type is a significant factor affecting hiding cover and forage for most species (Patton 1992). For instance, the early Gap Analysis Program in Idaho gathered occurrences for each species from the Idaho Conservation Data Center and the vegetation map that was compiled from existing large-scale vegetation maps and interpretation of Landsat images (Butterfield *et al.*, 1994). In addition, habitat preferences for each species were determined from several sources and subsequently a data file that assigned presence and absence of each vertebrate species to each vegetation type map was created. For each species composite polygons were created that are identified by country and vegetation type. Then, countries from the composite map layer where a species is coded as present were extracted. Species are modeled to be present in the entire vegetation polygons even when the species is not recorded.

However, the ecological literature demonstrates several limitations. The first limitation is the map scale or size of mapping unit. When we use small scale maps (1:500,000 – 1:1,000,000), each vegetation polygon, in fact, contains heterogeneous communities on the ground. Thus, it may be suitable to several species. Secondly, the vegetation map contains insufficient ecological information for some vegetation types. Vegetation types that have narrow shape or small remnants in the landscape (such as riparian habitats) but which are important for biodiversity are likely to be generalized with adjacent polygons. Therefore, species modelers should select an appropriate map scale for any particular species under investigation.

2.1.4 Inventory Data

The purpose of species collection by taxonomists is mainly for identification, thus only a limited number of specimens are collected during field surveys. In many cases, some species, such as palms, which have complex morphological structure, are rarely collected because it is inconvenient to install these in the herbarium. Therefore, a species modeler cannot rely only on specimens that are derived from herbarium or museum because they do not facilitate satisfactory appraisal of distribution. In such cases, incorporating data gathered from other sources, and in particular forest inventory activities may improve the distribution patterns of selected species.

Trisurat (2009) predicted changes in the distribution of tree species as a consequence of climate change in Thailand. Plant occurrence points (geo-referenced and labeled with the species name) were obtained from both the Forest Herbarium of the Department of National Park, Wildlife and Plant Conservation (formerly the Royal Forest Department) and from two forest inventory projects. As discussed in the previous section most specimens do not have geo-referenced locations. The Forest Resources Inventory and the Preparatory Studies to Install a Continuous Monitoring System for the Sustainable Management of Thailand's Forest Resources respectively established uniform fixed grids of 10 x 10 km and 20 x 20 km, over the entire country for gathering plant species (RFD/ITTO, 2002). In each sample plot, forestry officials record data on plot location (geo-referenced location, tree species, number of individuals per plot, etc.). The modelers cannot, however, check identifications of tree species collected from inventory data and possible confusions attached to closely allied species because most forest inventory projects do not keep herbarium specimens. In addition, most tree species that have limited economic value are not included in formal inventories.

2.1.5 Electronic Databases

Recently, advances in information technology (e.g. large-capacity electronic storage media, the Internet, the World Wide Web, distributional database technology) and in the policies of owners of primary data sources (e.g. large-scale digitization of data, creation of public-access databases) are providing new possibilities for the way that biodiversity information can be created, maintained, distributed and used (Bisby, 2000; Oliver *et al.*, 2000; Edwards *et al.*, 2000; Krishtalka & Humphrey, 2000; Krishtalka *et al.*, 2002), with the potential of much more to come (Godfray, 2002). Moreover, the amount, variety and resolution of spatially explicit electronic data that can be used to describe environments (e.g. RS data available via the Internet) are similarly growing (Sobero & Peterson, 2004).

Primary biodiversity data is now becoming accessible at an accelerated speed. Increasing numbers of museums and herbaria are computerizing data associated with natural history specimens (Krishtalka & Humphrey, 2000). In many cases, these datasets are being made available through the Internet (Sobero & Peterson, 2004). Excellent examples include: the New York Botanical Garden, the Museum of Vertebrate Zoology, the University of California at Berkeley, the Missouri Botanical Gardens and the Instituto Nacional de Biodiversidad, Costa Rica, but the list is growing very fast. Also, several centralized databases provide access to information held in jointly created specimen databases. Fishbase, for example (http://www.fishbase.org/home.htm) offers data from hundreds of thousands of specimens. This initial commitment to sharing data and providing open access to data is an important step towards greater information access in the biodiversity world (Sobero & Peterson, 2004).

More profoundly still, since 1998, several distributed biodiversity information networks have provided a new class of access to biodiversity information. In particular, two specialized search engines, The Species Analyst (http://speciesanalyst.net) and REMIB (http://www.conabio.gob.mx/remib/remib.html) have solved key problems that plagued earlier, single-database implementations. These facilities provide access to distributed databases, which means that the data remain at the institutions where the voucher specimens are housed, thus maintaining the connection between primary documentation (specimens) and the information product (the database) (Sobero & Peterson, 2004). Nevertheless, the contents of these dispersed databases are shared virtually via specialized Internet access engines. The Species Analyst and REMIB now connect databases of hundreds of collections, and serve data associated with millions of specimens. Still better access is now permitted by a next-generation integrating technology (DiGIR; http://digir.sourceforge.net/), which has now been implemented fully for the first time in the MaNIS project (The Mammal Networked Information System) (Stein & Wieczoreck, 2004) (http://elib.cs.berkeley.edu/manis/) and will become the standard protocol of the collections associated to the GBIF (Sobero & Peterson, 2004).

2.1.6 Field Observation

Among the six data sources, field observation is considered to be the most reliable and important source of data for species modeling. This is because it provides first-hand information on species distribution. The results will show both up-to-date information in the form of both presence and absence data. In addition, other environmental data at the site such as species number, vegetation type, topography, human threats, etc. can be recorded if required. However, compared to other approaches, field observation usually requires too much time and resources. Therefore, most scientists carefully set data sampling design and target species for collection prior to actual implementation. For example, the Western Forest Complex (WEFCOM) Project employed a rapid ecological assessment

(REA) (Sayre *et al.*, 2000) to assess wildlife occurrences in the WEFCOM area. Wildlife signs and visual sightings of eight target large mammals were recorded by approximately 50 park rangers who had been trained in map reading, acquisition of Global Positioning System (GPS) data and identification of wildlife tracks and signs prior to the surveys. Ground surveys were conducted in both wet and dry seasons in year 2001 along wildlife trails and patrolling routes across the WEFCOM landscape. The presence and absence of wildlife signs and visual sightings were detected and later these data were used in species distribution modeling (WEFCOM, 2004).

2.2 Geographic Information Systems

A geographic information system (GIS) is computer software that is capable of integrated handling and analysis of spatial data (Burrough, 1986). A GIS includes a sophisticated database that stores data on the geographical location of spatial entities and one or more of their attributes, and can test for relationships between different entities and attributes. Information is stored either in vector format or in raster format. Most GIS programs now allow easy interchange between these two formats.

Human beings are limited in their ability to interpret complex geographical information when multiple features are combined on the same printed map. A GIS simplifies the picture by mapping complex geographical features as points, lines and polygons because each feature is stored in a separate digital data layer. Thus where on a paper map these features are overlaid in an inflexible way, in a GIS they can be overlaid selectively and flexibly for easier interpretation.

The GIS approach is more than just an efficient method for presenting complex geographical information. It also offers a new way of looking at that information, by dividing it into major sets of individual features and allocating a separate digital map layer to each. A single feature can have multiple attribute layers associated with it. Advancements in computer technology, statistical modeling and GIS software allow the knowledge of species/habitat relationship to be used for prediction of the geographic distribution of individual populations of wildlife species (Yost *et al.*, 2008).

2.3 Representation of Species Features

When all available specimens and species observations have been located, their geographical coordinates are also needed. Species or records without the sources of information or without reference to administration boundary have less value for mapping. In early stages, surveyors did not have Global Positioning System (GPS) to locate geographical location of specimens. Their localities were usually defined by administration name, either province or district. For translating a place-name into geographical coordinates, taxonomists in Thailand and also in many other countries in the tropics normally either used a general administrative map or incorporated them with existing forest maps and topographic maps for identification of coordinates (Parnell *et al.*, 2003). However, place-name ambiguities often occur at different locations. In addition, place-names can change from time to time. Therefore, identification of the correct place-name requires research into the context of a report or collection.

In those studies in which the geographical location of specimens or records were obtained, species distributions frequently are represented on maps by either points on a map (Figure 1(a), continuous distributions over a large area (Figure 1(b)), or in terms of simple presence or absence of a species from a grid square, which may be as large as 1° x 1° (Figure 1(c); Granger *et al.*, 1995). Even if these maps were to reflect natural distributions they would be misleading, because many areas have not been properly surveyed and the apparent concentration of source species distribution may reflect collection bias (Nelson *et al.*, 1990).

2.4 Species Habitats and Ecological Niche

Habitat is the environment and the specific place where an organism lives. Patton (1992) defined habitat as all factors affecting an animal's chance to survive and reproduce in a specific place. These specific places are often described by a vegetation type or topographic feature (e.g., food, water, cover and space) and can be derived from maps. A food map should include shrub areas, location of big trees (*Ficus* sp.), concentrations of big trees, salt licks and natural openings, while a cover map should include plant canopy, location of caves and root trees, etc. A water map should show at least the stream network, water bodies and amphibian habitats. These maps can be made as overlays on topographic maps, either manually or automatically, to derive habitat suitability estimations for specific species (to be discussed in next section (Trisurat, 2009)).

Kimmins (2004) classified ecological niches according to three major concepts. Firstly, niche refers to the functional role of a species in an ecosystem. It stresses the entire complex of characteristics exhibited by the species, i.e., how the species fit into the complex functional processes of the ecosystem. Elton (1927) defined this term as the relationship of an animal to food and enemies. Secondly, niche refers to the habitat of a species: the range of environments in which it lives. Thus, the definition of species niche includes its adaptations to light, temperature, moisture, soil, fires, and amplitude of these factors to which it is exposed at various times. For instance, Trisurat *et al.* (2009) indicated that a lot of evergreen tree species in northern Thailand would lose their ecological niche due to future climate change. Thirdly, the definition of niche involves a statement of the geographic area or range over which a species is found. For example, the giant frog is found only in running water at high altitude in northern and western Thailand (WEFCOM, 2004).

Thus niche refers to the functional, adaptive and distributional characteristics of a species. The

Figure 1. The distribution of species in three common formats: A) Point map; B) Continuous map; and C) Grid type map. Source: Grainger et al. (1995)

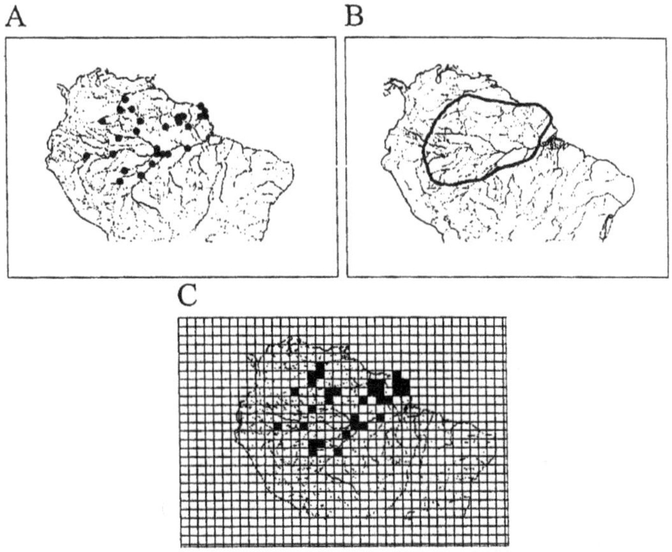

niches of a species have been likened to a volume within which the species is competitively supreme. In addition, the fundamental niche represents the maximum niche that species could occupy in the absence of competition from other species. Therefore, each species depends on the existence of a specific set of environmental conditions for its long-term survival (Hutchinson, 1957), they include not only the abiotic environment but also biotic factors of the respective ecosystem determining the abundance of resources as well as trophic chain interactions.

2.5 Species Modeling Methods

A model is an intellectual representation of a real-world situation. In developing a model, the modeler must identify a representation that accurately and comprehensively defines the problem (Patton, 1992). Species-distribution models are based on the assumption that the relationship between a given pattern of interest (e.g. species abundance or species occurrence) and a set of factors assumed to control it and can be quantified (Guisan & Zimmermann, 2000). Generally, there are three major steps involved with predictive modeling and mapping; (1) collect species level occurrence data and biophysical attributes of the landscape, (2) build the model to determine the best subset of predictors and their parameter coefficient and (3) application of the models to GIS data to predict probability of occurrence for un-sampled location (Corsi *et al.*, 2000; Yost *et al.*, 2008).

It should be noted, however, that modeling species abundance and distribution is not necessarily the same as modeling biodiversity. Most of the species models and maps are designed to depict the distribution of individual species and do not generate biodiversity as an integrated index such as Mean Species Abundance (MSA) (Chapter 8). Nevertheless, the modeling results are essential inputs for informed conversation planning, mapping patterns of biodiversity, detecting distributional changes from monitoring data and quantifying how variation in species performance relates to one or more controlling factors (Toxopeus *et al.*, 2007).

Basically, there are two approaches to developing species distribution maps: the deductive approach and the inductive approach, and the selection of these approaches are dependent on objectives and data availability (Stoms *et al.*, 1992). The deductive approach extrapolates known habitat requirements to the spatial distributions of habitat factors. If the habitat requirements are not well known, however, the habitat map can be derived from a sample of observations of the species locations to one or more habitat factors. This method is named inductive approach. Based on these two approaches, the existing species distribution models are categorized into three modeling methods, namely (1) cartographic overlay method, (2) species modeling using presence-absence data, and (3) species modeling using presence-only data. The elements of each approach are illustrated in sections below, with the occurrence of elephant in northeast Thailand as a study example.

3. MODELING ELEPHANT DISTRIBUTION

3.1 Study Area

Bun Tharik-Yod Mon, a proposed wildlife sanctuary, is part of the Pha Taem Forest Complex, which is comprised of five protected areas, namely Pha Taem National Park, Kaeng Tana National Park, Phu Jong-Na Yoi National Park and Yot Dom Wildlife Sanctuary, as well as Bun Tharik-Yot Mon, a newly proposed Wildlife Sanctuary (Figure 2). The complex is located between latitude 14° 12.5′- 15° 13.9′ N and longitude 104° 58.5′- 105° 8.5′ E in Northeast Thailand. All together, this complex constitutes 1,736 km^2 and its total perimeter is 730 km. Approximately 317 km or 43% of the total border length adjoins Laos (298 km or 40.96%) and Cambodia (18 km or 2.5%). The shape of Kaeng Tana and Yot Dom is relatively simple while the shapes of the remaining areas are complex. Currently, 18 ranger stations have been established and eight park officials as well as 350 temporary employees are deployed to safeguard the PPFC (Table 1). With financial and technical support, the Royal Thai Forest Department has initiated the transboundary biodiversity conservation between Thailand, Laos and Cambodia since 2001. Currently, the Project Phase II (2008-2009), which extends the scope of project area to cover nearby protected areas in neighboring countries (e.g., Phouxeingthong National Biodiversity Conservation Area (NBCA) (120,000 ha) in Laos, and Protected Forest for conservation of Genetic Resources of Plants and Wildlife in Preah Vihear Province (190,000 ha) in Cambodia, the so-called the Emerald Triangle Protected Forest Complex) aims to strengthen existing cooperation among the three countries, to enhance protection of biological resources along the tri-national borders, and to strengthen the involvement of local communities and stakeholders in sustainable use and management of natural resources in the buffer zone. The duration of project was two years (2008-2009) (RFD/ITTO, 2004).

The landscape of PPFC is flat to undulating terrain. Elevation is ranging from 100 m to 732 m above sea level and the dominant altitude is between 100-400 m. The terrain level in the west and northwest is relatively low and rises to the east and to the south of the complex, and immediately declines to Mae Khong River. Besides Mae Khong River, Nam Mun River also drains from the west to the east and passes through Kaeng Tana before joining Mae Khong River, while Lam Dom Noi and Lam Dom Yai Streams originating in Phu Jong - Na Yoi and Yot Dom, drain to the north. These rivers and streams are the main water resources for Pak Mun and Sirinthon hydropower reservoirs, respectively.

Three main vegetation types were described based on the interpretation of Landsat satellite images in 2002 (Figure 2): dry evergreen forest, mixed deciduous forest, and dry Dipterocarp forest. More than 288 tree species are identified and at least 49 mammal, 145 bird, 30 reptile and 13 amphibian species are recorded, but large wildlife species such as the Asian elephant, banteng, freshwater crocodile and tiger are observed only along the tri-national borders (Marod, 2003; Bhumpakphan, 2003). Biological features of protected areas in Cambodia and Laos were not

Table 1. Summary of key features of the Pha Taem Forest Complex

Name	Established [1/]	Area (km^2) [2/]	Perimeter (km) [2/]	Country bound. Km (%) [4/]	Shape Index [5/]	Ranger Station	Officials [6/]
Pha Taem NP	31 Dec 91	353.16	242.67	63.32 (27%)	3.64	5	3/100
Kaeng Tana NP	13 Jun 81	84.62	62.52	29.96 (48%)	1.92	4	2/90
Phu Jong Na Yoi NP	1 Jun 87	697.38	215.88	93.87 (43%)	2.31	4	1/90
Yot Dom WS	11 Oct 77	235.93	88.21	33.21 (37%)	1.62	4	1/60
Bun Tharik-Yot Mon	Proposed	365.86	186.15	96.40 (52%)	2.75	1	1/15
Total		1736.95	730.04 [3/]	316.76 (43%)		18	8/355

Notes: 1/ Royal Gazette, 2/ Calculated by GIS, 3/ Excluding shared border, 4/ Length of country boundary, 5/ Perimeter/2(π x a), 6/ Government Official/temporary employee

Figure 2. Location of Bun Tharik-Yod Mon proposed wildlife sanctuary, elephant observations and major land-use classes

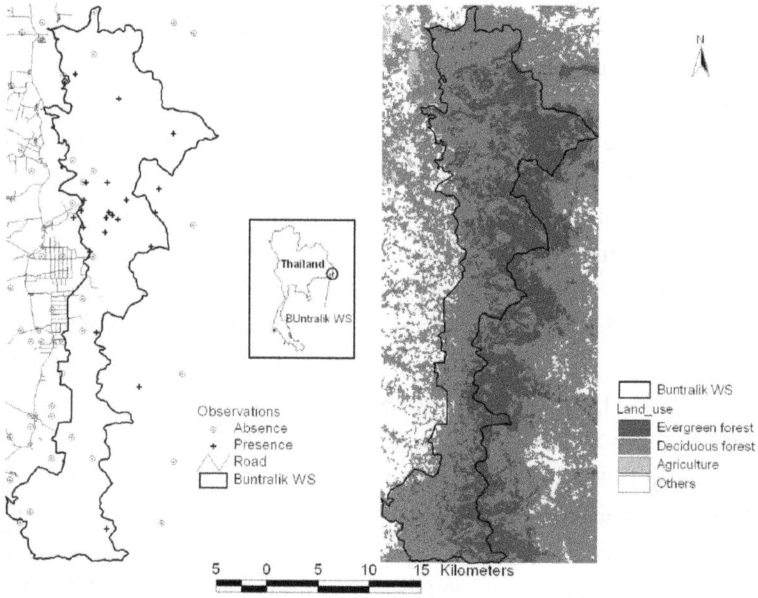

assessed during the Project Phase I. However, the on-going *UNDP/GEF Medium-size Project for the Northern Plain "Establishing Conservation Areas through Landscape Management" CALM* and on-going project phase II reveals an abundance of the populations of elephant, *Eld's* deer, Sarus crane and Giant Ibis inhabiting and breeding in Preah Vihear and areas adjoining Laos (Sonnon, 2003; Cheng, 2004). In addition, detailed assessment of biological features was conducted under the Project Phase II.

3.2 Wildlife Observation Data

In this chapter, the Asian elephant (*Elephas maximus*) was used as focal species to elaborate several species modeling methods that have been used for mapping distribution and conservation planning because elephant is classified an endangered species by the IUCN (2009) due primarily to poaching and habitat loss.

The Asian elephant has a large size, thick grey or grey-brown skin and a long trunk. The tracks of adults are 35–50 cm across with five toe marks on the front foot and four on the back. It is a landscape species and found in a wide variety of forested areas including monsoon forest, lowland rain forest, swamps and plantations. It is active by both day and night, although it prefers to remain in the shade during the heat of the day. Its diet comprises a wide variety of plants, including palms, bananas, twigs, barks and leaves from a range of trees and shrubs, and vines. The number of elephants is greatly reduced by hunting and forest loss.

Park ranger and wildlife experts conducted a field survey to record target wildlife species, including elephant in the study area and their geo-graphical locations were recorded in the WGS 1984 UTM zone 48N using Global Positioning System (GPS). The results reveals that all together, 32 sample sights of Asian elephant were recorded in the Bun Tharik-Yod Mon landscape. A few

Table 2. Explanatory variables used to generate the distribution models

Type	Predictor variable	Sources/techniques
Biophysical-factors	Land use/land cover	Image interpretation
	Accessibility to water (m)	Topographic map/image interpretation/surface analysis
	Elevation (digital elevation mode: DEM) (msl)	Topographic map/surface analysis
	Slope (%)	Topographic map/surface analysis
Anthropogenic factors	Distance to road (m)	Topographic map/surface analysis
	Distance to village (m)	GPS mapping
	Distance to ranger station (m)	GPS mapping

observations were recorded in Phu Jong Na Yoi National Park and Yot Dom wildlife sanctuary because rugged terrain along the national border forms biological and physical barriers to elephant movement. In addition, no single elephant sign was observed in Pha Taem and Kaeng Tana national parks because of severe human disturbance in these areas.

3.3 Habitat Factors for Elephant

Characteristics and habitat preferences for elephant were reviewed from a number of previous research studies that have been conducted to assess habitat utilization of this species, e.g. Wildlife Conservation Division and Forestry Research Center (1997, 1999) and Phrommakul (2003).

The review results show that four biophysical factors and three anthropogenic factors are important resource factors for determining habitat suitability for elephant (Trisurat, 2004). Bio-physical factors include land use/land cover, accessibility to permanent water, elevation and slope. Forest type is a significant factor to provide hiding cover and forage for herbivore species while water is a resource necessary for animals to survive, especially in the dry season, and to reproduce (Patton, 1992). In addition, elevation and slope are physical barriers to wildlife migration because most species prefer to inhabit lowland area rather than rugged terrain. Human factors that were identified were distance to road, distance to ranger station and distance to village.

The habitat factors used in the preparation of distribution models were obtained from different sources and spatial analysis techniques (Table 2). Cloud-free multi-temporal Landsat 5-TM digital imageries in 2002 were acquired from the Geo-informatics and Space Technology Development Agency. Then a sub-scene of the image covering the study area was extracted, mosaicked and rectified to the WGS 1984 UTM Zone 48N using ERDAS Imagine software. In addition, visual interpretation was employed to delineate land-use/land-cover into 10 classes: dry evergreen forest, mixed deciduous forest, dry Dipterocarp forest, secondary growth, scrub (degraded) forest, forest plantation, rubber plantation, agricultural area (paddy field and cash crops), settlement area, bare soil and water body. Key image features of the sampling vegetation type plots across the study area were developed to assist visual interpretation.

All roads (paved and unpaved road and trails) and water were digitized from topographic maps of 1:50,000 scale and satellite images. Locations of villages were both digitized from topographic maps and uploaded from GPS. Geo-referenced locations of villages and ranger stations obtained from the field survey were later converted to GIS. In addition, ArcView Spatial Analyst was used to generate layers of digital elevation model, slope, and distances to road, water, village and ranger station. All continuous data derived from spatial analyses and from the land use/land cover map were referenced to the same 100 m x 100 m resolution UTM grid square.

3.4 Cartographic Overlay

A Habitat Evaluation Procedure (HEP) to evaluate the requirements of selected species is one deductive approach that has been developed for use by the Fish and Wildlife Service (U.S. Department of Interior, 1980). Once the species has been selected a habitat suitability index based on research data and expert opinions is calculated. The assumption is that a Habitat Suitability Index (HIS), a numeric value summarizing habitat suitability based on habitat quality with a decimal value of 0.0-1.0, can be developed for the selected species.

There are a series of HIS equations as follows:

$$HIS = [V_1 + V_2 + V_3]/3 \text{ Compensatory Model}$$

$$HIS = [2V_1 + V_2 + V_3]/4 \text{ Weighted Mean}$$

$$HIS = [V_1 \text{ x } V_2 \text{ x } V_3]^{1/3} \text{ Geometric Mean}$$

Where HIS = habitat Index Suitability

Vi = habitat factors (e.g., food, cover, water, space).

A large number of habitat suitability indices have been developed for both terrestrial and aquatic habitats. In operation, a Geographic Information System (GIS) is used to create habitat factors, assign numeric values based on habitat quality, overlay all these layers and calculate suitability classes.

In this chapter, the modified Compensatory Model was used to evaluate habitat suitability of the Asian elephant at the Bun Tharik-Yod Mon proposed wildlife sanctuary in the Pha Taem Protected Forest Complex. First, all wildlife habitat factors were reclassified according to their attributes. Then, each attribute of habitat factor was ranked to determine its suitability for each species. The ranking scores were assigned as follows: 3 (suitable), 2 (moderate) and 1 (not suitable). For example, 10 classes of land use/land cover were broadly categorized into 3 classes: evergreen forest, deciduous forest and human-dominant land uses or non-forest. Based on literature reviews (Wildlife Conservation Division and Forestry Research Center 1997, 1999; Wildlife Conservation Division and Khon Kaen University, 2000; Phrommakul, 2003) dry evergreen forest is preferable for elephant; therefore this vegetation type is recorded as score 3. Meanwhile, score 2 is assigned for deciduous forest, scrub while score 1 is given to the remaining classes which are dominated by

Table 3. Spatial criteria and ranking scores for elephant habitat suitability

Theme	Classes	Ranking score
Land use	Evergreen Forest	3
	Deciduous Forest	2
	Human-dominant land uses or non-forest	1
Elevation (m)	0–600	3
	> 600	2
Slope (%)	0–20	3
	20–30	2
	> 30	1
Accessibility To permanent water (m)	<1,000	3
	1,000–3,000	2
	>3,000	1
Distance to main road (km)	< 5	1
	5–7	2
	> 7	3
Distance to ranger station (km)	< 1	1
	1–3	2
	>3	3
Distance to village (km)	< 5	1
	5–10	2
	> 10	3

human activities (Table 3). In addition, elephants normally avoid human activities, but roam in low altitude, flat terrain and prefer to remain close to water sources. Therefore, the attributes of the habitat factors were ranked accordingly.

All habitat factors were super-imposed on to one layer using raster-based GIS ArcView software. The output map contained accumulated scores of seven habitat factors. The accumulated scores ranked between 10-21 and mean value was 14.45. They were equally categorized into five classes to represent the habitat suitability index: (1) low, (2) relatively low, (3) moderate, (4) relatively high, and (5) high. The preliminary habitat suitability was masked by human settlement, agricultural area and water body because areas inside these regions did not inhabit wildlife. After masking, the draft suitability map was generalized by removing noise pixels for better visualization and for more practical use on the ground.

The result map (Figure 3) shows areas of each suitability class, where elephant is likely to be found in the Bun Tharik landscape. The likely habitats of elephant (moderate-high) cover approximately 68% of the Bun Tharik landscape. The predicted areas of low suitability, relatively low suitability, moderate suitability, relatively high suitability and high suitability for the elephant cover 31.2 <0.1, 27.2, 9.4 and 31.6% respectively. Relatively high to highly suitable habitats extend along the national border areas. In these areas, paved roads are few, human settlements are distant and dry evergreen forest is dominant. In addition, data from the field survey show a number of migratory routes that elephants from Laos use to travel to Thailand in the wet season and back to Laos in the dry season because of water availability. On the other hand, most areas situated in the west of Bun Tharik in Thailand, are predicted to have low to relatively low suitability for the elephant. These areas have been totally converted to farmlands or human settlements and a dense road network has been constructed.

Figure 3. Habitat suitability classes for elephant derived from the 3 model approaches

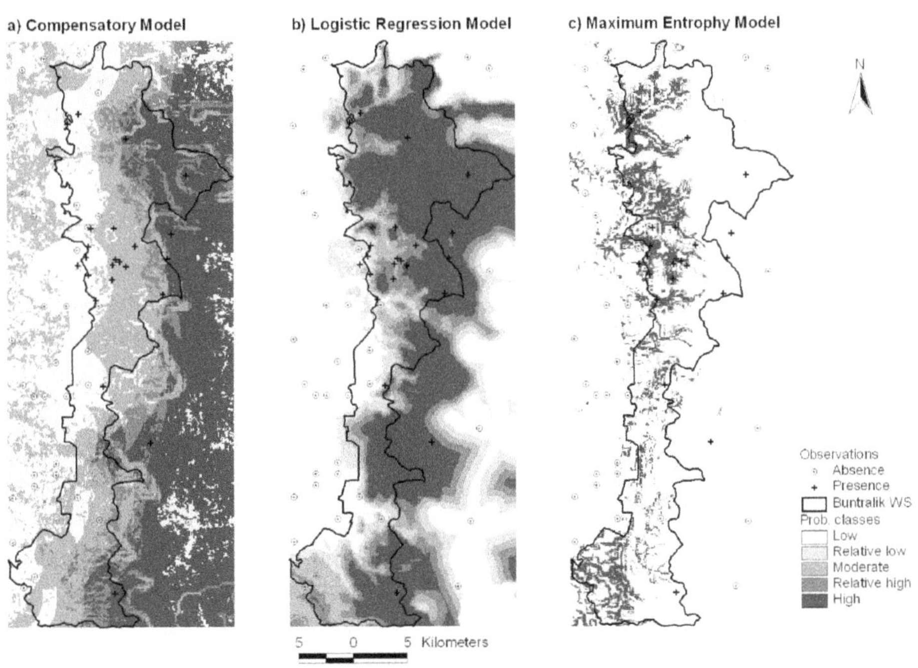

3.5 PRESENCE –ABSENCE DATA (LOGISTIC REGRESSION MODELING)

A range of species distribution models has been developed for binary response variables (presence/absence) such as Generalized Additive Models ((Hastie & Tibshirani, 1990) and Generalized Linear Models (GLM). Logistic regression modeling, a particular branch of GLM, is a multivariate statistical technique that is used to predict a binary dependent variable (presence or absence) from a set of variables (Atkinson & Massari, 1998). Thus, it is an inductive approach in which the result is derived from observation samples and related variables. The advantage of logistic regression is that the variables may be either continuous or discrete, or any combination of both types, and they do not necessarily have normal distributions. Therefore, it is not necessary to categorize explanatory factors before entering them in the model. The algorithm of logistic regression applies maximum likelihood estimation after transforming the dependent into a logit variable (the natural log of the odds of the dependent occurring or not). In this way, logistic regression estimates the probability of a certain event occurring (Atkinson & Massari 1998; Lee & Nelder, 1996). The logistic regression model is:

$$\text{Prob}_{event} = \frac{e^{Zi}}{1+e^{Zi}}$$

Where Z_i is the linear combination model of species I as follows:

$$Z = \beta_0 + \beta_1 X_1 + \beta_2 X_2 + \ldots + \beta_n X_n$$

βI = coefficient

Xi = independent variables (habitat factors)

The probability values derived from the regression models range from 0.0-1.0. The higher the value, the greater the likelihood of occupancy of the target species. A cut-off value of 0.5 was used for binary classification. Any pixel containing the probability values equal or greater than 0.5 was categorized as presence, while pixels with values lower than 0.5 were classified as absence. In some cases, the cut-off value for binary classification may be adjusted (0.4, 0.45, 0.50, 0.55 and 0.60) to maximize the fit for sample data based on a prior knowledge (Neter *et al.,* 1996; Trisurat *et al.*, 2010).

Based on a well established body of statistical theory, it is possible to do all of the following within the logistic regression model framework: (1) construct a parsimonious model that strikes a balance between bias and variance using criteria supported by an established body of statistical theory, (2) identify the relative importance of the predictor variables, (3) explore and interpret the response of the species to each predictor, (4) estimate the uncertainty associated with parameter estimates, (5) predict the probability of observing the species (rather than predicting binary presence-absence) and 6) explore spatially explicit patterns of uncertainty in predictions (Wisz & Guisan, 2009). Therefore, it remains the most widely used model to predict the potential distributions of species (Guisan & Zimmermann, 2000).

To illustrate this modeling method, the stepwise logistic multiple regression model was employed to generate the occupancy models for elephant in the PPFC. The presence/absence data were treated as dependent variables in the occupancy model. As mentioned above, 32 sample locations of Asian elephant were found in the Bun Tharik-Yod Mon landscape. Unfortunately, park rangers and wildlife experts did not record absence data, which is required for logistic modeling. This problem is quite common for species distribution models, thus the logistic regression model has been precluded from many studies of species distributions (Wisz & Guisan, 2009). In order to facilitate the use of the logistic regression model, when absence data were unavailable, a number of studies have used pseudo-absences in place of real absences. For instance, Wisz & Guisan (2009) proposed two strategies to select pseudo-absences: (1) randomly selecting from the background and (2) selecting from low suitability areas predicted by other species modeling methods, that did not require absence data. In this chapter, pseudo-absence was chosen from low suitability classes (low-moderate suitable) derived from the Compensatory Method or cartographic overlay.

In this study, the systematic sampling plots of 1-km separation were generated across the Bun Tharik-Yod Mon landscape. Then the balanced number of pseudo-absence data (the number of generated pseudo-absence was the same as the number of true presences - 32 samples) for elephant species was selected using stratified sampling method. It is essential to keep in mind that pseudo absence are never chosen in pixels having the same characteristic as presence, because these values are removed during the modeling (Enger *et al.*, 2004). Thus, the locations of pseudo-absence were selected from likely absence classes derived from the cartographic overlay technique (low, relatively low and moderate suitable classes). The number of pseudo absence was 4, 10, and 18 samples, respectively.

The independent variables included in the logistic model were the same as used for the Compensatory Model (four biophysical and three human disturbance factors). However, they were not categorized as they were calculated for the Compensatory Model. These factors were interpolated and recorded as continuous data in order to represent the real characteristics. In addition, land use/land cover (10 classes) was still recorded as discrete data. Independent variables were included in the logistic regression models if the p-value for the partial regression coefficient was less than 0.05.

The accuracy of the logistic regression model was assessed by using the area under curve (AUC) of a receiver operating characteristic (ROC) curve to assess the accuracy of each model (Hosmer & Lemeshow, 2000). The AUC provides a measure of discrimination between presence and absence. The interpretation of ROC as a general rule as below:

If ROC = 0.5: this suggests no discrimination

If 0.5≤ ROC < 0.7: this is considered poor discrimination

0.7≤ ROC < 0.8: this is considered acceptable discrimination

If 0.8≤ ROC < 0.9: this is considered excellent discrimination

If ROC ≥0.9: this is considered outstanding discrimination

The results of the logistic regressions indicated that elevation, distance to road, and distance to ranger station were significantly related to the distributions of elephant (Table 4). Slope, distance to stream and land use were excluded from the model because slope was highly correlated with elevation while distant to water had significant correlation with DEM, distance to road and ranger station.

The logistic regression model for elephants in Bun Tharik Yod Mon wildlife sanctuary is shown below.

$$\text{Prob}_{event} = \frac{e^{Zi}}{1+e^{Zi}}$$

Z = -4.86532 + 0.01884DEM + 0.00141Rd + 0.00062Vil

– 0.00063Rst ; AUC = 0.918

Where DEM = altitude (meter above mean sea level)

Table 4. Variables in the logistic regression equation for elephant

Independent variable		B	S.E.	Wald	df	Sig.	Exp(B)
	road_dist	.00141	.001	3.040	1	.050	1.001
	rang_dist	-.00063	.000	8.149	1	.004	.999
	vill_dist	.00062	.000	6.164	1	.013	1.001
	DEM	.01884	.007	6.827	1	.009	1.019
	Constant	-4.86532	1.711	8.084	1	.004	.008

Notes: road_dist = distance to road; rang_dist = distance to ranger station; Vill_dist = distance to village; DEM = digital elevation model or altitude in meter

Rd = distant to road (meter)

Vil = distant to main road (meter)

Rst = distant to ranger station (meter)

The logistic regression model indicated that elephants in Bun Tharik-Yod Mon proposed wildlife sanctuary are more likely to inhabit areas of high altitude, areas closer to ranger stations, and areas farther from villages and roads. This is due to that fact that most lowland area was converted to agriculture and human settlement. In contrast, it prefers to inhabit areas close to ranger stations because strict protection from disturbance is enforced.

In addition, the area under the ROC curve was 0.918 (Figure 4) which is considered as excellent discrimination between likely presence and likely absence. Therefore, the probability values of pixels in the Bun Tharik-Yod Mon proposed wildlife sanctuary were first reclassified into five suitability classes similar to the Modified Compensatory Method, for model comparison, as follows: 0.0-0.2 = low; 0.2-0.4 = relative low; 0.4-0.6 moderate; 0.6-0.8 relative high; and 0.8-1.0 = high values. Additionally, they were classified into two classes. Any pixel containing the probability values equal or greater than 0.5 was categorized as presence, while the values lower than 0.5 were classified as likely absence.

The probability of the elephant distributions derived from logistic regression is shown in Figure 3. Predicted areas of low suitability, relatively low suitability, moderate suitability, relatively high suitability and high suitability for the elephant cover 39.28, 11.38, 10.21, 10.13 and 28.97% respectively. By using the cut-off value of 0.5 for binary classification, the predicted distributions of elephants cover an area of 480 km^2 or 44.32% of the Bun Tharik Yod Mon landscape. Likely the habitat of elephants are located in the center, the north of Bun Tharik-Yod Mon and extends toward the east (Figure 2(b)). Elephants are unlikely to be present in the west and

Figure 4. The receiver operating characteristic (ROC) curve for elephant derived from logistic regression for elephant

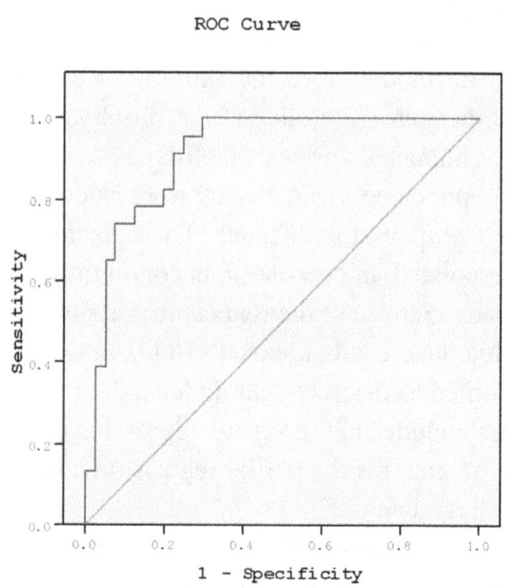

the middle part of the sanctuary where they are disturbed by human activities and deforestation.

3.5 Presence-Only Data (Maximum Entropy Model)

Reliable species distribution information on various scales is needed for both biogeographic and conservation purposes. Species distribution data from herbarium and museums, taxonomic literature, ecological communities, inventory data and field observations that were documented in databases and GIS can provide information relevant to the development of prediction maps (Dennis & Hardy, 1999; Chefaoui & Lobo, 2008). However, these heterogeneous data sources generally do not indicate the locations where the species have not been found after a sufficiently intense collection effort as pseudo-absences can decrease the reliability of prediction models (see Anderson, 2003; Loiselle *et al.*, 2003). Wisz & Guisan (2009) indicated that models built with true absences had the best predictive power and best discriminatory power, but models based on random pseudo-absences had among the lowest fit according to Akaike's information criterion (AIC).

There are many methods that use presence-only data for modeling species distributions. For instance, BIOCLIM predicts suitable conditions in a bioclimatic envelope, consisting of a rectilinear region in environmental space representing the range of observed presence values in each environmental dimension (Busby, 1986). In addition, DOMAIN predicts a suitability index by computing the minimum distance in environmental space to any presence record (Carpenter *et al.*, 1993) Environmental-Niche Factor Analysis (ENFA, Hirzel *et al.*, 2002) uses presence localities together with environmental data for the entire study area, without requiring a sample of the background to be treated like absence. It is similar to principle components analysis (Jolliffe, 2002), involving a linear transformation of the environmental space into orthogonal marginality and specialization factors. Then, environmental suitability is modeled as a Manhattan distance in the transformed space.

In this research, we used a niche-based model or the maximum entropy method (MAXENT) (Peterson *et al.*, 2001) to estimate the probability distribution of the elephants. This is because MAXENT was one of the strongest performing methods among four groups of modeling techniques (artificial neural networks, - BIOCLIM, classification and regression trees, - DOMAIN, generalized additive models,-GARP and generalized linear models, - MAXENT), particularly for species with a relatively low number of presence localities (Tognelli *et al.*, 2009).

MAXENT is a general-purpose method for making predictions or inferences from incomplete information. MAXENT uses entropy as the means to generalize specific observations of presence of a species, and does not require or even incorporate absence points within the theoretical framework. Presence-only points are observations of the presence of a species.

The idea of MAXENT is to estimate a target probability distribution by searching the probability distribution of maximum entropy (i.e., that is most spread out, or closest to uniform), subject to a set of constraints that represent our incomplete information about the target distribution. The information available about the target distribution often presents itself as a set of real-valued variables, called "features", and the constraints are that the expected value of each feature should match its empirical average (average value for a set of sample points taken from the target distribution). When MAXENT is applied to presence-only species distribution modeling, the pixels of the study area make up the space on which the MAXENT probability distribution is defined, pixels with known species occurrence records constitute the sample points, and the features can be climatic variables, elevation, soil category, vegetation type or any other environmental variables and functions thereof.

The advantages of MAXENT include the following: (1) it requires only presence data and environmental factors, (2) it can utilize both continuous and categorical variables, and (3) it is efficient at determining the algorithms for converging the optimal probability distribution (Philips *et al*., 2006). More detailed information on the theory of maximum entropy for modeling species distribution is available at Phillips *et al*. (2006). A tutorial on MAXENT and example data can be downloaded from www.cs.princeton.edu/~schapire/MAXENT.

To run MAXENT, two separate datasets are needed. The presence localities of species are in the file "samples\elephant.csv", the environmental layers or habitat factors are in the directory "layers", and the outputs will be stored in "outputs" directory. In this study, maximum entropy models were run with MAXENT (Phillips *et al*., 2006) using a convergence threshold of 10 with 1,000 iterations as an upper limit for each run. Elephant occurrence data was divided into two datasets. 75 percent of occurrences (24 presence points) was used to generate species distribution models (training data), while the remaining 25% (8 presence points) was kept as independent data (test data). The authors used the logistic threshold at maximum training sensitivity plus specificity for binary classification. This logistic threshold point has been used by many researchers (Cuesta-Camocho *et al*., 2006; Trisurat *et al*., 2009;). Similar to the logistic regression model, the area under the receiver operating characteristic (ROC) curve was used to test the model accuracy.

The results of MAXENT indicated that all seven predictor variables were included in the model. The relative contributions of environmental factors for the elephant distributions vary from variable to variable (Table 5). Among the seven factors, the models indicate that the contribution of 'Distant to ranger station' was the highest (46.3%) followed by Slope (26.4%). 'Altitude above mean sea level' had the lowest contribution (0.2%).

Table 5. Relative contribution factors to elephant distribution in Thailand

No	Prediction variable	% contribution
1	Distant to ranger station (m)	46.3
2	Slope (%)	26.4
3	Land use	8.4
4	Distant to stream (m)	7.9
5	Distant to road (m)	6.9
6	Distant to village (m)	3.9
7	Altitude above mean sea level (m)	0.2
	Total	100.0

The curves show how the logistic prediction changes as each environmental variable is varied, keeping all other environmental variables at their average sample value (Figure 5). For example, elephant prefers evergreen forest (code 1) and deciduous forests (codes 2 and 3) more than scrub (code 4) and non-forest classes (codes 5-10). In addition, elephants are unlikely to be found in areas more than 1000 m from a ranger station or a stream. Beyond that point, these two environmental variables have less effect on the logistic prediction and the response curve drops off sharply and reaches nearly 0 at the middle of the variable's range. The accuracies of the ecological niche model for both training and test data were very high (AUC = 0.968 for training data and AUC = 0.883 for test data) (Figure 6). Thus, the prediction models are considered excellent for the discrimination of presence and absence.

The probability of the occurrence of elephants was classified into five classes and the resulting map is shown in Figure 7 and Table 6. The predicted areas of low suitability, relatively low suitability, moderate suitability, relatively high suitability and high suitability for the elephant cover 79.25, 11.18, 4.12, 2.40 and 3.05% respectively. By using the logistic threshold at maximum training sensitivity plus specificity (0.15) for binary classification, the likely distributions for

Figure 5. Response curves showing how each environmental variable affects the MAXENT prediction

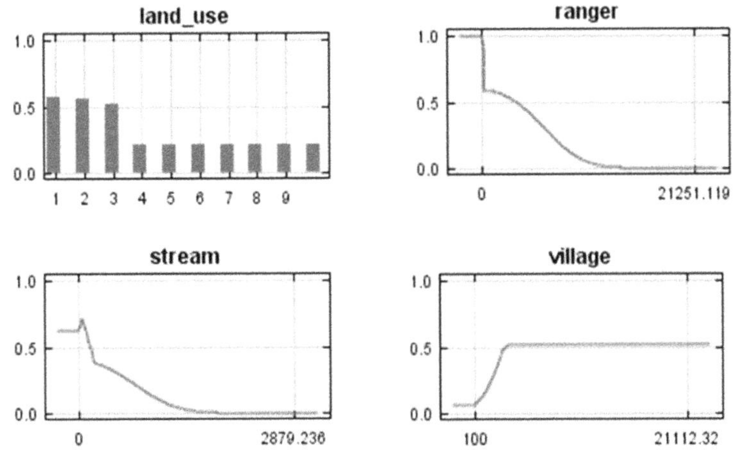

Figure 6. The receiver operating characteristic (ROC) curve for training and test data of elephant

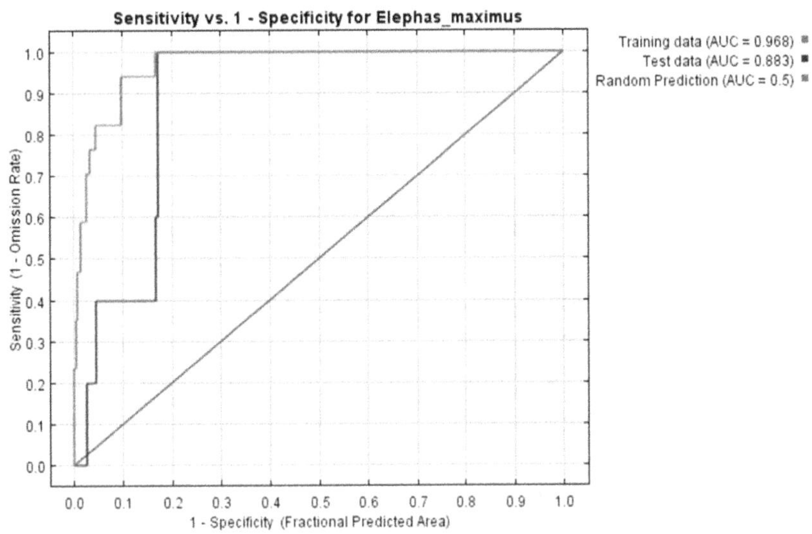

elephant cover an area of 99 km² or 9.14% of the Bun Tharik-Yod Mon landscape. Current suitable niches are located in the center and towards the west of the study area.

4. DISCUSSION AND CONCLUSION

In this chapter, three approaches for species modeling were used to predict species distribution implemented in Bun Tharik-Yod Mon landscape. These approaches are cartographic overlay, logistic multiple regression and the maximum entropy method. The advantages and disadvantages of each method are presented in Table 7.

The cartographic overlay or habitat suitability evaluation method is one of the deductive approaches for extrapolating from known habitat requirements to the spatial distributions of habitat factors. This approach is simple and it is appli-

Figure 7. Predicted presence and absence of elephant in Bun Tharik-Yod Mon landscape

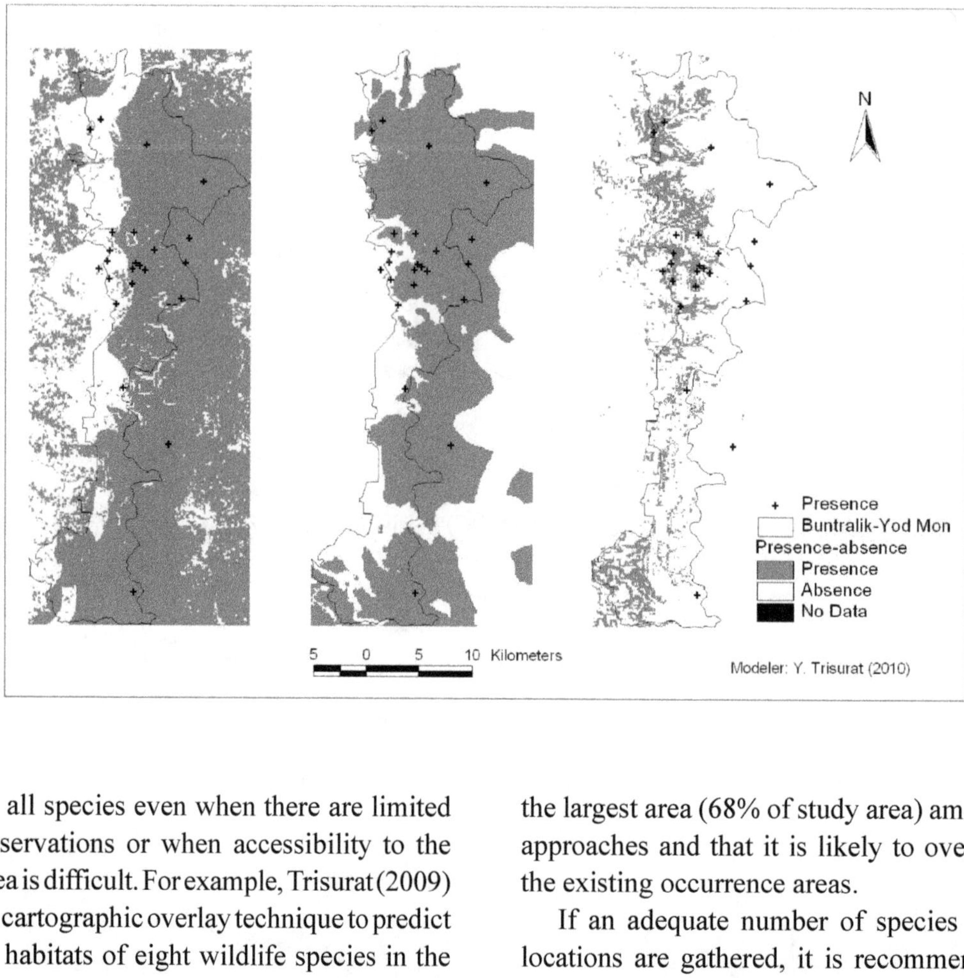

cable to all species even when there are limited field observations or when accessibility to the study area is difficult. For example, Trisurat (2009) used the cartographic overlay technique to predict suitable habitats of eight wildlife species in the Phataem Protected Forest Complex where thousands of landmines still exist. However, the results show that the potentially suitable habitats cover the largest area (68% of study area) among three approaches and that it is likely to overestimate the existing occurrence areas.

If an adequate number of species presence locations are gathered, it is recommended that an inductive approach such as logistic multiple regression or maximum entropy should be applied because these statistical techniques yield

Table 6. Percentage of predicted suitability for the area for elephant and predicted presence and absence, derived from the 3 models

Suitability classes	Cartography overlay	Logistic model	MAXENT
Low suitability	31.8	39.3	79.3
Relatively low	<0.1	11.4	11.2
Moderate	27.3	10.2	4.1
Relatively high	9.4	10.1	2.4
High	31.6	29.0	3.0
Presence	68.2	55.7	9.1
Absence	31.8	44.3	90.9

Table 7. Comparisons of spatial distribution models

Model	Advantages	Disadvantages
Cartography overlay	Simple to understand; Applicable for all species either with or without occurrence data	Normally overestimate the distribution range Largely depending on expert knowledge
Logistic regression model	The relative importance of different predictor variables in determining species distribution can be assessed	Requires large input dataset in order to obtain a meaningful model
Maximum entropy model	Use only presence data to run, easy to obtain; The relative importance of different predictor variables in determining species distribution can be assessed; Can effectively model species distribution from small dataset	Predicted distribution might be biased due to non-systematic samplings and can be overestimated or underestimated due to sampling scheme

more accurate results and provide an opportunity to test the accuracy of the model results (Elith, 2002; Phillips *et al.*, 2006). These methods use presence-absence data or presence-only data to create the model explaining relationship between independent predictor variables and species occurrence using the information-theoretic approach. The advantage of logistic regression and MAXENT methods is that the relative importance of different predictor variables determining the species distribution can be assessed (Margules & Sarkar, 2007). For instance, the results of the logistic regressions indicated that elevation, distance to road, and distance to ranger station were significantly related to the distributions of the elephants and logistic regression predicted approximately 56% of the area as suitable area which is 12% less than the cartographic overlay technique. However, the main weakness of this method is the requirement of large amounts of presence and absence data in order to obtain a meaningful model. Basically, true absence data are often not available for many cases, including this study. It is possible to create pseudo-absence data which are randomly selected from low suitable areas derived from the cartography overlay technique. Taking into account that an AUC value > 0.9 is qualified as outstanding (Hosmer & Lemeshow, 2000), the predicted species distribution is considered as an excellent discrimination between likely presence and likely absence and it is more realistic than the result derived from the cartographic overlay technique.

In addition, it is also possible to develop the species occurrence probability map using a niche-based model or the maximum entropy method (MAXENT) (Peterson *et al.*, 2001). This approach requires only presence data and environmental information and its performance is considered to be better than other methods using presence-only or presence-absence data (Elith, 2002; Phillips *et al.*, 2006). However, its performance is still in doubt due to the lack of accurate absence data. The predicted distributions therefore might be biased due to ad hoc or non-systematic sampling. This program might over-predict the distribution range (Margules & Sarkar, 2007) if species occurrence is spread all over the study area and possibly under-estimates the distribution range when the species has a clumped distribution or biased survey in particular areas, as is the case in this study due to landmines. The resulting map shows that approximately 9% of the study area is classified as presence. It should be noted that the predicted suitable areas cover the least area among the three approaches.

Although the results show a large difference in suitable areas among cartographic overlay, logistic multiple regression and maximum entropy models, it should not be concluded that one method always proves to be better than others. For example, Tognelli et al (2009) evaluated the performance of different techniques for modeling the distribution of Patagonian insects and the results indicated that the maximum entropy model was also one of the strongest performing methods, particularly for species sampled from a relatively low number of localities. In addition, Hirzel *et al.* (2002) used real presence and absence data to simulate species distribution, and the results showed that the species-niches model can provide higher accuracy than that obtained from the logistic regression method.

On the other hand, Wisz & Guisan (2009) indicated that model-results based on logistic regression with true absences data had the best predictive power and best discriminatory power, but that models based on random pseudo-absences had among the lowest fit. Models based on two-step approaches had intermediate fit and the lowest predictive power. This might be less likely to be the case for many rare and endemic species, which tend to occupy most of their potential habitats, as these species usually cover only a small proportion of the area taken into consideration. The results suggest that the performance of these methods depend on the type of organism being modeled (Engler *et al.*, 2004). In addition, the best option for improving the habitat suitability maps would certainly be to obtain additional data for the target species.

Species distribution not only depends on existing bio-physical but also on anthropologic factors. Future deforestation and other land-use changes and also climate change are critical threats to biodiversity in many countries in the tropics. Several studies have addressed these concerns. For instance, Trisurat *et al.* (2010) predicted land use change in northern Thailand and estimated the magnitude and extent of the impacts on wildlife distribution. In addition, Cuesta-Camocho *et al.* (2006) found that approximately 40% of native bird and plant species in the Northern Tropical Andes are predicted to be vulnerable or extinct in 2080. More cases are provided in other chapters. Thus, land use and climate changes should be included in future research.

REFERENCES

Anderson, R. P. (2003). Real vs. artifactual absences in species distributions: Tests for Oryzomys albigularis (Rodentia: Muridae) in Venezuela. *Journal of Biogeography, 30*, 591–605. doi:10.1046/j.1365-2699.2003.00867.x

Atkinson, P., & Massari, I. (1998). Generalized linear modeling of landslide susceptibility in the Central Apennines, Italy. *Computers & Geosciences, 24*, 373–385. doi:10.1016/S0098-3004(97)00117-9

Bhumpakphan, N. (2003). *Management of the Pha Taem protected forest complex to promote cooperation for trans-boundary biodiversity conservation between Thailand, Cambodia and Laos (Phase I): Wildlife ecology final report.* Bangkok: Faculty of Forestry, Kasetsart University.

Bisby, F. A. (2000). The quiet revolution: Biodiversity informatics and the Internet. *Science, 289*, 2309–2312. doi:10.1126/science.289.5488.2309

Burrough, P. A. (1986). *Principles of geographic information systems for land assessment.* New York, NY: Oxford University Press.

Busby, J. R. (1986). *Bioclimate prediction system (BIOCLIM). User's manual version 2.0.* Australian Biological Resources, Study Leaflet.

Butterfield, B. Csuti, B., & Scott, J. M. (1994). Modeling vertebrate distributions for gap analysis. In R. I. Miller (Ed.), *Mapping the diversity of nature* (pp. 53-68). London, UK: Chapman & Hall.

Carpenter, G., Gillison, A. N., & Winter, J. (1993). DOMAIN: A flexible modelling procedure for mapping potential distributions of plants and animals. *Biodiversity and Conservation, 2*, 667–680. doi:10.1007/BF00051966

Chefaoui, R. M., & Lobo, J. M. (2008). Assessing the effects of pseudo-absences on predictive distribution model performance. *Ecological Modelling, 210*, 478–486. doi:10.1016/j.ecolmodel.2007.08.010

Cheng, D. (2004). Monitoring large mammals of the northern plain landscape in Cambodia. *Paper presented at the 25th Annual Wildlife Seminar,* December 24-26, 2004. Faculty of Forestry, Kasetsart University, Bangkok, Thailand.

Corsi, F., De Leeuw, J., & Skidmore, A. K. (2000). Modeling species distribution with GIS. In Boitani, L., & Fuller, T. K. (Eds.), *Research techniques in animal ecology: Controversies and consequences* (pp. 389–434). New York, NY: Colombia University Press.

Cuesta-Camocho, F., Ganzenmuller, A., Peralvo, M., Novoa, J., & Riofrio, G. (2006). *Predicting species' niche distribution shifts and biodiversity change within climate change scenarios: A regional assessment for bird and plant species in the Northern Tropical Andes*. EcoCiencia, Peru: Biodiversity Monitoring Program.

Dennis, R. L. H., & Hardy, P. B. (1999). Targeting squares for survey: Predicting species richness and incidence of species for a butterfly atlas. *Global Ecology and Biogeography, 8*(6), 443–454. doi:10.1046/j.1365-2699.1999.00148.x

Edwards, J. L., Lane, M. A., & Nielsen, E. S. (2000). Interoperability of biodiversity databases: Biodiversity information on every desktop. *Science, 289*, 2312–2314. doi:10.1126/science.289.5488.2312

Elith, J. (2002). Quantitative methods for modeling species habitat: Comparative performance and an application to Australian plants. In Ferson, S., & Burgman, M. (Eds.), *Quantitative methods for conservation biology* (pp. 39–58). New York, NY: Springer-Verlag.

Elton, C. (1927). *Animal ecology*. London, UK: Sedgwick and Jackson.

Engler, R., Guisan, A., & Rechsteiner, L. (2004). An improved approach for predicting the distribution or rare and endangered species from occurrence and pseudo-absence data. *Journal of Applied Ecology, 41*, 263–274. doi:10.1111/j.0021-8901.2004.00881.x

Godfray, C. (2002). Challenges for taxonomy. *Nature, 417*, 17–19. doi:10.1038/417017a

Grainger, A., Rose, S., Trisurat, Y., & Brockelman, W. (1995). A GIS approach to mapping spatial variation in tropical rain forest biodiversity. In T. J. B. Boyle & B. Boontawee (Eds.), *Proceedings of the IUFRO Symposium on Measuring and Monitoring Biodiversity in Tropical and Temperate Forests* (pp. 335-354). Chiang Mai, Thailand: Royal Forest Department.

Guisan, A., & Zimmermann, N. E. (2000). Predictive habitat distribution models in ecology. *Ecological Modelling, 135*, 147–186. doi:10.1016/S0304-3800(00)00354-9

Haines-Young, R., Green, D. R., & Cousins, S. H. (1993). *Landscape ecology and GIS*. London, UK: Taylor and Francis.

Hall, J. B. (1984). Juniperus excela in Africa: A biogeographical study of an Afromontane tree. *Journal of Biogeography, 11*, 47–61. doi:10.2307/2844775

Hastie, T., & Tibshirani, R. (1990). *Generalized additive models*. London, UK: Chapman & Hall.

Hirzel, A., Hausser, H., Chessel, J. D., & Perrin, N. (2002). Ecological-niche factor analysis: How to compute habitat-suitability maps without absence data. *Ecology, 83*(7), 2027–2036. doi:10.1890/0012-9658(2002)083[2027:ENFAHT]2.0.CO;2

Hosmer, D. W., & Lemeshow, S. (2000). *Applied logistic regression*. New York, NY: John Wiley & Sons, Inc.doi:10.1002/0471722146

IUCN. (2009). *The IUCN red list of threatened species*. Gland, Switzerland: The World Conservation Union (IUCN).

Jolliffe, I. T. (2002). *Mathematics*. New York, NY: Springer-Verlag.

Kimmins, J. P. (2004). *Forest ecology: A foundation for sustainable forest management and environmental ethics in forestry* (3rd ed.). British Columbia, Canada: Prentice Hall.

Krishtalka, L., & Humphrey, P. S. (2000). Can natural history museums capture the future? *Bioscience, 50*, 611–617. doi:10.1641/0006-3568(2000)050[0611:CNHMCT]2.0.CO;2

Krishtalka, L., Peterson, A. T., Vieglais, D. A., Beach, J. H., & Wiley, E. O. (2002). The green Internet: A tool for conservation science. In Levitt, J. N. (Ed.), *Conservation in the Internet age: Strategic threats and opportunities* (pp. 143–164). Washington, DC: Island Press.

Larsen, K. (1979). Exploration of the flora of Thailand. In Larsen, K., & Holm-Nielsen, L. B. (Eds.), *Tropical botany* (pp. 125–333). London, UK: Academic Press.

Larsen, K. (1988). Botany in Aarhus 1963–1988. The first 25 years of the Botanical Institute, University of Aarhus. *Reports from the Botanical Institute. University of Aarhus, 17*, 1–92.

Larsen, K., & Warncke, E. (1966). Report on the flora of Thailand project – I. Expedition. *The Natural History Bulletin of the Siam Society, 21*, 251–262.

Lee, Y., & Nelder, J. A. (1996). Heirarchical generalized linear models (with discussion). *Journal of the Royal Statistical Society. Series B. Methodological, 58*, 619–678.

Loiselle, B. A., Howell, C. A., Graham, C. H., Goerck, J. M., Brooks, T., Smith, K. G., & Williams, P. H. (2003). Avoiding pitfalls of using species distribution models in conservation planning. *Conservation Biology, 17*, 1591–1600. doi:10.1111/j.1523-1739.2003.00233.x

Margules, C. R., & Sarkar, S. (2007). *Systematic conservation planning*. Cambridge, UK: Cambridge University Press.

Marod, D. (2003). *Management of the Pha Taem protected forest complex to promote cooperation for trans-boundary biodiversity conservation between Thailand, Cambodia and Laos (Phase I): Forest ecology final report*. Bangkok: Faculty of Forestry, Kasetsart University.

Miller, R. L. (Ed.). (1994). *Mapping the diversity of nature*. London, UK: Chapman & Hall.

Nelson, B. W., Ferreira, C. A. C., da Silva, M. F., & Kawasaki, M. L. (1990). Endemism centers, refugia and botanical collection density in Brazilian Amazonia. *Nature, 345*, 714–716. doi:10.1038/345714a0

Neter, J., Wasserman, W., Nachtsheim, C. J., & Kutner, M. H. (1996). *Applied linear regression models* (3rd ed.). Chicago, IL: Irwin.

Oliver, I., Pik, A., Britton, D., Dangerfield, J. M., Colwell, R. K., & Beattie, A. J. (2000). Virtual biodiversity assessment systems. *Bioscience, 50*, 441–449. doi:10.1641/0006-3568(2000)050[0441:VBAS]2.0.CO;2

Parnell, J. (2000). The conservation of biodiversity: Aspects of Ireland's role in the study of tropical plant diversity with particular reference to the study of the flora of Thailand and Syzygium. In Rushton, B. (Ed.), *Biodiversity: The Irish dimension* (pp. 205–216). Dublin, Ireland: Royal Irish Academy Special Publication.

Parnell, J., Simpson, D. A., Moat, J., Kirkup, D. W., Chantaranothai, P., & Boyce, P. C., & contributors. (2003). Plant collecting spread and densities: Their potential impact on biogeographical studies in Thailand. *Journal of Biogeography*, *30*, 193–209. doi:10.1046/j.1365-2699.2003.00828.x

Patton, D. R. (1992). *Wildlife-habitat relationships in forested ecosystems*. Oregon, USA: Timber Press Inc.

Peterson, A. T. (2001). Predicting species' geographical distributions based on ecological niche modeling. *The Condor*, *103*, 599–605. doi:10.1650/0010-5422(2001)103[0599:PSGDBO]2.0.CO;2

Phillips, S. J., Andersion, R. P., & Schapire, R. E. (2006). Maximum entropy modeling of species geographic distributions. *Ecological Modelling*, *190*, 231–239. doi:10.1016/j.ecolmodel.2005.03.026

Prommakul, P. (2003). *Habitat utilization and prey of the tiger (Panthera tigris) in the Eastern Thung Yai Naresuan Wildlife Sanctuary*. Unpublished M.S. thesis (in Thai). Kasetsart University, BangkoK, Thailand.

RFD/ITTO. (2002). *Management of the Pha Taem protected forest complex to promote cooperation for trans-boundary biodiversity conservation between Thailand, Cambodia and Laos (Phase I)*. Bangkok: International Tropical Timber Organization and Royal Forest Department.

RFD/ITTO. (2004). *Management of the Emerald Triangle protected forest complex to promote cooperation for trans-boundary biodiversity conservation between Thailand, Cambodia and Laos (Phase II)*. Bangkok: International Tropical Timber Organization and Royal Forest Department.

Santisuk, T., Smitinand, T., Hoamuangkaew, W., Ashton, P., Sohmer, S. H., & Vincent, J. R. (Eds.). (1991). *Plants for our future: Botanical research and conservation needs in Thailand.* Bangkok, Thailand: Royal Forest Department Thailand.

Sayre, R., Roca, E., Sedaghatkish, G., Young, B., Keel, S., Roca, R., & Sheppard, S. (2000). *Nature in focus: Rapid ecological assessment.* Washington D.C. and California: The Nature Conservancy, Island Press.

Smitinand, T., & Larsen, K. (1966). The flora of Thailand project. Report on the work in 1966. *The Natural History Bulletin of the Siam Society*, *21*, 341.

Sobero, J., & Peterson, A. T. (2004). Biodiversity informatics: Managing and applying primary biodiversity data. *Philosophical Transactions of the Royal Society of London. Series B, Biological Sciences*, *359*, 689–698. doi:10.1098/rstb.2003.1439

Sonna, P. (2003). *The status and distribution of Eld's deer cervus eldi siamensis in Preah Vihear Province. Wildlife protection Office*. Phnom Phen, Cambodia: Forest Administration.

Stein, B. R., & Wieczorek, J. (2004). Mammals of the world: MaNIS as an example of data integration in a distributed network environment. *Biodiversity Informatics*, *1*, 14–22.

Stoms, D. M. (1992). Effects of habitat map generalization in biodiversity assessment. *Photogrammetric Engineering and Remote Sensing*, *58*(11), 1587–1591.

Tognelli, M. F., Roig-Junent, S. A., Marvald, A. E., Flores, G. E., & Lobo, J. M. (2009). An evaluation of methods for modeling distribution of Patagonian insects. *Revista Chilena de Historia Natural (Valparaiso, Chile)*, *82*, 347–360. doi:10.4067/S0716-078X2009000300003

Toxopeus, A. G., Skidmore, A. K., de Bie, C. K., Venus, V., Marquez, J. R. G., & Pablos, N. S. … Lymbarakis, P. (2007). BIOFRAG-herpetological biodiversity and habitat fragmentation in the Mediterranean Basin. *First Mediterranean Herpetological Congress (CMH1), 16-20 April 2007, Marrakesh-Morocco.*

Trisurat, Y. (2004). *GIS database and its applications in ecosystem management*. Bangkok: Western Forest Complex Ecosystem Management Project, National Park, Wildlife and Plant Conservation.

Trisurat, Y. (2009). Application of geo-informatics for transboundary biodiversity conservation of the Pha Taem Protected Forest. *International Journal of Terrestrial Observation*, *1*(2), 17–29.

Trisurat, Y., Alkemade, R., & Arets, E. (2009). Projecting forest tree distributions and daptation to climate change in northern Thailand. *Journal of Ecology and Natural Environment*, *1*(3), 55–63.

Trisurat, Y., Alkeman, R., & Verburg, P. H. (2010). Projecting land-use change and its consequences for biodiversity in Northern Thailand. *Environmental Management*, *45*, 626–639. doi:10.1007/s00267-010-9438-x

Trisurat, Y., Pattanavibool, A., Gale, A. G., & Reed, D. (2010). Improving the viability of large mammal populations using landscape indices for conservation planning. *Wildlife Research*, *36*, 401–412. doi:10.1071/WR09110

U.S. Department of Interior. (1980). *Habitat as a basis for environmental assessment. ESM 101*. Washington, DC: Fish and Wildlife Service, Division of Ecological Service.

WEFCOM (Western Forest Complex for Ecosystem Management Project). (2004). *GIS database and its application for ecosystem management. Final Technical Report*. Bangkok: Royal Forest Department.

Wildlife Conservation Division and Forestry Research Center. (1997). *Application of remote sensing and GIS for monitoring forest land use in Huai Kha Khaeng Wildlife Sanctuary: Final report*. Bangkok, Thailand: Office of Natural Resources Conservation, Royal Forest Department/Faculty of Forestry, Kasetsart University.

Wildlife Conservation Division and Forestry Research Center. (1999). *Application of remote sensing and GIS for monitoring forest land use in Khao-Ang-Runai Wildlife Sanctuary: Final report*. Bangkok, Thailand: Office of Natural Resources Conservation, Royal Forest Department/Faculty of Forestry, Kasetsart University.

Wildlife Conservation Division and Khon Kaen University. (2000). *Application of remote sensing and GIS for monitoring forest land use in Phu Luan Wildlife Sanctuary: Final report*. Bangkok, Thailand: Office of Natural Resources Conservation, Royal Forest Department/Faculty of Forestry, Kasetsart University.

Wisz, M. S., & Guisan, A. (2009). Do pseudo-absence selection strategies influence species distribution models and their predictions? An information-theoretic approach based on simulated data. *BMZ Ecology*, *9*, 8. Retrieved February 20, 2009 from http://www.biomedcentral.com/1472-6785/9/8

WRI. IUCN, & UNEP. (1992). *Global biodiversity strategy: A policy-makers' guide*. A report produced in consultation with FAO and UNESCO. Washington, DC.

Yost, A. C., Petersen, S. L., Gregg, M., & Miller, R. (2008). Predictive modeling and mapping sage grouse (Centrocercus urophasianus) nesting habitat using maximum entropy and a long-term dataset from Southern Oregon. *Ecological Informatics*, *3*, 375–386. doi:10.1016/j.ecoinf.2008.08.004

Section 4
Case Studies

Chapter 10
Modeling Land-Use and Biodiversity in Northern Thailand

Yongyut Trisurat
Kasetsart University, Thailand

Rob Alkemade
PBL Netherlands Environmental Assessment Agency, The Netherlands

Peter H. Verburg
VU University Amsterdam, The Netherlands

ABSTRACT

Rapid deforestation has occurred in northern Thailand over the last few decades, and it is expected to continue. Besides deforestation, climate change has become a global threat to biodiversity in recent years and in the future. The government has implemented conservation policies aimed at maintaining a forest cover of 50% or more and has been promoting agribusiness, forestry, and tourism development in the region. The goal of this chapter was to analyze the likely effects of various directions of development on the region. Specific objectives were to: (1) forecast land-use change and land-use patterns across the region based on trend, integrated-management, and conservation-oriented scenarios, (2) analyze the consequences of deforestation and climate change for biodiversity, and (3) identify areas most susceptible to future deforestation and high biodiversity loss. The chapter combined a dynamic land-use change model (Dyna-CLUE) with a model for biodiversity assessment (GLOBIO3). The Dyna-CLUE model was used to determine the spatial patterns of land-use change for the three scenarios, viz trend, integrated management, and conservation oriented. The methodology developed for the Global Biodiversity Assessment Model framework (GLOBIO3) was used to estimate biodiversity intactness expressed as the remaining relative mean species abundance (MSA) of the original species relative to their abundance in the primary vegetation. The results revealed that forest cover in 2050 would mainly persist in the West and upper North of the region, which is rugged and not easily accessible. In contrast, the highest deforestation was expected to occur in the lower north. MSA values decreased from 0.52

DOI: 10.4018/978-1-60960-619-0.ch010

in 2002 to 0.45, 0.46 and 0.48, respectively, for the three scenarios in 2050. The expected MSA values were lower than the predefined target of 30% at outside protected areas for all land use scenarios. The lowest value is found for the trend scenario (20.8%). The expected MSA for trend scenario is below the predefined target of 70% due to high habitat loss and severe fragmentation from road development in the future. Nevertheless, the MSA values for integrated and conservation-oriented scenarios nearly meet the representation goal. Based on the model outcomes, conservation measures were recommended to minimize the impacts of deforestation on biodiversity. The model results indicated that only establishing a fixed percentage of forest was not efficient in conserving biodiversity. Measures aimed at the conservation of locations with high biodiversity values, limited fragmentation, and careful consideration of road expansion in pristine forest areas may be more efficient to achieve biodiversity conservation.

1. INTRODUCTION

Deforestation and land-use change are critical threats to biodiversity in Southeast Asia (Fox & Vogler, 2005). The Food and Agriculture Organization of the United Nations (FAO, 2010) recently announced that the world's total forest area is just over four billion hectares or 31% of the total land area. Deforestation, mainly the conversion of tropical forests to agricultural land, has decreased over the past ten years but continues at an alarmingly high rate in many countries. Globally, around 13 million hectares of forests were converted to other uses or lost through natural causes each year between 2000 and 2010 as compared to around 16 million hectares per year during the 1990s. South America and Africa had the highest net annual loss of forests in 2000-2010, with four and 3.4 million hectares respectively. Asia, on the other hand, registered a net gain of some 2.2 million hectares annually in the last decade, mainly because of large-scale afforestation programs in China and Vietnam, which have expanded their forest area by a total of close to four million hectares annually in the last five years. However, conversion of forested lands to other uses continued at high rates in many countries.

Forest loss in Thailand was ranked the highest of all countries in the Greater Mekong sub-region and as fourth in the top ten of tropical countries in terms of annual rate of loss in 1995 (CFAN, 2005). According to Charuphat (2000), forest cover in Thailand declined from 53% of the country's area in 1961 to approximately 25% in 1998. Deforestation in Thailand is mainly caused by commercial logging of primary forest, agribusiness, and urban development, driven by ongoing population growth (Panayotou & Sungsuwan, 1989) and the national development strategy (Delang, 2002) to gain foreign income. Cropper *et al.* (1996) indicated that road development and population growth explain about 70% of the deforestation that occurred in Thailand between 1976 and 1989. During this period, about 1.2 million new agricultural households and about 17,000 km of roads were added in northern and northeast Thailand.

Deforestation has been a concern for policy makers as it has been listed as the most important environmental issue in the Kingdom of Thailand in the last ten years (ONEP, 2006). In 1989, the Thai government declared the closure of commercial logging concessions as part of its change in strategy for national development. In addition, the Royal Thai Government (RTG) has implemented two measures to avoid further deforestation and increase forest cover, namely the establishment of a protected areas network and reforestation, respectively (Trisurat, 2007). Nevertheless, the latest assessments based on new and improved methods of measuring and classifying forest cover show that the remaining forest cover slightly decreased between 2000 and

2003 from 33.1% to 32.6% of the total land area (Royal Forest Department, 2007).

The impacts of deforestation are well known. Of primary concern are impacts on biodiversity (Redford & Richter, 1999) and the ability of biological systems to support human needs (Lambin *et al.*, 2003). Not only does deforestation causes habitat loss, but it also results in habitat fragmentation, diminishing patch size and core area, and isolation of suitable habitats (MacDonald, 2003). In addition, fragmentation provides opportunities for pioneer (light-demanding) species to invade natural habitat along the forest edge (Forman, 1995; McGarigal & Marks, 1995). Pattanavibool *et al.* (2004) found that the fragmented forest in the Mae Tuen Wildlife Sanctuary in northern Thailand contained lower densities of large mammals (e.g. Asian elephant, gaur) and hornbills compared to the relatively intact Om Koi Wildlife Sanctuary. In addition, recovery of degraded ecosystems to their original state is extremely difficult and time consuming. Fukushima *et al.* (2008) investigated the recovery of tree species composition in secondary forests in northern Thailand that had been abandoned after swidden cultivation for more than 20 years. The results indicated that native species in recently abandoned poppy fields were mostly absent and that it would take more than 50 years to reach climax species composition. In addition, Obserhauser (1997) also indicated that a high number of vascular species were observed and increasing numbers of animal species became established in the older plantations.

Models of land-use change can address two separate issues: where land-use changes are likely to take place (location of change) and at what rates changes are likely to progress (quantity of change). The first issue requires the identification of the natural and cultural landscape attributes that are the spatial determinants of change. The rate or quantity of change is driven by demands for land-based commodities and these demands are often described using economic models accounting for demand-supply relations and international trade (Verburg *et al.,* 2008). Land-use change models range from simple system representations including a few driving forces to simulation systems based on a profound understanding of situation-specific interactions among a large number of factors at different spatial and temporal scales, as well as environmental policies. Reviews of different land-use models have been provided by Verburg *et al.* (2004), Matthews *et al.* (2007), and Priess & Schaldach (2008).

Besides deforestation, climate change has become a global threat to biodiversity in recent years and in the future. According to several previous studies (Jorgensen & León-Yanez, 1999; Young *et al.*, 2002; Cuesta-Camach *et al.*, 2006;), climatic, soil, vegetation types and topographic variables are the most important environmental factors that determine plant species assemblage and their patterns of distribution that reflect both recent ecological conditions and processes and phylogeographic history (Avise, 2000). In addition, evidence from the fossil record and from recent observed trends shows that changes in global climate have a profound influence on species' range expansion and contraction. The Fourth IPCC Assessment Report indicated that mean temperature in Thailand will rise 2.0-5.5 °C under the HadCM3 A2 scenario (regionally-oriented economic development) (IPCC, 2007). Therefore, the predicted future climate change is expected to have a significant impact on the distribution of species (IPCC, 2001) in northern Thailand similar to several other countries.

There are a number of modeling methods for predicting the potential impacts of climate change on biodiversity. Recent advances employing geographic information system (GIS) technology allow correlative modeling of species' distributions for large geographic areas, especially when detailed information about the natural history of species is lacking (Peralvo, 2004; Anderson *et al.*, 2002). Species-distribution models (SDMs) are based on the assumption that the relationship between a given pattern of interest (e.g. species abundance or presence/absence) and a set of factors assumed to control it can be quantified (Guisan & Zimmermann, 2000). In addition, correlative approaches are capable of extrapolating from known sample points into unknown areas (Anderson *et al.*, 2003; Raxworthy *et al.*, 2003; Anderson & Martínez-Meyer, 2004).

This research focused on the northern region of Thailand, which contains the highest percentage of remaining forest cover and protected areas compared with other regions. Due to ongoing human population growth at 1.4% per annum expansion of agriculture (Land Development Department, 2003) and infrastructure development in the region, continuing deforestation and decreasing biodiversity can be expected. This chapter assessed the potential consequences of ongoing deforestation for biodiversity. Two spatial models were combined, namely the Dyna-CLUE (Conversion of Land Use and its Effects) model (Verburg *et al.*, 2002, Overmars *et al.*, 2007) and the Global Biodiversity Model framework (GLOBIO3; Alkemade *et al.*, 2009). The research aim was to analyze the likely effects on biodiversity of various conservation policy options in the region, with the combined use of these models. Specific objectives were to: (1) forecast land-use change and land-use patterns across the region based on three scenarios, (2) analyze the consequences of land-use and climate changes for biodiversity, and (3) identify areas most susceptible to future deforestation and high biodiversity loss.

2. MATERIALS AND METHODS

2.1 Study Area

Northern Thailand is situated between the northern latitudes of 14°56' 17" and 20°27' 5" and the eastern longitudes of 97° 20' 38" and 101° 47' 31", covering 17 provinces and encompassing an area of 172,277 km^2 or one-third of the country's land area (Figure 1). The dominant topography is mountainous, with a north-south orientation. The average annual temperature ranges from 20 to 34°C; the average annual rainfall ranges between 600 and 1,000 mm in low areas to more than 1,000 mm in mountainous areas. The rainy season is from May to October. The total population has been almost stable at 11 million over the last 10 years but the population is relatively different among provinces (Department of Local Administration, 2007). The growing population is leading to additional pressure on a limited land resource for the purposes of agricultural production and food self-sufficiency (Cropper *et al.*, 1996).

Northern Thailand was originally covered by dense forest. Dominant vegetation included dry dipterocarp and mixed deciduous forests at low and moderate altitudes, while pine forest, hill evergreen forest and tropical montane cloud forest dominated areas at high altitudes (Santisuk, 1988). The Land Development Department (LDD) indicated that forest cover in this region declined from 68% to 57% during 1961 to 2002 (Land Development Department, 2003). Except in protected areas, the lowland forests have been removed due to extensive logging in the past and the expansion of agricultural land. These areas are now extensively managed for agriculture, with rice on irrigated land, and vegetables and fruit trees (e.g. longan, lychee) elsewhere. Agriculture

Figure 1. Location of study area

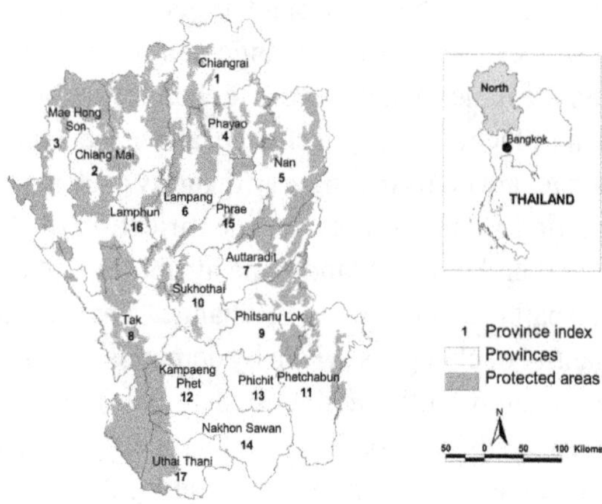

currently covers approximately 32% of the region. Secondary forest in mountainous northern Thailand has been the result mainly of swidden cultivation (Fukushima *et al.*, 2008). In addition, some swidden cultivation has been shortened in its cycle or changed to monoculture cash crops over the last 50 years (Schmidt-Vogt, 1999; Fox & Vogler, 2005). According to the Office of Agricultural Economics (OAE), approximately 50,000 ha of rubber plantations were planted in this region during 2004 to 2006 (Office of Agricultural Economics, 2007). The continuing increase in the rubber price in the last decade has stimulated a huge land demand for rubber plantations.

2.2 Land-Use Modeling and Scenario Development

The Dyna-CLUE model (Verburg *et al.*, 2002, Overmars *et al.,* 2007) was used to project land-use transitions for different scenarios during the period 2002-2050. The model has been used and validated in multiple case studies and has proven to be capable of simulating land-use dynamics in Southeast Asian mountain regions (Castella & Verburg, 2007; Pontius *et al.*, 2008) See details in Chapter 6. The model requires four inputs that together create a set of conditions and possibilities for which the model calculates the best solution by an iterative procedure. These inputs are: (1) land-use requirements (demand), (2) location characteristics, (3) spatial policies and restrictions, and (4) land-use type specific conversion settings. Land-use requirements and spatial policies are scenario specific, whereas the location characteristics and conversion settings are assumed equal for all scenarios.

Land-use requirements (demand) were calculated at the aggregate level as part of a specific scenario. Three land-demand scenarios for northern Thailand in 2050 were developed: (1) trend scenario, (2) integrated management scenario, and (3) conservation-oriented scenario. The characteristics of the three scenarios are shown in Figure 2 and descriptions of each scenario are presented below.

1) The trend scenario was based on a continuation of current trends of land use conversion of recent years (1998-2003; Office of Agricultural Economics, 2007). The annual loss of forest cover in this period was 0.6%. On the other hand, the increment rates of plantation, paddy, upland crop, tree crop, miscellaneous area and built-up area were 0.2, 0.8, 0.2, 2.5, 0.7 and 1.1%, respectively. If this trend is continued, natural forest cover in 2050 will increased to 45%, while agricultural area will increase from 33% to 44% (Figure 2).
2) The integrated scenario was derived from the long-term environmental policy (Office of

Figure 2. Land use requirements for different scenarios between 2002-2050

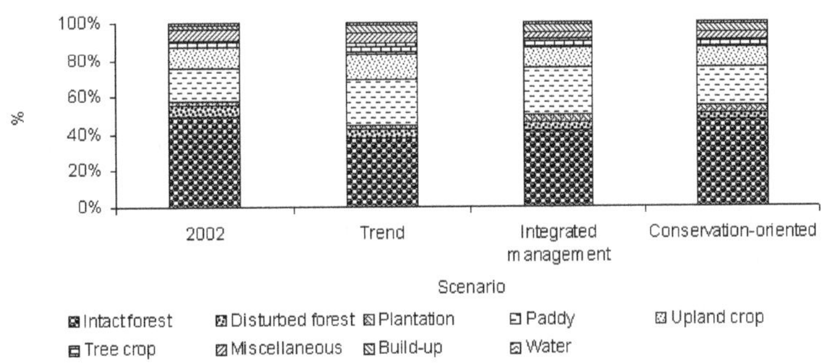

Environmental Policy and Planning, 1997) and aimed to maintain 50% forest cover at the national level. This figure is similar to the designated area of Class 1 Watershed, conservation forest and headwater sources (48%; Tangtham, 1992), and vulnerable land for agriculture due to soil erosion (leading to a total area of 51%; Land Development Department, 2001) in the region. In addition, this scenario will accommodate additional rubber plantation to cover an area of 480,000 ha (50% of the total suitable area) by 2050 (Department of Agriculture, 2003). Half of new rubber plantation will replace existing fruit tree plantations, which are facing low market price and the remaining areas will be situated in marginal upland cropland, idle land and partially in disturbed forest outside protected areas. The total agricultural area is expected to cover 30% of the region in 2050. Urban area will increase according to population growth but water body will remain stable. The increment of plantation forest is 4,800 ha/year, which is somewhat higher than previous years due to public concern on climate change and watershed degradation.

3) The conservation-oriented land use scenario aimed to maintain 55% of the region as forest cover and rehabilitate the degraded head watershed. Meanwhile, reforestation is preferable in the degraded Class 1 Watershed, which covers approximately 7,600 km^2 or 4.4% of the region. Additional rubber plantation will cover an area of 300,000 ha. The increment rate of urban area and the extent of water body will be the same as the integrated management scenario.

Location characteristics: The Dyna-CLUE model quantifies the location preferences of the different land uses based on logistic regression models. The coefficients (β) are estimated through logistic regression using the occurrence of the land uses in 2002 as the dependent variable. The goodness-of-fit of a logistic regression model is evaluated using the Receiver Operating Characteristic (ROC) (Swets, 1986). The value of the area under curve ranges between 0.5 (completely random) and 1.0 (perfect fit).

The original land-use classes derived from the 1:50,000 land-use map for 2002 (Land Development Department, 2003) were aggregated into nine classes to facilitate land-use simulations: (1) intact forest, (2) disturbed forest, (3) forest plantation, (4) paddy, (5) upland crop, (6) tree crop, (7) miscellaneous area (e.g., old clearing, wetland, rock outcrop), (8) built-up area, and (9) water body. In the current study, models were developed for seven land-use classes, as the 'intact forest' and 'water body' classes were assumed to remain unchanged during the simulation period.

The physical factors included topographic variables, annual precipitation, distance to available water, and soil texture. Some topographic factors (altitude, slope, and aspect) represented limiting factors for agriculture. Altitude, aspect, slope, distance to main roads, and distance to streams and rivers were extracted and/or interpolated from 1:50 000 topographic maps (Royal Thai Survey Department, 2002). A surface representing the spatial variation in annual precipitation was interpolated from rainfall data recorded at meteorological stations across northern Thailand, using universal kriging technique (Theobald, 2005). Soil types were derived from the 1:100,000 soil map (Land Development Department, 2001).

The socio-economic factors influencing deforestation included: distance to village, distance to city, distance to main road, and population density. Distance to village and population density were proxy indicators for local consumption, while the distance to road and to city parameters were important proxies for the costs of transporting agricultural commodities to market. Current population data were obtained from the Local Administration Department. Protected area coverage was digitized from the National Gazette map.

Figure 3. Land-use transition matrix for northern Thailand. Notes: 1- conversion is possible; 0 – conversion is not possible; 120, 130, 110 – conversion is possible after 20, 30 and 10 years, respectively from current situation

Future

Current

Current\Future	Intact forest	Dist. Forest	Plantation	Paddy	Upland	Tree crop	Misc.	Build-up	Water
Intact forest	1	1	0	1	1	1	1	1	0
Dist. Forest	120	1	1	1	1	1	1	1	0
Plantation	130	110	1	1	1	1	1	1	0
Paddy	0	0	1	1	1	1	1	1	0
Upland	0	0	1	1	1	1	1	1	0
Tree crop	0	0	0	0	0	1	1	1	0
Misc.	0	0	1	1	1	1	1	1	0
Build-up	0	0	0	0	0	0	0	1	0
Water	0	0	0	0	0	0	0	0	1

All the data were prepared at 500x500 m spatial resolution and spatial analyses were carried out using ArcGIS software.

Spatial policies and restrictions indicate areas where land-use changes are restricted through strict protection measures, such as protected areas. In the trend scenario, no spatial policies were implemented, thus forest encroachment could occur in protected areas if the location characteristics were favorable. In contrast, land-use policies were imposed for the integrated management and conservation-oriented scenarios. Under these latter scenarios, existing national parks and wildlife sanctuaries, which cover approximately 53,200 km^2 or 30% of the region (Royal Forest Department, 2007), were designated as *restricted areas*, so that no further encroachment was allowed in these areas and natural succession was possible.

Land-use type specific conversion settings influence the temporal dynamics of the simulations. Two sets of parameters are essential to characterize each land-use type: conversion elasticities and land-use transition settings. The elasticities were estimated based on capital investment, time, and energy costs and expert judgment, ranging from 0 (easy conversion) to 1 (irreversible change) (Verburg *et al.,* 2002). High values for this parameter were assigned to the primary forest, paddy, built-up area and water body classes, because these land-use types are not likely to be displaced because they involve a high commitment to investment or a large amount of time in the case of establishing primary forest. Medium values were given to disturbed forest and fruit trees. On the other hand, upland crop and miscellaneous areas are highly dynamic land uses, thus low values were assigned to them. In the land-use transition settings a minimum of 30 years was specified as a requirement for the natural succession of reforestation to primary forest and 20 years was specified for succession from disturbed forest back to primary forest, based on the work of Obserhauser (1997). Detailed land-use transition matrix is shown in Figure 3.

The Dyna-CLUE model uses all inputs to calculate the total probability for each grid cell of each land-use type based on the local suitability of the location derived from the logit model, the conversion elasticity, and the competitive strength of the land-use type (Verburg & Overmars, 2009). Where no constraints to a specific conversion were specified, the location was allocated to the land use with the highest total suitability. Using an iterative process, the competitive strength of the different land-use types was adapted until the total allocated area of each land use equaled the total land requirements specified in the scenario.

The resulting land-use patterns were analyzed with the help of the FRAGSTATS 3.0 software (McGarigal & Marks, 1995) to assess landscape structure change and fragmentation due to land-use changes. The following indices were calculated to imply forest fragmentation that may have a direct impact on biodiversity.

Area indices include (1) total area (TA): the total area that the land use class occupies in the study area. (2) number of patches (NP): the number of patches of a particular land use class. (3) mean patch size (MPZ): average area of a patch of a particular class. (4) largest patch index (LPI): the percentage of landscape area occupied by the largest patch of a particular land use class.

In addition, two *core area indices* were also analyzed. These indices include: (1) mean core area (MCA): the area within a fragment located beyond a specified edge distance (1 km for this study); (2) total core area (TCA): the sum of the total surface of all areas of a particular land use class (in hectares). A higher index of core areas indicates less fragmentation.

2.3 Calculation of Remaining Mean Species Abundance

The Global Biodiversity Model framework (GLOBIO3) was used to assess the consequences of different land-use scenarios for biodiversity (Alkemade *et al.,* 2009). The model was built on the simple cause-effect relationships between human-induced drivers and biodiversity impacts in the past, present and future, and it can be used at various scales. The relationships were derived from an extensive literature review and meta analyses and are described in Alkemade *et al.* (2009) and Chapter 8. The driving factors in the original GLOBIO3 model included: land use and land-use intensity (LU), infrastructure development (I), fragmentation (F), atmospheric nitrogen deposition () and climate change CC). This study did not include the last two drivers in the GLOBIO3 model because they were not applicable for local assessment.

The Mean Species Abundance (MSA) values range between 1.0 in an undisturbed or primary ecosystem and 0.0 in a completely destroyed ecosystem. The MSA values can be categorized into five classes: low (0.0-0.2); relatively low (0.2-0.4); medium (0.4-0.6); relatively high (0.6-0.8); and high (0.8-1.0). The original MSA_{LU} values as estimated by Alkemade *et al.* (2009) were adapted with slight modification to suit the local situation, using input obtained from national biodiversity experts and a literature review (Santisuk, 1988). The values used for intact forest, disturbed forest, forest plantation, secondary forest (miscellaneous), intensive agriculture (paddy), extensive agriculture (upland crop), perennial trees, and built-up area were estimated to be 1.0, 0.7, 0.4, 0.45, 0.1, 0.3, 0.2, and 0.05, respectively.

To estimate the effect of infrastructure, impact zones were calculated along road networks. Roads were buffered by different widths. The width of an impact zone depended on the land-use type, because the direct and indirect effects of roads on the neighboring plants and wildlife differ among ecosystems (United Nations Environmental Programme, 2001).

2.4 Evaluation of Effectiveness of Protected Area Network

Thailand's protected area system includes national parks (NP), wildlife sanctuaries (WS), forest parks (FP), non-hunting areas (NHA), conservation mangrove forests and Class 1 Watersheds, as well as botanical gardens and arboreta. In this study, only national parks and wildlife sanctuaries were considered as an effective protected area system for biodiversity conservation. The current protected area coverage derived from the National Gazette maps written in either A4 or A3 were readjusted to match the natural boundary based on the geo-referenced topographic map on 1:50,000 scales,

Table 1. Significant location factors related to each land use location

Variables	Disturbed forest	Plantation	Paddy	Upland crop	Tree crop	Misc.	Built-up
Altitude	+ [1/]	- [2/]	-	-	*n.s.* [3/]	+	+
Slope	*n.s.*	-	-	-	-	-	-
Soil texture [4/]							
• *Loam*	-	-	+	+	+	*n.s.*	*n.s.*
• *Sand*	*n.s.*	-	-	*n.s.*	+	*n.s.*	*n.s.*
• *Laterite*	+	-	*n.s.*	*n.s.*	+	+	-
• *Slope complex*	*n.s.*	*n.s.*	-	-	*n.s.*	+	-
• *Clay*	-	*n.s.*	+	*n.s.*	+		*n.s.*
• *Wetland*	*n.s.*	-	+	-	*n.s.*	+	*n.s.*
Distance to stream	+	-	-	+	-	-	-
Distance to village	*n.s.*	*n.s.*	-	-	-	-	*n.s.*
Distance to main road	-	-	*n.s.*	-	-	*n.s.*	-
Distance to city	-	-	-	+	-	+	-
Population density	-	-	*n.s.*	-	*n.s.*	-	+
ROC	0.68	0.71	0.93	0.83	0.80	0.66	0.88

Remarks: 1/ positive correlated; 2/ negative correlated; 3/ *n.s.*: not significant at 0.05 level; 4/ category variable

which were subsequently digitized. After completion of digitizing, the digital protected area maps were appended.

The purpose of this analysis was to evaluate the effectiveness of the protected area network designated to conserve biodiversity elements in the present and future to respond to the impacts of predicted land use and climate change. We modified the IUCN Red List B1 criterion (IUCN, 2004) and used it as representation goals for MSA. The representation goal was set at 30% in regional level, and 70% in protected area level. These outputs were evaluated by overlaying the protected area map with the MSA map. At regional level, we divided the accumulated MSA values with the accumulated MSA values for the entire northern Thailand. If the result is greater than 0.3 or 30%, it would imply protected areas are effective to conserve biodiversity in northern Thailand. In addition, if the average MSA value inside protected areas is greater than 0.7 or 70%, it would imply that protected areas are effective to conserve biodiversity in a protected area network. On the other hand, if the predicted MSA value inside protected areas is lower than the predefined target of 70% it means the protected areas are not effective. The representation goals were relatively high because the extent of occurrence was predicted from spatial distribution models and not direct observation.

3. RESULTS

3.1 Land-Use/Land-Cover Changes

The significant factors and coefficients derived from the logistic regression analysis that determined the location suitability of the seven land-use types are shown in Table 1. From this table, it can be seen that not all location factors were included in the regression models and each factor contributed differently depending on the land-use type. High altitude, lateritic soil and distance to available water were positively correlated with disturbed forest. These areas contained many limiting factors for agriculture, so further reclamation to agriculture was limited. In contrast, areas close to the stream network, situated in densely populated forested areas, accessible from main roads, at low altitude and on fertile soil (clay and loam) were identified as at risk of encroachment because they are prime targets for agriculture. In addition, rugged terrain, poor soil, and remoteness were limiting factors to future encroachment. The spatial distributions of the seven land-use types were explained moderately to well by the selected location factors, as indicated by the ROC values that measure the goodness-of-fit of the logistic regression models (Table 1). High ROC values were found for paddy (0.93) and built-up area (0.88). Relatively high-to-moderate values were found for upland crops (0.83), tree crops (0.80) and forest plantation (0.71). Low values were observed for disturbed forest (0.68) and miscellaneous land use (0.66). These differences in goodness-of-fit occurred because paddy requires specific land characteristics (e.g. poor drainage, clay texture) similar to a built-up area that forms in high population areas, close to roads and cities and at low altitude. Sand and lateritic soils are not systematic choices for upland crops. Disturbed forest and miscellaneous land use (including abandoned swidden cultivation) can be found in all altitude zones, often on soils vulnerable to soil erosion (Land Development Department, 2001; Thanapakpawin *et al.*, 2006). In addition, because these classes represented a wide range of different activities, lower ROC values were found.

Land-use/land-cover maps for 2002 and simulation results for 2050 for the three scenarios are shown in Figure 3. The results of the trend scenario without spatial policies and restrictions show that the highest rate of deforestation and a low percentage of forest cover were found in the four lower-north provinces of Phitsanulok, Sukhothai, Khamphaeng Phet and Nakhon Sawan (Figure 4). The dominant topography in these provinces is flat-to-gently-sloping terrain with alluvial deposits, which are highly suitable for agriculture and development. Forest areas have been converted to agriculture for several decades. In contrast, trends of deforestation in Chiang Mai, Phayao, Lampang, Lampun and Phetchabun provinces were expected to reverse. The annual deforestation rate would be less than 1% (previously 1.1-1.9%). The highest percentage of remaining forest cover in 2050 were found in the western provinces, such as Mae Hong Son (84%), followed by Chiang Mai (70%). Mae Hong Son province also showed the lowest deforestation rate (0.1% per annum) followed by Nan (0.2% per annum). These provinces have mostly shallow, erosive soils with low fertility and are located in sloping terrain, with very difficult access, so opportunities for agriculture are very restricted by natural barriers. Consequently, the population in Mae Hong Son was less than 300,000 and it had the lowest density (20 people/km^2) in Thailand. Although, the population in Nakhon Sawan was 1.1 million, the population density was 126 people/km^2 or 6-7 times that of Mae Hong Son province.

The integrated-management scenario, with protected areas and a slower rate of agricultural expansion, showed different land-use patterns. This scenario assumed less demand for agriculture and rubber plantations, leading to higher remaining forest cover and tree crops. Forest cover was expected to increase from the west to the upper north and along the eastern national border. Mae Hong Son and Chiang Mai would have more than 75% forest cover in 2050 (Figure 4), while Nan would have approximately 65%, which was similar to the conditions in 2002. Under this scenario, substantial increases in forest cover were also predicted for Chiang Rai, Nan, Tak and Phitsanulok provinces. For instance, Nan would gain approximately 2% forest cover in the next 48 years, due to the restriction of further encroachment into the reserves and the regrowth of natural vegetation in about 5,700 ha of abandoned agricultural areas.

Figure 4. Current and predicted land uses in 2050 under three different scenarios. Indices: 1= Chiang Rai; 2= Chiang Mai; 3= Mae Hong Son; 4= Phayao ; 5= Nan; 6= Lampang; 7= Uttaradit; 8= Tak; 9= Phitsanoluk; 10= Sukhothai; 11 =Phetchabun; 12= Kampaeng Phet; 13= Pichit; 14= Nakhon Sawan; 15= Phrae; 16= Lampun; 17= Uthai Thani

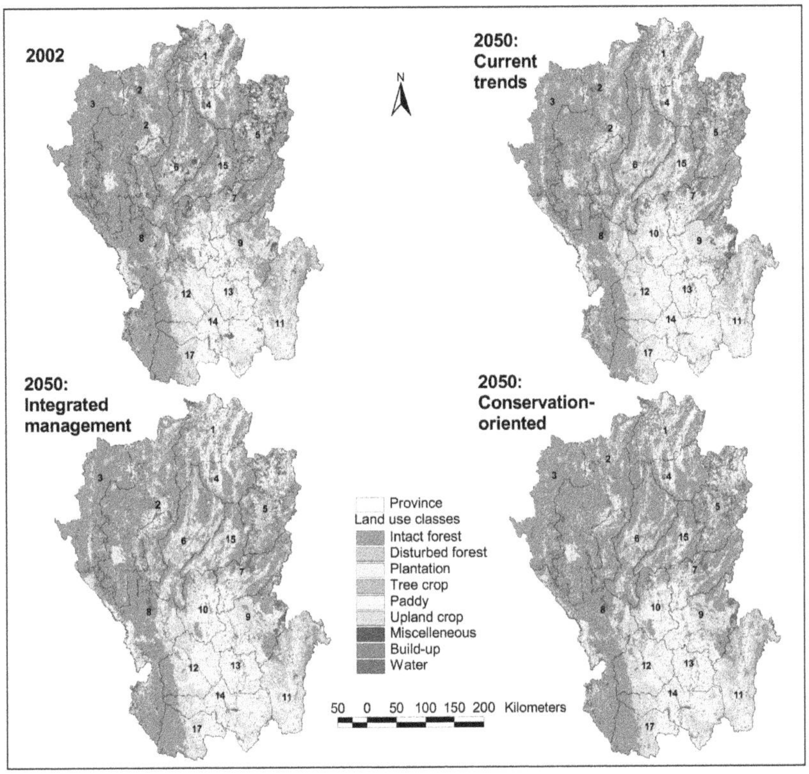

The results of the conservation-oriented scenario showed that the extent and distribution pattern of the remaining forest in 2050 were relatively similar to the conditions in 2002. Similar to other scenarios, high deforestation was found in Pichit, Sukhothai, Phetchabun, Kamphaeng Phet and Nakhon Sawan provinces. Mae Hong Son, Chiang Mai, Lampang, Tak and Phrae provinces still encompassed more than 70% forest cover (Figure 4). In addition, Nan province would gain approximately 5.6% forest cover, mainly converted from secondary forest and abandoned swidden cultivation at high elevation through reforestation.

The results of the Dyna-CLUE model revealed that the number of remaining forest patches increased in all scenarios over the 48 years of simulation. The number of patches increased from 1,250 in 2002 to values in 2050 of 1,783 for the trend scenario, 1,515 for the integrated-management scenario, and 1,321 patches for the conservation-oriented scenario. This index corresponds to mean patch size index, which revealed that the mean patch size of forests decreased from 7,930 ha in 2002 to 4,373 ha, 5,681 ha, and 7,214 ha for the trend, integrated-management, and conservation-oriented scenarios, respectively. In addition, the largest patch of forest cover substantially declined from 54% of the remaining forest cover in 2002 to 39% for the trend scenario, 44% for the integrated scenario and 50% for the conservation-oriented scenario due to fragmentation. Small, fragmented forest patches surrounded by agricultural land uses can be considered as disturbed forest or sink

Table 2. Landscape indices of remaining forest area and relative contribution of the different pressure indicators to reduced MSA in northern Thailand during 2002-2050

Landscape indices	2002	Trend scenario	Integrated scenario	Conservation-oriented scenario
Remaining intact forest (%)	49.2	37.2	41.6	47.0
Number of patches	1,254	1,783	1,515	1,312
Average mean patch size (ha)	7,930	4,373	5,681	7,214
Largest patch index	53.7	38.9	43.7	49.7
Total core area (km^2)	60,069	4,241	52,926	60,213
Mean core area (km^2)	48	24	35	45
Remaining MSA	0.52	0.45	0.46	0.48
Reduction by (%)				
Land use change	33.9	42.2	40.0	35.1
• Agriculture	26.2	34.7	33.0	28.2
• Forestry	3.9	3.6	4.1	4.0
• Urban	0.2	0.4	0.4	0.4
• Others	3.6	3.5	2.5	2.5
Infrastructure development	10.7	10.2	11.5	13.5
Forest fragmentation	3.2	3.0	3.0	3.3
Total	100.0	100.00	100.0	100.0

habitat (Forman, 1995), since the whole patch corresponds to a border area.

3.2 Remaining MSA

The remaining MSA for northern Thailand in 2002 was approximately 0.52, which was a decrease of 0.48 since human intervention first occurred (Table 2). The decrease was mainly caused by land-use change, especially agriculture (34%). The projected MSA value for the trend, integrated-management and conservation-oriented scenarios for 2050 was estimated at 0.45, 0.46 and 0.48, respectively (Table 2). The expansion of agriculture contributed to a future MSA loss of between 28 and 35%. The second highest impact factor was infrastructure development (road expansion), followed by forest management and fragmentation. The large effect of infrastructure development was a consequence of the assumption for all scenarios that the current unpaved roads would be developed to hard-surface standard within the next 48 years and would facilitate human access to remaining intact forest areas, thus leading to a reduction in MSA.

The highest MSA values were associated with high altitude and inaccessible areas in the west (Figure 5). Such areas received *de facto* natural protection except where communities had settled and were practicing agriculture. In addition, patchy areas of high MSA were scattered in remnant protected areas across the landscape. Thus, the persistence of forest cover and MSA in the future would be very likely to occur only in protected areas and on high slopes as indicated by Fox & Vogler (2005). Figure 5 also shows that the remaining MSA in the northwest (Chiang Mai province) would decline due to infrastructure development, particularly road construction, and fragmentation.

Figure 5. Remaining species abundance for northern Thailand under different land use scenarios in 2050 scenario 1) current trends, 2) integrated management, 3) conservation oriented

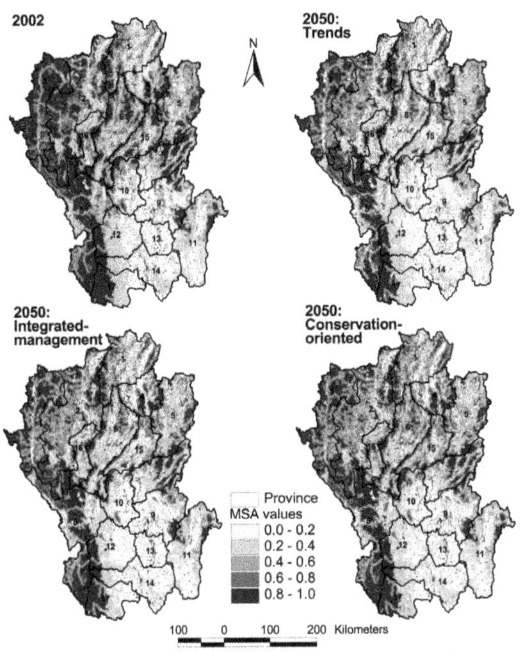

3.3 Effectiveness of Protected Areas

In this study we use the MSA as a surrogate of biodiversity at a coarse scale. In addition, the predefined representation goal is set at 30% in regional level and 70% in protected area level. The results of assessment show that the MSA values at regional level are lower than the expectation at present and in the future for all land use change scenarios. The lowest value is found for the trend scenario (MSA = 0.21) while the expected MSA for integrated and conservation-oriented scenarios are almost equal and marginally higher than the trend scenario (Table 3). However, the MSA is high within the protected area system, especially in 2002 which is approximately 0.75. The expected MSA for trend scenario is below the predefined target of 0.70 due to high habitat loss and severe fragmentation. Although, there will be no encroachment inside the protected area system in the future under the integrated and conservation-oriented scenarios, approximately 0.05% will be diminished as the result of infrastructure development and fragmentation by roads (Table 3). We assume that the unpaved roads will be developed in the future. Nevertheless, the MSA values for integrated and conservation-oriented scenarios nearly meet the representation goal.

4. DISCUSSION

Earlier studies implemented various land-use models in Thailand addressing various questions and had either focused on a specific case study or addressed one specific land-use sector (Watcharakitti, 1976; Rajan & Shibasaki, 1997; Trisurat, 1999; Barnaud *et al.*, 2006);. The modeling approach presented in this paper was chosen for the current study based on the available data, the

Table 3. Remaining MSA in northern Thailand and inside protected areas during 2002-2050

Landscape indices	2002	Trend	Integrated land use	Conservation-oriented
Total forested area (%)	57.0	45.0	50.0	55.0
Remaining MSA				
• Northern region	0.52	0.45	0.46	0.48
• Protected areas compared to the region	0.24	0.21	0.22	0.22
• Protected areas compared to their size	0.75	0.66	0.69	0.70

large spatial extent of the region and the complexity of the processes. In addition, a pixel size of 25 ha was used, which was considered to be an appropriate resolution for regional scale assessments and realistic in terms of the spatial scale of the different data used and the computational requirements of the modeling.

Alternative models, such as the Markov chain model, use previous land-use trends to envisage what will happen in the future, without considering the role of changes in the controlling natural, political and sociological factors, unlike the Dyna-CLUE model that explicitly addresses the dynamics of the different competing land uses. A dynamic approach that can account for competition between land uses is needed, where there are changing preferences for different land uses that have different environmental and geographic requirements. Therefore, Markov chain models would not be capable of addressing the different scenarios presented in this paper. Multi-agent models have high data requirements and so they are not often applied at a large regional scale, as was the case in the current study (Matthews *et al.*, 2007). In addition, logistic regression approaches in deforestation studies often focus on the identification of forest versus non-forest land use only, which is insufficient given the diverging characteristics of non-forest land uses.

The estimated MSA values were different from the biodiversity assessments based on the species-area relationship (SAR) concept, which estimates that 80-90% of the original species will remain if 30-40% of the area of any given terrestrial community or ecosystem can be conserved (Dobson, 1996). This is due to the SAR approach ignoring the variation of habitat quality and fragmentation effects and not including the species abundance (Gotelli, 2001). In addition, the SAR approach may underestimate the potential losses of MSA, especially when the remaining habitat is highly fragmented. For instance, Lomolimol (1982) found that not only the patch size, but also the distance between patches was a significant factor for terrestrial mammals. In addition, Delin & Andrén (2004) revealed that neither fragment size, nor the degree of isolation was significant for the distribution of Eurasian red squirrels (*Sciurus vulgaris*). The only factor that significantly influenced a density index was the proportion of spruce within a habitat fragment. Similar results were observed for terrestrial birds on British islands, where there was no significant relationship between the average number of visiting birds and island area or island distance, but habitat quality was significant (Stracey & Pimm 2009). The GLOBIO3 model incorporated habitat fragmentation and other human pressure factors making it an appropriate method for scenario analysis. Thus, it was an appropriate tool to assess biodiversity integrity in human-induced landscape and heterogeneous habitats.

The predicted forest cover in 2050 showed a similar pattern of forest distribution. High percentages of remaining forest cover in 2050 were predicted in the west and in the upper north provinces, since these regions have more protected areas compared to the lower north provinces (Figure 3). In addition, all scenarios indicated that agriculture was the largest contributor to biodiversity loss. The dominance of agriculture is not surprising, since agriculture covers vast areas in northern Thailand and will continue to increase substantially in the future, due to high demand for rubber and food. The result was in line with several reports (Office of Environmental Policy and Planning, 1997; Panayotou & Sungsuwan, 1989) indicating that expansion of agriculture had reduced significantly the amount of forest cover and biological resources over the last four decades.

Even though the predicted forest cover in 2050 was quite different among the three scenarios (45-55%), the overall MSA values showed less distinct differences. There are two reasons to explain this. First, the extent of forest cover was composed of intact forest, disturbed forest, and forest plantation. The estimated reforestation area under either the integrated-management scenario or conservation-oriented scenario was approximately 684,000 ha or 4.0% of the region, compared with 499,000 ha or 2.9% under the trend scenario. Based on calculations by Thai national experts, the MSA_{LU} value for plantation is 0.4 relative to undisturbed ecosystems, because forestry officials usually plant a single species, mainly indigenous *Pinus merkusi* or *P. kesiya* or the exotic *Eucalyptus camaldulensis*. Therefore, the increment of MSA is not proportional to the increase in forest cover given the change in forest types. Second, the GLOBIO3 model calculated MSA_F and MSA_I values only for natural areas, resulting in a lower contribution to forest fragmentation and infrastructure development to MSA in the trend scenario than in the integrated-management and the conservation-oriented scenarios, because of the smaller forest area and higher forest encroachment along road networks (Table 2). Road construction in densely forested areas in the northwest of the study area would facilitate human access to intact forest for agriculture, hunting and extraction of forest products, and, in the end, would lead to a reduction in biodiversity. Cropper *et al.* (1996) indicated that road development had caused significant loss of forest cover in the north of Thailand from 1976 to 1989. In addition, roads created forest edges that allowed increased light and wind penetration into core areas, forcing some species to move deeper into the forest, and in addition, roads formed physical barriers restricting wildlife movement (Allen *et al.,* 2001). The research findings could motivate policy makers to increase the existing 50% forest cover and strengthen land allocation policies (Office of Environmental Policy and Planning, 1997) by paying attention not only to the quantity of forest cover, but also to the configuration of the remaining forest and carefully monitoring tourism infrastructure development in pristine ecosystems. In addition, this research also identified hotspots of threats to biodiversity, where MSA was expected to decrease by 0.5 or more during 2002 to 2050. This value was chosen because it is similar to the quantitative criterion, E, of the IUCN Red List classification (IUCN, 2002). This criterion was derived from quantitative analysis identifying critically endangered species with a probability of extinction in the wild of least 50% over 10 years or three generations, as authorized agencies are able to allocate only limited resources to cope with deforestation and its consequences on biodiversity.

Despite the fact that the GLOBIO3 model provided a useful approximation of MSA and the relative contribution of pressure indicators on MSA, the existing model could be improved for more effective implementation at regional and local levels. Improvements could include the addition of other driving factors and validation. Firstly, only three pressure factors out of the five available in GLOBIO3 were used. The N-deposition and climate change are actually implemented in the model for global-level analysis, therefore these two factors should be calibrated before using at regional and local levels. Additional potential drivers, including poverty and forest fire, should be investigated in future research. In Thailand, at least five million forest dwellers live in reserve forests and depend on biological resources (Fox & Vogler, 2005; Royal Forest Department, 2007); most are living below the poverty line.

Secondly, validation of the accuracy of a prediction model is always important, in order to convince stakeholders and decision makers to accept the results. In this study, it was not possible to validate the predicted land-use map because land-use data beyond 2002 were not available. An absence of appropriate data for validation is a common problem in land-use modeling; only a few models and model applications are properly validated (Pontius *et al.*, 2008). The LDD produces a new land-use map when budget funding is available, which would provide validation of this application of the model. For GLOBIO3, one promising method for validation in the future would be to use actual species occurrences and species indices (e.g. species richness or species diversity) derived from long-term forest monitoring plots or species distribution models (Chapter 9) to test model accuracy.

Nevertheless, the combination of the Dyna-CLUE and GLOBIO3 models was very useful, not only to simulate land-use allocation, but also to visualize forest patterns in the landscape. In addition, the models identified 'hot zones' of deforestation and important areas for biodiversity conservation.

5. CONCLUSION

According to existing trends, forest cover loss in northern Thailand will continue unless strict protection measures are undertaken. The results of this study were:

1. The trend scenario was developed based on a continuation of the trends of land-use conversion of recent years. The existing forest cover of 57% of the region in 2002 was expected to decrease to 45% by 2050. However, forest loss was likely to vary strongly across the region. The remaining forest cover would be found mainly in the upper north and in the west where altitude is high and accessibility is low. The lowest loss and highest percentage of forest cover would be found in the northwest. In contrast, forest cover in the lower north provinces would be less than 20% by 2050. The estimated MSA value would decline from approximately 0.52 in 2002 to 0.45 in 2050. High threats to MSA would occur in areas covering approximately 4,910 km^2, mainly (located widespread) in the center of the region. Intensive and regular patrolling to minimize deforestation in protected areas is highly recommended because a lot of deforestation is predicted within the reserves.
2. The integrated-management scenario was directed by policies that aimed to maintain 50% forest cover. Under this scenario, much forested land in rugged terrain and protected areas remained intact due to the land not being suitable for agriculture and a restriction policy being undertaken in the existing protected area network, despite there being a high demand for rubber plantations. The estimated MSA value derived from the simulated land-use map in 2050 was 0.46. High threat areas covered approximately 2,719 km^2. Three conservation measures are recommended based on the results of this scenario: minimize future deforestation in protected areas and threats to MSA areas in the buffer zones, raise conservation awareness among local people and maintain ecosystem connectivity of fragmented protected areas.
3. The conservation-oriented scenario aimed at maintaining 55% of the region under forest cover. The results of model simulation showed that the extent and pattern of remaining forest cover in 2050 were relatively similar to the forest area in 2002, except in the lower north, which would have less forest cover. Nan province would gain substantial forest cover from secondary forest regeneration and abandoned swidden cultivation at high elevations. The estimated MSA value for this scenario was 0.48 of the original state. Threats to MSA covered an area of 2,225 km^2, with only 556 km^2 predicted inside protected areas. Besides protection and conservation awareness raising, it is recommended that forestry officials use many native species to rehabilitate degraded ecosystems inside protected areas, which will most likely improve MSA values.

The results of this research indicated that the '50% - 55% forest cover' policies for 2050 might not be the most efficient way to promote biodiversity conservation, and these ambitious targets might create significant confrontations over land demands between conservationists and landless farmers. The results suggest that it is more effective for authorized agencies to protect biodiversity by maintaining forest cover with high biodiversity (corresponding with high MSA) values. There is also a need to recognize that infrastructure development in dense forest may have a negative impact on biodiversity and the agencies may need to allocate resources to prevent future deforestation in risk areas.

ACKNOWLEDGMENT

The authors would like to thank the Kasetsart University Research and Development Institute (KURDI) and the Netherlands Environmental Assessment Agency (PBL) for financial support for this research project. In addition, gratitude is expressed to the Royal Forest Department, Department of National Park, Wildlife and Plant Conservation and Land Development Department for providing information. Nipon Tangtham and two anonymous reviewers provided valuable suggestions and comments during preparation of this chapter.

REFERENCES

Alkemade, R., van Oorschot, M., Nellemann, C., Miles, L., Bakkenes, M., & ten Brink, B. (2009). GLOBIO3: A framework to investigate options for reducing global terrestrial biodiversity loss. *Ecosystems, 12*(3), 349–359. doi:10.1007/s10021-009-9229-5

Allen, C. R., Pearlstine, L. G., & Kitchens, W. M. (2001). Modeling viable populations in gap analysis. *Biological Conservation, 99*(2), 135–144. doi:10.1016/S0006-3207(00)00084-7

Barnaud, C., Promburom, P., Bousquet, F., & Trébuil, G. (2006). Companion modelling to facilitate collective land management by Akha villagers in upper northern Thailand. *Journal of the World Association Soil & Water Conservation, 1*(4), 38–54.

Castella, J. C., & Verburg, P. H. (2007). Combination of process-oriented and pattern-oriented models of land use change in a mountain area of Vietnam. *Ecological Modelling, 202*(3-4), 410–420. doi:10.1016/j.ecolmodel.2006.11.011

CFAN (Forestry Advisers Network). (2005). *Deforestation: Tropical forests in decline*. CIDA Forestry Advisers Network. Retrieved February 2, 2008, from http://www.rcfa-cfan.org/english/issues.12-3.html

Charuphat, T. (2000). Remote sensing and GIS for tropical forest management. In *Proceedings of the Ninth Regional Seminar on Earth Observation for Tropical Ecosystem Management*, Khao Yai, Thailand, 20-24 November 2000. (pp. 42-49). The National Space Development Agency of Japan, Remote Sensing Technology Center of Japan, Royal Forest Department, and GIS Application Center/AIT, Khao Yai National Park Thailand.

Cropper, M., Griffiths, C., & Mani, M. (1996). *Roads, population pressures and deforestation in Thailand, 1976-1989*. New York, NY: The World Bank.

Delang, C. O. (2002). Deforestation in northern Thailand: The result of Hmong farming practices or Thai development strategies? *Society & Natural Resources, 15*, 483–501. doi:10.1080/08941920290069137

Delin, A. E., & Andren, H. (2004). Effects of habitat fragmentation on Eurasian red squired (*Sciurus vulgaris*) in a forest landscape. *Landscape Ecology, 14*, 62–72.

Department of Local Administration. (2007). *Population census in Thailand from 1994 to 2007*. Retrieved March 2, 2008, from http://www.dopa.go.th/ hpstat9/people2.htm

Dobson, A. P. (1996). *Conservation and biodiversity*. New York, NY: Scientific American Library.

mFood and Agriculture Organization of the United Nations. (2005). *Global forest assessment.* Retrieved March 10, 2008, from http://www.fao.org/forestry/site/fra/24690/en

Food and Agriculture Organization of the United Nations. (2010). *Global forest assessment 2010.* Rome: Food and Agriculture Organization of the United Nations.

Forman, R. T. T. (1995). *Land mosaics: The ecology of landscapes and regions.* Cambridge, UK: Cambridge University Press.

Fox, J., & Vogler, J. B. (2005). Land-use and land-cover change in montane mainland Southeast Asia. *Environmental Management, 36*(3), 394–403. doi:10.1007/s00267-003-0288-7

Fukushima, M., Kanzaki, M., Hara, M., Ohkubo, T., Preechapanya, P., & Chocharoen, C. (2008). Secondary forest succession after the cessation of swidden cultivation in the montane forest area in northern Thailand. *Forest Ecology and Management, 255*(5-6), 1994–2006. doi:10.1016/j.foreco.2007.12.022

Gotelli, N. J. (2001). *A primer of ecology.* Sunderland, MA: Sinauer Associates.

Lambin, E. F., Geist, J., & Lepers, E. (2003). Dynamics of land-use and land-cover change in tropical regions. *Annual Review of Environment and Resources, 28*, 205–241. doi:10.1146/annurev.energy.28.050302.105459

Land Development Department. (2001). *Soil erosion map.* Land Development Department, Ministry of Agriculture and Co-operatives, Bangkok, Thailand.

Land Development Department. (2003). *Annual statistics report year 2003.* Land Development Department, Ministry of Agriculture and Co-operatives, Bangkok, Thailand.

Lomolinol, M. V. (1982). Species-area and species-distance relationships of terrestrial mammals in the Thousand Island Region. *Journal Oecologia, 54*, 1432–1939.

MacDonald, G. (2003). *Biogeography: Introduction to space, time and life.* New York, NY: John Wiley & Sons, Inc.

Matthews, R., Gilbert, N., Roach, A., Polhill, J. G., & Gotts, N. M. (2007). Agent-based land-use models: A review of applications. *Landscape Ecology, 22*(10), 1447–1459. doi:10.1007/s10980-007-9135-1

McGarigal, K., & Marks, B. (1995). *FRAGSTATS: Spatial pattern analysis program for quantifying landscape structure.* (Gen. Tech. Rep. PNW-GTR-351). Portland, USA.

Oberhauser, U. (1997). Secondary forest regeneration beneath pine (*Pinus kesiya*) plantations in the northern Thai highlands: A chronosequence study. *Forest Ecology and Management, 99*(1-2), 171–183. doi:10.1016/S0378-1127(97)00203-X

Office of Agricultural Economics. (2007). *Agricultural statistics of Thailand 2004.* Ministry of Agriculture and Co-operatives, Bangkok, Thailand.

Office of Environmental Policy and Planning. (1997). *Thailand policy and perspective plan for enhancement and conservation of national environmental quality, 1997-2016.* Bangkok, Thailand: Ministry of Science, Technology and Environment.

Office of Natural Resources and Environmental Policy and Planning. (2006). *The state of the environment for the year 2005-2006.* Bangkok: Office of Natural Resource and Environmental Policy and Planning, Ministry of Natural Resources and Environment.

Overmars, K. P., Verburg, P. H., & Veldkamp, T. (2007). Comparison of a deductive and an inductive approach to specify land suitability in a spatially explicit land-use model. *Land Use Policy*, *24*(3), 584–599. doi:10.1016/j.landusepol.2005.09.008

Panayotou, T., & Sungsuwan, S. (1989). *An economic study of the causes of tropical deforestation: The case of northeast Thailand.* (Development Discussion Paper No. 284). Harvard University Institute of Economic Development, Harvard University, Massachusetts, USA.

Pattanavibool, A., Dearden, P., & Kutintara, U. (2004). Habitat fragmentation in north Thailand: A case study. *Bird Conservation International*, *14*, S13–S22. doi:10.1017/S0959270905000195

Pontius, R., Boersma, W., Castella, J.-C., Clarke, K., de Nijs, T., & Dietzel, C. (2008). Comparing the input, output, and validation maps for several models of land change. *The Annals of Regional Science*, *42*, 11–37. doi:10.1007/s00168-007-0138-2

Priess, J. A., & Schaldach, R. (2008). Integrated models of the land system: A review of modelling approaches on the regional to global scale. *Living Reviews in Landscape Research, 2*. Retrieved from http://www.livingreviews.org/lrlr-2008-1

Rajan, K. S., & Shibasaki, R. (1997). Estimation of agricultural productivity and its application to modelling the expansion of agricultural land in Thailand. *Journal of Agricultural Meteorology*, *52*(5), 815–818.

Redford, K. H., & Richter, B. D. (1999). Conservation of biodiversity in a world of use. *Conservation Biology*, *13*(6), 1246–1256. doi:10.1046/j.1523-1739.1999.97463.x

Royal Forest Department. (2007). *Forestry statistics year 2006*. Bangkok, Thailand: Ministry of Natural Resources and Environment.

Royal Thai Survey Department. (2002). *Topographic map scale 1: 50,000*. Ministry of Defense, Bangkok, Thailand.

Santisuk, T. (1988). *An account of the vegetation of northern Thailand. Geological Research, 5*. Stuttgart, Germany: Franz Steiner Verlag.

Schmidt-Vogt, D. (1999). *Swidden farming and fallow vegetation in northern Thailand. Geological Research, 8*. Stuttgart, Germany: Franz Steiner Verlag.

Stracey, C. M., & Pimm, S. L. (2009). Testing island biogeography theory with visitation rates of birds to British islands. *Journal of Biogeography*, *36*, 1532–1539. doi:10.1111/j.1365-2699.2009.02090.x

Swets, J. A. (1986). Measuring the accuracy of diagnostic systems. *Science*, *240*, 1285–1293. doi:10.1126/science.3287615

Thanapakpawin, P., Richey, J., Thomas, D., Rodda, S., Campbell, B., & Logsdon, M. (2006). Effects of land-use change on the hydrologic regimes of the Mae Chaem river basin, NW Thailand. *Journal of Hydrology (Amsterdam)*, *334*(1-2), 215–230. doi:10.1016/j.jhydrol.2006.10.012

Theobald, D. M. (2005). *GIS concepts and ArcGIS methods* (2nd ed.). Colorado: Colorado State University.

Trisurat, Y. (1999). *Land-use changes inside and around Srinakharin and Erawan National Parks: Fnal report submitted to the Royal Forest Department*. Bangkok, Thailand: Environ Planning, Inc.

Trisurat, Y. (2007). Applying gap analysis and a comparison index to assess protected areas in Thailand. *Environmental Management*, *39*(2), 235–245. doi:10.1007/s00267-005-0355-3

United Nations Environmental Programme. (2001). *GLOBIO: Global methodology for mapping human impacts on the biosphere*. Report UNEP/DEWA/TR 25. United Nations Environmental Programme, Nairobi.

Verburg, P., Eickhout, B., & van Meijl, H. (2008). A multi-scale, multi-model approach for analyzing the future dynamics of European land use. *The Annals of Regional Science*, *42*, 57–77. doi:10.1007/s00168-007-0136-4

Verburg, P., & Overmars, K. (2009). *Combining top-down and bottom-up dynamics in land use modeling: Exploring the future of abandoned farmlands in Europe with the Dyna-CLUE model*. Landsc Ecol. Retrieved August 15, 2009, from http://dx.doi.org/10.1007/s10980-009-9355-7

Verburg, P. H., Schot, P., Dijst, M. J., & Veldkamp, A. (2004). Land use change modelling: Current practice and research priorities. *GeoJournal*, *61*(4), 309–324. doi:10.1007/s10708-004-4946-y

Verburg, P. H., Soepboer, W., Limpiada, R., Espaldon, M. V. O., Sharifa, M., & Veldkamp, A. (2002). Land-use change modelling at the regional scale: The CLUE-S model. *Environmental Management*, *30*(3), 391–405. doi:10.1007/s00267-002-2630-x

Verburg, P. H., & Veldkamp, A. (2004). Projecting land use transitions at forest fringes in the Philippines at two spatial scales. *Landscape Ecology*, *19*(1), 77–98. doi:10.1023/B:LAND.0000018370.57457.58

Watcharakitti, S. (1976). *Tropical forest land-use evolution in northern Thailand. Forest Research Bulletin, 44*. Bangkok, Thailand: Faculty of Forestry, Kasetsart University.

Chapter 11
The Current and Future Status of Floristic Provinces in Thailand

P.C. van Welzen
Leiden University, The Netherlands

A. Madern
Leiden University, The Netherlands

N. Raes
Leiden University, The Netherlands

J.A.N. Parnell
Trinity College Dublin, Ireland

D.A. Simpson
Royal Botanical Gardens, UK

C. Byrne
Trinity College Dublin, Ireland

T. Curtis
Trinity College Dublin, Ireland

J. Macklin
Trinity College Dublin, Ireland

A. Trias-Blasi
Trinity College Dublin, Ireland

A. Prajaksood
Trinity College Dublin, Ireland

P. Bygrave
Royal Botanical Gardens, UK

S. Dransfield
Royal Botanical Gardens, UK

D.W. Kirkup
Royal Botanical Gardens, UK

J. Moat
Royal Botanical Gardens, UK

P. Wilkin
Royal Botanical Gardens, UK

C. Couch
Royal Botanical Gardens, UK

P.C. Boyce
Universiti Sains Malaysia, Malaysia

K. Chayamarit
Thailand Botanical Garden Association, Thailand

P. Chantaranothai
Khon Kaen University, Thailand

H-J. Esser
Botanische Staatssammlung München, Germany

M.H.P. Jebb
Ireland National Botanical Gardens, Ireland

DOI: 10.4018/978-1-60960-619-0.ch011

K. Larsen
University of Aarhus, Denmark

S.S. Larsen
University of Aarhus, Denmark

I. Nielsen
University of Aarhus, Denmark

C. Meade
National University of Ireland, Ireland

D.J. Middleton
Scotland Royal Botanical Garden, UK

C.A. Pendry
Scotland Royal Botanical Garden, UK

A.M. Muasya
University of Cape Town, South Africa

N. Pattharahirantricin
Thailand Department of National Parks, Thailand

R. Pooma
Thailand Department of National Parks, Thailand

S. Suddee
Thailand Department of National Parks, Thailand

G.W. Staples
Singapore Botanic Gardens, Singapore

S. Sungkaew
Kasetsart University, Thailand

A. Teerawatananon
Thailand National Science Museum, Thailand

ABSTRACT

Two databases containing distribution data of species and specimens show that within Thailand preferably four floristic or phytogeographical regions can be discriminated (areas with a typical, unique and distinct plant composition): the Southern, Northern, Eastern and Central Region. They differ from the seven regions used at present in the Flora of Thailand Project. Modelling the effects of slight climate changes due to global warming shows that the floristic regions will be different in 2050. Not only will the areas differ, but the numbers of species per area will decrease dramatically, although species from outside Thailand may migrate into Thailand. Predictions contain a high degree of uncertainty, and they may never come true as they are strongly influenced by small, currently unpredictable effects. Nevertheless, the loss of biodiversity and its consequences for climate, economies, health, et cetera, are already becoming noticeable. Therefore, the protection and improvement of biodiversity should become the main focus of attention for all governments in the region.

1. INTRODUCTION

Species are generally not randomly distributed. Plants and animals originate via evolution and this always happens in a geographically restricted area. Thus, it is not surprising that man has searched for patterns in these distributions, that is in the areas in which species are found. One means of finding these patterns is to examine if certain areas can be characterized by species which are in combination typical for the area. The resulting regions are called, in the case of plants, floristic or phytogeographical regions. Usually, a country/continent is completely subdivided into these areas and these are mutually exclusive.

Thailand has a species rich and complex biodiversity that differs in various parts of the country (MacKinnon, 1997; Wikramanayake *et al.*, 2002; Maxwell, 2004). Thailand harbours one of the 25 global hotspots of biodiversity (Myers *et al.*, 2000) known as the Indo-Burmese Region. Unfortunately, the biodiversity of Thailand is under severe threat (Stibig *et al.*, 2007). Indeed, the whole of Southeast Asia is on the verge of losing approximately three-quarters of its original forest cover by 2100, and up to 42% of its biodiversity (Sodhi *et al.*, 2004). In Thailand, clearance for agriculture and other uses has reduced forest cover to perhaps as low as 20% (Santisuk *et al.*, 1991), much of which may be degraded (Parnell *et al.*, 2003). According to Middleton (2003), forest cover has declined from 50% in the 1950's to 25% in 2000, as detected by Landsat-TM images, and this is one of the fastest rates of deforestation in the tropics (Middleton, 2003). Maxwell's (2004) view is even more dramatic as he estimates forest cover to be reduced to 15%. Currently, the 115 National Parks and Wildlife Sanctuaries in Thailand together cover 6.72 M ha and the forest therein is protected, which equates to 53% of the remaining forest area or 8% of the total land area (Middleton, 2003). Even though little remains of the original deciduous and evergreen forests of Thailand, it is still one of the biodiverse countries in Southeast Asia (estimates by Middleton, 2003, and Parnell, 2000, are that, 10,250 and 12,500 higher plant species, respectively are found). The reason for the high level of species richness in Thailand is that the country is situated on the borders or at the cross-roads between four major biogeographical regions: the Himalayas in the northwest, China in the north, Indochina in the east, and Sundaland in the south. The flora is therefore influenced by Indochinese, Indo-Burmese and Malesian elements.

Based on these influences Smitinand (1958) discriminated seven floristic regions in Thailand (Figure 1). His delimitation of these regions was based on a manuscript by Dr. A.F.G. Kerr, which is present in the Royal Botanic Gardens Kew, UK, but which we were unable to retrieve. Kerr distinguished six regions to which Smitinand added a seventh, the South-western region. The regions are still in use by botanists (e.g., Maxwell, 2004) and the Flora of Thailand Project (e.g., Santisuk & Larsen, 2009) today. The regions can be characterized as follows (sequences as used in the present floras; text largely after Smitinand, 1958):

Figure 1. Phytogeographical areas of Thailand as used in the Flora of Thailand project. Provincial borders are indicated. N = Northern (yellow), NE = North-eastern (dark green), E = Eastern (green), SW = South-western (middle blue), C = Central (light blue), SE = South-eastern (dark blue), P = Peninsular (red)

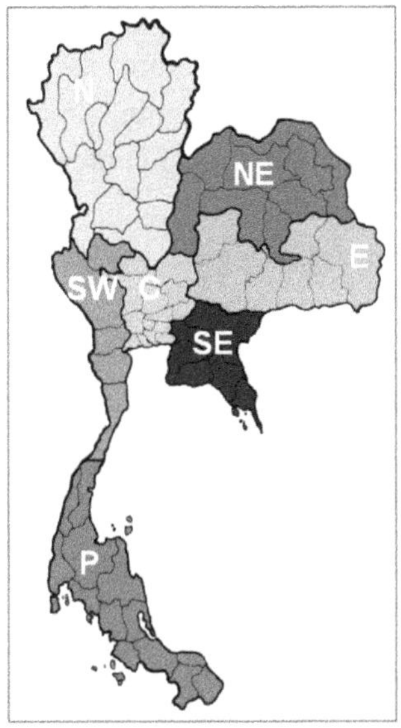

- The Northern Region (Figure 1: N, yellow) is influenced by Indo-Burmese floristic elements. Typical forests are dry deciduous forest, dry hill evergreen forest, and montane temperate forests. On higher elevations pines and rhododendrons are typical, on lower elevations Dipterocarpaceae. The area contains the highest mountains of Thailand.

- The North-Eastern Region (Figure 1: NE, dark green) is influenced by the Indo-Chinese Flora, but Indo-Burmese species are also found there, while there is also an affinity with South China. Vegetation types largely comprise dry deciduous to mixed deciduous forest with large tracts of dry evergreen forest or savannah in between. A large part of the area is dominated by table mountains. Again, pines are typical for higher elevations, Dipterocarpaceae for lower altitudes.
- The Eastern Region (Figure 1: E, green) is influenced by the Central and Southern Indo-Chinese Flora (Cambodia and S. Vietnam). Dry Dipterocarp forests are typical, as are extensive savannahs, and stands of (mainly) *Dipterocarpus obtusifolius* Teijsm., together with pines on periodically inundated soil. Again an area is dominant with many Dipterocarpaceae.
- The South-Western Region (Figure 1: SW, middle blue) corresponds to the Tenasserim or Lower Burmese Flora. Along the border evergreen forest may be found, but bamboo forests are very common on the plains (common species: *Thyrsostachys siamensis* Gamble, Poaceae); these gradually change into mixed deciduous and dry dipterocarp forests. Mangroves also occur in this region.
- The Central Region (Figure 1: C, light blue) is mainly cultivated with only small remnants of original forests. The trees which are present are mainly recruited from surrounding regions. Mangrove stands occur along the coast. There is no typical flora.
- The South-Eastern Region (Figure 1: SE, dark blue) mainly has Indochinese and Malesian floristic elements, but several Burmese species are also present. The most common vegetation type is tropical rain forest in which Dipterocarpaceae, *Lophopetalum* Wight ex Arn. (*Solenospermum* Zoll.; Celastraceae) and *Parkia* R.Br. (Fabaceae) dominate. Less common are periodically inundated savannahs with *Dipterocarpus obtusifolius* and *D. intricatus* Dyer, which are often subjected to fire during the dry season. Extensive mangrove swamps and tidal forests occur along the coast.
- The Peninsular Region (Figure 1: P, red) is mainly influenced by Malesian flora. The border of Malesia runs through the southern provinces (Van Steenis, 1950; Raes & Van Welzen, 2009). The northern part of the peninsula also contains Burmese elements. Tropical rain forest with Dipterocarpaceae are commonest, followed by mangrove forests and peat swamps.

Smitinand (1958) did not further characterize the regions. No analysis of indicator species exists, nor are the influences from other areas specified (e.g., there is no indication which species constitute a Malesian influence in the Peninsular region). Nevertheless, publications on Thai botany always use the seven regions. We wonder if this is justified. Unfortunately, investigating this is quite difficult, because Thailand is still heavily undercollected (Parnell *et al.*, 2003). Estimates for collection density range from 157,000 specimen in total for the whole country (Holmgren *et al.*, 1990) to a density of c. 0.5 specimens per km^2 (Parnell, 2000; Parnell *et al.*, 2003). Furthermore, collection localities are often more based on accessibility of areas, leading to biased sampling. The general pattern consists of decreasing collection numbers with increasing distance from populated places and roads. Most collection activity is within 8 km or less of a populated place (Parnell *et al.*, 2003; Reddy & Davalos, 2003; Loiselle *et al.*, 2008). Furthermore, collecting is excentric as islands, national parks and mountains are popular with collectors, while the cultivated lowlands are largely ignored (Santisuk *et al.*, 1991; Parnell *et al.*, 2003).

This means that the distributional data for most species are incomplete. Luckily, this problem can be overcome by the use of Species Distribution Models (SDMs). An SDM interpolates the relationships between species data (presence/absence or presence-only) in combination with the abiotic conditions at the species' collection localities and thereby predicts the potential presence of the species in non-sampled but environmentally suitable areas (Araujo & Guisan, 2006; Elith *et al.*, 2006; Peterson *et al.*, 2007; Franklin, 2009). This methodology allows us to predict the potential distribution of a species even for areas that suffer from incomplete and biased sampling, or for areas where no collections have been made. SDMs basically establish the dimensions of the ecological niche for a species using a provided set of abiotic parameters. An added advantage of these methods is that they allow the projection of the established niche dimension to future climatic conditions and thereby permit prediction of range shifts under global climate change scenario's (Bates *et al.*, 2008). This can be done for a vast number of species and their SDMs and allow assessment of whether the seven floristic regions of Smitinand (1958) are recognisable.

With this contribution we use various methods to investigate:

- whether or not the regions recognised by Smitinand (1958) have characteristic floral elements by which they can be recognized, or if not, then
- whether there are areas that can be recognized as floristic regions and to identify them
- whether the recognisable floristic areas are internally contiguous and can be characterised.
- whether these phytogeographical areas change in boundary or composition as climate changes.

We conclude by asking: will the Thai flora be threatened even more in the future than it is today?

2. MATERIAL AND METHODS

2.1 Databases

Two databases were compiled, which were analysed separately using different methodologies.

2.1.1 Database I

Database I has species information and contains presence/absence data per species per province for all species published in the Flora of Thailand project up to part 10.1. The 76 provinces, though not natural units and of variable size, are used because they are the basic and only distribution entities in the flora. The database comprises 3,187 indigenous species in 800 genera and 173 families; all cultivated, introduced and alien escapees (206) were omitted. These data will not be modelled. We used data from the Flora of Thailand for two reasons:

- The Flora of Thailand forms the first thorough inventory of all Thai species. The revisions provide the most reliable species delimitations and currently the best estimate of species distributions.
- The families in Flora of Thailand are published in the sequence in which they arrive, thus no classification is used to direct their publication. This means that the choice of families is biased towards the small families, but as far as distributions are concerned the selection is random and the data representative.

A Non-metric Multidimensional Scaling analysis (NMS; see Borg & Groenen, 2005) was performed on Database I, using Lance-Williams distance, 3-Dimensional solution with stress convergence and minimum stress set to 0.0005, 1,000 iterations and 500 random starts, yielded a normalised raw stress of 0.004, was performed with SPSS version 16.0. SPSS was chosen for this analysis as it is widely available and can perform an NMS on the very large data matrix we wished to analyse. For technical reasons, the SPSS algorithm is more efficient with dissimilarity/distance measures than with similarity/proximity measures and so requires distance matrices, not similarity matrices (Garson, 2009) – Sørensen's Coefficient is therefore not available as an option for NMS in the SPSS package. Of the measures available, the most appropriate (and most similar to Sørensen) is Lance-Williams distance – also known as Bray-Curtis distance, a long-established distance measure (Bray & Curtis, 1957; Legendre & Legendre, 1998). Several provinces, with very low amounts of collections, were omitted from the NMS analysis: Uttaradit, Mukdahan (united with Nakhon Phanom in Figure 3a, therefore not white), Khon Kaen, Kalasin, Maha Sarakham, Ratchaburi, and Lop Buri (69 provinces used in the analysis).

Database I was also analysed using Unweighted Pair Group Method with Arithmetic mean (UPGMA), with Sørensen's Coefficient, performed with the MultiVariate Statistical Package (MVSP) version 3.131, Kovach Computing Services, Anglesey, UK (for methodology see Sneath & Sokal, 1973). Sørensen's Coefficient (Sørensen, 1948) or Dice Coefficient (Dice, 1945) simply compares the similarity in presence of species between pairwise areas via the formula 2A/(2A+B+C), where A is the number of shared species, B the number of species only present in one area (absent in the other) and C the number of species only present in the other area (absent in the first area).

Two UPGMA cluster analyses were performed, one utilised all species in Database I, the other utilised these same data minus the species that occur in every province (these latter result in high clustering between provinces with low sampling intensity). In the Cluster Analyses several provinces were united. This was necessary, because some of the larger provinces, present when the first volumes of the Flora of Thailand were published, were later split-up. Therefore, there are relatively few records for these modern provinces as early published records only deal with the aggregated provinces. We therefore united: Phayao with Chiang Rai, Nong Bua Lam Phu with Udon Thani, Mukdahan with Nakhon Phanom, Yasothon and Amnat Charoen with Ubon Ratchathani, and Sa Kaeo with Prachin Buri. Thus the cluster analyses utilise 70 provinces not the present 76.

2.1.2 Database II

Database II was used for the species distribution modelling. The database is specimen based and contains collection data from herbarium databases and collecting trips. The collecting localities were georeferenced with online gazetteers (NGA GEOnet Names Server - http://earth-info.nga.mil/gns/html/index.html, Global Gazetteer - http://www.fallingrain.com/world/, Fuzzy Gazetteer - http://isodp.fh-hof.de/fuzzyg/query/, and Alexandria Digital Library -http://clients.alexandria.ucsb.edu/globetrotter/) and Jacobs (1962) for collections by Kerr and contemporaries. This database contains 32,254 records, divided over 237 families, 1,684 genera, and 6,029 species.

The application of SDMs requires environmental data in a spatial raster format (here we used grid cells of 5 arc-minutes or ca. 10 x 10 km resolution; see below). Only grid cells with data for all environmental data layers were retained, resulting in 24,070 grid cells of which 6,209 were within the boundaries of Thailand. Records close to the coast or near the shores of large lakes that just fell outside a grid cell were shifted to their nearest grid cell. To prevent overfitting of the models to certain environmental conditions we removed duplicate records from each grid cell, i.e. only one record per species was retained for every grid cell. Furthermore, because SDMs identify and interpolate relations between species presence records and environmental conditions we set the lower limit for species to be modelled at 5 unique records (i.e., presences in 5 different grid cells). These requirements were met by 1,399 species.

2.2 Environmental Predictors

To relate the current presence of species to climatic and edaphic conditions we used two datasets of environmental predictors. The first one is the WorldClim dataset (www.worldclim.org) that contains 19 bioclimatic predictors (1950-2000) and altitude (in m) at 5 arc-minute spatial raster resolution (Hijmans *et al.*, 2005). To prevent the border of Thailand from artificially influencing the analysis we used larger geographical dimensions than just Thailand (24,070 grid cells). The environmental predictors were clipped to latitude between 4 – 24°N, and longitude between 95 – 110°E. In total these data layers comprise 24,070 grid cells. To avoid problems with multi-collinearity (Graham, 2003), which can result in model over-fitting (Peterson *et al.*, 2007), we removed all predictors with Pearson's correlation coefficient $r > 0.71$. This procedure reduced the number of bioclimatic variables from 20 partly correlated to 7 uncorrelated variables. These variables are Bio01 – Annual mean temperature, Bio02 – Mean diurnal temperature range, Bio04 – Temperature seasonality, Bio07 – Temperature annual range, Bio12 – Annual precipitation, Bio17 – Precipitation of the driest quarter, and Bio18 – Precipitation of the warmest quarter.

The second dataset represented the edaphic conditions. For this dataset we used a selection of soil property values from the FAO Land and Water Digital Media series # 20 (FAO, 2002) with the same 5 arc-minute spatial resolution and geographic extent as the bioclimatic variables. The selection included the following six variables; CEC (Cation Exchange Capacity) of clay in the topsoil, Easy available water, Nitrogen percentage of the topsoil, Organic carbon pool, pH of the topsoil, and Textural class of topsoil. The FAO data do follow a logical increasing order, but are not fully continuous. Therefore, we used a Spearman rank correlation test to assess their independence. To prevent overweighting of over-represented combinations of edaphic conditions we performed the Spearman's rank correlation test on the 51 unique combinations of edaphic conditions. Maximum Spearman's rho for the six selected soil variables is 0.651. In total we used 7 bioclimatic and 6 edaphic uncorrelated variables to develop the SDMs.

To project the SDMs to future climatic conditions we also downloaded the bioclimatic variables of the CCM3 global climate change scenario for 2050 from the worldclim.org website (this scenario is currently no longer available). We selected the same seven bioclimatic variables for CCM3 scenario as were used to develop the SDMs under current climatic conditions at 5 arc-minute resolution and with the same geographical extent. The CCM3 models for 2050 predict for Thailand a maximum temperature rise of 1-1.2 °C (Peninsula, South-eastern, SE part of Eastern) to c. 1.7 °C (Chiang Rai, North-eastern, border Northern and South-western), and more rain for most of Thailand (up to 700 mm extra at the border between Northern and South-western, a part of the North-eastern, and Phangnga), and only less rain (up to 170 mm less) in the southern part of the Peninsula and around Bangkok. The data that represents the future edaphic conditions were identical with those used to model the distributions under current climatic conditions.

All data manipulations were performed with Manifold GIS (Manifold.net).

2.3 Species Distribution Modelling and Model Significance Testing

From the variety of available modelling techniques we selected Maxent (version 3.3.0; www.cs.princeton.edu/~schapire/maxent), the maximum-entropy approach for species habitat/distribution modelling (Phillips *et al.*, 2006; Phillips & Dudik, 2008). Maxent (with all default modelling rule settings) was chosen because it:

- is specifically developed to model species distributions with presence-only data,
- has been shown to outperform most other modelling applications (Elith et al. 2006),
- is least affected by georeferencing errors (Graham *et al.*, 2008),
- performs best when few presence records are available (Wisz *et al.*, 2008), and
- allows the projection of the identified niche dimensions to future (and past) climatic conditions.

To test the significance of the SDMs we used the bias corrected null-model methodology of Raes & ter Steege (2007). This method uses the threshold independent and prevalence insensitive Area Under the Curve (AUC) value as a measure of model accuracy (Fielding & Bell, 1997; Manel *et al.*, 2001; McPherson *et al.*, 2004). The method tests whether the AUC value of a species SDM deviates significantly from 999 AUC values of SDMs that were developed with equally many, but randomly drawn, records (Raes & ter Steege, 2007; Raes *et al.*, 2009). Instead of fully randomly drawing from the entire country (from all 6,209 grid cells), draws were only made from all 1,576 grid cells for which we have collection data. In this way we corrected for potential bias in environmental conditions represented by the collection localities. If the AUC value of a true SDM is larger than the 950th ranked AUC value of the 999 random models, it can be concluded that the chance that a random set of records generates an equally accurate model is less than 5% (p<0.05). We developed null-distributions and their 95% upper confidence interval (C.I.) limits (the 950th ranked values) for every number of records by which the 1,399 species were represented, i.e. for species represented by 5 records, 6 records, etc., till species represented by 89 records. Finally, we applied a curve-fit through the 95% C.I. values for the following three ranges. A linear fit for 5-9 records where Maxent only uses linear features (AUC= -0.013 × records + 1.019; R^2=0.99), a linear fit for 10-14 records where Maxent uses linear and quadratic features (AUC= -0.0018 × records + 0.931; R^2=0.79), and a power fit for ≥15 records where Maxent uses linear, quadratic and hinge features (AUC= 1.044 × records$^{-0.0353}$; R^2=0.98). The SDM AUC values were tested for significance against the fitted null-model values. In total 799 of the 1399 species represented by ≥ 5 records were significantly different from random expectation; 280 of these 799 species are shared with Database I.

As a final test we assessed whether certain parts of the country's environmental gradients were underrepresented by the 1,576 collection localities. We first divided each of the country's 13 environmental gradients (7 bioclimatic and 6 edaphic, see above) represented by 6,209 grid cells into 10 equally sized bins and developed 13 frequency distributions, one for each gradient. Secondly, we tested whether the frequency distribution of the environmental values of 1,576 collection localities deviated significantly from the frequency distribution of 6,209 grid cells covering the entire country's gradients with a Chi-square test (Loiselle *et al.*, 2008). This was done for each of the 13 variables. For none of the 13 environmental variables was the frequency distribution of the 1,576 collection localities different from the country's one, meaning that the 1,576 collection localities were not environmentally biased, thus the collection localities form a good representation for the rest of Thailand.

The continuous Maxent values were converted to discrete presence/absence values by applying the 10 percentile training presence threshold (Liu *et al.*, 2005). This threshold is a rather conservative value and omits 10% of the collection records from the predicted presence area. These are the areas with the lowest chances that the taxon concerned occurs. We applied this threshold to all 799 significant SDMs and developed a presence/absence matrix for the 6,209 grid cells falling within the borders of Thailand.

For all species with a significant SDM under current climatic conditions we also developed a presence/absence matrix from their projected future distributions. We used the same threshold values as were used for their present climatic conditions. These two presence/absence matrices are used to delimit the present and future species richness, weighted endemicity and the phytogeographic regions of Thailand.

2.4 Species Richness and Weighted Endemicity

To develop the species richness- and endemicity patterns the presence/absence matrices for both present and future climatic conditions were exported to Excel 2007. Summing the presences resulted in a value for the richness per grid cell. These values were geographically plotted with Manifold GIS.

The weighted endemicity values (Crisp *et al.*, 2001; Raes *et al.*, 2009) were derived from the species presence data. The contribution of each species to the total weighted endemicity value is proportional to the range over which it occurs (thus weighting is introduced). A species that is predicted to occur by its SDM in 100 grid cells has a weighted endemicity value of 1/100 for each grid cell in which it occurs; if the species only occurs in 2 grid cells, then for each of those cells the weighted endemicity value is 1/2. Thus, widespread species contribute little to the total weighted endemicity value of a grid cell and the opposite occurs for species with a limited distribution. The weighted endemicity values are summed for each grid cell to result in a map of weighted endemicity.

2.5 Phytogeographic Regions Based on SDMs

Phytogeographic regions were delimited from the two presence/absence matrices derived from the SDMs (current climatic and future climatic conditions) with the recommended hierarchical cluster analysis with a flexible beta (β = -0.25) linkage method and Sørensen distance measure (McCune & Grace, 2002; Raes, 2009) to cluster the 6,209 grid cells of Thailand with PC-ORD 5.32 (MjM Software Design, Gleneden Beach, Oregon). To prune the two cluster trees to their optimum number of cluster groups we used Indicator Species Analysis (ISA) (Dufrêne & Legendre, 1997). We ran the ISA for 2-20 cluster groups for both the present and future cluster trees, indicating the potential 2-20 phytogeographic regions under present and future climatic conditions. For each run ISA calculates an indicator value for each species. A perfect indicator means that presence of a species points to a particular group without error (McCune & Grace, 2002). The significance of the indicator values is tested with a montecarlo method (999 randomizations) and yields a *p*-value for each species. By summarizing the *p*-values for all species per ISA for 2-20 cluster groups, the number of cluster groups is revealed that has the most significant indicators and is therefore the optimal pruning point of the cluster tree. However, there was one problem with this analysis. Since our presence/absence matrices represent a spatially continuous grid monte carlo randomizations always resulted in maximum significant results. We therefore subsampled the 6,209 grid cells 5 times by drawing 621 random grid cells (10%). For each of the five subsets we ran the ISA and we report average *p*-values for 2-20 cluster groups, for both cluster trees. Finally, we report the indicator values for the optimal number of cluster groups based on the ISA of the entire matrix (6,209 grid cells).

Figure 2. (a) Species densities per province for all species described in the Flora of Thailand so far. Dark blue shows the lowest densities, red the highest; (b) Endemic species (i.e., those only occurring in a single province) densities per province for all species described in the Flora of Thailand; (c) Relationship between classes (see legends of a and b) of species richness and numbers of endemic species; the diamonds represent the provinces, quite a few provinces share the same diamond

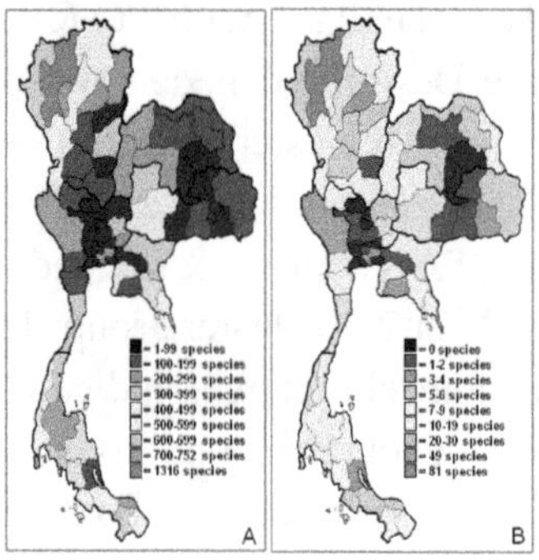

Figure 3. (a) Species densities per grid cell derived from the Species Distribution Models for the present day distributions (Database II); (b) Species weighted endemicity per grid cell derived from the Species Distribution Models for the present day distributions (Database II). (White and blue are low endemicity, red indicates high densities.)

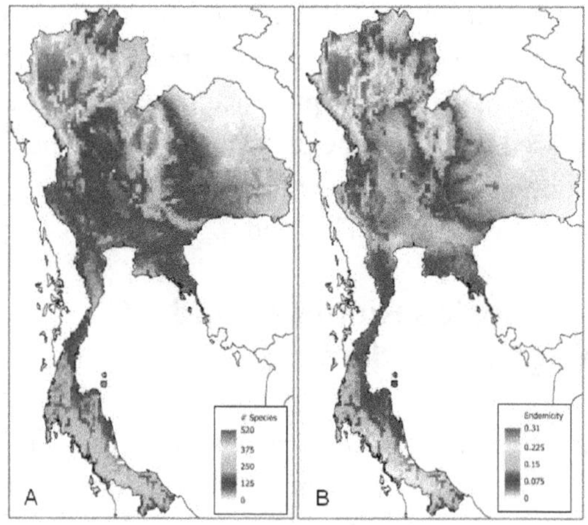

2.6 Remaining Forests and National Protected Areas

As final analysis we assessed which percentage of the potential forested extent is still covered with forest today. For this purpose we made use of the Global Land Cover 2000 dataset for South East Asia – v3 (http://bioval.jrc.ec.europa.eu/products/glc2000/data_access.php - last accessed 02/06/2010; Stibig *et al.*, 2004, 2007). From the 18 recognized landcover classes we selected class 1-5 (characterized as 'Tree cover') as natural forested vegetation. For each of the three florstic analyses based on SDMs we assessed their potential forested extent and the percentage thereof which remains forested today. To get an indication which area of the currently forested extent is protected by national parks we overlaid the three maps of remaining forest with the GIS data of National Protected Areas (data generously provided by Nantachai Pongpattananurak, see Acknowledgements).

3. RESULTS

Database I shows very variable numbers of species recorded per province (divided into nine classes in Figure 2a), varying between 1-99 species, in especially the central and eastern provinces, to 1,316 species in Chiang Mai. A more or less similar pattern of variation is present for the Thai endemic species per province (Figure 2b, also nine classes). There is an almost linear relationship between the classes of species richness per province and the classes of endemic species (Figure 2c). The summation of the 799 threshold SDMs based on Database II shows a different result. High concentrations of species (Figure 3a) and of weighted endemicity (Figure 3b) are present in the Northern region, the western part of the North-eastern region and to a lesser extent in the Peninsular part.

The NMS (Figure 4) performed on Database I shows six groups; of which the geographical representation is shown in Figure 5a. None of the

Figure 4. Result of Non-metric Multidimensional Scaling analysis of all data of Database I (presence/absence per province of species described in Flora of Thailand). The clustered provinces are depicted in Figure 5a

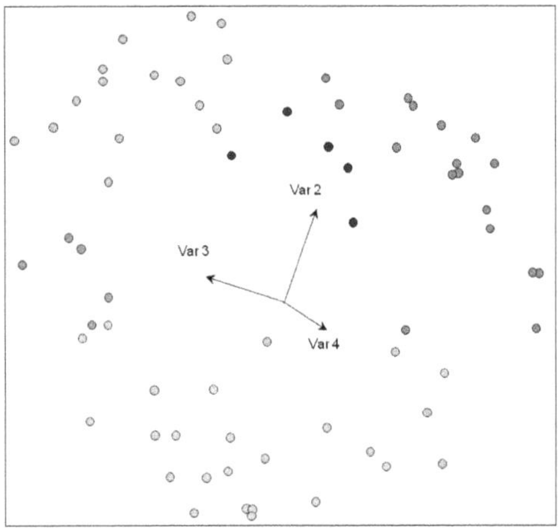

Figure 5. Maps showing clusters of provinces according to (a) Non-metric Multidimensional Scaling analysis (Figure 4); (b) Cluster Analysis of all species data (Figure 6a); (c) Cluster Analysis of all species minus those present in all provinces (Figure 6b). The colour schemes for all figures are comparable

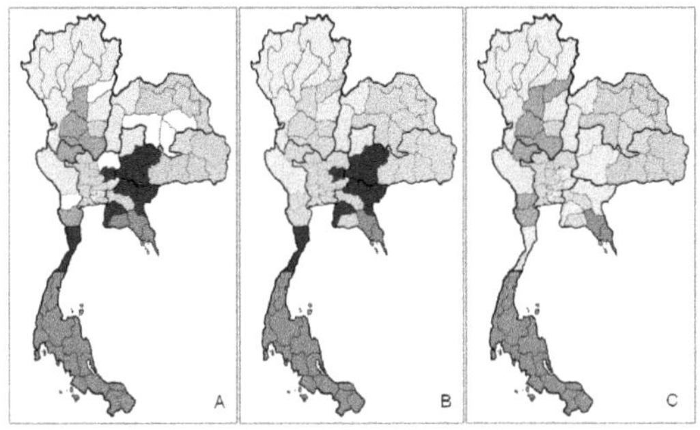

groups is distinctive, thus the placement of some of the dots within a particular group can be disputed. There is, however, considerable similarity between the patterns shown by NMS and that shown by the cluster analysis performed on all data minus the species present in all areas (compare Figures 5a and c). The Cluster Analyses were performed on all data of Database I (Figure 6a) and on all data minus the species present in all areas (Figure 6b). Note that all branches in the cluster analyses that point to the groups are very short (Figure 6a and 6b, which means that the overall similarity in species composition is low within each group. Figure 6a shows four distinct clusters, shown as provinces in Figure 5b. Figure 6b shows five distinct clusters, presented as prov-

Figure 6. Results of the Cluster Analyses. The colour schemes of the branches are as in the maps of Figure 5b and c, respectively, the colours of the province names correspond to those of the floral regions in Figure 1: (a) Cluster Analysis of all data of Database I; (b) Cluster Analysis of all species data of Database I minus the species present in all provinces

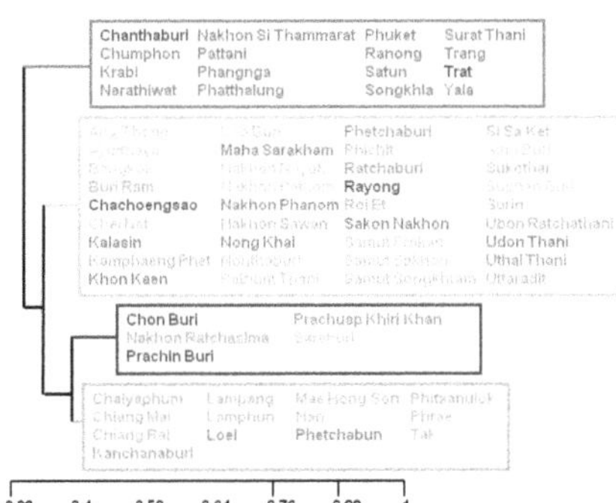

(a)

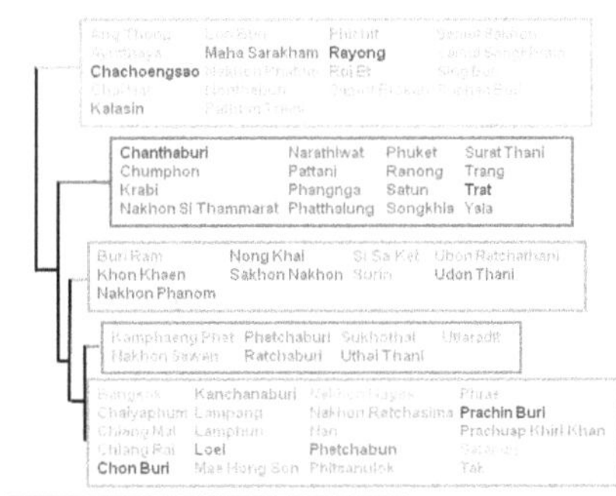

(b)

Figure 7. Summedp-values ofthe indicator values of 799 species for each number of clustergroups. The lowest values (red dots) indicate the optimal pruning point of the cluster tree, which is an indication for the number of floristic regions to be recognized. A. For the cluster analysis of SDMs under present climatic conditions and B. for the cluster analysis of SDMs projected to future climatic conditions of 2050

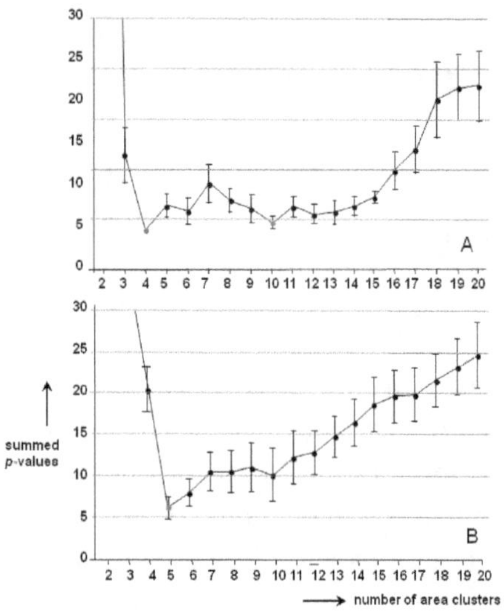

inces in Figure 5c. The main differences between both analyses are found in the light blue areas. In Figure 6a (Figure 5b) they are in the middle of the diagram, but in Figure 6b (Figure 5c) they are basal to the rest. All these light blue provinces have very few species (Figure 2a: ranging between 15 and 61 except for Rayong with 136 species) and, therefore, the light blue area cannot be regarded as a numerically well-defined floral region.

The analysis of the optimum number of clustergroups for the cluster analysis based on the 799 significant SDMs (Database II; Figure 7a) shows that Thailand can be divided into four and potentially ten floristic regions. The latter ten areas are a subdivision of the four areas (compare Figure 8a with 8b, the dotted line across the branches in Figure 8c indicates the division into four areas, the terminals the division into ten areas). Tables

Figure 8. Cluster analysis of the species compositions based on the present day SDM distributions per grid cell (C). (a) Optimisation for four phytogeographical areas; (b) Optimisation for ten areas (see Figure 7A); the ten areas are a further differentiation of the four areas; (c) Cluster diagram, the dotted line shows the division into four areas, also shown by the colours of the branches it transects; the triangles represent the united grid cells as in the ten area division

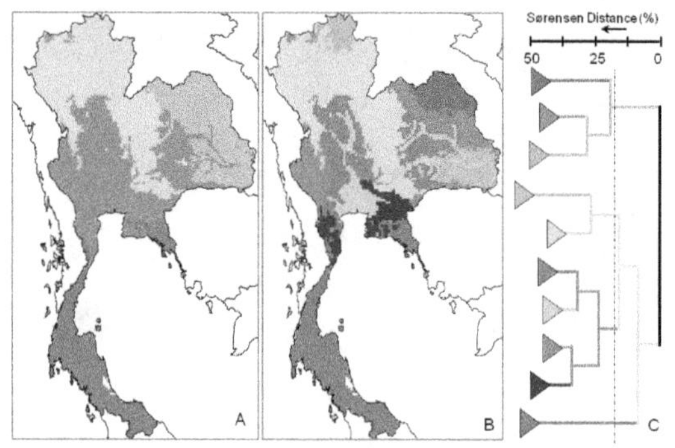

1 and 2 show the top ten indicator species typical for the four and ten floristic regions (no indicator species were found for the dark green area of Figure 8b). Only the indicator species for the Red area and the Yellow areas (Figures 8a and b) have high indicator values. When ten regions are distinguished then the yellow region (Figure 8a) is divided in a Yellow and Orange area (Figures 8b and c). The species with the highest indicator values for the Orange area are also represented in the Yellow area (both form a group in Figure 8a).

The cluster analysis of the projected SDMs for the year 2050 shows dramatic changes. The species richness values are shown in Figure 9a and the weighted endemicity values in Figure 9b. The highest number of species under projected future climatic conditions is 336 species, much lower than the value of 520 species found under the present conditions (Figure 3a). Subtraction of the presently predicted number of species (Figure 3a) from the predicted future values (Figure 9a)

Table 1. The top ten (when present) indicator values of species (numbers in bold in columns Red to Green) for the four phytogeographical areas under present climatic conditions (Figure 8a). The colors refer to those used in Figure 8a. Species names in bold are indicator species for the same area as at the present (based on ten areas) and in the future (Tables 2 and 3, respectively). The numbers are indicator values between 0 and 100

Family	Species	Area	Red	Blue	Yellow	Green
Santalaceae	*Dendrotrophe varians*	Red	93	0	0	0
Myrtaceae	*Syzygium polyanthum*	Red	92	1	0	0
Anacardiaceae	*Bouea oppositifolia*	Red	91	0	0	0
Dipterocarpaceae	*Shorea guiso*	Red	90	0	0	0
Phyllanthaceae	*Phyllanthus pulcher*	Red	89	1	0	0
Eriocaulaceae	*Eriocaulon willdenovianum*	Red	86	0	0	0
Euphorbiaceae	*Claoxylon longifolium*	Red	86	0	0	0
Violaceae	*Rinorea lanceolata*	Red	85	0	0	0
Cyperaceae	*Rhynchospora corymbosa*	Red	85	1	0	0
Cyperaceae	*Hypolytrum nemorum*	Red	85	1	0	0
Cyperaceae	*Schoenoplectus articulatus*	Blue	0	49	1	0
Euphorbiaceae	*Chrozophora rottleri*	Blue	2	45	0	1
Phyllanthaceae	*Bridelia ovata*	Blue	3	42	11	0
Rhamnaceae	*Ziziphus mauritiana*	Blue	1	40	2	3
Euphorbiaceae	*Mallotus spodocarpus*	Blue	0	40	0	0
Euphorbiaceae	*Acalypha indica*	Blue	1	38	8	3
Vitaceae	*Leea tetrasperma*	Blue	0	37	0	0
Concolvulaceae	*Operculina turpethum*	Blue	12	35	4	0
Sapindaceae	*Sisyrolepis muricata*	Blue	0	33	22	2
Euphorbiaceae	*Sampantaea amentiflora*	Blue	0	32	13	16
Urticaceae	*Boehmeria malabarica*	Yellow	0	0	77	0
Meliaceae	*Heynea trijuga*	Yellow	0	0	75	0
Urticaceae	*Debregeasia longifolia*	Yellow	0	0	75	0
Fabaceae	*Uraria campanulata*	Yellow	0	1	74	0
Acanthaceae	*Peristrophe lanceolaria*	Yellow	0	0	74	0
Cyperaceae	*Carex perakensis*	Yellow	0	0	74	0
Urticaceae	*Pouzolzia pentandra*	Yellow	0	4	73	0
Fabaceae	*Crotalaria ferruginea*	Yellow	0	0	73	0
Fabaceae	*Crotalaria assamica*	Yellow	0	0	73	0
Asteraceae	*Blumea lacera*	Yellow	0	0	73	0
Phyllanthaceae	*Breynia fruticosa*	Green	0	23	10	28
Apocynaceae	*Aganonerion polymorphum*	Green	0	10	4	24
Myrtaceae	*Syzygium ripicola*	Green	0	18	20	23

Table 2. The top ten (when present) indicator species for the ten phytogeographical areas under present climatic conditions (Figure 8b). The numbers above the columns refer to the colours as used in Figure 8b, the indicator species are in bold in column 1-10; Area 1 = Red, 2 = Dark Blue, 3 = Middle Blue, 4 = Light Blue, 5 = Purple, 6 = Yellow, 7 = Orange, 8 = Light Green, 9 = Middle Green, 10 = Dark Green. Species names in bold are indicator species for the same area as at the present (based on four areas) and in the future (Tables 1 and 3, respectively); in italics when indicative for different areas. The numbers are indicator values between 0 and 100

Family	Species	Area	1	2	3	4	5	6	7	8	9	10
Santalacae	**Dendrotrophe varians**	Red	**86**	1	0	0	0	0	0	0	0	0
Euphorbiaceae	*Claoxylon longifolium*	Red	**86**	0	0	0	0	0	0	0	0	0
Phyllanthaceae	*Aporosa aurea*	Red	**80**	0	0	0	0	0	0	0	0	0
Phyllanthaceae	*Baccaurea parviflora*	Red	**79**	0	0	0	0	0	0	0	0	0
Anacardiaceae	**Bouea oppositifolia**	Red	**77**	3	0	0	0	0	0	0	0	0
Pandaceae	*Galearia fulva*	Red	**75**	0	0	0	0	0	0	0	0	0
Violaceae	**Rinorea lanceolata**	Red	**74**	1	0	0	0	0	0	0	0	0
Euphorbiaceae	*Mallotus dispar*	Red	**74**	1	0	0	0	0	0	0	0	0
Phyllanthaceae	*Cleistanthus polyphyllus*	Red	**74**	0	0	0	0	0	0	0	0	0
Dipterocarpaceae	**Shorea guiso**	Red	**72**	3	0	1	0	0	0	0	0	0
Cyperaceae	*Fimbristylis dura*	Dark Blue	34	**38**	7	0	0	0	0	0	0	0
Anacardiaceae	*Buchanania sessilifolia*	Dark Blue	33	**36**	6	0	0	0	0	0	0	0
Euphorbiaceae	*Sumbaviopsis albicans*	Dark Blue	0	**32**	2	0	9	25	1	0	0	0
Annonaceae	*Rauwenhoffia siamensis*	Dark Blue	21	**31**	0	0	0	0	0	0	0	0
Convolvulaceae	*Tridynamia bialata*	Dark Blue	7	**28**	0	0	0	0	0	0	0	0
Rubiaceae	*Psilanthus merguensis*	Dark Blue	23	**28**	0	0	0	0	0	0	0	0
Convolvulaceae	*Argyreia collinsae*	Dark Blue	5	**28**	11	0	14	1	0	6	0	0
Euphorbiaceae	*Euphorbia atoto*	Dark Blue	13	**27**	0	0	4	0	0	0	0	0
Lauraceae	*Cinnamomum iners*	Dark Blue	15	**26**	0	6	2	4	0	0	0	0
Rhamnaceae	**Ziziphus cambodiana**	Dark Blue	0	**25**	17	0	8	3	0	0	0	0
Poaceae	*Cyathorhachis wallichiana*	Middle Blue	0	8	**48**	0	0	6	0	0	0	0
Vitaceae	*Tetrastigma assimile*	Middle Blue	0	2	**44**	0	0	3	0	0	0	0
Poaceae	*Dendrocalamus copelandii*	Middle Blue	0	1	**41**	0	0	8	0	0	0	0
Annonaceae	*Orophea brandisii*	Middle Blue	0	1	**35**	0	0	1	0	0	0	0
Rubiaceae	*Ixora cibdela*	Middle Blue	0	1	**35**	0	0	11	0	0	0	0
Araliaceae	*Schefflera pueckleri*	Middle Blue	0	1	**34**	0	0	21	4	0	0	0
Euphorbiaceae	*Macaranga siamensis*	Middle Blue	0	17	**34**	0	2	16	0	0	0	0
Euphorbiaceae	**Cleidion spiciflorum**	Middle Blue	0	2	**34**	0	0	32	0	0	0	0
Vitaceae	*Leea crispa*	Middle Blue	0	0	**33**	0	0	5	0	0	0	0
Anacardiaceae	*Buchanania latifolia*	Middle Blue	0	0	**32**	0	0	30	1	0	0	0
Vitaceae	*Leea tetrasperma*	Light Blue	0	2	0	**43**	14	0	0	0	0	0
Euphorbiaceae	*Euphorbia reniformis*	Light Blue	0	9	0	**40**	5	5	0	0	0	0
Phyllanthaceae	*Margaritaria indica*	Light Blue	0	0	0	**37**	4	15	0	0	0	0

continued on following page

Table 2. continued

Family	Species	Area	1	2	3	4	5	6	7	8	9	10
Cyperaceae	*Cyperus paniceus*	Light Blue	0	0	1	**37**	4	16	0	0	4	0
Lamiaceae	*Ocimum americanum*	Light Blue	0	7	1	**35**	5	13	0	0	0	0
Convolvulaceae	*Rivea ornata*	Light Blue	0	0	0	**33**	16	14	1	0	0	0
Euphorbiaceae	***Chrozophora rottleri***	Light Blue	1	11	1	**27**	22	0	0	3	0	0
Annonaceae	*Miliusa mollis*	Light Blue	0	10	11	**26**	4	15	0	0	1	0
Violaceae	*Rinorea bengalensis*	Light Blue	24	15	1	**24**	3	0	0	1	0	0
Polygalaceae	*Polygala triflora*	Light Blue	1	1	1	**22**	3	19	0	0	0	0
Anacardiaceae	*Spondias bipinnata*	Purple	0	6	2	0	**35**	0	0	11	0	0
Rhamnaceae	***Ziziphus mauritiana***	Purple	0	11	2	9	**29**	1	0	8	0	0
Cyperaceae	*Schoenoplectus articulatus*	Purple	0	6	0	28	**29**	0	0	0	0	0
Gesneriaceae	*Aeschynanthus parviflora*	Purple	0	8	4	3	**29**	17	0	1	0	0
Annonaceae	*Polyalthia cerasoides*	Purple	0	8	4	9	**27**	11	0	3	0	0
Annonaceae	***Melodorum siamense***	Purple	5	13	0	5	**27**	0	0	7	0	0
Putranjivaceae	*Drypetes roxburghii*	Purple	0	8	8	0	**26**	22	0	4	2	0
Euphorbiaceae	***Acalypha indica***	Purple	0	10	3	13	**26**	4	0	6	1	0
Anacardiaceae	*Mangifera altissima*	Purple	0	8	7	1	**25**	13	0	5	1	0
Euphorbiaceae	*Mallotus spodocarpus*	Purple	0	6	1	17	**25**	0	0	0	0	0
Fabaceae	*Tephrosia kerrii*	Yellow	0	0	0	0	0	**70**	0	0	0	0
Gesneriaceae	*Petrocosmea kerrii*	Yellow	0	0	0	0	0	**70**	0	0	0	0
Annonaceae	*Goniothalamus laoticus*	Yellow	0	0	1	0	0	**70**	0	0	0	0
Mimosaceae	*Cruddasia insignis*	Yellow	0	0	0	0	0	**70**	0	0	0	0
Buxaceae	*Sarcococca saligna*	Yellow	0	0	0	0	0	**69**	1	0	0	0
Poaceae	*Microstegium vagans*	Yellow	0	0	0	0	0	**68**	1	0	0	0
Meliaceae	*Heynea trijuga*	Yellow	0	0	0	1	1	**67**	0	0	0	0
Zingiberaceae	*Globba schomburgkii*	Yellow	0	0	0	0	0	**67**	0	0	0	0
Droseraceae	*Drosera peltata*	Yellow	0	0	0	0	0	**67**	1	0	0	0
Cyperaceae	*Carex cruciata*	Yellow	0	0	0	0	0	**67**	0	0	0	0
Polygalaceae	*Polygala persicariaefolia*	Orange	0	0	0	0	0	20	**61**	0	0	0
Poaceae	*Arundinella bengalensis*	Orange	0	0	0	0	0	14	**52**	0	0	0
Solanaceae	*Lycianthes macrodon*	Orange	0	0	0	0	0	32	**45**	0	0	0
Fabaceae	*Crotalaria ferruginea*	Orange	0	0	0	0	0	33	**45**	0	0	0
Urticaceae	*Boehmeria malabarica*	Orange	0	0	0	0	0	36	**41**	0	0	0
Urticaceae	*Debregeasia longifolia*	Orange	0	0	2	0	0	33	**40**	0	0	0
Fabaceae	*Dalbergia stipulacea*	Orange	0	0	0	0	0	31	**40**	0	0	0
Urticaceae	***Boehmeria clidemioides***	Orange	0	0	1	0	0	32	**40**	0	0	0
Actinidiaceae	*Saurauia petelotii*	Orange	0	0	5	1	1	23	**37**	2	0	0
Urticaceae	*Pouzolzia pentandra*	Orange	0	0	11	1	1	29	**35**	0	0	0
Clusiaceae	*Calophyllum calaba*	Light Green	22	4	0	0	1	0	0	**36**	4	0
Oleaceae	*Olea salicifolia*	Light Green	0	0	14	0	10	5	7	**32**	0	0
Phyllanthaceae	*Glochidion coccineum*	Light Green	0	0	12	1	9	2	4	**32**	0	0

continued on following page

Table 2. continued

Family	Species	Area	1	2	3	4	5	6	7	8	9	10
Euphorbiaceae	***Croton krabas***	Light Green	0	1	1	3	21	0	0	**27**	0	0
Phyllanthaceae	*Actephila collinsae*	Light Green	8	17	7	1	21	0	0	**22**	1	0
Phyllanthaceae	*Breynia fruticosa*	Light Green	0	5	5	3	17	4	0	**20**	14	1
Cyperaceae	*Eleocharis geniculata*	Light Green	18	15	3	10	15	0	0	**19**	0	0
Dioscoreaceae	*Dioscorea alata*	Light Green	0	0	14	0	6	10	0	**16**	8	0
Dipterocarpaceae	*Anisoptera costata*	Light Green	11	7	2	3	8	1	0	**13**	7	11
Cyperaceae	*Cyperus cuspidatus*	Blue Green	0	1	0	6	3	11	0	0	**28**	4
Cyperaceae	*Lipocarpha pygmaea*	Blue Green	0	0	0	12	22	13	0	0	**26**	1
Apocynaceae	*Aganonerion polymorphum*	Blue Green	0	0	2	6	6	2	0	11	**21**	0
Dipterocarpaceae	*Dipterocarpus obtusifolius*	Blue Green	1	3	2	7	13	14	0	7	**15**	10
Lamiaceae	*Orthosiphon rotundifolius*	Blue Green	0	0	6	1	4	5	0	0	**14**	0
Dioscoreaceae	*Dioscorea arachnida*	Blue Green	0	0	1	0	4	12	0	0	**14**	0
Euphorbiaceae	*Cladogynos orientalis*	Blue Green	11	4	3	2	1	5	0	0	**13**	0

shows a maximum reduction of 431 species in the north and a species gain of 108 species in the south (Figure 11a). At the same time the northern and the northeastern centres of diversity have shifted somewhat southwards.

The analysis of the optimum number of clustergroups for the cluster analysis on the 799 projected SDMs shows that the optimum pruning point is at five cluster groups (Figure 7b). The geographical location and cluster tree are shown in Figure 10. The position of the clusters supports a shift of the centre of species richness in northern Thailand (compare Figures 3 and 9). The Peninsula still forms one region. But in the North together with a part of the North-east two longitudinal parallel regions are formed, of which the Pink one is quite distinct. The Pink region corresponds to a long line in the cluster diagram (Figure 10b), indicating that many similar species present. The Pink area also corresponds with the Yellow northern area dilimited under present climatic conditions (Figure 8a). The future Yellow area (Figure 10a) is either a distinctly different floristic region or a subset of part of the Yellow region under present conditions (the future Yellow region forming the other subset; see also discussion). The central Thai regions are more or less split into two, whereby the two river systems and deltas of Central and Eastern Thailand form one

Figure 9. (a) Species densities per grid cell derived from the Species Distribution Models for the 2050 distributions (Database II); (b) Species weighted endemicity per grid cell derived from the Species Distribution Models for 2050 (Database II). In the centre of the circles there are red grid cells that show a very high weighted endemicity. (White and blue are low densities, red high densities)

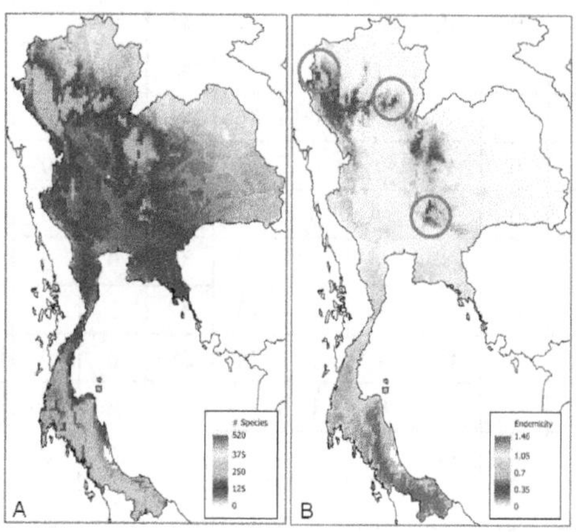

Figure 10. Cluster analysis of the species compositions based on the 2050 SDM distributions per grid cell (B). (a) Optimisation for five phytogeographical areas (see Figure 7b); (b) Cluster diagram; the triangles represent the united grid cells pertaining to the same floristic/phytogeographical region

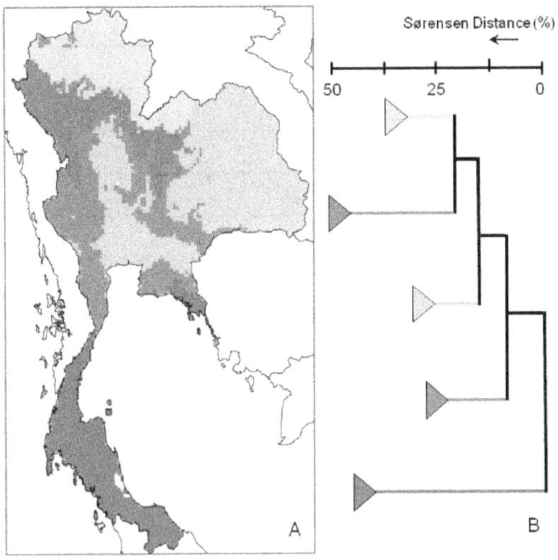

floral district (light blue) different from the other central Thailand lowland regions (dark blue). The indicator species for the five future floristic regions are listed in Table 3. The species indicative for the Peninsula (red in Figure 10a) and those of the pink region in the North and North-eastern have the highest indicator values; those of the other regions are also found in the neighboring regions.

When the floristic regions are overlaid with the Global Land Cover 2000 dataset it is shown that only 20.3% of Thailand was forested in 2000 (Figure 12). The current situation (year 2010) is probably worse. The percentages covers for forest for the present 4 and 10 floristic regions and the future 5 regions are geographically depicted in Figure 12 and given in Table 4. The hatched areas in Figure 12 indicated the National Protected Areas. For some regions the percentages of forest remaining are as low as 1% (Table 4).

Figure 11. (a) SDMs of 2050 with those of the present day subtracted. Red shows the areas where species densities become less in 2050, the higher the intensity the larger the loss of biodiversity. Green represents the grid cells with a gain of species, again the higher the intensity of green, the higher the increase in biodiversity; (b) Terrestrial Ecoregions of the Indo-Pacific for Thailand (Olson et al., 2001; Wikramanayake et al., 2002); downloaded from and after: http://www.worldwildlife.org/science/data/item1875.html

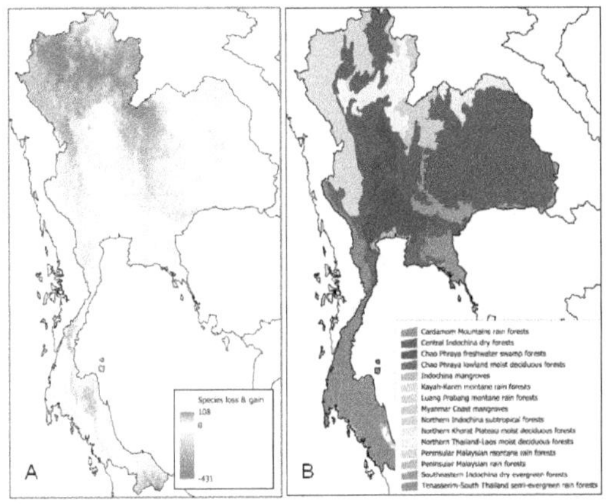

4. DISCUSSION

4.1 Reliability of Data

The flora of Thailand has an estimated 10,250—12,500 species (Parnell, 2000; Middleton, 2003) for which the two databases with samples of 3,187 and 1,399 species each (280 species in common) are considered to be representative, though this cannot be easily checked as a large part of the flora is still badly known, e.g., not yet published in the Flora of Thailand series. However, the results of the cluster and NMS analyses of both datasets are so highly similar (Figures 5 and 8), that we consider the two samples to be sufficiently large and diverse to enable firm conclusions to be drawn.

SDMs are models that only show the potential distribution ranges of species. Thus, it does not necessarily follow that species do occur in all

Table 3. The top ten (when present) indicator values of species (in bold in columns Red to Orange) for the five phytogeographical areas under future climatic conditions (Figure 10a). The colours refer to those used in Figure 10a. Species names in bold are indicator species for the same area if either four or ten areas phytogeographic areas are designated based on the present distributions (Tables 1 and 2, respectively); species names in italics are species indicative for different areas under present and future distributions. The numbers are indicator values between 0 and 100

Family	Species	Area	Red	Middle Blue	Light Blue	Pink	Yellow
Santalacae	***Dendrotrophe varians***	Red	**97**	0	0	0	0
Dipterocarpaceae	***Shorea guiso***	Red	**91**	0	0	0	0
Violaceae	***Rinorea lanceolata***	Red	**90**	1	0	0	0
Anacardiaceae	***Bouea oppositifolia***	Red	**90**	1	0	0	0
Cyperaceae	*Scleria ciliaris*	Red	**89**	0	0	0	0
Dipterocarpaceae	*Anisoptera curtisii*	Red	**89**	0	0	0	0
Myrtaceae	*Syzygium polyanthum*	Red	**88**	1	0	0	0
Cyperaceae	*Hypolytrum nemorum*	Red	**87**	1	0	0	0
Eriocaulaceae	*Eriocaulon willdenovianum*	Red	**87**	1	0	0	0
Phyllanthaceae	*Phyllanthus pulcher*	Red	**86**	2	0	0	0
Phyllanthaceae	*Phyllanthus collinsae*	Middle Blue	6	**32**	1	1	0
Rhamnaceae	*Ventilago calyculata*	Middle Blue	0	**31**	0	3	3
Euphorbiaceae	***Cleidion javanicum***	Middle Blue	14	**30**	7	29	0
Phyllanthaceae	*Bridelia ovata*	Middle Blue	8	**29**	20	8	0
Annonaceae	*Miliusa mollis*	Middle Blue	0	**28**	2	8	1
Euphorbiaceae	***Croton krabas***	Middle Blue	0	**27**	23	3	0
Phyllanthaceae	*Actephila collinsae*	Middle Blue	25	**27**	20	3	0
Euphorbiaceae	***Acalypha indica***	Middle Blue	1	**27**	13	0	0
Sapindaceae	*Zollingeria dongnaiensis*	Middle Blue	10	**25**	17	2	0
Rhamnaceae	***Ziziphus cambodiana***	Middle Blue	0	**24**	0	2	0
Euphorbiaceae	***Chrozophora rottleri***	Light Blue	14	19	**31**	1	0
Euphorbiaceae	*Agrostistachys indica*	Light Blue	28	15	**31**	3	1
Pteridaceae	*Acrostichum aureum*	Light Blue	24	16	**31**	3	2
Balsaminaceae	*Hydrocera triflora*	Light Blue	26	18	**30**	5	0
Polygalaceae	*Xanthophyllum lanceatum*	Light Blue	23	15	**29**	1	1
Sapotaceae	*Manilkara hexandra*	Light Blue	22	19	**28**	9	2
Rhamnaceae	***Ziziphus mauritiana***	Light Blue	7	23	**25**	1	0
Cyperaceae	*Scleria tonkinensis*	Light Blue	24	17	**25**	11	4
Annonaceae	***Melodorum siamense***	Light Blue	18	15	**25**	0	0
Euphorbiaceae	*Sampantaea amentiflora*	Light Blue	0	18	**24**	2	2
Asteraceae	*Blumea lacera*	Pink	0	0	0	**80**	0
Mimosaceae	*Xylia xylocarpa*	Pink	0	1	0	**79**	0
Fabaceae	*Crotalaria kurzii*	Pink	0	1	0	**78**	0
Magnoliaceae	*Michelia baillonii*	Pink	0	0	0	**75**	0

continued on following page

Table 3. continued

Family	Species	Area	Red	Middle Blue	Light Blue	Pink	Yellow
Urticaceae	***Boehmeria clidemioides***	Pink	0	0	0	**74**	0
Caesalpiniaceae	*Bauhinia viridescens*	Pink	0	0	0	**73**	1
Fabaceae	*Cajanus goensis*	Pink	0	0	0	**72**	0
Anacardiaceae	*Buchanania latifolia*	Pink	0	5	0	**72**	0
Fabaceae	*Uraria campanulata*	Pink	0	3	0	**71**	2
Poaceae	*Setaria palmifolia*	Pink	0	1	0	71	0
Dioscoreaceae	*Dioscorea glabra*	Orange	0	5	0	40	**43**
Dipterocarpaceae	*Dipterocarpus obtusifolius*	Orange	0	1	0	14	**38**
Dioscoreaceae	*Dioscorea decipiens*	Orange	0	6	0	31	**37**
Lentibulariaceae	*Utricularia scandens*	Orange	0	2	0	27	**36**
Phyllanthaceae	*Bridelia affinis*	Orange	0	0	0	5	**33**
Cyperaceae	*Lipocarpha pygmaea*	Orange	0	3	3	5	**21**
Convolvulaceae	*Rivea ornata*	Orange	0	1	4	4	**14**
Clusiaceae	*Hypericum japonicum*	Orange	8	0	0	3	**12**
Cyperaceae	*Cyperus cuspidatus*	Orange	0	0	0	0	**10**

Figure 12. Comparison between floristic regions, remaining forest (http://bioval.jrc.ec.europa.eu/products/glc2000/products.php, area 6) and areas protected by the National Park, Wildlife and Plant Conservation Department of Thailand. The maps only show the parts of the floristic regions that are still covered by forest (coloured parts), the protected areas are shaded. (a) Four present day floristic regions (Figure 8a); 1-4: four centres of plant diversity as indicated by Davis et al., 1995: 1= Doi Chiang Dao Wildlife Sanctuary, 2 = Doi Suthep-Pui National Park, 3 = Thung Yai-Huai Kha Khaeng World Heritage Site, 4 = Khao Yai National Park; (b) Ten present day floristic regions (Figure 8Bb; (c) Future floristic regions (Figure 10a)

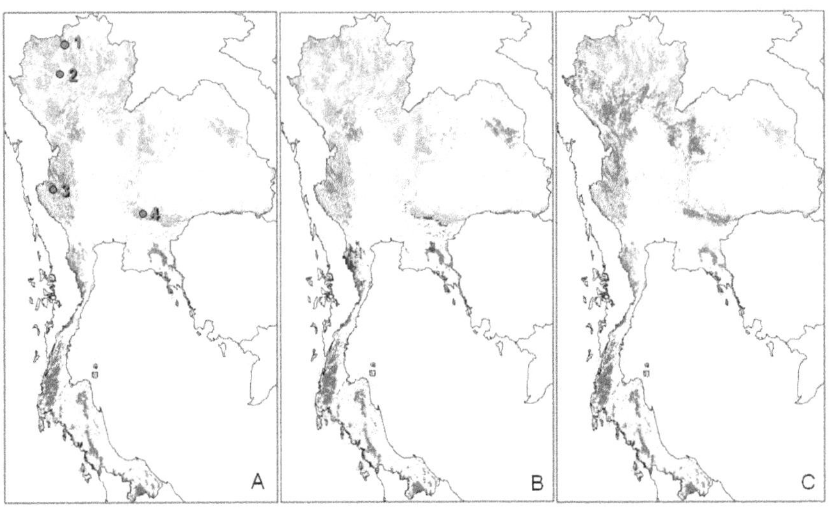

Table 4. The three sets of floral regions with their spatial coverage (percentage) of Thailand and the amount of forest left per floral region (e.g., the red area in Figure 8a occupies 20 % of Thailand and 26 % of the red area is still forested). The forest cover is based on http://bioval.jrc.ec.europa.eu/products/glc2000/products.php, area 6

	Region	% Surface of Thailand	Forest % left
Four Floral regions (Figure 8a)			
	Red	15	26
	Blue	35	12
	Yellow	32	35
	Green	17	5
Ten Floral regions (Figure 8b)			
	Red	15	26
	Dark blue	5	16
	Middle blue	8	31
	Light blue	8	2
	Purple	14	6
	Yellow	30	36
	Orange	3	18
	Light green	5	1
	Blue green	7	9
	Dark green	5	4
Five Future regions (Figure 10a)			
	Red	14	25
	Middle Blue	24	25
	Light Blue	37	5
	Pink	10	42
	Yellow	15	34

suitable grid cells, just that the potential exists. The use of large numbers of significant SDMs should diminish this problem. The distributions we modeled only used abiotic (climate and soil) variables to determine the distributions. Extending the distribution modeling with biotic variables in the future (e.g., dispersal capacity of seeds, pollination syndrome) will possibly increase the accuracy of the SDMs. Our data are on a scale of 5-arc minutes (ca. 10 by 10 km), which means that the effect of small areas (islands), size limited soil types (e.g., mangrove, swamp forests), coastal grid cells with more sea than land and fast changes in altitude (mountains) vanish as only mean values are used per grid cell.

4.2 Centres of Plant Diversity in Thailand

Davis *et al.* (1995) list four centres of plant diversity in Thailand, two in the Northern region: Doi Chiang Dao Wildlife Sanctuary and Doi Suthep-Pui National Park, Thung Yai-Huai Kha Khaeng World Heritage Site occurs in the South-western region, and Khao Yai National Park in the Central/Eastern region (see numbers 1-4 in Figure 12a). When we compare these centres with the centres of species richness and weighted endemicity as depicted in Figure 3A and B, then it is obvious that Khao Yai and Thung Yai-Huai Kha Khaeng mountains are not centres of plant diversity nor endemicity. Doi Chiang Dao and Doi Suthep (no. 1 and 2 Figure 12A) are at the edges of the northern centre of plant diversity (Figure 3a) and weighted endemism (Figure 3b). Doi Chiang Dao is at the northern edge and is not really in the red area indicative for high species richness, while Doi Suthep lies between two of the red areas. When compared with the projections for 2050 (Figure 9) are considered then, with the exception of Khao Yai, none of the plant diversity centres of Davis *et al.* (1995) form part of the centres of species richness or weighted endemism. The centres of species richness and weighted endemicity do not correspond now, and will never correspond, to those delimited on subjective expert opinion (Davis *et al.* 1995).

4.3 Present-Day Species Richness, Endemicity and Floristic Regions

The results of the analyses of Database I (NMS, two cluster analyses; Figure 5) and Database II (Figure 8) are very similar. In the following discussion one must keep in mind that the boundaries used in the analyses of Database I (Figure 5) are artificial, as they are province boundaries and these are political and not – at least to plants – natural boundaries. The grid cells in Figure 8 allow the depiction of more natural areas. But the question remains can phytogeographic regions be recognized? The answer seems to be affirmative, however, the number of floristic regions shown in Figs. 5 and 8 varies between four and ten floristic regions (Figure 7a). Thus, the next question is, which regions should be recognised?

Only two of the seven floristic regions phytogeographic regions as used in the Flora of Thailand (Smitinand, 1958; Figure 1) are confirmed as having reality by our analysis and these only in part (Peninsular and Northern regions). Instead, based on our analyses, we propose to recognise four areas (Figure 8a), which are all discontinuous, thereby more accurately depicting the natural borders of the areas and the presence of indicator species (see below).

4.3.1 Phytogeographic Area 1 (Figures 5, 8a and b: Red Area): Southern Region

This comprises the former Peninsular region together with part of the South-eastern region, mainly comprising the provinces Chanthaburi and Trat. The NMS analysis (Figure 5a) also adds the province Rayong (though this is not included with the cluster analyses, Figures 5b and c), an addition which is confirmed by the cluster analysis of the SDMs (Figure 8). In fact, only a part of Rayong seems to belong to the Peninsular Area (Red area in Figure 8a). In this figure the Southern Reigon extends into Prachuap Khiri Khan, thus further to the north than the NMS and other cluster analyses could/did show (Figure 5). The presence of the 'Peninsular' flora in the South-eastern area has never been recorded before. It is explicable climatically, as the Southwest monsoon, after having passed the narrow Peninsula, picks up humidity again when passing the Gulf of Thailand, after which it reaches Trat and Chanthaburi in the South-eastern and brings plenty of rain. Both provinces are among the wettest provinces of Thailand and are climatically similar to areas of the Peninsula. It is clear that the dark blue area forms a buffer area between it and the flora of the rest of the country (Figure 8b). Perhaps, the dark blue area indicates a transition between evergreen and deciduous forest exists in this region. Subdivision into 10 (Figure 8b) as opposed to 4 floristic (Figure 8a) regions does not alter the red area at all; suggesting that it is very clearly delimited.

Tables 1 and 2 show the top ten indicator species for this region. The southern limit of this region is artificial (border of Thailand), but mainly coincides with the Kangar (Malaysia)-

Pattani (Thailand) line, which denotes the border of the Malesian tropical everwet flora (Van Steenis 1950). This southern limit, which is not investigated further here, may denote a climate change between everwet forest in Malaysia and seasonal in the Peninsula, and may also be the result of two Mio/Pliocene (24-13 Ma and 5.5-4.5 Ma) seaways (Krabi to Surat Thani, Alor Setar to Songkhla; Woodruff 2003) traversing the Peninsula in a N-S direction when sea levels were even more than 150 m higher than present day. The northern boundary is present both in plant (Woodruff, 2003) and bird distributions (Hughes *et al.*, 2003). It represents the transition between seasonal evergreen rain forest (on the Peninsula) and mixed moist deciduous forest (to the north of the Peninsula). The boundary is skewed lying further to the north in the west than it does in the east. This latter effect is likely to be due to higher rainfall in the west, partly because of the winds, partly because of the mountain ranges in the Peninsula (Woodruff, 2003). In addition, the boundary itself is sharper in the east than in the west. However, the reason why the demarcation exists is still vague (Woodruff, 2003). From Figure 8b it is obvious that here too the dark blue area forms a buffer with more northern areas.

4.3.2 Phytogeographic Area 2 (Figures 5 and 8: Yellow Region): Northern Region

The second area, present in most analyses, is an expanded Northern region with extensions reaching into the North-eastern, Eastern, and South-western areas of Smitinand (1958; Figure 1). Typically this whole area contains discontinuous mountains. The cluster analysis based on the SDMs (Figure 8) shows a lesser extension into the South-western region (Kanchanaburi) than does the NMS (Figure 5a) and other cluster analyses (Figures 5b and c), but this is probably an artifact caused by the large size of Kanchanaburi and the use of provinces as units for the NMS and UPGMAs (Figure 5). Like the UPGMA (Figure 5c) the cluster analysis of the SDMs shows that the Northern region extends south to the former South-eastern region. The SDM division into 10 areas (Figure 8b) divides the Northern region into a Yellow and Orange area, which latter corresponds to most of Chiang Rai. Again, indicator species are present for the Yellow (Northern) and Orange (Chiang Rai) areas as shown in Table 1 (Northern only, incl. Chiang Rai) and in Table 2. Columns 6 and 7 of Table 2 show that there is no overlap of indicator species from the yellow Northern area into the Orange Chiang Rai, but there is overlap in the reverse direction as the indicator species for Chiang Rai are also present in the Northern region.

4.3.3 Phytogeographic Area 3 (Figures 5a, c, and 8a: Green Area): Eastern region.

A possible third floristic region is formed from the eastern part of the former North-eastern and Eastern regions extending westwards in the NMS and second UPGMA cluster analysis (Figures 5a and c) than in the cluster analysis based on the SDMs (Figure 8a). However, the cluster analysis of Figure 5b does not recognise this area. We believe that the results shown in Figure 8a are more realistic, accurate and precise than those of Figures 5a and c, because the grid cells represent more natural boundaries than do the provinces. From Figure 8 it is clear that the major river systems of the Eastern, draining into the Mekong, belong to a different floristic region (blue colour). When 10 floristic areas are recognized (Figure 8b) as opposed to 4, then the Green Area (Figure 8a) falls apart into three sub-areas. However, it should be noted that the most northern of these (Dark Green) lacks indicator species (Table 2, area 10). Therefore, we do not believe that such further subdivision is appropriate and we prefer to recognise only the green area of Figure 8a.

4.3.4 Phytogeographic Area 4 (Figures 5a, c, and 8a: Blue area): Central region.

The last region we recognise comprises the Central Lowlands. This is the least continuous area of all four phytogeographic areas and the one for which the results of our various analyses differ the most (Figures 5 and 8). Several of the differences shown in Figure 5 can be explained as a further subdivision of the Central Blue Area (Figure 8b). The river systems (light blue Figure 8b) in the central and former North-eastern/Eastern regions of Thailand form one subdivision, and these may be also indicated at a gross level by the light blue areas shown in Figures 5a and c. That these two river system constitute a floristic subregion is quite remarkable, because they are independent and their physiography is different. The central river system drains into the Gulf of Thailand, flows mainly through lowland, is broad and forms the delta in which Bangkok is situated. The rivers in the North-eastern/Eastern flow to the Mekong river, run through aras of higher altitude and the rivers are narrow in comparison with the central river system. Indicator species for the various Central Lowland (Blue) areas can be found in Tables 1 (Area Blue) and 2 (areas 2, 3, 4, and 5).

4.3 Comparison of Floristic Regions with WWF Ecoregions

In the WWF Ecoregions for Thailand (Olson *et al.*, 2001; Wikramanayake *et al.*, 2002) 15 different floristic regions (Figure 11b) are delimited. When compared with the floristic regions derived from the SDMs in Figure 8a and b many differences are obvious.

The mangroves (two areas in Figure 11b, Myanmar Coast mangroves, light blue-green, ecoregion IM1404, and the Indochina Mangroves, light middle blue, ecoregion IM1402) are lacking in the floristic regions based on the SDMs, because the mangrove areas are too small to be distinguished in the grid cells as separate soil types. Clearly, therefore our analysis cannot disprove the existence of these areas and these two floristic regions should be recognized.

Another difference is that the WWF recognized two additional areas (Figure 11b) for the Peninsular region, the Peninsular Malaysian montane rain forest, yellow, ecoregion IM0144, and the Peninsular Malaysian lowland rain forest, orange, ecoregion IM0146, which were not recognized by any of our analyses (Figures 5 and 8). For the remainder of the Peninsular region (Figure 11b), the delineation of the Tenasserim-South Thailand semi-evergreen rain forests (Figure 11b, red, ecoregion IM0163) is very similar to our SDM cluster results (Figure 8), differing only by extending a bit further to the north (Kanchanaburi province). Where our cluster analysis also recognizes the South-eastern provinces as part of the Peninsular region, the WWF recognizes this as a separate area, the Cardamom Mountains rain forest (Figure 11b, pink, ecoregion IM0106).

Especially in the North the WWF recognizes many more floristic- or ecoregions. The Luang Prabang montane rain forest (green area in Figure 11b; ecoregion IM0121) falls into two parts and more or less corresponds with much of the orange Chiang Rai area in Figure 8b. Otherwise the areas delimited by the WWF in the North are quite different to ours. Where our analysis recognizes only the Northern region (Figure 8a, yellow) and at best two regions (Figure 8b, yellow & orange) the WWF recognizes five different ecoregions: i) the Kayah-Karen montane rain forests (Figure 11b, light blue, ecoregion IM0119) in the north-west, ii) the Northern Thailand-Laos moist deciduous forests (Figure 11b, orange-yellow, ecoregion IM0139) in the east of the North Thailand, iii) the north-eastern Luang Prabang montane rain forests (Figure 11b, blue-green, ecoregion IM0121), iv) the Northern Indochina subtropical forests (Figure 11b, green, ecoregion IM0137). Furthermore, a large part of the Northern Thailand is traversed by v) the Central Indochina dry forests (Figure

11b, purple, ecoregion IM0202), which in our analysis mainly corresponds with the cultivated areas in central and eastern Thailand (Blue and Green in Figure 8a).

The WWF central Chao Phraya freshwater swamp forests (Figure 11b, dark blue, ecoregion IM0107) corresponds with our light blue, Central Lowlands area in Figure 8b. In the east of Thailand the WWF recognizes the Northern Khorat Plateau moist deciduous forests (Figure 11b, yellow-green, ecoregion IM0138), a region which partly corresponds with our dark green region (Figure 8b). Our analysis (Figure 8) was unable to delineate a region that corresponds with the WWF South-eastern Indochina dry evergreen forests (Figure 11b, orange-red, ecoregion IM0210) that is located along the southern part of east Thailand.

The differences between both maps can be explained by the fact that Olson *et al.* (2001) and Wikramanayake *et al.* (2002) use only few species and their maps use both zoological and botanical information with the emphasis on the zoological differences between areas. Our Figure 8 is based on an as large as possible botanical dataset only.

4.5 Future Floristic Regions in 2050

The number of species per grid cell (Figure 9) diminishes strongly in the Northern and North-eastern when compared with the present species numbers (Figure 3), from 520 species for many grid cells in Figure 3a to 336 in Figure 9a. Also, the spatial extent of the northern/northeastern area with higher biodiversity decreases strongly when compared to the present situation. The Peninsular region appears to be relatively stable, though endemicity even seems to increase (compare Figures 3b and 9b, note the different scales!). The difference in the number of species per grid cell between the present and the future is shown in Figure 11a. Red areas indicate loss of species, which is mainly the case in the north and central Thailand, and some parts of the Peninsular region. On the other hand, the Eastern region and parts of the Peninsular region show an increase in the number of species (green). The pattern in Figure 11a might indicate that dispersal of the flora from the red areas in the west to the green areas in the east could occur.

In 2050 the centres of weighted endemicity are reduced to a few red grid cells (circled in Figure 9b), and though these contain very high numbers of endemics, they are generally very small, perhaps too small to maintain such a rich endemic flora.

Table 5 shows which species will become extinct in Thailand in 2050 (i.e., their projected SDMs are blank). The total numbers of species drop far more (compare Figure 9a with Figure 3a, both figures have the same scale). In the North the drop of c. 200 species is caused by species that become absent in most but not all grid cells (i.e., their SDMs show far less grid cells in 2050 than at present).

The ISA (Figure 7b) shows that the optimal pruning point of the cluster tree based on projected SDMs is at five phytogeographical areas (Figure 10a); one more than recognized under the present climatic conditions. The Southern Area (Figure 10a, red) becomes slightly smaller. The Northern region (Figure 8, yellow) mainly subtracts to the Pink Area in Figure 10a and the yellow area in Figure 10a may become a region with a new floristic composition (see next paragraph). The eastern and central regions are subdivided differently. The Central river system and delta together with the eastern sandstone shield (light blue in Figure 10a) becomes one region, intersected and flanked by a middle blue region.

The floral change that Thailand faces according to the models is very dramatic, not only is there a substantial reduction in the number of species, but also the floristic regions differ, at least in part. However, the predictions for 2050 are premature. It is still questionable if the climatic change models used are sufficiently accurate, e.g., a lesser rise in temperature will have a drastically different influence. Also, we can only use species that are currently present to be projected according to the

Table 5. Species of Database II, which become extinct in 2050 according to the predictions of the SDMs

Family	Species
Gesneriaceae	*Aeschynanthus hosseusii*
Convolvulaceae	*Argyreia siamensis*
Poaceae	*Arthraxon hispidus*
Poaceae	*Arthraxon lancifolius*
Caesalpiniaceae	*Caesalpinia digyna*
Fabaceae	*Clitoria macrophylla*
Mimosaceae	*Cochlianthus gracilis*
Fabaceae	*Desmodium pulchellum*
Fabaceae	*Desmodium triflorum*
Lamiaceae	*Elshotzia winitiana*
Convolvulaceae	*Evolvulus nummularius*
Cyperaceae	*Fimbristylis yunnanensis*
Lamiaceae	*Geniosporum siamensis*
Orchidaceae	*Geodorum recurvum*
Poaceae	*Isachne albens*
Lamiaceae	*Isodon hispidus*
Acanthaceae	*Justicia grossa*
Scrophulariaceae	*Limnophila chinensis*
Orchidaceae	*Liparis paradoxa*
Lauraceae	*Litsea cubeba*
Molluginaceae	*Mollugo pentaphylla*
Cyperaceae	*Pycreus sanguinolentus*
Poaceae	*Sporobolus kerrii*
Poaceae	*Sporobolus tetragonous*
Fabaceae	*Stylosanthes sundaica*
Myrtaceae	*Syzygium tetragonum*
Poaceae	*Themeda arundinacea*
Poaceae	*Tripogon trifidus*
Fabaceae	*Vigna dalzelliana*
Asteraceae	*Wedelia montana*

climate senario of 2050 and we cannot take into account the flora of adjacent countries, which is much less well-known than that of Thailand. It is very likely that deciduous species from China, Laos, Burma, etc. will migrate to the Northern region, so that the loss in flora will in effect be less. Trisurat *et al.* (2009) showed for northern Thailand that deciduous species, especially, will gain importance, and these are exactly the species that might migrate from more northern areas to Thailand (i.e., they will probably populate the northern yellow area in Figure 10A). Thus, the heavy biodiversity loss as predicted by the SDMs for the northern half of north Thailand may be compensated for.

4.6 Remaining Forest and Protected Areas

A concern of modelling is that the models only show areas with currently suitable environmental conditions and in 2050. The models do not consider human influence. Figure 12 shows which parts of the floristic regions are still forested (the remaining coloured parts), this is also expressed in percentage terms in Table 5. The maps (four present-day floristic areas, Figure 12a, ten present-day floristic areas, Figure 12b, five future floristic regions, Figure 12c) are overlain with the presently protected areas. Another concern is the granularity of the model used to delimit the floristic regions as it makes a substantive difference both to the amount of forest in each floristic region and to the amount protected therein. So, if we group the present day flora into four floristic areas (Figure 8a) we find that each is relatively well covered with forest, though the coverage in the east is minimal (Figure 12a; Table 5: green area: 5% forested). If instead, we form ten instead of four regions (Figure 8b) then Chiang Rai (orange region) has little forest left (18%) and has hardly any coverage by protected areas. Likewise, the purple and light blue areas in the centre of Thailand (Figure 8b) have almost no forest left (6% and 2%, respectively) and none of it is protected. The dark blue area (Figure 8b) contains some protected areas in the former South-western region (Kaeng Krachan, Maenam Phachi) and Khao Ang Ru Nai in the former South-eastern region (9% forest left). The three green regions in the east almost completely lack forest. Best represented is the blue-green middle region with 9% forest left, which is protected by areas like Phu Phan, Huai Huat and Phu-si-tan. The dark green area (upper region, 4% forest) has one tiny protected area, Phu Wua. The worst forested and protected area is in light green (lower region, 1% forest) and the only protection is in the far east along the Cambodian border (Pha Tam, Kaeng Tana, Phu Chong Na Yoi).

If deforestation and land use change will completely stop immediately, then there will be some forested part left in each of the five recognized future floristic regions (Figure 12c), with the light blue area in the east least covered by the present day forest remnants (5% cover, Table 5). The predicted movement of species in the Southern region (Figure 11a) from the western towards the eastern part of the Peninsula and the predicted influx of species from across the Malaysian border, is luckily covered by already existing protected areas. Thus, it is evident that the areas that are currently protected are and will remain crucial to protect as much botanical diversity as possible. Increasing the size and number of the protected areas and creating dispersal links between them (as the Forestry Department and the Forest Division of Kasetsart University are planning to do) should remain items that are high on the political agendas. Further fragmentation of biodiversity rich areas (road expansion, shifting cultivation, population pressure, etc.) should thereby be limited as much as possible (Trisurat *et al.*, 2010).

5. CONCLUSION

The distinction of floristic regions or phytogeographic areas is based on the co-occurrence of plant species. Co-occurrence is generally indicative of interaction between the species, not only the plant species used in our samples, but all plants and animals in a region. Therefore, the term ecoregion is also often used. Smitinand (1958) recognized seven floristic regions in Thailand, which continue to be used. Only two of these are more or less present in the results of our analyses (Peninsular part and Northern). We would like to replace them with a Peninsular Province (extending in the South-eastern), a Northern Province (with extensions into the South-western and South-eastern), an Eastern Province and the Central Lowlands. All of them are more or less continuous areas, with at least ten species that can be used as indicator species.

We also demonstrate that a small change in climate may have severely disruptive effects on the ecosystems. Many species may disappear, interactions will probably change, species will migrate, but most important is the imminent diversity loss, amplified because migration is not always possible due to large scale human agricultural or other activities. The Species Distribution Models (SDMs) and climate scenarios for 2050 may be criticized because of the uncertainty in the models. Nevertheless, the effects of climate change are already noticeable. Loss of biodiversity and disintegration of ecosystems are already resulting in loss of ecosystem services such as scarcity of fresh water, timber, fisheries, genetic resources, climate regulation, protection from natural hazards, erosion control, etc. Other problems, like desertification, air pollution, chronic diseases, slowing economies, etc., wil undoubtedly increase. Discussions within the World Economic Forum (www.weforum.org) are attempting to address these problems and link them to economic losses that already amount to tens of billions of US dollars per year. Thus, preventing biodiversity loss is globally important: forewarned is forearmed, and it should become the main political focus of every country.

ACKNOWLEDGMENT

Nantachai Pongpattananurak (Kasetsart University, Faculty of Forestry, Department of Conservation) is thanked for his mapped information about the protected areas in Thailand. Yongyut Trisurat and Sarawood Sungkaew (Kasetsart University, Faculty of Forestry) are thanked for their invitation to contribute to this book.

REFERENCES

Anderson, R. P., & Peterson, A. T., & Gomez—Laverde, M. (2002). Using niche—based GIS modeling to test geographic predictions of competitive exclusion and competitive release in South American pocket mice. *Oikos*, *98*, 3–16. doi:10.1034/j.1600-0706.2002.t01-1-980116.x

Araujo, M. B., & Guisan, A. (2006). Five (or so) challenges for species distribution modelling. *Journal of Biogeography*, *33*, 1677–1688. doi:10.1111/j.1365-2699.2006.01584.x

Bates, B. C., Kundzewicz, Z. W., Wu, S., & Palutikof, J. P. (2008). *Climate change and water*. Geneva: Technical Paper of the Intergovernmental Panel on Climate Change.

Crisp, M. D., Laffan, S., Linder, H. P., & Monro, A. (2001). Endemism in the Australian flora. *Journal of Biogeography*, *28*, 183–198. doi:10.1046/j.1365-2699.2001.00524.x

Davis, S. D., Heywood, V. H., & Hamilton, A. C. (1995). *Centres of plant diversity 2*. Cambridge, U.K.: WWF and IUCN Publications Unit.

Dufrêne, M., & Legendre, P. (1997). Species assemblages and indicator species: The need for a flexible asymmetrical approach. *Ecological Monographs*, *67*, 345–366.

Elith, J., Graham, C. H., Anderson, R. P., Dudik, M., Ferrier, S., & Guisan, A. (2006). Novel methods improve prediction of species' distributions from occurrence data. *Ecography*, *29*, 129–151. doi:10.1111/j.2006.0906-7590.04596.x

FAO. (2002). *TerraStat: Global land resources GIS models and databases for poverty and food insecurity mapping*. Land and Water Digital Media Services.

Fielding, A. H., & Bell, J. F. (1997). A review of methods for the assessment of prediction errors in conservation presence/absence models. *Environmental Conservation*, *24*, 38–49. doi:10.1017/S0376892997000088

Franklin, J. (2009). *Mapping species distributions: Spatial inference and prediction*. Cambridge, UK: Cambridge University Press.

Graham, C. H., Elith, J., Hijmans, R. J., Guisan, A., Peterson, A. T., & Loiselle, B. A. (2008). The influence of spatial errors in species occurrence data used in distribution models. *Journal of Applied Ecology*, *45*, 239–247. doi:10.1111/j.1365-2664.2007.01408.x

Hijmans, R. J., Cameron, S. E., Parra, J. L., Jones, P. G., & Jarvis, A. (2005). Very high resolution interpolated climate surfaces for global land areas. *International Journal of Climatology*, *25*, 1965–1978. doi:10.1002/joc.1276

Holmgren, P. K., Holmgren, N. H., & Barnett, L. C. (1990). *Index Herbariorum*. New York: International Association of Plant Taxonomists.

Hughes, J. B., Round, P. D., & Woodruff, S. D. (2003). The Indochinese-Sundaic faunal transition at the Isthmus of Kra: An analysis of resident forest bird species distributions. *Journal of Biogeography*, *30*, 569–580. doi:10.1046/j.1365-2699.2003.00847.x

Jacobs, M. (1962). Reliquiae Kerrianae. *Blumea*, *11*, 427–493.

Liu, C., Berry, P. M., Dawson, T. P., & Pearson, R. G. (2005). Selecting tresholds of occurrence in the prediction of species distributions. *Ecography*, *28*, 385–393. doi:10.1111/j.0906-7590.2005.03957.x

Loiselle, B. A., Jorgensen, P. M., Consiglio, T., Jimenez, I., Blake, J. G., Lohmann, L. G., & Montiel, O. M. (2008). Predicting species distributions from herbarium collections: Does climate bias in collection sampling influence model outcomes? *Journal of Biogeography, 35*, 105–116.

MacKinnon, J. (1997). *Protected areas systems review of the Indo-Malayan realm*. ABS/WCMW/World Bank, Canterbury.

Manel, S., Williams, H. C., & Ormerod, S. J. (2001). Evaluating presence—absence models in ecology: The need to account for prevalence. *Journal of Applied Ecology, 38*, 921–931. doi:10.1046/j.1365-2664.2001.00647.x

Manifold System. (2008). *Manifold® System 8.0 Universal Edition*. Nevada: Manifold Net Ltd.

Maxwell, J. F. (2004). A synopsis of the vegetation of Thailand. *The Natural History Journal of Chulalongkorn University, 4*, 19–29.

McCune, B., & Grace, J. B. (2002). *Analysis of ecological communities*. Gleneden Beach, OR: MjM Software Design.

McPherson, J. M., Jetz, W., & Rogers, D. J. (2004). The effects of species' range sizes on the accuracy of distribution models: Ecological phenomenon or statistical artefact? *Journal of Applied Ecology, 41*, 811–823. doi:10.1111/j.0021-8901.2004.00943.x

Middleton, D. J. (2003). Progress on the flora of Thailand. *Telopea, 10*, 33–42.

Myers, N., Mittermeier, R. A., Mittermeier, C. G., da Fonseca, G. A. B., & Kent, J. (2000). Biodiversity hotspots for conservation priorities. *Nature, 403*, 853–858. doi:10.1038/35002501

Olson, D. M., Dinerstein, E., Wikramanayake, E. D., Burgess, N. D., Powell, G. V. N., & Underwood, E. C. (2001). Terrestrial ecoregions of the world: A new map of life on Earth. *Bioscience, 51*, 933–938. doi:10.1641/0006-3568(2001)051[0933:TEOTWA]2.0.CO;2

Parnell, J. A. N. (2000). The conservation of biodiversity: Aspects of Ireland's role in the study of tropical plant diversity with particular reference to the study of the flora of Thailand and *Syzygium*. In Rushton, B. (Ed.), *Biodiversity: The Irish dimension* (pp. 205–216). Dublin, Ireland: Royal Irish Academy Special Publication.

Parnell, J. A. N., Simpson, D. A., Moat, J., Kirkup, D. W., Chantaranothai, P., & Boyce, P. C. (2003). Plant collecting spread and densities: Their potential impact on biogeographical studies in Thailand. *Journal of Biogeography, 30*, 193–209. doi:10.1046/j.1365-2699.2003.00828.x

Peterson, A. T., Papes, M., & Eaton, M. (2007). Transferability and model evaluation in ecological niche modeling: A comparison of GARP and Maxent. *Ecography, 30*, 550–560.

Phillips, S. J., Anderson, R. P., & Schapire, R. E. (2006). Maximum entropy modeling of species geographic distributions. *Ecological Modelling, 190*, 231–259. doi:10.1016/j.ecolmodel.2005.03.026

Phillips, S. J., & Dudik, M. (2008). Modeling of species distributions with Maxent: New extensions and a comprehensive evaluation. *Ecography, 31*, 161–175. doi:10.1111/j.0906-7590.2008.5203.x

Raes, N. (2009). *Borneo: A quantitative analysis of botanical richness, endemicity and floristic regions based on herbarium records*. Unpublished PhD thesis, National Herbarium of the Netherlands, Leiden University Branch, Leiden.

Raes, N., Roos, M. C., Slik, J. W. F., van Loon, E. E., & Ter Steege, H. (2009). Botanical richness and endemicity patterns of Borneo derived from species distribution models. *Ecography, 32*, 180–192. doi:10.1111/j.1600-0587.2009.05800.x

Raes, N., & Ter Steege, H. (2007). A null-model for significance testing of presence-only species distribution models. *Ecography, 30*, 727–736. doi:10.1111/j.2007.0906-7590.05041.x

Raes, N., & Van Welzen, P. C. (2009). The demarcation and internal division of Flora Malesiana: 1857 – Present. *Blumea, 54*, 6–8. doi:10.3767/000651909X475888

Reddy, S., & Davalos, L. M. (2003). Geographical sampling bias and its implications for conservation priorities in Africa. *Journal of Biogeography, 30*, 1719–1727. doi:10.1046/j.1365-2699.2003.00946.x

Santisuk, T., & Larsen, S. (Eds.), *Flora of Thailand* (*Vol. 6*). Bangkok, Thailand: The Diamond Printing Co.

Santisuk, T., Smitinand, T., Hoamuangkaew, W., Ashton, P., Sohmer, S. H., & Vincent, J. R. (1991). *Plants for our future: Botanical research and conservation needs in Thailand.* Bangkok: Research Publication by USAID/RFD/WWF-US.

Smitinand, T. (1958). The genus Dipterocarpus Gaertn.f. in Thailand. *Thai Forest Bulletin (Botany)*, *4*, 1–50.

Sodhi, N. S., Koh, L. P., Brook, B. W., & Ng, P. K. L. (2004). Southeast Asian biodiversity: An impending disaster. *Trends in Ecology & Evolution, 19*, 654–660. doi:10.1016/j.tree.2004.09.006

Stibig, H. J., Belward, A. S., Roy, P. S., Rosalina-Wasrin, U., Agrawal, S., & Joshi, P. K., Hildanus, Beuchle, R., Fritz, S., Mubareka, S., & Giri, C. (2007). A land-cover map for South and Southeast Asia derived from SPOT-vegetation data. *Journal of Biogeography, 34*, 625–637. doi:10.1111/j.1365-2699.2006.01637.x

Trisurat, Y., Alkemade, R., & Arets, E. (2009). Projecting forest tree distributions and adaptation to climate change in northern Thailand. *Journal of Ecology and Natural Environment, 1*, 55–63.

Trisurat, Y., Alkemade, R., & Verburg, P. H. (2010). Projecting land-use change and its consequences for biodiversity in Northern Thailand. *Environmental Management, 45*, 626–639. doi:10.1007/s00267-010-9438-x

Van Steenis, C. G. G. J. (1950). The delimitation of Malaysia and its main plant geographical divisions. In C. G. G. J. van Steenis (Ed.), *Flora Malesiana* Ser. 1, 1. (pp. lxx—lxxv). Noordhoff-Kolff n.v., Djakarta.

Wikramanayake, E. D., Dinerstein, E., Loucks, C. J., Olson, D. M., Morrison, J., & Lamoreux, J. ... Hedao, P. (2002). *Terrestrial ecoregions of the Indo-Pacific: A conservation assessment.* Washington, DC: Covelo and London: Island Press

Wisz, M. S., Hijmans, R. J., Li, J., Peterson, A. T., Graham, C. H., & Guisan, A. (2008). Effects of sample size on the performance of species distribution models. *Diversity & Distributions, 14*, 763–773. doi:10.1111/j.1472-4642.2008.00482.x

Woodruff, D. S. (2003). Neogene marine transgressions, palaeogeography and biogeographic transitions on the Thai-Malay Peninsula. *Journal of Biogeography, 30*, 551–567. doi:10.1046/j.1365-2699.2003.00846.x

ENDNOTE

1 Colour plates (ThaiPhytogeogrAreasFigures(2).pdf) can be downloaded from ftp://mail.nnm.nl (in Windows Explorer: Click Page and Open FTP Site in Windows Explorer).

Chapter 12
Biodiversity Modelling Experiences in Ukraine

Vasyl Prydatko
International Association Ukrainian Land and Resource Management Center, Ukraine

Grygoriy Kolomytsev
I.I.Schmalhausen Institute of Zoology of National Academy of Sciences of Ukraine, Ukraine

ABSTRACT

Biodiversity modeling in Ukraine was recently developed in order to support policy making and for providing information to e.g. the reporting to the UN Convention of Biological Diversity. This is the first and highly ambitious study on biodiversity and its conditions in Ukraine and some surrounding countries. It includes four different methods to assess and project biodiversity changes: the indicative-index approach, the GLOBIO Mean Species Abundance (MSA) and two species based approaches, one using habitat changes as driving factor (EEBIO) and the other includes climate change (SDM_GLM). The indicative-index methodology dealt with 128 species and demonstrated low impact of climate change from 1950-2002, and is presented in a special Web-agro-biodiversity-searchable 'BINU' system for the users in Ukraine. It contains 96 agro-biodiversity indicators-indices. The EEBIO approach links species distribution maps, compiled from different sources to habitat change maps, resulting in a series of 800 GIS maps. The MSA-approach gives a general view of the intactness of biodiversity and shows a low impact of climate change by 2002 and a high impact due to habitat loss. A training package for educational purposes is derived from the analyses. The SDM-GLM-approach provided detailed species-based maps of the expected changes in habitats condition caused by land use change and climate change. Finally, the selected 54 indicator species (vascular plants, insects, amphibians, birds and mammals) demonstrated a surprising diversity of SDM-GLM-trends by 2030-2050. It proved that expected climate change, together with land-use change would provoke numerous expected and unexpected species-habitat alterations. If the final model is correct, then in the near future in Ukraine in particular, scientists and decision makers will by 2050 find about 4% of new species or will lose up to 13% of existing species.

DOI: 10.4018/978-1-60960-619-0.ch012

1. INTRODUCTION

In Ukraine until 2003, climate change, land use change and biodiversity were mainly discussed as philosophical issues in scientific publications and no attention was given to the evidences on changes in biodiversity resulting from pressures like climate change. In 2003-2005, the UNEP-GEF funded Biodiversity Indicators for National Use (BINU) project proposed the indicative-index approach and demonstrated possible impact of land use change (LUC) and climate change (CC) on agrobiodiversity of Ukraine (Sozinov *et al.*, 2005a, 2005b). In 2007, the internationally oriented Ukrainian Land and Resources Management Centre (ULRMC) jointly with the Netherlands Environmental Assessment Agency (PBL) carried out an application of a pressure based biodiversity model at national and regional level. Based on that study a book, 'Landscape Ecology' was published for educational purposes (Prydatko *et al.*, 2008a, 2008b). In June 2008, the partners completed the second project on biodiversity modelling, i.e. the 'Projection of Species- and Species-Climate Based Models' and scenario development using the GLOBIO approach for the Ukraine Region, which was mainly focused on Ukraine and neighboring countries like Belarus, and Moldova. At the same time, the methodology used required a much larger geographical space for better simulation. It also required a broader set of species including rare and 'red-data-book' species as well as alien species. In 2008, the geographical space for the species-based-models was extended to twelve Eastern European countries, which we called the EEBIO region. The final modelling has been applied for projections from 2000 towards 2030 and 2050. This paper summarizes and compares the different modeling approaches and discusses them at the conceptual level and in their possible applicability for the Ukraine region

2. REGIONAL BIODIVERSITY MODELLING HISTORY

The regional history of biodiversity modelling only started in 2005 with a serious attempt of digitizing biodiversity distribution maps. Unlike other European countries, Ukraine has demonstrated slow progress of biodiversity modelling (at least for applications at the level of decision makers) during 1990's and 2000's. This is in contrast to well known opinions about many successes in biodiversity conservation during 1992-1998. However these attempts were more virtual instead of evidence based studies of its natural analogy as stated by Prydatko (2000).

The first location-based evaluation of the performance of Ukraine's commitments under Convention on Biological Diversity (CBD) was done and summarized four years ago (Sozinov & Prydatko, 2006). It reported both satisfactory and unsatisfactory indexes of Ukraine's 14 years of membership to the Convention (since the Convention was signed). During this period, Ukraine was placed before Congo and after Togo on the basis of efforts devoted to preserve biological diversity (in percentage to the GDP). At the same time, over 200 legislative documents were issued (and approximately 13 normative documents developed per year), which directly or indirectly facilitated the preservation of biological diversity and the active development of cooperation in this subject. Regardless of 14 years of experience as a member to the Convention, only 8% of the documents issued ensured direct application of the articles and decisions of the Convention on Biological Diversity, which might be considered as the documents of practical CBD-directives. During 14 years the reporting of Ukraine remained unsatisfactory as only 15% of the obligatory reports were submitted. According to the selective data, the reporting activity placed Ukraine on the same level with Uganda and lower than Armenian and Uzbek.

This contributed to low assessment scores, given by the public during the All-Ukrainian survey in 2005 – Ukrainian biological diversity, i.e. 2.7 out of 5 (Prydatko *et al.*, 2006). The authors recommended that in order to ensure the fulfillment of the most actual tasks and commitments under the CBD, including the 2010 Target, Strategic Plan of the Convention, New Millennium Targets, and repayment of informational debts, the national team will have to organize intensive activity and requires to consolidate efforts of governmental and non-governmental environmental organizations, as well as of separate experts. Unfortunately, these recommendations have remained only on paper and were not implemented.

The above mentioned case explains why scientific advance did not take place with respect to biodiversity on the national level. It shows that the importance of the modern biodiversity modelling in Ukraine is based not only on national, but also on international Species Distribution Model (SDM) examples. (The combination forms basis for many new outcomes and maps).

Thus, the latest GLOBIO approaches and results, including the backcasted results from the IMAGE-GLOBIO methodology of 1970, were not known in Ukraine. The first Ukrainian publication on the GLOBIO appeared in 2003 (Tekelenburg *et al.*, 2003), but in 2006 the list of publications increased considerably due to targeted funding of EEBIO project run by ULMRC. During those years the lead participants developed informative atlases of animals and plants distribution of which became available at a searchable system (ULRMC, 2010a, web). The first MSA-map for Ukraine has been developed by one of the authors in Enschede (the Netherlands) during the special training course 'Regional and National Biodiversity Modelling and Analysis', 2006.

The projects focused on national and regional application of GLOBIO (see other chapters in this book) facilitated the acceptance of GLOBIO considerably. These projects translated the GLOBIO approach for the global level. GLOBIO for global purposes uses a coarse 50 by 50 km grid, global data on land use and climate changes and a specific method to link the model compartments (ARISFLOW). In the projects a GIS based approach was developed, which enabled the use of data at different and more detailed levels, including statistics at administrative units and detailed gridded data (chapter 19).

In addition, participation in the GLOBIO sound activity gives the country scientific chance to observe the biodiversity modelling area with series of the tested scales and allows for considering the neighboring areas' models. It is an important advantage, which can improve local research level. (For example, as for today the Red Data Book in Ukraine deals with map-schemes of habitats, but not with its digital analogues).

Big sized grid modelling can save time and money for potential researches and decision makers. We calculated that traditional modelling with field survey requires much more investment than RS-GIS index. (The target was habitat changes research near the big cities for 20 bird species for the period of 13 years based on remote sensing data). The local experiment of 2005 demonstrated that human activity provoked ±25% of changes. It took us six months to implement the RS-GIS. But what will happen if the area is bigger and the modelling instruments list will be longer? and How should we measure 'climate change' and 'land use change' joined impact on wildlife?

All our examples here in the article (about indicators, indices, modern digital maps, MSA, SDM-GLM) are rather exceptions than the common day-to-day results.

3. MODELLING APPROACHES

Since the last decade of the 20th century, the effects of global climate change have been observed all over the world. Nowadays, this is the secondary type of pressure, after land use change that has impact on biodiversity. Nevertheless, in the course of time, this process will become more and more dominant. Climate change affects species by transformation of their habitats. Some species will be driven out of their old habitats and will have to find new habitats that are optimal for them because the precipitation pattern will change as well as their distribution within biomes during a year etc. It is observed that some species 'move' into the high-level land areas, adapting to different climate conditions and humidity. At present, a well-known fact is the critical state of many amphibian populations. The regional modelling history, as mentioned above, suggested that a large-scale study of the impact of climate change on plant and animal species in Ukraine, was not scientifically correct and only took place three to five years ago.

Today this story on modelling knowledge does not look so strange in the light of some realities concerning biodiversity research in Ukraine during the last ten years (Sozinov & Prydatko, 2006). Currently, the potential institutional network for biodiversity modelling assessment in Ukraine for example, includes, ULRMC, Institute of Zoology NASU, National University of Life and Environmental Sciences of Ukraine, and some other GO's and NGO's. Other potential stakeholders includes ministries, committees and scientific institutes of Ukraine i.e Ministry of Emergencies; Ministry for Environmental Protection; State Committee on Land Resources; State Committee on Water Management; State Committee on Forestry; State Service of Geodesy, Cartography and Cadastre; State Service of Reserves; State administrations of Zakarpattia, Mykolaiv, Kherson and Kyiv Oblasts; National Academy of Sciences of Ukraine institutes and branch institutes and international organizations, and international projects.

As we have summarized in our book 'Landscape Ecology' published in 2008: we witnessed the end of the disputes era when scientists discussed on 'who loves landscape ecosystems and biodiversity most', and have moved to the new era of simulations and competition of models.

Four different approaches to describe the effects of land use change and climate change on biodiversity have been developed for the Ukraine and surrounding countries. These are the Indicator-Index approach, the EEBIO mapping approach, the MSA approach and the SDM-GLM approach. We will describe and discuss them briefly.

3.1 Indicator-Index Approach

In 2003-2005, new recommendations of CBD were taken into account in Ukraine and in other partner countries of the UNEP-GEF Biodiversity Indicators for National Use (BINU) Project to better manage the process of indicators and indices development. In 2004, the BINU Project integrated all the key questions and joined indicators into the bilingual BINU searchable list (ULRMC, 2010b). There were about two thousand visits per months registered at the BINU web in autumn of 2009.

It was a useful, pre-modelling period of research of different pressure factors important for agro-biodiversity in Ukraine. The factor groups were named as: driving forces (D), pressure (P), state (S), impact (I), response (R). As a result, an extensive starting list of 64 key questions was reduced to 5 questions to be used then in the selection of indicators of national importance. Local experts selected 128 wild species (34% birds, 23% mammals, 23% vascular plants and 20% invertebrates) that served as indicators for the assessment of biodiversity in agro-landscapes. The data permitted a preliminary assessment of impacts on biodiversity and pressures in major natural-agricultural zones, i.e. Forest, Forest-Steppe, Steppe, the Carpathian Mountains, and the Crimean Mountains.

The final important conclusion was made that during 1950-2002 agro-biodiversity was affected more by the 'land use changes' pressure than was 'climate change'. This development process was important, in particular for the national ecological network purposes, in which agro-landscapes played a significant role (Sozinov *et al.*, 2005a, 2005b).

3.2 Mean Species Abundance approach (MSA)

The Mean Species Abundance (*MSA*) is defined as: the relative means of species abundance of originally occurring species and can be calculated out of the mean trends in population size of a representative cross section of species. This trend is expressed as a relative change of the original, pristine population. *MSA* addresses the homogenization process by dealing only with original species in a particular area, and omits increase in the opportunistic species which may mask the loss in the original species if looking only at total species richness (Alkemade *et al.*, 2009). This measure of the mean species abundance is similar to the Biodiversity Intactness Index (Scholes & Biggs, 2005) and can be considered as a proxy for CBD indicators.

Ukraine occupies the largest part of the GLOBIO Ukraine Region. Therefore, it serves as an example for *MSA*. The first regional calculation was done in 2007. The steps followed for development of climate-change-part for the final *MSA* map have been briefly described below.

In accordance with *MSA*-methodology requirements (Alkemade *et al.*, 2009) the drivers or pressure factors considered by GLOBIO3 include: *land-cover change* (to be received with the help of IMAGE), *land-use intensity* (to be received partly with the help of IMAGE), *atmospheric nitrogen (N) deposition* (from IMAGE), *infrastructure development* (as was applied in GLOBIO-2 version), as well as *fragmentation of habitats* and *climate change* (both to be derived from the IMAGE model and the GLC 2000 land cover map).

GLOBIO3 calculates the overall *MSA* by multiplying *MSA* values for each driver and for each $IMAGE_i$ 0.5*0.5 degree grid cell:

$$MSA_{Xi}=MSA_{LUi} \cdot MSA_{Ni} \cdot MSA_{Ii} \cdot MSA_{Fi} \cdot MSA_{CCi}$$

Basic GLOBIO-3 methodology can be applied from global to national scales. The national-scaled GLOBIO-3 application was created successfully using grids of 1x1 km size or better. It should be noted that on a topographic map a grid is not a regular quadrangle but a trapezium. For example, maps of Ukraine have 1:100,000 scales and grid size of 20x30'. There is a manual that describes the methodology of the model and allows for flexible use of software during model run stage (Rooij, 2006, see also Chapter 19).

For calculation of MSA_I, MSA_F and MSA_N values (biodiversity loss due to negative impact of such factors as 'infrastructure', 'fragmentation', 'atmospheric N deposition') easily, the Faculty of Geo-Information Science and Earth Observation (ITC) of the University of Twente (the Netherlands) and the Netherlands Environmental Assessment Agency (PBL) developed an instrument for the automatic process i.e. *GLOBIO Toolbox*. At the present time it is compatible with *ArcGIS 9.2* only.

So, today the following software is necessary for the modelling process: *ArcInfo* version 9.2, using *GLOBIO Toolbox, ArcInfo* version 9.1 and lower (without using the automatic instruments), as well as Excel (or its analogue that is distributed free of charge as part of the office package *Open Office*).

3.3 EEBIO Mapping Approach

The EEBIO-model abbreviation stands for 'the Eastern European GLOBIO v 1.0'. 'Eastern Europe' is defined as the territory, approximately between 25^0 E and 70^0 E longitude, and part of the eleven countries of the former Soviet Union. In GIS, this territory is limited by the 70^0 E longitude, for what the *WGS1984* frame of reference was chosen. For species that occupy the territory beyond the Polar Circle, the *North Pole Lambert Azimuthal Equal Area* frame of reference was used.

For the purpose of Brown Bear (*Ursus arctos*) habitats modelling in GIS environment, ULRMC used its own processed remote sensing data (MODIS-2000, MODIS-2002, SPOT-2004, Landsat-4 TM, Landsat-5 TM, Landsat-7 ETM+, Terra Aster 2000, Terra Aster 2002), and some data of ten regional GISs. For the cases of 'traditional' analysis of habitat changes, ULRMC used the climate data derived from the IMAGE model (MNP, 2006) to fit the Species Distribution Model (SDM). That outcome was very close to IMAGE model output for the two scenarios that have been recently used in the OECD Environmental Outlook by 2030.

ULRMC had the technical possibility to apply 'future land-use' with application of available GIS-examples for the land-cover-classes: 'forest' (through year-steps of 1928, 1973, 1980, 1990, 2000, 2002, 2005), and was applied successfully for several land-use-sensitive species. For example: Wood Grouse *(Tetrao urogallus),* Black Grouse *(Lyrurus tetrix),* European Elk (*Alces alces*), and was really successfully for Brown Bear.

On July 7, 2008 the models were demonstrated at the Ministry of the Environment and were well received. Our list of outcome scenarios included: (1) 20 animated SDM-GLM- and (2) GIS-map-scenarios at the BioModel web with the usage of GIF-animation (BioModel, 2010). Some results were published in the handbook (Prydatko *et al.*, 2008a), and used in local universities syllabi in Kyiv, i.e. 'Terrestrial Ecology and Biological Indicator Methods', 'Landscape Ecology', 'Applied Ecology'.

3.4 SDM Approach and GLM Application

Generalized Linear Modeling (GLM) is a flexible generalization of ordinary least squares regression, as described by John Nelder and Robert Wedderburn in 1970s. It was selected by environmentalists for development of Species Distribution Modelling (SDM). For performing GLM analysis, the statistical environment 'R' can be used as a language and environment for statistical computing and graphics, an integrated suite of software facilities for data manipulation, calculation and graphical display.

In 2007-2008, ULRMC together with PBL researched GLM application method applying the R 2.6.2 environment scripts for SDM purposes (Prydatko & Kolomytsev, 2008). The starting package of files was tested on the plant species Pedunculate Oak – *Quercus robur*.

The SDM is now a routine procedure. In 2008, we spent about 10-14 step-algorithms per one species GLM-view, and required one day for a GIS operator to process data for a pair.

The team faced difficulties as many available forest maps of the former USSR had no mosaic, which was very important for appropriate transformation of scanned pictures to shape files. ULRMC used the Digital Ukraine product of 1997-2001 to imitate needed forest coverage of 1970s and 1980s in the basic GIS. It was obtained through our technical experiments by means of 'Erase' and 'Clip' options of the ArcToolbox (ArcMap). It was a long-term technical task since all respective layers in the older ArcView 8.0 format dealt with several thousand elements together with huge joined attributive tables. The erase- and clip- basic files (settlements and roads) had several thousand elements too. Eventually the planned shape files were produced and used successfully for modeling of all forests-dependent birds and mammals habitats (*Tetrao urogallus, Lyrurus tetrix, Sus scrofa, Felis lynx, Felis sylvestris*) and vascular plants (*Anemone nemorosa, Anemone ranunculoides, Anemone silvestris*), etc.

For some plants and animals ULRMC produced unique 'forest edges' shape files, which were available for 1990 and 2001. The idea of those habitat imitations required technically some additional steps, such as development of special buffers (0.5, 1, 3 or 5 km) around all forest aggregations for the selected species preferences.

We provided additional historical analysis of archived data of 1927-2004 for 22 plant species, which were new for the EEBIO approach list: *Adonis aestivalis, Adonis annua, Aegilops cylin-*

drica, Bifora radians, Acroptilon picris, Cerastium arvense, Cerastium holosteoides, Cynodon dactylon, Consolida orientalis, Lactuca tatarica, Linaria genistifolia, Lycopsis arvensis, Lycopsis orientalis, Elilotus wolgicus, Ranunculus oxyspermus, Ranunculus sardous, Scleranthus perennis, Spergula arvensis, Anemone ranunculoides and Anemone sylvestris. It was done because there is a need to learn more about real changes of habitats across the natural zones in Ukraine. For the SDM-GLM experiments, *Anemone ranunculoides and Anemone sylvestris* were selected.

The team carried out 19 supportive species-based approximations (logarithmic, polynomial, power associatively and linear filtering) for illustrating local trends in a better manner. The preferences have been given to trends with trusting level (R^2) higher than 0.5...0.7. Some approximations demonstrated the trusting level up to 0.8. The supportive information explained some GLM-predictions better and, possibly the first species to be extinct (migration) from GLOBIO Ukraine Region by 2050, i.e. *A.alces, O.leucoicephala, L.tetrix* and some other species. For *Spermophilius pigmaeus*, we built up the model and the approximation on the dependence of the mammal on human tillage activity and that was based on some published historical data of 1810-2005. It allowed us to build more realistic power approximations.

Key questions served as a good methodological tool that always helped with acceleration of future modelling concept and final outcome. For this purpose Bio-Model group stated the following key questions:

- What are the main causes of biodiversity change in the GLOBIO Ukraine Region and what was the role (%) of land use change (LUC), climate change (CC) and other factors?
- How to separate 'land use change' from 'climate change' effects and which key factors have to be analyzed in the project first of all?
- What species can be selected primarily for expected 'species envelopes', i.e. LUC-dependent and CC-dependent?
- How to improve the database 'to the past' and 'to the future' (in scope of 'positive' and 'negative' data)?
- How to better demonstrate results to customers and then to potential users?

Finally, new regional SDM-GLM scenarios for vascular plants and vertebrate animals have been developed in Ukraine.

4. RESULTS

4.1 Indicator-Index Approach Outcome Set

During 2003-2005, the UNEP-GEF BINU Project (Ukraine) conducted a research on 'impact of humans on agro-biodiversity' in Ukraine during 1950-2002. The summary of the research is as follows: land use change (37%), inadequate environmental management (16%), habitat fragmentation (7%), exploitation (9%), toxification (7%), disturbance (6%), and others (18%). The most important conclusion was that 'land use changes' dominated in the region and summed up to 37%, but 'climate changes' were less than 1% (Prydatko, 2005). According to the experts' assessment during 1950-2003, the number of such agricultural dependent wild species of different taxonomic groups in Ukraine had declined at first and then stabilized, rose or continued declining (Sozinov *et al.*, 2005a).

The Bio-model's follow-up studies with usage of more powerful SDM-GLM scenarios demonstrated that the climate impact on habitats would be larger and wider and it has to be addressed to a longer list of species.

4.2 MSA-Approach Outcome Set

In our *MSA*-experiment, the entry data from public sources were included into eight key groups. The 4th group of 8 was the climate change, i.e. aspect of temperature increase since 1970 according to the IMAGE-model. Scientists from all over the world are trying to reflect these trends, evaluate the changes and upload them in to the model. The PBL's researchers retrieved the first simple regression equations, where the *MSA* value depends on average global temperature change (Δt, C^0). Resulting 'mean biodiversity' in this case has a well known mark as MSA_{CC}. In the GLOBIO model, these regression equations were used in order to find out possible theoretical *MSA* values in different biomes of the dry land (Alkemade *et al.*, 2009). The biome sensitivity values are derived from IMAGE and EuroMove. In the GLOBIO-3, only their final combination was used. Climate change impact is calculated assigning MSA_{CC} values for each of biomes and accounting CC-response of the biome. The resulting is a raster map, where each pixel or a group of pixels indicates the expected impact of climate change on biodiversity.

ArcGIS produces MSA_{CC} on the map. In our case, a land cover of a typological unit of natural-agricultural zoning of Ukraine is taken as a land cover of the biome. The input data for MSA_{CC} modeling include: natural-agricultural zoning of Ukraine; degree of biomes sensitivity: and a table of annual temperature changes. The calculated total *MSA* for Ukraine territory was 32.4%. The calculated MSA_{CC} and the map are shown in Figure 1.

4.3 EEBIO Approach Outcome Set

In 2007 our team built 54 EEBIO spatial models, which included 24 species of the climate-change-pressure-group and 30 species of the land-use-change-pressure-group (ULRMC, 2010a). The species group selection exercise was based on our previous experience and many regional scientific publications.

In this article, we focus on the Brown Bear (*Ursus arctos*) example only, because the animal depends on a combination of factors. In the chapter below readers can compare EEBIO-sounded and SDM-GLM-sounded results. About four year ago, the given point of view (Figure 2) was expressed

Figure 1. The summarized map of climate change impacts on biodiversity in Ukraine, MSA aspect

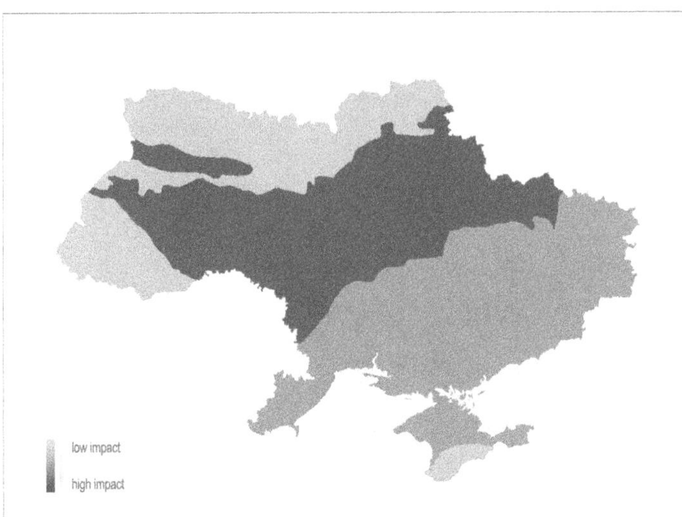

by us for the first time in the region. The model was actually for the period of 2000-2001. The creation of a digital thematic basis with resolution up to 200 m and better was the key purpose of that detailed EEBIO-sounded modelling. In addition, as you can see below, the Brown Bear's digital layers were the basis for embedding habitat in R-environment for construction of SDM-GLM-scenario, which took into account climate change (Figure 3), (but not only) land-use change (Figure 2). So, during the three years after EEBIO project we brought together the forest-change-data and climate-change-data in the one model.

For the creation of Figure 2, we used eight mapping resources, several common publications and ULRMC's remote sensing data archive for 1998-2005. Our key steps are as follows: (1) construction of a map basis regarding forest-covered areas on the basis of ULRMC's archives, (2) marking the entire territory where 'percentage of forest land' was not less than 40% and (3) removal from the digital map of the territories, which were, theoretically, left out by species. For example, they were all settlements with a buffer zone of about 500-meter around them and sometimes highways, etc. In total, 6561 settlements have been taken into account for the model. As a result, the mapping model demonstrates the trend of species areal fragmentation in the Carpathian region and location of most the risky existing or occurring discontinuities, that were getting larger and could create problems for migration of Brown Bear between forest-covered islands. The size of the area, calculated in ArcMAP 9x was 9,124 km^2, with a density of 0.32 individuals per km^2 (according to the data from the annual forest assessment where species population was recorded as 0,3 thousand).

The EEBIO project forecast was that in the near future similar discontinuities would occur on the two lineaments of settlements i.e.: '*Svalyava→ Olenyovo→Pavlovo→Polyana*' and '*Mizhgirya →Kolochava→Krasna*', a (special peculiar) isolation area, where it is difficult to expect positive results in registering this species has already been created in the northern-east part of the area, on the territory of Ukraine. An example of interim map and legends were accessed via the Internet (ULRMC, 2010a).

4.4 SDM-GLM Outcome Set

The SDM-GLM scenarios in our experiment were a new phase followed by the period of revision and updating of EEBIO project data. During 2008, the team built 46 examples of SDM-GLM. Some descriptions of them are available on the Internet (BioModel, 2010). Before are two examples only,

Figure 2. Historical and current habitat changes for Brown Bear, Ursus arctos, in the Western Ukraine

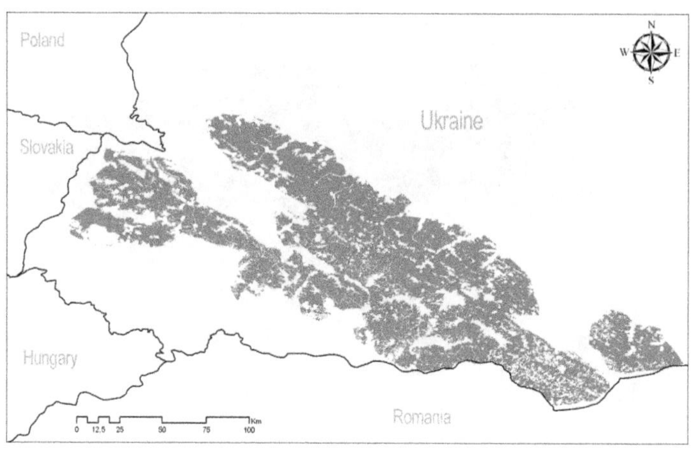

Figure 3. Historical and predicted habitat changes for European Elk (Alces alces) in EEBIO Project Region by 2050

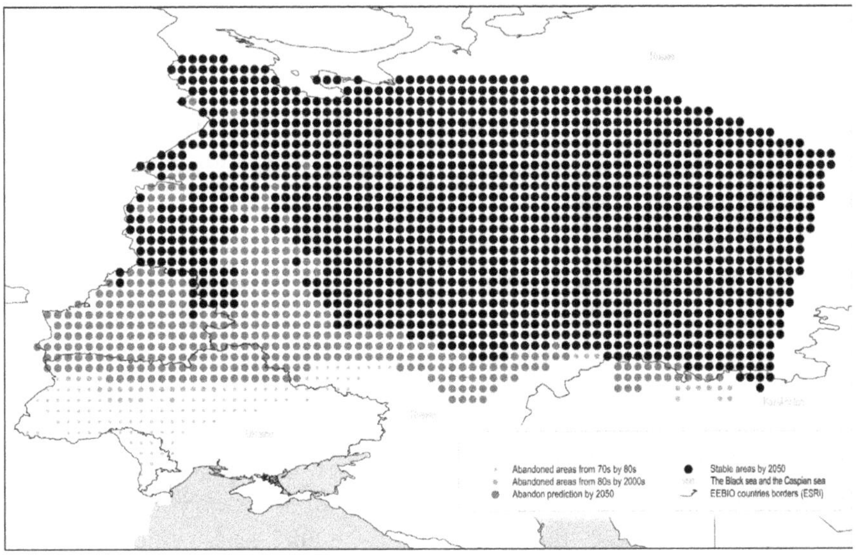

which show historical and predicted future shift of distribution of European Elk and Brown Bear (Figure 3, Figure 4).

Table 1 and Figure 5 give demonstrate a summary of species habitat changes and trends by 2050 in GLOBIO Ukraine Region. Some trends are as follows: seems to be extinct (Dniprov's Birch, Klokovii Birch); decreasing locally (Alpine Grizzled Skipper, Fire Salamander, Alpine Newt, Carpathian Newt, Wild Cat); decreasing-disappearing (Chestnut); shifting east (*Anabasis aphylla*); increasing (Corncrake); increasing lo-

Figure 4. Predicted habitat changes for Brown Bear (Ursus arctos) in EEBIO Project Region by 2050

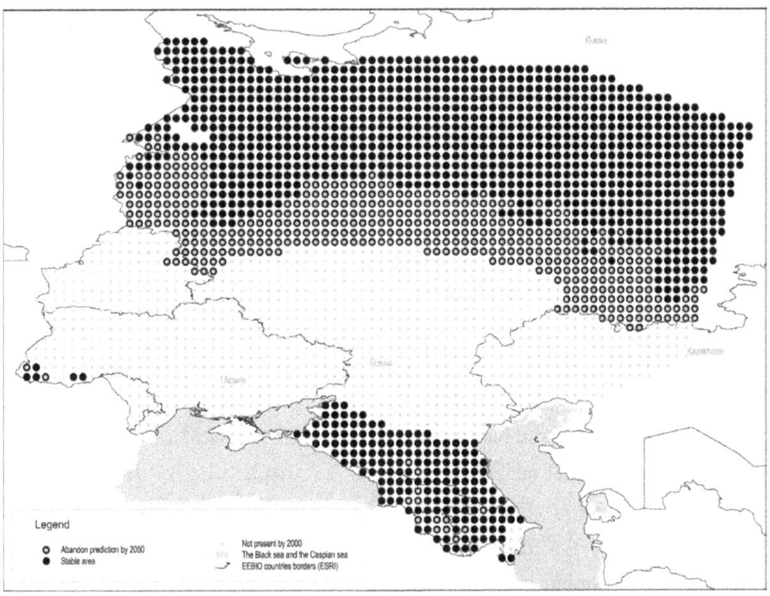

cally, in Russia (Slender-billed Curlew); shifting north (European Elk, Downy Birch, Small-leaved Lime, Natterjack Toad, European Tree Frog, and mainly all birch species, i.e. Elegant Birch, River Birch, Silver Birch. It is predicted that Red Squirrel habitats will move of north and/or to south to Caucasia. Lesser Noctule demonstrates a 'north and decreasing trend locally.

So, the modelling shows that habitats of predominant species show trends of shifting to 'north'. The secondary dominant trend is shift of species towards 'north-east' and rarely 'north-west': European Polecat, Aquatic Warbler, European Turtle Dove, Yellow Water-lily, Pedunculate Oak, Wood Grouse, Lynx, Blue Hare, Noctule, Lesser Horseshoe Bat, Greater Noctule, Steppe Polecat, Imperial Eagle, White-headed Duck, European Licorice, Great Bustard, Marbled Polecat, Great Jerboa, Little Ground Squirrel.

It is also important (necessary) to know which species will move to the southern part of the Region in the 2050-future despite of global climate change. The SDM-GLM scenarios showed that some species will have very complicated trends, i.e. north-west-and-south-east (European Rabbit, some birches, which could disappear in the region); north-east-and-south-east (Black Grouse, Wild Boar, Brown Bear).

Indeed, brown bear has no chance to return to the Crimea Peninsula where it was in the historical past, but its respective habitats can (potentially suitable territories in terms of climate are not included in the Figure 4).

The list of species, whose habitats would not undergo any serious changes, is small and includes shrubby birch and jackal. Consequently, present local notes about increased observations of jackal in south-western Ukraine are not a proof that it will migrate soon to the Carpathian Region and further. The local Birch species will move to south. Then, only four species (from the list of 54) will possibly migrate with its habitats to the south-east: Yellow Anemone, Snowdrop Anemone, and Scotch Pine, Norway Spruce. At least, habitats of only two plant species demonstrated a tendency to migrate to the western part of the region, i.e. Wood Anemone and Tauric Wormwood.

5. SOLUTIONS AND RECOMMENDATIONS

1. During the 2003-2010 period four new approaches on modelling of pressure-based biodiversity changes (indicator-index, EEBIO mapping, MSA and SDM-GLM) were applied in the Ukraine Region.
2. The indicative-index methodology dealt with 128 species and demonstrated low impact of climate change on wildlife during 1950-2002, and the level was less than 1% of the total pressures. The land use change pressure however was dominant. The supportive RS-GIS index demonstrated that during 10 years humans could (create) about 25% of enlargement of the total territory for some species, but simultaneously decrease the total territory (27%) for some others. The changes would occur without climate change pressure.
3. In 2006, the map-based EEBIO project dealt with 800 maps for distribution of 130 species and demonstrated that diversity of trend examples were dependent not only on the land-use-change and climate-change, but also on awareness of scientists. We found out that debating vision of scientists concerning the areal changes of the White-headed Duck, (*Oxiura leucocephala),* migrated to the north-west direction in accordance with the climate-change trend. So, the 'mistakes' of environmentalists have a tendency to migrate synchronously with a climate-change. The total of Polar Bear (*Ursus maritimus*) habitats in the EEBIO GIS environment depended on vision of specific author: the gap was about 17 thousand km^2 (Prydatko *et al.,* 2008b). In the same time, the climate

Table 1. BioModel summary of species-habitat-change-trends in GLOBIO Ukraine Region. Source: MNP-ULRMC Join Project 'Projection of Species- and Species-Climate Based Models on to the GLOBIO Ukraine Region, and Scenarios Development', E/555050/01/MO (2006)

Species name (Latin)	Species name (English)	Species name (Ukrainian)	GIS-class	Dominated pressure	Predicted SDM-GLM trend by 2050
Acrocephalus paludicola	Aquatic Warbler	Ocheretyanka Prudka	bird	CC	NE
Alces alces	European Elk	Los'	mamm	CC	N
Allactaga major	Great Jerboa	Tushkan Velyky	mamm	LUC	NW, NWW
Anabasis aphylla	Anabasis	Anabasis Bezlysty	plant	CC	E
Anemone nemorosa	Wood Anemone	Anemona Lisova	plant	CC	W
Anemone ranunculoides	Yellow Anemone	Anemona Zhovta	plant	CC	SE
Anemone sylvestris	Snowdrop anemone	Anemona Snigova	plant	CC	SW
Aquila heliaca	Imperial Eagle	Orel Mogylnyk	bird	CC	NW
Artemisia taurica	Tauric wormwood	Polyn Tavryysky	plant	CC	W
Betula borysthenica	Dniprov's Birch	Bereza Dniprovska	plant	CC	(seems to extinct)
Betula humilis	Shrubby Birch	Bereza Nyz'ka	plant	CC	no changes
Betula klokovii	Klokovii Birch	Bereza Klokova	plant	LUC	(seems to be extinct)
Betula microlepis	Elegant Birch	(Bereza Elegantna)	plant	LUC	N
Betula obscura	River Birch	Bereza Temna	plant	LUC	N
Betula pendula	Silver Birch	Bereza Korelska	plant	LUC	N
Betula pubescens	Downy Birch	Bereza Pukhnasta	plant	CC	N
Betula sp. (7 species)	Betula sp. (7 species)	Bereza (All 7 Species)	plant	LUC	N
Bufo calamita	Natterjack Toad	Ropukha Ocheretyana	amphib	LUC	N
Canis aureus	Jackal	Shakal	mamm	LUC	no changes
Castanea sativa	Chestnut	Kashtan Yistivny	plant	CC	decreasing, disappearing
Crex crex	Corncrake	Derkach	bird	LUC	increasing
Fagus sylvatica	Beech	Buk (Zvychainy)	plant	CC	S
Felis lynx	Lynx	Rus'	mamm	LUC	NE
Felis sylvestris	Wild Cat	Kit Lisovyi	mamm	LUC	decreasing locally
Glycyrrhyza glabra	European Licorice	Solodka Ghola	plant	CC	NW
Hyla arborea	European Tree Frog	Zvychayna Kvaksha	amphib	LUC	N
Lepus timidus	Blue Hare	Zayats Bilyak	mamm	LUC	NE
Lyrurus tetrix	Black Grouse	Teteruk	bird	LUC	NE, SE
Mustela eversmanni	Steppe Polecat	Tkhir Stepovy	mamm	LUC	N-NW
Mustela putorius	European Polecat	Tkhir Chorny	mamm	LUC	N, NE
Numenius tenuirostris	Slender-billed Curlew	Kulion Tonkodz'oby	bird	CC	increasing locally
Nuphae lutea	Yellow Water-lily	Latattya Zhovte	plant	CC	NE
Nyctalus lasiopterus	Greater Noctule	Vechernytsya Veletens'ka	mamm	LUC	NE, NW
Nyctalus leisleri	Lesser Noctule	Vetchernytsya Mala	mamm	LUC	N, decreasing locally
Nyctalus noctula	Noctule	Vetchernytsya Dozirna	mamm	LUC	NE
Oryctolagus cuniculus	European Rabbit	Dyke Krolia	mamm	LUC	NW-SE

continued on following page

Table 1. continued

Species name (Latin)	Species name (English)	Species name (Ukrainian)	GIS-class	Dominated pressure	Predicted SDM-GLM trend by 2050
Otis tarda	Great Bustard	Drohva	bird	LUC	NW
Oxyura leucocephala	White-headed Duck	Savka	bird	CC	NW
Picea abies	Norway Spruce	Yalyna Yyevropeiska	plant	CC	SW
Pinus silvestris	Scotch Pine	Sosna Zvychaina	plant	CC	SE
Pyrgus andromedae	Alpine Grizzled Skipper	(MetelykAndromeda)	insect	CC	decreasing locally
Quercus robur	Pedunculate Oak	Dub Zvychainy	plant	CC	NE
Rhinolophus hipposideros	Lesser Horseshoe Bat	Pidkovonos Maly	mamm	LUC	NE and decreasing locally
Salamandra salamandra	Fire Salamander	SalamandraPlyamysta	amphib	LUC	decreasing locally
Sciurus vulgaris	Red Squirrel	Vyvirka	mamm	LUC	N, and to Caucasia
Spermophilus pygmaeus	Little Ground Squirrel	Chovrakh Maly	mamm	LUC	NW, NWW
Sstreptopelia turtur	European Turtle Dove	Gorlytsya Zvychaina	bird	CC	NE
Sus scrofa	Wild Boar	Kaban Dyky	mamm	LUC	NE,SE
Tetrao urogallus	Wood Grouse	Glushets	bird	LUC	NE
Tilia cordata	Small-leaved Lime	Lypa Dribnolysta	plant	CC	N
Triturus alpestris	Alpine Newt	Tryton Alpiysky	amphib	LUC	decreasing locally
Triturus montandoni	Carpathian Newt	Tryton Karpatskyi	amphib	LUC	decreasing locally
Ursus arctos	Brown Bear	Vedmid' Bury	mamm	CC	NE,SE
Vormela peregusna	Marbled Polecat	Pereguznya	mamm	LUC	NW

Note: LUC- land use change, CC – climate change, mamm – mammals, amphib - amphibians.

Figure 5. Summary: SDM-GLM trends for 54 indicator species by 2050

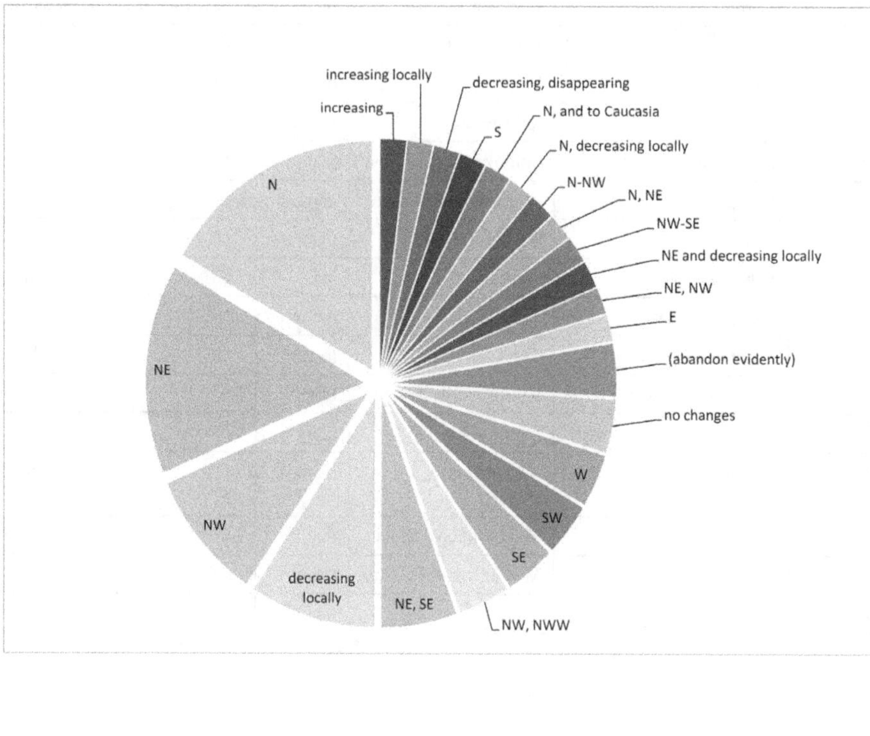

change aureoles on the map were combined with habitats change aureoles.

4. In 2007, the MSA-approach demonstrated the low impact of the climate change by 2002, while the land-use-change pressure dominated.
5. The SDM-GLM approach provided modern 2D (gif-animated) maps and demonstrated a surprising diversity of habitat change trends by 2030-2050. It proved that expected climate change (together with well-known land-use change) would provoke numerous non-simplified and unexpected habitat changes. The scales could be so big that expected habitat changes of European Elk would be equal to 20.4% of its current area and for Brown Bear it would be more than 36%. If the proposed 'climate change' model is correct, in the near future, Ukraine will have about 4% of new species, but lose about 13% of existing species by 2050.
6. The calculated human impact on the wildlife by land-use-change near the Ukrainian capital Kyiv could be about 27% during 10 years (regarding of the total loss). The calculated climate change impact on wildlife in the region might be about 36% during 50 years (regarding of the total change habitats for some big mammals).
7. The modelers of climate-change impacts on the wildlife have to plan development of more than one scenario of habitat changes and use more than one methodology. Only then the results will be more reliable for science and decision making.

6. CONCLUSION

Unfortunately, our region, society falls behind about 15 years in applying new modelling approaches i.e. IMAGE, MSA, SDM-GLM etc. At the same time much has been achieved during the last decade, although it remains difficult to translate the findings to the level of decision makers. This is also because of difficulties in communicating the findings in a proper way, and many regional scientists tend to discuss the details instead of the broad figures.

It's apparently, (continuously) easier to explain a decreasing of space research by the lack of required resources. But, there are much more profitable ways to see further and to know more, especially when the task comes to agricultural plants, invasive species, and rare protected species. We can also assume and hope that very hot summer of 2010 (at least in Ukraine), will push the decision makers to pay more attention to biodiversity changes, forecasts, and related modelling. Moreover, much has been changed in the GLOBIO Ukraine region and the outside world. There is in the capital of Ukraine, Kyiv, of the increase an 'Italian sky', which can be described as cloudless with the sun as hot as in Australia. This summer's thunderstorms do not bring much relief, and the humid stuffiness is reminiscent of Malaysia's stifling heat. The tropical contrasts are even more visible in the wild nature, most notably, the Ukrainian Carpathians. Unprecedentedly, the high Hogweed (*Heracléum* sp.) has become as common an element of the landscape as our picturesque stacks and wooden churches. (Its juice contains hazardous substances that provoke severe sunburns under ultraviolet light). In Russia, the situation is similar. This July, a regional TV channel reported weekly cases of sunburns from this plant, which was found in fields and gardens. Already in Southern Ukraine, no summer passes without new cases of the biting Kara Kurt *(Latrodectus tredecimguttatus*), which is more typical in the semi-deserts of Central Asia. Almost no one in the Ukrainian steppes had heard about this species until 1970. There is another exotic example found in the Carpathian Mountains. This summer, one of the article's authors repeatedly found octopus stinkhorn or Archer's mushroom (*Clathrus archeri*) near Shajan. It came to Europe from maintained Australia and Tasmania.

These regional stories continue, and our students discuss it with great interest. This helps them understand the reasons for modelling. Perhaps the students of this new generation are special. They are involved in modelling the habitat changes and biodiversity; they are the future managers, which could support the research discussed in this book.

Some of our efforts from 2005-2009 have been crowned with success (i.e the adaptation of knowledge about GLOBIO, as well as the modelling techniques developed in the Netherlands for the Ukrainian higher education requirements). At least, some teachers at the National University of Biological Resources and Life Sciences (NUBRLS) in Kyiv, the National Taras Shevchenko University of Kyiv, the Uzhgorod National University (Uzhgorod), and the National University of Kyiv-Mohyla Academy (Kyiv) expressed their interest in our textbooks, which explain the usage of new biodiversity-sound-indicators and indices and in particular, the MSA. In the last few years, several syllabi from NUBRLS reflected all three approaches to the contemporary evaluation of biodiversity, for example, the aforementioned indicator-index, EEBIO-based, and SDM. Our work with NUBRLS students showed that they were able to independently perform many tasks within the disciplines, simply referred to today as biodiversity and modelling. The work included Landscape Ecology, Applied Ecology, Terrestrial Ecosystems, and Biological Monitoring Methods. (The obstacles to this development were merely a lack of technical training and skills, and/or inability of some students to deal with GIS).

Another one of our attempts to implement the modelling knowledge at the postgraduate education level (i.e. in an institute for advanced studies of ecologists and managers under the umbrella of the environmental ministry) had no success. Apparently, the breakdown can be found in such a case where a barrier forms because the decision makers have low awareness and technical training in these situations. In any case, without holding allowances, posters, maps, and other supporting materials it would be impossible to initiate learning of GLOBIO at the tertiary level. Our obvious advantages were the published textbooks on biodiversity and its evaluation and modelling (Sozonov *et al.*, 2005a,2005b; Prydatko *et al.*, 2008a, 2008b).

The authors believe that critical observations and remarks will be met with understanding, as the main objective of the research is to unite the efforts to define key questions and take into account the received lessons, to improve the quality of biodiversity modelling, applying the synergic approach, using the powerful intellectual and organizational potential Ukraine possesses, ant to consolidate the activity of government and non-government organizations.

The proper development of pressure-based-biodiversity modelling is of critical importance for the GLOBIO Ukraine region. It could be an important part of the Biotic Geo-information Science Modelling (BGSM). The new measures to reduce key negative causes of climate change and land use change should be considered by decision-makers. At this stage of research, the attention has to be focused on diversity of SDMs approaches as a supportive part of the GLOBIO methodology.

ACKNOWLEDGMENT

The Authors express their gratitude to Mr. Eric Arets (Wageningen UR) and Mr. Wilbert van Rooij (PBL) for their kind assistance in the projects research, implementation and useful consultancy, as well as to PBL for financing our first textbook edition.

REFERENCES

Alkemade, R., van Oorschot, M., Miles, L., Nellemann, C., Bakkenes, M., & ten Brink, B. (2009). GLOBIO3: A framework to investigate options for reducing global terrestrial biodiversity loss. *Ecosystems, 12*, 374–390. doi:10.1007/s10021-009-9229-5

Bakkenes, M., Eickhout, B., & Alkemade, R. (2006). Impacts of different climate stabilisation scenarios on plant species in Europe. *Global Environmental Change, 16*(1), 19–28. doi:10.1016/j.gloenvcha.2005.11.001

BioModel. (2010). *ULRMC BioticGIS Modelling Group*. Retrieved April 6, 2010, from http://biomodel.org.ua/

Bouwman, A., Kram, T., & Goldewijk, K. (Eds.). (2006). *Integrated modeling of global environmental change. An overview of IMAGE 2.4. Netherlands Environmental Assessment Agency*. Bilthoven, The Netherlands: MNP.

CLUE. (2008). *Land use and land cover change model*. Retrieved February 1, 2009, from http://www.cluemodel.nl/

GLOBIO. (2007). *Modelling human impacts on biodiversity*. Retrieved February 1, 2009, from http://www.globio.info/

NEEA. (2007). *The International Biodiversity Project: Understanding biodiversity, ecosystem services and poverty in order to support policy makers*. The Netherlands Environmental Assessment Agency. PBL website. Retrieved November 15, 2007, from http://www.pbl.nl/en/publications/2007/WorldwideBiodiversityLossAndPoverty_HowDoTheyRelate.html

Prydatko, V. (2000). *Biodiversity and bioresources of Ukraine: Review of SoE publications*. (1992-1998). *Re-evaluation of trends* (1966-1999). *The environment and resources*. Kyiv: ERRIU. ISBN 966-95141-1-6

Prydatko, V., & Kolomytsev, G. (2008). Climate and biodiversity changes by 2050 in GLOBIO Ukraine region. *Ukrainian-American Environmental Association Newsletter, 7*(5), 34.

Prydatko, V., Pristinska, G., Panina, O., & Vasylkivsky, B. (2006). *Textbook on collection and processing of the information for national reports of Ukraine on implementation of the Convention on Biological Diversity*. Kyiv, Ukraine: EcoPravo.

Prydatko, V. I. (2005). Indication and indicators: Development and use of it for purpose to evaluate state of agrobiodiversity in Ukraine. In Sozinov, O. O. (Eds.), *Agrobiodiversity of Ukraine: Theory, methodology, indicators, examples. Book 1* (pp. 94–113). Kyiv, Ukraine: Nichlava.

Prydatko, V. I., Kolomytsev, G. O., Burda, R. I., & Chumachenko, S. M. (2008). *Landscape ecology: Textbook on application of pressure-based biodiversity modelling for national and regional educational purposes. Book 1 & 2*. Kyiv, Ukraine: NAU.

Rooij, W. van., & Tekelenburg, T. (2007). Land use modelling focusing on the impact on biodiversity. Retrieved October 20, 2009, from www.fao.org/forestry/foris/ppt/outlook2020/land-use-modelling.pdf

Scientific modelling. (n.d.). *Wikipedia*. Retrieved March 27, 2010, from http://en.wikipedia.org/wiki/Scientific_modelling

Sozinov, O., & Prydatko, V. (2006). *Basic report on the implementation of the Convention on Biological Diversity in Ukraine*. Retrieved December 12, 2006, from http://www.undp.org.ua/i/files/Bio_base.pdf

Sozinov, O. O., Prydatko, V. I., Tarariko, O. H., & Shtepa, Y. N. (2005a). *Agrobiodiversity of Ukraine: Theory, methodology, indicators, examples. Book 1 &2*. Kyiv, Ukraine: Nichlava.

Tekelenburg, A., Prydatko, V., Alkemade, J. R. M., Schaub, D., Luhmann, E., & Meijer, J. R. (2003). *Assessment of wild biodiversity in agricultural land use. First design and perspectives of a pressure-based Global Biodiversity Model.* In 6th Annual Ukraine's ESRI User Conference Geoinformation Technologies in the Management of Territorial Development. (pp. 184-196). Simpheropol, Ukraine: TNU. Retrieved from http://www.ulrmc.org.ua/publication/envmanag/globio%20rivm%20ulrmc_ru.pdf

ULRMC, Ukrainian Land and Resource Management Center. (2010a). *EEBIO searchable service: Maps, species based models, habitats, pressures*. Retrieved April 6, 2010, from http://www.ulrmc.org.ua/services/eebio/is/index.asp?lang=EN

ULRMC, Ukrainian Land and Resource Management Center. (2010b). *BINU searchable list of agrobiodiversity indicators*. Retrieved April 6, 2010, from www.ulrmc.org.ua/services/binu/is/index.asp?lang=EN

van Rooij, W. (2006). *Manual: Using GLOBIO-3 for determining the current level of biodiversity on a national/regional scale for the course*. Enschede: Regional and National Biodiversity Modelling and Analysis. ITC.

Wikipedia (n.d.). *Decline in amphibian populations.* Retrieved July 11, 2010, from http://en.wikipedia.org/wiki/Decline_in_amphibian_populations

Yatsyk, A. V., & Shenchuk, V. Y. (2006). *Encyclopedia on water industry, nature management, nature procreation, sustainable development*. Kyiv, Ukraine: Geneza.

Chapter 13
Regional Scenarios of Biodiversity State in the Tropical Andes

Carolina Tovar
Universidad Nacional Agraria La Molina, Peru

Carlos Alberto Arnillas
Universidad Nacional Agraria La Molina, Peru

Manuel Peralvo
CONDESAN, Ecuador

Gustavo Galindo
Instituto de Recursos Biológicos "Alexander von Humboldt", Colombia

ABSTRACT

Biodiversity assessment represents a baseline for developing conservation strategies, but the assessment of future impact of some policies also requires the development of scenarios. These assessments are particularly important and difficult in areas with high biodiversity such as the Tropical Andes. Therefore three countries were analyzed: Colombia, Ecuador and Peru using the framework of GLOBIO3 to assess the remaining biodiversity for 2000 and for two 2030 scenarios: market forces and policy reforms. The purpose was to identify the most vulnerable areas to biodiversity loss, the most important drivers and the implications of such losses for conservation. Detailed information for each country was used to build the drivers of biodiversity loss (land use/land cover, infrastructure, fragmentation and climate change). The authors discuss the use of this methodology for Andean countries, how the results can be useful for policy and decision makers, and provide suggestions to improve GLOBIO3 at national scales.

1. INTRODUCTION

Global environmental change processes due to human intervention are generating observable effects in different regions of the world. Humans have influenced land cover, atmospheric composition and even soil composition. We are just beginning to understand the complex ways in which these changes affect biodiversity, ecosystem services

DOI: 10.4018/978-1-60960-619-0.ch013

and the goods they provide for human survival. Nevertheless, the changes are expected to persist or even expand because of the continuous resource demand to satisfy human necessities. In this context, several scientists have started research processes to understand the potential impacts on the environment and the drivers of these changes. The most important and known group, the Intergovernmental Panel on Climate Change (IPCC), focus on a global scale providing valuable scientific data about the likely consequences of Climate Change (IPCC, 2007). Despite the advances to understand how global policy decisions can impact ecosystems (IPCC, 2007; UNEP, 2007; OECD, 2008), there are still several steps pending.

Most of the drivers of environmental change at global level are similar to those affecting the regional level. Regional scenarios are important to support decisions at more relevant scales for countries and for the territorial planning within countries. These scenarios are progressively more important in a context of increasing connections between local and global communities where global processes influenced local decisions. Fortunately, biodiversity, mainly the services that it provides, is increasingly been recognized as a key factor to secure the survival of human societies, next to having a value by itself. A first problem to deal with scenarios is the definition of biodiversity indexes, as biodiversity is a complex multilevel concept that includes genetic diversity, species, communities and ecosystems (Millennium Ecosystem Assessment, 2005). Moreover, biodiversity needs to be related to the environmental services that ecosystem processes provide. To attain this, not one, but several indexes that can provide information about biodiversity could be needed. Therefore, the Convention on Biological Diversity (CBD) suggested developing a series of indicators to describe biodiversity changes (CBD, 2006).

Due to the previous considerations, we were in this analysis, interested in a biodiversity index that can provide information to national and regional policy makers about the impact of policy decisions on biodiversity. Our interest area was to investigate three Andean countries, Colombia, Ecuador and Peru. The topography and the presence of several atmospheric conditions define a large environmental heterogeneity, which covers the most arid areas and the wettest areas in the world. We used a regional version of the GLOBIO3 framework to answer two questions: (1) What would be the more vulnerable areas in the Tropical Andean countries given projected changes in main drivers of biodiversity loss? (2) What are the most important drivers of biodiversity loss and what will be the implications for conservation? In the process, we evaluated the applicability of the framework to the environmental regional context and with that experience, discussed the applicability of such framework to regional policy context.

2. CURRENT APPROACHES

The current conscience about global environmental change requires tools to evaluate the impact of human activities, and hence, policy decisions on environment. In order to do this, scientists have developed different approaches to assess current biodiversity state. On the one hand, extinction rates give first approximations at the species level (e.g. Thomas et al., 2004). However, these calculations present drawbacks related to overestimation, uncertainty or lack of sensitivity to short-term change (Balmford, Green, & Jenkins, 2003). Other studies have focused on population size and the reduction of these populations (e.g. Houlahan et al., 2000), or the reduction of habitat range for specific taxa (e.g. Ceballos & Ehrlich, 2002). On the other hand, there are number of studies that have approached the problem considering directly the impact of human induced pressures on biodiversity. The habitat index is one of them, that considers the percentage of undisturbed vegetation based on population density and land use (Hannah, Lohse, Hutchinson, Carr,

& Lankerani, 1994) or the human footprint that ranks different degrees of impact of population density, land transformation, accessibility and electricity power infrastructure (Sanderson et al., 2002). These approaches share the use of rankings based on expert opinion, which is not always ideal in terms of consistency, and the generation of biodiversity assessments at finer resolutions.

A different set of approaches implements theoretically and empirically-based links between the state of biodiversity and a set of drivers that affect it. This is the case of the biodiversity intactness index (BII), which considers the "abundance of a large and diverse set of organisms in a given geographical area, relative to their reference populations" (Scholes & Biggs, 2005). The estimation of the impacts of land use classes is based on the area affected by each land use class (protected, moderate use, degraded, cultivated, plantation, urban) and considers the remaining fraction of pre-colonial population on each species. Similarly, the relative Mean Species Abundance of originally occurring species (MSA) is another biodiversity index, which analyzes remaining biodiversity under the framework of the Global Biodiversity Model (GLOBIO3) (Alkemade et al., 2009). The drivers considered in GLOBIO3 are land use, fragmentation, climate change, atmospheric nitrogen deposition and infrastructure development. Although no index can synthesize the complexity of the relationships between biodiversity and human influences on landscapes, they provide a valuable platform to generate data about the spatial variability of these relationships.

In addition to current biodiversity assessments, a second step to improve the understanding of environmental impact of policies is the development of future scenarios of biodiversity loss. These scenarios, where the effect of different management decisions can be tested, constitute a useful tool for politicians and policy makers (Cumming, 2007). Future scenarios can also help to locate areas where the combination of different drivers can potentially affect biodiversity. Examples of future global biodiversity scenarios are: projections of biome change for 2100 (Sala et al., 2000) based on expert opinion, changes in number of threatened species of mammals and birds for 2050 based on population growth and species richness (McKee, Sciullli, Fooce, & Waite, 2003) or the global projections for 2050 of the MSA values (Alkemade et al., 2009). These analyses, implemented for large areas, provide insights at global scales. Therefore, these exercises need to simplify important processes that operate at regional levels. For instance, global climate change models (GCM) do not capture the complex climatic gradients of mountainous regions (Urrutia & Vuille, 2009). In addition, global land use/cover change (LUCC) models normally conflate drivers at continental levels homogenizing complex socio-economic and biophysical characteristics of different regions and countries (Lambin et al., 2001).

Regional models of biodiversity are necessary especially in areas of recognized biological importance. The tropical Andes are considered one of the 25 hotspots for their concentration of endemic species (Myers, R. Mittermeier, C. Mittermeier, Fonseca, & Kent, 2000) but at the same time, it is a highly vulnerable region having irreplaceable characteristics (Brooks et al., 2006). The environmental heterogeneity, mainly defined by the Andes, a mountainous chain along the north-south axis, is mirrored by social and economic diversity and the existence of a complex set of production systems. These systems reflect different patterns of adaptation to local resource bases, historical trajectories of change and development, and inter and intra regional processes of migration. All of these factors contribute to the fragility of biodiversity in the tropical Andes.

So far, in South America, assessments of possible future states of biodiversity have used climate change as the main driver (Nogué, Rull, & Vegas-Vilarrúbia, 2009) or climate change and land use as main drivers, but not in an integrated way (Higgins, 2007; Feeley & Silman, 2010). However, more work is needed regarding a synthetic biodiversity index that allows an assessment considering more drivers and their direct and indirect effects. For example, LUCC dynamics related to the construction of infrastructure (e.g. roads) affect biodiversity both directly (e.g. through improved patterns of accessibility and further LUCC) and indirectly through the resulting patterns of fragmentation.

3. ESTIMATING BIODIVERSITY TRENDS IN TROPICAL ANDEAN COUNTRIES

As we mentioned earlier, we used a regional version of the GLOBIO 3 framework to solve two questions: a) What would be the most vulnerable areas in the Tropical Andean countries given projected changes in main drivers of biodiversity loss? b) What would be the most important drivers in biodiversity loss and what will be the implications for conservation? The areas of potential biodiversity change were identified through the comparison of the remaining biodiversity for the year 2000 and two scenarios of future remaining biodiversity for the year 2030.

3.1 Study Area

The study area comprises the continental territories of Colombia, Ecuador and Peru with an approximate combined area of about 2,700,000 km^2 (Figure 1). These three countries harbor one of the most notable concentrations of biological diversity on Earth with high levels of diversity and endemism (R. Mittermeier & C. Mittermeier, 2005). At the same time, the region experiences varied levels of human impacts related to different land use patterns of human use of the territory. For example, Josse et al. (2009) reports that 59%, 43% and 12% of the Andean region within Colombia, Ecuador and Peru, respectively, correspond to anthropogenic landscapes.

From a historical viewpoint, Colombia's economy is based on agriculture. Gradually there was a shift to the construction, mining, commerce, industrial, transport and financial sectors that have gained more importance (DANE, 2008). Policies have been centered toward the implementation of free trade treaties as a way to incorporate the country in a global economy. The governments have supported the construction and actualization of the transportation network, proposing important highways, maritime and river ports and other nationwide infrastructure projects that complement the country's policies. The progressive liberalization of the economy has lead to an intense structural shift in the agriculture sector. The production of annual crops that were commonly subsidized (e.g. rice, sorghum and cotton) has been in crisis while extensive and intensive livestock grazing and permanent crops have increased. Bio-fuel crops have been stimulated by benefits from credits and commercial policies, given their apparent advantages in the domestic (and international) markets. These crops have been developed by large-scale organized enterprises. On the other hand, coffee, the most important export product during the last century and pillar of a smallholding rural economy, has decreased in area and in production (DANE, 2007). This decrease responds to the low prices in the international markets and to the growing opportunities for more profitable economic activities. Illegal crops have had fluctuations in area during the last decade (UNODC, 2008). Even though the government has implemented an aggressive eradication campaign, these crops continue to play an important role in the rural economy, mainly for the remote areas of the country.

Ecuador presents three well-defined regions that organize the composition and structure of landscapes at a macro – level. The Pacific Coast of Ecuador represents a transition between the hyper arid conditions found in the Peruvian Desert and the hyper humid climate of the Choco region (M. Munday & G. Munday, 1992; Davis,

Figure 1. Location of study area

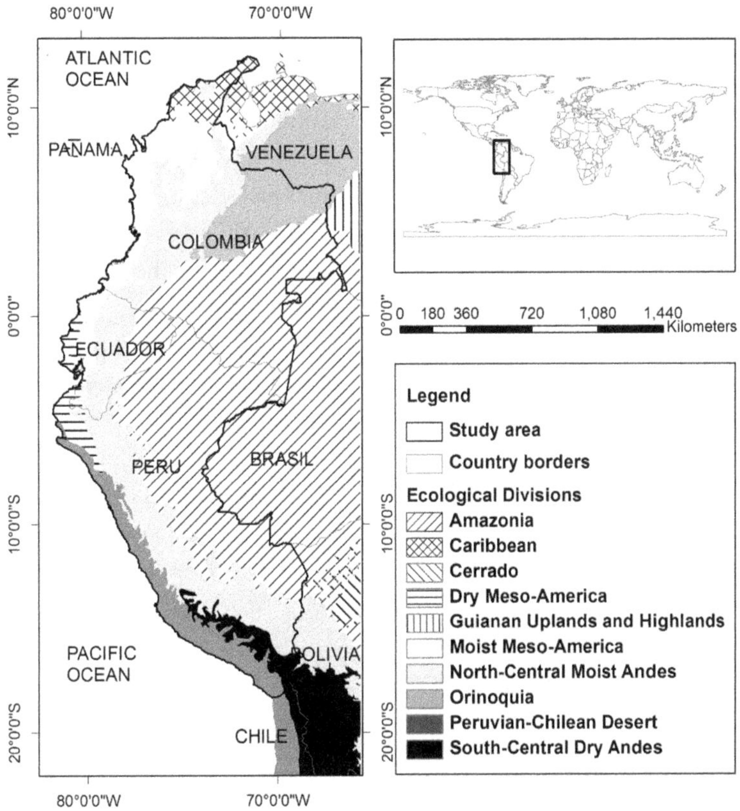

Heywood, & Hamilton, 1997). Historically, the coastal region has experienced the development of highly intensive productive systems. The main products are linked to international markets such as banana plantations, rice and sugar cane especially in the central and southern portions of the Coast. In contrast, the moister region in the north of the Coast has less intensive agriculture. The most important land use systems are associated with the extraction of tropical hardwoods by a complex set of actors that include smallholders, wood exporting companies, and middlemen (Sierra & Stallings, 1998; Sierra, 1999). The Ecuadorian Andes have experienced extensive patterns of agricultural land uses, mainly distributed in the valleys. Even after two processes of agrarian reform, land tenure remains unequal. The valley bottoms concentrate market oriented and capital intensive agricultural operations while in the versants of the Andean ranges prevails a mixed market and subsistence smallholder agriculture (Caviedes & Knapp, 1995). The Ecuadorian Amazon region experienced radical changes in land use patterns in the second half of the past century. The construction of a road, important for the exploitation of the oil fields, promoted a migration wave form the Coast and Andean regions to the Amazon region (Walsh, Messina, Crews-Meyer, Bilsborrow, & Pan, 2002). The northern portion of the Ecuadorian Amazon has witnessed widespread processes of forest loss and fragmentation, associated with the expansion of cattle ranching and agro industrial monocrops (e.g. Oil palm). The central and southeastern portion of the region has been less affected by extensive ecosystem conversion and harbors lower population densities associated with indigenous groups. Subsistence agriculture associated with slash and burn practices still dominate land use dynamics in these areas.

In Peru as in Ecuador, there are also three main regions: Coast, Andes and Amazonia. In the Coastal region several coastal valleys divide the Sechura desert, where the 54.6% of the Peruvian population lives (Walsh et al., 2002). Only the northern part of the Coastal region, given the influence of El Niño Current, has dry forest vegetation where the main agricultural crops are rice, cotton, lemon and mango (MINAG, 2007). In the Andean region, traditionally, agriculture has been the main activity especially based on the cultivation of corn and potatoes. In addition, large extensive livestock in the Andean grasslands constitute the source of income for many families. Finally, mining is a very important activity, but problems regarding land resources tenure (Bury, 2005), ecological consequences and human health problems in some localities have lead to conflicts between population and mining companies. The Amazonian region remains with low population density and the main activities are the extraction of natural resources such as rubber and timber. Agriculture areas began to increase in the last fifty years, partly due to the migration from Andean areas and population and colonization policies. In recent years, several projects such as dams for hydroelectric generation, roads, bio-fuel crops and oil and gas exploitation have started and some of the impacts are already noticeable. The current social situation in Peru is characterized by enormous internal changes, influenced by the set of relatively recent public policies applied for the last governments which have promoted the signature of free trade agreements with other countries such as United States and China.

3.2 Modeling Biodiversity: the GLOBIO Framework

The biodiversity assessment was performed using the mean species abundance (MSA), an index developed by the Netherlands Environmental Assessment Agency (PBL), together with UNEP-WCMC, UNEP-GRID-Arendal under the framework of the GLOBIO3 (Alkemade et al., 2006). This index is not based on the reduction of species richness but on the alteration of the mean abundances per species. The MSA calculates the remaining biodiversity for a specific year after accounting for the effect of different drivers using the following formula for each cell in the study area:

$$MSA_i = MSAL_{LU\,i} * MSA_{I\,i} * MSA_{F\,i} * MSA_{CC\,i} * MSA_{N\,i}$$

where *i* refers to each pixel on the map, MSA_{LU}, MSA_I, MSA_F, MSA_N and MSA_{CC} are the remaining MSA after considering the effect of land use, infrastructure, fragmentation, nitrogen deposition, and climate change, respectively. The values of each MSA range from 0, where no original biodiversity is found, to 1, where biodiversity is pristine. The design of GLOBIO3 allows the estimation of each of the drivers impacts on biodiversity from a small set of land maps: land use, transportation (roads and rails), biome distribution, and nitrogen deposition density maps. The construction of the empirical relationships between the MSA and each driver lies on a meta-analyses of peer-reviewed literature (Alkemade et al., 2009).

Land Use

We calculated the MSA_{LU} according to the relation established between land use classes and empirical remaining biodiversity (Table 1) based on Alkemade et al. (2009).

Table 1. Values for MSA_{LU}, Alfa and sigma according to land cover and land use class (Source: Adapted from Alkemade et al. (2006))

Land cover / Land Use	Description	MSA_{LU} (%)	Alfa	Sigma
Tropical forest	Primary forest with little or none human influence	100	0.1658	0.6214
Shrublands	Shrublands with little or none human influence		0.1614	0.8134
Grasslands	Grasslands where domestic cattle could have partially replaced native species of ruminants.	100	0.1614	0.8134
Desert	Desert areas	100	0.1614	0.8134
Glaciers	Glaciers	100	0.1614	0.8134
Secondary Forest	Forest succession in deforested areas.	50	0.1614	0.8134
Forest Plantations	Planted trees, predominantly homogenous monospecific systems for timber production. The species can be exotic or native.	20	0.1614	0.8134
Biofuels		20	0.1614	0.8134
Perennial crops	Planted trees to produce fruit, coffee, cocoa, and so on. The operation means that the soil is left untreated for long periods of time.	20	0.1614	0.8134
Extensive agriculture	Agricultural areas where the use of fertilizers and pesticides is limited. The production is predominantly for subsistence.	30	0.1614	0.8134
Commercial intensive agriculture	Agricultural areas, with high use of fertilizers and pesticides. The production is predominantly commercial.	10	0.1614	0.8134
Fully managed irrigated agriculture	Irrigated agricultural areas, intensively managed. High levels of fertilizers and pesticides. The production is predominantly commercial.	5	0.1614	0.8134
Mining areas		5	0.0001	1
Artificial Grass	Forests converted to pasture for cattle grazing.	10	0.1614	0.8134
Urban Areas	Areas with high density of artificial structures (e.g. Cities, suburban areas, roads, airports, etc.).	5	0.0001	1

Infrastructure

Values for MSA_I are higher in areas further away from roads and decrease with distance to the closest road. Additionally the calculation considered infrastructure impact would be different according to the biome or the land use. The MSA_I was calculated using the following equation:

$$MSA_I = \alpha * \ln(0.001*(dist + 10)) + \delta$$

where α and δ are specific parameters for the types of land use and land cover defined in Figure 3, Chapter 8, *dist* represents the distance in meters to the infrastructure roads. The calculation of MSA_I only considered the area with natural vegetation. For areas with intense human activity such as urban areas, the effect of infrastructure is already included in the land use impact, therefore, in these areas, the infrastructure impact was not considered as a separate factor.

Fragmentation

The MSA associated with fragmentation takes into account the patch size of natural vegetation in which the unit of analysis is located. Larger areas would have higher values of remaining biodiversity. MSA_F values were assigned by natural vegetation patch size, patches between 0 and 1 km^2 have a value of 0.55 and the following values are shown in Figure 2 and Figure 4, chapter 8. Processes are different between different ecosystems; therefore, patch size was calculated independently for the following biomes: forest, grassland-shrubland, desert and glaciers. Areas with human activities were assigned a MSA_F of 1 as the fragmentation itself does not generate any biodiversity loss in these anthropogenic areas.

Nitrogen Deposition

This factor considers the effect of nitrogen accumulation due to the use of fertilizers (Figure 2, chapter 8). For this case study, there was no reliable information for South America, and this factor was therefore excluded from the analysis.

Climate Change

The effect of climate change on biodiversity is analyzed from the perspective of potential impacts on variation in habitat extension. Species distribution would be affected by changes in temperature and local patterns of precipitation. In general, climate change scenarios predict an increase of temperature (IPCC, 2007), which may lead to any of the 3 following situations: some species or biomes distribution areas may disappear, ranges may be displaced and in other areas, biomes or species may remain stable. Considering this, the MSA_{CC} is calculated as the ratio representing the stable area over the original area. This index is calculated using the following equation:

$$MSA_{CC} = 1 - Slope * \Delta\, Temperature$$

The impact of temperature change on biodiversity was calculated for each biome type (Table 2). The slope represents the sensitivity of each biome to the temperature changes and it is specific for each biome.

3.3 Current and Future Datasets

The land use map for the year 2000 was based on available information for each country. We up-scaled these maps from their original resolution to 1 Km pixel size and adjusted the land use classes to fit each legend to the GLOBIO3 legend. The biome information was constructed mixing information from several sources, since currently, there is no map that provides such information in a medium resolution scale (100 m). The biomes required for the model were desert, grassland/shrubland, savannas, and forest. We obtained the high Andes grasslands map from Josse et al. (2009), the Peruvian desert was recovered from Veliz et al. (2008), the Colombian desert from the GLC2000 (Eva et al., 2003) and the Orinoquia region from the WWF ecoregion map (Olson et

Figure 2. MSA_F values for different natural areas of different patch size

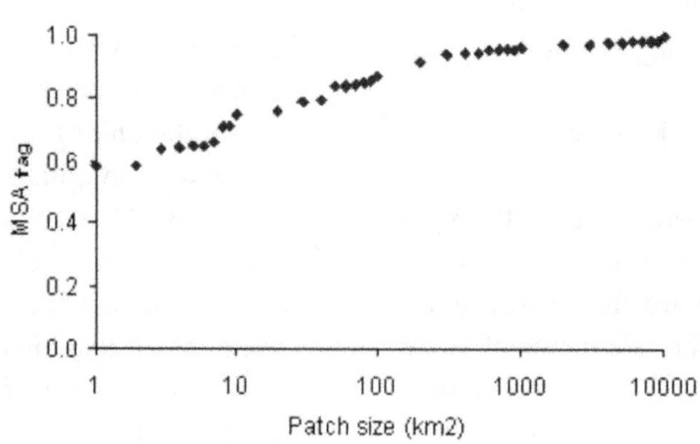

Table 2. Sensibility value and MSA_{CC} for different biomes for years 2000 and 2030 (Source: Alkemade et al. (2006))

Biome	Slope or Sensibility ($°C^{-1}$)	MSA_{cc} 2000 ΔT°C = 0,569	MSA_{cc} 2030 ΔT°C = 1,298
Shrubs	0,129	0,9266	0,8326
Natural Grasslands and steppes	0,098	0,9442	0,8728
Desert	0,036	0,9795	0,9533
Tropical Forest	0,034	0,9807	0,9559

al., 2001). The road maps for each country were, in order to assess infrastructure and fragmentation impacts, systematized from the national sources of official cartography. No information about the construction of future roads infrastructure was included due to inconsistencies in the quality of national databases.

We worked with two scenarios based on the Third Global Environmental Outlook (Raskin & Kemp-Benedict, 2002). This document presents projections for scenarios of Market forces and Policy reform, consistent with scenarios A1 and B1 of the IPCC, respectively, and consistent with the projections made by FAO for the years 2015 and 2030 (Bruinsma, 2003). According to this information, the Market Forces scenario considers that agricultural areas and man-made pastures for livestock will grow at the current rate until 2030. The Policy reform scenario considers that some international policies and agreements will reduce by half the agriculture and man-made pastures growth rate from 2015 onwards. Both scenarios were used as a guide, being later adjusted for the current characteristics of each country, and where information was available, with the time trend of recent years (Colombia: DANE, 2007, 2008; Peru: MINAG, 2007).

Under these scenarios, two future land use maps were calculated using the CLUE model (Conversion of Land Use and its Effects; Verburg et al., 2002; Verburg & A. Veldkamp, 2004). These future land use maps constituted the main input for modeling future remaining biodiversity using GLOBIO 3. The CLUE model allocates the land demand to each landscape unit (pixel), using deterministic rules. These allocation rules consist of the generation of probability maps for each land use class, the identification of the potential changes between land use classes (matrix of change), and the estimations of how easy it is for a pixel of one class to change into another class (elasticity) (Verburg et al., 2002).

The probability maps were built using a backward stepwise regression considering the following variables: climatic (annual mean temperature, annual total precipitation, annual ombrothermic index and ombrothermic index for the driest trimester) obtained from Rivas-Martínez, Sánchez-Mata, & Costa (1999); topographical (elevation, slope, total curvature, terrain convergence index, exposure topographic index: smoothed and non-smoothed); accessibility (access time to market, modified from Jarvis et al., 2006) and restriction variables (legal protection system: natural protected areas).

The matrix of change was slightly different for each country because of the differences between the land use systems of each country (Details in Arnillas et al., 2008). The land use classes with the highest elasticity are those in which the probability of occurrence in one year is not affected because the previous year has had the same use. The completely inelastic classes (with more inertia) are those that once established are very unlikely to change. Among the less elastic land use classes are mining, urban areas and areas where significant investments in infrastructure have been done. Extensive agriculture, intensive agriculture and livestock were assigned intermediate levels of elasticity, depending on the characteristics of each one.

3.4 Biodiversity Change Assessment

The biodiversity change assessment was based on the comparison of the MSA values between the 2030 scenario and the year 2000 by subtraction of the values cell-by-cell. We analyzed the changes considering countries and the ecological divisions developed by Josse et al. (2003). The contribution of each driver was calculated following the MSA methodology (Alkemade et al., 2006, 2009).

4. RESULTS

4.1 Current Biodiversity State

The remaining MSA estimated for the year 2000 is 75.9% for the whole study area, but with important patterns of spatial variability among and within countries (Figure 3). At the country level, the Andean region of Colombia and Ecuador showed extensive areas with low values of MSA, including the Pacific coast of Ecuador. Overall, Peru presented smaller disturbed areas. The areas with lower MSA were located around the most important cities in Colombia (i.e. Bogota, Medellin and Cali) and Ecuador (Guayaquil and Quito). In Peru, the most affected areas were around Tarapoto and Yurimaguas, located in the rainforest. In terms of conservation there is a generalized correspondence between protected areas with areas of high remaining biodiversity (high MSA, Figure 3).

Land use was the most important driver of biodiversity loss, being responsible for the loss of 16.3%. Fragmentation was the second in importance representing a loss of 3.3%. Land use impact was greater in Ecuador, almost five times the Peruvian land use impact. The driver Climate change showed little differences among countries but the highest value was registered in Peru (2.8%, Figure 4).

In the analysis per ecological division, the Caribbean division, located in north Colombia (Figure 1) had the lowest MSA value of all the divisions (37.2%, Figure 5). The Moist Meso-America division situated in the Colombian and Ecuadorian coast only kept half of its biodiversity. On the contrary, the Amazonia region of the three countries and the South-Central dry Andes in south Peru had the largest values of MSA (92.3% and 85.7% respectively). The largest land use effect was registered in the Caribbean division while the South-Central dry Andes had the smallest land use impact. Accordingly, this last division showed the highest climate change impact (5.2%, Figure 5). Infrastructure impact was greater for the Dry meso America division located in the north Peruvian coast and south Ecuadorian coast.

4.2 Future Biodiversity State

As expected from the future land use demands, both 2030 scenarios projected a reduction in the remaining biodiversity for the whole study area. The Market Forces scenario projects a loss of 4.7% while Policy Reform scenario projects 4.2% (Figure 4). The MSA change maps reflected not only areas with a reduction of the MSA values but also with an increase of MSA values (green values in the Figure 3). The areas that showed some recovery are larger in the Policy Reform scenario than in the Market Forces scenario, especially in north Colombia. However, this might be an artifact of the model implementation allowing an easy migration of pixels from one class to another.

There are more areas with higher biodiversity loss (above 50%) in the Market Forces scenario than in the Policy Reform scenario (Figure 3). Areas with losses between 5 and 50% appeared principally in the Colombian eastern savannas of the Orinoco and the Peruvian Andean region

Figure 3. Biodiversity assessment for 2000 (a) and changes in MSA value for the 2030 Market Forces (b) and Policy Reform Scenarios (c)

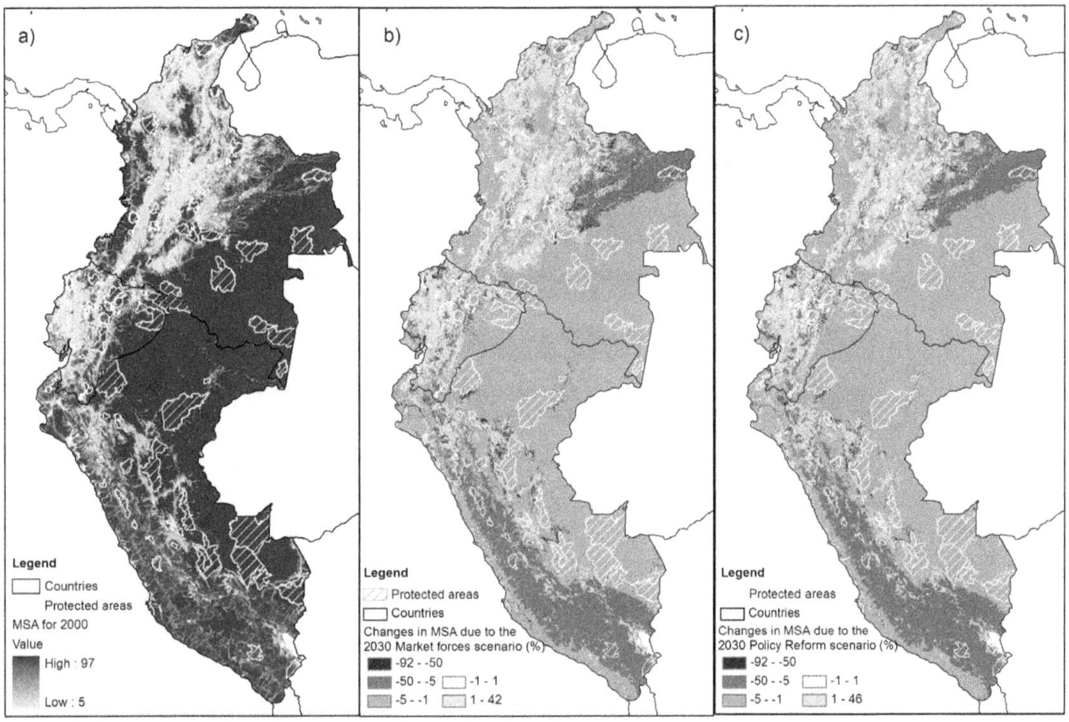

(Figure 3 and Figure 5). This last region, associated with grasslands, would show the highest impact of climate change, according to the model (see Table 2).

At the country level Ecuador would have the lowest values of remaining MSA for 2030, followed by Colombia and finally by Peru for both scenarios (Figure 4). In a comparison with the values of the year 2000, Ecuador also showed the highest biodiversity loss (5.7% for Market Forces scenario), but Peru is the second one while Colombia would have a loss of 3.9% for the Market Forces scenario (Figure 6). The drivers of biodiversity loss of Ecuador maintain their place in importance for the 2030 scenario; the most important one was land use, then infrastructure, fragmentation and finally climate change (Figure 4). In Colombia, the effect of infrastructure was slightly greater than climate change for 2000, while in the 2030 projections, climate change doubled the effect of infrastructure. The same pattern was observed for Peru.

In Colombia, the Orinoco region showed the largest areas of MSA loss in the 2030 scenarios (Figure 3). Here, there is less representation of protected areas in the national system. The Amazon and Pacific regions have small decreases in MSA while there is some recovery of natural vegetation in the eastern and western slopes of the Andes. In Ecuador, areas with greater potential biodiversity loss correspond to the areas of intensive human use on inter – Andean valleys,

Figure 4. Remaining biodiversity and biodiversity loss per driver for each country. 2000 assessment, Market Forces scenario (MF) and Policy Reform scenario (PR)

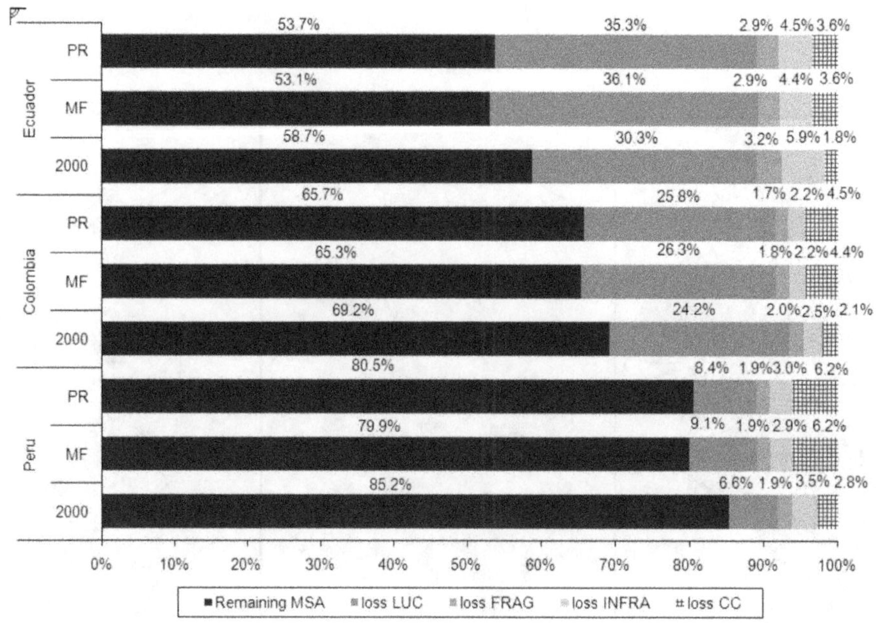

Figure 5. Remaining biodiversity and biodiversity loss per driver for each Ecological Division. 2000 assessment, Market Forces scenario (MF) and Policy Reform scenario (PR)

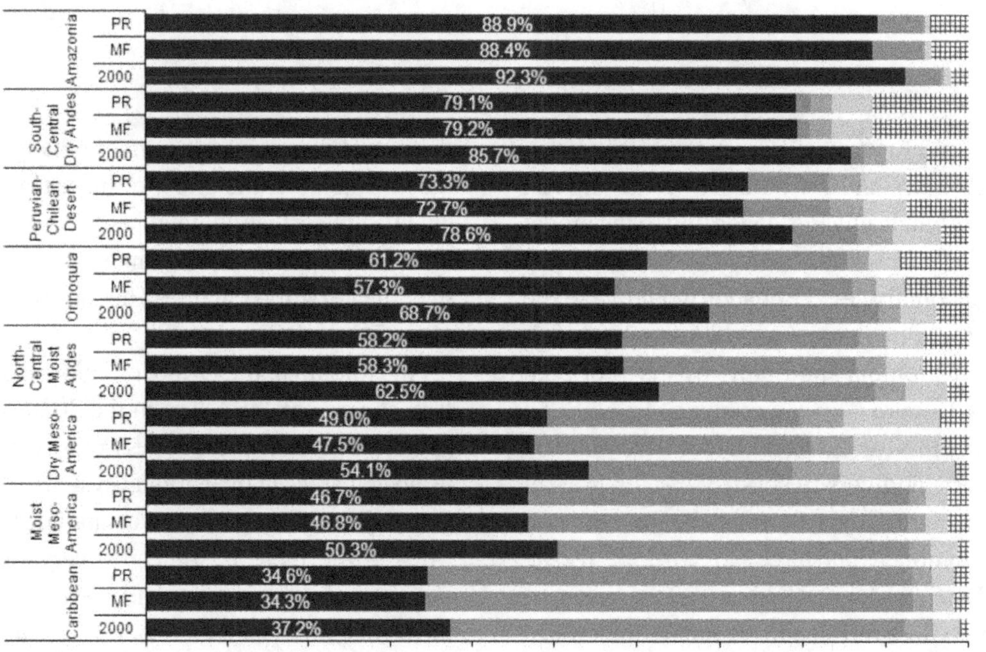

Figure 6. Biodiversity loss for the period 2000-2030 according to the Market Forces scenario (MF) and the Policy Reform (PR) scenario per country

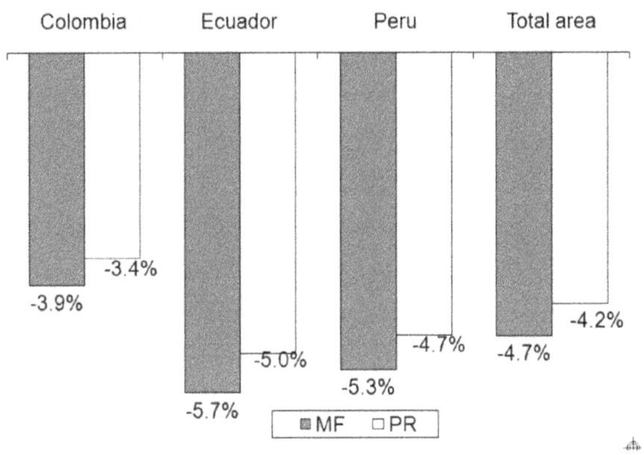

and the deforestation fronts in western Amazonia. In contrast, areas with low decreases of remaining biodiversity were observed in the Amazonia and the east side of the Andes. In Peru, the areas at most risk are in the Amazonian. The expansion of agriculture developed in rainforests will tend to increase.

In the analysis per ecological division, the Caribbean division was the only one with less than half the original biodiversity in 2000. However, for the 2030 projections two other ecological divisions showed less than 50% as remaining biodiversity: Moist Meso-America and Dry Meso-America (Figure 5). Despite the Caribbean division has the lowest MSA values for 2030, the Orinoquia division is the one that showed the highest biodiversity loss for the period 2000-2030 (Figure 7). The Caribbean, Moist Meso-America, Dry Meso-America, North-Central Moist Andes and Orinoquia divisions had land use as a main driver of biodiversity loss with a value twice the next important driver, which reflects the land use threat to these areas. In the Peruvian-Chilean desert and the Amazonia, land use impact is also the largest one but the difference in climate change is not as high as it was in 2000. Finally, the South-Central Dry Andes is still the only region where the climate change effect is higher than the land use effect. Indeed most of the biodiversity loss for 2030 is due to this driver, 11.8% for both scenarios in comparison to the 5.2% found for 2000.

Figure 7. Biodiversity loss for the period 2000-2030 according to the Market Forces scenario (MF) and the Policy Reform (PR) scenario per Ecological Division

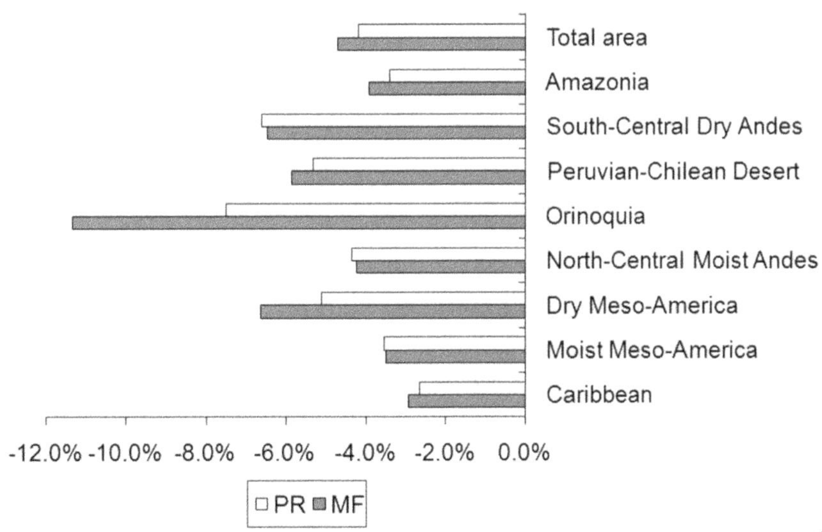

5. DISCUSSION

5.1 Vulnerable Areas in Tropical Andean Countries

According to the results, the ecological divisions that showed the less remaining biodiversity for the present and the future scenarios are located in the coastal areas of Colombia, Ecuador and north Peru i.e. The Caribbean, the Moist Meso-America and the Dry Meso-America (Figure 5). These results contrast with the global results obtained by Alkemade et al. (2009), which present the Orinoquia as the region with the lowest remaining biodiversity followed by the Andean regions of Ecuador and Colombia, but also including the coastal areas and the Peruvian-Chilean Desert. Even though there is some agreement between our results and those of Alkemade et al. (2009) the use of national land use maps with higher resolution in the present study allows for a more accurate evaluation.

The Caribbean division is principally composed of mangrove forest, dry forest and shrublands in north Colombia. A great part of the Colombian population is concentrated here with a large history of landscape transformation and a variety of land uses. Therefore, this area currently shows high degree of transformation and fragmentation. The main activity is large-scale cattle raising, however there is an expansion of mining and biofuel crops.

The Moist Meso-America division is composed of the Choco (mainly in the Colombian Pacific coast), the Magdalena Humid forest of Colombia and the Western Ecuador moist forest. Climatically, high levels of precipitation and humidity characterized this region. Socioeconomically, it shows high levels of poverty and the presence of indigenous and afro-descendant communities, especially in Colombia. In Ecuador, the Coast has experienced high levels of human land use and population densities due to the existence of fertile soils and seasonally dry weather, especially in the central and southern areas (Murphy & Lugo, 1995). These areas are dominated by market-oriented agricultural systems, many of which are destined to international markets (e.g. bananas). The great disturbance of the Ecuadorian part of this division and the Magdalena Humid forest of Colombia is mainly responsible for the low value of remaining biodiversity. On the other hand, the Choco Region is rather well conserved (Figure 3).

The Dry Meso-America division consists of the coastal dry forest of Ecuador and Peru. Here, the Ecuadorian part shows the most affected areas even though land use patterns in this part of the country are less intensive than the agro-export region of the lower portion of the Guayas river basin. In Peru, these areas are dedicated to agricultural activities, replacing the dry forest for areas for the cultivation of rice, cotton, lemon, mango and some others crops.

Despite the three divisions mentioned above are the ones with less biodiversity for 2000 and for 2030, they are not necessarily the most vulnerable ones. Other areas show a greater biodiversity loss for the period 2000-2030. For example, the Orinoquia Division composed of tropical gallery forests and savannas, would have more reduction in MSA values for the period 2000-2030 (Figure 7). Most of the human population is concentrated in the piedmont part of the region where the expansion of agriculture and petroleum activities is evident. Extensive cattle raising activities characterize the eastern savannas. This region may become the next "colonization frontier" of Colombia due to the high impact projects that are being planned. The projections of future biodiversity of this study corroborate this trend. Many of the biofuel and permanent crops are being harvested at the cost of savannas, gallery and dry forests. In addition, there is 4% of difference in biodiversity loss between the Market Forces scenario and the Policy Reform (Figure 7). This is explained by the extensive areas of this region dedicated to livestock activity that will be favored under the Market Forces scenarios.

The policies affecting land use might be different for the three countries, and therefore, the ecological divisions can show different patterns for each country. In this sense, the North-Central moist Andes, above the 3000 masl, that includes the Paramo (Colombia, Ecuador and north Peru) and the Puna (Central and South Peru) ecosystems show different degrees of biodiversity loss along their distribution. For the year 2000, the Colombian Paramos, showed very low values of remaining biodiversity, while Peruvian Punas had medium to high values mostly due to the infrastructure effect (Figure 3). However, in the future scenarios, the Peruvian Puna would show a biodiversity loss range between 5% and 50% while the Colombian and the Ecuadorian paramos did not show much change (Figure 3). This may reflect a relative stabilization of the already widespread distribution of agricultural areas in the Colombian and Ecuadorian Andes. On the other hand, the Peruvian Puna has extensive livestock activities distributed over most of the natural grasslands. This use has some impact on biodiversity that was not possible to take into account with the available information. Therefore, there might be an overestimation of the MSA. Nevertheless, this effect is small compared to other types of land uses.

5.2 Most Important Drivers in Biodiversity Loss and Implications for Conservation

Land use is the most important driver of biodiversity loss at the country and at the ecological division level and for both 2000 and 2030. This driver may be slightly underestimated due to the absence of maps identifying activities such as selective logging or extensive grazing in natural grasslands. Although the effect of these activities may be partially incorporated in the infrastructure impact, it is possible that some sectors remain with higher MSA values than they should have. For that reason, values of MSA should be interpreted as moderate overestimation especially in forests and the Peruvian mountain natural grasslands. The analysis of the other drivers should consider the increment of land use impact. For example, the calculation of fragmentation only considers natural areas, therefore, since more areas have human activities, these are not considered for the fragmentation effect causing the decrease of its value. The same applies for the projected impact of climate change for both scenarios. Therefore, the impact of the other drivers in future scenarios should be estimated more accurately with specific models that can provide independent information to MSA.

The explicit inclusion of several drivers in the biodiversity model allows analyzing the biodiversity conservation process in a more comprehensive way. For example, it would be possible to support territorial zoning and other management options to locate areas where agricultural expansion is less detrimental (e.g, causing minor fragmentation, reducing the impact of infrastructure or evaluating more carefully the impact of climate change). Likewise, the increase in productivity would also be an option that avoids the expansion of agricultural frontier and even continue fulfilling the demand of food for the population by 2030.

Our results were evaluated at two different levels of agregation, one at the country level and the other one at the ecological division level. One of the most important advantages of GLOBIO3 is its flexibility to identify spatial patterns of change in multiple management units. The use of GLOBIO3 easily translates into maps the degree of threats that can be used for example for territorial planning. While it is still necessary to refine national information (such as trends in land use based on more recent agricultural census) we believe that the results draw attention to the most vulnerable areas today. It also allows the identification of more important areas for biodiversity conservation (Trisurat, Alkemade, & Verburg, 2010). However, the appropriate scale for which the methodology provides an appropriate platform for decision-making is a function of the quality and level of detail of the data used.

The outputs of both scenarios are a diagnosis of possible future biodiversity based on a combination of local and international trends. However, it is possible to propose different scenarios. Therefore, both, the central government and local governments, by means of using GLOBIO3-CLUE methodology, have an alternative to assess the impacts of their decisions. Possible outcomes before they take decisions on agricultural policy, territorial zoning, environmental policies, and new road infrastructure, among others are some examples of this. In this way, the evaluation of desired or undesired outputs as a result of a policy give tools to politicians, policy makers and managers, as they are the ones deciding finally (Cumming, 2007). Nevertheless, it is also important to highlight the uncertainties of these models. The uncertainties for climate change projections have been evaluated for Ecuador (Buytaert, Célleri, & Timbe, 2009) and this is one of the drivers of GLOBIO. Therefore, we should also contemplate the uncertainties of the other drivers, in order to give policy makers a complete tool for making decisions.

5.3 Future Research Directions in Tropical Andes

In this exercise, the mean species abundance (MSA) is the indicator for biodiversity assessment, when analyzing how the species abundance changes in comparison to a referent. This kind of indexes that integrate information at species and ecosystem levels has been recognized as important for their contribution to understanding trends and for being easy to communicate (Pereira & Cooper, 2006). However, most of the reliability depends on good input data. In this sense, the relationships established between MSA and LUCC classes, infrastructure, fragmentation and climate change needs to be further validated for tropical countries. Specific experiments that can control several variables at the same time, during long temporal periods might also be necessary. We should also consider that the relation between MSA and the drivers at the global approach could be modified due to the change in the scale.

Some improvement can be made regarding some of the drivers, for example to, the second most important driver, infrastructure. The information of roads that we incorporated into the GLOBIO3 model was static, assuming that there will be no change until the year 2030. In the future, this information should be incorporated in the model dynamically to highlight the impacts of infrastructure projects over biodiversity and land use. In the same way, the use of population density in future projections will improve the predictions. This is especially important when there is evindece of the relation between human population growth and threatened species of birds and mammals for example (McKee et al., 2003). However, to include both drivers in a dynamical way, we need to reinforce the model with socioeconomic information. There is also a need of new models, that can provide information about specific processes such as the socioeconomic or political conditions that promote the creation of roads or that modifies the people flux along the roads. Another important aspect is that, a similar impact to infrastructure would be found near croplands. Around these agricultural areas, there is higher possibility to access resources than further away. Specially in forests, people usually hunt, gather resources and the agriculture activities can facilitate the occurrence of introduced species. This process could be easily included in the MSA algorithm, but must be calibrated first with existing data.

The effect of climate change could be improved by the use of more detailed climatic information. Some regional climate models (RCM) have been applied or tested in the South American Region (Urrutia & Vuille, 2009; Buytaert et al., 2010). Even though they remain coarse for analyses in the montanious region, this is the best source of information we have at the moment. The calculation of the MSA_{CC} in GLOBIO includes variation of the effect according to biomes (Table 2), however a better approximation should include a more detailed ecosystem classification. A better resolution would allow the distinction between montane forest and dry forest, for instance. A first approximation could be used for example for the Ecoregions of WWF (Olson et al., 2001). Some inconsistencies such as a low climate change impact on tropical glaciers would thereby be avoided.

Fragmentation should also be revised for the finer scale analysis. For example, national borders should not create false fragmentation. The analysis of the current and future scenarios can be complemented with land use change outputs of global models that would provide the border conditions to assess the MSA. Another option, that would require further discussion, is the use of main rivers as boundaries of fragments. Even though several animals and plants can cross the river; this might not be the case for the smaller ones. The main rivers such as the Amazonas and some of their tributaries are important barriers, mainly for sub-population dynamics. Following this line of reasoning, vegetation might be naturally fragmented into small patches, not necessarily caused by human intervention. Therefore, it would be worthwhile to explore changes in patch size as an indicator of fragmentation rather than current patch sizes. This can be done using the ratio between current patch size and original patch sizes. The procedure could use a species-area approach, similar to those applied by Thomas et al. (2004) and Hubbell (2001).

Other drivers such as nitrogen deposition could not be tested for our study area due to the absence of reliable input data for the South American region. Similar problems was faced by Trisurat et al. (2010) when analyzing the case of Thailand at the regional level. This driver was possible to apply at global level, but not at regional level. In addittion, there is still some other drivers that would need to be tested such as the selective extraction and hunting. Even though the difficulties of quantifying these drivers, they are important in the Amazonian regions of the tropical countries. Some studies have pointed out that in hunting areas, hunting intensity would be a better predictor of species diversity than vegetation disturbances (Naughton-Treves et al., 2003). In this regard, drivers should be incorporated according to the different biodiversity threats of each specific region. However, problems incorporating drivers that have high correlation with other factors need to be solved first.

Finally, some issues should be addressed to improve the communication between scientists and decision makers. A first key issue is the more explicit inclusion of the relationship between biodiversity and environmental services. Since decision makers have to respond to socioeconomic demands, the valuation of biodiversity in terms of environmental services would be more tangible. A model considering this would help to support or reject policies considering both biodiversity itself and the economic impact. The second issue relates to finding new ways of explaining the importance of biodiversity. In fact, the link with environmental services might be useful as well. However, to be able to support a discussion about policy decisions, a definition of a threshold for a "significant change" in MSA would be needed. A first step towards this target should be the inclusion of uncertainty measures in the current MSA estimation. Nevertheless, more research and conceptual support would allow answering some questions. For example: Is a 0.1 loss in MSA an affordable reduction? Which is the lower boundary to avoid serious damage to the ecosystem at this landscape scale?

6. CONCLUSION

The evaluation of biodiversity is not a simple task, nevertheless, the MSA represents a simple index to understand biodiversity changes, a necessary quality to communicate information to decision makers and general public (Pereira & Cooper, 2006). The use of high-resolution inputs such as the national land use maps allowed the identification of areas with the lowest remaining biodiversity values (Caribbean) and the area with the highest biodiversity loss for the period 2000-2030 (Orinoquia). However, without more detailed experiments and models, the improvement in data quality obtained from national data would be lost. In addition, we should be aware that other aspects such as, for example, genetic diversity, are hard to consider (Cumming, 2007). Indeed, MSA does not consider this. Moreover, some other characteristics have also been highlighted as important when discussing future biodiversity modeling such as dispersal capabilities, reproductive potential, and biotic interactions (Ibáñez et al., 2006). Despite all this, MSA shows a more integrative approach for developing future scenarios under different drivers that considers more than climate change (a very popular topic nowadays). Moreover, our results suggest that land use change would have a greater impact than climate change in a near future. The inclusion of drivers integrated in the "boundaries" of national policy actions, is another important topic, since the framework can provide tools to evaluate national policies that can affect land use changes. This kind of models require a way to translate policy decisions into land use demand, and in this topic, further research is needed by adopting an interdisciplinary approach.

ACKNOWLEDGMENT

We would like to thank the Netherlands Environmental Agency (PBL) for financing this project and the Instituto Alexander Von Humboldt for providing the information of Colombia.

REFERENCES

Alkemade, R., Bakkenes, M., Bobbink, R., Miles, L., Nellemann, C., Simons, H., & Tekelenburg, T. (2006). GLOBIO 3: Framework for the assessment of global terrestrial biodiversity. In A. F. Bouwman, T. Kram, & Goldewijk (Eds.), *Integrated modelling of global environmental change. An overview of IMAGE 2.4* (pp. 171-186). Bilthoven, Netherlands: Netherlands Environmental Assessment Agency (MNP).

Alkemade, R., van Oorschot, M., Miles, L., Nellemann, C., Bakkenes, M., & ten Brink, B. (2009). GLOBIO3: A framework to investigate options for reducing global terrestrial biodiversity loss. *Ecosystems*, *12*, 374–390. doi:10.1007/s10021-009-9229-5

Arnillas, C. A., Galindo, G., Peralvo, M., & Tovar, C. (2008). *Validation and diffusion of the GLOBIO methodology in the Andean region. Informe Final*. Lima, Peru: PBL.

Balmford, A., Green, R. E., & Jenkins, M. (2003). Measuring the changing state of nature. *Trends in Ecology & Evolution*, *18*(7), 326–330. doi:10.1016/S0169-5347(03)00067-3

Brooks, T. M., Mittermeier, R., da Fonseca, G. A. B., Gerlach, J., Hoffmann, M., & Lamoreux, J. F. (2006). Global biodiversity conservation priorities. *Science*, *313*, 58–61. doi:10.1126/science.1127609

Bruinsma, J. (Ed.). (2003). *World agriculture: Towards 2015/2030. An FAO perspective*. London: Earthscan Publications Ltd.

Bury, J. (2005). Mining mountains: neoliberalism, land tenure, livelihoods, and the new Peruvian mining industry in Cajamarca. *Environment & Planning A*, *37*, 221–239. doi:10.1068/a371

Buytaert, W., Célleri, R., & Timbe, L. (2009). Predicting climate change impacts on water resources in the tropical Andes: the effects of GCM uncertainty. *Geophysical Research Letters, 36*, L07406. doi:10.1029/2008GL037048

Buytaert, W., Vuille, M., Dewulf, A., Urrutia, R., Karmalkar, A., & Célleri, R. (2010). Uncertainties in climate change projections and regional downscaling: implications for water resources management. *Hydrology andl Earth System Science, 14*, 1247-1258. doi:javascript:openWin(this);window.document.basic1.submit();

Caviedes, C., & Knapp, G. (1995). *South America*. Englewood Cliffs: Prentice Hall.

Ceballos, G., & Ehrlich, P. R. (2002). Mammal Population Losses and the Extinction Crisis. *Science, 296*, 904–907. doi:10.1126/science.1069349

Cumming, G. S. (2007). Global biodiversity scenarios and landscape ecology. *Landscape Ecology, 22*, 671–685. doi:10.1007/s10980-006-9057-3

DANE. (2007). Colombia: Departamento Administrativo Nacional de Estadística, Indicadores de producción agropecuaria. Retrieved from http://www.dane.gov.co/index.php? option=com_content&task=category&sectionid=18&id=41&Itemid=152

DANE. (2008). Producto interno Bruto. Segundo trimestre de 2008 Base 2000. Departamento Nacional de estadística. Retrieved from http://www.dane.gov.co/files/ investigaciones/boletines/pib/presen_PIB_IItrim08.pdf

Davis, S., Heywood, V. H., & Hamilton (Eds.). (1997). *The Americas*. Centres of plant diversity (IUCN.). Gland.

Eva, H., de Miranda, E., Di Bella, C., Gond, V., Huber, O., Sgrenzaroli, M., et al. (2003). The Land Cover Map for South America in the Year 2000. European Commision Joint Research Centre. Retrieved from http://www-gem.jrc.it/glc2000

Feeley, K. J., & Silman, M. R. (2010). Land-use and climate change effects on population size and extinction risk of Andean plants. *Global change biology, Accepted article*. doi:10.1111/j.1365-2486.2010.02197.x

Hannah, L., Lohse, D., Hutchinson, C., Carr, J. L., & Lankerani, A. (1994). A preliminary inventory of human disturbance of world ecosystems. *Ambio, 23*(4/5), 246–250.

Higgins, P. A. T. (2007). Biodiversity loss under existing land use and climate change: an illustration using northern South America. *Global Ecology and Biogeography, 16*, 197–204. doi:10.1111/j.1466-8238.2006.00278.x

Houlahan, J. E., Findlay, C. S., Schmidt, B. R., Meyer, A. H., & Kuzmin, S. L. (2000). Quantitative evidence for global amphibian population declines. *Nature, 404*, 752–755. doi:10.1038/35008052

Hubbell, S. (2001). *The Unified Neutral Theory of Biodiversity and Biogeography*. Princeton University Press.

Ibáñez, I., Clarck, J. S., Dietze, M. C., Feeley, K. J., Hersh, M., & LaDeau, S. (2006). Predicting biodiversity change: outside the climate envelope, beyond the species–area curve. *Ecology, 87*(8), 1896–1906. doi:10.1890/0012-9658(2006)87[1896:PBCOTC]2.0.CO;2

IPCC (2007). *Climate Change 2007: Synthesis Report*. Contribution of Working Groups I, II and III to the Fourth Assessment Report of the Intergovernmental Panel on Climate Change [Core Writing Team, Pachauri, R.K and Reisinger, A. (eds.)]. IPCC, Geneva, Switzerland, 104 pp.

Jarvis, A., Castano, S., Hyman, G., Gebhardt, S., Guevara, E., Castro, M., et al. (2006). TNC Threats Assessment Version 1.1. CIAT. Retrieved from http://conserveonline.org/workspaces/ersm.pilots/pilot/SACRThreats/view.html

Josse, C., Cuesta, F., Barrena, V., Cabrera, E., Chacón-Moreno, E., Ferreira, W., et al. (2009). *Ecosistemas de los Andes del Norte y Centro. Bolivia, Colombia, Ecuador, Perú y Venezuela* (Secretaría General de la Comunidad Andina, Programa Regional ECOBONA – Intercooperation, CONDESAN-Proyecto Páramo Andino, Programa BioAndes, EcoCiencia, NatureServe, IAvH, LTA-UNALM, ICAE-ULA, CDC-UNALM, RUMBOL SRL.). Lima.

Josse, C., Navarro, G., Comer, P., Evans, R., Faber-Langendoen, D., & Fellows, M. (2003). *Ecological systems of Latin America and the Caribbean. A working classification of terrestrial systems*. NatureServe.

Lambin, E., Turner, B. L., Geist, H., Agbola, S. B., Angelsen, A., & Bruce, J. W. (2001). The causes of land-use and land-cover change: moving beyond the myths. *Global Environmental Change, 11*, 261–269. doi:10.1016/S0959-3780(01)00007-3

McKee, J. K., Sciullli, P. W., Fooce, C. D., & Waite, T. A. (2003). Forecasting global biodiversity threats associated with human population growth. *Biological Conservation, 115*, 161–164. doi:10.1016/S0006-3207(03)00099-5

Millennium Ecosystem Assessment. (2005). *Ecosystems and Human Well-being: Biodiversity Synthesis*. Washington, D.C.: World Resources Institute.

(2007). *MINAG* (p. 180). Lima, Peru: Estadística Agraria Mensual. Julio.

Mittermeier, R., & Mittermeier, C. (2005). *Megadiversity: Earth's Biologically Wealthiest Nations*. Mexico: Cemex.

Munday, M., & Munday, G. (1992). The climate of south–west Ecuador. In Best, B. (Ed.), *The threatened forests of south-west Ecuador* (pp. 7–59). Leeds: Biosphere Publications.

Murphy, P. G., & Lugo, A. (1995). Dry forests of Central America and the Caribbean. In Bullock, S., Mooney, H. A., & Medina, E. (Eds.), *Seasonally Dry Tropical Forests* (pp. 9–34). Cambridge University Press. doi:10.1017/CBO9780511753398.002

Myers, N., Mittermeier, R., Mittermeier, C., Fonseca, G., & Kent, J. (2000). Biodiversity hotspots for conservation priorities. *Nature, 403*, 853–858. doi:10.1038/35002501

Naughton-Treves, L., Mena, J. L., Treves, A., Alvarez, N., & Radeloff, V. C. (2003). Wildlife Survival Beyond Park Boundaries: the Impact of Slash-and-Burn Agriculture and Hunting on Mammals in Tambopata, Peru. *Conservation Biology, 17*(4), 1106–1117. doi:10.1046/j.1523-1739.2003.02045.x

Nogué, S., Rull, V., & Vegas-Vilarrúbia, T. (2009). Modeling biodiversity loss by global warming on Pantepui, northern South America: projected upward migration and potential habitat loss. *Climatic Change, 94*, 77–85. doi:10.1007/s10584-009-9554-x

OECD. (2008). OECD Environmental. *Outlook to*, 2030.

Olson, D. M., Dinerstein, E., Wikramanayake, E. D., Burgess, N. D., Powell, G. V. N., & Underwood, E. C. (2001). Te rrestrial Ecoregions of the Wo rld: A New Map of Life on Earth. *Bioscience, 51*(11), 933–938. doi:10.1641/0006-3568(2001)051[0933:TEOTWA]2.0.CO;2

Pereira, H. M., & Cooper, H. D. (2006). Towards the global monitoring of biodiversity change. *Trends in Ecology & Evolution, 21*(3), 123–129. doi:10.1016/j.tree.2005.10.015

Raskin, P. D., & Kemp-Benedict, E. (2002). *Global Environmental Outlook Scenario Framework. Background Paper for UNEP's Third Global Environmental Outlook Report* (p. 102). Stockholm Environment Institute–Boston Center.

Rivas-Martínez, S., Sánchez-Mata, D., & Costa, M. (1999). North American boreal and western temperate forest vegetation. Retrieved from http://www.ucm.es/info/cif/book/namerica2/namerica_02_1.htm

Sala, O. E., Chapin, F. S. III, Armesto, J. J., Berlow, E., Bloomfield, J., & Dirzo, R. (2000). Global Biodiversity Scenarios for the Year 2100. *Science*, *287*, 1770–1774. doi:10.1126/science.287.5459.1770

Sanderson, E. W., Jaiteh, M., Levy, M. A., Redford, K. H., Wannebo, A. V., & Woolmer, G. (2002). The human footprint and the last of the wild. *Bioscience*, *52*(10), 891–904. doi:10.1641/0006-3568(2002)052[0891:THFATL]2.0.CO;2

Scholes, R. J., & Biggs, R. (2005). A biodiversity intactness index. *Nature*, *434*, 45–49. doi:10.1038/nature03289

Sierra, R. (1999). Traditional resource-use systems and tropical deforestation in a multiethnic region in North-west Ecuador. *Environmental Conservation*, *26*(2), 136–145. doi:10.1017/S0376892999000181

Sierra, R., & Stallings, J. (1998). The Dynamics and Social Organization of Tropical Deforestation in Northwest Ecuador, 1983-1995. *Human Ecology*, *26*(1), 135–161. doi:10.1023/A:1018753018631

Thomas, C. D., Cameron, A., Green, R. E., Bakkenes, M., Beaumont, L. J., & Collingham, Y. C. (2004). Extinction risk from climate change. *Nature*, *427*(8), 145–148. doi:10.1038/nature02121

Trisurat, Y., Alkemade, R., & Verburg, P. (2010). Projecting land-use change and its consequences for biodiversity in northern Thailand. *Environmental Management*, *45*, 626–639..doi:10.1007/s00267-010-9438-x

UNEP. (2007). *Global Environment Outlook GEO4, envieonment for development*. Malta: PROGRESS PRESS LTD.

UNODC. (2008). Colombia Coca Cultivation Survey. United Nations Office on Drugs and Crime (UNODC) and Government of Colombia. Retrieved from http://www.unodc.org/unodc/en/crop-monitoring/index.html

Urrutia, R., & Vuille, M. (2009). Climate change projections for the tropical Andes using a regional climate model: Temperature and precipitation simulations for the end of the 21st century. *Journal of Geophysical Research*, *14*, D02108. doi:10.1029/2008JD011021

Veliz, C., Tovar, A., Tovar, C., Regal, F., & Vásquez, P. (2008). ¿Qué áreas conservar en nuestras Zonas Áridas? Seleccionando sitios prioritarios para la conservación en la Ecorregión Desierto de Sechura - Perú. *Zonas Áridas*, *12*(1), 36–59.

Verburg, P., Soepboer, W., Veldkamp, A., Limpiada, R., Espaldon, V., & Mastura, S. S. A. (2002). Modeling the spatial dynamics of regional land use: the CLUE-S model. *Environmental Management*, *30*(3), 391–405. doi:10.1007/s00267-002-2630-x

Verburg, P., & Veldkamp, A. (2004). Projecting land use transitions at forest fringes in the Philippines at two spatial scales. *Landscape Ecology*, *19*, 77–98. doi:10.1023/B:LAND.0000018370.57457.58

Walsh, S., Messina, J., Crews-Meyer, K., Bilsborrow, R., & Pan, W. (2002). Characterizing and modeling patterns of deforestation and agricultural extensification in the Ecuadorian Amazon. In Crews-Meyer, K., & Walsh, S. (Eds.), *Linking people, place and policy: a GIS science approach* (pp. 187–214). Boston: Kluwer Academic Publishers.

Chapter 14
The Influence of Changing Conservation Paradigms on Identifying Priority Protected Area Locations

Alan Grainger
University of Leeds, UK

ABSTRACT

Conservation planning for climate change adaptation is only one in a long sequence of conservation paradigms. To identify priority locations for protected areas it must compete with three other contemporary paradigms: conservation of ecosystem services, optimizing conservation of ecosystem services and poverty alleviation, and reducing carbon emissions from deforestation and forest degradation. This chapter shows how conservation paradigms evolved, discusses the merits of different approaches to modelling potential impacts of climate change on biodiversity, and describes the hybrid BIOCLIMA model and its application to Amazonia. It then discusses conservation planning applications of the three other contemporary paradigms, illustrated by examples from Amazonia and Kenya. It finds that rapid paradigm evolution is not a handicap if earlier paradigms can be nested within later ones. But more sophisticated planning tools are needed to identify optimal locations of protected areas when climate is changing, and to use protection to mitigate climate change. These should encompass the complex interactions between biodiversity, hydrological services, carbon cycling services, climate change, and human systems.

DOI: 10.4018/978-1-60960-619-0.ch014

1. INTRODUCTION

The potential impacts of global climate change on biodiversity could be catastrophic. According to one early estimate, up to 35% of all species could be committed to extinction by 2050, six times the impact if habitat loss continues at current rates (Thomas *et al.*, 2004). More detailed studies are needed at national and regional scales to assist pro-active conservation planning to ameliorate harmful impacts (Thuiller *et al.*, 2008). The need for preventative measures is particularly urgent in the humid tropics, which contains half of all species in the world (Pimm, 2001).

But how can we identify the optimal locations for protected areas when climate is changing, and also use this protection to help in mitigating climate change? One answer may lie in the fact that facilitating adaptation of biodiversity to global climate change is only one of a sequence of paradigms that have emerged during the evolution of conservation planning. Despite its recent origin it must compete for research funding and attention from policy makers with three alternative paradigms. These are: (1) conservation of ecosystem services; (2) optimizing conservation of ecosystem services and poverty alleviation; and (3) reducing carbon emissions from deforestation and forest degradation - the REDD scheme of the UN Framework Convention on Climate Change.

This chapter reaches two main conclusions. First, the rapid evolution of conservation paradigms is not a handicap if earlier paradigms can be nested within later ones. Second, more sophisticated tools are required to indicate how conservation planning should encompass the complex interactions between ecosystem services, biodiversity, climate change and human systems.

2. CONSERVATION PLANNING AND BIODIVERSITY

Conservation planning paradigms in the tropics have evolved rapidly in the last 50 years, superseding the 'recreational' paradigm that dominated conservation since its birth in temperate countries in the nineteenth century (Adams, 2008).

Conservation in the 1960s generally followed the 'reservation' paradigm. The boundaries of national parks and other protected areas were legally identified, local people were excluded from using them, and the general public were allowed varying degrees of access for recreation. Criteria used to select where to site protected areas included size, remoteness, natural beauty and the need to maintain the habitats of threatened animal species. The latter was assisted by regular publication of ('Red') lists of threatened species by the International Union for the Conservation of Nature (IUCN) (Vie *et al.*, 2008).

In the 1970s, a new 'conservation and development' paradigm became widely accepted, since social exclusion had not improved protection on the ground. In 'biosphere reserves', established within the Man and Biosphere Programme (MAB) of the UN Educational, Scientific and Cultural Organization, core protected areas were surrounded by 'buffer zones' of strictly limited use, negotiated with local people (Batisse, 1990). Funds were also made available to improve the productivity of adjacent areas designated for sustainable use.

In 1980 a 'globally structured' paradigm was advocated by the IUCN World Conservation Strategy (WCS). Each country was now advised to protect representative areas of key ecosystems and concentrations of endemic species, to ensure that globally important species and all of the world's major ecosystem types would be sustained in some form (IUCN, 1980).

The 'globally structured' paradigm marked the switch from paradigms initiated solely by practitioners to those which also draw on scientific knowledge. This trend continued with the 'hot spots' paradigm, which recognized that it was not possible to give equal priority to every area of high conservation values. Funds were limited and, given the perceived speed of tropical deforestation (Grainger, 1980), time was short. So it was important to focus on 'hot spots' - areas with high densities of endemic or rare species that were severely threatened by habitat (usually forest) destruction (Myers, 1988). This was, unusually, a hybrid paradigm which combined conservation value and rate of habitat destruction.

In the 1990s the 'biological diversity' (or biodiversity) paradigm emerged and has since co-existed with the 'hot spots' paradigm. It initially referred only to conserving species diversity and how to prevent extinctions (Wilson, 1988), but was later elaborated to refer to conserving the joint diversity of ecosystems, species and genes (McNeely *et al.*, 1990). This effectively combined previously parallel lines of action and research on conserving representative ecosystem types (as in the WCS), endemic and threatened species (as in the 'hotspots' and earlier paradigms), and plant genetic resources. Work on the latter began in the late 1960s with a focus on crop plants and led to the formation in 1974 of the International Board for Plant Genetic Resources (Jackson & Ford-Lloyd, 1990). Since measuring biodiversity is a great challenge (Boyle & Boontawee, 1995; Rose & Grainger, 2002) this paradigm is still to be properly implemented. Scientists participated in devising indicators to monitor progress in realizing the UN Convention on Biological Diversity's Target 2010 plan (Balmford *et al.*, 2005), to reduce the rate of biodiversity decline, but have asked for future targets and indicators to be clearer (Mace *et al.*, 2010).

Since the turn of the Millennium, conservation planning has been refined yet again to incorporate adaptation to the effects of global climate change on biodiversity (Thuiller *et al.*, 2008). The novelty of the paradigm is evident in the Third and Fourth Assessment Reports of Working Group II of the Intergovernmental Panel on Climate Change (IPCC). The Third Assessment Report refers to impacts on biodiversity in only general terms (McCarthy *et al.*, 2001). The Fourth Assessment Report devotes considerable space to the detailed research carried out since 2001 in simulating the possible impacts of climate change on species distributions (Parry *et al.*, 2007).

Whereas earlier approaches to protecting concentrations of biodiversity assumed unchanging habitats in a static climate, such assumptions are no longer tenable in an era where climate is changing and habitats will shift with it. Greater flexibility in siting protected areas is required to ensure the resilience of biodiversity as a whole. So protected area boundaries must not only contain the present distributions of ecosystems, species and genes but where they might shift to as climate changes. Again, pragmatism is important, and in view of other pressures on land use, compact 'migration corridors' between large protected areas may be a practical way to facilitate biotic transfers (Malhi *et al.*, 2008). Work on assessing the biotic impacts of climate change effectively began with a conference on plant genetic resources, held in Birmingham, UK in 1989 (Jackson *et al.*, 1990). This had dual importance, as it considered not only the impacts of climate change, but the ability of humanity to respond to this by breeding new crop strains suited to changed climates. The latter required retaining the widest possible diversity of the genetic resources of the world's principal crop plants. But it was the development of models that could simulate the potential impact of climate change on the distributions of many individual species (e.g. Thomas *et al.*, 2004) which brought this topic to the international conservation policy agenda.

This new 'climate change adaptation' paradigm is consistent with all those that preceded it except one. The 'conservation and development' paradigm needs modification in the climate change era, since MAB 'buffer zones' of stable land use, intended to support survival within the core protected areas they surround, could undermine species resilience by blocking migration.

3. MODELLING GLOBAL CLIMATE CHANGE IMPACTS ON BIODIVERSITY

We now examine the different approaches to modelling the potential impacts of global climate change on biodiversity. These may be divided into four main categories: biome models; dynamic global vegetation models; climate envelope models; and process models (Morin *et al.*, 2008).

3.1 Biome Models

Biome models predict present and future distributions of major ecosystem types based on spatial variation in climatic and other environmental variables. They are still valuable at large scale despite their coarse resolution (Malcolm *et al.*, 2006).

3.2 Dynamic Global Vegetation Models

Dynamic global vegetation models are the heirs of biome models. They still do not model species individually, but achieve higher resolution by aggregating species in groups of 'plant functional types' with similar physical characteristics (Higgins, 2007).

3.3 Climate Envelope Models

Climate envelope models explain current and future distributions of individual species in terms of envelopes provided by correlations between the presence of a species and climatic and other environmental variables. They are widely used for continental simulations (e.g. Thomas *et al.*, 2004) but exclude interactions between species; factors that determine if a species flourishes in a set of environmental conditions; and migation mechanisms when conditions change (Araújo & Luoto, 2007). They also assume that a species will occupy the same niche when climate changes (Dormann, 2007).

3.4 Process Models

Process models explain how species presence is affected by population dynamics, e.g. in terms of growth, reproduction, mortality, dispersal etc. (Keith *et al.*, 2008). Empirical calibration increases their explanatory power and enhances their scientific credibility, but makes them data intensive and limits their spatial scope. This has restricted applications at continental scale (Chuine & Beaubien, 2001) and in data-poor developing countries.

3.5 The Merits of Climate Envelope and Process Approaches

The two most favoured approaches at present are climate envelope models and process models. Each has its advantages and disadvantages.

Climate envelope models have proven useful for large-scale modelling but are criticized by those who favour process models. According to their detractors they do not incorporate the processes underlying the dynamics of populations of individual species, or interactions between the populations of difference species. They therefore involve considerable uncertainty (Araújo *et al.*, 2005). Critics also question the assumption that a species will occupy the same niche in future climates as it does at present (Dormann, 2007).

Process based models are favoured by those requiring good empirical calibration for scientific credibility, but their complexity and data intensity limits large-scale applications (Morin *et al.*, 2008). So quantification is often possible for data-rich developed countries but not for data-poor developing countries.

Pimm (2007) believes that the case for climate envelope models is compelling. He also finds credible the estimate by Thomas *et al.* (2004) that the threat posed by climate change far exceeds that due to habitat loss. Thuiller *et al.* (2008) point to the reality of trade-offs between complexity and tractability, and argue that there is still no agreement on whether more complex models have greater accuracy than simple ones. Morin *et al.* (2008) suggest that climate envelope models could overestimate the potential for extinction, but suggest that both they and process models have their place. Ideally, therefore, both types of models should be simulated so their results can be compared.

Jeschke & Strayer (2008) admit that climate envelope models often explain present species distributions, despite their simplicity, but claim that their predictions of trends are rarely tested. This point has been answered by Green *et al.* (2008) who found that this type of model could explain trends in how the populations of 42 British bird species varied with climate change over a 25 year period.

3.6 Hybrid Models

Keith *et al.* (2008) make a convincing case for hybrid models, arguing that climate envelope models and process models "in isolation. fail to deal with interactions and dependencies between small-scale population processes and landscape scale habitat change. Coupled together, they are more likely to produce projections that are both realistic and robust to uncertainty."

4. THE BIOCLIMA MODEL

To see how climate-change biodiversity models can assist conservation planning under the new 'climate change adaptation' paradigm we look in detail at the BIOCLIMA model (Miles *et al.*, 2004). This was an early example of a hybrid model. For while its outputs were combined by Thomas *et al.* (2004) with others obtained with climate envelope models, it integrated the climate envelope and process approaches in a manner suited to data-poor tropical countries.

In contrast to other models, which simulate possible trends in an arbitrary sample of species (e.g. Morin *et al.*, 2008), BIOCLIMA simulated trends in a sample chosen to represent regional plant biodiversity. For a single main ecosystem type - lowland tropical rain forest - a sample of tree species was selected by taxonomic family and then by plant functional types containing the five traits that directly influence the processes determining a plant's response to climate change, i.e. reproductive rate, dispersal mechanisms and pre-adaptations to expected stresses. The traits were life-form; height class; deciduousness; pollination mode; and dispersal mode.

BIOCLIMA followed climate envelope models in requiring for each species a map of its *potential distribution*, based on its requirements for actual evapotranspiration (AET), annual moisture deficit (MD), and seasonality of moisture availability (SMA). This was corrected to give a more reliable map of its actual or *realized distribution,* to allow for heterogeneity in environmental conditions, e.g. the presence of mountains that are barriers to dispersal.

But BIOCLIMA also followed process models in simulating a size-structured population for each 1° grid cell in the realized distribution of a species from 1900 to 1990, when it was assumed to be in equilibrium with its environment. Future population change was then simulated until 2095. This was based on how survival, growth and reproduction were affected by changes in AET, MD and SMA when climate changed according to a suitable scenario of the Hadley Centre Global Climate Model HADCM2, built and run by the UK Meteorological Office. Simulation was then continued for another 100 years to test the long-term viability of the surviving population.

Species viability was evaluated by analysing changes in its population and the overlap between its realized distribution and the new projected potential distribution. The weaker the overlap, and the poorer the fit between these two distributions and protected areas, the more likely a species would be "committed to extinction", and the more urgent the need for conservation measures. Migration to new projected distribution areas was not included in BIOCLIMA.

The original computer model accessed plant and climate data stored in a Microsoft Access database. Species response to climate change was simulated by algorithms coded in Microsoft Visual Basic. Spatial outputs were represented graphically by IDRISI Geographical Information System software.

5. APPLYING THE BIOCLIMA MODEL TO AMAZONIA

BIOCLIMA was devised to simulate climate change impacts on biodiversity in Amazonia. This section summarizes findings described by Miles *et al.* (2004).

5.1 Climate Change Scenarios

Two HADCM2 global climate model scenarios were used. The 'worst case' scenario, called the Standard Impact (SI) scenario, used the outputs of bioclimate variables in the IS92a 'business as usual' scenario of HADCM2. This was the standard scenario for low levels of international action on climate change, and assumes an approximate doubling of atmospheric CO_2 concentration. In the 'best case' - or Reduced Impact (RI), scenario - selected values for climate parameter changes were half those in the SI scenario.

5.2 Species Selection

Selection of species in lowland Amazonia began with 228 families of angiosperms thought to occur in Amazonia. This was cut to 132 after some families were excluded for lack of strong evidence of occurrence or because species distributions were concentrated in montane or arid regions. The 132 families were classified into cladistic groups and representatives of each group selected, with preference given to endemic families and those with many genera. An iterative process of selection, taking into account the existence of good quality data on distributions, led to a final sample of 14 families. One hundred and ninety-three species from these families were evaluated for their ability to represent all plant functional types. A final sample of 69 species for which good distributional data were available was chosen for simulation.

5.3 Results

Simulations between 1990 and 2095 showed that actual evapotranspiration (AET) increased in west Amazonia and decreased in the northeast. The greatest rise in moisture deficit (MD) was in the northeast, but areas of high MD encroached into central Amazonia during the simulation period. MD remained relatively low in northwest Amazonia for much of the period, so it could serve as a refuge for moist forest species. Seasonality of moisture availability (SMA) remained highest in the east, while the northeast acquired an aseasonal moisture deficient climate at the end of the simulation period. SMA increased throughout most of Amazonia, indicating rising moisture stress, even in previously aseasonal areas, but remained low in the aseasonal northwest.

For most species the simulations showed no significant change in their *realized distributions* between 1990 and 2095. No species became extinct over more than one third of its estimated range, though many declined to a very low population density which would make them highly vulnerable to extinction.

Yet significant changes occurred in the *potential distributions* of all species, leaving many populations as non-viable relicts. By 2095, 28 of the 69 species in the sample (41%) were under greatest threat in the SI scenario, compared with 14 species (21%) in the RI scenario (Table 1). *Pouteria reticula* was close to joining the 28 threatened species in the SI scenario, with 98% of all cells non-viable, compared with 60% in the RI scenario (Figure 1).

By 2095, 20 species in the SI scenario (29% of the sample) and 9 species in the RI scenario (13%) either had no potential distribution at all, or their original distribution was so far from their potential distribution that they had no realistic chance of reaching it by dispersal. The other non-viable species did have areas of potential distribution near their current realized distribution and so could disperse into the newly suitable cells.

Changes in the spatial distributions of moisture deficits and seasonality also affected species distributions over the simulation period. In both scenarios the most favourable habitats for moist forest species in 2095 were in the more aseasonal western Amazonia, and in high altitude areas, which are also concentrated in the west. In the SI scenario many species gained new potential distributions along the western edge of their current simulated range.

Northeast Amazonia underwent the most profound long-term change in species density and composition. All modelled populations lost viability in the SI scenario, and only a small proportion remained viable in the RI scenario. Initial spatial distribution appeared to be the most decisive factor in explaining change response: species that were widely distributed throughout Amazonia showed the greatest resistance to change; and species with narrow ranges or poor tolerance of

Table 1. Numbers of species becoming non-viable by 2095 in the Standard Impact (SI) and Reduced Impact (RI) scenarios (Source: Miles et al. (2004))

	SI		RI	
	No	%	No	%
Non-viable species	28	41	14	21
No potential distribution or possibility of reaching one	20	29	9	13
Of which endemics	7	10	0	0

Figure 1. Distribution of Pouteria reticulata (Sapotaceae) in Amazonia in 2095, simulated in the BIOCLIMA Standard Impact scenario (SIS) and Reduced Impact scenario (RIS). Viable cells are shaded blue, green or yellow; non-viable cells are shaded red, orange or purple; unoccupied potential distribution cells are shaded grey (1990), black (2095) or white (1990 and 2095). Source: Miles (2002).

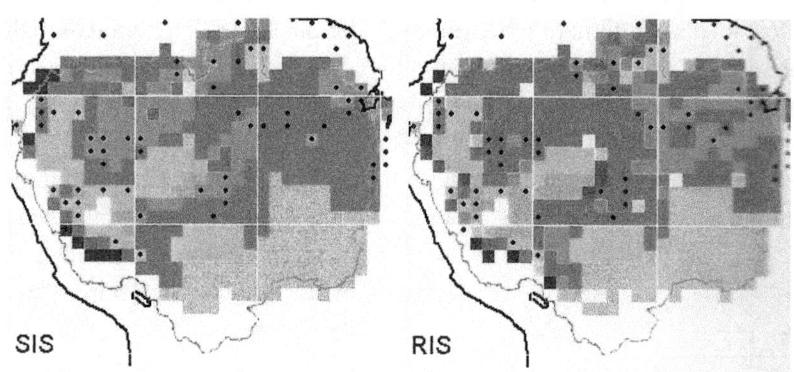

moisture deficits underwent the greatest changes in population density.

5.4 Recommendations for Conservation Planning

Simulations with BIOCLIMA suggested that the best remaining refugia for lowland moist forest species were in the far west of Amazonia, near the Andean highlands. Some species might even expand their altitudinal range, displacing existing montane forest flora, which would in turn migrate themselves. For this reason the focus of conservation actions to increase resilience should not be confined to the far west Amazonian lowlands but encompass a wider altitudinal range. In view of the uncertainty associated with the impacts of climate change, existing protected areas in lowland Amazonia should not be discarded. Ideally, as many reserves as possible would include both lowland and montane forests or migration corridors between these.

6. PLANNING APPLICATIONS OF OTHER CONSERVATION PARADIGMS

Biodiversity, however, is just one of a number of ecosystem services, and while it does play a fundamental role its primacy is contested, as shown in this section. So countries also face the challenge of identifying the optimal locations of protected areas when climate is changing, and finding ways to use protection to mitigate that climate change, without impacting adversely on multiple ecosystem services. Insights into how to accomplish these multiple goals can be gained by examining three other conservation paradigms that are now in vogue.

6.1 Conserving Multiple Ecosystem Services

The concept of ecosystem services has become increasingly important in the scientific literature since the year 2000 (De Groot *et al.*, 2002). Following its promotion in the Millennium Ecosystem Assessment (MEA, 2005) it has also been widely adopted in conservation and development planning. The conventional classification (MEA, 2005) divides ecosystem services into:

1. Production services, which supply goods such as food and raw materials.
2. Regulation services, which regulate the flows of gases, water and wastes.
3. Cultural services, which provide recreational and other opportunities.
4. Supporting services, which provide resilience and cycle nutrients.

It regards biodiversity as underpinning all four services.

This classification has several disadvantages which could impede practical application. First, the outcomes of production services were previously represented by the concept of 'ecosystem goods'. Including them within the same classification may seem scientifically neater, but it removes the easily grasped dichotomy between 'harvesting ecosystem goods' and 'conserving ecosystem services', and this could constrain conservation rather than improving it. Second, placing all key regulation services in a single category could also limit their profile to planners. Third, omitting biodiversity from the main classification system could inhibit comprehensive planning.

Working Group II of the Intergovernmental Panel on Climate Change (IPCC), which deals with impacts, vulnerability and adaptation, has referred to 'ecosystem goods and services' in both the Third Assessment Report (TAR) (McCarthy *et al.*, 2001) and the Fourth Assessment Report (FAR) (Parry *et al.*, 2007). In both cases hydrological services were discussed in a separate chapter from other ecosystem services and biodiversity. While the TAR only referred to ecosystem goods and services in general terms, the FAR used a specific classification. This resembled that listed above but with biodiversity included as one of the supporting services. We shall use this approach here.

Research into the link between biodiversity and other ecosystem services is still embryonic (Mooney, 2010). Only a partial overlap between them has been found at global scale (Naidoo *et al.*, 2008), though synergies are likely to be highest in tropical forests (Turner *et al.*, 2007).

Kenya is a good place to test the multiple ecosystem services concept. It is currently experiencing severe problems arising from the lack of sustainable hydrological services from its five main river catchments: the Mau Escarpment, Mount Elgon, Mount Kenya, the Aberdare Range and the Cherangani Hills (Figure 2). These regulate annual flows of water resulting from highly seasonal precipitation (Krhoda, 1988).

The Mau Escarpment is the catchment of all but one of the main rivers of the Rift Valley. These include: the Nzoia, Yala, Nyando, Sondu, Mara and Kerio, which flow into Lake Victoria; the Molo, which flows into Lake Baringo; the Ewaso Nyiro, which flows into Lake Natron; and the Njoro, Nderit, Makalia and Naishi, which flow into Lake Nakuru. Lake Victoria is shared with Uganda and Tanzania and is the source of River Nile. Lake Turkana is shared with Ethiopia, and Lake Natron with Tanzania.

Forest covers at most 6% of Kenya (FAO, 2006) and is dispersed in 26 blocs. Mau Forest is the largest bloc, being found in 403,775 ha of forest reserves. Its actual present area is not recorded, but 39,000 ha were cleared from 1986 to 2000 (Baldyga *et al.*, 2007) and 16,897 ha from 2000 to 2005 (Akotsi *et al.*, 2006).

Deforestation has had major impacts on Kenya's hydrological services (Baldyga et al., 2004), affecting boreholes, rivers and lakes. Continuity of river flows is important since 70% of Kenya's electricity is generated by hydroelectric power plants. Wildlife depending on rivers and lakes have also been affected: the migration of wildebeest has been impeded by lack of water in the Mara River, and flamingo breeding has been affected by the drying of Lake Natron (Swallow, 2006). The consequent reduction in livestock pastures has exacerbated pre-existing conflicts over land between nomadic pastoralists and farmers. Discontinuities in access to water are likely to get worse as global climate change continues (Awuor, 2007).

Deforestation in the Mau Forest is not purely the result of unplanned forest encroachment. An investigation by the Government of Kenya has revealed that earlier administrations gave land titles in Mau Forest to their supporters in contravention of the law. This began in the 1990s and in October 2001 the government alienated another 15% of Mau (Republic of Kenya, 2004).

In September 2009 a new Kenyan administration decided to take drastic action to allow forest

Figure 2. The principal forests of Kenya, with the five main catchment forests labelled in upper case. Source: based on Sayer et al. (1992).

to regenerate to restore hydrological services. It began to evict large numbers of illegal settlers from the Mau Forest. By February 2010 most settlers with no formal land titles had been evicted, and there were plans to remove the remaining 18,649 families with illegal land titles.

While the evictions should help to restore hydrological services they could have a detrimental effect on other ecosystem services and biodiversity. Since poor rural people must generally grow food to survive, controlling deforestation in one part of a country could cause it to 'leak' to forests elsewhere. Mau and the four other large mountainous areas that constitute Kenya's main river catchments are of recent volcanic origin, so their density of endemic species and species generally is low. The highest biodiversity forests are actually elsewhere (Figure 2): near the coast; in the far west, e.g. the Kakamega Forest is an outlier of the Congo Basin forests; and the Taita and Taveta Hills in the south, which are the northern tip of the Eastern Arc mountains centred in Tanzania (Sayer *et al.*, 1992). These latter forests are a tropical biodiversity 'hot-spot' (Myers *et al.*, 2000). So high biodiversity forests could be vulnerable to deforestation leakage if forests that convey most hydrological services are preferentially protected.

Kenya's experience suggests that treating biodiversity as underpinning all ecosystem services may not provide the best framework for studying ecosystem services or for conservation planning. The implication that the magnitudes of ecosystem services and biodiversity are correlated is not empirically valid in Kenya. High hydrological services do not generally coincide with high biodiversity there. So preferentially protecting forests that convey high hydrological services will not automatically protect high biodiversity forests too, and may even put the latter at greater risk of deforestation through leakage. It therefore seems best to follow the IPCC in treating biodiversity as a supporting service which can be managed separately from other ecosystem services, while also allowing for its collective impacts.

6.2 Optimizing Ecosystem Services Conservation and Poverty Alleviation

Many countries are committed to fulfilling the eight Millennium Development Goals. The first is to eradicate poverty and hunger and the seventh is to ensure environmental sustainability (UN, 2000). British scientists are being encouraged to connect the Millennium Development Goals and Millennium Ecosystem Assessment by devising techniques to conserve ecosystem services and alleviate poverty (NERC/ESRC/DFID, 2008).

This can be seen as an elaboration of the earlier 'conservation and development' paradigm. Yet limited results of earlier integrated strategies (Adams *et al.*, 2004) suggest that such "win-win" outcomes are not inevitable. Only 16% of a sample of World Bank projects with poverty alleviation and conservation goals recorded net gains in both goals (Tallis *et al.*, 2008). Barrett *et al.* (2005) advise "scholars and practitioners to guard against wishful thinking that synergies naturally emerge" and claim that "trade-offs are the norm."

The difficulties involved in implementing this paradigm are evident in Western Amazonia, near the Andean highlands. Simulations with the BIOCLIMA model, summarized above, suggested that the best remaining refugia for lowland moist forest species would be in this region. Keeping forests intact was also vital to give lowland species the flexibility to migrate to montane areas in case the westernmost lowlands do not provide suitable habitats. Yet this transitional zone between Amazonia and the Andes also has very fertile soils. These could be the basis for sustainable and productive farming systems that would reduce the poverty of those who practise them (Porro *et al.*, 2008). So the governments of Bolivia and Peru would face a challenge if they tried to implement this paradigm by aiming to conserve biodiversity and cut poverty in the same area. Careful planning is needed to provide refugia and migration corridors while allocating some lands to productive and sustainable farming.

Another example of the potential to conserve ecosystem services and alleviate poverty is the Mau Forest in Kenya, discussed above. Kenya was ranked 147th in 2007 out of 182 countries by the UN Human Development Index and 92nd out of 135 countries by the Human Poverty Index (UNDP, 2009).

Poverty is still high in many areas, including western Kenya where the main river catchments are located. It has been exacerbated by the deterioration in hydrological services. Hydroelectric power generates 70% of all electricity, but a new plant in Nyanza province, completed in 2007 and expected to raise national power supply by 5%, is still not operating, as the level of the Sondu Miriu River is too low. So the resulting lack of energy has limited the country's ability to generate income. Falling levels of rivers and lakes have driven down wildlife populations, and this has affected tourist numbers and income, not only in Kenya but in neighbouring Tanzania too (Gereti *et al.*, 2003).

Mau itself is the home of the Ogiek. With a population of 20,000 they are one of the country's last remaining forest dwelling tribes (African Union, 2003). They live in communally held pieces of land administered by Councils of Elders according to clans and family units. Until the 1950s they relied on Mau Forest's production services, hunting wild game, gathering fruits, collecting herbs and cultivating bees. But their cultures, traditions and territory have since been eroded, so many Ogiek now practise small-scale cropping and livestock raising.

The poverty of the Ogiek could be reduced if they received payments to improve the sustainability and productivity of their livelihoods while acting as 'ecosystem stewards' to protect the forests on the Mau catchment (Swallow *et al.*, 2009). This in turn would reduce the external costs resulting from irregular water supplies that are undermining the livelihoods of many lowland 'environmental services beneficiaries' groups, ranging from farmers to the tourist industry.

Two challenges remain, however. First, people evicted from Mau Forest must be resettled in ways that alleviate their poverty but do not impact negatively on ecosystem services elsewhere in the country, e.g. by causing deforestation to leak to high biodiversity forests. Second, reforestation must be sufficient to ensure resilience to extra variability in rainfall caused by global climate change.

6.3 Conserving Forest Carbon Cycling Services

Unlike the other paradigms considered here, the Reduced Emissions from Deforestation and Degradation (REDD) scheme planned by the UN Framework Convention on Climate Change (UNFCCC) has not been devised specifically to conserve species, or biodiversity in the broader sense. Instead it focuses on conserving carbon cycling ecosystem services (and hence forest carbon stocks), to mitigate global climate change and prevent some of the impacts predicted earlier.

Biodiversity conservation has, however, been implicitly assumed to be a co-benefit of REDD, on the basis that any forest that is high in carbon density should be high in biodiversity too. As tropical forests contain half of all carbon stored in terrestrial vegetation (Watson *et al.*, 2000) and at least half of all the species on the planet (Pimm, 2001) this might seem a reasonable assumption. High carbon density is indeed positively correlated with high species diversity in grasslands (Tilman *et al.*, 2008) but there is no evidence to show that this is replicated in tropical forests. Concerns were first raised by Miles & Kapos (2008), and then restated by the Marburg Declaration of the Association for Tropical Biology and Conservation and Society for Tropical Ecology (ATBC/STE, 2009), and a team of leading tropical forest scientists (Grainger *et al.*, 2009).

Early estimates suggest that only half of high biodiversity forests, i.e. those featuring in at least four hot spots lists, also have high carbon density

(Kapos *et al.*, 2008). So preferentially protecting high carbon density forests to comply with commitments to REDD could by default allow deforestation to 'leak' to forests that are high in biodiversity but are not as strongly protected since they are low in carbon density. This provides further support for the IPCC approach of including biodiversity as one of a number of supporting ecosystem services.

There are two main reasons why this potential hazard was not spotted earlier. First, political compartmentalization. Responsibility for climate change and biodiversity is divided between the UNFCCC and the UN Convention on Biological Diversity (UNCBD) respectively. Both were agreed at the UN Conference on Environment and Development in 1992, but there has long been concern about their different goals (Kim, 2004). This is exacerbated by their different approaches to implementation: the UNFCCC has developed an increasingly complex set of rules while the UNCBD adopts a 'softer' regulatory stance (Caparrós & Jacquemont, 2003). The two conventions communicate via a Joint Liaison Group and in 2004 agreed a Joint Work Programme. However, UNCBD rules are too weak to require Parties to forego any UNFCCC actions that could damage biodiversity (Caparrós & Jacquemont, 2003).

The second reason is linked to divisions among biological scientists, some of whom focus on forest carbon while others study biodiversity. These divisions are perpetuated by the IPCC. In the latest report of Working Group II, responsible for impacts, vulnerability and adaptation, a chapter on ecosystems discussed the impacts of climate change on biodiversity (Fischlin & Midgley, 2007). But there was no corresponding chapter in the report of Working Group III, which deals with mitigation. Biodiversity impacts of mitigation rated just three paragraphs in its forestry chapter, and only concerned forest plantations (Metz *et al.*, 2007). An earlier IPCC study on climate change and biodiversity dealt mostly with impacts of climate change on biodiversity. Potential impacts of reducing deforestation merited just half a page and were generally seen as positive, though the need for more research was recognized (Gitay *et al.*, 2002).

While REDD is not strictly a conservation paradigm, its influence on tropical forest conservation will be at least equivalent to that of preceding paradigms, as it will have the full weight of the UNFCCC behind it. So when implementing REDD it is vital to take an approach that is fully inclusive of all ecosystem services. REDD rules are expected to recognize the need to recognize the rights and welfare needs of indigenous peoples (UNFCCC, 2009), so if this integrated approach is adopted it would conform to the British government's aim of conserving ecosystem services and alleviating poverty.

7. TOWARDS A COMPREHENSIVE PLANNING TOOL

This study suggests that to identify the optimal locations of protected areas when climate is changing, and to use protection to mitigate climate change in a sustainable way that does not undermine biodiversity by deforestation, will require a more sophisticated conservation planning tool than currently exists.

Such a tool could use the ecosystem services concept as a framework for encompassing the complex interactions between biodiversity, hydrological services, carbon cycling services, climate change - and human systems too.

However, it is notable that one sub-discipline which might have been expected to contribute to the development of such an integrated tool, ecological economics, has until recently neglected ecosystem services (Perrings, 2006).

8. CONCLUSION

Conservation paradigms have evolved rapidly over the last fifty years. This reflects our earlier ignorance and ability to learn from experience in implementing relatively simplistic paradigms.

Rapid evolution is not a handicap if earlier paradigms can be nested within later ones. Usually this has occurred, and the growing complexity of paradigms has not been at the expense of consistency. However, the REDD scheme, which has been developed within the carbon policy and scientific communities, rather than in the conservation policy and scientific communities, is an exception to this. So conserving forests solely for carbon benefits will have some positive biodiversity impacts but some negative impacts too (Grainger *et al.*, 2009).

The growing complexity of paradigms is entirely justified, because the problems of balancing the economic, social and environmental dimensions of human welfare are becoming more complex with every year that passes. The notion that the global climate change can have a dramatic effect on forest biodiversity has only become widely recognized within the last 10 years. This chapter has argued that to meet present needs we need a far more sophisticated conservation planning tool than currently exists. It should help planners to select forests for protection to reduce the carbon emissions that contribute to climate change and maximize the resilience of remaining forests to climate change. At the same time it should avoid negative impacts on biodiversity or other ecosystem services and, by alleviating the poverty of local people, give them an incentive to respect the boundaries of protected areas.

Ultimately, it is for every government - with the support of non-governmental organizations, local communities and international donors - to ensure that comprehensive and integrated plans are in place to optimize economic, social and environmental benefits in this way. Once the REDD scheme becomes operational it is likely that an increasing number of countries will establish their own national forest carbon monitoring systems (Baker *et al.*, 2010). It would be excellent if these systems could also have: (a) an integrated biodiversity component so that potential negative impacts of REDD protection on biodiversity can be averted; and (b) a land use component, so that future agricultural expansion can be planned to benefit people in particular localities and in the country as a whole, while minimizing negative environmental impacts.

The international scientific community has a major role to play in this. First, by continuing to develop more sophisticated tools to support conservation and development planning by using the information generated by these new national carbon and biodiversity monitoring systems. Second, by reducing the huge deficit of global information on the spatial distributions of forests and their attributes, such as carbon density and biodiversity, through better global monitoring.

REFERENCES

Adams, W. M. (2008). *Green development* (3rd ed.). London, UK: Routledge.

Adams, W. M., Aveling, R., Brockington, D., Dickson, B., Elliott, J., & Hutton, J. (2004). Biodiversity conservation and the eradication of poverty. *Science*, *306*, 1146–1149. doi:10.1126/science.1097920

African Union. (2003). *Report of the African Commission Working Group of experts on indigenous populations/communities*. Banjul: African Union.

Akotsi, E., Ndirangu, J. K., & Gachanja, M. (2006). *Changes in forest cover in Kenya's five water towers 2003-2005*. Department of Resource Surveys and Remote Sensing/Kenya Forests Working Group, Kilmani.

Araújo, M. B., & Luoto, M. (2007). The importance of biotic interactions for modelling species distributions under climate change. *Global Ecology and Biogeography, 16*, 743–753. doi:10.1111/j.1466-8238.2007.00359.x

Araújo, M. B., Whittaker, R. J., Ladle, R. J., & Erhard, M. (2005). Reducing uncertainty in projections of extinction risk from climate change. *Global Ecology and Biogeography, 14*, 529–538. doi:10.1111/j.1466-822X.2005.00182.x

ATBC/STE. (2009). *The Marburg declaration, Marburg, Germany, July 2009*. Frankfurt: Association for Tropical Biology and Conservation and the Society for Tropical Ecology.

Awuor, C. (2007). *Assessing policy influences on people's relationship to water ecosystem services: The Kenyan experience*. London, UK: International Institute for Environment and Development.

Baker, D. J., Richards, G., Grainger, A., Brown, S., DeFries, R., & Gonzalez, P. (in press). Achieving forest carbon information with higher certainty: A five step strategy. *Environmental Science & Policy*.

Baldyga, T. J., Miller, S. N., Direse, K. L., & Gichaba, C. M. (2007). Assessing land cover change in Kenya's Mau Forest region using remotely sensed data. *African Journal of Ecology, 46*, 46–54. doi:10.1111/j.1365-2028.2007.00806.x

Baldyga, T. J., Miller, S. N., Shivoga, W., & Gichaba, M. (2004). *Assessing impact of land cover change in Kenya*. Paper presented at the Annual Conference of the American Society for Photogrammetry and Remote Sensing, May 2004, Denver, Colorado.

Balmford, A., Crane, P., Dobson, A., Green, R. E., & Mace, G. M. (2005). The 2010 challenge: Data availability, information needs and extraterrestrial insights. *Philosophical Transactions of the Royal Society B, 360*, 221–228. doi:10.1098/rstb.2004.1599

Barrett, C. B., Lee, D. R., & McPeak, J. G. (2005). Institutional arrangements for rural poverty reduction and resource conservation. *World Development, 33*, 193–197. doi:10.1016/j.worlddev.2004.07.008

Batisse, M. (1990). Development and implications of the biosphere reserve concept and applicability to coastal regions. *Environmental Conservation, 17*, 111–116. doi:10.1017/S0376892900031878

Boyle, T. J. B., & Boontawee, B. (Eds.). (1995). *Measuring and monitoring biodiversity in tropical and temperate forests*. Bogor: Centre for International Forestry Research.

Caparrós, A., & Jacquemont, F. (2003). Conflicts between biodiversity and carbon sequestration programs: Economic and legal implications. *Ecological Economics, 46*, 143–157. doi:10.1016/S0921-8009(03)00138-1

Chuine, I., & Beaubien, E. (2001). Phenology is a major determinant of tree species range. *Ecology Letters, 4*, 500–510. doi:10.1046/j.1461-0248.2001.00261.x

De Groot, R. S., Wilson, M. A., & Boumans, R. M. J. (2002). A typology for the classification, description and valuation of ecosystem functions, goods and services. *Ecological Economics, 41*, 393–408. doi:10.1016/S0921-8009(02)00089-7

Dormann, C. F. (2007). Promising the future? Global change projections of species distributions. *Basic and Applied Ecology, 8*, 387–397. doi:10.1016/j.baae.2006.11.001

FAO. (2006). *Global forest resources assessment 2005. (FAO Forestry Paper No. 147)*. Rome: UN Food and Agriculture Organization.

Gereti, E. J., Wolanski, E., & Chiombola, E. A. T. (2003). *Assessment of the environmental, social and economic impacts on the Serengeti ecosystem of the developments in the Mara River Catchment in Kenya*. Dar Es Salaam: Tanzania National Parks.

Gitay, H., Suárez, A., Watson, R. T., & Dokken, D. J. (Eds.). (2002). *Climate change and biodiversity*. Geneva: IPCC Technical Paper V. Intergovernmental Panel on Climate Change.

Grainger, A. (1980). The state of the world's tropical forests. *The Ecologist, 10*, 6–54.

Grainger, A., Boucher, D. H., Frumhoff, P. C., Laurance, W. F., Lovejoy, T., & McNeely, J. (2009). Biodiversity and REDD at Copenhagen. *Current Biology, 19*, R974–R976. doi:10.1016/j.cub.2009.10.001

Green, R., Collingham, Y., Willis, S., Gregory, R., Willis, S. G., & Gregory, R. D. (2008). Performance of climate envelope models in retrodicting recent changes in bird population size from observed climatic change. *Biology Letters, 4*, 599–602. doi:10.1098/rsbl.2008.0052

Higgins, P. A. T. (2007). Biodiversity loss under existing land use and climate change: An illustration using northern South America. *Global Ecology and Biogeography, 16*, 197–204. doi:10.1111/j.1466-8238.2006.00278.x

IUCN. (1980). *World conservation strategy*. International Union for the Conservation of Nature and Natural Resources, Gland.

Jackson, M. T., & Ford-Lloyd, B. V. (1990). Plant genetic resources - a perspective. In Jackson, M. T., Ford-Lloyd, B. V., & Parry, M. L. (Eds.), *Climatic change and plant genetic resources* (pp. 1–17). London, UK: Belhaven Press.

Jackson, M. T., Ford-Lloyd, B. V., & Parry, M. L. (Eds.). (1990). *Climatic change and plant genetic resources*. London, UK: Belhaven Press.

Jeschke, J. M., & Strayer, D. L. (2008). Usefulness of bioclimatic models for studying climate change and invasive species. *Annals of the New York Academy of Sciences, 1134*, 1–24. doi:10.1196/annals.1439.002

Kapos, V., Ravilious, C., Campbell, A., Dickson, B., Gibbs, H. K., & Hansen, M. C. … Trumper, K. C. (2008). *Carbon and biodiversity, a demonstration atlas*. UNEP World Conservation Monitoring Centre, Cambridge.

Keith, D. A., Akçakaya, H. R., Thuiller, W., Midgley, G. F., Pearson, R. G., & Phillips, S. G. (2008). Predicting extinction risks under climate change: Coupling stochastic population models with dynamic bioclimatic habitat models. *Biology Letters, 4*, 560–563. doi:10.1098/rsbl.2008.0049

Kim, J. (2004). Regime interplay: The case of biodiversity and climate change. *Global Environmental Change, 14*, 315–324. doi:10.1016/j.gloenvcha.2004.04.001

Krhoda, G. O. (1988). The impact of resource utilization on the hydrology of the Mau Hills Forest in Kenya. *Mountain Research and Development, 8*, 193–200. doi:10.2307/3673447

Mace, G. M., Cramer, W., Diaz, S., Faith, D. P., Larigauderie, A., & La Prestre, P. (2010). Biodiversity targets after 2010. *Current Opinion in Environmental Sustainability, 2*, 3–8. doi:10.1016/j.cosust.2010.03.003

Malcolm, J. R., Liu, C., Neilson, R. P., Hansen, L., & Hannah, L. (2006). Global warming and extinctions of endemic species from biodiversity hotspots. *Conservation Biology, 20*, 538–548. doi:10.1111/j.1523-1739.2006.00364.x

Malhi, Y., Roberts, J. T., Betts, R. A., Killeen, T. J., Li, W., & Nobre, C. A. (2008). Climate change, deforestation and the fate of the Amazon. *Science, 319*, 169–172. doi:10.1126/science.1146961

McCarthy, J. J., Canziani, O. F., Leary, N. A., Dokken, D. J., & White, K. S. (Eds.). (2001). *Climate change 2001. Working group II: Impacts, adaptation and vulnerability. Intergovernmental Panel on Climate Change*. Cambridge, UK: Cambridge University Press.

McNeely, J. A., Miller, K. R., Reid, W. V., Mittermeier, R. W., & Werner, T. B. (1990). *Conserving the world's biological diversity*. International Union for the Conservation of Nature, Gland.

MEA. (2005). *Millennium Ecosystem Assessment: Ecosystems and human well-being*. Washington, DC: Island Press.

Metz, B., Davidson, O., Bosch, P., Dave, R., & Meyer, L. (Eds.). (2007). *Climate change 2007. Working group III: mitigation. Intergovernmental Panel on Climate Change*. Cambridge, UK: Cambridge University Press.

Miles, L. (2002). *The impact of global climate change on tropical forest biodiversity in Amazonia*. Unpublished doctoral dissertation, University of Leeds, Leeds.

Miles, L., Grainger, A., & Phillips, O. L. (2004). Impact of global climate change on tropical forest biodiversity in Amazonia. *Global Ecology and Biogeography*, *13*, 553–565. doi:10.1111/j.1466-822X.2004.00105.x

Miles, L., & Kapos, V. (2008). Reducing greenhouse gas emissions from deforestation and forest degradation, global land-use implications. *Science*, *320*, 1454–1455. doi:10.1126/science.1155358

Mooney, H. A. (2010). The ecosystem-service chain and the biological diversity crisis. *Philosophical Transactions of the Royal Society B*, *365*, 31–39. doi:10.1098/rstb.2009.0223

Morin, X., Viner, D., & Chuine, I. (2008). Tree species range shifts at a continental scale: New predictive insights from a process-based model. *Journal of Ecology*, *96*, 784–794. doi:10.1111/j.1365-2745.2008.01369.x

Myers, N. (1988). Threatened biotas: Hot spots in tropical forests. *The Environmentalist*, *8*, 1–20. doi:10.1007/BF02240252

Myers, N., Mittermeier, R. A., Mittermeier, C. G., Fonseca, G. A. B., & Kent, J. (2000). Biodiversity hotspots for conservation priorities. *Nature*, *403*, 853–858. doi:10.1038/35002501

Naidoo, R., Balmford, A., Costanza, R., Fisher, B., & Green, R. E. Lehner. B.,... Ricketts, T. H. (2008). Global mapping of ecosystem services and conservation priorities. *Proceedings of the National Academy of Sciences USA*, *105*, 9495-9500.

NERC/ESRC/DFID. (2008). *Ecosystem services for poverty alleviation: Strengthening research capacity in developing countries*. Swindon: Natural Environment Research Council.

Parry, M. L., Canziani, O. F., Palutikof, J. P., Van Der Linden, P. J., & Hanson, C. E. (Eds.). (2007). *Climate change 2007. Working group II: Impacts, adaptation and vulnerability*. Cambridge, UK: Cambridge University Press/ Intergovernmental Panel on Climate Change.

Perrings, C. (2006). Ecological economics after the millennium assessment. *International Journal of Ecological Economics and Statistics*, *6*(FO6), 8–22.

Pimm, S. L. (2001). *The world according to Pimm: A scientist audits the Earth*. New York, NY: McGraw Hill.

Pimm, S. L. (2007). Climate disruption and biodiversity. *Current Biology*, *183*, R117–R119.

Porro, R., Benitez, S., Rubiano, J., Mulligan, M., Naranjo, L. G., & Jarvis, A. (2008). *Securing biostability in the Amazon/Andes: A situation analysis. Report to NERC/ESRC/DFID*. London, UK: Kings College, University of London.

Republic of Kenya. (2004). *Report of the Commission of Inquiry into illegal/ irregular allocation of public land*. Nairobi, Kenya: Government Printer.

Rose, S., & Grainger, A. (2002). Multivariate mapping of spatial variation in biodiversity in Peruvian Amazonia. *Diversity & Distributions*, *9*, 237–250. doi:10.1046/j.1472-4642.2003.00019.x

Sayer, J., Harcourt, C., & Collins, N. M. (1992). *Conservation atlas of tropical forests: Africa*. London, UK: Macmillan.

Swallow, B. M. (2006). Forests, flowers or flamingos. *ETFRN News*, *45*, 52.

Swallow, B. M., Kallesoe, M. F., Iftikhar, U. A., Van Noordwijk, M., Bracer, C., & Scherr, S. J. (2009). Compensation and rewards for environmental services in the developing world: Framing pan-tropical analysis and comparison. *Ecology and Society*, *14*(2), 26.

Tallis, H., Kareiva, P., Marvier, M., & Chang, A. (2008). An ecosystem services framework to support both practical conservation and economic development. *Proceedings of the National Academy of Sciences of the United States of America*, *105*, 9457–9464. doi:10.1073/pnas.0705797105

Thomas, C. D., Cameron, C., Green, R. E., Bakkenes, M., Beaumont, L. B., & Collingham, Y. C. (2004). Extinction risk from climate change. *Nature*, *247*, 145–148. doi:10.1038/nature02121

Thuiller, W., Alberta, C., Araujo, M. B., Berry, P. M., Cabeza, M., & Guisane, A. (2008). Predicting global change impacts on plant species distributions. *Perspectives in Plant Ecology, Evolution and Systematics*, *9*, 137–152. doi:10.1016/j.ppees.2007.09.004

Tilman, D., Reich, P. B., Knops, J., Wedin, D., Mielke, T., & Lehman, C. (2008). Diversity and productivity in a long-term grassland experiment. *Science*, *294*, 843–845. doi:10.1126/science.1060391

Turner, W. R., Brandon, K., Brooks, T. M., Costanza, R., Da Fonseca, G. A. B., & Portela, R. (2007). Global conservation of biodiversity and ecosystem services. *Bioscience*, *57*, 868–873. doi:10.1641/B571009

UN. (2000). *Millennium declaration*. New York, NY: United Nations.

UNDP. (2009). *Human development report (2009)*. New York, NY: Palgrave Macmillan/UN Development Programme.

UNFCCC. (2009). *Revised negotiating text. Ad hoc Working Group on long-term cooperative action under the convention. FCCC/AWGLCA/2009/INF.1*. Bonn: Framework Convention on Climate Change.

Vié, J.-C., Hilton-Taylor, C., Pollock, C., Ragle, J., Smart, J., Stuart, S. N., & Tong, R. (2008). The IUCN red list: a key conservation tool. In Vié, J.-C., Hilton-Taylor, C., & Stuart, S. N. (Eds.), *The 2008 review of the IUCN red lst of threatened species*. International Union for the Conservation of Nature and Natural Resources, Gland.

Watson, R. T., Noble, I. R., Bolin, B., Ravindrath, N. H., Verardo, D. J., & Dokken, D. J. (Eds.). (2000). *Land use, land-use change and forestry*. Cambridge, UK: Cambridge University Press.

Wilson, E. O. (Ed.). (1988). *Biodiversity*. Washington: National Academy Press.

Chapter 15
Land Degradation and Biodiversity Loss in Southeast Asia

Rajendra P. Shrestha
Asian Institute of Technology, Thailand

ABSTRACT

Land degradation and biodiversity loss are important global change issues because of their enormous effect on the functioning of ecosystem. Despite the fact that there have been tremendous concerns on land degradation and biodiversity loss for nearly two decades, there is still the need of having a sound data and information base, specifically in developing countries. The need has been more pronounced in the face of climate change as these three issues are intricately interlinked. Southeast Asia is an important geographic region from all these perspectives, as it has high biodiversity on the verge of rapid loss, continuing rapid land degradation due to desire of higher economic development, and of climate change importance with a large tract of forest areas in the region. This chapter, first of all, examines general status of land degradation and biodiversity in the region and goes on presenting two case studies. The first case study, based on secondary data, presents land degradation assessment in the Lower Mekong Basin demonstrating the use of spatial data and technologies and various land degradation indicators. The second case study specifically documents plant diversity and examines the relationship of plant diversity with biomass and soil erosion by making use of field surveyed primary data. Both studies aim at producing additional information which can help make better landuse allocation and planning for ecosystem maintenance without compromising much on regional or local livelihood through production.

DOI: 10.4018/978-1-60960-619-0.ch015

1. LAND DEGRATION

Given the fact that only 11 percent of the global land surface can be considered as prime or Class I land to feed burgeoning global population, the issues of land degradation is of major significance for world food security and the quality of the environment. Hence, land degradation will remain high on the international agenda in the 21st century. Land degradation refers to land, which due to natural processes or human activity is no longer able to sustain properly an economic function and/or the original natural ecological function (GEF, 1999) arising from the causes, like deforestation, inappropriate agricultural practices, overgrazing. Land degradation involves two interlocking and complex systems: the natural ecosystem and the human social system. Land degradation can take various processes and forms, such as soil erosion due to water and wind, physical deterioration (compaction, sealing), chemical deterioration (soil fertility decline, salinization, acidification), vegetation degradation. The term 'desertification' is found widely used to indicate land degradation in arid, semi-arid and dry sub-humid areas resulting from various factors, including climatic variations and human activities. Causes of degradation and desertification are several and complex. These include socioeconomic factor (e.g. land tenure, marketing, institutional support, income and human health), political (e.g. incentives, political stability), landuse patterns and practices. Globalization phenomena is also adding to land degradation.

Land degradation has multiple and complex impacts on ecosystem functions and services of environment through a range of direct and indirect processes. This ultimately impacts peoples' livelihood through reduced ecosystem functions (e.g. reduced productivity, flooding, and sedimentation) and/or through pollution of ecosystem (e.g. pollution of soil and water). There is clear inter-linkage between land degradation, biodiversity and climate change. While linkage between land degradation and climate change can be of significant importance at global level, the linkage between land degradation and biodiversity can be of significance both at global to local level. In Figure 1, land degradation interrupts the regulating and provisioning services of ecosystems, in particular to reduced primary production and nutrient cycling and reduced carbon sequestration into above- and below-ground carbon reserves (MEA, 2005). Similarly, land degradation also affects biodiversity through loss of nutrients and soil moisture. There are also internal feed-back mechanisms that affect the process and state of these important issues of global importance.

Southeast Asia typically includes the following countries, namely Brunei, Cambodia, East Timor, Indonesia, Lao s, Malaysia, Myanmar, Philippines, Singapore, Thailand, and Vietnam. The issue of land degradation is of global importance, as we increasingly face the challenge of food insecurity and declining ecosystem services, particularly in Southeast Asia, which is highly populated and experiencing a rapid pace of both economic development and ecosystem degradation. Increasing yield to supply more food is almost impossible as almost all cultivable lands are under cultivation. Land degradation can further aggravate the situation (FAO, 2002) since the region as a whole may already have passed the safe limits for agricultural expansion (Eswaran *et al.*, 2001).

According to FAO's (2003) TERRSTAT database, in general, all Southeast Asian countries but Laos suffers from land degradation. The proportion, degraded area of the country's total area, ranges from 36% in Myanmar to 100% in Brunei Darussalam (Table 1). Significant proportions of the rest of the areas of the countries are also still affected by some less severe form of degradation. In majority of case, soil erosion due to water is the main cause of degradation, as the region gets relatively higher rainfall. This form of erosion is accelerated by deforestation and inappropriate agricultural management practices. Chemical

Figure 1 Interlinkage between land degradation, biodiversity loss and climate change. Source: adopted/ modified from MEA (2005)

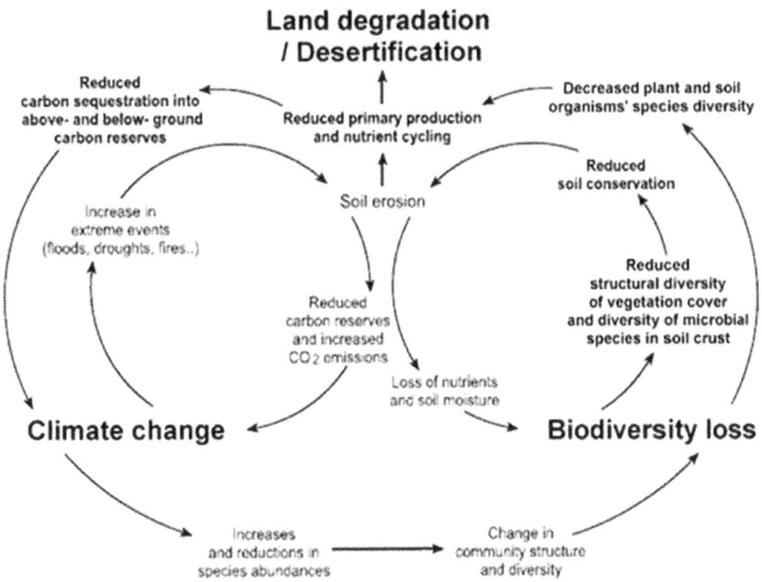

deterioration particularly in cultivated landscape due to continuous and higher use of chemical, e.g. fertilizers, herbicides, is the second major land degradation problem in the region. It is reported that there are also instances of physical deterioration in some countries.

2. BIODIVERSITY LOSS

Biodiversity (also called as Biological diversity) refers to both the genetic variability among individuals of a species and the abundance of individuals within a species (Cutter & Renwick, 1999). Biodiversity is defined as the variability among living organisms from all sources, including, inter alia, terrestrial, marine and other aquatic ecosystems and the ecological complexes of which they are part; this includes diversity within species, between species, and of ecosystems (Heywood & Bates, 1995). Biodiversity provides a foundation for the continued existence of a healthy planet and our own well-being by enhancing ecosystem resilience therefore able to recover more readily from stresses such as drought or human induced habitat degradation. Variety of biological resources also provides opportunities for adapting to change. The unknown potential of genes, species and ecosystems is enormous and of inestimable. Takacs (1996) argues the biodiversity to have scientific, ecological, economic, social amenity, biophilic, transformative, intrinsic, spiritual, and aesthetic value. More discussion on the biodiversity and its value are provided in chapters 8, 9, 12 of this volume.

According to UN ESCAP (2010), Asia and the Pacific region had the world's highest number of threatened species in the year 2008, with almost one third of all threatened plants, and over one third of all threatened animal species. According to International Union for the Conservation of Nature and Natural Resources (IUCN) red list of threatened species, Southeast Asia, particularly in Asia and the Pacific, is one of the global hotspots of threatened species of both animals and plants despite the region is home to half of world's

Table 1. Land degradation extent, type and their causes in selected Southeast Asia Countries

	Total area		Degradation type (% of total area)				% Area	Cause	Type
Country	**000 km²**	**None**	**Light**	**Moderate**	**Severe**	**Very severe**	**degraded**		
Brunei Darussalam	6	0	0	0	100	0	100	A,D	P,C,W
Cambodia	183	13	2	36	27	22	49	D	W
Indonesia	1 898	1	35	26	32	6	38	D,A	W,C
Laos	232	0	16	83	0	1	1	D	W
Malaysia	329	0	0	17	83	0	83	D,A	W,C
Myanmar	668	1	0	63	35	1	36	D,A	W,C
Philippines	295	3	0	18	76	3	79	D	W
Thailand	516	0	2	20	28	50	78	D,A	W,C
Viet Nam	330	0	0	22	29	49	78	D,A	W,C
Cause: A = agriculture; D = deforestation									
Type: W = water erosion; C = chemical deterioration; P = physical deterioration Source: compiled from FAO (2003)									

terrestrial and marine biodiversity. With regard to terrestrial biodiversity, the Southeast Asian countries, namely Indonesia, Malaysia, Philippines, Thailand, Vietnam, top the chart among the countries in Asia and the Pacific having highest threatened plant and animals species. Besides threatened species, IUCN (2003) has indeed reported that three plant and eight animal species have been listed as 'extinct' in Southeast Asia.

According to UN ESCAP (2010), the proportion of protected land areas in Southeast Asian countries ranges from 2.7% (in Philippines, Singapore) to 32.4% (in Brunei Darussalam) of the country's area (Table 2). There is however a positive trend of increase in the protected land areas in nearly last two decades. With regard to number of threatened animal and plant species, Indonesia has the highest threatened mammal species (183), bird species (115), Reptiles (27), where as it is Malaysia which has highest threatened amphibian species (47) and plant species (686) among the countries in the region. Timor-Leste, in general, has less number of both threatened animal and plant species but it is also the fact that there is no complete information on the threatened species in this country. In general, in case of all the countries, the numbers of threatened animal and plant species have increased in last two decades.

Southeast Asia is also equally rich on marine biodiversity. Malaysia, Indonesia and Philippines are mega diversity countries with 80 per cent of the global biological diversity. Some of the last remaining intact expanses of mangroves occur in South-East Asia and around 30 per cent of the world's coral reefs are situated in the region. Despite this fact, the proportion of marine protected areas in Asia-pacific, such area is generally low 10% of less of country's territorial water except for Australia. Among Southeast Asian countries, such area ranges from less than 1% (for Cambodia) to about 5% (Malaysia).

The threats to biodiversity include landuse change (forest conversion, agricultural expansion mono-cropping), forest fire, hunting for wildlife trade, fragmentation due to infrastructure and agricultural development. Deforestation is the major threat to accelerating loss of biodiversity. Southeast Asia now has the highest deforestation rate in Asia-Pacific and the region could lose three quarters of its original forests by 2100

Table 2. Protected areas, threatened animals and plant species in Southeast Asian countries

	PLA (% of surface area)		T Mammals (no. of sp.)		T Birds (no. of sp.)		T Reptiles (no. of sp.)		T Amphibians (no. of sp.)		T Plants (no. of sp.)	
	1990	2008	1990	2008	1990	2008	1996	2008	1996	2008	1993	2008
Brunei Darussalam	27.7	32.4	n.a.	35	n.a.	21	4	5	0	3	n.a.	99
Cambodia	0.1	21.9	21	37	13	25	9	12	1	3	8	31
Indonesia	3.5	5.7	49	183	135	115	19	27	0	33	283	386
Lao PDR	0.8	15.9	30	46	18	23	7	11	0	5	6	21
Malaysia	13.0	14.1	4	70	23	42	14	21	0	47	490	686
Myanmar	2.4	5.4	n.a.	45	42	41	20	22	0	n.a.	29	38
Philippines	2.0	2.7	12	39	39	67	7	9	2	48	198	216
Singapore	2.2	2.7	n.a.	12	5	14	1	4	0	n.a.	15	54
Thailand	11.2	16.7	26	57	34	44	16	22	0	4	382	86
Timor-Leste	n.a.	7.3	n.a.	4	n.a.	5	n.a.	1	n.a.	n.a.	n.a.	n.a.
Viet Nam	0.9	3.0	28	54	34	39	12	27	1	17	350	147

PLA= Protected Land Areas, T= Threatened, n.a. = not available
Source: compiled from UN ESCAP, 2010.

and up to 42% of its biodiversity (Sodhi *et al.*, 2003). Brook *et al.* (2002) projected that habitat loss through continuing deforestation would lead to extinction of 21-48% mammals in the region by 2100. Forest fire which has occurred all the times in the region is also an important cause of biodiversity loss besides other impacts of forest fire on people's health, navigation, and damage to properties. Southeast Asia is also considered as a major hub of wildlife trade both of illegal and legal types.

Conservation of biodiversity is a complex and challenging issue. This requires an understanding of direct and indirect causes of biodiversity loss and developing appropriate strategic solutions to address those causes. There should also be every effort from all levels in all kind of landscapes to not only arrest the further loss of these important biological resources but also increase the biodiversity. With increasing globalization along with quest for economic gain, agricultural expansion by forest conversation and mono-cropping of annuals, e.g. such as sugarcane, cassava and of perennials, e.g. rubber, oil palm, eucalyptus, have remained one of the major cause of biodiversity loss in several countries of South Asia. Since majority of land areas are cultivated in the region, the need of enhancing and conserving biodiversity in cultivated landscape, e.g. agricultural area, becomes equally important.

3. ASSESSESSMENT OF LAND DEGRADATION AND BIODIVERSITY LOSS

Despite the global importance of land degradation, the available data on the extent of land degradation in dry lands are limited (FAO, 2005) and this stands true with areas other than dry lands. TERRSTAT database on land resource potential and constraints statistics at country and regional level, which was presented above are the estimates based on various small scale maps and inventories that were not always up to date, reliable or both (FAO, 2003). The other best known land degradation data are of the Global Assessment of Human Induced Soil Degradation (GLASOD) study at the global scale (1:10 million scale) and Soil Degradation in South and Southeast Asia (ASSOD) for the regional scale (1:5 million scale). These data are compiled from variety of available data of various types and scales, and information gathered through expert opinion. ASSOD data classifies various land degradation types according to five impact categories. Besides being coarse resolution data, both GLASOD and ASSOD intend to reflect the actual situation in the field by expressing soil degradation in qualitative terms as impact on productivity, the spatial extent classified under each impact category vary tremendously between GLASOD and ASSOD (Oldeman & van Lynden, 1998). This suggests that relatively precise data on the spatial extent of degradation is still lacking in the region. Since the earlier studies completed long ago are often based on the expert judgments, there is the need of updating land degradation database using current information. Such information can serve as quantified baseline information on land degradation for the regional level planning and strategies formulation for natural and land resources conservation. In this study, we use geospatial data available from secondary sources to assess the land degradation based on four major indicators in the Greater Mekong Subregion (GMS) in order to produce quantified information on land degradation.

This is same in case of biodiversity assessment, which perhaps is even more challenging. The current knowledge of biodiversity is based on what we know about the nature and ecosystem and what we tend to know is perhaps a small proportion of what the planet has to offer. In most cases, species richness per se is one of the most widely used indicators for the assessment of biodiversity (Clergue *et al*., 2005). More to that, it is important to understand how such species richness can be maintained and what component of ecosystem and human activities are responsible factors.

On this premises, we present here two case studies: land degradation assessment at the regional scale and biodiversity assessment at watershed scale. These studies demonstrate how such assessment could be carried out, which can serve as an important information and knowledgebase for planning of a sustainable practice for maintaining a health ecosystem and eventually reducing climate change impact.

4. CASE STUDY I: LAND DEGRADATION ASSESSMENT IN THE LOWER MEKONG BASIN

4.1 Introduction

The study area, Lower Mekong Basin (LMB), cuts across four countries, including most of Cambodia and Lao PDR and substantial portions of Thailand and Vietnam covering area about 606,000 sq. km. (Hook *et. al*., 2003). The climate is governed by monsoons-steady winds that blow alternately from the northeast and the southwest, each for about half of the year. The regional rainfall varies significantly from the driest region in the basin (Northeast Thailand) where annual rainfall is mostly between 1,000-1,600 mm, to the wettest regions (Northern and Eastern Highlands) with 2,000-3,000 mm annually. Maximum temperatures range from 30°C up to 38°C while minimum temperatures range from 15°C in lower plain to subzero in the higher altitude during winter.

Approximately 53 million people lived in the LMB. Rice cultivation is the main agricultural activity while other crops are also grown in upland in different part of the basin. Inappropriate agricultural practices are highly prevalent in the study area thus contributing to several types of land degradation.

Assessment of land degradation to map and quantify their severity is important for the purpose of rehabilitating the degraded areas and identifying the potential areas for intensive cultivation. As mentioned above, while TERRASTAT, GLASOD and ASSOD land degradation data are available at global and regional level, these studies relied heavily on expert opinion (FAO, 2005) and quantified information are still lacking in the developing world (Symeonakis & Drake, 2004). There is no regional or basin specific data available for use or even for validation. Using spatial data, this study demonstrates assessment of land degradation status based on few land degradation indicators in the Lower Mekong Basin (LMB).

4.2 Methodology

The basic data used were:

- Remote Sensing Data
    - MODIS data: NDVI Data (Normalized Difference Vegetation Index) for the year 2002 downloaded from the University of Tokyo at http://webmodis.iis.u-tokyo.ac.jp/Asia/ (Wataru, 2001).
    - NOAA-AVHRR data covering study area was used to prepare land cover map of the area. The scene was a 1.1 km resolution composite image of 21-30 September 2000 and was obtained from the Asian Center for Research on Remote Sensing, Bangkok.
- Soil Data: FAO-UNESCO soil map (1:5M) was downloaded from Food and Agriculture Organization (FAO) website at http://www.fao.org/ag/agl/agll/wrb/soilres.stm#down (FAO, 1979)
- Rainfall: the monthly rainfall data over several stations in the Mekong region for the period of 1999-2002 downloaded from the website of the University of Tokyo at http://hydro.iis.u-tokyo.ac.jp/GAME-T/GAIN-T/updates.html (University of Tokyo, 2005).
- SRTM GTOPO30 elevation data (1-km resolution) was used to prepare slope map. The data was downloaded from the Global Land Cover Facility website at http://glcf.umiacs.umd.edu/data/srtm/ (USGS, 2004).

In this study, four major indicators of degradation, namely vegetation cover, runoff, rain use efficiency (RUE) and soil loss, were computed for assessing the composite land degradation status (Figure 2). Digital classification of remote sensing data was carried out using ERDAS IMAGINE™ software and all overlay analyses ware carried out in Geographic Information System (GIS) environment using ARCMAP™ 9.0 and ArcView™ 3.3 software.

4.2.1 Estimation of Vegetation Cover

Vegetation cover is an important indicator of land degradation (Rubio & Bochet, 1998) because of

Figure 2. Research methodology

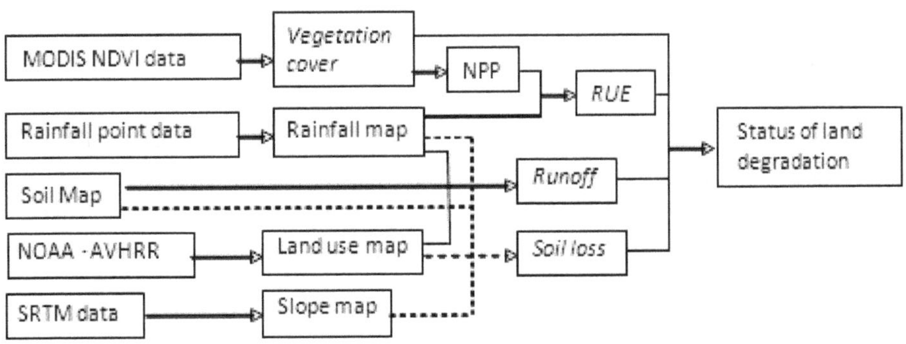

its important role in protecting the land surface from direct action of casual factors of degradation as well as in maintaining the physical condition of soils. This indicator is expressed in various forms, one being percentage cover of vegetation. Most often, it is found that NDVI from remote sensing data are used to compute vegetation cover. In this study, we used MODIS NDVI data to compute the annual integrated NDVI and subsequently the vegetation cover (V_c) using the equation suggested as follow (Zhang *et al.*, 1999).

$$V_c = 1.333 + 131.877(NDVI)$$

4.2.2 Estimation of Rain Use Efficiency

RUE is an important indicator of land degradation (Le Houerou, 1984; Prince *et al.*, 1998). RUE can be considered as the ratio of net primary productivity (NPP) to precipitation. NPP can be used to quantify the net carbon absorption rate by living plants (NRC, 2006) and thus serves as a measure of dryland condition (WRI, 2002). Since direct observations of NPP are not available in many instances, it can be estimated with other available techniques and models. NPP can be calculated using NDVI. MODIS NDVI was used to calculate NPP using the regression model suggested and calibrated as (Prince *et al.*, 1998).

$$NPP\ (Mg\ ha^{-1}a^{-1}) = 3.139 \sum NDVI - 3.852$$

4.2.3 Estimation of Runoff

Runoff can be defined as the movement of water, usually from precipitation, across the earth's surface towards streams, channels, lakes, oceans, depressions or low points in the earth's surface. Runoff is considered an important indicator of land desertification (Sharma, 1998). The USDA Soil Conservation Service runoff curve number (CN) procedure is the best known and widely used model of this type (SCS, 1972).

$$Q = (P-0.2S)^2/(P+0.8S)$$

where, Q = runoff (mm); P = rainfall (mm); S = amount of rainfall (mm) which can soak into the soil during the storm, given by S = (25400/CN) – 254 (when water depths expressed in mm), CN is a fraction of soil hydrologic conditions, landuse and infiltration characteristics of soils, it can be extracted from published tables and is dependent on landuse, crop, management and hydrological soil group. Landuse map was obtained by employing digital classification of NOAA-AVHRR data. Landuse and soil map were used to compute CN value. Rainfall map was used for annual rainfall.

4.2.4 Estimation of Soil Loss

Soil loss is important indicator of desertification. Water erosion is prominent in the study area, soil loss due to water erosion was estimated in this study using Universal Soil Loss Equation (USLE) (Wischmeier & Smith, 1978). The equation can be expressed as:

$$E = RKLSCP$$

where, E is the mean annual soil loss (ton/ha/yr); R is the rainfall erosivity factor; K is the soil erodibility factor; L is the slope length factor; S is the slope steepness factor, C is the crop management factor; and P is the erosion control practice factor. The derivation of individual USLE factor was carefully done drawing upon the findings of previous studies (Shrestha *et al.*, 1996). R-factor was encoded in rainfall layer; K in FAO-UNESCO soil map; S in slope map derived GTOPO-30 DEM data; L estimated from land cover map; and CP in land cover map of 2000 interpreted from NOAA-AVHRR data. Finally, all the GIS layer containing USLE factors were overlaid to compute average annual soil loss.

4.2.5 Assessment of Land Degradation

Above four indicators were combined to come up with a composite land degradation status. This was achieved by combining all four indicators and selecting worst to best in each category based on the range of the computed values to come up with severity classes of land degradation, using Boolean logic in GIS environment.

4.3 RESULTS AND DISCUSSION

The distribution of degradation indicators of the study area is presented in Figure 3 and the proportion of area under each class of individual indicator is presented in Table 3. Below is brief discussion on each of them.

4.3.1 Vegetation Cover

The computed maximum vegetation cover was about 84%. The area with 60-80% vegetation

Figure 3. Spatial distribution of different land degradation indicators.(a) vegetation cover, (b) rain use efficiency, (c) runoff, (d) soil loss

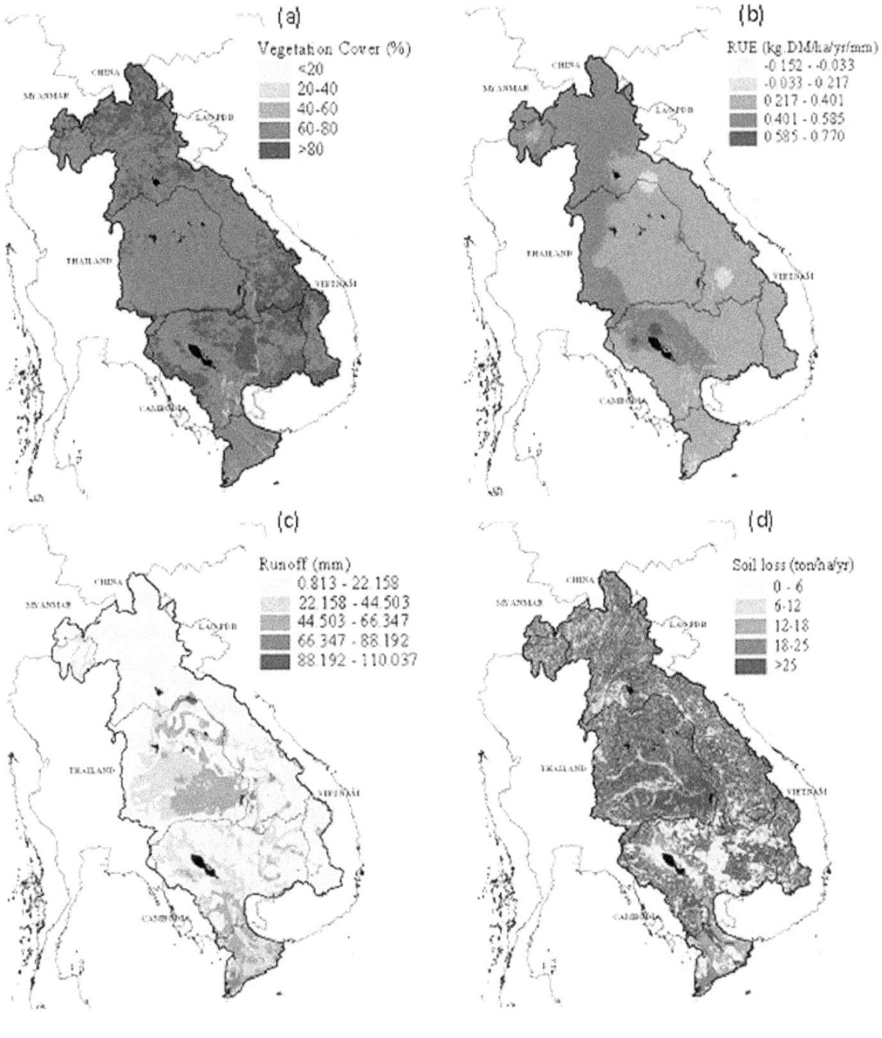

Table 3. Area distribution of different degradation indicators.

Vegetation cover (%)	% Area	RUE (kg.DM/ha/yr/mm)	% Area	Runoff (mm)	% Area	Soil loss (ton/ha/yr)	% Area
<20	0.82	-0.151 - -0.033	0.79	88.192-110.037	0.12	>25	44.97
20 - 40	0.10	-0.033 – 0.217	2.03	66.347–88.192	1.03	18-25	16.27
40 - 60	0.88	0.217 – 0.402	62.71	44.503–66.347	12.36	12-18	12.21
60 - 80	70.36	0.402 – 0.586	33.84	22. 158–44.503	20.17	6-12	13.05
>80	27.84	0.586 – 0.770	0.63	0.813–22.158	66.32	0-6	13.49

Total area: 606,000 km^2

cover was about 70% of the total area of the LMB. A quarter of the study area has the vegetation cover more than 80%. This is similar as reported by (Ogawa et al., 2005) in their study conducted at the Mekong basin using MODIS data, where the vegetation density was higher than 60% of the areas.

4.3.2 Rain Use Efficiency

The observed RUE value ranged from -0.15 to 0.77 kg.DM/ha/yr/mm with the average about 0.36 kg. DM/ha/yr/mm. More than 80% of the areas are having RUE above the average; there is only 2% of the total land area having RUE below the average. These areas included *Se Don* and *Se Kong* watershed in Vietnam and small part of northeastern of Thailand.

4.3.3 Runoff

Runoff is usually higher in the area with less vegetation cover and higher runoff can cause higher land degradation compared to the areas with low runoff due to higher soil erosion. The computed runoff values ranged from nearly zero (0.81) to a maximum about 110 mm. The average runoff is about 22.86% for the study area. More than 80% of the total area was observed to have runoff below the average. There is only 1.03% of the total area with high runoff.

4.3.4 Soil Loss

Generally, 12 tons/ha/yr is considered as the maximum permissible soil loss; however, the limit might differ from place to place given the respective emphasis on soil conservation. The higher amount of soil loss indicates higher degradation severity with declining productivity.

In the study area, soil loss ranged from 0 to 99 tons/ha/yr. More than 80% of the total area of LMB was observed to have soil loss rate exceeding 12 tons/ha/yr. The distribution of soil loss was random in general, however, Nam Mun watershed, northeastern of Thailand had higher soil loss rate compared to other parts of the basin. By and large, most of the lower part of basin had lower rate of soil loss.

4.3.5 Status of Land Degradation

In this study, based on the four indicators, viz vegetation cover, runoff, rain use efficiency, and soil loss, about 8.62% area of the LMB were classified as having very severe land degradation, 18.34% as severe, 38.80% as moderate, 28.04% as slight and 6.19% as no degradation (Table 4). The distribution of different degradation severity

Table 4. Status of degradation in the Lower Mekong Basin

Degradation status	% Area
Very severe	8.62
Severe	18.34
Moderate	38.80
Slight	28.04
No degradation	6.19

Total area: 606,000 sq. km.

is presented in Figure 4. The north eastern region of Thailand is having large proportion of land degradation compared to others. The findings of the study indicated that a substantial amount of area is degraded in the LMB, calling for the need of appropriate land conservation measures in different areas.

4.4 Conclusion Remarks

There is increasing concern on the seriousness of land degradation as it directly affects the land productivity and thus food security. This is important to conduct land degradation studies for better planning and management of our valuable land resources in the areas like LMB with higher population but decreasing availability of arable land. Hence, it is important to examine the degradation severity to understand the potential of improving land quality so that production could be increased by conserving such degraded land meanwhile aiding towards better environmental conservation.

The study suggests that about one quarters of LMB land area is severely degraded and another three quarters are moderately, slightly and no degradation with their area distribution in the LMB. The results are however based on available quantified data; it should be treated as a guide to prioritize the conservation efforts by developing appropriate landuse measures within each country of the LMB. Since land degradation is a complex interplay of several factors, inclusion of other indicators in the analysis would help to improve the precision of the analysis of such regional level studies.

5. CASE STUDY II: LINKAGE OF PLANT DIVERSITY WITH BIOMASS AND SOIL EROISON IN AGRICULTURAL LANDSCAPE OF KHLONG YAI WATERSHED, THAILAND

5.1 Introduction

Biodiversity has been recognized as an important factor in maintaining or enhancing agricultural sustainability (Brookfield *et al.*, 2002). Agricultural biodiversity, also called as 'agro-biodiversity' is now an established term in its own right and is defined as 'the variety and variability of animals, plants and microorganisms that are used directly or indirectly for food and agriculture, including crops, livestock, forestry and fisheries' (FAO, 1999). Agricultural biodiversity has multiple functions, such as contributing to food and livelihood security, to production and environmental sustainability, and to rural productivity (FAO, 1997) besides controlling land degradation (Stocking, 2002) and increasing nutrient use efficiency. Evidence from experimental and intercropping systems has demonstrated that higher species richness can be associated with increased productivity (Tilman, 1996), which is probably due to the differences in nutrient cycling characteristics that can regulate soil fertility (Hooper & Vitousek, 1998).

Biodiversity conservation in agricultural landuses is one of the greatest challenges, especially in the tropics (NEMA, 2001), e.g. in Thailand, where new opportunities have arisen through national and international market demand for industrial crops and commercial vegetable crops, has been undergoing rapid changes in land-use pattern. Such commercialized landuses mostly focus on single species and result in losses of local varieties from farmers' fields (Rerkasem & Rerkasem,

Figure 4. Land degradation status in the LMB

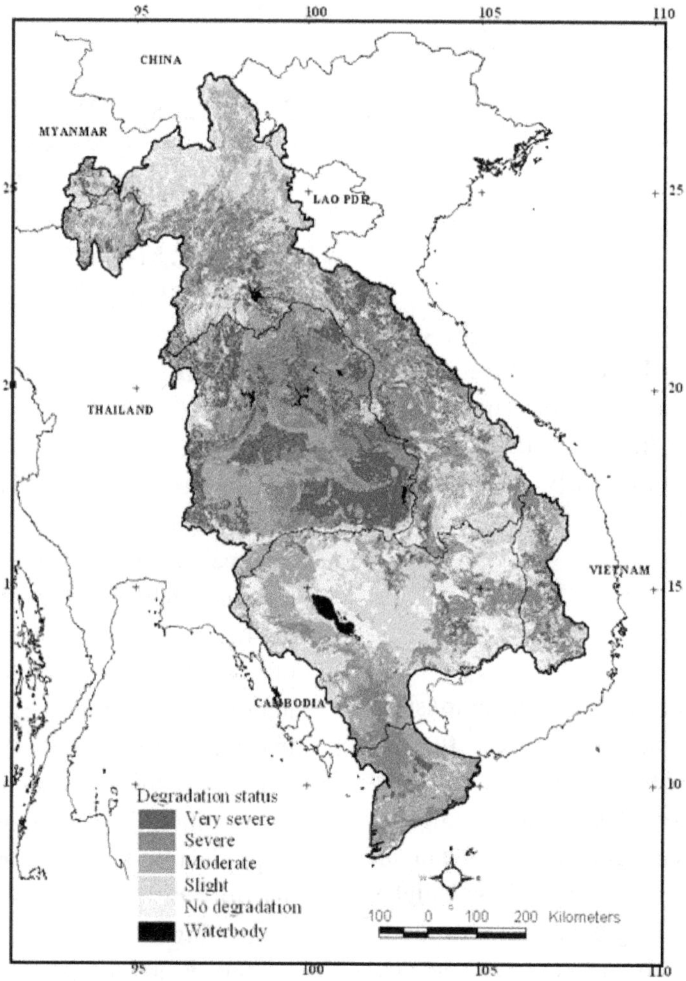

2000). Due to difficulties in gathering the data required for assessing agro-biodiversity, various surrogates can be developed using information on land-use dynamics (Dumanski & Pieri, 2000).

Plants diversity may increase total resource capture and thus have a higher net primary production (Hooper, 1998). Such an increase in net primary production with increasing plant diversity is mainly attributed to increased nutrient and water uptake due to different depths of root systems (Berendse, 1979), increased leaf area index and light capture due to differences in shoot architectures (Tilman, 1996), and increased efficiency of resource capture over time due to differences in phenology (Gulmon *et al.*, 1983). Soil erosion causes losses in soil productivity, degradation of water quality, and loss of organic carbon (Lal *et al.*, 1998). Plant diversity increases soil respiration and microbial biomass because of increased net primary productivity and therefore greater C input (Feike *et al.*, 2005). Monocultures promote soil erosion (Power & Follett, 1987), and the loss of plant diversity can alter the susceptibility to soil erosion (Korner, 1999).

Commercialized agriculture in the form of mono-cropping of both annuals and perennials, which results in low plant diversity is common practice due to several reasons. This study carried out in Khlong Yai watershed of Thailand focuses on plant diversity or plant species richness, including both crops and spontaneous vegetation, as it depends directly on the management practices in the cultivated landscape. Hence, assessment of plant diversity, which affects primary production, of different agricultural landuses can give useful baseline information to plan sustainable landuse management.

The study area, *Khlong Yai* watershed, is situated between 12° 65' to 13^0 14' N latitudes and 101^0 03' to 101^0 44' E in the eastern seaboard region of Thailand covering 170,175 ha. The climate is tropical monsoon and rainy season that extends from May to October, caused by the south west monsoon. The average annual rainfall is 1383 mm in 120 average annual rainy days. The average annual temperature is 28.3 °C. More than 75% of the sub-watershed has flat or gently undulating topography making the area good for upland cultivation. Almost area (80%) of the watershed is cultivated and major landuses are different shrub mono-crops, mixed orchards, tree mono-crops, and tree shrub inter crops, of para rubber (*Hevea brasiliensis)*, mixed orchards, pineapple and cassava.

5.2 Methodology

5.2.1 Sampling Design

The study was conducted at the landuse stage and field types, particularly in referring dominant vegetation or crop type, in line with the suggested methodology described in Zarin *et al.* (2002). These were identified as three basic vegetation layers, namely tree layer, shrub and herb layer. Fields of mixed orchard, para rubber, eucalyptus and coconut were considered as tree crop landuses (tree layer) as these crops grow to a height of more than two meters. Land-uses with pineapple, cassava and sugarcane are referred to as shrub crop landuse (shrub layer) because their height ranges between 0.5–2 meter. Paddy is considered as herb layer as its stem does not produce woody, persistent tissue and generally dies back at the end of each growing season.

Hence, the sampling was designed in such a way that the information on all these three layers could be collected during the field survey, which was conducted in early 2007. We used nested plot/sub plot design (Avery & Burkhart, 1983) containing 20x20, 10x10 m, 5x5 m and 1x1 m quadrats, nested within each other, as sampling units for plant diversity and biomass. 20x20 m quadrats were basically used for morphometric measurements of the tree layer for biomass estimation, tree species identification and counting of tree numbers in mixed orchards as the most diverse type of landuse. 10x10 m quadrats were used for tree crop landuses with a greater uniformity of species. 5x5 m quadrats were used for measurements in the shrub layer. Sampling of biomass of herbaceous species and grasses for bio mass estimation, and for counting of herbaceous species, and of number of individuals within species was done in the 1x1 m quadrats. A stratified sampling design using landuse as strata with number of sampling sites, proportional to size of area covered by each landuse class was employed as a sampling framework. The total number of sampling quadrats was 75, ranging from 4-12 quadrats in each landuse category. Soil type was also considered while selecting sample quadrats to ensure the dominant soils in terms of areal coverage in the study area are represented. Farm household owning the field with sample quadrats were identified and interviewed to collect relevant information including utility value of species and yield data.

5.2.2 Assessment of Plant Diversity

For the assessment of biodiversity or, more narrowly, plant diversity, various methods and indices are available. In this study, species richness, Shannon index, Simpson index, and species utility index (Zarin *et al.*, 2002) were used to estimate plant diversity for each land-use type. The indices were calculated separately for the different vegetation layers, herb layer, shrub layer and tree layer. In addition to that, the number of layers was also considered in order to incorporate the vertical aspect of diversity. This is important because of the hypothesis that productivity, which is mainly based on resource utilization, will be high in diverse landuses with different vertical layers (Hooper & Vitousek, 1998). The standardized methodology of linear scaling was adopted in order to combine different indices, so that a single index could be derived to rank different landuses in terms plant diversity.

Species richness is a simple numerical count of the number of species found in a given sampling unit (Maggurran, 1988), the quadrat in our case. The Shannon's index is a measure of the average degree of uncertainty in predicting to what species an individual chosen at random from a sample will belong to. The average uncertainty increases as the number of species increases and as the distribution of individuals among species becomes even. The Shannon's diversity index was calculated by multiplying species proportional abundance by its natural log.

$$H = -\sum_{i=1}^{N} p_i \ln p_i$$

Where p_i is the proportion of individuals found in the species *i*. Simpson's index (Simpson, 1949) gives the probability of two randomly chosen individuals drawn from a population belonging to the same species. Simpson's index was calculated by adding the sum of squares of proportional abundance of each species identified in the sampling quadrats. The higher the probability that individuals belong to the same species, the lower would be the diversity. The index was converted to (1-D) for easy interpretation, because a higher value of (1-D) also indicates a higher diversity.

$$D = \sum pi^2$$

The species utility index was calculated by dividing the number of species identified as useful by the farmers by the total number of identified species. Utility index was calculated by combining the species in all three layers. In addition, the number of layers was also taken as one index in order to incorporate the vertical aspect of diversity and to avoid bias due to richness of only one layer.

5.2.3 Ranking of Plant Diversity

The landuse types in the study were heterogeneous in terms of type of crops grown, management, and number of layers of crops and plants. As plant diversity is meaningful when considered both the horizontal and the vertical dimension, it is therefore essential to combine all the indices into one index to compare plant diversity of respective landuses. In this view, linear scaling of the different indices in different layers is suggested to obtain a single index for ranking plant diversity of the different agricultural landuses. This will help to choose better landuses for maintaining plant diversity in future. Linear scaling was done using the equation given below.

$$R = [(Yi - Y_{min})/(Y_{max} - Y_{min})] * 10$$

Where, R is rescaled diversity index. Yi, Y_{max}, and Y_{min} stand for i^{th} diversity index, maximum value of ith diversity index among landuses, and minimum value of ith diversity index among landuses, respectively. All the calculated indices were linear scaled at a range of 10 and averaged to get a single plant diversity index. Landuses were then ranked according to the calculated single plant diversity index.

5.2.4 Estimation of Biomass

Biomass of each landuse was estimated using regression equations suggested by FAO (1997) for the data collected from the nested sampling quadrates for tree species and spontaneous shrubs. Yield statistics from the household survey and harvest index from secondary sources were combined to estimate biomass of shrub crops. Herbaceous biomass of each landuse was measured by sampling the herbaceous layer of 1x1m quadrat. Finally, the total biomass per hectare was calculated by adding up tree, shrub and herb biomass. The average of total biomass estimates of all quadrate sites within a land-use represents the value for that land-use. The detailed methodology of biomass estimation has been described in Gnanavelrajah *et al.*, (2008).

5.2.5 Estimation of Erosion

The universal soil loss equation (USLE) given by Wischmeir & Smith (1978) was used to model soil erosion. The equation estimates the mean annual soil erosion in tons/ha/yr, resulting from the multiplication of six factors of soil erosion: rainfall erosivity soil erodibility, slope length, slope steepness, crop management and erosion control practice. Each factor was computed using the appropriate method suitable for local conditions as described in Gnanavelrajah (2007), which were encoded in GIS thematic layers before computing the mean annual soil erosion.

5.3 RESULTS AND DISCUSSION

5.3.1 Plant Diversity in Herb, Shrub and Tree Layer of Different Land-Uses

The plant species collected as described in sampling design section were identified. Altogether 46 herbaceous species, 22 Shrub species and 19 tree species were identified in all the landuse types in the study area. The most common vegetation layer in the observed landuse types was herbaceous species which were found in almost all landuses. Figure 5 presents the Shannon index, Simpson index, and Species richness of each herbaceous, shrub and tree layers for 11 landuse

Figure 5. Shannon index, Simpson index, and Species richness of herbaceous, shrub and tree layers

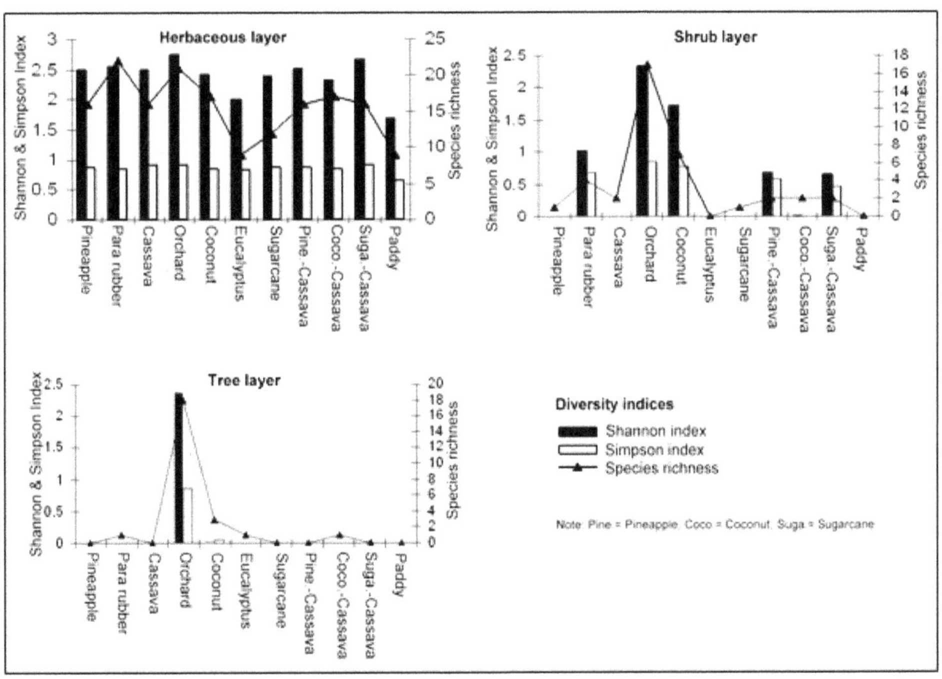

types. The highest species richness of 22 was recorded in the herbaceous layer of para rubber landuse followed by 21 in orchards. Eucalyptus and paddy had the lowest species count of 9 each. The computed index shows that orchard landuse had the highest Shannon index of 2.756, while paddy had the lowest index value of 1.690. The computed Simpson index was found highest (0.909) in Sugarcane–cassava landuse followed by orchard (0.907). Paddy scored the lowest Simpson index (0.659). This suggests that orchard landuse, which is mostly a mixed species orchard, has comparatively greater diversity in the herbaceous layer than other landuses. The area gets relatively higher rainfall and thus supporting different kinds of fruit trees to grow.

Eucalyptus and paddy did not have any species in the shrub layer, whereas pineapple and sugarcane had only one species each, hence no Shannon or Simpson indices were calculated. Orchard layer scored the highest diversity with corresponding computed indices of 2.336, 0.853 and 17 for Shannon index, Simpson index, and species richness, respectively for shrub layer (Figure 6). Coconut with indices of 1.306,.692, and para rubber with 1.023,.688, 4 values scored second and third rank, respectively, with regard to plant diversity in case of shrub layer. In addition to spontaneous species, orchard and coconut had also useful and cultivated species in this layer. Very young para rubber plantations also had cultivated species (pineapple) in their shrub layer. In other land-uses, the presence of only one single species in the shrub layer at a particular time contributed to lower diversity.

Most of the landuses, namely pineapple, cassava, sugarcane, sugarcane-cassava, pineapple-cassava and paddy did not have any tree species in the fields. Landuses coconut-cassava, eucalyptus and para rubber had only one tree species, coconut had three tree species, and orchard landuse had a variety of tree species as shown by the higher species richness of 18, Shannon index of 2.369 and Simpson index of 0.873.

5.3.2 Utility Index

The utility index was determined based on the farmers' opinion, however no in-depth study was made on how these species are used. In Table 5, the highest utility index of 61% was found in case of orchard landuse containing many cultivated species which are to be of useful for farmers and the lowest for eucalyptus (9%). Landuses, like co-

Figure 6. Biomass vs Plant diversity

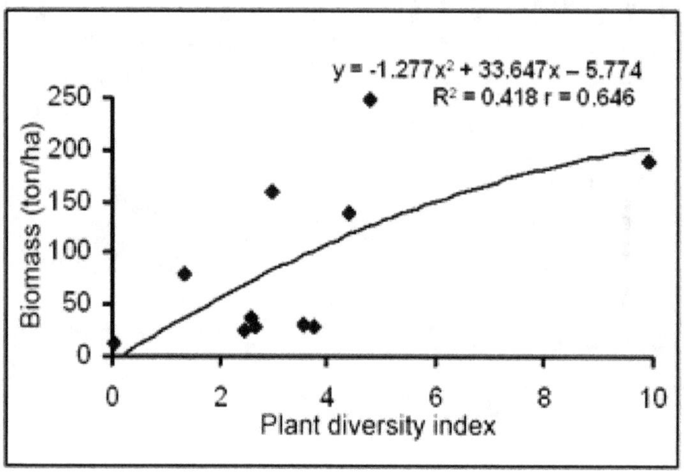

conut, coconut-cassava and sugarcane-cassava had utility indices of 29%, 24% and 23%, respectively. Coconut plots had useful species, both cultivated and spontaneous growing, in addition to the main field crop as coconut. The coconut-cassava and sugarcane-cassava landuses had more useful spontaneous species, which contributed to a higher utility index when compared to other landuses.

5.3.3 Plant Diversity in Different Land-Uses

The plant diversity in terms of number of different plant species was presented at the landuse level. The comparison of plant diversity between the landuses revealed that the orchard landuse had the highest plant diversity and the paddy landuse had the lowest plant diversity as seen in the above Figure 6. The cropping pattern of orchard was mixed cropping with a variety of crops in all three layers where as paddy field was mono-cropping with intensive weed management, which also contributes to less plant diversity. Hence, it is largely due to the selective use of herbicides and other chemicals in the paddy field for controlling weeds of any kind, where as in the mixed orchard fields it is rarely practiced, however farmers do apply some insecticides but not herbicides or weedicides. Paddy field being, a lowland landuse, might contain number of other flora including aquatic but this was not considered in this study.

Table 5. Species utility index of land uses

Land-use	Species utility index (%)
Pineapple	18
Para rubber	21
Cassava	11
Orchard	61
Coconut	24
Eucalyptus	9
Sugarcane	15
Pineapple-cassava	17
Coconut-cassava	31
Sugarcane-cassava	22
Paddy	11

Coconut and para rubber landuses were ranked at the second and third place, respectively, with regard to plant diversity. Even though these are mono-cropping landuses, the fact that these being perennial crops, the farmers do not practice intensive management against weeds compared to the annual crops, resulting into higher number of plant species grown naturally. Mixed landuses, such as sugarcane-cassava and pineapple-cassava ranked fourth and fifth, respectively. These two landuses are mono-crop rotations and therefore had a higher diversity than single mono-crop landuses, such as cassava, pineapple or sugarcane. Coconut– cassava, which is an intercrop landuse ranked sixth. In comparison to coconut monocrop, coconut-cassava intercrop was found to have less diversity due to cassava an annual crop whose field preparation needs complete tilling of the field thus reducing the scope for spontaneous vegetation growth.

Ranks seven, eight and nine were all landuses with mono-cropping of shrubs and recorded lower plant diversity. Even though eucalyptus is a tree layer mono-crop with low management intensity, its diversity was even lower than those of shrub mono-crops. Similar findings of low plant species diversity in eucalyptus plantations as compared to native species plantation have been reported (Sangha & Jalota, 2005). This is probably due to the allelopathy effects of eucalyptus species (Verma & Totey, 1999) and/or to the toxic effects of allelochemicals on soil micro-organisms (Chander, 1995).

5.3.4 Biomass of Landuses

Landuse under para rubber had the highest average total biomass of 247.89 tons/ha while landuse under paddy had the lowest of 12.87 tons/ha (Table 6). Total biomass of mixed orchard was slightly lower (189.43 tons/ha) but was not significantly different from that of para rubber. Other landuses having lower total biomass were those with no tree layers in the fields, such as pineapple, cassava, pineapple–cassava rotation, sugarcane, and sugarcane–cassava rotation. Among the landuses having tree crops, coconut, coconut-cassava and eucalyptus had less total biomass compared to mixed orchard and para rubber because of high plant spacing and less intense management of coconut and eucalyptus plantations. Mixed orchard was highly variable in terms of type of plants, age and management.

Shrub biomass was highest in sugarcane (28.59 tons/ha) being a C_4 plant which produces biomass efficiently. All other shrub crops ranked second in shrub biomass. All perennial landuses except coconut-cassava intercrop scored the lowest shrub biomass group. All shrub crops had lower herb biomass compared to tree crops. Among shrub crops sugarcane and pineapple had lowest herb biomass because of intense weed management practice and the closed spacing and canopy structure of such plants. Landuse under cassava, pineapple-cassava rotation and coconut-cassava intercrop had higher herb biomass compared to other shrub landuses because of less intense management of cassava crop leading to high weed growth.

Landuse under perennial trees had higher herb biomass compared to landuse under shrub type vegetation because of less competition and less intense weed management. As there are no trees in this category, no tree biomass was computed. Landuses under eucalyptus, coconut and coconut-cassava have lower tree biomass compared to mixed orchard and para rubber due to lesser per plant biomass and also higher plant spacing. The biomass of sugarcane (37.79 tons/ha) in the study area is comparable to that of the reported value of 39.71 tons/ha by Rahman *et al.* (1992). However, De Silva & De Costa (2004) reported much higher biomass values (46.32 – 63.25 tons/ha) for sugarcane. All tree crop land-uses had higher biomass/ha compared to the shrub crop species indicating the important ecological function, such as carbon sequestration, of tree crop species. It is also to note that the landuse under intercrop, coconut-cassava, had higher biomass than the mono-crop, cassava or coconut.

Table 6. Average biomass of land uses

Land use	Above ground			Below ground biomass	Total biomass
	Tree biomass	Shrub biomass	Herb biomass		
	Tons/ha				
Pineapple	0	18.50[b]	0.85[a]	5.8[a]	25.17[a]
Para rubber	187.53[c]	1.39[a]	1.75[c]	57.20[d]	247.89[d]
Cassava	0	20.36[b]	1.86[c]	6.66[a]	28.89[a]
Orchard	141.76[bc]	1.31[a]	2.63[d]	43.71[cd]	189.43[cd]
Coconut	100.70[ab]	4.81[a]	1.51[bc]	32.10[bc]	139.17[bc]
Eucalyptus	60.14[a]	0	1.80[c]	18.58[b]	80.52[bc]
Sugarcane	0	28.59[c]	0.47[a]	8.72[a]	37.79[a]
Pineapple-cassava	0	22.71[b]	1.25[b]	7.19[a]	31.15[a]
Coconut-cassava	100.72[ab]	20.43[b]	1.20[b]	36.71[bc]	159.07[bc]
Sugarcane-cassava	0	21.36[b]	1.47[bc]	6.85[a]	29.69[a]
Paddy	0	9.13[a]	0.77[a]	2.97[a]	12.87 [a]

5.3.5 Relationship between Plant Diversity and Biomass

The biomass, as an increasing function of plant diversity has been reported in the work of Tilman *et al.* (1997). In this study, while considering all landuses and the biomass from all layers, a significant positive correlation (Pearson correlation, r= 0.646) was found between average plant diversity and biomass of observed landuses in the study area (Figure 3) reaffirming the findings of previous reported works. When plant diversity and biomass are compared at individual landuse level, the overall trend of relationship looks satisfactory (Figure 7), however this is not in complete consistency that the highest the plant diversity the highest the biomass or vice versa. This could be explained by the relative efficiencies of individual species in converting resources into biomass and degree of complementary and competitive interaction among species (Hooper, 1998). In case of shrub crop landuses, there is a trend of increase in biomass with increasing plant diversity among landuses such as pineapple, cassava, pineapple-cassava and sugarcane-cassava. Even though sugarcane had the highest biomass among shrub crop land-uses, its plant diversity was rather low. The analysis at the individual landuse level suggests that even though there is an overall positive linkage between biomass and plant diversity, higher plant diversity does not necessarily imply higher biomass in all cases as reported in other studies of Hooper (1998) and Hooper & Vitousek (1997).

However, using species richness as a simple measure of biological diversity does not provide enough explanatory power, as ecosystem processes are affected by functional characteristics of organisms involved rather than by taxonomic identity. Moreover, the observation that increasing species diversity leads to increasing functional group diversity (Schmid *et al.*, 2001) in most natural ecosystems does not need to be always true for agricultural systems. This can serve as an explanation for landuses with low plant diversity yielding higher biomass in this study. Sugarcane, for instance, which has a higher biomass than all other shrub crop landuses, namely cassava, pineapple, paddy, sugarcane-cassava and pineapple-cassava, was characterized by low plant diversity, which was higher only then that of paddy. Higher plant diversity of other shrub crop landuses did not increase the number of C_4 plants, hence no increase in biomass is observed. Similarly, while coconut-cassava had a higher biomass than coconut, plant diversity was higher in coconut than in coconut-cassava. As cassava is a root crop, its biomass production is higher in comparison to that of the spontaneous shrub species found in landuse under coconut.

5.3.6 Soil Erosion

The potential soil erosion from different landuses was assessed. 84% of the study area have potential erosion rate of 2 tons/ha/yr or below. About 6 and 7% of the total area were found to have 2 to 4 and 4-12 tons of erosion/ha/yr, respectively. It is obvious that the landform of the study area is flat to undulating, which are less prone to soil erosion. Only 3% of the study area was found to have erosion rate exceeding the maximum permissible limit of 12 tons/ha/yr, particularly in those landuses with upland crops, such as cassava, sugarcane-cassava, the fields of young para rubber.

5.3.7 Relationship between Plant Diversity and Soil Erosion

Soil erosion though computed at the minimum mapping unit of spatial scale, it was aggregated at the landuse level to examine the relationship between plant diversity and soil erosion. In general, a negative correlation was observed between average soil erosion of land-uses and their respective plant diversity (Figure 8). However, the correlation was not significant. As soil erosion is a function of rainfall, soil, topography, vegetation type and

Figure 7. Landuse-wise biomass and plant diversity

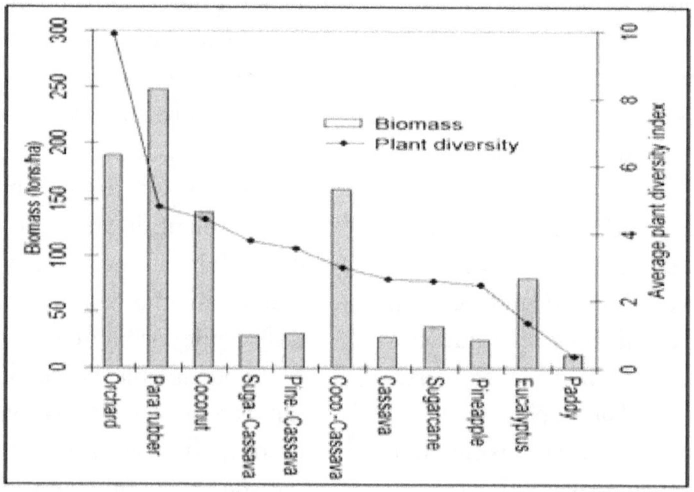

land management practices, it was difficult to get a clear and unambiguous correlation in this study. However, comparison of individual land-uses yielded some interesting information.

Higher average potential soil erosion was observed in landuse under sugarcane and eucalyptus, which had low plant diversity. Low soil erosion was observed in landuses with higher plant diversity, for example mixed orchard, except coconut, which has relatively higher plant diversity and also higher soil erosion. This may be due to the effect of coconut canopy which do not protect soil enough form being eroded. The lower average erosion in para rubber and pineapple may be attributed to the dense canopy structure, which can effectively reduce rainfall erosivity. In case of paddy, lowest plant diversity was associated with lowest average soil erosion. This is mainly due to the fact that paddy cultiva-

Figure 8. Landuse-wise soil erosion and plant diversity

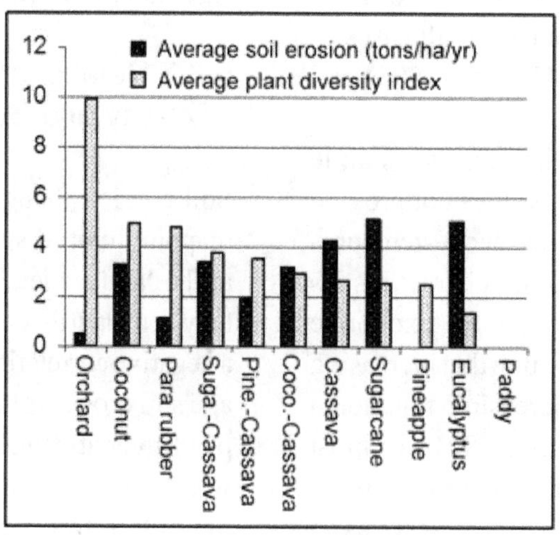

tion is practiced in flat terrain in the study area, where soil erosion is naturally of less significance.

5.4 Concluding Remarks

Given that substantial land areas on the earth are cultivated, the cultivated landscape has an enormous scope to conserve biodiversity. The study found that the landuse under orchard had the highest, and the paddy had the lowest plant diversity. Mono-crop of shrub type crops, such as cassava, pineapple and sugarcane had lower plant diversity than all mono-crop of tree crops with the exception of eucalyptus. But rotational mono-crop, such as pineapple-cassava and sugarcane-cassava, or intercrop, such as coconut-cassava had higher plant diversity than shrub mono-crop. These findings are in agreement with other observations that mono cropping in the case of shrub crops reduces biodiversity (Thrupp, 1998). Tree mono-crops, on the other hand, had higher plant diversity than shrub crop rotations or tree shrub intercrops. A significant positive correlation was observed between biomass and average plant diversity of landuses. However, when landuses were compared individually, higher biomass of land-uses did not always correspond to higher plant diversity and vice versa. With respect to soil erosion, plant diversity is negatively correlated, however, no strong evidence was demonstrated as the correlation was non-significant. In-depth studies with field measurements would help better to examine the relation between plant diversity and erosion.

It can be concluded from the findings of this study that the trend towards mono-cropping of shrubs, which can be expected to accelerate in Thailand due to the prioritization of export crops and more recently bio-fuels, will lead to a further reduction in plant diversity on a landscape level. Such information on the relationship between plant diversity, biomass and soil erosion can help for taking better decision in relation to landuse allocation to achieve various goals, such as soil conservation, nutrient recycling and carbon sequestration. Conservation or promotion of landuses, including the management of perennial trees should be based on the satisfaction of farmers expressed by the utility index. Choice of landuse should be determined by the capacity of landuses to fulfill more than only one function over a long time perspective.

ACKNOWLEDGMENT

Thanks are due to the University of Tokyo, Asian Center for Research on Remote Sensing, Global Land Cover Facility, Food and Agriculture Organization for allowing to freely download MODIS NDVI, rainfall, Shuttle Radar Topography Mission, NOAA-AVHRR, and Soil data. Research grant provided by the Asian Institute of Technology to carry out plant diversity study is highly appreciated. Several persons helped at different stages of the two studies including field works. Their help is also duly acknowledged.

REFERENCES

Avery, T. E., & Burkhart, H. E. (1983). *Forest measurements*. New York, NY: McGraw-Hill.

Berendse, F. (1979). Competition between plant populations with different rooting depths: Theoretical considerations. *Oecologia, 43*, 19–26. doi:10.1007/BF00346669

Brook, B. W., & Sodhi, N. S., Ng, & P. K. L. (2003). Catastrophic extinctions follow deforestation in Singapore. *Nature, 424*, 420–423. doi:10.1038/nature01795

Brookfield, H., Padoch, C., Parsons, H., & Stocking, M. (2002). *Cultivating biodiversity: Understanding, analyzing and using agricultural diversity*. London, UK: ITDG Publishing.

Chander, K., Goyal, S., & Kapoor, K. (1995). Microbial biomass dynamics during decomposition of leaf litter of poplar and eucalypt in a sandy loam. *Biology and Fertility of Soils, 19*, 357–361. doi:10.1007/BF00336107

Clergue, B., Amiaud, B., Pervanchon, F., Lasserre-Joulin, F., & Plantureux, S. (2005). Biodiversity: Function and assessment in agricultural areas, a review. *Agronomy for Sustainable Development, 25*, 1–15. doi:10.1051/agro:2004049

Cutter, S. L., & Renwick, W. H. (1999). *Exploitation, conservation, preservation: A geographic perspective on natural resource use*. New Jersey: John Wiley & Sons, Inc.

De Silva, A. L. C., & De Costa, W. A. J. M. (2004). Varietal variation in growth, physiology and yield of sugarcane under two contrasting water regimes. *Tropical Agricultural Research, 16*, 1–12.

Dumanski, J., & Pieri, C. (2000). Land quality indicators: Research plan. *Agriculture Ecosystems & Environment, 81*, 93–102. doi:10.1016/S0167-8809(00)00183-3

Eswaran, H., Lal, R., & Reich, P. F. (2001). Land degradation: An overview. In E. M. Bridges, I. D. Hannam, L. D. Oldeman, F. W. T. Pening de Vries, S. J. Scherr & S. Sompatpanit (Eds.). *Responses to land degradation. Proceedings of the 2nd International Conference on Land Degradation and Desertification*, Khon Kaen, Thailand.

FAO. (1979). *FAO-UNESCO soil map of the world: Southeast Asia*. Paris: Food and Agricultural Organization of the United Nations.

FAO. (1997). *Estimating biomass and biomass change of tropical forests: A primer. (Forestry Paper No. 134)*. Rome: FAO.

FAO. (1999). *Agricultural biodiversity.* Multifunctional Character of Agriculture and Land Conference. Background paper 1. Maastricht Netherlands.

FAO. (2002). *World agriculture: Towards 2015/2030. Summary report*. Rome: Food and Agriculture Organization.

FAO. (2003). *TERRSTAT: Land resource potential and constraints statistics at country and regional level*. Retrieved May 15, 2010, from http://www.fao.org/ag/agl/agll/terrastat/

FAO. (2005). *Properties and management of drylands*, Retrieved June 12, 2006, from http://www.fao.org/ag/agl/agll/drylands/assessment.htm

Feike, A. D., Sarah, E. H., Peter, B. R., & Johannes, M. H. K. (2005). Divergent effects of elevated CO_2, N fertilization, and plant diversity on soil C and N dynamics in a grassland field experiment. *Plant and Soil, 272*, 41–52. doi:10.1007/s11104-004-3848-6

GEF. (1999). *Report of the STAP Expert Group Workshop on Land Degradation.* (GEF/C.14/Inf. 15). Global Environment Facility.

Gnanavelrajah, N. (2007). *Sustainable land-use options for enhancing carbon sequestration, plant diversity and productivity: A case of Khlong Yai Sub-watershed, Thailand.* Unpublished PhD dissertation, Asian Institute of Technology, Bangkok.

Gnanavelrajah, N., Shrestha, R. P., Schmidt-Vogt, D., & Samarakoon, L. (2008). Assessment of carbon stock and soil carbon management in agricultural landuses in Thailand. *Land Degradation and Development, 19*(3), 242–256. doi:10.1002/ldr.838

Gulmon, S. L., Chiariello, N. R., Mooney, H. A., & Chu, C. C. (1983). Phenology and resource use in three co-occuring grassland annuals. *Oecologia, 58*, 33–42. doi:10.1007/BF00384539

Heywood, V. H., & Bates, I. (1995). Introduction. In Heywood, V. H., & Watson, R. T. (Eds.), *Global biodiversity assessment*. Cambridge, UK: Cambridge University Press.

Hook, J., Susan, N., & Robyn, J. (2003). *Social atlas of the Lower Mekong Basin*. Phnom Penh: Mekong River Commission.

Hooper, D. U. (1998). The role of complementarity and competition in ecosystem responses to variation in plant diversity. *Ecology*, *79*(2), 704–719. doi:10.1890/0012-9658(1998)079[0704:TROCAC]2.0.CO;2

Hooper, D. U., & Vitousek, P. M. (1998). Effect of plant composition and diversity on nutrient cycling. *Ecological Monographs*, *68*, 121–149. doi:10.1890/0012-9615(1998)068[0121:EOPCAD]2.0.CO;2

IUCN. (2003). *2003 IUCN red list of threatened species*. Retrieved May 10, 2009, from http://www.redlist.org

Korner, C. (1999). *Alpine plant life*. Berlin, Germany: Springer.

Lal, R., Kimble, J. M., Follet, R. F., & Cole, C. V. (1998). *The potential of US croplands sequester carbon and mitigate the green house effect*. Chelsea, MI: Ann Arbor Press.

Le Houerou, H. N. (1984). Rain-use efficiency: A unifying concept in arid-land ecology. *Journal of Arid Environments*, *7*, 213–247.

Magguran, A. E. (1998). *Ecological diversity and its measurement*. Princeton, NJ: Princeton University Press.

Millennium Ecosystem Assessment. (2005). *Ecosystems and human well-being: Synthesis*. Washington, DC: Island Press.

NEMA. (2001). *State of the environment report for Uganda*. Kampala: National Environment Management Authority, Ministry of Water Lands and Environment.

NRC. (2006). *Net primary productivity*. Retrieved June 2, 2006, from http://ccrs.nrcan.gc.ca/optic/coarse/beps/npp_e.php

Ogawa, S., Rikimaru, A., Masumoto, T., Tanji, H., & Maharaxay, M. (2005). Landuse and water management of agriculture in Mekong River basin. *Journal of Agricultural Meteorology*, *60*(5), 371–374.

Oldeman, L. R., & van Lynden, G. W. J. (1998). Revisiting the GLASOD methodology. In Lal, R., Blum, W. H., Valentine, C., & Stewart, B. A. (Eds.), *Methods for assessment of soil degradation* (pp. 423–440). Boca Raton, FL: CRC Press.

Power, J. F., & Follett, R. F. (1987). Monocultures: Advantages, limitations, and alternatives. *Scientific American*, *256*(3), 78–86. doi:10.1038/scientificamerican0387-78

Prince, S. D., Brown de Colstoun, E., & Kravitz, L. L. (1998). Evidence from rain-use efficiency does not indicate extensive Sahelian desertification. *Global Change Biology*, *4*, 359–379. doi:10.1046/j.1365-2486.1998.00158.x

Rahman, M. H., Pal, S. K., & Alam, F. (1992). Effect of nitrogen, phosphorus, potassium, sulphur, zinc and manganese nutrients on yield and sucrose content of sugarcane (Saccharum officinarum) in floodplain soils of Bangladesh. *Indian Journal of Agricultural Sciences*, *62*, 450–455.

Rerkasem, B., & Rerkasem, K. (2000). Agrodiversity for in situ conservation of Thailand's native rice germplasm. *CMU Journal*, *1*(2), 15–23.

Rubio, J. L., & Bochet, E. (1998). Desertification indicators as diagnosis criteria for desertification risk assessment in Europe. *Journal of Arid Environments*, *39*, 113–120. doi:10.1006/jare.1998.0402

Sangha, K. K., & Jalota, R. J. (2005). Value of ecological services of exotic Eucalyptus tereticornis and native albergia sissoo tree plantations of North-Western India. *Conservation & Society*, *3*(1), 92–105.

Schmid, B., Joshi, J., & Schlapfer, F. (2001). Empirical evidence for biodiversity-ecosystem functioning relationships. In Kinzing, A. P., Pacala, S. W., & Tilman, D. (Eds.), *The functional consequences of biodiversity: Empirical progress and theoretical extension. Monographs in Population Biology* (pp. 120–150). Princeton, NJ: Princeton University.

SCS. (1972). *Hydrology sect. 4, SCS national engineering handbook.* Washington, DC: USDA-SCS.

Sharma, K. D. (1998). The hydrological indicators of desertification. *Journal of Arid Environments, 39*, 121–132. doi:10.1006/jare.1998.0403

Shrestha, R. P., Eiumnoh, A., & Baimoung, S. (1996). *Soil erosion assessment and its policy implications: A case study of RS and GIS applications in Uthai Thani, Thailand.* Retrieved January 28, 2010, from http://www.gisdevelopment.net/aars/acrs/1996/ss/ss1004.shtml

Simpson, E. H. (1949). Measurement of diversity. *Nature, 163*, 688. doi:10.1038/163688a0

Sodhi, S. N., Koh, L. P., Brook, B. W., & Ng, P. K. L. (2003). Southeast Asian biodiversity: An impending disaster. *Trends in Ecology & Evolution, 19*(12), 654–660. doi:10.1016/j.tree.2004.09.006

Stocking, M. (2002). Diversity: A new strategic direction for soil conservation. Sustainable utilization of global soil and water resources. *Proceedings of the 12th International Soil Conservation Conference* (pp. 53-58). Beijing, China: Tsinghua University Press.

Symeonakis, E., & Drake, N. (2004). Monitoring desertification and land degradation over sub-Saharan Africa. *International Journal of Remote Sensing, 25*, 573–592. doi:10.1080/0143116031000095998

Takacs, D. (1996). *Philosophies of paradise.* Retrieved May 16, 2010, from http://www.dhushara.com/book/diversit/restor/takacs.htm#anchor370984

Thrupp, L. (1998). *Cultivating diversity: Agrobiodiversity and food security.* Washington, DC: World Resources Institute.

Tilman, D. (1996). Biodiversity: Population versus ecosystem stability. *Ecology, 77*, 350–363. doi:10.2307/2265614

Tilman, D., Knops, J., Wedin, D., Reich, P., Ritchie, M., & Siemann, E. (1997). The influence of functional diversity and composition on ecosystem process. *Science, 277*, 1300–1302. doi:10.1126/science.277.5330.1300

UN ESCAP. (2010). *Statistical yearbook for Asia and the Pacific 2009.* Retrieved June 8, 2010, from http://www.unescap.org/stat/data/syb2009/

University of Tokyo. (2005) *Rainfall data.* GAME-T2 Data Center. Retrieved September 16, 2005, from http://hydro.iis.u-tokyo.ac.jp/GAME-T/GAIN-T/updates.html

USGS. (2004). *Shuttle radar topography mission.* United States Geological Survey. Retrieved October 20, 2005, from fttp://ftp.glcf.umiacs.umd.edu/glcf/srtm

Verma, R. K., & Totey, N. G. (1999). Biological diversity, medicinal potential of ground flora and improvement in soil quality under plantations raised on degraded bhata land. *Advances in Forestry Research in India, 20*, 37–69.

Wataru, K. (2001). *Aqua/Terra MODIS bi-weekly composite database (Asia).* Institute of Industrial Science, University of Tokyo. Retrieved October 20, 2005, from http://webmodis.iis.u-tokyo.ac.jp/Asia/

Wischmeier, W. H., & Smith, D. D. (1978). *Predicting rainfall erosion losses: A guide to conservation planning*. U.S. Department of Agriculture, Agricultural handbook No 537.

WRI. (2002). *Drylands, people, and ecosystem goods and services: A Web-based geospatial analysis*. World Resource Institute. Retrieved June 11, 2006, from http://www.wri.org

Zarin, D. J., Huijun, G., & Enu-kwesi, L. (2002). Guidelines on the assessment of plant species diversity in agricultural landscapes. In Brookfield, H., Padoch, C., Parsons, H., & Stocking, M. (Eds.), *Cultivating biodiversity: Understanding, analyzing and using agricultural diversity*. London, UK: ITDG Publishing.

Zhang, X., Drake, N. A., Wainwright, J., & Mulligan, N. (1999). Comparison of slope estimates from low resolution DEMs: Scaling issues and fractal method for their solution. *Earth Surface Processes and Landforms*, *24*, 763–779. doi:10.1002/(SICI)1096-9837(199908)24:9<763::AID-ESP9>3.0.CO;2-J

Chapter 16
Sustainable Land Use and Watershed Management in Response to Climate Change Impacts:
Overview and Proposed Research Techniques

Nguyen Kim Loi
Nong Lam University, Vietnam

ABSTRACT

With the changes in climatic, biophysical, socio-cultural, economic, and technological components, paradigm shifts in natural resources management are unavoidably and have to be adapted/modified to harmonize with the global changes and the local communities' needs. This chapter focuses on sustainable land use and watershed management in response to climate change impacts. The first part covers some definitions and background on sustainable land use, watershed management approach, and sustainable watershed management. The second part describes the use of the Geographic Information System (GIS) and Spatial Decision Support System (SDSS) model focusing on the framework for a planning and decision making, computer-based system for supporting spatial decisions. The mathematical programming has been reviewed focusing on optimization algorithms that include optimization modeling and simulation modeling for decision making. Finally, the example of methodology development for sustainable land use and watershed management in response to climate change in Dong Nai watershed, Vietnam is presented.

DOI: 10.4018/978-1-60960-619-0.ch016

1. INTRODUCTION

Current climate change estimates indicate that major environmental changes are likely to occur due to climate change, in practically every part of the world, with a majority of these changes being felt through modification of the hydrological cycle, e.g. floods, droughts and storms. Climate change impacts are also estimated to be particularly severe in many developing countries of the world and especially in Vietnam. The recent studies (Dasgupta *et al.*, 2007; IPCC, 2007) have concurred that Viet Nam will be one of most vulnerable countries to climate change in the world. Gradual changes, such as sea level rises and higher temperatures, more extreme weather such as drought, and more intense typhoons are all on the horizon and will have a potentially devastating impact on the country's people and economy. According to the latter study, 10.8% of Vietnam's population, mostly those people living in the two river deltas (Red & Mekong river deltas), would be affected by sea level rise (SLR) of just 1 meter (Dasgupta *et al.*, 2007). According to the IPCC (2007), a 1 meter SLR in Vietnam would lead to flooding of up to 20,000 km^2 of the Mekong River delta and 5,000 km^2 of the Red River delta. In the Mekong River delta alone, more than 1 million people would be affected directly.

The above statements do not only reflect the importance of watershed resources in natural resources management but also imply for the integrated management, which all stakeholders must consider where developing management activities from the beginning of a project establishment. Hence, this chapter focuses on sustainable land use and watershed management in response to climate change impacts. A broad understanding of various topics in sustainable land use and watershed science and modeling technology is required to complete the studies presented in this chapter and it is important to thoroughly review each of them. The first part covers some definition and background to sustainable land use, watershed management approach, and sustainable watershed management. The second part is concerned with the Geographic Information System (GIS) and Spatial Decision Support System (SDSS) model focusing on the framework for planning and decision making, and the computer-based system for supporting spatial decisions. The mathematical programming system has been reviewed focusing on optimization algorithms that include optimization modeling, and simulation modeling for decision making. Finally, an example of methodology development for sustainable land use and watershed management in response to climate change in Dong Nai watershed, Vietnam is presented.

2. DEFINITION AND BACKGROUND OF WATERSHED MANAGEMENT APPROACH

If one asks why we need to manage natural resources based on watershed boundary, the answer would be to recognize that sustained land or resource based development depends on the interaction of all the activities that take place in the watershed. Uplands and lowlands are physically linked in a watershed via the hydrologic cycle. Upstream activities affect downstream opportunities and problems by influencing the flow of water, sediments and other waterborne materials through the system. For recognizing this fact, one has to merely look at the numerous examples where poor upstream land use practices result in disaster downstream. Upland erosion not only leads to long-term losses of upland productivity, but also loss of storage capacity in reservoirs which in turn leads to loss of hydropower production, increased flooding, or loss of irrigation capacity downstream. Soil loss brings adverse downstream impacts even when reservoirs are not present. More frequent over bank flows and flood damages will likely result. In addition, lack of adequate water to dilute wastes and general water quality deterioration from uplands results in more serious pollution, including public health problems.

It is essential for the success of watershed management that, there is a clear understanding of some of its basic underlying concepts. The below section endeavors to define key terms and principles that are relevant to watershed management.

2.1 Watershed

Brooks *et al*.(1992) described that watershed is a topographically delineated area of landform where rainwater can drain as surface runoff through a river system with a common outlet, which could be a dam, irrigation or domestic water supply off-take point or where the river discharges into a larger river, lake or sea. A watershed is part of a larger system stretched across the Earth's surface, with adjacent watersheds separated by boundaries or divides.

The term "Watershed" is synonymous with "river basin", "drainage area" and "catchment". The term "river basin" is often used in reference to large watersheds (usually over 100,000 ha). In contrast, "catchment" usually refers to small watersheds (ranging from less than 1,000 ha to 100,000 ha). A watershed is a self-contained system consisting of intricately interacting biotic and abiotic components and often of several linked ecosystems or portions of a number of ecosystems.

A watershed is not necessarily an upland or mountainous landform; it may occur in a lowland setting and the land surface may be a major site for residential, commercial, industrial, agricultural, educational, experimental, environmental and forest land uses, which are often conflicting and competing with each other for limited watershed land resources. Watersheds are a major source of nutrients and pollutants, which are deposited in lakes, coastal areas, and rivers.

2.2 Watershed Management Approach

Brooks *et al.* (1992) defined watershed management as the process of guiding and organizing land and other resource uses in a watershed to provide desired goods and services without adversely affecting soil and water resource. It is also defined as the application of business methods and technical principles for manipulating and controlling watershed resources to achieve a desired set of objectives, such as maximum supply of usable water, minimization of soil erosion and sedimentation problems and reduction of flood and drought occurrences.

Planning and implementation of both technical and policy initiatives are necessary to enable the natural and human resources of individual watersheds to contribute to one or more of the following development aims:

Improved rainwater management within individual watersheds so as providing quality water from both surface and groundwater sources on a sustainable basis to meet the needs of different water users (human settlement, lowland farmland/irrigation systems, power and transport infrastructure, fish ponds and coral reefs/coastal resources) within and downstream of the watershed, and increased protection from flood and sedimentation damage for the downstream area of the watershed; Improved standard of living, through the maintenance, enhancement and development of existing and new sustainable livelihood opportunities for those individuals, households and communities whose welfare needs are met wholly, or in part, by the utilization of watershed resources; Improved maintenance, enhancement and protection of those areas that are important for bio-diversity conservation; Improved care and management of the natural resources within watersheds, thereby enabling them to be used for economically productive purposes (water, forestry, agriculture, tourism, power generation, etc.) on a sustained basis while maintaining and enhancing their social and environmental service functions.

In the context of limited natural resources and rapid population growth, the concepts of multiple use and sustainable management have been established to cope with the need for long term social stability of future generations. Watershed management involves the integrated management of all the natural resources of a drainage basin, in order to protect, maintain or improve water yields. It requires synthetic approach, integrating the various aspects of hydrology, ecology, soils, physical climatology and other sciences to provide the scientific basis of management. Then, to develop from this basis, rational procedures of applying this information to achieve desired results and to derive guidelines for choosing acceptable management alternatives within the scope of social wants and needs (Satterlund & Adams, 1992).

Watershed management is a term mainly used by foresters and soil conservationists. The holistic approach which includes all facets of complex interactions among bio-technical, social, economic, institutional and political factors is taken into consideration to ensure that all resources development activities are implemented in concert with one another to achieve a variety of objectives successfully. It can be summarized as a part of natural resources which composes of three main principles: (1) Land use planning in terms of land capability and suitability, (2) Resource utilization and conservation which depend on natural resources characteristics, and (3) Pollution control in terms of erosion, floods, protection of aesthetic values and others mitigation impact planning (Hewlett & Nutter, 1969; Jermar, 1987; & Chunkao, 1981, 1992).

In addition, the most significant outcome of the United Nations Conference on Environment and Development "The Earth Summit" held in Rio de Janeiro in June 1992 was Agenda 21. This document is set to guide and drive action towards sustainable development as a key text for all concerned with policy and practice. The Local Government Management Board in the UK has produced a simplified guide to Agenda 21, (Gardiner, 1994) by the following:

"Watershed resources must be planed and managed in an integrated and holistic way to prevent shortage of water, or pollution of water sources, from impeding development. Satisfaction for basic human needs and preservation of ecosystem must be the priorities; after these, water users should be charged appropriately. By the year 2000, all states should have national action programs for water management based on catchment basins or sub–basins and efficient water–use programs. These could include integration of water resource with land–use planning and other development and conservation activities, demand management through pricing or regulation, conservation, re-use and recycling of water".

Gardiner (1994) also insisted that in order to become an executable sustainable development action plan, it must satisfy criteria in at least three major dimensions: Ecological–Social–Economic (ESE). Although the general aspiration or criteria of social and economic are almost easy to identify, the process of ecological phenomena requires better understanding for setting up the appropriate criteria of the system functions.

2.3 Watershed Functions

The interactions between the structures, biotic and abiotic, which function mainly in terms of hydrological process, nutrient and food chains in the watershed ecosystem are extremely close. The overexploitation of some resources will have impact on their natural relationships and always contribute to undesirable outcomes for human beings. However, humans are inextricably bound to their ecosystem and function only as consumers.

2.4 Hydrologic Processes

The hydrologic balance or water budget is a fundamental concept of hydrology and a useful method for the study of the hydrologic cycle. The hydrologic cycle represents the processes and pathways involved in the circulation of water from land and water bodies, to the atmosphere and back again. The cycle is complex and dynamic but can be simplified if we categorize components into input, output or storages.

The hydrologic processes of the biosphere and the effects of vegetation and soils on these processes such as precipitation, infiltration, percolation, evaporation, transpiration, surface runoff, subsurface flow, and groundwater flow can all be affected by land management activities. Likewise, man can alter the magnitude of various storage components including soil water, snow packs, lakes, reservoirs, and rivers. With a water budget, we can examine existing watershed systems, quantify the effects of management impacts on the hydrologic cycle and in some cases predict or estimate the hydrologic consequences of proposed or future activities.

2.5 Sustainable Watershed Management

Sustainability involves ensuring a long term supply of water of adequate quality for all designated purposes, for which an area is intrinsically suitable while minimizing adverse economic, social and ecological impacts and maintaining the structure and function of the natural system (Diane, 2002). Sustainable watershed management involves informed decision-making in a complex array of biophysical, social and economic environments made up of processes and interactions between ecosystems, their components and between human groups intervening in such ecosystems. Decisions involve the allocation of resources, formulation of policies, interventions and manipulations of natural resources present in the naturally-defined area confined by a watershed or hydrological basin. Due to the complexity of issues involved in watershed management, this requires a multi-disciplinary, holistic and integrated approach. An ecological approach for managing watersheds recognizes the interconnectedness and relationships of mutual dependence between the ecosystems and the degree in which manipulations of the structure and functions of one ecosystem may result in inputs and changes to the structure and functions of other related ecosystems.

2.6 Ecosystem Management

Ecosystem based management as an approach to resolve the fragmented management of terrestrial and aquatic resources and for achieving the goals of sustainable development and biodiversity and ecosystem integrity has been increasingly significant in managing the vast expanse of natural resources in the Asia-Pacific region.

Agenda 21 reflects a global consensus and commitment at the highest political level on how to make development socially, economically, and environmentally sustainable. It includes protecting the atmosphere, an integrated approach to planning for and managing land resources, combating deforestation, managing fragile ecosystems: combating desertification and drought and sustainable mountain development, promoting sustainable agriculture and rural development, conservation of biodiversity, managing biotechnology, protecting and managing the ocean, protecting and managing fresh water, and supporting science for sustainable development (Shlaepfer, 1997)

The term *"ecosystem management"* has been defined by several ecologists, like Gordon (1993), Grambine (1994), Christenson *et.al.* (1996), and Thomas & Huke (1996) cited by Baker *et al.* (1996) in the EPA report. Based on the definition given by these authors, the more likely appropriate definition for the present situation could be modified as:

"Ecosystem management is a goal-driven approach to restore and sustaining ecosystem structure andfunctions and value using the best science available together with local wisdom. It entails working collaboratively with central government, tribal, and local government, community group, private landowners, and other stakeholders to develop a vision of desire future ecosystem condition. This vision integrates ecological, economic, and social factors affecting the management unit defined by ecological, not political boundaries. The goal is to restore and maintain the health of ecosystems while supporting economies and socio-culture of communities as well as whole society"

Recently the ecosystem management concept has been applied in natural resources management in both the developed and developing countries and not only for terrestrial ecosystem but also for aquatic and coastal ecosystem management (Vallega, 2002). Moreover, watershed management has been used as a planning and implementing unit for sustainable ecosystem management.

3. GEOGRAPHIC INFORMATION SYSTEM AND ITS APPLICATIONS

GIS is a system, including hardware and software, designed for data collection, analysis and retrieving results of data in different positions on earth for solving the complex planning problems. Tomlinson (1985) defined GIS as the information filled in the map for decision making. Geographic Information System (

GIS) helps to increase the efficiency of geographic data collection and analysis, including the results from the real world information or fact. By geo-referencing with a coordinate system, any real world feature can be represented in the GIS system, including its descriptive data or picture (Ounon, 1990; Maguire, 1991).

GIS are gaining importance and widespread acceptance as a tool for decision support in land, infrastructure, resources, environmental management and spatial analysis and in urban and regional development planning. With the development of GIS, environmental and natural resource managers have information systems at their disposal, which data are readily accessible, easily combined and flexibly modified for decision making in environmental and natural resource management. It is thus reasonable to expect a better informed and more explicitly reasoned decision-making process. But despite the proliferation of GIS software systems and the surge of public interest in the application of the system to resolve the real world problems, the technology is commonly seen as complex, inaccessible, and alienating to the decision makers (Fedra, 1993, 1994; Maidment, 1993; Geertman & Stillewl, 2002).

The reasons for this estrangement are varied. In part the early development and commercial success of GIS were fuelled more by the need for efficient spatial inventory rather than decision support systems. As a result, even today only few systems provide any explicit decision analysis tools. In addition, the technology is built upon a very broad base of scientific disciplines, ranging from cartography to remote sensing, computer science, statistics etc. This implies that to become broadly involved in GIS use, an extensive background in the digital data management, mapping sciences and information technology is required. Furthermore, the technology has strong elements of modernity and scientific rigor that is strongly cultivated by vendors, consultants, and other advocates. As a result, GIS has become a field requiring a host of intermediaries between the end user and the data provider: technicians, system managers, analysts, user interfaces, query languages and so on. In addition to these are the institutional and organizational issues of the technology transfer. Although, recent development in GIS software's and Web Technology has made GIS more user-friendly, therefore usable and accessible to more users (Geertman & Stillwel, 2002). Information technology may either democratize information by making it more equitably accessible, or it may have the opposite effects of disproportionately empowering a selected sector of society. The lack of analytical tools to aid decision evaluation and policy formulation efficiently and the continuing mystification of the field have unfortunately often led to the latter in GIS (Fox, 1991; Geertman & Stillwel, 2002). In many cases GIS has become a rifting technology, tending to divert the process of decision making away from decision makers and into the hands of GIS analyst and host of other highly trained technological intercessors (Eastman *et al*., 1995). For alleviating the above problems, GIS should be upgraded by decision support system (DSS) functionality in a user friendly environment. However, there is a trade-off between the efficiency and ease of use, and the flexibility of the system. When more options are predetermined and available from the menu of choices, the more defaults are provided; the easier it becomes to use a system for an increasingly small set of tasks. There is also trade-off between the ease of understanding and the precision of the results. Providing visual or symbolic presentation will change the quality of the information in the course of transformation from quantitative to qualitative data sets. Finally, the easier the system the harder it is to make and maintain.

GIS may record and demonstrate in two types: raster or grid format and vector format. Grid format or pixel is referred to the coordinating system; details of spatial data may record variance in the grid size. The dominant information of spatial is manipulation of data and reversible to transform into digital data. Vector format may be used for the continuation of spot and coordination to allocate the objects or interest. The advantage of vector format is the storage data area, which is not large, and symbols of data may be similar to the real data, however, it is difficult to perform operation requiring calculation (Ongsomwang, 1995).

Spatial area is the important database. GIS system relates to the database and conjugates data in map and ground check because it is very important for analysis. Therefore, the structure of the map and coordinates are important for the accuracy of the facts and analysis of data.

4. COMPUTER-BASED SYSTEM FOR SUPPORTING SPATIAL DECISION

Based on the classification of decision problems and corresponding computer systems, four types of systems for supporting spatial decisions can be distinguished: Spatial data processing systems (SDPSs), Spatial decision support system (SDSS), Spatial expert systems (SESs), and spatial expert support system (SESSs). SDPSs and SDSS are briefly discussed below.

4.1 Spatial Data Processing Systems (SDPSs)

Spatial data processing systems (SDPSs) are applicable in a decision making situation where all four components of problem solving activities are structured; that is, all required data are available, there is a well-defined set of evaluation criteria and constraints, the problem can be solved by standard procedures, and there is no need for a complex strategy for generating and evaluating alternatives. The assumption behind this type of system is that the problem is solved by a computer. There is no need to involve decision makers in solving process activities. Central to SDPSs for solving decision problems is the ability to: (1) to incorporate all elements of the decision problems into a model representing the problems, and (2) to use systematized techniques or algorithms for analysis of the model. The model is the description of a decision making situation, while an algorithm analyzes that description to generate a solution to the problem. A map of a subway system – a model – is not the same as using that map to find a route between two stations – an algorithm. Several different algorithms can be used to analyze the same model. For example, there are usually a number of algorithms to solve the same location – allocation model (Ghosh & Rushton, 1986).

4.2 Spatial Decision Support System (SDSS)

The concept of spatial decision support system (SDSS) has evolved in parallel with Decision Support System-DSS (Densham & Rushton, 1987; Densham & Goodchild, 1989; Densham 1991; Crossland *et al.*, 1995). Based on a generic definition of DSS (Keen & Scott-Morton, 1978), SDSS can be defined as an interactive, computer-based system designed to support a user or group of users in achieving a higher effectiveness of decision making, while solving a semi-structured spatial decision problem.

The structure of SDSS can be described by identifying the major components or subsystems of the system. An SDSS typically contains three generic components: a database management system (DBMS) and geographical database, a model-based management system (MBMS) and model base, and a dialogue generation and management system (DGMS) as shown in Figure 1. The data subsystem performs all data related tasks; that is, it stores, maintains, and retrieves data from database, extracts data from various resources (Table 1). It provides access to data as well as controls the program necessary to get those data in a appropriate form for a particular decision making problem. The MBMS component provides links between different models so that the output of one model is the input for another model. The importance of the dialogue subsystem cannot be overemphasized since all the capabilities of the SDSS must be articulated and implemented through it. In addition, the decision maker or user is considered to be a part of the system. As mentioned earlier, the unique contributions of DSS are derived from the interaction between computer and user.

4.3 Mathematical Modeling for Decision Making with GIS

There are two major thrusts in mathematical modeling within GIS environment: Optimization and simulation (Fotheringham & Rogers, 1994; Steyaert & Goodchild, 1994). Each represents a fundamentally different approach to problem solving. Broadly speaking, the output of optimization models is a prescription of strategy. Simulation, on the other hand, is a descriptive approach.

Figure 1. Components of Spatial decision support system Source: Malczewski (1999)

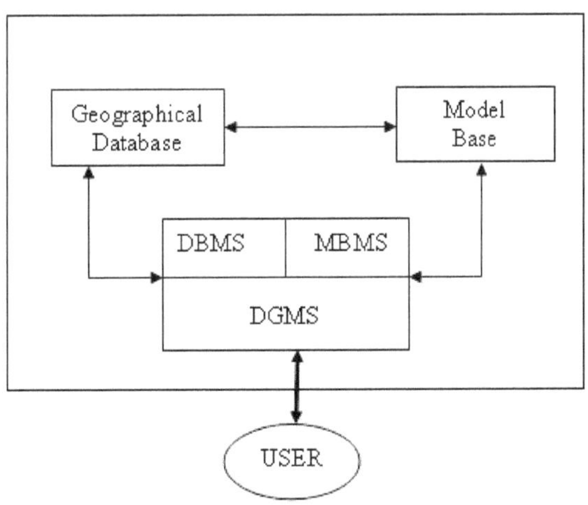

Table 1. Function of Decision Support Systems (DSS)

Components	Functions
Data Base and Management (DBM)	**Types of data** locational (e.g. coordinates) topological (e.g. points, lines, polygons and relationships between them) attributes (e.g. geology, elevation, transportation network)
	Logical Data Views relational DBMS hierarchical DBMS network DBMS object-oriented DBMS
	Management of Internal and External Databases acquisition storage retrieval manipulation directory queries integration
Model Base and Management (MBM)	**Analysis** goal seeking optimization simulation what-if
	Statistics and forecasting exploratory spatial data analysis confirmatory spatial data analysis time series geostatistics
	Modeling decision maker's preference value structure hierarchical structure of goals, evaluation criteria, objectives and attributes pairwise comparison multiattribute value/utility consensus modeling
	Modeling uncertainty data uncertainty decision rule uncertainty sensitivity analysis error propagation analysis
Dialog Management (DM)	**User friendliness** consistent, natural language comments help and error messages novice and expert mode
	Variety of dialog styles command lines pull-down menus dialogue boxes graphical user interfaces
	Graphical and tabular display visualization in the decision space (high-resolution cartographic displays) visualization in the decision outcome space (e.g. two and three-dimensional scatter plots and graphs, tabular rapports)

Source: Malczewski (1999)

4.4 Optimization Modeling

Optimization is a normative approach to identify the best solution for a given decision problem (Wilson *et. al.*, 1981; Thomas & Huggett, 1980). An optimization method is a modeling method that seeks to find the best (maximum or minimum) solution to a well-defined management problem. A well-defined problem is one, which has been structured in a way that the optimization method can utilize. Common to all optimization models is a quantity to be minimized or maximized. The quantity is often termed the objective or criterion function. The constraints define the set of feasible solutions. The solution to an optimization problem determines the values of decision variables subjective to a set of constraints. Thus, in the most general term an optimization model can be written as follows:

```
Minimize or maximize f(x)
Subject to x ∈ X
```

Where f(x) is a criterion function, x is a set of decision variables, and X is a set of feasible alternatives. If the problem involves a single criterion function, the problem is referred to as a single-criterion decision model. When more than one criterion function is to be optimized simultaneously, the model is called a multi-criteria problem.

4.5 Simulation Modeling for Decision Making

In the broad sense, simulation is a methodology for performing experiments using a model of the real-world system (Rubinstein, 1981; Mather, 1991; Englund, 1993). The primary difference between optimization and simulation is their starting point. Optimization procedures start with a definition of the system objectives and specify the actions that will satisfy those objectives at the optimum level. Once the optimum conditions are established, the vicinity of the optimal points is analyzed to determine the effect of variations in the system. Simulation modeling starts with the actions and studies their effects on the overall system objectives by testing different policies under various external conditions. Simulation is the exploratory approach to decision problems. It either reproduces a process or obtains a sample of many possible outcomes. Components of a system being simulated are defined mathematically and related to each other in a series of functional relationships. The result is a mathematical description of the complete decision process. The model is solved repeatedly using different parameters and different decision variables every time. As those values are changed, a range of solutions is obtained for the problem and the best solution is chosen from that range. This approach is similar in philosophy to post-optimality analysis, except that it is not restricted to the neighborhood of the optimum point.

Given that simulation is based on a mathematical model, two classifications of simulation approaches can be identified: static versus dynamic and deterministic versus stochastic (Rubinstein, 1981). A static simulation is one in which experiments are performed on a model having variables and parameters that are not time dependent. A dynamic simulation includes systems that change over time. Deterministic simulations involve variables and parameters that are fixed and known with certainty, whereas stochastic simulations assign probability distributions to some or all of the variables and parameters. This type of simulation provides a powerful tool in solving probabilistic problems, where the distribution of the final results is more important that a point estimate for the result. Such simulations are also sometimes referred to as Monte Carlo simulation because of their use of random variables (Openshaw & Whitehead, 1985; Openshaw, 1991; Fisher, 1991).

5. AN EXAMPLE OF METHODOLOGY DEVELOPMENT FOR SUSTAINABLE LAND USE AND WATERSHED MANAGEMENT IN RESPONSE TO CLIMATE CHANGE IN DONG NAI WATERSHED, VIETNAM

5.1 Dong Nai Watershed

The Dong Nai watershed of Vietnam is the largest national river basin and the economic center of the country in southern Vietnam. The watershed includes lowland areas that are subjected to annual flooding in the wet season and salinity intrusion in the dry season as well as mountainous highland areas of up to 1,600 m elevation. In addition, for administrative and planning purposes, a series of several smaller coastal basins are combined with the Dong Nai basin adding to a total surface area of 48,471 km^2 within Vietnam, or about 15 percent of the country's land surface area.

The Dong Nai watershed includes 11 provinces and Ho Chi Minh City (Figure 2). The Dong Nai watershed is highly developed, with a relatively low share of agricultural GDP and high income per capita and high population density, compared with other regions in Vietnam.

In Dong Nai watershed, large forest area has been replaced by the expansion of agricultural area, for food subsistence and then, for cash crop production, especially since the beginning of the "open economy" in 1980s. Traditional management systems for forest, land and water have been replaced by subsidiary state-run enterprises and agencies, which were not well motivated to enforce formal regulations and avoid it from becoming an open-access situation.

In reality, the rate of land use changes and forest resource depletion in upland ecosystems of Dong Nai watershed is alarming. The upland area has witnessed a rapid increase in population, resulting from massive immigration since the end of the war in 1975. In the beginning, poor landless

Figure 2. Dong Nai watershed map

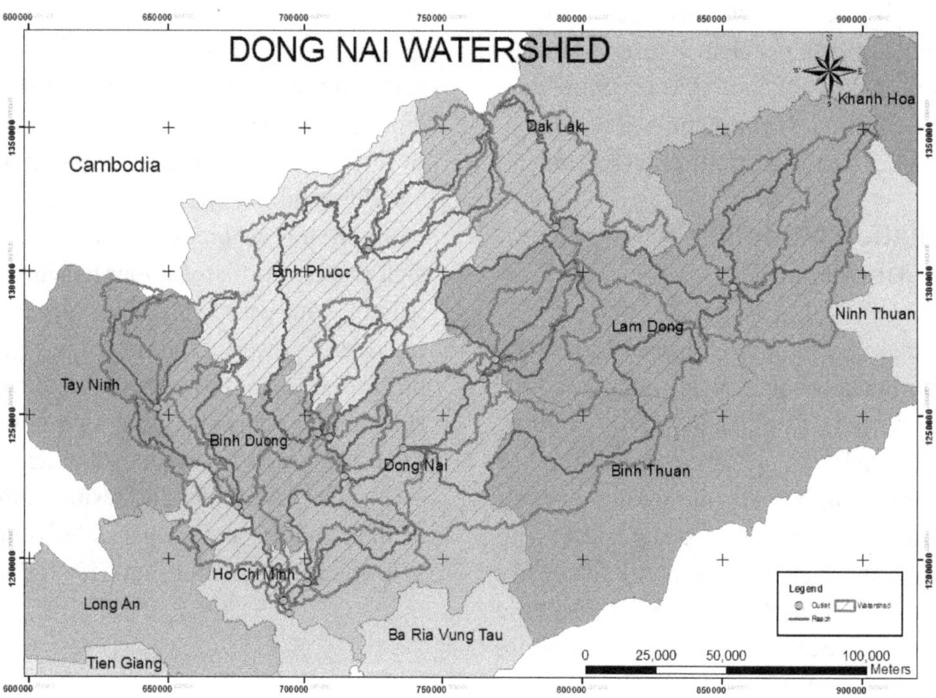

farmers from densely populated provinces came to the upland seeking livelihood alternatives under the national program with the creation of new economic zones (NEZ), or to work as hired labors in state-run forestry and agriculture enterprises.

In practice, the resource utilization in the watershed involves multiple objectives, many of which are incompatible or conflicting. Watershed management mechanisms therefore need to be analyzed in depth by considering economic, social, and environmental goals. Under the constraints of watershed resources and permissible ecological impacts, the effective and harmonious watershed management policies are needed to satisfy needs of both local communities and the national – regional governments. Development of such multi-objective plans requires the formulation of a mathematical programming technique or quantitative management approach, capable of quantifying the degree to which any proposed management meets objectives such as: (1) satisfactory net income, (2) desirable agricultural products, and (3) permissible soil loss and runoff.

5.2 Objectives

In order to formulate a sustainable land use and watershed management plan in response to climate change in Dong Nai watershed, the specific objectives of this study are as follows:

1. To assess land use/land cover change in Dong Nai watershed during the period from 1999 to 2009;
2. To determine the decision variable coefficients for sustainable land use;
3. To apply Linear Programming (LP) technique for optimizing land use allocation in Dong Nai watershed under the criteria of multiple objectives, limited resources, and permissible impacts to the water yield; and
4. To apply GIS and SDSS techniques for relocating the optimal land uses in response to climate change.

5.3 Methods

5.3.1 Approach to Problem Identification and Solution

Emerging water problems threaten the livelihood of local people and the sustainability of the whole watershed ecosystems in Vietnam. Coffee growers in the Western Highlands extracted water exhaustedly to save their trees in the El Ninõ year of 1997, the people in Central Coastal provinces lost their lives, livestock and household properties in the historical flood of 1999 and 2006; farmers in the Mekong Delta suffered a long lasting flood in 2000; these all are examples of recent water problems in Vietnam. The direct causes of these problems vary between locations and scale of the analysis but the common cause has been closely related to the development approach and the imbalance of trade-offs between conservation and development in watershed areas.

The dynamics of current development in Vietnam are results of Government interventions over the past 25 years. Uplands in Vietnam have been considered by Government authorities as a strategic region of the country and policies for their development have been implemented in two different directions. In one perspective, it is a region of watershed forests that need to be protected to ensure environmental security for the lowland area; typical example of policies under this context is the national program for fixed agriculture and resettlement, implemented since 1968 to stop swidden agriculture and to demarcate forest and agricultural land. In the other direction, the upland is regarded as "a sleeping princess to be awaked" and heavy investments have been made for redistribution of population and boosting agricultural production since 1975. As the result of these policies, uplands in Vietnam have undergone remarkable socio-economic and environment transitions.

The population has increased and the social structure has changed rapidly; demographic pressure of immigrations has usually been described as the motivation for this dynamics. Large forest area has been replaced by the expansion of agricultural area, for food subsistence and then, for cash crop production, especially since the beginning of the "open economy" in 1990s. Traditional management systems for forest, land and water were replaced by subsidiary state-run enterprises and agencies, which were not well motivated to enforce formal regulations that could to prevent them from becoming open-access resources.

At present in most parts of the country, similar scenarios of deforestation are unfolding. Forests are cleared; hillsides are planted. This has resulted in significant environmental degradation on a local, regional and eventually contributing at the global scale. This study concentrated on the abovementioned problems from the view point of physical disturbance due to resource degradation.

5.3.2 The Scenario Planning Process

The general methodology applied to this case study of Dong Nai watershed is shown in Figure 3. The principal planning task is aiming at the efficient planning of the future of watershed resources. The objective of each plan is to assist in deciding upon the socio-economic, physical and environmental data that are required in formulating the different planning scenarios. The derived objective is also used later in the methodology to evaluate the efficiency of each proposed planning scenario.

A number of socio-economic, physical and environmental data inputs are required to drive the land-use planning scenarios. The core socio-economic data inputs include: population, birth rate, death rate, immigration rate, and migration rate. The main physical and environmental data inputs include: water flow, land-use (forest land, agricultural land, special land, bare land, and

Figure 3. General methodology in developing SDSS for sustainable land use and watershed management in response to climate change impacts

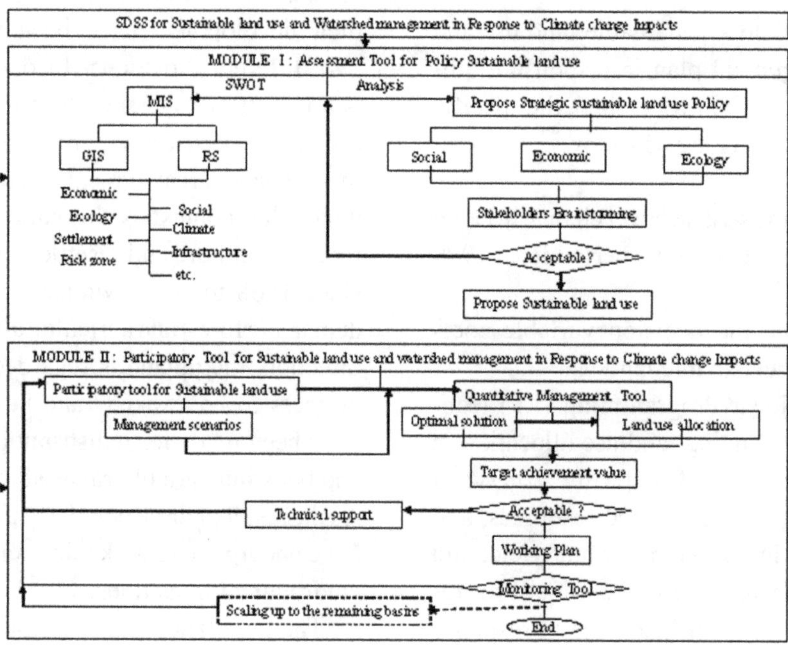

urban), and soil erosion, sediment, water discharge.

The next step of the planning process is to formulate possible land-use scenarios. Three land-use planning scenarios under climate change scenarios are formulated for Dong Nai watershed.

- Scenario A: "future trends" is based on existing socio-economic trends;
- Scenario B: "land allocation for maximizing economic" by using optimization modeling of land valuation data;
- Scenario C: "land allocation for sustainable land use in response to climate change" was derived using a number of environmental layers and assigning weighting scores to each layer by using a Multiple Criteria Analysis (MCA) approach.

As mentioned earlier, evaluation of each of the three land-use scenarios is undertaken using the mathematical/quantitative planning strategies. The objectives and policies contained within these strategies are used in evaluating the efficiency of each proposed land-use scenario, through the use of a Goals Achievement Matrix (GAM). The process of evaluation is iterated until a convincing desire is reached on the target. Finally, the end result of the scenario planning approach is the formulation of a final plan, to be reviewed accordingly.

5.3.3 Models for Predicting Annual Land-Use Changes

In order to obtain year-by-year land use changes, the Markov's Chain model was applied to determine probability of land use change based on the land use evolution between two given periods. The general form of the model to predict land use change from 1st date (year) to the 2nd date (year) is expressed in Figure 4.

Where γ_{ij}: is probability of change determined from overlaying of two different periods of land use map. This can be transformed (backwards) in general matrix multiplication as shown in Figure 5.

In this study, land use and land cover changes in Dong Nai watershed were carried out using modeling techniques recommended by Chunkao

Figure 4. Pathway of land-use change

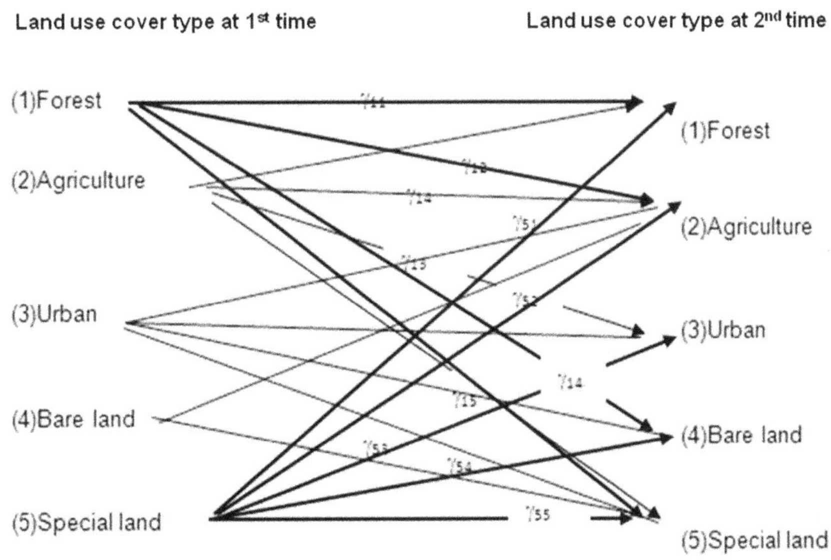

Figure 5. Markov's Chain Model

$$[V_1, V_2, \ldots, V_5]_1 * \begin{pmatrix} \gamma_{11}, \gamma_{12}, \gamma_{13}, \ldots \gamma_{15} \\ \gamma_{21}, \gamma_{22}, \gamma_{23}, \ldots \gamma_{25} \\ \vdots \\ \gamma_{51}, \gamma_{52}, \gamma_{53}, \ldots \gamma_{55} \end{pmatrix} = [V_1, V_2, \ldots, V_5]_2$$

(Proportion of land use of the first date) * (Matrix of probability of land use change) = (Proportion of land use of the second date)

(1992) and modified by Loi (2002, 2005). The steps in deriving year-by-year land use proportion are:

Land use /cover unit design

The term "Patch" (P), which is used to represent the homogenous appearance of plant community in the landscape that appears uniformly, was initially designed here in as:

P_1 = Forest land
P_2 = Agricultural land
P_3 = Settlement/ Urban
P_4 = Bare land
P_5 = Special land

Rule for Change Between Periods

Changes in land use and land cover in each Patch at any given time vary according to interaction between population, technology, education, economic and policy. In this study, at time t_1, area of each Patch is a function of a coefficient (c_i) at t_1 and the patch area (AP_i) at time t_o which can be simply written as:

$$(AP_1)_{t1} = c_1\ AP_{1(to)}$$
$$(AP_2)_{t1} = c_2\ AP_{2(to)}$$
$$(AP_3)_{t1} = c_3\ AP_{3(to)}$$
$$(AP_4)_{t1} = c_4\ AP_{4(to)}$$
$$(AP_5)_{t1} = c_5\ AP_{5(to)}$$

Where c_1 to c_5 is land use and land cover change coefficients; t = time;

AP_1 to AP_5 is area for P_1 to P_5.

Thus, the above equations can be generally re-written as:

$$(AP_n)_{(t+1)} = c_n\ AP_{n\ (t)}$$

For the year 1999, 2004 and 2009, the size of area under investigation is considered as a function of human activities which can be expressed as:

1990: $A_{(t1)} = AP_{1\ (t1)} + AP_{2\ (t1)} + AP_{3\ (t1)} + AP_{4\ (t1)} + AP_{5\ (t1)}$
1995: $A_{(t2)} = AP_{1\ (t2)} + AP_{2\ (t2)} + AP_{3\ (t2)} + AP_{4\ (t2)} + AP_{5\ (t2)}$
2000: $A_{(t3)} = AP_{1\ (t3)} + AP_{2\ (t3)} + AP_{3\ (t3)} + AP_{4\ (t3)} + AP_{5\ (t3)}$

$A_{(t1)} = A_{(t2)} = A_{(t3)}$ = total study area (Dong Nai Watershed area) from time t_1 to t_3.

Estimating Annual Change of Land-Use Units

Change in land use and land cover in each Patch at any given time (t_1) varies according to the change (Δ) of population, technology, education, economic and policy among the time interval ($t_o - t_1$). For this study, between the time $t_o - t_1$, the changes between different patches are expressed in Table 2.

Change of patch P_1, between $t_o - t_1$, to other land use can be logically expressed as:

$$(AP_1)_{t1} = c_1AP_{1(to)} = AP_{1(to)} - \gamma_{12}AP_{1(to)} - \gamma_{13}AP_{1\,(to)} - \gamma_{14}AP_{1\,(to)} - \gamma_{15}AP_{1(to)} + \gamma_{21}AP_{2\,(to)} + \gamma_{31}AP_{3\,(to)} + \gamma_{41}AP_{4\,(to)} + \gamma_{51}AP_{5\,(to)} \quad (*)$$

In the same manner, change of patch P_2, P_3, P_4, P_5, between t_o- t_1, to the others land use patches can be expressed as:

$$(AP_2)_{t1} = c_2\,AP_{2(to)} = AP_{2(to))} - \gamma_{21}AP_{2(to)} - \gamma_{23}AP_{2\,(to)} - \gamma_{24}AP_{2\,(to)} - \gamma_{25}AP_{2(to)} + \gamma_{12}AP_{1\,(to)} + \gamma_{32}AP_{3(to)} + \gamma_{42}AP_{4\,(to)} + \gamma_{52}AP_{5\,(to)}$$

$$(AP_3)_{t1} = c_3AP_{3(to)} = AP_{3(to)} - \gamma_{31}AP_{3(to)} - \gamma_{32}AP_{3\,(to)} - \gamma_{34}AP_{3\,(to)} - \gamma_{35}AP_{3(to)} + \gamma_{13}AP_{1\,(to)} + \gamma_{23}AP_{2\,(to)} + \gamma_{43}AP_{4\,(to)} + \gamma_{53}AP_{5\,(to)}$$

$$(AP_4)_{t1} = c_4AP_{4(to)} = AP_{4(to)} - \gamma_{41}AP_{4(to)} - \gamma_{42}AP_{4\,(to)} - \gamma_{43}AP_{4\,(to)} - \gamma_{45}AP_{4(to)} + \gamma_{14}AP_{1\,(to)} + \gamma_{24}AP_{2\,(to)} + \gamma_{34}AP_{3\,(to)} + \gamma_{54}AP_{5\,(to)}$$

$$(AP_5)_{t1} = c_5AP_{5(to)} = AP_{5(to)} - \gamma_{51}AP_{5(to)} - \gamma_{52}AP_{5\,(to)} - \gamma_{53}AP_{5\,(to)} - \gamma_{54}AP_{5(to)} + \gamma_{15}AP_{1\,(to)} + \gamma_{25}AP_{2\,(to)} + \gamma_{35}AP_{3\,(to)} + \gamma_{45}AP_{4\,(to)}$$

$(AP_1)_{t1}$ = Area of patch P_1 at time t_1
$(AP_1)_{to}$ = Area of patch P_1 at time t_0
c_1 = Coefficient of change for patch P_1 which is implicitly caused by human dimension in the study area during period t_o to t_1.
γ_{ij} = Coefficient indicating, probability of land use change from patch P_i to patch P_j.

In above equation (*) "plus (+)" indicates the transformation from Patch "P_2, P_3, P_4, P_5" to Patch P_1, and "minus (-)" indicates the conversion from Patch P_1 to Patch "P_2, P_3, P_4, P_5". In other equations, the same pattern of "plus (+)" and "minus (-) explain the transformation according to γ_{ij} and AP_i.

5.3.4 Estimating Soil Loss, Sediment Yield and Water Using SWAT Model

The Soil and Water Assessment Tool (SWAT) has been widely applied for modeling watershed hydrology and simulating the movement of non-point source pollution. The SWAT is a physically – based continuous time hydrologic model with Arcview GIS interface developed by the

Table 2. Matrix coefficient land use and land cover change between time t_o to t_1.

to \ t_1	P_1	P_2	P_3	P_4	P_5
P_1	γ_{11}	γ_{12}	γ_{13}	γ_{14}	γ_{15}
P_2	γ_{21}	γ_{22}	γ_{23}	γ_{24}	γ_{25}
P_3	γ_{31}	γ_{32}	γ_{33}	γ_{34}	γ_{35}
P_4	γ_{41}	γ_{42}	γ_{43}	γ_{44}	γ_{45}
P_5	γ_{51}	γ_{52}	γ_{53}	γ_{54}	γ_{55}

Blackland Research and Extension Center and the USDA-ARS (Arnold et al., 1998) to predict the impact of land management practices on water, sediment, and agricultural chemical yields in large complex basins with varying soil type, land use and management conditions over long periods of time. The main driving force behind the SWAT is the hydrological component. The hydrological processes are divided into two phases, the land phase, which control amount of water, sediment and nutrient loading in receiving waters, and the water routing phase which simulates movement through the channel network. The SWAT considers both nature sources (e.g. mineralization of organic matter and N-fixation) and anthropogenic contributions (fertilizers, manures and point sources) as nutrient inputs (Somura *et al.*, 2009). The SWAT is expected to provide useful information across a range of timescales, i.e. hourly, daily, monthly, and yearly time-steps (Neitsch *et al.*, 2002).

The SWAT model approach applied to the case study area of Dong Nai watershed is shown in Figure 6. The principal planning task is aimed at the efficient planning of Dong Nai watershed in future. The objectives of each plan are to assist in deciding the socio-economic, physical and environmental data that are required for formulating the different planning scenarios. The derived objectives are also used later in the methodology to evaluate the efficiency of each proposed planning scenario.

Impact assessment of changes in land use practices and human practices in Dong Nai watershed on soil loss, sediment yield and water yield to the Tri An reservoir during the period from 1999 – 2009 were conducted.

The SWAT model requires methodological data such as daily precipitation, maximum and minimum air temperature, wind speed, relative humidity, and solar radiation data. Spatial data sets including digital parameter layers such as parameters (R, K, C and P) and topography (LS) was digitized from the associated maps. LS factor of the watershed is derived from digital elevation model (DEM) obtained from topography.

5.4 Land Use Allocation Mapping

The main aim of this research is to, based on results from the three scenarios (Scenario A: future trends scenario, scenario B:land allocation for maximizing economic scenario, and scenario C: land allocation for sustainable land use in response to climate change), map the location of land uses, allocated by different scenarios using GIS techniques with the given criteria.

The linear programming (LP) and goal programming (GP) do not provide a spatial representation for the suggested land use allocations on how many hectares of each land use should be changed, and also do not indicate which specific hectares should be altered. Therefore, two approaches were employed for mapping new locations of allocated land uses.

In order to obtain the necessary information for setting up the above criteria, some of the land use types needed to be further analyzed and combined involving many parameters. This situation is well suited to the use of a GIS-technique. The ArcGIS program provides a broad set of functions to fulfill the requirements to this problem. After the analysis is performed, the program provides a value for the area which meets all the criteria.

Auxiliary variables are used to locate each land use change. According to linear programming and goal programming we know exactly how many hectares of each land use changes should be located. In other words, which grid cell should be selected to transform to other land use classes. The selection of these cells is performed using three criteria variables: slope, soil depth, and rainfall. Besides three criteria variables, we can set up a fourth criterion that is distance from the existing forest land to new forest land. This was used if above four variables could not meet the new forest land area we need. The selection of transition cells from any land use to a new

Figure 6. Application of SWAT model in Dong Nai watershed

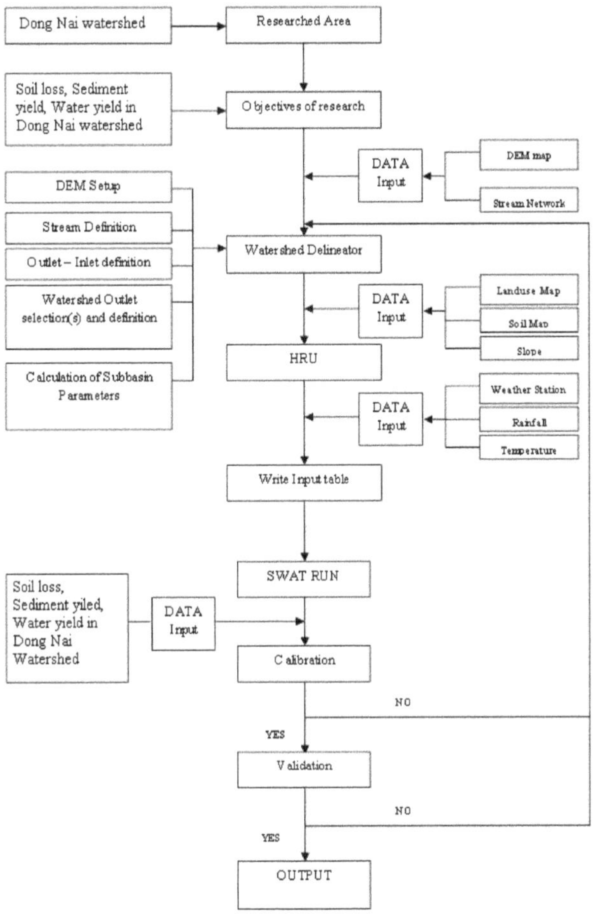

land use is performed in a similar manner. For example in the new additional forest land could be found out by setting the criteria: (1) minimal percent slope is 15%; (2) Soil depth should be in level 1 or level 2; and (3) Rainfall should be less than 1500 mm/yr.

6. CONCLUSION

The concept of sustainable land use and watershed management has always been important in natural resources management and still holds relevance in the context of climate change. Along with other environmental degradation problems, floods, drought and sea-level rises are the emerging hydrological problems in Vietnam due to climate change. The geographic information system in combination with the spatial decision support system can be useful for analyzing environmental degradation problem and determining suitable measures for sustainable land use and watershed management to cope with the impacts of climate change. The general methodology proposed for the study in Dong Nai watershed as implementing unit include formulation and evaluation of three land use scenarios using socio-economic and environmental data and mathematical/quantitative planning tools, so that land use maps for allocation can be generated. Later, the method can be used in other sites with modifications as needed.

REFERENCES

Arnold, J. G., Srinivasan, R., Muttiah, R. S., & Williams, J. R. (1998). Large area hydrologic modeling and assessment part I: Model development. *Journal of the American Water Resources Association, 34*, 73–89. doi:10.1111/j.1752-1688.1998.tb05961.x

Baker, J., Shepherd, S., & Edyvane, K. (1996). The use of marine fishery reserves to manage benthic fisheries, with emphasis on the South Australian abalone fishery. In R. Thackway. (Ed.), *Developing Australia's representative system of marine protected areas. Criteria and guidelines for identification and selection* (pp. 103–113). Proceedings of a technical meeting 22–23 April 1996, Adelaide, South Australia, Australia. Department of the Environment, Sports and Territories, Canberra, Australian Capital Territory, Australia.

Brooks, K. N., Peter, F. F., Hans, M. G., & John, L. T. (1992). *Hydrology and the management of watershed.* Iowa, USA: Iowa State University Press.

Christenson, N. L., Bartuska, A. M., & Brown, J. H. (1996). The report of the Ecological Society of America: Committee on the Scientific Basis for Ecosystem Management. *Ecological Applications, 6*(3), 665–691. doi:10.2307/2269460

Chunkao, K. (1981). *Introduction to watershed resources management in humid tropics. Regional training course vol 1-24. Watershed resources management and environmental monitoring in humid and tropical ecosystems.* UNESCO, USAID, MAB (USA), and NEB (Thailand).

Chunkao, K. (1992). *Watershed management.* A paper presented at the training program on natural resources management and conservation watershed (Applied Remote Sensing/GIS short course) at Asian Institute of Technology, Bangkok, September 7-11, December 1992.

Crossland, M. D., Perkins, W. C., & Wynne, B. E. (1995). Spatial decision support systems: An overview of technology and a test efficiency. *Decision Support Systems, 14*(3), 219–235. doi:10.1016/0167-9236(94)00018-N

Dasgupta, S. Laplante. B., Meisner, C., Wheeler, D., & Yan, J. (2007). *The impact of sea level rise on developing countries: A comparative analysis.* World Bank Policy Research Working Paper 4136, February 2007.

Densham, P., & Rushton, G. (1987). Decision support systems for locational planning. In Golledge, R. G., & Timmermans, H. (Eds.), *Behavioral modeling in geography and planning* (pp. 56–90). New York, NY: Croom Heelm.

Densham, P. J. (1991). Spatial decision support systems. In Maguire, D. J., Goodchild, M. F., & Rhind, D. W. (Eds.), *Geographical Information Systems: Principles and applications* (pp. 403–421). Harlow, Essex, UK: Longman.

Densham, P. J., & Goodchild, M. F. (1989). Spatial decision support systems: A research agenda. *Processing of GIS/LIS '89* (pp. 707-716). Orlando, FL.

Diane, M. (2002). *Community approaches to watershed management.* Centre for Rural Studies and Enrichment. St. Peter's College, Muenster. SK. Retrieved from http://www.saskriverbasin.ca/Conference/2002/Presentations/Diane%20Martz.ppt

Eastman, J. R., Jin, W., Kyem, A. K., & Toledano, J. (1995). Raster procedures for multi-criteria/multi-objective decisions. *Photogrammetric Engineering and Remote Sensing, 61*(5), 539–547.

Englund, E. J. (1993). Spatial simulation: Environmental applications. In Goodchild, M. F., Steyaert, L. T., & Parks, B. O. (Eds.), *GIS and environmental modeling: Process and research issues* (pp. 432–437). Fort Collins, CO: GIS World, Inc.

Fedra, K. (1993). GIS and environmental modeling. In Goodchild, M. F., Parks, B. O., & Steyaert, L. T. (Eds.), *Environmental modeling with GIS* (pp. 35–50). New York, NY: Oxford University Press.

Fedra, K. (1994). Model-based environment information and decision support systems. In Hilty, J. (Ed.), *Informatik für den Umweltschutz* (pp. 37–58). Metropolis Verlag, Marburg.

Fisher, P. F. (1991). Modeling soil map-unit inclusions by Monte Carlo simulation. *International Journal of Geographical Information Systems*, *5*(2), 193–208. doi:10.1080/02693799108927843

Fotheringham, S., & Rogers, P. (1994). *Spatial analysis and GIS*. London, UK: Taylor & Francis.

Fox, J. M. (1991). Spatial information resource management in Asia: A review of institutional issues. *International Journal of Geographical Information Systems*, *5*(1), 59–72. doi:10.1080/02693799108927831

Gardiner, J. L. (1994). Sustainable development for river catchments. *Journal of the Institution of Water and Environmental Management*, *8*, 308–319. doi:10.1111/j.1747-6593.1994.tb01109.x

Geertman, S., & Stillwell, J. (2002). *Planning support systems in practice: Advances in spatial science*. New York, NY: Springer Publishers.

Ghosh, A., & Rushton, G. (1986). *Spatial analysis and location-allocation models*. New York, NY: Van Nostrand Reinhold.

Gordon, J. (1993). Ecosystem management: An idiosyncratic overview. In Aplet, G. H., Johnson, N., Olson, J. T., & Sample, V. A. (Eds.), *Defining sustainable forestry* (pp. 240–247). Washington, DC: Island Press.

Grambine, R. E. (1994). What is ecosystem management? *Conservation Biology*, *8*, 27–38. doi:10.1046/j.1523-1739.1994.08010027.x

Hewlett, J. D., & Nutter, W. L. (1969). *An outline of forest hydrology. School of Forest Resources*. Georgia: University of Georgia Press.

IPCC. (2007). *Climate change 2007: Impacts, adaptation, and vulnerability. The fourth assessment report of the Intergovernmental Panel of Climate Change*. Cambridge, UK: Cambridge University Press.

Jermar, M. K. (1987). *Water resources and management*. Amsterdam, The Netherlands: Elsevier Science, Publishing Company Inc.

Keen, P. G. W., & Scott-Morton, M. S. (1978). *Decision support systems: An organizational perspective*. Reading, MA: Addison-Wesley.

Loi, N. K. *(2002)*. Effect of land use/land cover changes and practices on sediment contribution to The Tri An Reservoir of Dong Nai Watershed, Vietnam. *Unpublished M.Sc. thesis, Graduate School, Kasetsart University, Bangkok, Thailand.*

Loi, N. K. (2005). *Decision support system (DSS) for sustainable watershed management in Dong Nai watershed – Vietnam.* Unpublished Ph.D thesis, The Graduate School, Kasetsart University, Bangkok, Thailand.

Maguire, D. I. (1991). An overview and definitions of GIS. In Maguire, D. J., Goodchild, M. F., & Rhind, D. W. (Eds.), *Geographical information systems* (pp. 9–20). London, UK: Longman.

Maidment, D. R. (1993). GIS and hydrologic modeling. In Goodchild, M. F., Parks, B. O., & Steyaert, L. T. (Eds.), *Environmental modeling with GIS*. New York, NY: Oxford University Press.

Malczewski, J. (1999). *GIS and multi-criteria decision analysis*. New York, NY: John Wiley & Sons, Inc.

Mather, P. M. (1976). *Computation methods of multivariate analysis in physical geography*. Chichester, West Sussex, UK: Wiley.

Neitsch, S. L., Arnold, J. G., Kiniry, J. R., Srinivasan, R., & Williams, J. R. (2002). *Soil and water assessment tool: User's manual* (Version 2000). (GSWRL Report 02-02, BRC Report 2-06). Temple, Texas, USA.

Ongsomwang, S. (1995). *Fundamental of GIS*. Bangkok: Forest Resources Analysis Division, Forestry Academic Office, Royal Forest Department.

Openshaw, S. (1991). Developing appropriate spatial analysis method for GIS. In Maguire, D. J., Goodchild, M. F., & Rhind, D. W. (Eds.), *Geographical information systems* (pp. 389–402). London, UK: Longman.

Openshaw, S., & Whitehead, P. (1985). A Monte Carlo simulation approach to solving multi-criteria optimization problems related to plan making, evaluation, and monitoring in local planning. *Environment and Planning*, *12*(4), 321–334.

Ounon, T. (1990). The way to GIS development in academic institute. *Proceedings of the Conference Decumbent Entitled Application of Remote Sensing Information and GIS for Development and Management of Natural Resource* (pp. 10-17). Division of Natural Resource Assessment by Satellite, National Research Council, Bangkok.

Rubinstein, R. Y. (1981). *Simulation and the Monte Carlo method*. New York, NY: Wiley.

Satterlund, D. R., & Adams, P. W. (1992). *Wildland watershed management* (2nd ed.). New York, NY: John Wiley & Sons, Inc.

Somura, H., Hoffman, D., Arnold, J. G., Takeda, I., & Mori, Y. (2009). *Application of the SWAT Model to the Hii River Basin, Shimane Prefecture, Japan. Soil and Water Assessment Tool (SWAT): Global Applications*. World Association of Soil and Water Conservation. Special Pub. No.4.

Steyaert, L. T., & Goodchild, M. F. (1994). Integrating geographic information systems and environmental simulation models: A status review. In Michener, W. K., Brunt, J. W., & Stafford, S. G. (Eds.), *Environmental information management and analysis: Ecosystem to global scales* (pp. 333–355). London, UK: Taylor & Francis.

Thomas, J. W., & Huke, S. (1996). The forest service approach to healthy ecosystems. *Journal of Forestry*, *94*(8), 14–18.

Thomas, R. H., & Huggett, R. J. (1980). *Modelling in geography: Mathematical approach*. London, UK: Harper & Row.

Tomlinson, R. F. (1985). *An introduction to geographic information system*. A paper presented at the U.N. Seminar on the Role of Surveying. Mapping and Charting in Country Development Programming Aylmer, Quebec.

Vallega, A. (2002). The regional approach to the ocean, the ocean regions, and ocean regionalization – A postmodern dilemma. *Ocean and Coastal Management*, *45*, 753. doi:10.1016/S0964-5691(02)00104-7

Wilson, A. G., Coelho, J. D., MacGill, S. M., & Williams, H. C. W. L. (1981). *Optimization in location and transport analysis*. Chichester, West Sussex, UK: Wiley.

Chapter 17
Modeling of Current and Future State of Biodiversity in Central America Using GLOBIO3 Methodology

Denisse McLean R.
Biodiversity Modeling Project, IRBIO, Honduras

ABSTRACT

The modeling of the state of biodiversity in Central America using GLOBIO3 methodology was carried out by the Regional Biodiversity Institute for the Central American Commission on Environment and Development. For each country, current and future states of biodiversity under three socio-economic scenarios were explored. The country results were integrated into one regional assessment. The aim of this chapter is to explain how GLOBIO3 was adapted to the national scale. The main issues and the approaches adopted to solve them are described. The results from the Central American experience are presented followed by a discussion on main model limitations and derived recommendations. Finally, the challenges countries face to integrate the results into their government agendas are analyzed. This chapter is expected to be helpful for potential users of GLOBIO3 who are interested in the application of the methodology on a national and sub regional scale.

1. INTRODUCTION

Earth biodiversity is experiencing a series of accelerated deterioration, mainly due to human influence. Evidence shows that rates of extinctions have risen to historical levels and ecosystems ability to supply goods and services has been substantially reduced (United Nations [UN], 1993).

The most affected by biodiversity degradation are the people with fewer resources. These people usually depend directly on biodiversity and ecosystem services for their subsistence and they have fewer resources to deal with unfavorable environmental condition. As a result, bio-

DOI: 10.4018/978-1-60960-619-0.ch017

diversity degradation represents a major barrier for the achievement of the international objective of poverty reduction (Mertz *et al.*, 2007; World Conservation Union [IUCN], 2010; UN, 2010). In this context, the countries that integrate the Convention on Biological Diversity (CBD) agreed on 2002 to significantly reduce the rate of biodiversity loss at global, regional and national level (CBD, 2006).

To achieve this goal, a broader understanding of the effects of biodiversity loss is needed. Existing knowledge on biodiversity composition and functioning is only partial, same with our understanding on the mechanisms through which humans affect biodiversity and the consequences of the effects. Methods are needed to evaluate the state of biodiversity, to estimate future trends, and to evaluate potential intervention strategies considered by decision makers to achieve conservation goals. In particular, countries need to identify where they stand regarding the CBD target and which are the most efficient ways to achieve it.

This type of evaluation may represent a challenge for many countries. While having the political will to do it, they may lack the financial resources, technical capacity or the relevant information required. This is frequently the case of developing countries, where monitoring systems are weak, resources are scarce and there is still a lack of support for timely decision making. While many developing countries have high levels of biodiversity associated to their geographical location and to their areas of undisturbed ecosystems (United Nation Environmental Program [UNEP] & Central American Commission on Environment and Development [CCAD], 2006), both the population's pressing needs and the economic growth planned in government agendas are expected to be achieved at the cost of natural resource depletion. These conditions make fast, effective, and affordable approaches to biodiversity assessment in developing countries even more a priority.

The case study presented in this chapter is an example of such an approach. The modeling of the current and future state of biodiversity in Central America was carried out in 2009 and 2010 by the Regional Biodiversity Institute (IRBIO) for the Central American Commission on Environment and Development (CCAD). CCAD requested assistance from the Netherlands Environmental Assessment Agency (PBL) and UNEP to support the development of scientifically sound policy support tools as part of their commitment with the CBD. PBL extended its support by assisting in the implementation of a regional biodiversity assessment based on individual country models. GLOBIO3 methodology was used to evaluate the current state of biodiversity and to compare with the future state according to: (1) a baseline scenario with the projection of ongoing growth trends, (2) a policy option of the implementation of the Alliance for the Sustainable Development of Central America (ALIDES) and (3) a policy option of trade liberalization.

The aim of this chapter is to describe the application of the GLOBIO3 methodology in the Central American context. The specific objectives are to describe the adaptation process and to identify key success factors and main constrains. The downscaled methodology is explained and the main issues faced during the modeling experience and the approaches adopted to solve them are described. The results are presented, followed by a discussion on main modeling limitations and recommendations. Finally, the challenges countries face to update the model and integrate the results into their government agendas are analyzed. The chapter is expected to help potential users of GLOBIO3 methodology, such as government agencies, NGOs and institutions who are interested in the application of GLOBIO3 on a national or sub regional scale.

2. BACKGROUND

2.1 The Use of the Models

The dynamic and complex nature of biodiversity implies that for efficient conservation, current state assessments should be complemented with methods capable of capturing the underlying dynamics processes involved. In order to design adequate intervention strategies, decision makers have to rely on models to predict the potential reaction of biodiversity to the drivers of change.

Biodiversity models have been developed on the basis of different paradigms for different research objectives (Rounsevell *et al.*, 2006). The objectives depend on the spatial and temporal scope at which results will be applied and on the level of decision making concerning output users. Common research priorities in modeling have included the study of trends, evaluation of potential conservation scenarios and identification of key areas of degradation (Parks *et al,* 2004).

Models, by definition, focus on limited aspects of biodiversity components. Aspects such as species distribution, richness, abundance or ecosystem extent and quality are evaluated under certain conditions to characterize and compare between occurring and hypothesized situations (Spangenberg, 2007). Biological surrogates are commonly used as proxies of larger groups of elements (species, ecosystems or others). Each model has to be specific about the aspect of biodiversity that is being addressed and about the scope and limitations of selected indicators (Haynes-Young, 2009). Although not exhaustively, models have provided reasonable outputs that support decision makers on the development and selection of management practices that are suitable for their conservation priorities.

Still, the levels of decision making of national, regional and sub regional authorities –instead of, for instance, protected area or wildlife managers- require more general indicators and analytical frameworks. These instances deal with practical issues such as environmental laws, regulations and funding priorities. As so, they could benefit from models with more straightforward application and interpretation methods.

Integrated modeling frameworks have been developed to assess the overall state of biodiversity in terms of simple indicators. Among these are the Biodiversity Integrity Index (Majer & Beeston, 1996), the Biodiversity Intactness Index (BII) (Scholes & Biggs, 2005), the Living Planet Index (LPI) (World Wildlife Found [WWF], 1998) and the Mean Species Abundance index applied in this case study through the GLOBIO3 methodology.

2.2 GLOBIO3

The GLOBIO3 biodiversity model was developed by PBL and the GRID-Arendal center of UNEP. The methodology estimates the impact of human drivers on biodiversity in terms of the single indicator of Mean Species Abundance or MSA. The MSA represents the mean abundance of original species of an area relative to its abundance in pristine or undisturbed ecosystems (Alkemade *et al.*, 2009). It is a measure of the intactness or naturalness of a location (GLOBIO, 2010), as it is expressed as a proportion (0 to 1) or a percentage (0% to 100%) of remaining species abundance from the abundance in original state.

The MSA is affected by the combination of a set of selected drivers. Five human induced drivers of biodiversity degradation have been identified and included in the methodology: land use change and intensity, road infrastructure developments, natural area fragmentation, climate change and atmospheric nitrogen deposition. See details in Chapter 8.

2.3 CLUE

When modeling the future state of biodiversity, once the scenarios have been characterized, Pressure – State – Response (PSR) framework considerations are integrated into a driver model to obtain quantitative estimates of the GLOBIO3 inputs in the future. In the global methodology, the IMAGE model (Integrated Model to Assess the Global Environment) is used (Bouwman *et al.,*2006). For the land use driver, estimated changes need to be spatially allocated before integration. For this purpose the CLUE (Changes in Land Use and its Effects, Verburg *et al.*, 2002) modeling framework is used.

The CLUE model combines the principles of local suitability and dynamic competition between land uses to allocate the land use demands estimated in scenarios (see more details in chapter 6). Previous work with CLUE model has been done in Costa Rica, Ecuador, Java, China, Honduras (Kok & Winograd, 2002) and in the Mesoamerican Reef (MAR) region (Luijten *et al.*, 2006). Although the objectives of these studies have not always been linked to biodiversity -they have included, for instance, evaluation of land use dynamics or land use change impacts in hydrological models- these studies have validated the application of the CLUE model at various scales (Verburg *et al.*, 2006).

2.4 Model Advantages

The advantages of GLOBIO3 framework in combination with the CLUE model include its transparency, ease of replication, low input requirement (Verboom *et al.*, 2007) and most importantly its focus on the drivers and pressures of biodiversity loss and on the forces that shape those drivers. Instead of focusing on how biodiversity is being affected in certain components, the methodology focuses on explaining why biodiversity is changing in terms of the main drivers of that change and enables the assessment of the relative contribution of each driver to the total loss (Alkemade *et al.*, 2009). Spangenberg (2007) explicitly suggested a pressured based approach in order to provide relevant information to decision makers to prevent further biodiversity losses. Drivers and pressures are easier to measure than the biodiversity components themselves and they can serve as the interface between the socioeconomic forces behind them and the impact they exert.

Drivers included in GLOBIO3 have been identified as significant causes of biodiversity degradation in numerous studies. The consequences of land use intensity have been extensively documented (Reidsma et al., 2006; Haines-Young, 2009); mainly the effects of conversion of natural forest covers into cultivated or urbanized lands (Fahring, 2003; Lambin *et al.*, 2003; Chazal & Rounsevell, 2009). Climate change and eutrophication were recognized after land use as the most important causes of biodiversity degradation in the Millennium Assessment (Eickhout *et al.*, 2007). Young *et al.* (2005) identified resource overexploitation and agricultural intensification as main threats for biodiversity. Moreover, Spangenberg (2007) identified intensive land use, high energy consumption (with resulting climate change) and habitat fragmentation by new road infrastructure driven by economic expansion, as the primary driving forces affecting biodiversity in all its components.

In addition, GLOBIO3 offers an integrated framework to study land use and climate change pressures in biodiversity assessment. Chazal & Rounsevell (2009) explained that the characterization of the relationship between these two drivers and biodiversity is currently limited by the lack of "process understanding, data availability and inherent scenarios uncertainties". It was suggested that any assessment on biodiversity state should at least include land use and climate change drivers as determinants.

Applications of the GLOBIO3 framework have been completed at the global (sCBD, 2006; sCBD & PBL, 2007; Alkemade *et al.*, 2009) and regional scale (Verboom *et al.*, 2007). These applications have helped to raise awareness on the consequences of development pathways on biodiversity at the international level. A review by Leemans *et al.* (2007) validated the scientific soundness of the methodology and determined it was suitable to provide information on depictions of current and future biodiversity trends at global and regional scales. Nonetheless, attention has been given to downscaling the methodology in order to assist other levels of decision making (Verburg *et al.,* 2006).

There is an imperative need to integrate biodiversity conservation as a transversal issue in a wide range of policy domains if substantial progress in conservation and management is to be achieved (Spangenberg, 2007). Leve & Mounolou (2003) stated that biodiversity should be "a framework for considering the whole range of questions raised by human relationships with other species and the environment, a mediator between ecological and social systems". Since it is human needs and interventions that shape the drivers of biodiversity loss (O'Rourke, 2006; Young *et al.*, 2005), it is essential that the policy domains defining these interventions consider biodiversity as a variable. It has been suggested to integrate the biodiversity debate into sectors such as agriculture and land use in order to search for the alternative long term policies that will be compatible with biodiversity conservation (Mattison & Norris, 2005; Poschlod, Bakker & Kahmen, 2005; Rounsevell *et al.*, 2006).

2.5 Modeling on National Scale

In that sense, downscaling the global GLOBIO3 methodology represents four major contributions: First, a finer spatial resolution improves the understanding of processes that are location specific and more dependent on the spatial pattern of land use, such as connectivity between natural areas and land use transition sequences (Verburg *et al.*, 2006). These processes are poorly assessed when addressed at the global scale, where their relationship with biodiversity degradation could render weaker or even non-existent (Kok & Veldkamp, 2001; Chazal & Rounsevell, 2009). Second, the downscaling procedure makes the integration of country specific information on scenarios and drivers possible, which renders outputs that are more applicable for national policy considerations. Third, a reduced spatial extent, taking the country level as the unit of analysis, is more suitable to explore the specific relationships between drivers that vary between contexts. Studies have found that these relationships can vary significantly between administrative units (Kok & Veldcamp, 2001; de Koning *et al.*, 1998; Verburg & Chen 2000; Verburg *et al.*, 2006). In global and regional assessments, unrealistic interactions between drivers emerge from the aggregation of spatial units. This reduces the applicability of global assessment results for national policy support. Finally, stakeholder involvement could also be better addressed at the country level, since national administrative units are the relevant framework for the definition of policy priorities and for deliberate application of transversal policy instruments. Leemans *et al.* (2007) and Mertz *et al.* (2007) stated that stakeholder involvement in policy valuation and monitoring should be strengthened, as a key element in the process of biodiversity conservation.

In this study GLOBIO3 methodology has been downscaled for application at a sub regional scale build upon national scale assessment for Central America. Other downscaled national and sub regional GLOBIO3 applications have been carried out in Nicaragua (Lezama-López, 2007), Vietnam (Than So, 2008), Mexico (CCAD & IRBIO, 2009) and Thailand (Trisurat *et al.*, 2010). The methodology has proved to be adequate for context with limited data, time and resource availability.

3. APPLYING GLOBIO3 AT THE CENTRAL AMERICAN CONTEXT

3.1 Study Area

The Central American region is comprised by seven countries -Guatemala, Belize, Honduras, El Salvador, Nicaragua, Costa Rica and Panama- within a relatively small extension of approximately 524,000 km^2. The countries share several geographic characteristics such as climate, topography, soil, vegetation, economic conditions and population dynamics. Along with the southern part of Mexico, the region comprises the Mesoamerican biodiversity hotspot. It concentrates a great number of species including 17,000 plant, 440 mammal, 1,100 bird, 700 reptile and 550 amphibian species in 0.35% of the world territory. The region has high levels of endemism as a result of its transitional location between the North and South America continental masses (Conservation International [CI], 2007).

Central America has experienced an accelerated process of resource consumption and environmental degradation that started with the establishment and expansion of cities after the colonization period. The process continued with the intensification of agriculture and the establishment of coffee, bananas and palm plantations through the western Pacific plains in the 1900 and was followed by the boost of timber extraction from tropical and subtropical forests from the Atlantic region in the XX century (CI, 2007). Population expansion and inequalities in the distribution of assets have intensified the situation. Central American countries are currently concentrating efforts on evaluating the state of their resources and the potential pathways to achieve a significant level of conservation and sustainable management. A main priority in this sense is to give more relevance to biodiversity on environmental and other sectors' policy discussion.

3.2 Methodology

The case study application was based on the GLOBIO3 and CLUE-s frameworks described by Alkemade *et al.* (2009) and Verburg *et al.* (2002), respectively. According to the methodology, each of the drivers included in GLOBIO3 is related to independent variables that correspond to human pressures on biodiversity.

For downscaled studies, the methodology consists in collecting spatial information on the independent variables from the area under study, reclassifying their local values into the GLOBIO3 impact categories, and assigning the corresponding MSA impact values to drivers. National land use/land cover, road infrastructure, ecosystem and population density maps were used to estimate the impact of land use, infrastructure and fragmentation drivers. For climate change and nitrogen deposition no local information was available and data had to be derived from global models.

To estimate the pattern of land use distribution in 2030 with the CLUE model, specific inputs had to be prepared. The most important inputs were the demand tables and the land use suitability or location factors maps.

A demand table expresses the estimated amount of area to be occupied by each land use category of the region under study in each year of the simulation. In other words, it represents the distribution of the area of a region between its different land use classes. Demand tables were built from the qualitative storylines described in scenarios.

Land use suitability maps were obtained from a regression analysis. The factors that are expected to influence the occurrence of land use classes were evaluated for each country. Soil (chemical and physical properties), geographic (altitude, rainfall, temperature, slope), socioeconomic (distance to towns, distance to rivers) and demographic (population density) factors were included. Previous studies by Kok & Veldkamp (2001), Kok (2001) and Kok & Winograd (2002) in the Central American region have shown that these factor categories are determining and equally important to explain the occurrence of land use classes and recommended their integration into national assessments. The factors with significant beta coefficients at 95% confidence level were used as model inputs to characterize location suitabilities for the different land use types. To verify the prediction capacity of factors a Relative Operator Characteristic (ROC) analysis was done. Pontius & Schneider (2001) validated that ROC of 65 to 70% were significantly better than random and were considered acceptable. In the downscaled procedure a modified version of the model adapted for small regional extents –the CLUE-S (Verburg, 2006) - was used. Future land use maps derived from the CLUE-S modeling were used as inputs in GLOBIO3 to estimate the future impact of land use, infrastructure and fragmentation drivers on biodiversity.

Inputs were analyzed using ArcGIS software in a 1*1km spatial resolution. O'Rourke (2006) stated that biodiversity measures tend to be strongly dependent on the spatial and temporal scales chosen since every phenomenon has its own emerging properties at different levels. Analyzing effects at various scales simultaneously was suggested. This was not possible in this case study due to time and resource constraints. However, while finer resolutions have been used in previous land use change studies (Lujten *et al.*, 2006), the 1km resolution was considered adequately. This resolution was considerably finer than the 0.5*0.5 degrees applied in global modeling and it has been a resolution frequently applied in this type of downscaled assessments (GLOBIO, 2010). The studies by Kok & Veldamp (2001) verified that in the Central American context no significant changes take place in the selection of the determinant factors of land use occurrence when coarsening resolution. Furthermore, they verified that a coarser resolution increased model performance and gave more explanatory power to poor data quality, as long as uniform administrative units remained the largest extent of analysis.

For each country, the modeling of the current state was done for the year of the latest land use/ land cover map available. A time horizon of approximately 30 years (up to 2030) was selected for future state modeling to adequately assess the impact of the different drivers without introducing high levels of uncertainty to results.

3.3 Main Issue in Case Study

The main issue in this case study was how to adapt the GLOBIO3 methodology to the Central American context. While region members have common goals and expectations regarding the assessment, each country is still an independent administrative unit with its own laws, regulation and jurisdiction. As explained above, the national level is the most suitable unit of analysis in environmental and socio economic downscaled assessments. For that reason, it was chosen to perform individual country evaluations and to integrate them afterwards into one regional analysis.

Adjustments were needed to achieve this purpose. Seven major adaptation areas were identified.

1. Land use MSA values revision: Existing scientific literature findings tends to be insufficient or unrepresentative for regions where little research and monitoring has been done (Leemans *et al.*, 2007). This is usually the case for developing countries. Consequently, the general MSA values of the global land use impact categories were not always applicable for particular classes found in national land use maps. For instance, some land use classes did not fit into any of the global GLOBIO3 land use categories; other classes did fit but understanding of prevailing local conditions revealed that corresponding MSA value was not suitable; in other cases, the map classification was too general and classes with different MSA values were aggregated into a single class. Therefore, MSA values had to be revised by local experts on land use effect on biodiversity.
2. Scenario design: In global applications, scenarios are based on IMAGE model outputs. IMAGE model implementation requires great amount of input including demographic, economic and technological developments as major driving forces shaping scenarios (Leemans, 1999). These inputs are usually not available in developing countries. For the downscaled procedure, simpler approaches for scenario building

have been suggested (Luijten *et al.*, 2006). These approaches take advantage of already developed local expertise by developing qualitative scenario storylines based on expert knowledge and available time series data which are later transformed into the quantitative inputs for CLUE.

3. Demand table building: Taking the previously mentioned approach for scenario building represented further complexities for the constructions of CLUE's demand tables. Figures supporting storylines were not as categorical as they would be in stricter scenario building approaches. There was a lack of consistent information on land use history and projections available for all countries and all land use classes. Baseline information had to be obtained from different sources, each using its own assumptions for determining expected changes. Moreover, baseline information on land use classes' areas not always matched land use maps figures, a condition that had to be observed when applying percentages of change to base areas to estimate land use projections.
4. Integration of country assessments: For each country, maps and aggregated figures of remaining MSA and biodiversity loss per pressures were obtained as outputs. The modeling team had to decide on how to integrate these results into the regional assessment. This was the first time a regional evaluation was carried out based on individual country studies.
5. Team organization: All the above considerations meant that the implementation of the methodology required a combination of expertise from a range of disciplines. Geographic information system knowledge was required for spatially explicit procedures of drivers' layer combination, CLUE input preparation and model output presentation. Thoughtful knowledge on the biodiversity of the region and the characteristics of land use systems was needed for the MSA value revision. An understanding of the dynamics of socioeconomic and political forces and the mechanisms through which forces affect biodiversity was needed for storyline building. In addition, knowledge of each country was needed to interpret and validate the results.
6. Stakeholder involvement: Direct stakeholder involvement in the assessment process was crucial to effectively transfer model results and effectively integrate them in national and regional policy discussion. It was also important to gather the relevant country specific information needed for each assessment.
7. Capacity building: There was also a need to develop scientific and technical capacity, both in the team in charge of the first modeling approach and in the technical staff in charge of model updating and continuation in each of the countries.

3.4 Solution Approaches

To deal with issues described above, the following approaches were adopted:

3.4.1 MSA Value Revision

A list of the land use classes contained in national land use maps was delivered to a scientific team. This team was integrated by experienced biologists, with demonstrated knowledge of the land use systems of the region. The experts reviewed previous scientific studies containing information on local species abundance and ecosystem characteristics. The information was complemented with their empirical knowledge.

Land use map classes were reclassified according to GLOBIO3 land use driver impact categories. For classes that did not fit any impact categories, different MSA values were assigned according to their relative intensity compared to the global categories. For example, the class know as "tacotal" or "charral" in Nicaragua and Costa Rica land use maps was left as a separate class known as "agro silvopastoral" -a land use system that combines fallow, low input agriculture, extensive livestock grazing and wood production- and an MSA value of 0.5 was assigned. When classes fitted a category but were known to have a different MSA value, a new value was assigned, as in Guatemalan forest plantations that received the MSA value of 0.4 instead of the standard 0.2 due to their known management system. If classes were too general, an appropriate MSA value was estimated by analyzing class description and the location and area extension the class occupied in the map. This was the case of Honduran land use map that had a single grassland class and did not distinguish between cultivated and natural grasslands. The class was assigned an MSA of 0.7 and was finally classified as "Livestock grazing".

Experts assigned MSA values to all original land use maps classes, combining only repetitive categories, as for example citric/banana/oil palm plantations under the "perennials and bio-fuels" category. However, although experts had MSA values for all land use classes, it was impossible to integrate them all in future biodiversity modeling due to limitations faced in scenarios and demand table building. Since current and future state assessments had to be comparable within each country, certain classes had to be regrouped. To avoid the loss of the revised MSA values, area weighted averages of the MSA values of aggregated classes were calculated (Table 1).

3.4.2 Scenario Building

A team of regional socioeconomics and policy experts was asigned the task of developing scenario storylines and collecting the available quantitative information to support them.

Scenarios are systematically organized perception about alternative future settings in which present situation may unfold (Agriculture Ecosystem & Environment, 2006). They are created as internally consistent storylines (Shearer, 2005) that are latter transformed into quantitative variations of a selected set of parameters, generally by using subjective expert judgments (Abildtrup *et al.*, 2006). In the context of GLOBIO3, the main objectives of scenarios were to identify key factors of biodiversity degradation, to evaluate potential courses of intervention (Verburg *et al.*, 2006) and to integrate these findings in the modeling process in the form of CLUE input demand tables.

Table 1. MSA values for the land uses of the Central American countries

Land Use Class	Guatemala	Belize	Honduras	El Salvador	Nicaragua	Costa Rica	Panamá
Primary forest	1.00	1.00	1.00	-	1.00	1.00	1.00
Forest plantation	0.40	-	-	0.20	0.31	-	0.20
Secondary forest	0.50	-	0.70	-	0.40	0.64	0.49
Light used forest	-	-	-	0.70	-	-	0.70
Agro forestry	0.20	-	-	0.40	0.50	0.40	-
Extensive agriculture	0.30	0.30	0.30	0.30	0.20	-	0.30
Intensive agriculture	0.10	-	0.10	0.10	0.10	0.17	-
Natural grasslands	0.40	-	-	-	-	-	-
Cultivated grasslands	0.05	0.10	-	0.10	0.10	0.10	0.10
Livestock grazing	-	-	0.70	0.70	-	-	-
Agrosilvo pastoral	-	-	-	-	0.50	0.50	-
Others	x	x	X	x	x	x	x

In this study, a development scenario and two policy options were explored: (1) Baseline scenario represented the persistence of the ongoing growth pattern mainly related to population and economic dynamics. (2) ALIDES option represented the implementation of the Alliance for the Sustainable Development of Central America, an agreement between countries to foster sustainable growth through regional cooperation and incentive mechanisms. (3) Trade liberalization option represented the implementation of the Central American Free Trade Agreements.

The drivers of biodiversity degradation in each scenario were determined using the PSR framework. The storylines were supported with quantitative information on drivers' past behavior –for Scenario 1- or with projections of future drivers' evolution –for Policy options 2 and 3-. Information was obtained from regional socioeconomic reports, environmental outlooks, statistical databases, central bank registries, journals, magazine articles, press releases and original agreement documents. The sources mostly contained information on how macroeconomic changes modified the area demands of different land use types. No agent based method to characterize storylines and determining demands was explored. The storylines and main underlying assumptions were specified for each scenario (Table 2).

Not every aspect of scenarios original storylines was integrated in the quantitative estimates. In the context of the methodology, scenarios were required to focus on the variation of region's land uses. The aspects that did not influence land use variation directly had to be disregarded. For example: strengthened democracy in ALIDES scenario. Others aspects' indirect influence had to be translated in variation of a land use category. For example: population growth interpreted as an increased demand for food, and consequently for agricultural land uses.

Given the varied character of storylines and data sources, quantitative support information was not presented uniformly. Baseline scenario figures were expressed as yearly variation of the area occupied by existing land use classes. The other policy options expressed variations as a relative deviation from the baseline projections. This assumption contributed to reduce scenarios inconsistencies by restricting changes to reasonable estimates. The differences in results would reflect differences in scenarios instead of differences in underlying databases. But both policy options were independent from one another and disregarded each other's effect completely. In addition, all three scenarios assumed undisturbed development of events. No political disturbance or natural event occurrence was considered.

3.4.3 Transforming Scenario Figures into CLUE's Demand Tables

Instead of using the IMAGE model, the tables were built directly through the application of scenario variation figures to the areas of the original classes from national land use maps. For this purpose, a third team was organized. The modeling team was integrated by geographic information system specialists who had the task of organizing input data, preparing it according to methodology, executing the models and presenting the results.

Demand table building did not follow a strict methodology. Instead, the process had to be highly flexible for two main reasons: (1) the variable nature of input data and (2) the lack of coincidence between scenario and country maps' base areas. The official figures of extension area occupied by land uses are generally reported only for macro economically important classes, such as agriculture and livestock production. Land uses classes with less macroeconomic relevance (e.g. subsistence agriculture, fallow lands) tend to have no records, or records tend to be incomplete, since not all the areas dedicated to those land uses are reported. When scenario figures were compared to the areas occupied by the same classes in country land use maps, the figures differed significantly.

Table 2. Scenarios of the Central American case study with main assumptions and input data display

	Baseline Scenario 2030	ALIDES 2030	Trade Liberalization 2030
Main assumptions in storylines	Population growth. Increased migration.	Primary sector growth through to the transformation of traditional agriculture and grasslands into complex multi level production systems.	Increased access to markets for goods and services generated in the region.
			Export growth and diversification.
	Remittances, tourism & clothes assembly factories become main sources of income.	Promotion of reforestation and sustainable forestry production. Secondary sector growth from the processing of primary sector production. Sector becomes main source of non agricultural income.	Increased access to goods and services manufactured outside the region. Import and foreign investment growth.
	GDP and imports growth	Sustainable growth of tertiary sector (ecotourism and environmental services).	Laws and regulations to incentive this dynamic.
	Primary sector is reduced.		Exports: agriculture, bio-fuels and intensive silviculture. / Hydroelectrically, tourism and other services. / Clothes assembly industries.
	Agricultural land is destined to export crops.	Increasing effects along time.	Imports: Food, fossil fuels. Investments: infrastructure, urban expansion, communications.
	Permanent crops and pastures remain relatively constant.Fixed relative effects along time,		Decreasing effects along time.
Input data display	Fixed % yearly area variation per country for:	Variable % yearly increase of	Fixed % period variation per country (variable between periods) for: Primary forest,
	Forest cover, Forest plantations, Extensive agriculture, Intensive agriculture & Grasslands.% yearly variation is applied to area from each previous year.	Secondary forest for the entire region. % yearly variations are applied to Base Line scenario projections.	Secondary forest, Forest plantations, Extensive agriculture, Intensive agriculture & Grasslands.
	Land use maps areas are used as base (year 0).	Secondary forest increases are compensated with grasslands or intensive agriculture reduction.	Three periods: year 0-2015, 2016-2020, 2021-2030. % period variations are applied to Base Line scenario projections.

Demand tables were built, following a series of general steps: First, a revision of both land use maps and scenario figure classification system was done. This implied identifying any classification premises or assumptions undertaken in scenarios. For example, in Baseline scenario there was no distinction between primary and secondary forest, and in Trade liberalization scenario grassland variations applied only to cultivated grasslands. It also implied identifying limitations in national land use map classification systems. For instance, the "agriculture/livestock" class in the map from Panama combines classes with different corresponding MSA.

Then, a comparison of the areas occupied by each land use class in both systems was done. If possible, the areas were matched by reallocating certain classes in a different category. For example, some classes from El Salvador map such as "Agriculture and grassland mosaic" or "Agriculture, grassland and vegetation mosaic" did not clearly belong to a category. So they were reallocated according to the areas of extensive agriculture and grasslands found in the scenario figures.

The next step was to aggregate minor classes of small extension, classes that experienced little variation and classes for which no scenario figures were available. For example, the "perennials and bio-fuels" category was combined with "extensive agriculture", since no specific information of its variation was found. Water bodies, natural bare areas and build up areas were aggregated under the "Other" category. In the CLUE model, they are considered to remain constant throughout the simulation. The considerations on each reclassification varied for every country, depending on the level of detail of the land use maps and the differences between the map and scenario figures (see example on Table 3).

Finally, the percentages of change from scenarios were applied to the CLUE categories. The tables were fitted so that for each year described in the table, the sum of the area of all land uses was equal to the total country area. In most of cases, the sum exceeded the total area. It was assumed that this surplus was a consequence of a reduction in natural areas. Thus, the numbers were subtracted from forest or natural grassland categories.

It was important to maintain, in as much as possible, the MSA values assigned to the original land use categories. Since the "Other" category remained constant in CLUE, the category was latter disaggregated into its components to assign corresponding MSA values.

3.4.4 Integrating Country Results into the Regional Assessment

A simple approach was adopted: A regional map was built by combining country maps in a single layer. Maps were combined for display but country information was not merged. Regional figures were estimated by integrating the quantitative results of remaining MSA and shares of biodiversity loss of the seven countries (derived from the Access query) in each scenario through area weighted addition. So that:

$$\text{MSA or Drivers' share}_{\text{Region}} =$$
$$\sum\left[\text{MSA or Drivers' share}_{\text{Country}} *(\text{Area}_{\text{Country}}/\text{Area}_{\text{Region}})\right]$$

3.4.5 Coordinating Team Work

Three teams with different areas of expertise were working simultaneously to complete the assessment: the MSA team, the scenario experts and the modeling team. Many tasks were fulfilled by independent consultants. For best model implementation and output interpretation team members came from and were located in different countries of the region.

This working scheme represented one of the main challenges for sharing and processing information. Regular contact was maintained by electronic mail, periodic team meetings were held to update team members on modeling progress. Still, difficulties emerged when partial results had to be discussed, particularly if assumptions adopted had not been made explicit.

3.4.6 Establishing Stakeholder's Involvement

Stakeholder's involvement was well established from the beginning. Countries' national environmental authorities through their regional Environmental Commission were interested in the project initiative. Countries had a strong interest in modeling results to assess their progress towards biodiversity conservation goals, mainly since the modeling exercise was a first approach in the implementation of a scientifically sound evaluation methodology in the region. Once the modeling phase concluded, countries had the in-

Table 3. Land use reclassification for the GLOBIO3 downscaled methodology. Example from Nicaragua.

Original Land Use Map Classes	Area (Km2)	GLOBIO3 Classes	MSA	CLUE Classes
Brushwoods and grasses	10509	Cultivated grasslands	0.70	Cultivated grasslands
Open broadleaf forest	9591	Primary forests	1.00	Primary forests
Cultivated grasslands	11497	Cultivated grasslands	0.10	Cultivated grasslands
Tacotal (Fallow lands)	33239	Agro silvopastoral	0.50	Agro silvopastoral
Yearly crops	4128	Extensive agriculture	0.20	Extensive agriculture
Flood lands	3451	Primary forests	1.00	Primary forests
Inhabited areas	330	Built up areas	0.05	Others
Water	673	Water bodies	1.00	Others
Forest/Palm	726	Primary forests	1.00	Primary forests
Closed broadleaf forest	30196	Primary forests	1.00	Primary forests
Cities and towns	222	Built up areas	0.05	Others
Citrics	30	Extensive agriculture	0.20	Extensive agriculture
Bamboo	52	Primary forests	1.00	Primary forests
Mangroves	712	Primary forests	1.00	Primary forests
Mixed forest	221	Primary forests	1.00	Primary forests
Closed pine forest	1013	Primary forests	1.00	Primary forests
Open pine forest	5134	Secondary forests	0.40	Secondary forests
Bare ground	354	Natural bare & rock	1.00	Others
Forest plantations (pine)	11	Forest plantations	0.31	Forest plantations
Cacao/Banana	15	Extensive agriculture	0.20	Extensive agriculture
Coffee (shade)	2045	Agro forestry	0.50	Agro forestry
Sugar cane	487	Intensive agriculture	0.10	Intensive agriculture
Shrubs	3469	Primary forests	1.00	Primary forests
Irrigated yearly crops	273	Intensive agriculture	0.10	Intensive agriculture
Coffee (no shade)	200	Forest plantations	0.31	Forest plantations
Orchards	163	Intensive agriculture	0.10	Intensive agriculture
Forest plantations	18	Forest plantations	0.31	Forest plantations
Volcanic areas	69	Natural bare & rock	1.00	Others
Banana plantations	83	Extensive agriculture	0.20	Extensive agriculture
Beaches	15	Natural bare & rock	1.00	Others
Shrimp farms	72	Water bodies	1.00	Others
Gullies with vegetation	12	Natural bare & rock	1.00	Others
Tobacco	16	Intensive agriculture	0.10	Intensive agriculture
Rocks	8	Natural bare & rock	1.00	Others

tention to integrate the models to their permanent assessment strategies and systems. The authorities collaborated with the project by providing the necessary information.

3.4.7 Transferring Results

To successfully transfer results to the countries PBL collaborated with capacity building workshops. Meetings were held to organize MSA, scenarios and modeling teams and to train each team in its corresponding part of the methodology. Later, two workshops were held to train selected officers from the environmental authorities –specialized in biology and geographic information system- in the GLOBIO3 and CLUE methodology. The members of the modeling team trained previously assisted in the guidance of these workshops. The trained officers received the model outputs from the assessment team and should be in charge of integrating the results to their instances' information systems and updating, rerunning and improving the models according to upcoming needs and capabilities.

4. RESULTS

4.1 Land Use Modeling

CLUE model provided the expected land use change distribution in three scenarios (Figure 1). Current state map is a combination of the countries' original land use maps with classes aggregated in a single classification. While it is presented as a single picture, the information corresponds to different years depending on the latest available map for each country: 2000 for Nicaragua and Costa Rica, 2002 for Honduras and El Salvador, 2005 for Guatemala, 2008 for Panama and a combination of three sources from 2000, 2006 and 2008 for Belize. The 2030 maps are the CLUE outputs.

Figure 1. Current and future land use modeling result maps for the Central American region

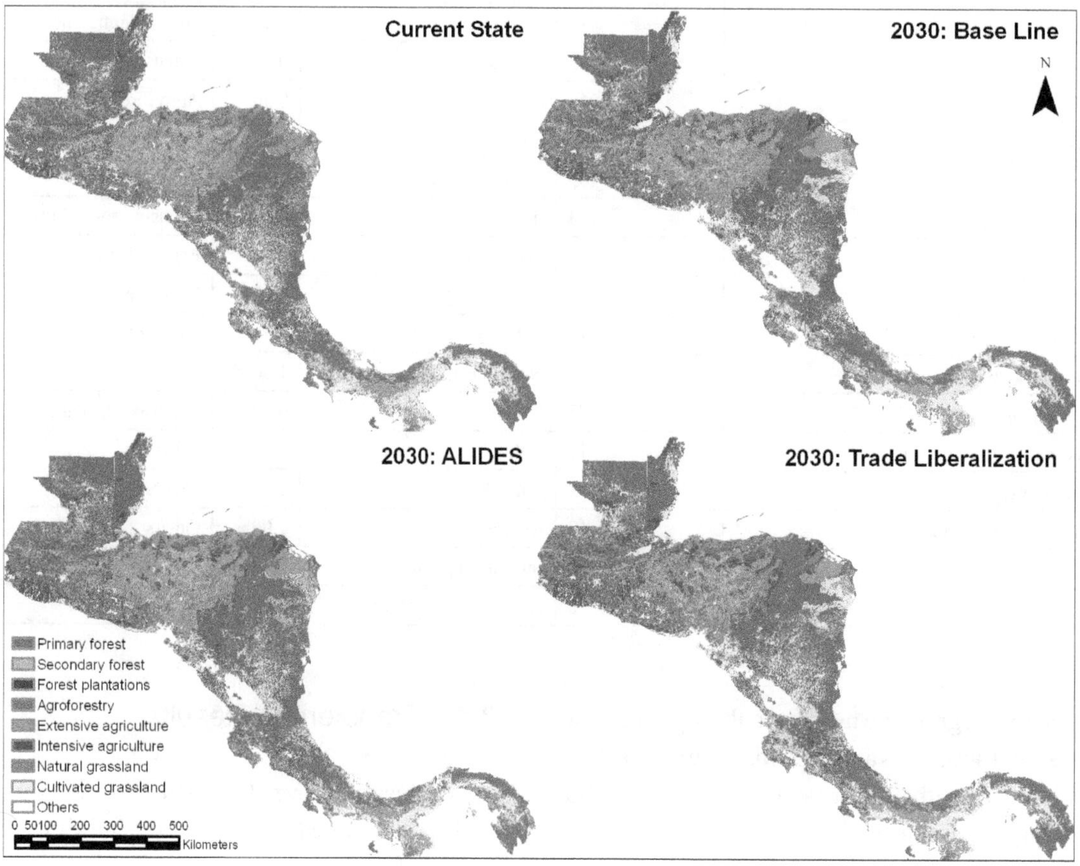

Table 4. Current land use and future land use modeling results for the Central American region.

	Current State	Baseline Scenario 2030		ALIDES 2030		Trade Liberalization 2030	
		%	Var.	%	Var.	%	Var.
Guatemala (20.98% of total area)							
Primary forest	38.23%	30.01%	-8.22	29.96%	-8.26	27.94%	-10.28
Secondary forest	21.29%	16.95%	-4.34	19.49%	-1.80	14.41%	-6.88
Forest plantation	0.26%	0.83%	0.57	0.83%	0.57	0.56%	0.30
Agro forestry	7.87%	5.55%	-2.32	11.79%	3.92	5.56%	-2.31
Extensive Agriculture	0.54%	7.14%	6.61	2.69%	2.16	2.38%	1.84
Intensive Agriculture	15.32%	23.38%	8.06	19.08%	3.76	30.54%	15.22
Natural grasslands	8.90%	8.46%	-0.44	8.46%	-0.44	8.46%	-0.44
Cultivated grasslands	4.12%	4.20%	0.08	4.22%	0.10	6.68%	2.56
Others	3.47%	3.47%	0.00	3.47%	0.00	3.47%	0.00
Belize (4.22% of total area)							
Primary forest	66.92%	60.68%	-6.24	78.50%	11.58	60.68%	-6.24
Grasslands	9.24%	11.76%	2.52	5.68%	-3.56	11.76%	2.52
Agriculture	17.73%	21.39%	3.66	9.64%	-8.08	21.39%	3.66
Others	6.11%	6.17%	0.06	6.17%	0.06	6.17%	0.06
Honduras (21.43% of total area)							
Primary forest	24.39%	13.69%	-10.69	13.65%	-10.74	13.42%	-10.96
Secondary forest	24.96%	14.02%	-10.94	18.32%	-6.64	15.28%	-9.68
Extensive Agriculture	38.08%	48.77%	10.69	48.70%	10.61	28.05%	-10.03
Intensive Agriculture	1.75%	13.25%	11.50	13.28%	11.53	18.68%	16.93
Livestock grazing	9.50%	8.95%	-0.55	4.74%	-4.76	23.24%	13.74
Others	1.33%	1.33%	0.00	1.33%	0.00	1.33%	0.00
El Salvador (4.05% of total area)							
Light used forest	15.47%	11.75%	-3.73	11.75%	-3.73	12.86%	-2.61
Agro forestry	22.76%	22.75%	0.00	24.21%	1.45	22.76%	0.00
Forest plantation	0.32%	0.32%	0.00	0.32%	0.00	0.36%	0.04
Extensive Agriculture	13.04%	7.47%	-5.56	7.48%	-5.56	9.56%	-3.48
Intensive Agriculture	22.97%	25.49%	2.52	25.49%	2.52	27.99%	5.01
Livestock grazing	15.78%	21.56%	5.78	21.56%	5.78	16.34%	0.57
Cultivated grasslands	2.73%	3.73%	1.00	2.26%	-0.47	3.20%	0.47
Others	6.93%	6.93%	0.00	6.93%	0.00	6.93%	0.00
Nicaragua (25.10% of total area)							
Primary forest	37.87%	25.79%	-12.08	25.78%	-12.08	26.47%	-11.40
Secondary forest	3.88%	2.64%	-1.24	8.54%	4.66	2.30%	-1.58
Forest plantation	0.17%	0.20%	0.03	0.20%	0.03	0.30%	0.13
Agro forestry	1.59%	1.66%	0.07	1.70%	0.10	1.66%	0.07
Extensive Agriculture	3.30%	3.00%	-0.29	3.00%	-0.30	2.01%	-1.29
Intensive Agriculture	0.74%	1.29%	0.55	1.29%	0.55	1.59%	0.85
Cultivated grasslands	16.73%	27.51%	10.77	15.16%	-1.58	27.76%	11.03
Agrosilvo pastoral	25.13%	27.31%	2.18	33.75%	8.62	27.31%	2.19

continued on following page

Table 4. continued

	Current State	Baseline Scenario 2030		ALIDES 2030		Trade Liberalization 2030	
Others	10.59%	10.59%	0.00	10.59%	0.00	10.59%	0.00
Costa Rica (9.82% of total area)							
Primary forest	35.23%	36.31%	1.08	36.25%	1.02	24.93%	-10.30
Secondary forest	15.44%	15.74%	0.30	17.41%	1.97	12.01%	-3.43
Agro forestry	7.81%	7.12%	-0.69	7.09%	-0.72	7.12%	-0.69
Intensive Agriculture	13.39%	13.77%	0.38	13.67%	0.28	24.25%	10.86
Agrosilvo pastoral	5.95%	5.65%	-0.30	11.79%	5.84	5.62%	-0.33
Cultivated grasslands	15.85%	15.09%	-0.77	7.47%	-8.39	19.75%	3.90
Others	6.32%	6.32%	0.00	6.32%	0.00	6.32%	0.00
Panama (14.40% of total area)							
Primary forest	42.10%	40.91%	-1.19	35.05%	-7.05	42.07%	-0.03
Secondary forest	0.15%	0.29%	0.13	0.29%	0.14	0.05%	-0.10
Light used forest	11.13%	11.08%	-0.05	11.87%	0.73	10.80%	-0.34
Forest plantation	0.54%	0.20%	-0.34	0.14%	-0.40	0.03%	-0.50
Extensive Agriculture	21.01%	22.54%	1.52	30.99%	9.97	24.92%	3.91
Cultivated grasslands	22.64%	22.56%	-0.07	19.24%	-3.40	19.70%	-2.93
Others	2.43%	2.43%	0.00	2.43%	0.00	2.43%	0.00
Regional							
Primary forest	35.10%	27.72%	-7.38	27.60%	-7.49	26.45%	-8.65
Forest plantation	0.19%	0.27%	0.08	0.26%	0.07	0.21%	0.02
Secondary forest	12.33%	8.81%	-3.52	11.91%	-0.42	8.06%	-4.27
Light used forest	2.23%	2.07%	-0.16	2.18%	-0.05	2.08%	-0.15
Agro forestry	3.74%	3.20%	-0.54	4.58%	0.84	3.20%	-0.54
Extensive Agriculture	12.65%	16.25%	3.60	16.52%	3.86	10.99%	-1.66
Intensive Agriculture	6.77%	11.36%	4.59	9.95%	3.19	15.23%	8.46
Natural grasslands	4.54%	4.57%	0.02	3.66%	-0.88	7.42%	2.87
Cultivated grasslands*	10.38%	13.16%	2.78	8.52%	-1.86	13.77%	3.39
Agrosilvo pastoral	6.89%	7.41%	0.52	9.63%	2.74	7.41%	0.52
Others	5.18%	5.18%	0.00	5.18%	0.00	5.18%	0.00

Allocated results differ from the land use demand tables by a specified level of acceptable error or deviation. This level was determined in CLUE parameters. For each country, the minimum error level for which the model was able to find a solution was used. The error levels were higher when the country had one or more classes occupying a small proportion of area. For this reason, it is suggested to aggregate these classes into other categories, as long as that does not imply a significant loss of MSA differentiation. Table 4 presents the distribution of country and region's area between their respective land use classes in the current state and in the three modeled scenarios. Variations with respect to current state are also presented. Country areas as percentage of region area are expressed in parenthesis.

The results show that in the Baseline scenario, the region experiences a reduction of its forested areas, mainly due to an increase in agricultural lands and cultivated grasslands. This corresponds to the persistence of the economic growth pattern of the region in the last decades, characterized by agricultural expansion and intensification, timber extraction and land abandonment.

In the ALIDES option, this trend will be compensated by a mitigation of the loss of secondary forest; as a result of the policies. The scenario stimulates the transformation of traditional agriculture and grassland activities into sustainable production systems that eventually become secondary forest. Meanwhile, in Trade Liberalization option baseline trends will be intensified because of the expected increase in demands for primary production export goods associated with the implementation of the Free Trade Agreements.

These general tendencies were observed in all countries. Not all land use categories were present in every country as a result of the level of detail of the available information.

4.2 Biodiversity Modeling

GLOBIO3 provided the regional MSA maps in the current state and three future scenarios (Figure 17.2). Map displayed the remaining biodiversity of the region in terms of the Mean Species Abundance. Dark areas correspond to high biodiversity values, meaning that their biodiversity is relatively undisturbed or intact. Light areas correspond to low biodiversity values, meaning that their biodiversity has been severely disturbed by human influence. Water bodies such as lakes and rivers were not evaluated in the methodology and appear as clear areas in the maps.

The aggregated figures of the modeling with the remaining MSA values and biodiversity loss per driver for each country and for the region are presented in Table 5.

Figure 2. Current and future biodiversity modeling result maps for the Central American region

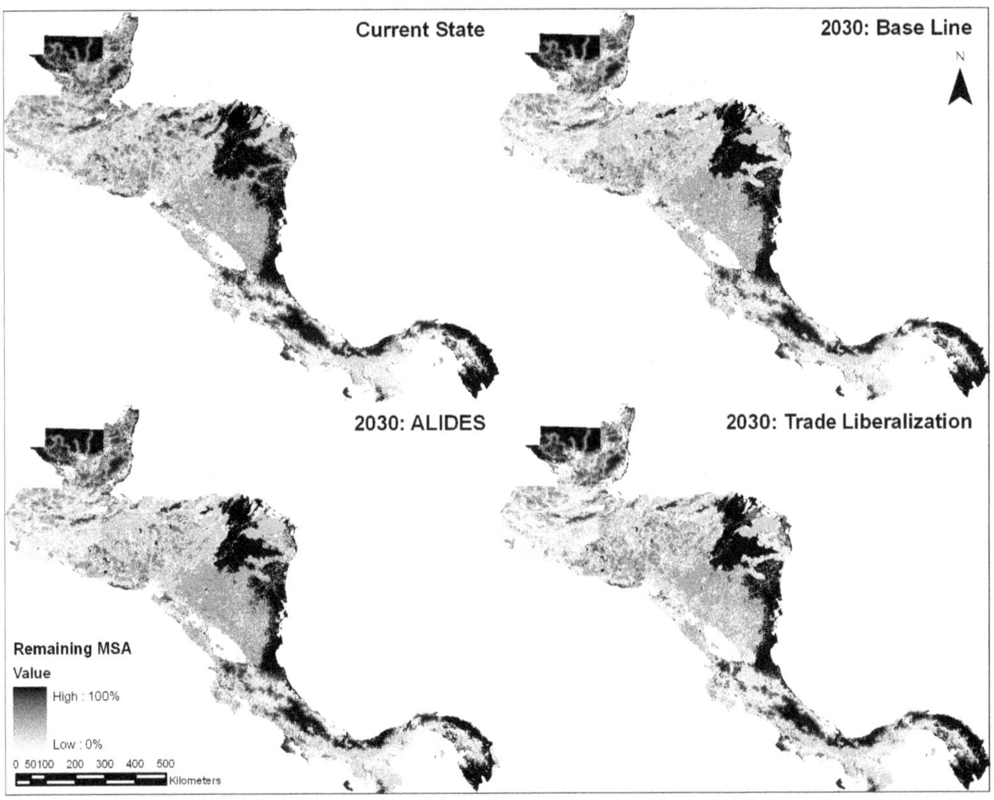

Table 5. Current and future biodiversity modeling results for the Central American region

	Current State	Baseline Scenario 2030		ALIDES 2030		Trade Liberalization 2030	
		%	Var.	%	Var.	%	Var.
Guatemala							
Remaining MSA	39.44%	33.61%	-5.83	34.03%	-5.41	31.33%	-8.11
Infrastructure	13.85%	13.28%	-0.57	13.34%	-0.51	13.08%	-0.77
Fragmentation	14.83%	14.21%	-0.62	14.27%	-0.56	14.00%	-0.83
N Deposition	0.00%	0.00%	0.00	0.00%	0.00	0.00%	0.00
Climate Change	4.01%	3.84%	-0.17	3.86%	-0.15	3.78%	-0.23
Land Use	27.88%	35.06%	7.18	34.50%	6.62	37.80%	9.92
Belize							
Remaining MSA	54.27%	49.78%	-4.49	57.59%	3.32	47.05%	-7.22
Infrastructure	25.29%	23.96%	-1.33	25.18%	-0.11	23.80%	-1.49
Fragmentation	1.93%	1.84%	-0.09	1.94%	0.01	1.83%	-0.10
N Deposition	0.00%	0.00%	0.00	0.00%	0.00	0.00%	0.00
Climate Change	3.08%	2.90%	-0.18	3.05%	-0.03	2.88%	-0.20
Land Use	15.43%	21.52%	6.09	12.25%	-3.18	24.43%	9.00
Honduras							
Remaining MSA	46.37%	36.73%	-9.64	37.03%	-9.34	39.17%	-7.20
Infrastructure	11.13%	6.15%	-4.98	6.09%	-5.04	8.73%	-2.40
Fragmentation	4.11%	2.93%	-1.18	2.65%	-1.46	3.11%	-1.00
N Deposition	0.00%	0.00%	0.00	0.00%	0.00	0.00%	0.00
Climate Change	1.88%	4.31%	2.43	4.32%	2.44	4.25%	2.37
Land Use	36.50%	49.87%	13.37	49.91%	13.41	44.75%	8.25
El Salvador							
Remaining MSA	30.88%	28.54%	-2.34	28.97%	-1.91	27.60%	-3.27
Infrastructure	6.35%	7.91%	1.56	7.95%	1.60	7.04%	0.69
Fragmentation	4.20%	4.30%	0.10	4.23%	0.04	3.89%	-0.31
N Deposition	0.00%	0.00%	0.00	0.00%	0.00	0.00%	0.00
Climate Change	2.02%	4.45%	2.44	4.46%	2.44	4.49%	2.47
Land Use	56.56%	54.80%	-1.76	54.39%	-2.17	56.98%	0.42
Nicaragua							
Remaining MSA	58.05%	50.42%	-7.63	54.19%	-3.86	47.13%	-10.92
Infrastructure	5.18%	0.53%	-4.65	1.14%	-4.04	0.53%	-4.65
Fragmentation	2.66%	0.79%	-1.87	0.91%	-1.75	0.79%	-1.87
N Deposition	0.00%	0.00%	0.00	0.00%	0.00	0.00%	0.00
Climate Change	1.96%	4.48%	2.52	4.45%	2.49	4.48%	2.52
Land Use	32.14%	43.77%	11.63	39.30%	7.16	47.07%	14.93
Costa Rica							
Remaining MSA	45.74%	41.48%	-4.26	42.87%	-2.87	36.27%	-9.47
Infrastructure	9.21%	14.07%	4.86	16.86%	7.66	8.91%	-0.30

continued on following page

Table 5. continued

	Current State	Baseline Scenario 2030		ALIDES 2030		Trade Liberalization 2030	
Fragmentation	4.16%	3.38%	-0.78	3.45%	-0.71	3.25%	-0.91
N Deposition	0.00%	0.00%	0.00	0.00%	0.00	0.00%	0.00
Climate Change	1.88%	4.16%	2.29	4.08%	2.21	1.88%	0.00
Land Use	39.02%	36.91%	-2.11	32.73%	-6.29	49.70%	10.68
Panama							
Remaining MSA	52.22%	48.23%	-3.99	45.59%	-6.62	50.25%	-1.97
Infrastructure	3.71%	5.46%	1.75	4.90%	1.19	4.38%	0.67
Fragmentation	2.67%	1.88%	-0.78	1.99%	-0.68	1.97%	-0.69
N Deposition	0.00%	0.00%	0.00	0.00%	0.00	0.00%	0.00
Climate Change	2.66%	4.16%	1.50	4.15%	1.49	4.18%	1.52
Land Use	38.74%	40.27%	1.52	43.36%	4.62	39.21%	0.47
Regional							
Remaining MSA	48.09%	41.64%	-6.45	42.77%	-5.32	40.56%	-7.53
Infrastructure	9.47%	7.90%	-1.57	8.29%	-1.19	7.73%	-1.74
Fragmentation	5.80%	4.78%	-1.02	4.78%	-1.02	4.76%	-1.05
N Deposition	0.00%	0.00%	0.00	0.00%	0.00	0.00%	0.00
Climate Change	2.54%	4.16%	1.62	4.15%	1.62	3.91%	1.38
Land Use	34.10%	41.52%	7.42	40.01%	5.91	43.05%	8.95

4.3 Current State

The results show that in the current state, region has a remaining MSA of 48%, meaning that 52% of MSA has been lost due to human induced pressures. According to current state map (Figure 17.2), most of this remaining is concentrated in the eastern regions of the Atlantic coast. This area is where undisturbed natural tropical and subtropical forests are located. The largest protected areas are also located in this region. By contrast, western lands in the Pacific have lower levels of remaining biodiversity. The Pacific coast is occupied by more intensive land uses, such as agricultural and livestock systems and by the remnants of dry forest and shrub lands. Population and road infrastructure are concentrated there, where most of the intense economic activity of the region takes place.

The main source of biodiversity loss was identified as the land use driver, which accounted for a loss of 34.1% of MSA. Infrastructure, fragmentation and climate change drivers accounted for 9.47%, 5.8% and 2.54% of the loss respectively.

For nitrogen deposition, data available from the IMAGE model at 0.5*0.5 degree resolution showed that the region did not experience any excess nitrogen, since nitrogen deposition did not exceed ecosystem's critical load under any of the explored scenarios.

4.4 Scenarios and Policy Options

In the Baseline scenario the region experiences the loss of an additional 6.45% of its biodiversity compared to the current situation. This loss is attributed to increased effects of the land use driver, which would account for the loss of an additional 7.42% of MSA. The region losses approximately of 10% of its total area in forest cover due to a corresponding increase in agricultural and grassland uses (Table 17.4).

Infrastructure and fragmentation would reduce their impact share on MSA by 1.57% and 1.02%, respectively. In the methodology, the effect of these drivers is only estimated for natural areas, since in non natural (human intervened) areas the effect of infrastructure and fragmentation on biodiversity has already been accounted for in the land use impact. As a result, if natural areas decrease, shares of these drivers will also decrease. Climate change driver increases its effect on MSA by an additional 1.62%.

In the ALIDES policy options, the effects are less severe. The region experiences the loss of an additional 5.32% of biodiversity compared to current state. This is 1.13% less than in Baseline scenario. The additional effect of the land use driver is smaller: an additional loss of MSA by 5.91% than compared to current state (1.51% less than in Baseline scenario), because of the regrowth of secondary forest (Table 17.4). Other drivers' influence remains similar to Baseline scenario results.

Alternatively, effects are more severe in Trade Liberalization policy option. The region experiences additional loss of MSA by 7.53% compared to current state, which is 1.08% more than in Baseline scenario. This is explained by the greater effect of the land use driver. Land use accounts for a loss of MSA by 8.95% compared to current state (1.53% more than in Baseline scenario), because of the increases in the area for agriculture and livestock production. Other drivers' influence is similar to the previous scenarios results, although infrastructure reduces its share of the loss because natural areas in this scenario are further reduced.

In future state maps, the effects of the three scenarios seem to affect the entire region (Figure 17.2). The areas with high MSA from the Atlantic will experience degradation, but main degradation will remain concentrated in the Pacific, where most of the land use intensification will take place.

The individual country results show remaining MSA values of 39.44% for Guatemala, 54.27% for Belize, 46.37% for Honduras, 30.88% for El Salvador, 58.05% for Nicaragua, 45.74% for Costa Rica and 52.22% for Panama in current situation. Land use was the most significant driver of biodiversity loss in all cases. The future state results in the countries followed the tendencies described in regional results, with Baseline scenario biodiversity losses being reduced in the ALIDES policy option or intensified in the Trade Liberalization option.

5. DISCUSSION

Overall, results show that state of biodiversity in the region has been significantly affected and will experience further degradation under all the evaluated scenarios. The differences between scenarios and current state are greater than the differences between scenarios themselves. This implies that the effects of the policy options evaluated do not modify the baseline tendencies significantly.

The main driver of biodiversity loss is land use intensity. In future state, land use changes towards more intense uses will account for most of the additional loss. Infrastructure and fragmentation impacts are less significant and are reduced due to decrease in the extension of natural areas. However, their effect on biodiversity should not be neglected, particularly since these drivers are the precursors of the land use influence. The areas intervened and fragmented by road infrastructure for natural resource extraction are later occupied by shifting cultivation farmers, who foster the land use intensification process. The contribution of climate change to biodiversity degradation increased in the future, but was relatively constant between scenarios. This is attributed to the virtual inertia of the climate change phenomenon, compared to variations in local drivers' conditions (Eickhout *et al.*, 2007). The reliance on global models to derive data will remain important in the integration of drivers such as climate change in downscaled procedures.

Differences in results between the countries can be attributed to real differences in the state of biodiversity, but also to differences between the land use classification systems used to assign MSA impact values. The extent to which differences are attributed to one cause or another is unknown. Only general assumptions can be drawn from land use maps. For example, in the case of Panama there was no intensive agriculture class. Agriculture increases from Trade Liberalization scenario had to be considered extensive, which implies that Trade Liberalization effects were underestimated in this case. In the case of Honduras, there was only one grassland class with an intermediate MSA value. This limitation had a ambiguous effect on scenario results: grassland conversions to extensive/integrated agriculture systems and later to secondary forest contemplated in ALIDES scenario resulted in a biodiversity loss instead, while Trade Liberalization grassland increase seemed to have contributed to biodiversity conservation. This uncertainty implies that results are not entirely comparable between the countries and considerations based on each map reclassification criteria should be made explicit when interpreting model results.

5.1 Output Interpretation

Future scenarios and policy options did not intended to be exact calculations of changes in MSA, but projections based on the selected assumptions. While scenarios' quantitative estimates facilitated the visualization of the qualitative storylines and helped to manage uncertainties, they were fundamentally subjective and lack the accuracy to be use as the single decision making tool (Shearer, 2005). Hence, conclusions of the modeling exercise should not be taken from absolute figures. Instead, model output interpretation should focus on discussing the differences between scenarios and the revealed trends in general (Verboom *et al.*, 2007). Output comparisons can be considered to be robust since they are based on a same set of main assumptions adopted (Verboom *et al.*, 2007).

Given that scenarios and policy options made conservative assumptions on growth and stability, it is likely that the impacts on biodiversity were underestimated. Since the inputs were represented at the country extent, it is at this extent that model outputs are valid. The result interpretations for specific sub-regions are not accurate. For instance, when in country reports delivered to CCAD, outputs were analyzed for protected areas; the results were significant for the group of protected areas of a country but not necessarily for each of the individual areas on their own. The smaller extents can be analyzed with GLOBIO3 methodology, but inputs should have the relevant scale and detail level.

5.2 Limitations of the Methodology

There are advantages and limitations of varied nature in the application of GLOBIO3 methodology. For instance, the selected indicator of Mean Species Abundance can be considered as an application of the CBD indicator of abundance and distribution of selected species. Thus, its evaluation is in line with the achievement of the CBD target (Leemans *et al.*, 2007). However, the MSA does not cover the entire biodiversity concept. An assessment of the MSA indicator is limited to the average response of a representative set of species under given driver conditions. No individual species response is evaluated. It is recommended that, when used in extensive biodiversity assessments, complimentary indicators should be included (www.globio.info, 2010). By using original (undisturbed) state as the baseline, MSA has the advantage of disregarding the misleading effect that invasive species have with other indicators. But this also means that equal weight is assigned to any undisturbed ecosystem. When aggregating the MSA values of a region, equal weight is assigned to each area unit disregarding its original richness (Leemans *et al.* 2005). This may lead to an underestimation of biodiversity losses when richness varies significantly across the evaluated region. These aspects require additional considerations when interpreting the results.

The CLUE model is able to predict evolution of the most likely landscape. This enables decision makers to analyze how scenario assumptions on local and regional land use, economic and environmental policies can influence the area under study. Still, the limitations to allow feedbacks between (local and regional) scales (Turpin *et al.*, 2009), between impacts and between drivers in the sequential modeling process (Verburg, 2006) are drawbacks of the framework. A dynamic approach, where local policies progressively influence land use demands and new occurring characteristics of landscape influence local suitability could be more accurate. The model is rather static and currently explores only path dependency relations in locating land uses (Verburg, 2006).

The GLOBIO3 methodology itself has limitations regarding its core structure and underlying assumptions. According to Leemans *et al.* (2005), two of the main assumptions of the model are that local changes in species richness are equal to regional changes in mean abundance and the correlation between them is a constant. These assumptions could not be valid, if the correspondence of these factors changes between the scale at which local changes were estimated and the scale at which regional results are being extrapolated, or if the regional pattern of species abundance and distribution implies that the correlation with local species richness is not constant. A limitation of the model structure is that it only applies to terrestrial ecosystems (Alkemade *et al.*, 2009). As a result, no marine or freshwater body could be evaluated in this study. However, an inland aquatic module of the model has been developed that will complement the biodiversity assessments with GLOBIO3.

There were additional issues regarding modeling uncertainties. Chazal & Rounsevell (2009) identified the following uncertainties in land use and biodiversity analysis: quality of input data, understanding of relevant processes, capacity of the model to represent processes and range of plausible outcomes. Inherent uncertainties related to scenario building described by Rounsevell et al. (2006) included the subjective nature of storyline interpretation, the assumptions behind the land use change models, quality of baseline data and errors with statistical downscaling techniques.

In the case study, uncertainties were introduced mainly in scenario development, demand tables building and through the consecutive reclassifications of national maps into GLOBIO3 impact categories. The fact that for every country the combination and sources of inputs was different represented an additional source of uncertainty in overall regional result interpretation. Also, given that processes were only assessed at a defined spatial extent and resolution, variables that exerted influence at different scales were considered exogenous and were derived from other sources (Rounsevell *et al.*, 2006). This coupling of modeling frameworks introduced uncertainties since different spatial and temporal scales involved might not have been fully compatible and feedback considerations were limited (Agriculture Ecosystem & Environment, 2006).

Not all of the sources of uncertainties were possible to control. As Rounsevell *et al.* (2006) stated that uncertainties are an inherent part of any modeling process and are acceptable to an extent. But, model maker and users should have full understanding on which assumptions were taken and what are the consequences of those uncertainties in output interpretation.

5.3 Limitations of Results

The main limitations faced in this study were: (1) the lack of appropriate information, (2) the differences in outputs between countries and (3) the difficulties to validate results. Data needed for MSA revision, scenario construction and model execution was incomplete and uneven between countries. The scientific studies undertaken in the region to explore relationships between local land use systems and species abundance are limited and provide only a general insight. Some key socioeconomic datasets were not accessible, and available time series data were not consistent for all seven countries. In some cases, countries had difficulties to supply required data inputs at the appropriate scale.

The different levels of detail of country land use maps constrained the possibilities of adequately applying GLOBIO3 impact categories to outputs. In some cases land use maps were essentially land cover maps, which distinguished between types of vegetation but not between degrees of use intensity. Land use model validations are usually difficult to complete since time series on land use distribution are not registered systematically and are not always available (Pontius & Schneider, 2001; Chazal & Rounsevell, 2009). No validation possibility was explored in this study because of information and time constraints (see Pontious *et al.*, 2004 for validation alternatives).

Regarding the validation of model progress and preliminary results, the complexity involved with model execution and output presentation to stakeholders within the initial phases of capacity building made validations difficult. The preliminary results were reviewed during the second training workshop. The adjustments were done mostly to land use MSA values. A meeting of scenario and modeling teams was held to validate scenario figures and to integrate them into the demand tables. However, validation of the finals results is pending. The adjustments derived from these discussions will be object of future research.

All the same, the GLOBIO3 methodology represented a practical framework to integrate available information and by complementing it with other sources to obtain a valuable tool for policy makers.

6. FUTURE RESEARCH DIRECTIONS AND MODEL TRANSFERRING

Future steps to implement GLOBIO3 in Central America deal with the transfer of models from the assessment team to the countries' environmental instances. Once the models were delivered to authorities, technical officers trained in the project started to face the challenge of model updating, improvement and continuation.

The main research directions will be related to improving available information, updating scenarios according to new findings, emerging interest, opportunities and needs, maintaining the acquired technical capacity and fully integrating modeling results as a tool for policy support. First of all, environmental authorities or corresponding instances should concentrate efforts on filling the identified information gaps with updated and more detailed land use maps, spatial biophysical, demographic and socioeconomic information at adequate scale and more importantly, improved knowledge on species response to local land uses and broader, consistent time series data on land uses occupation. This would enable a better classification of land uses with adequate MSA impact values, making model more suitable for specific context (Haines-Young, 2009). Comprehensive data series would allow for better model calibration and further validation of results. Moreover, the combination of these improvements on information availability would render more consistent results between units of study. Additional information on future estimates of other drivers would improve model performance as well. For instance, local data on nitrogen deposition would reveal nitrogen excess levels that were not evident at the coarse scale.

The next step should be updating the scenarios.. The explored scenarios and policy options could be enriched with more information on land use history or expected results from policy implementation. In addition, if the region or the countries want to evaluate further development pathways new storylines could be developed. Technical teams could work in standardizing scenario building techniques according to their capacities, so that assessment could be executed more rapidly. Such an approach would enable officers to revise models regularly. If feedbacks are included, the application could become an established mechanism for biodiversity conservation monitoring in the region.

Technical capacity should be maintained in environmental instances. At the moment, each country has two trained technicians fully capable of model execution. If these technicians are replaced, authorities should make sure that new technicians are trained to continue with the modeling. Additionally, framework developers are working to update the methodology by including more drivers and improving the quality of quantitative relationships based on new scientific findings (GLOBIO, 2010). Authorities should be aware of these updates and should prepare their technicians when new tools become available.

Finally, to actually influence biodiversity conservation, attention should be given to the full integration of model results into political decision making. So far, results have been used by countries in the reports on the progresses on the achievement of the CBD target. Authorities should embed the outputs and foster discussion with other decision making instances. Now that the drivers that significantly influence biodiversity degradation have been demonstrated, the mechanisms to deal with these drivers have to be considered. Any measure to control the influence of drivers will contribute to the efforts of biodiversity conservation and sustainable management.

8. CONCLUSION

The modeling of current and future state of biodiversity in the Central American region showed that biodiversity has been significantly affected and this trend is likely to continue under the scenarios considered. The GLOBIO3 methodology provided insight on the effects of alternative scenario and policy options on biodiversity conservation. It offered decision makers a suitable tool for national policy support, especially to stimulate policy discussion and to integrate the topic of biodiversity into various policy domains. If models are updated properly, the continuous evaluation of potential political interventions and pathways could be possible for Central American countries. While the presented case study is a preliminary application of the downscaled methodology, it represents a substantial scientific and technical achievement for the region. Considerations revised in this case study should facilitate the implementation of future modeling initiatives in similar contexts.

ACKNOWLEDGMENT

The Biodiversity Modeling Project was commissioned to the IRBIO by the Central American Commission on Environment and Development. Author would like to thank the Netherlands Environmental Assessment Agency for the financial and scientific support to this project. Special thanks to Wilbert van Rooij for his valuable collaboration, remarks and suggestions through the project execution and article review.

REFERENCES

Abildtrup, J., Rosato, P., Gylling, M., Fekete-Farkas, M., Audsley, E., Giupponi, C., & Rounsevell, M. D. A. (2006). Socio-economic scenario development for the assessment of climate change impacts on agricultural land-use: A pairwise comparison approach. *Environmental Science & Policy*, *9*(2), 101–115..doi:10.1016/j.envsci.2005.11.002

Alkemade, R., van Oorschot, M., Miles, L., Nellemann, C., Bakkenes, M., & ten Brink, B. (2009). GLOBIO3: A framework to investigate options for reducing global terrestrial biodiversity loss. *Ecosystems*, *12*(3), 374–390. doi:10.1007/s10021-009-9229-5

Bouwman, A. F., Kram, T., & Goldewijk, K. K. (Eds.). (2006). *Integrated modeling of global environmental change. An overview of IMAGE 2.4*. Bilthoven, The Netherlands: Netherlands Environmental Assessment Agency.

Central American Commission on Environment and Development & Regional Biodiversity Institute. (2009). *Reporte preliminar del modelo y análisis del estado actual y futuro de la biodiversidad para México con un enfoque de modelación en los estados fronterizos con Guatemala*. Manuscript submitted for publication.

Chazal, J., & de,., & Rounsevell, M. D. A. (2009). Land-use and climate change within assessment of biodiversity change: A review. *Global Environmental Change*, *19*, 306–315..doi:10.1016/j.gloenvcha.2008.09.007

Conservation International. (2007). *Biodiversity hotspots. Mesoamerica*. Retrieved from http://www.biodiversityhotspots.org /xp/hotspots/Mesoamerica /Pages/default.aspx

Eickhout, B., van Meijl, H., Tabeau, A., & van Rheenen, T. (2007). Economic and ecological consequences of four European land use scenarios. *Land Use Policy*, *24*, 562–575..doi:10.1016/j.landusepol.2006.01.004

GLOBIO Consortium. (2010). *What is GLOBIO?* Retrieved from http://www.globio.info/ what-is-globio

Haines-Young, R. (2009). Land use and biodiversity relationships. *Land Use Policy*, *26S*, S178–S186..doi:10.1016/j.landusepol.2009.08.009

Kok, K., & Veldkamp, A. (2001). Evaluating impact of spatial scales on land use pattern analysis in Central America. *Agriculture Ecosystems & Environment*, *85*, 205–221. doi:10.1016/S0167-8809(01)00185-2

Kok, K., & Winograd, M. (2002). Modelling land-use change for Central America, with special reference to the impact of hurricane Mitch. *Ecological Modelling*, *149*, 53–69. doi:10.1016/S0304-3800(01)00514-2

Leemans, R. (1999). Modelling for species and habitats: New opportunities for problem solving. *The Science of the Total Environment*, *240*, 51–73. doi:10.1016/S0048-9697(99)00320-4

Leemans, R., Gaston, K. J., van Jaarsveld, A. S., Dixon, J., Harrison, J., & Cheatle, M. E. (2007). *International review of the GLOBIO model version 3. (MNP Report No 555050002/2007)*. Bilthoven: Netherlands Environmental Assessment Agency.

Lezama-López, M. (2007). *El Índice de Capital Natural como instrumento de análisis de pérdida de biodiversidad en Nicaragua*. Technical Report. Netherlands Environmental Assessment Agency. Retrieved from http://www.globio.info/assessments-with-globio

Lujten, J., Miles, L., & Cherrington, E. (2006). *Land use change modelling for three scenarios for the MAR region*. Technical Report. UNEP World Conservation Monitoring Center.

Majer, J. D., & Beeston, G. (1996). The biodiversity integrity index: An illustration using ants in Western Australia. *Conservation Biology*, *10*(1), 65–73. doi:10.1046/j.1523-1739.1996.10010065.x

Mattison, E. H. A., & Norris, K. (2005). Bridging the gaps between agricultural policy, land-use and biodiversity. *Trends in Ecology & Evolution*, *20*(11), 610–616..doi:10.1016/j.tree.2005.08.011

Mertz, O., Ravnborg, H. M., Lövei, G., Nielsen, I., & Konijnendijk, C. C. (2007). Ecosystem services and biodiversity in developing countries. *Biodiversity and Conservation, 16*, 2729–2737.. doi:10.1007/s10531-007-9216-0

O'Rourke, E. (2006). Biodiversity and land use change on the Causse Méjan, France. *Biodiversity and Conservation, 15*, 2611–2626..doi:10.1007/s10531-005-5402-0

Parks, B. O., Fornwall, M. D., & Quinn, J. F. (Eds.). (2004). *Proceedings from NBII: First NBII Biodiversity Modeling Workshop: Results and Recommendations.* U.S. Geological Survey, Center for Bioinformatics, Denver, CO.

Pontius, R. G. Jr, Huffaker, D., & Denman, K. (2004). Useful techniques of validation for spatially explicit land-change models. *Ecological Modelling, 179*, 445–461..doi:10.1016/j.ecolmodel.2004.05.010

Pontius, R. G. Jr, & Schneider, L. C. (2001). Land-cover change model validation by an ROC method for the Ipswich watershed, Massachusetts, USA. *Agriculture Ecosystems & Environment, 85*, 239–248. doi:10.1016/S0167-8809(01)00187-6

Poschlod, P., Bakker, J. P., & Kahmen, S. (2005). Changing land use and its impact on biodiversity. *Basic and Applied Ecology, 6*, 93–98.. doi:10.1016/j.baae.2004.12.001

Reidsma, P., Tekelenburg, T., Van den Berg, M., & Alkemade, R. (2006). Impacts of land-use change on biodiversity: An assessment of agricultural biodiversity in the European Union. *Agriculture Ecosystems & Environment, 114*, 86–102.. doi:10.1016/j.agee.2005.11.026

Rounsevell, M. D. A., Berry, P. M., & Harrison, P. A. (2006). Future environmental change impacts on rural land use and biodiversity: A synthesis of the ACCELERATES project. *Environmental Science & Policy, 9*, 93–100..doi:10.1016/j.envsci.2005.11.001

Rounsevell, M. D. A., Reginster, I., Araújo, M. B., Carter, T. R., Dendoncker, N., & Ewert, F. (2006). A coherent set of future land use change scenarios in Europe. *Agriculture Ecosystems & Environment, 114*, 57–68..doi:10.1016/j.agee.2005.11.027

Scholes, R. J., & Biggs, R. (2005). A biodiversity intactness index. *Nature, 434*, 45–49..doi:10.1038/nature03289

Secretariat of the Convention on Biological Diversity. (2006). *Global biodiversity outlook 2*. Montreal.

Secretariat of the Convention on Biological Diversity & Netherlands Environmental Assessment Agency. (2007). *Cross-roads of planet Earth's life. Exploring means to meet the 2010 biodiversity target*. Montreal.

Shearer, A. W. (2005). Approaching scenario-based studies: Three perceptions about the future and considerations for landscape planning. *Environment and Planning. B, Planning & Design, 32*(1), 67–87..doi:10.1068/b3116

Spangenberg, J. H. (2007). Biodiversity pressure and the driving force behind. *Ecological Economics, 61*, 146–158. doi:10.1016/j.ecolecon.2006.02.021

Thanh, S. V. (2008). *Biodiversity assessment and modeling: Review and potential application in Vietnam.* Paper presented at International Symposium on Geoinformatics for Spatial Infrastructure Development in Earth and Allied Sciences 2008. Retrieved from http://www.globio.info/assessments-with-globio/ national-sub-regional-assessments

Trisurat, Y., Alkemade, R., & Verburg, P. H. (2010). Projecting land-use change and its consequences for biodiversity in northern Thailand. *Environmental Management, 45*, 626–639..doi:10.1007/s00267-010-9438-x

Turpin, N., Dupraz, P., Thenail, C., Joannon, A., Baudry, J., Herviou, S., & Verburg, P. (2009). Shaping the landscape: Agricultural policies and local biodiversity schemes. *Land Use Policy*, *26*, 273–283..doi:10.1016/j.landusepol.2008.03.004

United Nation Environmental Program. (2001). UNEP/DEWA/TR.01-3. In C. Nellemann, L. Kullerud, I. Vistnes, B., C. Forbes, E. Husby, G. P. Kofinas, B. P. Kaltenborn, et al (Eds.), *GLOBIO. Global methodology for mapping human impacts on the biosphere.*

United Nations. (1993). *Multilateral Convention on biological diversity.* United Nation Treaty Series. 1760, I 31619.

United Nations. (2010) *Goal 7: Ensure environmental sustainability.* Retrieved from http://www.un.org/millenniumgoals /environ.shtml

Verboom, J., Alkemade, R., Klijn, J., Metzger, M. J., & Reijnen, R. (2007). Combining biodiversity modeling with political and economic development scenarios for 25 EU countries. *Ecological Economics*, *62*, 267–276..doi:10.1016/j.ecolecon.2006.04.009

Verburg, P. H. (2006). Simulating feedbacks in land use and land cover change models. *Landscape Ecology*, *21*, 1171–1183..doi:10.1007/s10980-006-0029-4

Verburg, P. H., Schulp, C. J. E., Witte, N., & Veldkamp, A. (2006). Downscaling of land use change scenarios to assess the dynamics of European landscapes. *Agriculture Ecosystems & Environment*, *114*, 39–56..doi:10.1016/j.agee.2005.11.024

Verburg, P. H., Soepboer, W., Veldkamp, A., Limpiada, R., Espaldon, V., & Mastura, S. S. A. (2002). Modelling the spatial dynamics of regional land use: The CLUE-S model. *Environmental Management*, *30*(3), 391–405. doi:10.1007/s00267-002-2630-x

World Conservation Union. (2010). *About biodiversity. The biodiversity challenge.* Retrieved from http://www.iucn.org/what /tpas/biodiversity/about/

World Wildlife Fund. (1998). *Living planet report 1998*. Gland, Switzerland: World Wildlife Fund.

Young, J., Watt, A., Nowicki, P., Alard, D., Clitherow, J., & Henle, K. (2005). Towards sustainable land use: Identifying and managing the conflicts between human activities and biodiversity conservation in Europe. *Biodiversity and Conservation*, *14*, 1641–1661..doi:10.1007/s10531-004-0536-z

Chapter 18
Spatial Model Approach for Deforestation:
Case Study in Java Island, Indonesia

Lilik B. Prasetyo
Bogor Agriculture University, Indonesia

Chandra Irawadi Wijaya
Bogor Agriculture University, Indonesia

Yudi Setiawan
Bogor Agriculture University, Indonesia

ABSTRACT

Java is very densely populated since it is inhabited by more than 60% of the total population of Indonesia. Based on data from the Ministry of Forestry, forest loss between 2000-2005 in Java was about 800,000 hectares. Regardless of the debate on whether the different methodologies of forest inventory applied in 2005 have resulted in an underestimation of the figure of forest loss or not, the decrease of forest cover in Java is obvious and needs immediate response. Spatial modeling of the deforestation will assist the policy makers in understanding this process and in taking it into consideration, when decisions are made on the issue. Moreover, the results can be used as data input to solve environmental problems resulting from deforestation. The authors of this chapter modeled the deforestation in Java by using logistic regression. Percentage of deforested area was considered as the response variable, whilst biophysical and socioeconomic factors, that explain the current spatial pattern in deforestation, were assigned as explanatory variables. Furthermore, the authors predicted the future deforestation process, and then, for the case of Java, it was validated with the actual deforestation derived from MODIS satellite imageries from 2000 to 2008. Results of the study showed that the impacts of population density, road density, and slope are significant. Population density and road density have negative impacts on deforestation, while slope has positive impact. Deforestation on Java Island tends to occur in remote areas with limited access, low density population and relatively steep slopes. Implication of the model is that the government should pay more attention to remote rural areas and develop good access to accelerate and create alternative non agricultural jobs in order to reduce pressure on the forest.

DOI: 10.4018/978-1-60960-619-0.ch018

1. INTRODUCTION

Like other developing countries, most of government and community income in Indonesia still depend on natural resources. As a result natural forest resources have been under great pressure of conversion. Based on FAO (2005)the rate of natural forest conversion in Indonesia is about 1.2% per year. This figure is higher than deforestation in Brazil (0.4%) and RD Congo (0.4%). Based on MOF RI (2007) total conversion of forest in the 5 biggest islands of Indonesia was, during 2000 – 2005, about 1.9 million hectares and Sumatra & Kalimantan were the biggest contributors. Deforestation relates to many factors, e.g. population growth (Palo, 1994), forest logging (Kummer, 1991), shifting cultivation (Thapa & Weber, 1990; MOF RI, 2007), illegal logging (MOF RI, 2007), resettlement (Hurst, 1990), road construction (Hirsch, 1987; Geist & Lambin, 2001) and Krutilla *et al.* (1996), international debt (Kahn & McDonald, 1994), and policy failure by government (Repetto & Gillis, 1988).

There are many publications pointing out that population increase will affect land use changes (Ramankutty et al., 2002). Angelsen & Kaimowitz (1999) argued that increased population growth leads to increase of demand for forest land and resources, and furthermore, the high rates of deforestation will drive to poverty. The connection between population growth and the rate of deforestation is also pointed out by Zhang et al. (2000). He stated that population growth in China is the main factor contributing to the loss of natural forests. Studies from Brazil (Andersen, 1996), Mexico (Barbier & Burgess, 1996), and Thailand (Cropper *et al.*, 1997) also gave similar results. However, Sunderlin & Resosudarmo (1996) pointed out that the impact of human populations on the deforestation in Indonesia is site-specific.

So far, analyses of deforestation are based more on numerical statistical data and less on consideration of spatial context, whilst, in fact, it is very important to assist policy makers in understanding the process and take it into consideration, when decisions are made. Important data on the rate and spatial distribution of deforestation have been provided by the analysis of remote sensing images (DeFries *et al.*, 2000). Furthermore, Lambin (2001) and Angelsen & Kaimowitz (1999) pointed out, that other researchers had studied deforestation at detailed scales by identifying the causes and underlying driving factors of the processes leading to deforestation. These models make an important contribution to the integrated analysis of the different deforestation trajectories in their environmental and socio-economic context.

Land-use and land cover change analysis in Java have been investigated by Verburg *et al.* (1999). They have predicted that land use change will especially occur in the lowland areas, either directly through construction or indirectly through the demand for higher value crops. The upland areas will stay primarily rural. The models were developed based on rough grid spatial data equal to 40 km x 40 km (1,600 km2) derived from agricultural surveys by the Central Bureau of Statistics and coupled with provincial forest cover data.

The objective of this study is to illustrate possible application of spatial modeling for deforestation in Java by using available forest cover data derived from remote sensing data and social economical data derived from village surveys (Potensi Desa/PODES), which were mapped on 10 km x 10 km grid spatial data.

2. METHODOLOGY

2.1 Datasets, Data Preparation and Statistical Analysis

In order to analyze spatial patterns of deforestation and make the prediction on deforested areas with a probability of conversion in the future, several datasets were used (Table 1). The information on forest cover in Java & Sulawesi was obtained from datasets of the land use map of Department of

Table 1. Data used, assumption and criteria in the deforestation model

Data	Source	Assumption	Criteria
Deforestation	Analysis from Land use map by Department of Forestry (LU 2000, 2005) and Ministry of Environment (LU 2005)	Analyzed from forest cover change from 2000 to 2005, and the ideal threshold was a half of grid size ($100 km^2$). But, since that threshold was not significant, $20 km^2$ was used as a threshold for deforestation	Deforestation > $20 km^2$ = 1 Deforestation < $20 km^2$ = 0
Slope (X1) (c_slope)	Generated from SRTM DEM USGS (2004), Shuttle Radar Topography Mission 90 x 90m, Global Land Cover Facility, University of Maryland, College Park, Maryland, February 2000.	Slope data was stretched into 8 Bit Data (0 – 255)	Min Value = 0 Max Value = 255
Population density (X2) (c_popdens)	Analyzed from BPS-Statistics Indonesia, data PODES 2000 and 2005	Population density data was stretched into 8 Bit Data (0 – 255)	Min Value = 0 Max Value = 255
Elevation (X3) (c_elev)	Generated from SRTM DEM USGS (2004), Shuttle Radar Topography Mission 90 x 90m, Global Land Cover Facility, University of Maryland, College Park, Maryland, February 2000.	Elevation data was stretched into 8 Bit Data (0 – 255)	Min Value = 0 Max Value = 255
Road density (X4) (c_road)	Extracted from Base and Topographic map Scale 1:25.000 by National Coordinating Agency for Surveys and Mapping, Indonesia (1999)	Road density data was stretched into 8 Bit Data (0 – 255)	Min Value = 0 Max Value = 255
Population having agricultural sector source income (X5) (c_ptdens)	Analyzed from BPS-Statistics Indonesia, data PODES 2000 and 2005	Population having agricultural sector source income was stretched into 8 Bit Data (0 – 255)	Min Value = 0 Max Value = 255
Population having non-agricultural sector source income (X6) (c_nptdens)	Analyzed from BPS-Statistics Indonesia, data PODES 2000 and 2005	Population having non-agricultural sector source income was stretched into 8 Bit Data (0 – 255)	Min Value = 0 Max Value = 255

Forestry in 2000, and a land use map of Ministry of the Environment in 2005. Both data sets were vector format data (Esri shape file).

First we synchronized the datasets with the same definition of forest cover, and then forested areas were separated from non-forested area. The pattern of forest cover represented in the deforested map was the result of the history of deforestation events from 2000 to 2005. Based on the deforestation map we developed a binary grid (10 x 10 km) map of deforestation, where value 1 represented deforested area and 0 represented non-deforested area. The grid was assigned as 1 (deforested area) if about 20 km2 forest within one grid is converted to other land cover (Figure 1). Further, each data parameter of independent

Figure 1. Deforestation in 2000-2005

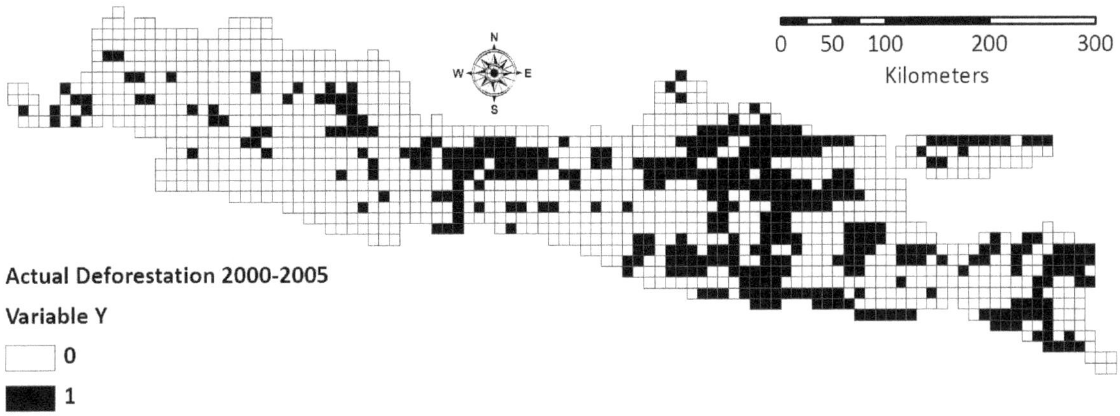

variables were re-sampled in a 10 km grid as a unit analysis in the model (Table 1, Figures 2-7). Vector grid data of 10 km were made by creating fishnet command in ArcGIS. Grid attributing process for vector data was conducted by Hawth Tools, free add-on extension in ArcGIS version 9.2 (http://www.spatialecology.com/htools) and raster data by ERDAS Imagine 9.1.

As explained above the population growth is expected to be potentially the major driver of deforestation. A map of population density from 2000 to 2005 was generated at the village-level using national census data (*Potensi Desa/* PODES). The population growth is continuously changing in time and space; therefore, simulations were made in this study. Two scenarios were used, namely an increase of those independent variables as high as 1.2% for normal/moderate scenario and an increase of 2.4% of those variables for extreme scenario. Decision of using normal/moderate scenario is based on recent data of population increase in Java and the extreme scenario was assumed as two times as high as the normal scenario.

In order to quantitatively validate our predictions of deforestation, we used MODIS satellite images in 250 m resolution with 16-day composite which were acquired in February 2000, February

Figure 2. Slope of the study area

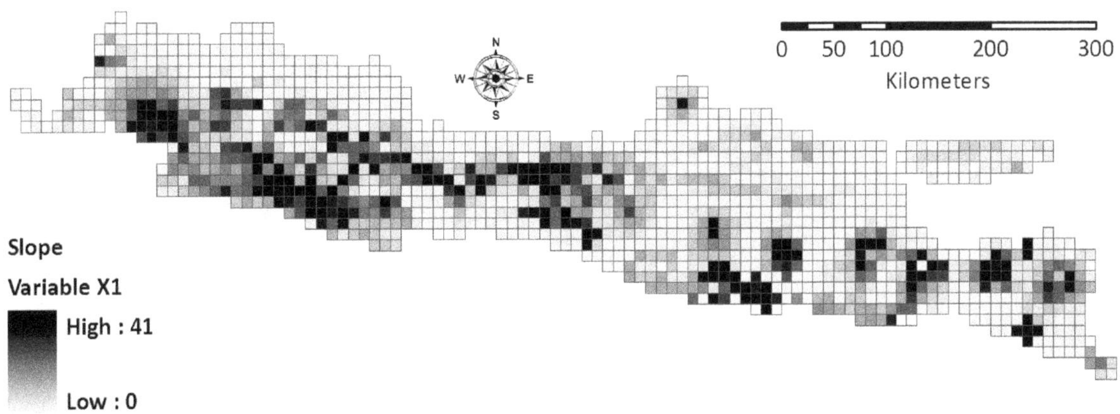

Figure 3. Population density of the study area

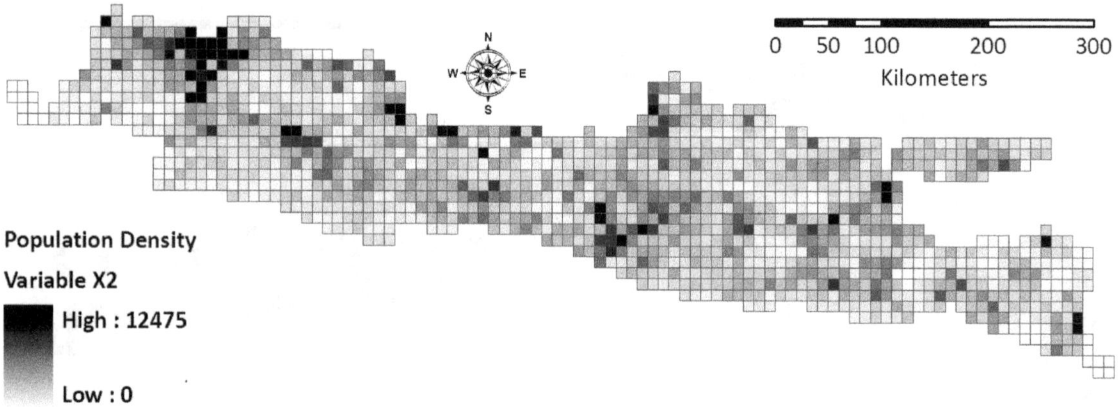

Figure 4.

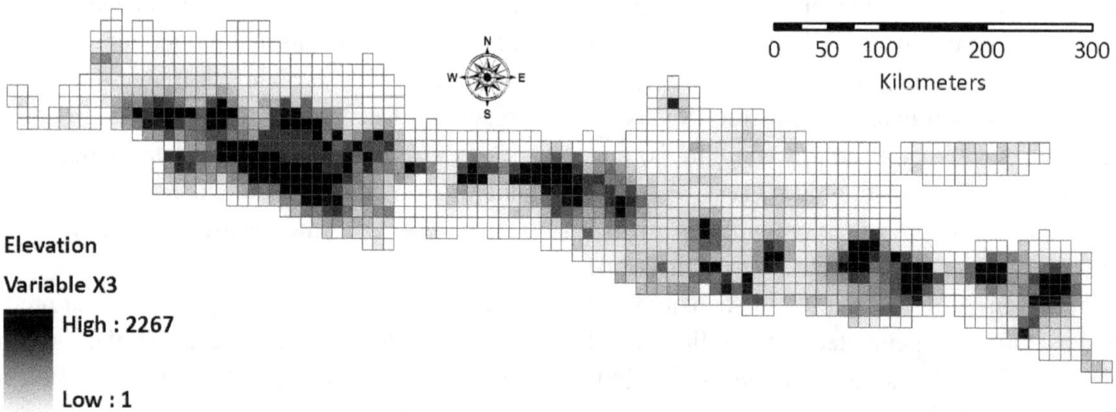

Figure 5. Elevation of the study area

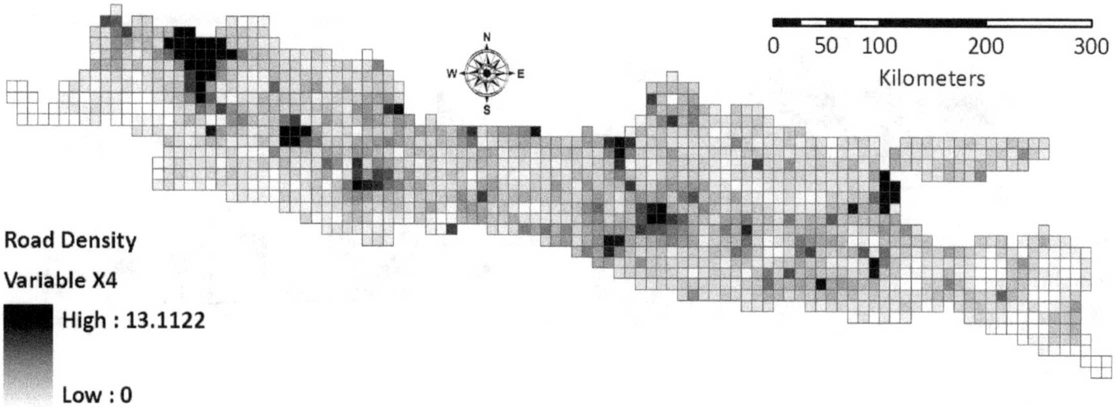

Figure 6. Road density of the study area

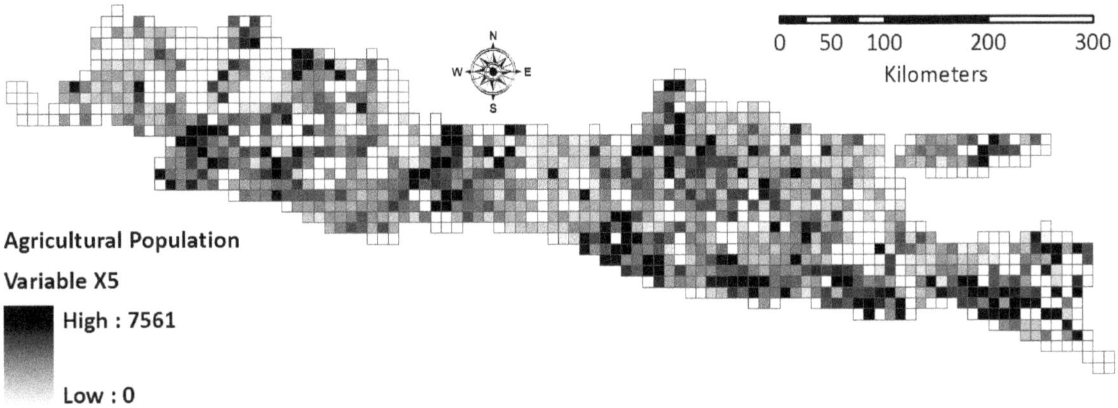

2008, August 2000 and August 2008. The image MODIS was obtained from Land Processes Distributed Active Archive Center, U.S. Geological Survey, http://lpdaac.usgs.gov/datapool/datapool.asp.

Pixels forest value of MODIS was identified and classified from MODIS datasets in different season data in order to get annual forest and non-forest coverage. Then, a forest-non forest map was re-sampled to 10 km grid size. The flow of the study is presented in Figure 8.

2.2 Statistical Modeling

As explained in Table 1, the six independent variables were used as predictors in the analysis. Logistic regression as statistical modeling was employed for estimating event probabilities of the occurrence of the deforestation as a dichotomous dependent variable. The regression coefficients obtained were used for integrating the spatial layers and the result was aggregated using a logit transformation [P = {exp(a+BX..)/1+(exp(a+BX..)}] by using IBM SPSS Statistic ver.19, to obtain the probabilistic map of deforestation.

Figure 7. Household engaged in non-agriculture sector

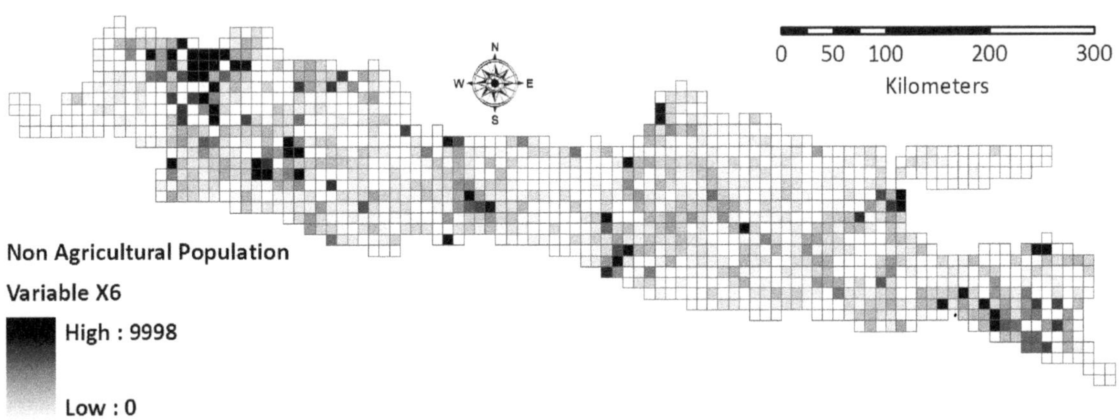

Figure 8. Flow of the research

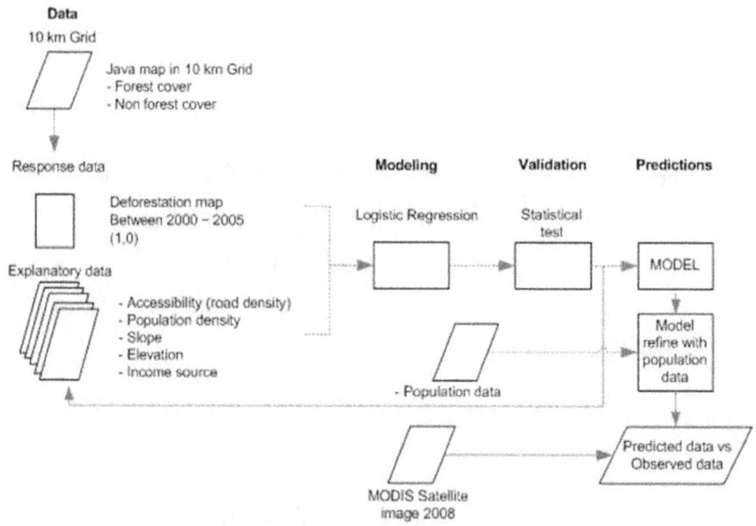

The initial specification of the model, based on theoretical considerations and data availability, is shown in Box 1. Where:

P: probability of the occurrence of deforestation; a: intercept; β: coefficient of parameter, c_slope ; slope ; c_popdens: population density; c_elev: elevation; c_road: road density; c_ptdens: percentage of population having agricultural sectors source of income; c_nptdens: percentage of population having non-agricultural sectors source of income

Spatial modeling was done using logistic regression to predict the future spatial location of forest conversion, whereby the predictions using two kinds of population growth rate were 1.2% (normal/moderate scenario) and 2.4% (extreme scenario).

Results of logistic regression models are often judged as successful if predicted probabilities, i.e. $P > 0.5$ correspond with the observed occurrence and value $P < 0.5$ with the absence of occurrence. Finally, we validated the deforestation map predicted in 2008 as a result of deforestation modeling with observed data of cleared forest/non-forest areas, which was interpreted from MODIS satellite imagery. Our aim was to validate only the approximate location of predicted forest conversion, and not to quantify the change. Then, the model was used to predict the occurrence of deforestation in 2020.

3. RESULTS AND DISCUSSION

3.1 Logistic Regression Equation

The result of logistic regression of Java is presented in the following equation model 1, shown in Box 2. Where:

Box 1.

$$1 + e^{-(a+\beta_1(c\_slope)+\beta_2 c\_popdens)+\beta_3(c\_elev)+\beta_4(c\_road)+\beta_5(c\_ptdens)+\beta_6(c\_nptdens)}$$

P: probability of the occurrence of deforestation; population density; c_elev: elevation; c_road: road density; c_ptdens: percentage of population having agricultural sectors source of income; c_nptdens: percentage of population having non-agricultural sectors source of income

Results of the Hosmer and Lemeshow test showed, that the model fits the data, which is indicated by the significance level being not less than 0.05. It means that the model can be applied for further analysis. Result of Nagelkerke R Square of the Logistic regression is 0.534. It means that variables, that were used only can explain 53.4% of the deforestation probability and the rest was influenced by other variables, that were not considered in the model. Significance tests of each variable are presented in Table 2. They showed that slope, population density and road density were significant in predicting the deforestation process, while variables derived from village survey data

Box 2.

$$p = \frac{e^{(1.480+0.024(c\_slope)-0.144(c\_popdens)+0.014(c\_elev)-0.109(c\_road)+0.003(Ptdens)-0.002(c\_nptdens)}}{1+e^{(1.480+0.024(c\_slope)-0.144(c\_popdens)+0.014(c-elev)-0.109(c\_road)+0.003(ptdens)-0.002(c\_nptdens)}} \quad (1)$$

Table 2. Coefficient of the equation

Factors	B	S.E.	Wald	df	Sig.	Exp(B)
Constant	1.480	0.243	37.131	1	0.000	4.393
c_slope	0.024	0.007	12.445	1	0.000	1.025
c_popdens	-0.144	0.021	45.498	1	0.000	0.866
c_elev	0.014	0.006	5.466	1	0.019	1.014
c_road	-0.109	0.017	39.308	1	0.000	0.897
c_ptdens	0.003	0.004	0.614	1	0.433	1.003
c_nptdens	-0.002	0.011	0.019	1	0.890	0.998

B: estimated logit coefficient, S.E: Standard Error of the coefficient, Wald = $[B/S.E]^2$, df: degree of freedom, Sig: significance level of the coefficient, Exp(B): is the odds ratio of the individual coefficient

Figure 9. Deforestation derived from MODIS during 2000-2008

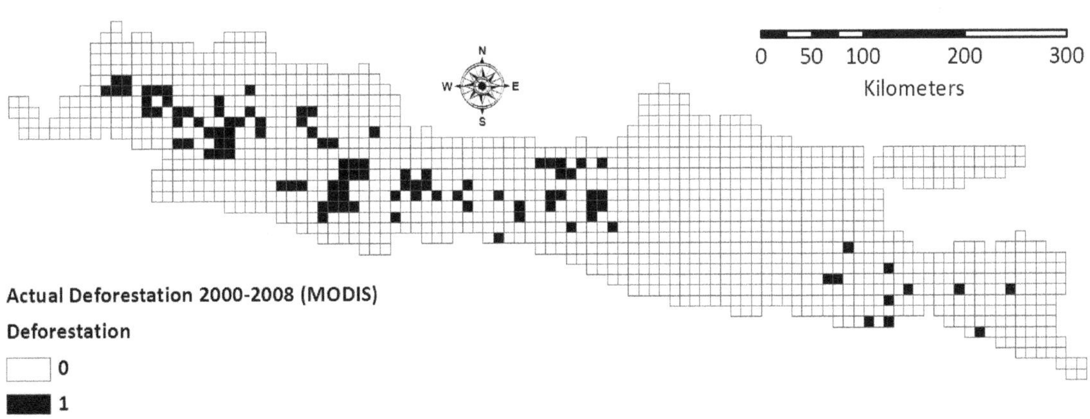

were not significant. Road and population density were having negative impact on deforestation. It means that as population density and road density increase, deforestation probability will decrease. This fact is opposite to findings of other researchers. This is due to the fact that most of remaining forests are distributed in rural areas with limited access and lower population density. Meanwhile slopes were having positive impact. It means that deforestation now tends to occur at steeper slope areas.

3.2 Model Validation & Prediction

The logistic regression model was also used to predict the deforestation in 2008, and was validated using observed deforestation data derived from MODIS satellite imagery taken in 2000 and 2008 (Figure 9). The validation result showed that the overall accuracy of the model is equal to about 66.67% both for moderate & extreme scenarios. In case of moderate scenario, the producer accuracy and user accuracy for un-deforested area were 72.73% and 11.43% respectively. Meanwhile, the producer accuracy and user accuracy for deforested area were 53.66% and 94.29%, respectively. In the case of extreme scenario, the producer accuracy and user accuracy for un-deforested area were 55.00% and 15.71% respectively. Meanwhile, the producer accuracy and user accuracy for deforested area were 55.05% and 85.71%, respectively. Overlays of actual and predicted deforestation are presented in Figures 10 and 11.

For deforestation predictions in 2020, a moderate scenario was selected. Results of the predictions are presented in Figure 12. The figure shows some deforested areas occur in remote rural areas and some of them are in National Parks, such as Halimun Salak National Park, Ujung Kulon National Park and protected forests in the southern part of Bandung, in West Java, as well as in Alas Purwo National Park in East Java.

Policy implication of the result model prediction is that the government should pay more attention to rural areas and has to develop good access and to create non-agricultural sectors jobs in order to reduce pressure on forest, especially in districts, that will face serious deforestation. Un-resolved forest border conflicts between communities and the government, as an underlying factor of state forest (government forest area) encroachment (Prasetyo *et al.*, 2008), should be mediated. This is due to the fact, that there are some villages, that are situated within and surrounding national state forest areas. Percentages of the village areas within and surrounding state forests in West Java, Banten, and East Java are 26.37%, 21.09%, and 35.96% of the total state forest area, respectively (BPS, 2007).

4. CONCLUSION

This study showed the utility of a combination of a statistical modeling approach and spatial analysis for analyzing and predicting deforestation. Population density, road density and slope were found to be the important variables in the model for explaining the pattern of deforestation observed in Java. Meanwhile variables related to economic factors, derived from village level data, were not significant. Deforestation in 2020 will tend to occur in remote rural areas, especially in national park areas. Government should therefore pay more attention to rural/remote areas and create non-agricultural sectors jobs in order to reduce the pressure on forests.

ACKNOWLEDGMENT

We would like to express our gratitude to the Coordinating Ministry for Economic Affairs, Republic of Indonesia for their support.

Figure 10. Deforestation derived from moderate scenario

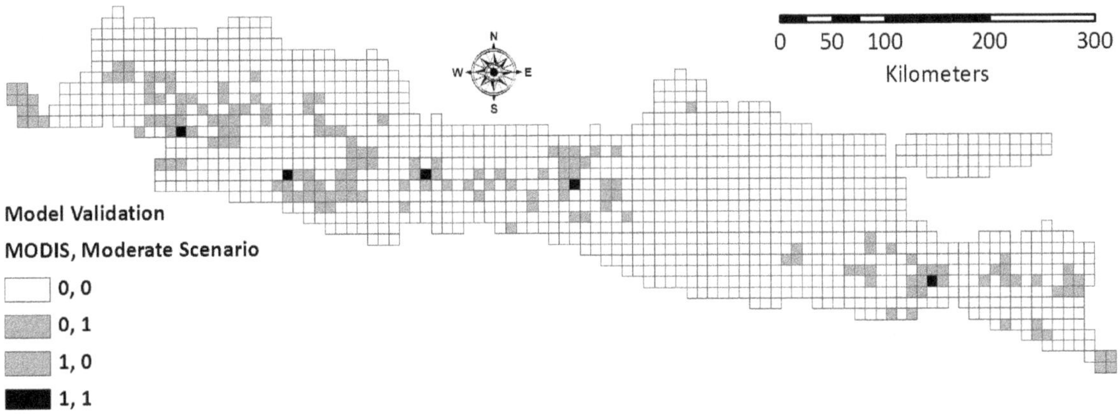

Figure 11. Deforestation derived from extreme scenario

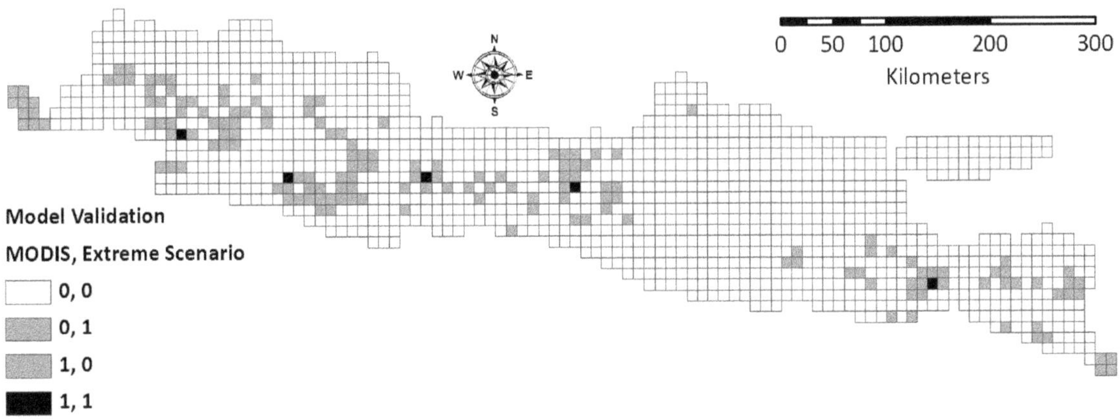

Figure 12. Deforestation prediction in 2020

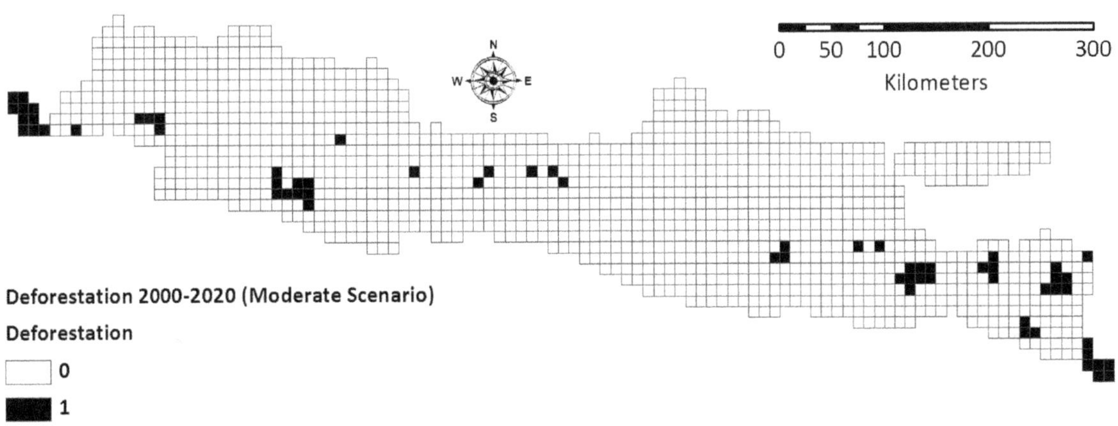

REFERENCES

Andersen, L. E. (1996). The causes of deforestation in the Brazilian Amazon. *Journal of Environment & Development*, *5*(3), 309–328. doi:10.1177/107049659600500304

Angelsen, A., & Kaimowitz, D. (1999). Rethinking the causes of deforestation: Lessons from economic models. *The World Bank Research Observer*, *14*(1), 73–98.

Barbier, E. B., & Burgess, J. C. (1996). Economic analysis of deforestation in Mexico. *Environment and Development Economics*, *1*(2), 203–239. doi:10.1017/S1355770X00000590

BPS (Biro Pusat Statistik). (2009). *Identifikasi Desa Dalam Kawasan Hutan. Kerja sama Pusat Rencana dan Statistik Kehutanan, Departemen Kehutanan dengan Direktorat Statistik Pertanian*. Jakarta: Badan Pusat Statistik. (in Indonesian)

Cropper, M., Griffiths, C. W., & Mani, M. (1997). *Roads, population pressures, and deforestation in Thailand, 1976-89*. (Policy Research Working Paper 1726). World Bank, Policy Research Department, Washington, D.C.

De Fries, R. S., Hansen, M. C., & Townshend, J. R. G. (2000). Global continuous fields of vegetation characteristics: A linear mixture model applied to multi-year 8km AVHRR data. *International Journal of Remote Sensing*, *21*, 1389–1414. doi:10.1080/014311600210236

FAO. *(2005)*. Working paper 18: Global forest resources assessment update 2005, specification of national reporting tables for FRA 2005. *Rome*

Geist, H. J., & Lambin, E. F. (2002). Proximate causes and underlying driving force of tropical deforestation. *Bioscience*, *52*(2), 143–150. doi:10.1641/0006-3568(2002)052[0143:PCAUDF]2.0.CO;2

Hirsch, P. (1987). Deforestation and development in Thailand and Singapore. *The Journal of Tropical Geography*, *8*(2), 129–138. doi:10.1111/j.1467-9493.1987.tb00190.x

Hurst, P. (1990). *Rainforest politics: Ecological destruction in Southeast Asia*. London, UK: Zed Books Ltd.

Kahn, J., & McDonald, J. (1994). International debt and deforestation. In K. Brown. & D.W. Pearce (Eds.), *The causes of tropical deforestation*, (pp. 55-106). Berkeley, CA: University of California Press.

Krutilla, K., Hyde, W. F., & Barnes, D. (1995). Peri-urban deforestation in developing countries. *Forest Ecology and Management*, *74*(2), 181–195. doi:10.1016/0378-1127(94)03474-B

Kummer, D. M. (1991). *Deforestation in the postwar Philippines*. The University of Chicago and London.

Lambin, E. F., Turner, B. L., Geist, H. J., Agbola, S. B., Angelsen, A., & Bruce, J. W. (2001). The causes of land-use and land-cover change: Moving beyond the myths. *Global Environmental Change*, *11*(4), 261–269. doi:10.1016/S0959-3780(01)00007-3

MOF RI. (2007). Statistik kehutanan [Departemen Kehutanan Indonesia] [In Indonesia]. *Indonesia*, 2007.

Palo, M. (1994). Population and deforestation. In Brown, K., & Pearce, D. W. (Eds.), *The causes of tropical deforestation* (pp. 55–106). London, UK: UCL Press.

Prasetyo, L. B., Wibowo, S. A., Kartodihardjo, H., & Tonny, F., Haryanto, Sonaji, R., & Setiawan, Y. (2008). Land use and land-cover changes of conservation area during transition to regional autonomy: Case study of Balairaja Wildlife Reserve in Riau Province, Indonesia. *Tropics*, *17*(2), 99–108. doi:10.3759/tropics.17.99

Ramankutty, N., Foley, J. A., & Olejniczak, N. J. (2002). People on the land: Changes in global population and croplands during the 20th century. *Ambio*, *31*, 251–257.

Repetto, R., & Gillis, M. (1988). *Public policies and misuse of forest resources*. Cambridge, UK: Cambridge University Press. doi:10.1017/CBO9780511601125

Sunderlin & Resosudarmo. (1996). *Rates and causes of deforestation in Indonesia: Towards a resolution of the ambiguities*. CIFOR.

Thapa, G. B., & Weber, K. E. (1990). Actors and factors of deforestation in tropical Asia. *Environmental Conservation*, *17*(1), 19–27. doi:10.1017/S0376892900017252

Verburg, P. H., Veldkamp, A., & Bouma, J. (1999). Land use under condition of high population pressure: The case of Java. *Global Environmental Change*, *9*, 303–312. doi:10.1016/S0959-3780(99)00175-2

Zhang, Y., Uusivuori, J., & Kuuluvainen, J. (2000). Econometric analysis of the causes of forestland use/cover change in Hainan, China. *Canadian Journal of Forest Research*, *30*, 1913–1921. doi:10.1139/cjfr-30-12-1913

Chapter 19
Embedding Biodiversity Modelling in the Policy Process

Nguyen Dieu Trinh
Ministry of Planning and Investment, Vietnam

Wilbert van Rooij
AIDEnvironment, The Netherlands

ABSTRACT

Biodiversity modeling for supporting policy processes is a relatively new field. Models can help policy makers to get a quick assessment of biodiversity and provide them with answers to some of their key questions on biodiversity. Models also allow them to evaluate the effects of proposed environmental policies on biodiversity and whether the policies are likely to meet their environmental targets and thus allow policies to be revised accordingly to meet the targets. In order to use modeling as a standard tool to support policy makers, it should be embedded in a policy process. The Strategic Environmental Assessment (SEA) is such a process that is well suited to include biodiversity modeling. Besides, it is forward-looking, has proper scale and timing components, and it needs an integrated approach to link social consequences on land use change and impacts on biodiversity. The modeled impacts on biodiversity can be used in SEA to guide the decision process. The results of the GLOBIO3 application at national level in Vietnam were considered useful for policymakers; however, the tools are not yet properly embedded in a policy context requiring number of conditions to be met to deliver appropriate information to the policy makers.

1. INTRODUCTION

Biodiversity is declining rapidly in many places and ecosystems. Without some promising measures to counteract the current development process, biodiversity decline will continue globally (sCBD, 2010). Land-use change is the major driver of biodiversity loss, but other drivers like pollution, fragmentation and climate change play an increasing role. Policy makers are increasingly aware

DOI: 10.4018/978-1-60960-619-0.ch019

of the risks of biodiversity loss, as biodiversity supports many ecosystem services and benefits to human well-being (MA, 2005). Therefore, policy makers need and want to be well informed on expected biodiversity loss. They need information on current and expected trends in biodiversity, and insights into the main driving forces, such as land-use change, which is duly recognized in by Convention on Biological Diversity (UNEP, 2007). The information can be used to adjust proposed policies and design alternative options in such a way, that environmental objectives can be realized. This chapter outlines the background and applicability of a national biodiversity model aimed at informing policy makers on biodiversity impacts taking a case of Vietnam.

Biodiversity modelling adds to existing methodologies of biodiversity assessment by bridging the information gaps since many countries in the world have not yet established an ecological network, that monitors all species groups by frequent inventories throughout the entire country. Biodiversity modelling thus can assess whether policies meet environmental objectives for biodiversity and helps to answer some key questions related to biodiversity, such as:

- What is changing? (indicators and monitoring)
- How is it changing? (modelling)
- What can we do about it? (assessment of drivers)
- What is the impact of policies? (assessments of policy options)

Biodiversity mainly depends on changing environmental factors therefore modelling of biodiversity focuses on the relationship between drivers and their impacts. It can be done either by relating species occurrences with environmental drivers (species modelling) or by directly relating a biodiversity indicator to these drivers (pressure based modelling).

The GLOBIO3 biodiversity model belongs to the latter approach. GLOBIO3 uses the mean species abundance (MSA) of originally occurring species relative to their abundance in undisturbed ecosystems. It describes the 'intactness' of an area so that, e.g. primary forests have a maximum possible MSA value and asphalted parking places have otherwise. The Convention on Biological Diversity (CBD) proposes five types of indicators to assess the status of biodiversity:

1. Trends in the extent of selected biomes, ecosystems and habitats;
2. Trends in abundance and distribution of selected species
3. Change in status of threatened species: Red list index
4. Trends in genetic diversity of domesticated animals, cultivated plants, and fish species of major socioeconomic importance.
5. Coverage of protected areas

MSA belongs to the second group. The extent of ecosystems is also derived from GLOBIO3. GLOBIO3 has been developed to assess effects of environmental change on biodiversity. The model can assess past, present and future biodiversity, expressed in a limited set of indicators, at national, regional and global scales. The model is built on simple cause-effect relationships between driving forces and biodiversity impacts in terms of MSA. These relations are derived from extensive literature research. Using these general relationships allows assessments in cases where limited field data are available. This makes the assessments time and cost-effective. Drivers are land-cover change, land-use intensity, fragmentation, climate change, atmospheric nitrogen deposition, and infrastructure development. Input data from these drivers are derived from available statistical data, spatial maps, other models and expert knowledge (see details in Chapter 8). GLOBIO3 has been used successfully in several integrated national (Trisurat *et al.*, 2010), regional and global assessments (Alkemade *et al.*, 2009).

It can generate the following direct and indirect outputs:

- the impacts of environmental drivers on MSA and their relative importance;
- expected biodiversity trends under various future scenarios; and
- the likely effects of various policy options.

For the implementation of the GLOBIO3 model at national scale, the original global model is downscaled to a spatial resolution of 1 * 1 km^2 and uses more detailed national datasets and local expert knowledge. The national application of the GLOBIO3 model can be integrated with the CLUE land use model in order to assess the impact of land use change on future biodiversity. CLUE (*Conversion of Land Use and its Effects*) is a model that is used to carry out a regional analysis of land use change as land-use change is the most important driver of biodiversity change (see details Chapter 6).

2. CAPCITY BUILDING

Since 2002 the Dutch government supported the International Biodiversity Project, which was carried out by the Netherlands Environmental Assessment Agency (PBL). The goal of this ongoing project is to build tools and institutional capacity for continuous support of biodiversity related policies, including the Convention on Biological Diversity (CBD). The current GLOBIO3 model is one of the biodiversity assessment tools that have been developed by the International Biodiversity Project. An international collaboration network with government organisations, NGOs and research institutes has been established to support the implementation of the tools for supporting policy makers at national, regional and global scale.

The GLOBIO3 model (http://www.globio.info) has been applied on a global scale for biodiversity assessments in the UNEP Global Biodiversity Outlook-2 (CBD, 2006), Geographical Environmental Outlooks (UNEP, 2007) and for FAO's Agricultural Assessments (FAO, 2006). At the regional scale the model is used for the Global Deserts Outlook (UNEP, 2006), the Fall of the Water (Nelleman,. 2004) EU-ruralis (Rienks, 2008) and for the Strategic Environment Framework for the Greater Mekong Subregion in Southeast Asia (GMS-EOC, 2008). At the national scale it has been implemented in the Ukraine, Colombia, Ecuador, Peru, Mexico, Nicaragua, Guatemala, Belize, Honduras, El Salvador, Panama, Kenya, Mozambique, Zambia, Cambodia, Laos, Myanmar, Thailand and Vietnam.

Capacity building activities of the project includes organization of training courses on land use and biodiversity modelling in several countries. It is important for optimal result of modelling exercise by combining bottom up technical training with a top down policy approach. Modelling intended to be embedded by the Agenda21 program as part of the sustainability indicator toolset, which is used for the development of national socio-economic development plans in Vietnam was such an effort of embedding modelling into policy. However, this intention has not been achieved yet. The reasons for this are explained in the next section of this chapter. At present, there is another project in Vietnam in which modelling is tried to be embedded in a policy context i.e. the 'Biodiversity modelling Inclusive Strategic Environmental Assessment Project' for the Quang Nam Land Use Plan for 2011-2020. Based on the experience of this project general guidelines will be developed for the integration of biodiversity modelling with Strategic Environmental Assessments (SEA).

3. BIODIVERSITY ASSESSMENT AND POVERTY LINKAGES PROJECT IN VIETNAM

3.1 Project Background

Vietnam joined the Convention on Biological Diversity (CBD) in 1994 and is one of 25 biodiversity hot spots in the world. Like in the rest of the world, economic development is occurring and putting pressure on the environment and on biodiversity. Adequate policies are required to lower these pressures. Biodiversity in Vietnam is declining and this may have negative consequences for the provision of many goods and services from ecosystems.

Recently, the Vietnamese government has paid attention to environmental protection with general assessment tools, but the limited knowledge on biodiversity and its linkages with poverty hampers balanced decision making and is thus a major obstacle for sustainable development. The information on impacts of policy decisions on biodiversity and ecosystem services is thus very limited.

The Ministry of Planning and Investment (MPI) is the focal point for preparing, monitoring evaluating and reporting on the whole development progress of the country, to the National Assembly and the Government. Besides its other assigned responsibilities by the Government, the MPI now needs to:

- pilot assessments of environmental issues in general and biodiversity in particular, as well as assessment of poverty reduction and the biodiversity – poverty linkages;
- prepare for improved information on biodiversity support for the inputs of Social and Economic Development Plans (SEDP)

Existing planning tools and environmental assessment tools appear to be inadequate and not easy to use. The indicator set of sustainable development currently proposed by the Ministry of Planning and Investment includes two indicators that are supposed to indicate the biodiversity status. These are percent forest cover and proportion of protected areas related to natural areas. But these two indicators do not represent the quality of forest and biodiversity values satisfactory. It becomes apparent that at the present time there is a need to develop new methodologies to assess forest quality and/or biodiversity value and link these with socio-economic development and poverty issues. In order to tackle this problem, the MPI carried out a cooperation on "Biodiversity Assessment and Poverty Linkage" in cooperation with PBL of Netherlands and scientists and directors from different research and management institutions. The main objective of the Dutch-Vietnamese cooperation is to support MPI in planning with thematic monitoring and evaluation of the biodiversity in Vietnam, by developing practical indicators and models that can be applied at national scale. In context of development strategy, plans and programs, these outputs will help policy makers to:

- Get information on past, present and future scenarios of status of biodiversity in Vietnam in a more convincing manner and with quantitative measures;
- Increase awareness towards the importance of biodiversity contributing to sustainable development;
- Investigate the potential negative impacts of development plans on biodiversity;
- Explore linkages between poverty and biodiversity;
- Access biodiversity trends and possible impacts on the poor/poverty;
- Avoid policy decisions that lead to lose – lose situation between biodiversity and poverty;
- Define ways to integrate biodiversity factors into the planning process at an early stage, to ensure minimizing the above mentioned negative impacts and gaining

Figure 1. Diagram of cooperation between PBL - MPI - SNV and other stakeholders

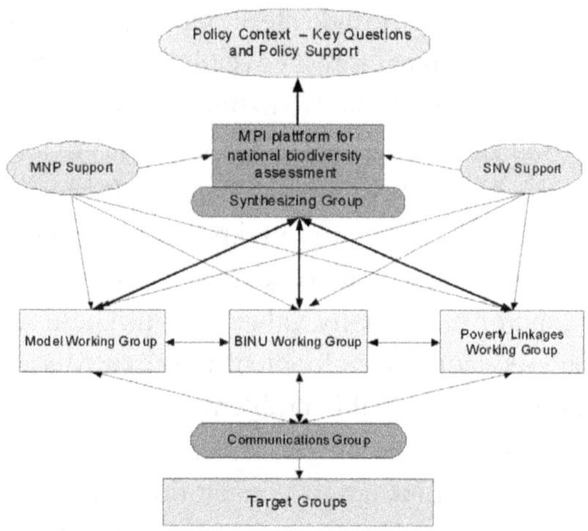

more results from poverty alleviation programs and SEDPs; and provide monitoring and evaluation tools for government and other related parties (bi-lateral evaluation for SEDP implementation).

3.2 General Approach

Over the last three years biodiversity modelling has been introduced in Vietnam to support national and regional policy makers with a tool for assessments of biodiversity through the International Biodiversity Project, a collaboration between MPI, PBL and the Netherlands Development Agency SNV (Figure 1). Two explorative case studies (Mangrove and Upland Case Studies by CRES in 2003-2004) two biodiversity modelling courses and various workshops were conducted in Hanoi. For an efficient implementation of the activities three working groups were established in the first stage of the project:

1. Biodiversity Indicators for National Use (BINU) group;
2. Biodiversity – Poverty linkages group;
3. Biodiversity Modelling group.

Table 1. Overall methodology of the project for 3 scientific working groups

Group	Past	Present	Future
Modelling	(+)	(++)	(++) Baseline scenario/ Policy option/CLUE methodology
Biological Indicator	(+)	(++)	(--) not applicable
Poverty & Biodiversity Linkage	(--) no information	(++)	(+) trend/vision/political plans

Notes: (++): Good information and results
(+): Moderate information and results
(--): No information or not applicable

In the second stage, a synthesizing group and a communication group were added that integrated the work from the groups and disseminated the findings to communities via leaflets, newsletters and by organising a puzzle contest at a secondary level school on biodiversity, environment and sustainable development for future citizens. To ensure scientific involvement, competent research institutes – CRES, the Forestry Department - FIPI, and other consultants were involved. The cooperation started in January 2006 with a round table meeting, followed by several technical meeting/workshops with participants from line ministries, local authorities, NGOs, INGOs, research institutes.

In Table 1 the overall methodology is illustrated based on the overview of existing research, data availability, compatibility and reliability concerning the biodiversity assessment and poverty linkages.

In general, there is relatively good information available that describes the present situation for the Biodiversity Indicator and Poverty group studies, but less detailed information for the past, and nearly no information for the future. Hence, three working groups were involved to collect, analyze and assess existing data at different temporal and spatial scales related to fields of biodiversity, poverty and their linkages. In the study, assessments have been carried out for the years 1993, 2000, 2020 and 2050. These years were selected because the first three years are time marks in socio-economic plans and strategies in Vietnam and the year 2050 for long-term forecasting. The applied spatial scales of the assessment are national, regional, and local. The national scale is important for MPI in the planning, monitoring and evaluation and integration process. The regional scale refers to the regional planning process driven by factors and might help to understand the situation in each particular region; and the local scale is important for obtaining understanding and implementing at the local level.

3.3 Modelling Process and Results

In this study, the GLOBIO3 model was not only used for assessing of biodiversity in Vietnam but the outcomes were also used for studying the relationship between biodiversity and poverty. For the implementation of GLOBIO3 in Vietnam, the model was first downscaled to national scale. This was realized by using national datasets, the integration of the model with the CLUE land allocation model and by making use of local expert knowledge. The 2000 land use map by the Forestry Department (FIPI) was used to derive the current biodiversity impact by land use. For deriving the impact of infrastructure and fragmentation, both the FIPI land use map and national road map were used. For the impact caused by nitrogen deposition and climate change, output of the Integrated Model to Assess the Global Environment (IMAGE) model were used (Chapter 5; Bouwman *et al*., 2006).

The land use map was grouped into 34 land use classes by specialists of the biodiversity Indicator working group. All land use classes were ranked according to their biodiversity level and compared with the biodiversity values for the generic GLOBIO3 classes, as used in the global application. The specialists assigned a biodiversity value to each unique land use class by interpolation of the MSA values of the generic GLOBIO3 forest classes.

In the expert interpolation method, the land use MSA values are not determined by extensive research, but by interpolation of known MSA values for generic land use classes. The best results would be achieved if the actual MSA value per land use class would be determined by extensive field work. Since inventories of species abundance for species groups per land use type are lacking in Vietnam and most other countries in the world, the interpolation method is used to adjust the generic relations for the local conditions. During the land use biodiversity valuation process, the local experts did not only look at the generic biodiversity values, but also compared the biodiversity of the unique land use type with that of the original vegetation at each location. The latter was derived from the global WWF Eco-region map.

For each pressure type MSA_pressure grid maps were calculated, representing the biodiversity loss in terms of MSA per pressure, per grid cell of 1*1km. In Figure 2 an overview is given of these MSA_pressure maps. The light color stands biodiversity loss and the dark color for low biodiversity loss.

Each MSA pressure map has grid cells with a value between 1 (high biodiversity) and 0 (low biodiversity). The overall MSA_total impact map was generated by a multiplication of all maps in a GIS grid calculation using the following formula:

$$MSA_{tot} = MSA_{lu} * MSA_{infra} * MSA_{frag} * MSA_{nitr} * MSA_{clim}$$

The overall biodiversity loss has been analysed for eight administrative regions and per protected area.

Figure 3 shows the overall MSA and biodiversity loss map for the year 2000. The two large areas with low biodiversity in the north and south of the map are respectively the Red River and Mekong river deltas with a high intensive rice

Figure 2. MSA_pressure maps generated with the GLOBIO3 model. From left to right: MSA_Land use, MSA_Infrastructure, MSA_Fragmentation MSA_Nitrogen deposition and MSA_Climate change

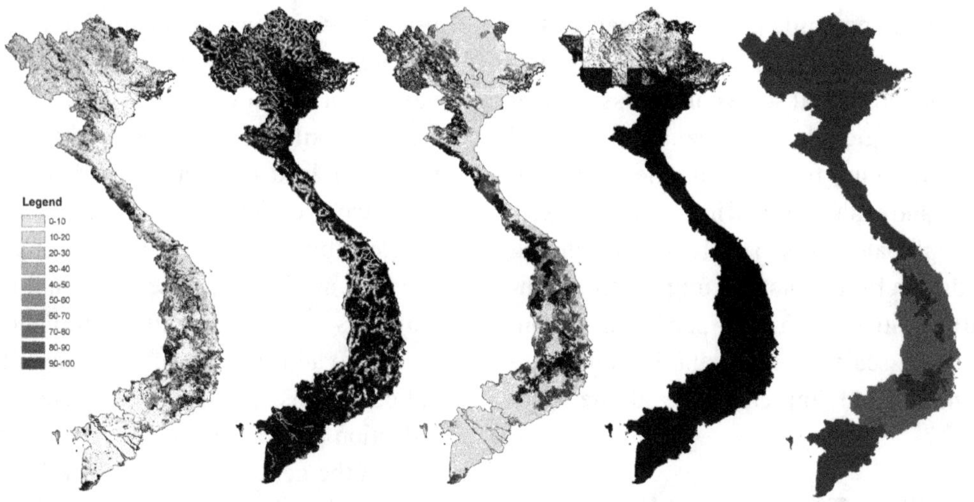

production. The pie chart shows the distribution of biodiversity loss per pressure type. The remaining biodiversity in Vietnam in year 2000 has a MSA of 26%. Seventy-four percent is lost because of the pressures specially land use, which is the largest cause being responsible for a biodiversity loss i.e. 56%.

The distribution of biodiversity loss per pressure per region shows that the Central Highland has the highest remaining biodiversity and the

Figure 3. Overall MSA loss in 2000 in Vietnam with distribution per pressure and per region.

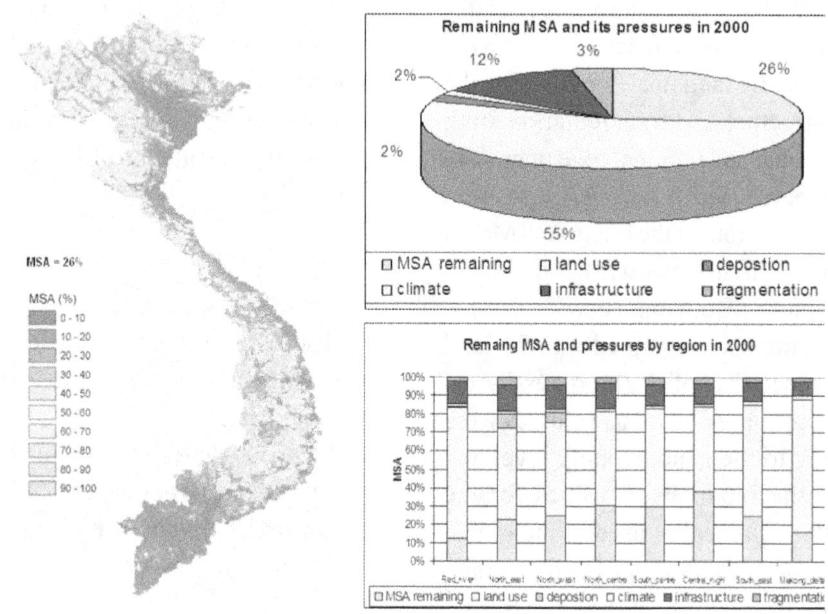

Table 2. Vietnam baseline scenario for the periods 2000-2020 and 2020-2050

Criteria	Rate of change	
	Period 2000-2020	Period 2020-2050
Agricultural land	- 0.5% /year	- 0.1% /year
Plantation	+ 500 km^2/year	+ 400 km^2/year
Primary forest (rich and medium)	+ 0.2% /year	+ 0.1% /year
Slightly disturbed forest	+ 2% /year	+ 0.2% /year
Heavily disturbed forest	+ 2% /year	+ 0.2% / year
Regrowth shrub and bushes	+ 3.5% /year	+ 0.35% /year
Shifting cultivation land	- 5% /year	- 7% /year
Residential land, urban land	+ 2% /year	+ 1.5% /year

Figure 4. Projected land use map Vietnam 2020, baseline scenario

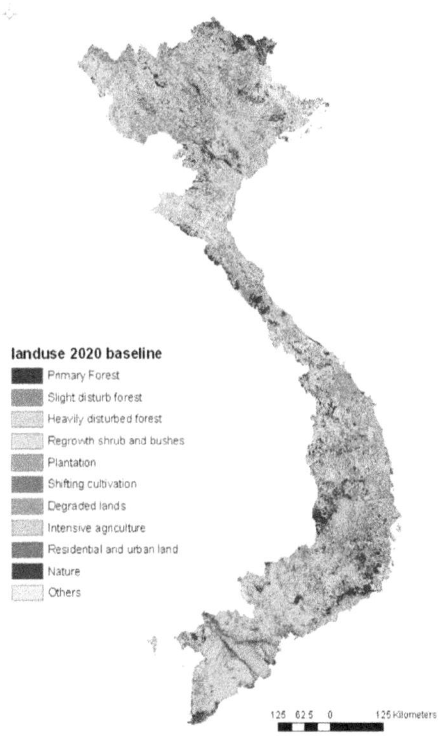

two river deltas have the lowest. The relative high losses in the two Northern provinces caused by Nitrogen deposition are the result of the use of fertilisers and air pollution in the neighbouring region in China. The northerly winds, drift Nitrogen southwards to Vietnam where it is deposited on the soil. Differences can also be seen in the impact of infrastructure and fragmentation per region. The latter is related to infrastructure, i.e. roads dissecting natural areas. The influence of climate change is still relatively small, but this effect will become larger in the near future.

The assessments have also been carried out for the years 1993, 2020 and 2050. With respect to the historical analysis of the 1993 land use data it appeared that the FIPI 1993 map had inconsistencies in relation to forest cover compared to the FIPI 2000 map. This was due to the fact that land use classes for the 2000 map were classified based on different criteria than those used for the 1993 map. For this reason the land use map of 1993 was not used in the analysis of biodiversity trends. The future maps were calculated with the CLUE model based on a land use scenario with two policy options produced by the Model and Indicator groups. The baseline scenario was based on information from national socio-economic development plans, the strategy for conservation and development, action plans and on the tendency of economical development and land use change by local experts. In case the information was lacking, the historical trends based on census date were extrapolated. In order to keep results consistent, the different land use maps were first aggregated. A summary of the expected rates of change for the base line scenario is shown in Table 2.

Before the CLUE model could be used to calculate the future land use maps, based on the scenario, demand table model parameters were set. A land use conversion matrix was made that indicated which land use class is allowed to change into another land use class. Location maps that have been used in CLUE to determine the regression equations to allocate land use change to new locations with the highest probability on maps

Table 3. Results of GLOBIO3 modelling for the baseline scenario in Vietnam

No	Remaining MSA and its pressure factors	2000 (%)	2020 (%)	2050 (%)
A	Remaining MSA	26.34	27.01	24.75
B	Pressure factors (Causes of biodiversity loss)			
	1. Land use change	54.11	47.07	43.60
	2. Infrastructure development	12.11	17.66	21.87
	3. Fragmentation	3.33	3.07	2.85
	4. Climate change	1.67	2.75	4.74
	5. Nitrogen composition (pollution)	2.43	2.44	2.19

are: digital elevation, slope, precipitation, population density, cost-distance to town, soil texture and soil depth and fertility. Figure 4 shows the resulting CLUE output map, i.e. the 2020 land use map for the baseline scenario.

With respect to the future impact of the other pressures a slight increased impact of the infrastructure has been applied with a correction for the expected population increase. No future road map was available. The future fragmentation map is constructed by analysing the remaining patch size areas of nature, based on the future land use map, dissected by the existing infrastructure map. The future nitrogen deposition and climate change maps have been derived from the Image model. The GIS analysis of the future MSA_pressure map shows that the MSA indicator first increases by 0.67% for the period 2000-2020, and then decreases by 2.26% from 2020-2050 (Table 3).

According to the results of the GLOBIO 3 model, the MSA values in all regions of the world, including South East Asia, have a tendency to decline from 2000 to 2050 (Alkemade *et al.*, 2009). However, in the case of Vietnam, the MSA has a tendency to slightly increase from 2000 to 2020. This remarkable deviation is explained by the fact that the Vietnamese efforts of reforestation have increased the forest cover. The new forest cover contributed to a higher MSA value than the existing degraded lands. The increase caused by this positive land use change and resulting lower fragmentation, is slightly higher than the decrease caused by infrastructure development, climate change and pollution. However, since there was no forest expert in the working group, the planted forest might be a bit overvalued while the relative large degraded land use class with varying vegetation patterns might be undervalued.

Figure 5. MSA trends in different region of Vietnam

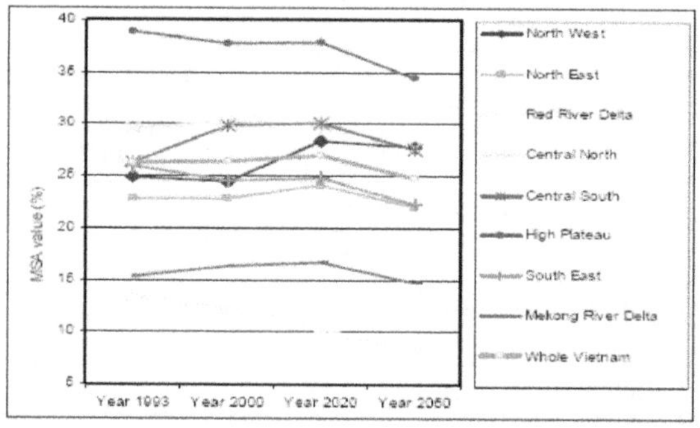

From 2020 onwards, the reforestation program is reduced from 500 km^2 per year to an estimated 400km^2 per year. The decrease of the MSA from 2020 to 2050 can be explained as inability of reduced reforestation to compensate for the growing losses caused by the increasing pressures. However, other pressure factors may play a more important role in future. Infrastructure development, in combination with population density effect, for example, is projected to increase from 12.11% in 2000 to 21.87% in 2050.

The biodiversity change over the years is not the same in all regions in Vietnam. Figure 5 shows the trends for the eight different regions of Vietnam. Disregarding the 1993 data because of its data inconsistencies, the largest differences can be seen for the two Northern provinces with an increase of MSA between 2000 and 2020, while in the same period the biodiversity declines in the Red River Delta. The differences in MSA per region, for the long term projection are less and show a general decline of MSA between 2020 and 2050 in the entire country.

In addition to the baseline scenario for 2020 and 2050 the project working groups also designed a policy option in which policy makers plan special measures to conserve biodiversity. The baseline scenario assumed that the current land use tendency, related to biodiversity conservation, would continue and applied it to the future, while the policy option scenario assumed a strong application of forestry/biodiversity conservation policy measures.

The assumptions for the conservation option were:

- Total forest cover (plantation + primary) in 2030 will reach 40% of the country's land area;
- Protected areas (PAs) increase from 7% to 10% of the land cover;
    - Existing parks and primary forests above 1000 m increase; and
    - Strict law enforcement; no land use change inside PAs.

The conservation policy option is calculated for its effects on biodiversity with GLOBIO3 and CLUE and showed an overall biodiversity increase of 1.37%. Although the overall increase appears small, the policy has a relative large effect on conservation of the national parks. The ongoing biodiversity loss in the parks as assumed for the baseline scenario is halted, while the existing parks in mountainous areas increase in extent.

3.3 Biodiversity Poverty Linkages

Four case studies were selected, each with a special theme: shifting cultivation, migration, hydropower and the construction of roads and infrastructure to examine the current state of poverty and biodiversity and their linkage. Since a poverty assessment is beyond the scope of this book only the findings of linkages will be described in brief. Poverty and biodiversity degradation are often interlinked for the reasons that (1) Poverty makes the poor communities dependent on small natural resources in local areas, and therefore makes them easily affected by natural and social changes and (2) Poverty results in lack of investment capital for production, infrastructure development, culture, education and environment improving projects. The lack of production and infrastructure development could well be positive for the biodiversity. However, biodiversity itself cannot be considered as a causal factor of poverty as exhibited by a map overlay of biodiversity with poverty which showed only a slight correlation between high biodiversity and poverty with no clear spatial linkage between them. In areas with a high remaining biodiversity/MSA low production opportunities exist because of: remoteness, low productive soils, low productivity, high transportation costs, no governmental control and no support of traditional livelihoods. Migration also produces new groups of poor people (40%) in illegal situations and with lack of support.

3.4 Role of Modelling Output in Policy Process

The models are useful and their methodologies are in line with international standards to quantify biodiversity objectives. However, the models are generic and it is a challenge to adjust them further to local conditions in Vietnam. Some of the key questions related to biodiversity could be answered with help of the models were that biodiversity is changing as shown by computed individual or overall impact of MSA through five major pressures. This suggests that using the contribution of each pressure to biodiversity loss, mitigation measures could be developed that could reduce the impact of individual pressures. With an additional assessment of drivers that increase the pressures on biodiversity, policy makers will have a better understanding of the underlying factors that cause biodiversity loss. It also allows to calculate the impact of different policy options and scenarios.

After one year of implementation and nearly three years of dissemination of the results from the cooperation, we see that bringing biodiversity into the planning process is a continuous process. The achievement of the project includes technical assessment process gradually emerging into policy-oriented activities, however a lot of activities and challenges still need to be addressed in terms of institutional and technical aspects via support from the government as well as bilateral and multilateral development partners in aspects of technical assistance and finance for continuing the work. Immediate benefits from the project were:

- An effective introduction of a new indicator and biodiversity assessment method that has potential for application in Vietnam
- A strengthened network and cooperation among various groups (modellers, researchers, NGOs, local communities, and policy-makers).

One of the prerequisites to get modelling output being used in a policy process is the need to communicate model results and their added value to policy-makers. Thus, the outputs need to be policy-relevant information on past and current environmental performances. The strategies to communicate the model results to policy makers are as below.

First of all, the target audience, that needs to be informed of the results of the study reports, should be identified. In this case, the key target group is policy-makers who tend to require distilled information with clear messages and implications for their policies. This could be done with a range of communication tools beyond the level of scientific reports, such as summary reports, interviews, brochures, presentations and Q&A tools.

Secondly, the message needs to be communicative attractive and meaningful. Language plays a vital role. The complex technical academic words have to be translated into simple language, that is understandable and digestible to both public and policy makers. The information and messages should appeal to the needs of the people to influence them.

The information delivery strategy in Vietnam is illustrated in Figure 6.

Figure 6. Delivery strategy in Vietnam.

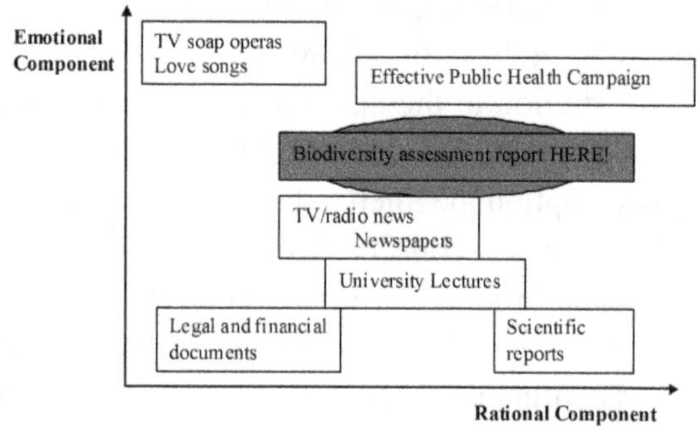

4. BIODIVERSITY MODELLING IN STRATEGIC ENVIRONMENTAL ASSESSMENTS

Land use and biodiversity modelling can be a valuable tool for policy makers who want to integrate environmental aspects in their decision making process. But in order to make it a part of a standard policy toolkit, it should preferably be embedded in a political process that directly benefits from it. The Integrated Assessment (IA) approach, the sustainability development processes and more specific the Strategic Environmental Assessment (SEA) methodology appear to be well suited for embedding modelling.

IA is defined as 'a participatory process of combining, interpreting and communicating knowledge from various disciplines in such a way that a cause-effect chain - involving environmental, social and economic factors – associated with a proposed public policy plan or programme can be assessed to inform decision makers' (UNEP,

Table 4. Building blocks for Integrated Assessment (UNEP, 2009)

Integrated Assessment building blocks		
A. Process	B. Policy institutional context	C. Analytical Contents
A1: Process design and links	B1: Institutional analysis and change	C1: Strategic framework and identification of key sustainability issues
A2: Policy-making decision windows	B2: IA team organizational model	C2: Trends and scenarios
A3: Communication strategy	B3: Stakeholder engagement and strengthening civil society	C3: Identification of opportunities and formulating alternative policy options
	B4: Evaluation and learning	C4: Assessment of impacts / risks and benefits.
		C5: Monitoring and evaluation

2009). The methodology intends to highlight connections between policies that strive to increase human well being and environmental sustainability. An overview of the IA methodology is shown in Table 4.

The modelling perfectly suits the Analytical contents building block since it provides information for key questions in relation to biodiversity, integrates the impact of major drivers that lead to biodiversity loss, calculates the biodiversity status for the past, present and future, helps in building land use scenarios and is able to calculate biodiversity and land use trends. The modelling tool is well suited for monitoring and evaluation of existing or planned policies.

The SEA is probably the best political instrument for which modelling can be used. Although voluntarily, it is gradually implemented in more and more countries to integrate environmental aspects with the planning process. The SEA is closely related to IA but focuses more on the environment. It is one of the outcomes of the eighth Conferences of the Parties meeting (COPVIII) for the Convention of Biological Diversity (CBD) with the endorsement of voluntary guidelines on biodiversity including environmental impact assessments (UNEP/CBD, 2006). SEA can be defined as 'the formalized, systematic and comprehensive process of identifying and evaluating the environmental consequences of proposed policies, plans or programmes to ensure that they are fully included and appropriately addressed at the earliest possible stage of decision making and on par with economical and social considerations' (Sadler & Verheem, 1996). An SEA might be applied for an entire sector or to a geographical area. In contrary to most Environmental Impact Assessments (EIA), SEA is in general applied *before* political decisions are taken for implementation. SEA is therefore proactive and sustainability driven whilst EIA is largely reactive (Slootweg *et al.*, 2006). With the help of modelling, policy decisions can first be evaluated on their general biodiversity impact, before decisions are made that might have a large impact on the environment.

Both modelling and the SEA methodology have a lot in common e.g.:

- **Integrated**: Address interrelationships of biophysical and social aspects. Both make use of the conceptual framework used by the Millennium Ecosystem Assessment which addresses linkages between direct and indirect drivers of change, ecosystem services and human well-being.
- **Participatory**: Both IA and EIA are participatory approaches that involve the participation of stakeholders throughout the entire decision making process. Also, for the modelling with CLUE and GLOBIO3, experts from different institutions are brought together to work on ecological (land use valuation), social (scenario building) and spatial data (GIS analysis).
- **Focussed**: Concentrate on key issues of sustainable development and provide usable information for development planning and decision making

- **Scale:** The sectoral or geographical implementations are often at regional or national scale. The impact of pressures is analysed separately and mutually for the study area or for the administrative region within the area.
- **Cost and time effective**: Make use of existing information, can be conducted in a relative short and cost effective period of time.
- **Iterative:** Ensure the assessment results in an early stage of the policy process, enabling adjusting decision making. Intended strategic policies can be evaluated on their impacts.

Another big advantage of using land use and biodiversity modelling within the SEA context is the quantification aspect. Although there are many reports about SEA, they lack information on how to derive quantitative information on the biodiversity status in a country or region. Existing information on biodiversity is in general scattered and not complete. A lot of information is available for some hot spot areas, but is often lacking for large human influenced areas. Additional inventories could provide such information, but generally they are too costly and very time consuming. The GLOBIO3 model allows generating information on biodiversity indirectly and uses available information on other indicators. The biodiversity status is modelled via generic relationships of other indicators with biodiversity. With the help of this methodology, information on biodiversity can be quantified and converted into trends, which are an important contribution to an SEA.

6. CONCLUSION

The Generic models have been used for a long time in economical assessments, but the introduction of models in national or regional biodiversity assessments is a relative new. It is very likely that they will become part of the standard policy tool set for biodiversity assessments in the near future as the modelling intends to add valuable information to the existing biodiversity assessment methodologies. The indicator that is calculated by the model, such as MSA, is complementary to other biodiversity indicators and together they aim to give a representative indication of the biodiversity.

GLOBIO3 and CLUE can be a powerful policy support tool, especially within the Strategic Environmental Assessment process because of its timing and scale, limited data needs and ability to calculate impact of different scenarios, and support analysis whether the political targets will be met. The models are basic and intended to be developed further for local conditions. Additional research, support, institutionalisation, policy embedding and further implementations will improve their acceptance as policy tools.

With regard to future work, it is important to have additional validation for local conditions although models like GLOBIO3 and CLUE have been tested in several countries. It is also desirable to have local implementation of the models in the context of SEA and IA along with further capacity development specific to technological knowhow.

REFERENCES

Alkemade, R., van Oorschot, M., Miles, L., Nellemann, C., Bakkenes, M., & ten Brink, B. (2009). GLOBIO3: A framework to investigate options for reducing global terrestrial biodiversity loss. *Ecosystems, 12*(3), 374–390. doi:10.1007/s10021-009-9229-5

Bouwman, A. F., Kram, T., & Goldewijk, K. (Eds.). (2006). *Integrated modelling of global environmental change*. An overview of IMAGE 2.4. (Report no. 500110002).

CBD. (2006). *Global biodiversity outlook 2*. Montreal.

FAO. (2006). *Global forest resources assessments 2005. Progress towards sustainable forest management. FAO forestry paper*. Rome: FAO.

GMS-EOC. (2008). *Subregional environmental performance assessment (EPA) report. National performance assessment and subregional strategic environment framework for the greater Mekong Subregion. TA No. 6069*. Bangkok, Thailand: GMS Environmental Operations Center.

Nelleman, C. (Ed.). (2004). *The fall of the water*. Arenda, Norway: United Nations Environmental Programme – GRID.

Rienks, W. A. (Ed.). (2008). *The future of rural Europe*. Wageningen, The Netherlands: Wageningen University Research and Netherlands Environmental Assessment Agency.

Sadler, B., & Verheem, R. (1996). *Strategic environmental assessment: Status, challenges and future directions*. The Netherlands: Ministry of Housing, Spatial Planning and the Environment.

Slootweg, R., Kolhoff, A., Verheem, R., & Höft, R. (2006). *Biodiversity in EIA and SEA*. Background document to CBD Decision VIII/28: Voluntary Guidelines on Biodiversity-Inclusive Impact Assessment. The Netherlands.

Trisurat, Y., Alkemade, R., & Verburg, P. (2010). Projecting land use change and its consequences for biodiversity in northern Thailand. *Environmental Management*, *45*, 626–639. doi:10.1007/s00267-010-9438-x

UNEP. (2006). *Global deserts outlook. Division of early warning and assessment*. Nairobi, Kenya: United Nations Environmental Programme.

UNEP. (2007). *Global environmental outlook 4: Environment for development*. Nairobi, Kenya: United Nations Environmental Programme.

UNEP. (2009). *Integrated assessment: Mainstreaming sustainability into policymaking. A guidance manual*. Nairobi, Kenya: United Nations Environmental Programme.

UNEP/CBD. (2006). *Decision adopted by the conference of the parties to the convention on biological diversity at its eighth meeting VIII/28. Impact assessment: Voluntary guidelines on biodiversity-inclusive impact assessment*. Curitiba, Brazil, 20-31 March 2006.

Section 5
Conclusion

Chapter 20
Conclusion and Recommendations

Yongyut Trisurat
Kasetsart University, Thailand

Rob Alkemade
PBL Netherlands Environmental Assessment Agency, Netherlands

Rajendra P. Shrestha
Asian Institute of Technology, Thailand

ABSTRACT

This chapter summarizes key findings of all the chapters contained in the book and presents analytical views on how modeling of land use and climate change and the consequent biodiversity change may potentially be used to assess past, current, and future threats to biodiversity and livelihoods of people at local and regional levels. In addition, this chapter identifies some key results, future innovations and research needs, e.g., accurate land use prediction, downscaling world climate data to local condition, and biodiversity/species distribution model. It also includes how to effectively implement the model results for conservation of land and biodiversity such as protected area system plan, optimal land use policy, environmental impact assessment, and strategic environmental assessment.

1. INTRODUCTION

The various chapters in this book describe how modeling of land use and climate change and the consequent biodiversity change may potentially be used to assess past, current and future threats to biodiversity and livelihoods of people at local and regional levels. Some chapters also point at the use of new fields like land use and biodiversity informatics. In addition, embedding the model results into policy support and implementation was discussed. This chapter identifies some key results, future innovations and research needs, as well as effective implementation of the model results.

DOI: 10.4018/978-1-60960-619-0.ch020

2. BIODIVERSITY AND ECOSYSTEM SERVICES

Biodiversity is defined as the variability among living organisms from all sources including, inter alia, terrestrial, marine and other aquatic ecosystems and the ecological complexes of which they are part; this includes diversity within species, between species and of ecosystems (Secretariat of the Convention on Biological Diversity, 2006). Biodiversity is also a valuable resource for humans. These values of biological resources are classified into two broad categories: direct values and indirect values (Mc Neely, 1998), which are similar to the concept of ecosystem services (MA, 2005). Ecosystem services are divided into four broad categories: provisioning, such as the production of food and water; regulating, such as the control of climate and disease; supporting, such as nutrient cycles and crop pollination; and cultural, such as spiritual and recreational benefits. They are considered of importance as the resource-base for many people, especially the rural poor. Protecting ecosystem services from being degraded may help eradicate poverty at local, national and international levels. Sustainable management of agricultural land and forests may be targeted as the protection of these ecosystem services. In Thailand, the National Economic and Social Development Plan (2008-2011) aims to develop the value of biodiversity and local wisdom for improving the livelihoods of local communities and eradicating local poverty (NESDB, 2008). Meanwhile, the Millennium Development Goals (MDGs) were formally established by the United Nations General Assembly. The MDGs targets for 2015 also address issues of poverty eradication and sustainable development using biological resources as resource-base (MA, 2005).

3. LAND USE, CLIMATE AND BIODIVERSITY MODELS

Deforestation causes a number of effects on biological and physical environment, such as habitat loss, habitat fragmentation, species extinction, deterioration of soil properties, drought, flooding, especially if the resulting cleared land is not managed sustainably. Increased fragmentation often results in the subdivision of the natural environment into isolated patches of different sizes and shapes (Turner and Corlett, 1996) and diminish species distribution and gene flow (Raabova et al., 2007), as well as favors species adapted to edge habitats, but prevents species living in core areas (Yahner, 1988). Section 2 of this book provides general information on the consequence of deforestation and climate change on biodiversity, and shows how Geo-informatics tools to monitor and assess biodiversity and land use change.

Besides deforestation, climate change is one of the greatest challenges of the 21st century for biodiversity conservation. Based on the future development scenarios, especially A1F1 (business-as-usual), temperature would increase by some 2.4 to 6.4 degrees Celsius and the sea would rise some 26 to 59 centimeters at the end of the century, potentially flooding large coastal zones and numerous islands, if no adaptation measures are taken (Secretariat of the Convention on Biological Diversity, 2003). Changes in climate have the potential to directly and indirectly affect individuals, populations and species, ecosystems, and the geographic location of ecological systems. Examples of effects include extinction of wildlife populations, change in phenology and hatching, and immigration of species, disrupted plant communities, species and ecosystems are projected to be impacted by extreme climatic events. In Section 3, several examples of projected impacts of climate change on biodiversity are described

3.1 Land Use Models

Projections of land use change are derived from socio-economic models, yielding claims for agricultural land and land dedicated for forestry depending on assumptions of productivity changes. A range of models is developed to make these projections more geographically explicit by allocating land-use categories, meeting land needs, and to better assess and project the future role of land-use and land-cover change in the functioning of the earth system (Veldkamp & Lambin, 2001).

The consequent impacts of land use change are presented in several case studies chapters. We found that Geo-informatics, which comprise Geographic Information System (GIS), Global Positioning System (GPS) and Remote Sensing (RS) are important tools for land use and biodiversity studies.

As of now a number of land use models are being used to predict future land-use change. In this book, two simple and classical models, including Markov Chain Model and logistic regression are included. In addition, several advanced modeling approaches for a complex, dynamic and spatial problem that combine system model and future demands were developed in recent years. In this book, we present the Integrated modeling of global environmental change (IMAGE), which proved its value at global and regional levels in many environmental assessments and Dyna-CLUE, mostly used at (sub-) national levels. The data and information derived from the land use model are input for broader policy-exploring tools, for example for both biodiversity models and comprehensive climate mitigation strategies and regimes.

3.2 Biodiversity Modeling

Exploring the relationship between species and the habitat and other features provided by ecosystems is fundamental in conservation and biodiversity management. Sustaining biodiversity requires knowledge about its geographical distribution and pattern, as well as an understanding of the processes that drives biodiversity at different scales (Skidmore et al., 2006). Decision makers and resource managers need to have a clear and reliable view of the distribution of species and their abundance in the landscape as well as knowledge of relative suitability of habitats for a given species. Predictive modeling and mapping based on these relationships, forms an analytical foundation for informed conservation planning. Advancements in computer technology, statistical modeling and Geographic Information System (GIS) software allow the knowledge of species/habitat relationship to be used for prediction of the geographic distribution of individual population of wildlife species (Yost et al., 2008).

The modeling of biodiversity can be approached from two different angles: An aggregated biodiversity index is compiled from data and the index is related to changes of environmental pressures (pressure based model); and individual species are related to environmental variables and the model results are combined into aggregated indices (species based model). An example of the former is GLOBIO3, which addresses (1) the impacts of environmental drivers on mean species abundance of originally occurring species (MSA) and their relative importance; (2) expected trends under various future scenarios; and (3) the likely effects of various policy response options, examples of the latter are the many climate envelope models developed of which examples are described in many case studies. It should be noted that most biodiversity pressure models describe impacts on terrestrial ecosystems. The recently developed GLOBIO aquatic model is based on a similar approach but it needs further improvement.

3.3 Species Modeling

It should be noted, however, that modeling species abundance and distribution is not the same as modeling biodiversity pressure. Most of the species models and maps are designed to depict the distribution of individual species. They do not generate biodiversity as a biodiversity proxy (e.g. mean species abundance) *per se*. In some cases, however, certain indicator species can be representative of biodiversity in general, but this difference should be clearly noted and accounted for.

Predictive species mapping is founded in the ecological niche theory and predictor analysis and rests on the premise that species distribution can be predicted from the spatial distribution of environmental variables that are correlated with or control the occurrence of a species (Yost et al., 2008; Phillips et al., 2006). To predict species potential distribution, a range of models has been developed. While major differences exist regarding the statistical algorithms used and their species occurrence data type requirements, all models generate predictions in multidimensional ecological space. Species distribution models therefore do not predict species geographic occurrences as such, but produce a spatially explicit probability surface (sometimes converted to binary output only) that represents habitat suitability in ecological hyperspace after factoring in some specified constraints (sometimes including variable interactions) (Herkt, 2007). In this book, there are four cases studies related to species modeling. Chapter 7 elaborates on the concepts of species modeling and presents three popular techniques to generate species distribution: including cartographic overlay, binary response (presence/absence) prediction model, and presence-only data model. Models described in chapters 11, 12, and 13 use presence-only data (MAXENT) to predict present and future distributions of plants and animals.

4. EFFECTS OF LAND USE AND CLIMATE CHANGE ON BIODIVERSITY

All case studies conducted in various regions (east to west, north to south) and multi-scales (global to local) across the globe revealed similar results about the negative impacts of future land use and climate change on biodiversity. Deforestation causes a number of consequent effects on the biological and physical environment, such as habitat loss, habitat fragmentation, species extinction, deterioration of soil properties, drought, flooding, etc. Fragmentation occurs in conjunction with loss of area and includes changes in composition, shape and configuration of resulting patches.

Future land use change is a dominant driver among several human-induced activities. At global and regional level, the main drivers for terrestrial systems defined in the GLOBIO model include land use; nitrogen deposition, fragmentation, infrastructure and climate change. Meanwhile, freshwater biodiversity is declining due to many interacting drivers, such as constructions of dams and other structures, wetland conversion, pollution, overexploitation and invasive species (MA, 2005; Revenga *et al.*, 2005). GLOBIO aquatic currently describes the impacts of land use and pollution by nutrients and the impact of the flow regime (due to dams and canalization and water abstraction), and climate change effects on the regime. Among the above mentioned biodiversity drivers, land use change is considered the most important contributor followed by fragmentation and others.

The results of a case study in northern Thailand revealed that only establishing a fixed percentage of forest or habitat sizes as described in the species-area relationship concept (Dobson, 1996), was not efficient in conserving biodiversity. This is due to the species-area relationship approach ignoring the variation of habitat quality and fragmentation effects and not including the species abundance (Gotelli, 2001). Measures aimed at the conservation of locations with high biodiversity values, limited fragmentation and proper land allocation policies are needed to achieve biodiversity conservation. For example, the Market forces scenario projects more biodiversity loss than the policy reform scenario in the Tropical Andean countries. Similarly, the model results in Central

America indicated that in the Baseline scenario, the region experiences a high reduction of its forested areas and biodiversity, mainly due to an increase in agricultural land and cultivated grassland. In the Alliance for the Sustainable Development of Central America (ALIDES), the effects are less severe; as a result of the policies to stimulate the transformation of traditional agriculture and grassland activities into sustainable production systems. Meanwhile, in the Trade Liberalization option baseline trends will be intensified because of the expected increase in demands for primary production export goods associated with the implementation of the Free Trade Agreements.

Climate change is an emerging threat to biodiversity and stimulate effect to an additional loss of biodiversity beside land use change. The expected negative impacts include extinction of wildlife populations, change in phenology, hatching and immigration of species, disrupted plant communities, change in species distribution moving poleward or upward, increase the rate of loss of species and create opportunities for the establishment of new species (Secretariat of the Convention on Biological Diversity, 2003). Thus, climate change in combination with habitat destruction, degradation and fragmentation may lead to new waves of species extinctions in the near future as species are set on the move but are unable to reach cooler refuges due to altered, obstructing landscapes. The research results in Ukraine confirmed that expected climate change together with land-use change would provoke numerous non-simplified and unexpected habitat changes. However, climate change impacts on biodiversity vary among regions, countries and altitude. The case study in Central America showed little differences among countries, but the lowest impacts was found in Belize and the highest value was registered in El Salvador and Nicaragua.

The research results in the Amazone revealed that changes in the spatial distributions of moisture deficits and seasonality also affected species distributions over the simulation period. In the Standard Impact (SI) and Reduced Impact (RI) scenarios. the most favorable habitats for moist forest species in 2095 were in the more a-seasonal western Amazonia, and in high altitude areas, which are also concentrated in the west. In the SIS scenario many species gained new potential distributions along the western edge of their current simulated range. In addition, northeast Amazonia underwent the most profound long-term change in species density and composition. Yet significant changes occurred in the potential distributions of all species, leaving many populations as non-viable relicts. By 2095, approximately 41% of sample species were under greatest threat in the SIS, compared with 21% in the RI scenario.

5. EMBEDDING THE MODEL RESULTS INTO LAND USE PLANNING

There have been several attempts to integrate and/or embed the results of land use, climate change and biodiversity analyses into conservation planning at various scales. These attempts can be done by three approaches: (1) identify optimal locations of protected areas and using protection to mitigate climate change in a sustainable way that does not undermine biodiversity by deforestation; (2) formulate an optimal land use policy; and (3) formulate legal and practical framework of environmental impact assessment (EIA), and strategic environmental assessment (SEA).

5.1 Protected Area System

The results of simulations derived in many case studies suggest that to identify the optimal locations of protected areas while climate is changing, will require a more sophisticated conservation planning tool than currently exists. Conservation planning will become even more complex if protection will be used to mitigate climate change in a sustainable way, not undermining biodiversity by deforestation, The results of the model implemented in northern Thailand indicated that only establishing a fixed percentage of forest (40-50%) was not efficient in conserving biodiversity due to the high biodiversity that exists in the protected area system. Measures aimed at the conservation of locations with high biodiversity values, limited fragmentation and careful consideration of road expansion in pristine forest areas, especially in existing protected areas are more efficient for achieving biodiversity conservation goals. It was thought that authorized agencies to establish more protected areas to include both lowland and montane forests or migration corridors between these in order to protect the best remaining lowland moist forest species and montane forest flora. These recommendations are similar to the model outputs/outcomes of other parts of the globe (e.g., Tropical Andes, Central America and Ukraine).

5.2 Optimal Land Use Policy

All results obtained from GLOBIO3, SWAT, the hybrid BIOCLIMA model, and logistic regression models implemented in many countries and regions show the similar results that deforestation and land use change are critical threats and from main drivers to biodiversity loss in the past, present and future. Not only does deforestation cause habitat loss, but it also results in habitat fragmentation by diminishing patch size and core area, and isolation of suitable habitats (MacDonald, 2003). In addition, fragmentation provides opportunities for pioneer (light-demanding) species to invade the natural habitat along the forest.

GLOBIO3 and Dyna-CLUE are powerful policy support tools, especially within Strategic Environmental Assessment process in Vietnam because of their scale, limited data needs, ability to calculate impact of different scenarios, and support analysis if political targets will be met. The models are basic and adopted to integrate into policy processes for national and local conditions. The GLOBIO3 methodology provided insight into the effects of alternative scenario and policy options on biodiversity conservation. It offered decision makers a suitable tool for national policy support, especially to stimulate policy discussion and to integrate biodiversity into other policy domains. For example, the modeling of current and future state of biodiversity in the Central American region showed that biodiversity has been significantly affected and that this trend is likely to continue under the options considered.

The SWAT for modeling watershed hydrology and simulating the movement of non-point source pollution applied in Dong Nai watershed in Vietnam provided efficient planning under different scenario developments on sustainable land use and watershed management in response to climate change impacts. In addition, the spatial modeling (logistic regression model) of the deforestation in Java, Indonesia will assist the policy makers to understand the process and to take into consideration how to effectively solve environmental problems resulting from deforestation. The result of the study showed that, in order to solve these problems the Government should pay more attention to population control especially in rural areas and to create alternative non agricultural jobs, as well as to reduce road construction in remaining forest.

Besides the impact on biodiversity loss, deforestation and land use change also causes impact on other ecosystem services which are important for human livelihoods, This issue has been raised in the scientific literature and the Millennium Ecosystem Assessment (MEA, 2005). Therefore, the current development and implications of model results are also providing information for policies targeted at poverty eradication and ensuring environmental sustainability. Kenya is a good place to test the multiple ecosystem services concept. It is currently experiencing severe problems arising from the lack of sustainable hydrological services from its five main river catchments. However, Biodiversity loss itself is not mentioned as a causal factor for poverty. A map overlay of biodiversity with poverty was carried out and showed a slight correlation between high biodiversity and poverty but the spatial linkage is not clear. In areas with a high remaining biodiversity/MSA low production opportunities exist because of remoteness, low productive soils, low productivity, high transportation costs, no governmental control, no support, traditional livelihoods and migration of poor people.

5.3 Environmental Impact Assessments (EIA) and Strategic Environmental Assessments (SEA)

SEA is defined as 'the formalized, systematic and comprehensive process of identifying and evaluating the environmental consequences of proposed policies, plans or programmes to ensure that they are fully included before political decisions are taken for implementation'. Therefore, SEA is proactive and sustainability driven whilst EIA is largely reactive. The SEA is closely related to integrated assessment, but focuses more on the environment. It is highly relevant to the guidelines on biodiversity including the environmental impact assessments that were approved at the 8th Conferences of the Parties meeting for the Convention of Biological Diversity (CBD) (UNEP/CBD, 2006). The case study in Vietnam both at national and local levels was a good example to show the potential uses of biodiversity modelling into policy processes. This collaborative project introduces an effective introduction of a new indicator and assessment method that is potential for application in Vietnam in terms of real need, especially in the planning process to balance the fast economic growth and quality of growth in observation towards biodiversity conservation. Right now, the biodiversity modelling process and results are endorsed by the Ministry of Environment and embedded in the national policy processes of Vietnam.

6. FUTURE DIRECTIONS AND RESEARCH

6.1 Accurate Land Use Map

All studies presented point at land use change as the dominant factor affecting biodiversity. The consequent fragmentation further impacts biodiversity. The existing IMAGE model generates land-use/land-cover map with a resolution of 0.5 by 0.5 degrees. This spatial resolution is appropriate for global and regional assessment; however it is not suitable for national (small extent) and local levels. Some studies used Dyna-CLUE to replace IMAGE model for local and national levels (Verburg & Overmars, 2009). However, the model is rather static and currently explores only path dependency relations in local land uses. Another limitation of the CLUE model is that it does not allow feedbacks between local and regional scales, and between impacts and drivers in the sequential modeling process.

Therefore, an improved and dynamic land-allocation model is essential. An improved model should integrate assessment of global and regional environmental sustainability. Moreover, it should provide appropriate spatial resolution and reflect a country-by-country representation of drivers and parameters of future land-use determinants. Possibly, the model should integrate satellite-based data and high resolution soil characteristics and derived soil properties.

6.2 Downscaled Climate Data

The estimation of the effect of climate change on biodiversity is of growing concern. Biodiversity models can be improved by the use of more detailed climatic information. Some regional climate models (RCM) have been applied or tested for the South American Region. Even thought they remain coarse for analyzing the montanious region, this is the best source of information we have at the moment. The calculation of the climate effects in

GLOBIO includes variation of the effect at the level of biomes. A better approximation should include a more detailed ecosystem classification. A better resolution would, for instance, allow the distinction between montane forest and dry forest. A first approximation could be used, for example using the Ecoregions of WWF. In addition, Hutchinson (2000) developed the *ANUSPLIN, ver 4.1* software to downscale global climate data to approximately 1-km resolution using topographic and geographical locations. The calibrated data provides a more accurate fit to location conditions. This process would avoid some inconsistencies such as the low climate change impact on tropical glaciers.

6.3 Biodiversity/Species Distribution Models

The two approaches for modeling biodiversity have progressed from simple approaches directly derived from land use and climate patterns (e.g. Huntley et al., 1995) or using expert opinion (Sala et al., 2000) to more sophisticated statistical approaches (e.g. Araújo & New, 2006; Alkemade et al., 2009). The advantage of the pressure based approach is the possibility to include a multitude of different pressures and the simple interpretation of the biodiversity index chosen. The need for projections of different biodiversity indices in policy and conservation planning urges the development of new models focusing on indices coping for the broad definition of biodiversity, including species richness, species extinctions and ecosystem services.

The species based approach is mostly developed for describing species distribution changes resulting from projected climate change. The predicted impacts are still premature as the accuracy of the current climatic change models is still debated, especially if applied at local and regional levels. A lesser rise in temperature in some region may have a drastically different influence on species distribution. Furthermore only species already present nowadays are used for the predictions in the future, as the species of adjacent countries are normally not taken into consideration. It is very likely that species from nearby will possibly migrate to the study area, so that the loss in flora will in effect be much less. Climate change thus also create opportunities for many species to increase there ranges. The different possible statistical techniques are now exhaustively explored and it may be time for including multiple drivers, such as land use change and fragmentation, and the notion of dispersal, into the models coping for the different scales at which these drivers are acting.

As these developments are merely a broadening of current approaches, the real future of biodiversity modeling lays, in our opinion, in the inclusion of dynamic models. One of the possibilities is the inclusion of biodiversity, in terms of species or traits, into the functional types modeled in dynamic vegetation models. Dynamic wildlife models are now only designed to predict a few species, but extension of these approaches to more species including a simple food web, may lead to a more dynamic biodiversity modeling.

Comparison of these approaches, allowing for describing biodiversity and ecosystem services using different metrics, yield a clear statement on uncertainty of biodiversity projections, which is indispensable for good policy support. The latter approach is successfully adopted by the IPCC and may be one of the key challenges for the recently launched Intergovernmental Panel for Biodiversity and Ecosystem Services (IPBES)

6.4 Conservation Planning and Policy Process

Biodiversity is just one of a number of ecosystem services, and while it does play a fundamental role its primacy is contested, as shown in many cases studies. Many countries, especially in the tropics, face the challenge of identifying the optimal locations of protected areas when climate and land use is changing, and finding ways to use protection to mitigate those predicted impacts, in addition the also face the challenge to accomplish these multiple goals by examining three other conservation paradigms that are now in vogue: conservation of ecosystem services; optimizing conservation of ecosystem services and poverty alleviation; and reducing carbon emissions from deforestation and forest degradation (REDD).

Conservation responses to land use and climate change are more complicate than the existing conservation strategies such as protected areas, sustainable land use planning and connectivity between conservation areas. More sophisticated planning tools and mechanisms are needed to respond to the challenges of dynamic land use and climate change.

The current protected area network does not represent all ecosystem and species, particularly in the Tropics (Trisurat, 2007). The Parties to the Convention on Biological Diversity (COPs) adopted the 2010 biodiversity targets, which specify that by 2010 at least 10% of each of the world's ecological regions should be effectively conserved in protected areas systems (target 1.1), the rate of loss and degradation of natural habitats decreased (target 5.1); and maintained and enhanced resilience of the components of biodiversity to adapt to climate change (target 7.1), etc (Secretariat of the Convention on Biological Diversity, 2006). The COPs encourages each member country to complete gap analysis and protected area system plan and establish biodiversity connectivity between conservation areas. However, the funding deficiencies are greatest in most tropical countries. In addition, connectivity forests along remnant mountain ecosystems may not be as effective as connecting lowland and uplands because species will move upslope with warming, so connecting fragmented ecosystems has relatively less benefit than connecting lowlands and uplands. Therefore, the most cost-effective approach is to extend existing protected areas to potentially inhabit the projected future range of species and protect risk areas for deforestation (Trisurat et al, 2010). Such implementation offers the opportunity to strengthen both biodiversity representation in protected areas and their potential future representation.

Recently, the UN Framework Convention on Climate Change (UNFCCC) has initiated the Reduced Emissions from Deforestation and Degradation (REDD) scheme, which focuses on conserving carbon cycling ecosystem services (and hence forest carbon stocks), to mitigate global climate change and prevent some of the impacts predicted earlier. Basically, it has not been devised specifically to conserve species, or biodiversity in the broader sense. It is recommended that any country participating in the REDD scheme should have: (1) an integrated biodiversity component so that potential negative impacts of REDD protection on biodiversity can be averted; and (2) include a land use component, so that future agricultural expansion can be planned to benefit people in particular localities and in the country as a whole, while minimizing negative environmental impacts.

Besides protected areas, connectivity and REDD, regional and global coordination and modified mechanisms will be important in dealing with land use and climate change. These include Strategic Environmental Assessment (SEA). The SEA allows environmental planners and policy maker to check whether the impact of environmental policies on biodiversity is likely to meet their environmental sustainability, which is more effective that Environmental Impact Assessment (EIA). The case study in Vietnam clearly shows that biodiversity modeling was effectively embedded in a right scale and timing of the political process.

Beyond technical and conservation, coordination is also needed to ensure that national, regional and international strategies work in concert in response to climate change. The Group on Earth Observations (GEO) was established to improve the coordination of existing Earth observation data sets, implement new observations and promote the generation of Earth observation products. GEO oversees a Global Earth Observation System of Systems (GEOSS) as the mechanism to achieve these goals. A Biodiversity Observation Network, or GEO BON, is one of the GEOSS (GEO-BON, 2008).

GEO BON will provide to users information documenting and interpreting changes in biodiversity. This information will form the basis for future assessments by the envisaged IPBES (Intergovernmental Platform for science-policy on Biodiversity and Ecosystem Services). It is anticipated that this international coordination will strengthen biodiversity conservation at all levels to deal with biodiversity threats, especially future land use and climate change.

REFERENCES

Alkemade, R., van Oorschot, M., Miles, L., Nellemann, C., Bakkenes, M., & ten Brink, B. (2009). GLOBIO3: A framework to investigate options for reducing global terrestrial biodiversity loss. *Ecosystems*, *12*(3), 374–390. doi:10.1007/s10021-009-9229-5

Bouwman, A. F., Kram, T., & Goldewijk, K. (Eds.). (2006). *Integrated modelling of global environmental change. An overview of IMAGE 2.4*. The Netherlands Environmental Assessment Agency. (Report no. 500110002).

Dobson, A. P. (1996). *Conservation and biodiversity*. New York, NY: Scientific American Library.

GEO-BON. (2008). *The geo biodiversity observation network: Concept document*. Postdam, Germany: Groups on Earth Observation.

Gotelli, N. J. (2001). *A primer of ecology*. Sunderland, MA: Sinauer Associates.

Herkt, M. (2007). *Modelling habitat suitability to predict the potential distribution of Erhard's Wall Lizard Podarcis erhardii on Crete*. The Netherlands: ITC.

Hutchinson, M. F. (2000). *ANUSPLIN, ver 4.1. user guide. Centre for Resource and Environmental Studies*. Canberra, Australia: The Australian National University.

MacDonald, G. (2003). *Biogeography: Introduction to space, time and life*. New York, NY: John Wiley & Sons, Inc.

McNeely, J. A. (1998). *Economics and biological diversity: Developing and using economic incentives to conserve biological resources*. Gland, Switzerland: IUCN.

Millennium Ecosystem Assessment-MA. (2005). *Ecosystems and human well-being: synthesis note: The designation of areas undergoing land-cover change resulting from degradation in drylands has been omitted*. Washington, DC: Island Press.

National Economic and Social Development Board- NESDB. (2008). *The 10th national economic and social development plan (2008-2011)*. Bangkok: Prime Minister Office.

Phillips, S. J., Andersion, R. P., & Schapire, R. E. (2006). Maximum entropy modeling of species geographic distributions. *Ecological Modelling*, *190*, 231–239. doi:10.1016/j.ecolmodel.2005.03.026

Raabová, J., Münzbergová, Z., & Fischer, M. (2007). Ecological rather than geographic or genetic distance affects local adaptation of the rare perennial herb, Aster amellus. *Biological Conservation*, *139*, 348–357. doi:10.1016/j.biocon.2007.07.007

Revenga, C., Campbell, I., Abell, R., de Villiers, P., & Bryer, M. (2005). Prospects for monitoring freshwater ecosystems towards the 2010 targets. *Philosophical Transactions of the Royal Society Bulletin*, *360*, 397–413. doi:10.1098/rstb.2004.1595

Sala, O. E., Chapin, F. S. III, Armesto, J. J., Berlow, E., Bloomfield, J., & Irzo, R.... Wall, D. H. (2000). Global biodiversity scenarios for the year 2100. *Science, 287*, 1770–1774. doi:10.1126/science.287.5459.1770

Secretariat of the Convention on Biological Diversity. (2003). *Interlinakages between biological diversity and climate change. Advise on the integration of biodiversity considerations into the implementation of the United Nations Framework Convention on Climate Change and Its Kyoto protocol.* Montreal.

Secretariat of the Convention on Biological Diversity. (2006). *Global biodiversity outlook* 2. Montreal.

Skidmore, A. K., Toxopeus, A. G., De Bie, C. A. J. M., Corsi, F., & Venus, V. (2006). *Herpetological species mapping for the Mediterranean.* Unpublished manuscript presented at the ISPRS mid-term Symposium 2006 Remote Sensing: From pixels to processes. ITC, Enschede, the Netherlands.

Trisurat, Y. (2007). Applying gap analysis and a comparison index to assess protected areas in Thailand. *Environmental Management, 39*, 235–245. doi:10.1007/s00267-005-0355-3

Trisurat, Y. Bhumpakphan, N., Dechyosdee, U., Kanchanasakha, B., & Tanhikorn, S. (2010). Identifying priority areas for biodiversity conservation in northern Thailand: Land use changes and species modeling approaches. In V. Rescigno & S. Maletta (Eds.), Biodiversity hotspots. (pp. 81-103). New York, NY: Nova Science Publishers.

Turner, I. M., & Corlett, R. T. (1996). The conservation value of small isolated fragments of lowland tropical rain forest. *Trends in Ecology & Evolution, 11*, 330–333. doi:10.1016/0169-5347(96)10046-X

UNEP/CBD. (2006). *Decision adopted by the conference of the parties to the convention on biological diversity at its eighth meeting VIII/28.* Impact assessment: Voluntary guidelines on biodiversity-inclusive impact assessment. Curitiba, Brazil.

Veldkamp, A., & Lambin, E. F. (2001). Editorial: Predicting land-use change. *Agriculture Ecosystems & Environment, 85*, 1–6. doi:10.1016/S0167-8809(01)00199-2

Verburg, P., & Overmars, K. (2009). Combining top-down and bottom-up dynamics in land use modeling: Exploring the future of abandoned farmlands in Europe with the Dyna-CLUE model. *Landscape Ecology, 24*, 1167–1181. doi:10.1007/s10980-009-9355-7

Yahner, R. H. (1988). Changes in wildlife communities near edges. *Conservation Biology, 2*(4), 333–339..doi:10.1111/j.1523-1739.1988.tb00197.x

Yost, A. C., Petersen, S. L., Gregg, M., & Miller, R. (2008). Predictive modelling and mapping sage grouse (*Centrocercus urophasianus*) nesting habitat using Maximum Entropy and a long-term dataset from Southern Oregon. *Ecological Informatics, 3*, 375–386. doi:10.1016/j.ecoinf.2008.08.004

Compilation of References

Abildtrup, J., Rosato, P., Gylling, M., Fekete-Farkas, M., Audsley, E., Giupponi, C., & Rounsevell, M. D. A. (2006). Socio-economic scenario development for the assessment of climate change impacts on agricultural land-use: A pairwise comparison approach. *Environmental Science & Policy*, *9*(2), 101–115..doi:10.1016/j.envsci.2005.11.002

Abram, N. J., Gagan, M. K., McCulloch, M. T., Chappell, J., & Hantoro, W. S. (2003). Coral reef death during the 1997 Indian Ocean dipole linked to Indonesian wildfires. *Science*, *301*, 952–955. doi:10.1126/science.1083841

Achard, F., Eva, H. D., Stibig, H. J., Mayaux, P., Gallego, J., Richards, T., & Malingreau, J. P. (2002). Determination of deforestation rates of the world's humid tropical forests. *Science*, *297*, 999–1002. doi:10.1126/science.1070656

Adams, D. M., Alig, R. J., & McCarl, B. A. (1996). An analysis of the impacts of public timber harvest policies on private forest management in the U.S. *Forest Science*, *42*(3), 343–358.

Adams, W. M. (2008). *Green development* (3rd ed.). London, UK: Routledge.

Adams, W. M., Aveling, R., Brockington, D., Dickson, B., Elliott, J., & Hutton, J. (2004). Biodiversity conservation and the eradication of poverty. *Science*, *306*, 1146–1149. doi:10.1126/science.1097920

African Union. (2003). *Report of the African Commission Working Group of experts on indigenous populations/communities*. Banjul: African Union.

Agarwal, S., Joshi, P. K., Shukla, Y., & Roy, P. S. (2003). SPOT VEGETATION multitemporal data for classifying vegetation in south central Asia. *Current Science*, *84*(11), 1440–1448.

Agrawala, S., Moehner, A., Hemp, A., van Aalst, M., Hitz, S., & Smith, J. ... Mwaipopo, O. U. (2003). *Development and climate change in Tanzania: Focus on Mount Kilimanjaro.* Organisation for Economic Co-operation and Development, Paris.

Akotsi, E., Ndirangu, J. K., & Gachanja, M. (2006). *Changes in forest cover in Kenya's five water towers 2003-2005*. Department of Resource Surveys and Remote Sensing/Kenya Forests Working Group, Kilmani.

Alcamo, J. (Ed.). (1994). *IMAGE 2.0: Integrated modeling of global change*. Dordrecht, The Netherlands: Kluwer Academic Publishers.

Alcamo, J., Leemans, R., & Kreileman, E. (Eds.). (1998). *Global change scenarios of the 21st century. Results from the IMAGE 2.1 model*. Oxford, UK: Elsevier Science.

Alkemade, R., vanOorschot, M., Miles, L., Nellemann, C., Bakkenes, M., & ten Brink, B. (2009). GLOBIO3: A framework to investigate options for reducing global terrestrial biodiversity loss. *Ecosystems*, *12*(3), 374–390. doi:10.1007/s10021-009-9229-5

Alkemade, R., Bakkenes, M., & Eickhout, B. (2011). Towards a general relationship between climate change and biodiversity: an example for plant species in Europe. *Regional Environmental Change*, *11*(1), S143–S150. doi:10.1007/s10113-010-0161-1

Alkemade, R., Bakkenes, M., Bobbink, R., Miles, L., Nellemann, C., Simons, H., & Tekelenburg, T. (2006). GLOBIO 3: Framework for the assessment of global terrestrial biodiversity. In A. F. Bouwman, T. Kram, & Goldewijk (Eds.), *Integrated modelling of global environmental change. An overview of IMAGE 2.4* (pp. 171-186). Bilthoven, Netherlands: Netherlands Environmental Assessment Agency (MNP).

Allan, B. F., Keesing, F., & Ostfeld, R. S. (2003). Effect of forest fragmentation on Lyme disease risk. *Conservation Biology*, *17*, 267–272. doi:10.1046/j.1523-1739.2003.01260.x

Allen, C. R., Pearlstine, L. G., & Kitchens, W. M. (2001). Modeling viable populations in gap analysis. *Biological Conservation*, *99*(2), 135–144. doi:10.1016/S0006-3207(00)00084-7

Alongi, D. M. (2008). Mangrove forests: Resilience, protection from tsunamis, and responses to global climate change. *Estuarine, Coastal and Shelf Science*, *76*, 1–13. doi:10.1016/j.ecss.2007.08.024

Anbar, A., Duan, Y., Lyons, T., Arnold, G., Kendall, B., & Creaser, R. (1903–1906). Buick, R. (2007). A whiff of oxygen before the great oxidation event? *Science*, *317*(5846).

Anderegg, W. R. L., Prall, J. W., Harold, J., & Schneider, S. H. (2010). Expert credibility in climate change. *Proceedings of the National Academy of Sciences of the United States of America*, *107*(27), 12107–12109. doi:10.1073/pnas.1003187107

Andersen, L. E. (1996). The causes of deforestation in the Brazilian Amazon. *Journal of Environment & Development*, *5*(3), 309–328. doi:10.1177/107049659600500304

Anderson, J. R., Hardy, E. E., Roach, J. T., & Witmer, R. E. (1976). *A land use and land cover classification system for use with remote sensor data, (US Geological Survey Professional Paper, No. 964)*. Washington, DC: USGS.

Anderson, R. P. (2003). Real vs. artifactual absences in species distributions: Tests for Oryzomys albigularis (Rodentia: Muridae) in Venezuela. *Journal of Biogeography*, *30*, 591–605. doi:10.1046/j.1365-2699.2003.00867.x

Anderson, R. P., & Peterson, A. T., & Gomez—Laverde, M. (2002). Using niche—based GIS modeling to test geographic predictions of competitive exclusion and competitive release in South American pocket mice. *Oikos*, *98*, 3–16. doi:10.1034/j.1600-0706.2002.t01-1-980116.x

Anderson, J. R., Hardy, E. E., Roach, J. T., & Witmer, R. E. (1976). *Land use and land cover classification system for use with remote sensor data*. U.S. Geological Survey professional paper (964).

Angelsen, A., & Kaimowitz, D. (1999). Rethinking the causes of deforestation: Lessons from economic models. *The World Bank Research Observer*, *14*(1), 73–98.

Angelson, A., & Kaimowitz, D. (Eds.). (2001). *Agricultural technologies and tropical deforestation*. New York, NY: CABI Publishing. doi:10.1079/9780851994512.0000

Araujo, M. B., & Guisan, A. (2006). Five (or so) challenges for species distribution modelling. *Journal of Biogeography*, *33*, 1677–1688. doi:10.1111/j.1365-2699.2006.01584.x

Araújo, M. B., & New, M. (2006). Ensemble forecasting of species distributions. *Trends in Ecology & Evolution*, *22*(1), 42–47. doi:10.1016/j.tree.2006.09.010

Araújo, M. B., & Luoto, M. (2007). The importance of biotic interactions for modelling species distributions under climate change. *Global Ecology and Biogeography*, *16*, 743–753. doi:10.1111/j.1466-8238.2007.00359.x

Araújo, M. B., Whittaker, R. J., Ladle, R. J., & Erhard, M. (2005). Reducing uncertainty in projections of extinction risk from climate change. *Global Ecology and Biogeography*, *14*, 529–538. doi:10.1111/j.1466-822X.2005.00182.x

Arino, O., Bicheron, P., Achard, F., Latham, J., Witt, R., & Weber, J. L. (2008). GlobCover: The most detailed portrait of Earth. *European Space Agency Bulletin*, *136*, 24–31.

Arnell, N. W. (2000). Thresholds and response to climate change forcing: The water sector. *Climatic Change*, *46*, 305–316. doi:10.1023/A:1005699210660

Arnillas, C. A., Galindo, G., Peralvo, M., & Tovar, C. (2008). *Validation and diffusion of the GLOBIO methodology in the Andean region. Informe Final*. Lima, Peru: PBL.

Arnold, J. G., Srinivasan, R., Muttiah, R. S., & Williams, J. R. (1998). Large area hydrologic modeling and assessment part I: Model development. *Journal of the American Water Resources Association*, *34*, 73–89. doi:10.1111/j.1752-1688.1998.tb05961.x

Asner, G. P., Broadbent, E. N., Oliveira, P. J. C., Keller, M., Knapp, D. E., & Sylva, J. N. M. (2006). Condition and fate of logged forests in the Brazilian Amazon. *Proceedings of the National Academy of Sciences of the United States of America*, *103*, 12947–12950. doi:10.1073/pnas.0604093103

Asner, G. P., Jones, M. O., Martin, R. E., Knapp, D. E., & Hughes, R. F. (2008). Remote sensing of native and invasive species in Hawaiian forests. *Remote Sensing of Environment, 112*(5), 1912–1926. doi:10.1016/j.rse.2007.02.043

Asner, G. P., Rudel, T. K., Aide, T. M., DeFries, R., & Emerson, R. (2009). A contemporary assessment of change in humid tropical forest. *Conservation Biology, 23*(6), 1386–1395. doi:10.1111/j.1523-1739.2009.01333.x

ATBC/STE. (2009). *The Marburg declaration, Marburg, Germany, July 2009*. Frankfurt: Association for Tropical Biology and Conservation and the Society for Tropical Ecology.

Atkinson, P., & Massari, I. (1998). Generalized linear modeling of landslide susceptibility in the Central Apennines, Italy. *Computers & Geosciences, 24*, 373–385. doi:10.1016/S0098-3004(97)00117-9

Attorre, F., Francesconi, F., Taleb, N., Scholte, P., Saed, A., Alfo, M., & Bruno, F. (2007). Will dragonblood survive the next period of climate change? Current and future potential distribution of Dracaena cinnabari (Socotra, Yemen). *Biological Conservation, 138*, 430–439. doi:10.1016/j.biocon.2007.05.009

Auch, R., Taylor, J., & Acevado, W. (2004). *Urban growth in American cities: Glimpses of U.S. urbanization*. U.S. Geological Survey Circular 1252. U.S. Geological Survey, Sioux Falls, SD.

Austin, M. P. (2002). Spatial prediction of species distribution: An interface between ecological theory and statistical modelling. *Ecological Modelling, 157*(2-3), 101–118. doi:10.1016/S0304-3800(02)00205-3

Avery, T. E., & Burkhart, H. E. (1983). *Forest measurements*. New York, NY: McGraw-Hill.

Awuor, C. (2007). *Assessing policy influences on people's relationship to water ecosystem services: The Kenyan experience*. London, UK: International Institute for Environment and Development.

Baker, D. J., Richards, G., Grainger, A., Brown, S., DeFries, R., & Gonzalez, P. (in press). Achieving forest carbon information with higher certainty: A five step strategy. *Environmental Science & Policy*.

Baker, J., Shepherd, S., & Edyvane, K. (1996). The use of marine fishery reserves to manage benthic fisheries, with emphasis on the South Australian abalone fishery. In R. Thackway. (Ed.), *Developing Australia's representative system of marine protected areas. Criteria and guidelines for identification and selection* (pp. 103–113). Proceedings of a technical meeting 22–23 April 1996, Adelaide, South Australia, Australia. Department of the Environment, Sports and Territories, Canberra, Australian Capital Territory, Australia.

Bakkenes, M., Alkemade, J. R. M., Ihle, F., Leemans, R., & Latour, J. B. (2002). Assessing effects of forecasted climate change on the diversity and distribution of European higher plants for 2050. *Global Change Biology, 8*, 390–407. doi:10.1046/j.1354-1013.2001.00467.x

Bakkenes, M., Eickhout, B., & Alkemade, R. (2006). Impacts of different climate stabilisation scenarios on plant species in Europe. *Global Environmental Change, 16*(1), 19–28. doi:10.1016/j.gloenvcha.2005.11.001

Baldyga, T. J., Miller, S. N., Direse, K. L., & Gichaba, C. M. (2007). Assessing land cover change in Kenya's Mau Forest region using remotely sensed data. *African Journal of Ecology, 46*, 46–54. doi:10.1111/j.1365-2028.2007.00806.x

Baldyga, T. J., Miller, S. N., Shivoga, W., & Gichaba, M. (2004). *Assessing impact of land cover change in Kenya*. Paper presented at the Annual Conference of the American Society for Photogrammetry and Remote Sensing, May 2004, Denver, Colorado.

Balmford, A., Green, R. E., & Jenkins, M. (2003). Measuring the changing state of nature. *Trends in Ecology & Evolution, 18*(7), 326–330. doi:10.1016/S0169-5347(03)00067-3

Balmford, A., Crane, P., Dobson, A., Green, R. E., & Mace, G. M. (2005). The 2010 challenge: Data availability, information needs and extraterrestrial insights. *Philosophical Transactions of the Royal Society B, 360*, 221–228. doi:10.1098/rstb.2004.1599

Barbier, E. B., & Burgess, J. C. (1996). Economic analysis of deforestation in Mexico. *Environment and Development Economics, 1*(2), 203–239. doi:10.1017/S1355770X00000590

Barnaud, C., Promburom, P., Bousquet, F., & Trébuil, G. (2006). Companion modelling to facilitate collective land management by Akha villagers in upper northern Thailand. *Journal of the World Association Soil & Water Conservation, 1*(4), 38–54.

Barrett, C. B., Lee, D. R., & McPeak, J. G. (2005). Institutional arrangements for rural poverty reduction and resource conservation. *World Development, 33*, 193–197. doi:10.1016/j.worlddev.2004.07.008

Barron, E., & Sorooshian, S. (1997). *Assessing the impacts of climate on regional water resources*. Paper presented at the Investigators Working Group (IWG) of the Mission to Planet Earth/Earth Observing System *(MTPE/EOS*), San Diego, California.

Bartholome, E., Belward, A., Beuchle, R., Eva, H., Fritz, S., & Hartley, A. (2004). *Global land cover for the year 2000*. European Commission.

Bartholomé, E., Belward, A. S., Achard, F., Bartalev, S., Carmona-Moreno, C., & Eva, H. (2002). *Global land cover mapping for the year 2000—project status November 2002. Publications of the European Communities, EUR 20524 EN*. Luxembourg: European Commission.

Bartholomé, E., & Belward, A. S. (2005). GLC2000: A new approach to global land cover mapping from earth observation data. *International Journal of Remote Sensing, 26*(9), 1959–1977. doi:10.1080/01431160412331291297

Bates, B. C., Kundzewicz, Z. W., Wu, S., & Palutikof, J. P. (2008). *Climate change and water*. Geneva: Technical Paper of the Intergovernmental Panel on Climate Change.

Batisse, M. (1990). Development and implications of the biosphere reserve concept and applicability to coastal regions. *Environmental Conservation, 17*, 111–116. doi:10.1017/S0376892900031878

Battrick, B. (Ed.). (2005). *Global Earth Observation System of Systems GEOSS. 10-Year implementation plan reference document. Group on Earth Observations. GEO 1000R / ESA SP – 1284, ESA Publication Division*. Noordwijk, the Netherlands: ESTEC.

Battrick, B. (Ed.). (2006). *The changing Earth. New scientific challenges for ESA's living planet programme. SP-1304, ESA Publication Division*. Noordwijk, the Netherlands: ESTEC.

Baudry, J., & Thenail, C. (2004). Interaction between farming systems, riparian zones, and landscape patterns: A case study in western France. *Landscape and Urban Planning, 67*, 121–129. doi:10.1016/S0169-2046(03)00033-1

Bawa, K. S. (1990). Plant-pollinator interactions in tropical rainforests. *Annual Review of Ecology and Systematics, 21*, 399–422. doi:10.1146/annurev.es.21.110190.002151

Baydack, R. K., & Campa, H. III. (1999). Setting the context. In Baydack, R. K., Campa, H. III, & Haufler, J. B. (Eds.), *Practical approaches to the conservation of biological diversity*. Washington, DC: Island Press.

Beaman, J. H., & Burley, J. S. (2003). Progress in the floristic inventory of Borneo. In Padoch, C., & Peluso, N. L. (Eds.), *Borneo in transition: People, forests, conservation and development* (pp. 93–113). Oxford, UK: Oxford University Press.

Belward, A. S. (1999). International co-operation in satellite sensor calibration: The role of the CEOS Working Group on calibration and validation. *Advances in Space Research, 23*(8), 1443–1448. doi:10.1016/S0273-1177(99)00296-3

Belward, A., & Loveland, T. (1995). The IGBPDIS 1 km land cover project. In Curran, P. J., & Robertson, C. (Eds.), *Remote sensing in action* (pp. 1099–1106). Southampton, UK: University of Southampton.

Bender, D. J., Contreras, T. A., & Fahrig, L. (1998). Habitat loss and population decline: A meta-analysis of the patch size effect. *Ecology, 79*, 517–533. doi:10.1890/0012-9658(1998)079[0517:HLAPDA]2.0.CO;2

Benitéz-Lopéz, A., Alkemade, R., & Verweij, P. A. (2010). The impacts of roads and other infrastructure on mammal and bird populations: A meta-analysis. *Biological Conservation, 143*, 1307–1316. doi:10.1016/j.biocon.2010.02.009

Bennet, A. F., & Saunders, D. A. (2010). Habitat fragmentation and landscape change. In Sodhi, N. S., & Ehrlich, P. R. (Eds.), *Conservation Biology for all* (pp. 88–106). Oxford, UK: Oxford University Press. doi:10.1093/acprof:oso/9780199554232.003.0006

Berbet, M. L. C., & Costa, M. H. (2003). Climate change after tropical deforestation: Seasonal variability of surface albedo and its effects on precipitation change. *Journal of Climate, 16*, 2099–2104. doi:10.1175/1520-0442(2003)016<2099:CCATDS>2.0.CO;2

Berendse, F. (1979). Competition between plant populations with different rooting depths: Theoretical considerations. *Oecologia, 43*, 19–26. doi:10.1007/BF00346669

Berry, M., Flamm, R., Hazen, B., & Macintyre, R. (1996). LUCAS: A system for modeling land use change. *IEEE Computational Science & Engineering, 3*, 24–35. doi:10.1109/99.486758

Bhumpakphan, N. (2003). *Management of the Pha Taem protected forest complex to promote cooperation for trans-boundary biodiversity conservation between Thailand, Cambodia and Laos (Phase I): Wildlife ecology final report.* Bangkok: Faculty of Forestry, Kasetsart University.

Biemans, H., Haddeland, I., Kabat, P., Ludwig, F., Hutjes, R. W. A., & Heinke, J. (2011). (in press). Impact of reservoirs on river discharge and irrigation water supply during the 20th century. *Water Resources Research.* doi:10.1029/2009WR008929

BioModel. (2010). *ULRMC BioticGIS Modelling Group.* Retrieved April 6, 2010, from http://biomodel.org.ua/

Bisby, F. A. (2000). The quiet revolution: Biodiversity informatics and the Internet. *Science, 289*, 2309–2312. doi:10.1126/science.289.5488.2309

Blaschke, T. (2006). The role of the spatial dimension within the framework of sustainable landscapes and natural capital. *Landscape and Urban Planning, 75*(3-4), 198–226. doi:10.1016/j.landurbplan.2005.02.013

Bobbink, R., Hicks, K., Galloway, J., Spranger, T., Alkemade, R., & Ashmore, M. (2010). Global assessment of nitrogen deposition effects on terrestrial plant diversity: A synthesis. *Ecological Applications, 20*, 30–59. doi:10.1890/08-1140.1

Bobbink, R., Ashmore, M., Braun, S., Fluckiger, W., & Van den Wyngaert, I. J. J. (2003). *Empirical nitrogen critical loads for natural and semi-natural ecosystems: 2002 update.*

Böhner, J., & Lehmkuhl, F. (2005). Environmental change modelling for Central and High Asia: Pleistocene, present and future scenarios. *Boreas, 34*(2), 220–231. doi:10.1080/03009480510012917

Bonan, G. B. (2008). Forests and climate change: Forcings, feedbacks, and the climate benefits of forests. *Science, 320*(5882), 1444–1449. doi:10.1126/science.1155121

Bouwman, A. F., & Beusen, A. H. W. (2009). Human alteration of the global nitrogen and phosphorus soil balances for the period 1970-2050. *Global Biogeochemical Cycles, 23*(4), 26–42.

Bouwman, A. F., Kram, T., & Klein Goldewijk, K. (Eds.). (2006). *Integrated modelling of global environmental change: An overview of IMAGE 2.4. Netherlands Environmental Assessment Agency.* Bilthoven, The Netherlands: MNP.

Bouwman, A. F., Van Drecht, G., Knoop, J. M., Beusen, A. H. W., & Meinardi, C. R. (2005). Exploring changes in river nitrogen export to the world's oceans. *Global Biogeochemical Cycles, 19*, 1–14. doi:10.1029/2004GB002314

Bouwman, A. F., Van Vuuren, D. P., Derwent, R. G., & Posch, M. (2002). A global analysis of acidification and eutrophication of terrestrial ecosystems. *Water, Air, and Soil Pollution, 141*, 349–382. doi:10.1023/A:1021398008726

Bouwman, A., Kram, T., & Goldewijk, K. (Eds.). (2006). *Integrated modeling of global environmental change. An overview of IMAGE 2.4. Netherlands Environmental Assessment Agency.* Bilthoven, The Netherlands: MNP.

Bouwman, A. F., Kram, T., & Goldewijk, K. K. (Eds.). (2006). *Integrated modeling of global environmental change. An overview of IMAGE 2.4.* Bilthoven, The Netherlands: Netherlands Environmental Assessment Agency.

Bouwman, A. F., Van Der Hoek, K. W., Van Drecht, G., & Eickhout, B. (2006b). World livestock and crop production systems, land use and environment between 1970 and 2030. In Brouwer, F., & McCarl, B. (Eds.), *Rural lands, agriculture and climate beyond 2015: A new perspective on future land use patterns* (pp. 75–89). Dordrecht, The Netherlands: Springer.

Bouwman, A. F., Kram, T., & Klein Goldewijk, K. (Eds.). (2006a). *Integrated modelling of global environmental change. An overview of IMAGE 2.4*. Bilthoven, the Netherlands: Netherlands Environmental Assessment Agency (MNP).

Bouwman, A. F., Kram, T., & Goldewijk, K. (Eds.). (2006). *Integrated modelling of global environmental change. An overview of IMAGE 2.4*. The Netherlands Environmental Assessment Agency. (Report no. 500110002).

Bowman, D. M. J. S. (1993). Biodiversity much more than biology inventory. In Gaston, K. J. (Ed.), *Biodiversity: A Biology of numbers and difference*. Oxford, UK: Blackwell Science.

Boyle, T. J. B., & Boontawee, B. (Eds.). (1995). *Measuring and monitoring biodiversity in tropical and temperate forests*. Bogor: Centre for International Forestry Research.

BPS (Biro Pusat Statistik). (2009). *Identifikasi Desa Dalam Kawasan Hutan. Kerja sama Pusat Rencana dan Statistik Kehutanan, Departemen Kehutanan dengan Direktorat Statistik Pertanian*. Jakarta: Badan Pusat Statistik. (in Indonesian)

Brady, M. A. (1997). Effects of vegetation changes on organic matter dynamics in three coastal peat deposits in Sumatra, Indonesia. In Rieley, J. O., & Page, S. E. (Eds.), *Biodiversity and sustainability of tropical peatlands* (pp. 113–134). Cardigan, UK: Samara Publishing Limited.

Brook, B. W., & Sodhi, N. S., Ng, & P. K. L. (2003). Catastrophic extinctions follow deforestation in Singapore. *Nature, 424*, 420–423. doi:10.1038/nature01795

Brookfield, H., Padoch, C., Parsons, H., & Stocking, M. (2002). *Cultivating biodiversity: Understanding, analyzing and using agricultural diversity*. London, UK: ITDG Publishing.

Brooks, T. M., Pimm, S. L., Kapos, V., & Ravilious, C. (1999). Threat from deforestation to montane and lowland birds and mammals in insular South-east Asia. *Journal of Animal Ecology, 68*(8), 1061–1078. doi:10.1046/j.1365-2656.1999.00353.x

Brooks, T. M., Mittermeier, R., da Fonseca, G. A. B., Gerlach, J., Hoffmann, M., & Lamoreux, J. F. (2006). Global biodiversity conservation priorities. *Science, 313*, 58–61. doi:10.1126/science.1127609

Brooks, K. N., Peter, F. F., Hans, M. G., & John, L. T. (1992). *Hydrology and the management of watershed*. Iowa, USA: Iowa State University Press.

Brown, D. G., Pijanowski, B. C., & Duh, J. D. (2000). Modeling the relationships between land-use and land-cover on private lands in the Upper Midwest, USA. *Journal of Environmental Management, 59*, 247–263. doi:10.1006/jema.2000.0369

Brown, I. (2006). Modelling future landscape change on coastal floodplains using a rule-based GIS. *Environmental Modelling & Software, 21*(10), 1479–1490. doi:10.1016/j.envsoft.2005.07.011

Brown, K. S., & Hutchings, R. W. (1997). Disturbance, fragmentation and the dynamics of diversity in Amazonian forest butterflies: Implications for conservation. In Laurance, W. F., & Bierregaard, R. O. (Eds.), *Tropical forest remnants: Ecology, management and conservation of fragmented communities* (pp. 91–110). Chicago, IL: University of Chicago Press.

Brown, C. (2000). *The global outlook for future wood supply from forest plantations*. Working Paper No GFPOS/WP/03 (p 141). FAO, Rome.

Bruinsma, J. (Ed.). (2003). *World agriculture: Towards 2015/2030. An FAO perspective*. London: Earthscan Publications Ltd.

Bruna, E. (1999). Seed germination in rainforest fragments. *Nature, 402*, 139. doi:10.1038/45963

Brussaard, L. (1997). Biodiversity and ecosystem functioning in soils. *Ambio, 26*(8), 563–570.

Bugden, J. L., Andrey, J., & Howarth, P. J. (2004). A SAR process model for land-cover mapping. *Canadian Journal of Remote Sensing, 30*(2), 195–204.

Bunce, R. G. H., Metzger, M. J., Jongman, R. H. G., Brandt, J., De Blust, G., & Elena-Rossello, R. (2008). A standardized procedure for surveillance and monitoring European habitats and provision of spatial data. *Landscape Ecology, 23*(1), 11–25. doi:10.1007/s10980-007-9173-8

Bunker, D. E., DeClerck, F., Bradford, J. C., Colwell, R. K., Perfecto, I., & Phillips, O. L. (2005). Species loss and aboveground carbon storage in a tropical forest. *Science, 310*, 1029–1031. doi:10.1126/science.1117682

Burel, F., & Baudry, J. (1995). Species biodiversity in changing agricultural landscapes: A case study in the Pays d'Auge, France. *Agriculture Ecosystems & Environment*, *55*, 193–200. doi:10.1016/0167-8809(95)00614-X

Burnett, C., & Blaschke, T. (2003). A multi-scale segmentation/object relationship modelling methodology for landscape analysis. *Ecological Modelling*, *168*(3), 233–249. doi:10.1016/S0304-3800(03)00139-X

Burrough, P. A. (2000). *Principles of Geographical Information systems*. London, UK: Oxford University Press.

Burrough, P. A., & McDonnell, R. A. (1998). *Principles of Geographical Information Systems*. Oxford, UK: Oxford University Press.

Burrough, P. A. (1986). *Principles of geographic information systems for land assessment*. New York, NY: Oxford University Press.

Bury, J. (2005). Mining mountains: neoliberalism, land tenure, livelihoods, and the new Peruvian mining industry in Cajamarca. *Environment & Planning A*, *37*, 221–239. doi:10.1068/a371

Busby, J. R. (1986). *Bioclimate prediction system (BIOCLIM). User's manual version 2.0*. Australian Biological Resources, Study Leaflet.

Butterfield, B. Csuti, B., & Scott, J. M. (1994). Modeling vertebrate distributions for gap analysis. In R. I. Miller (Ed.), *Mapping the diversity of nature* (pp. 53-68). London, UK: Chapman & Hall.

Büttner, G., & Maucha, G. (2006). *The thematic accuracy of CORINE Land Cover 2000*. Assessment using LUCAS (land use/cover area frame statistical survey). (EEA Technical report no.7/2006). Copenhagen, Denmark.

Büttner, G., Feranec, J., & Jaffrain, G. (2002). *Corine land cover update 2000: Technical guidelines*. (Technical report 89). EEA, Copenhagen.

Buytaert, W., Célleri, R., & Timbe, L. (2009). Predicting climate change impacts on water resources in the tropical Andes: the effects of GCM uncertainty. *Geophysical Research Letters*, *36*, L07406. doi:10.1029/2008GL037048

Buytaert, W., Vuille, M., Dewulf, A., Urrutia, R., Karmalkar, A., & Célleri, R. (2010). Uncertainties in climate change projections and regional downscaling: implications for water resources management. *Hydrology andl Earth System Science*, *14*, 1247-1258. doi:javascript:openWin(this);window.document.basic1.submit();

Caparrós, A., & Jacquemont, F. (2003). Conflicts between biodiversity and carbon sequestration programs: Economic and legal implications. *Ecological Economics*, *46*, 143–157. doi:10.1016/S0921-8009(03)00138-1

Carpenter, S. R., Pingali, P. L., Bennett, E. M., & Zurek, M. B. (Eds.). (2005b). *Ecosystems and human well-being: Scenarios* (*Vol. 2*). Washington, DC: Island Press.

Carpenter, G., Gillison, A. N., & Winter, J. (1993). DOMAIN: A flexible modelling procedure for mapping potential distributions of plants and animals. *Biodiversity and Conservation*, *2*, 667–680. doi:10.1007/BF00051966

Castella, J. C., Kam, S. P., Quang, D. D., Verburg, P., & Hoanh, C. T. (2007). Combining top-down and bottom-up modelling approaches of land use/cover change to support public policies: application to sustainable management of natural resources in northern Vietnam. *Land Use Policy*, *24*, 531–545. doi:10.1016/j.landusepol.2005.09.009

Castella, J. C., Pheng Kam, S., Dinh Quang, D., Verburg, P. H., & Thai Hoanh, C. (2007). Combining top-down and bottom-up modelling approaches of land use/cover change to support public policies: Application to sustainable management of natural resources in northern Vietnam. *Land Use Policy*, *24*, 531–545. doi:10.1016/j.landusepol.2005.09.009

Castella, J. C., & Verburg, P. H. (2007). Combination of process-oriented and pattern-oriented models of land-use change in a mountain area of Vietnam. *Ecological Modelling*, *202*, 410-420.

Castella, J. C., & Verburg, P. H. (2007). Combination of process-oriented and pattern-oriented models of land use change in a mountain area of Vietnam. *Ecological Modelling*, *202*(3-4), 410–420. doi:10.1016/j.ecolmodel.2006.11.011

Caviedes, C., & Knapp, G. (1995). *South America*. Englewood Cliffs: Prentice Hall.

CBD. (1992). *The United Nations Convention on Biological Diversity*. Retrieved December 29, 2009, from http://www.cbd.int/

CBD. (2002). *The United Nations Convention on Biological Diversity*. 2010 Biodiversity Target. Retrieved December 28, 2009, from http://www.cbd.int/2010-target/

CBD. (2006). *Global biodiversity outlook 2*. Montreal.

Ceballos, G., & Ehrlich, P. R. (2009). Discoveries of new mammal species and their implications for conservation and ecosystem services. *Proceedings of the National Academy of Sciences of the United States of America*, *105*, 11505–11511.

Ceballos, G., & Ehrlich, P. R. (2002). Mammal Population Losses and the Extinction Crisis. *Science*, *296*, 904–907. doi:10.1126/science.1069349

CEC. (1994). *CORINE land cover technical guide*. European Union, Directorate-General Environment, Nuclear Safety and Civil Protection, Office for Official Publications of the European Communities, Luxembourg. EUR, ISSN 1018-5593, ISBN 92-826-2578-8.

Central American Commission on Environment and Development & Regional Biodiversity Institute. (2009). *Reporte preliminar del modelo y análisis del estado actual y futuro de la biodiversidad para México con un enfoque de modelación en los estados fronterizos con Guatemala*. Manuscript submitted for publication.

CFAN (Forestry Advisers Network). (2005). *Deforestation: Tropical forests in decline*. CIDA Forestry Advisers Network. Retrieved February 2, 2008, from http://www.rcfa-cfan.org/english/ issues.12-3.html

Chander, K., Goyal, S., & Kapoor, K. (1995). Microbial biomass dynamics during decomposition of leaf litter of poplar and eucalypt in a sandy loam. *Biology and Fertility of Soils*, *19*, 357–361. doi:10.1007/BF00336107

Chapman, C. A., Chapman, L. J., Struhsaker, T. T., Zanne, A. E., Clark, C. J., & Poulsen, J. R. (2005). A long-term evaluation of fruiting phenology: Importance of climate change. *Journal of Tropical Ecology*, *21*, 31–45. doi:10.1017/S0266467404001993

Chapman, C. A., Chapman, L. J., Vulinec, K., Zanne, A., & Lawes, M. J. (2003). Fragmentation and alteration of seed dispersal processes: An initial evaluation of dung beetles, seed fate, and seedling diversity. *Biotropica*, *35*, 382–393.

Charuphat, T. (2000). Remote sensing and GIS for tropical forest management. In *Proceedings of the Ninth Regional Seminar on Earth Observation for Tropical Ecosystem Management*, Khao Yai, Thailand, 20-24 November 2000. (pp. 42-49). The National Space Development Agency of Japan, Remote Sensing Technology Center of Japan, Royal Forest Department, and GIS Application Center/AIT, Khao Yai National Park Thailand.

Chaurasia, R., Loshali, D. C., Dhaliwal, S. S., Shrma, M., Kudrat, P. K., & Tiwari, A. K. (1996). Land use change analysis for agriculture management – a case study of tehsil Talwandi Sabo, Punjab. Photonirvachk. *Journal of the Indian Society of Remote Sensing*, *24*(2), 89–98. doi:10.1007/BF03016124

Chazal, J., & de,., & Rounsevell, M. D. A. (2009). Land-use and climate change within assessment of biodiversity change: A review. *Global Environmental Change*, *19*, 306–315..doi:10.1016/j.gloenvcha.2008.09.007

Chefaoui, R. M., & Lobo, J. M. (2008). Assessing the effects of pseudo-absences on predictive distribution model performance. *Ecological Modelling*, *210*, 478–486. doi:10.1016/j.ecolmodel.2007.08.010

Chen, C. F., & Chen, M. H. (2005). Generalization of GIS polygon data using curvature-based approach. [IGARSS]. *International Geoscience and Remote Sensing*, *5*, 3498–3501.

Chen, J. (2005). Quality evaluation of topographic data from SRTM3 and GTOPO30. *Geomatics and Information Sciences*, *30*(11), 941–944.

Chen, J. M., Liu, J., Leblanc, S. G., Lacaze, R., & Roujean, J. L. (2003). Multi-angular optical remote sensing for assessing vegetation structure and carbon absorption. *Remote Sensing of Environment*, *84*(4), 516–525. doi:10.1016/S0034-4257(02)00150-5

Cheng, D. (2004). Monitoring large mammals of the northern plain landscape in Cambodia. *Paper presented at the 25th Annual Wildlife Seminar*, December 24-26, 2004. Faculty of Forestry, Kasetsart University, Bangkok, Thailand.

Chomitz, K. M., & Gray, D. A. (1996). Roads, land use, and deforestation: A spatial model applied to Belize. *The World Bank Economic Review*, *10*(3), 487–512.

Christensen, J. H., & Christensen, O. B. (2003). Climate modeling Sever summertime flooding in Europe. *Nature*, *421*, 805–806. doi:10.1038/421805a

Christenson, N. L., Bartuska, A. M., & Brown, J. H. (1996). The report of the Ecological Society of America: Committee on the Scientific Basis for Ecosystem Management. *Ecological Applications*, *6*(3), 665–691. doi:10.2307/2269460

Chuine, I., & Beaubien, E. (2001). Phenology is a major determinant of tree species range. *Ecology Letters*, *4*, 500–510. doi:10.1046/j.1461-0248.2001.00261.x

Chunkao, K. (1981). *Introduction to watershed resources management in humid tropics. Regional training course vol 1-24. Watershed resources management and environmental monitoring in humid and tropical ecosystems.* UNESCO, USAID, MAB (USA), and NEB (Thailand).

Chunkao, K. (1992). *Watershed management.* A paper presented at the training program on natural resources management and conservation watershed (Applied Remote Sensing/GIS short course) at Asian Institute of Technology, Bangkok, September 7-11, December 1992.

Clarke, K. C., Hoppen, S., & Gaydos, L. J. (1998). Loose-coupling a cellular automata model and GIS: Long term urban growth prediction for San Francisco and Washington, Baltimore. *International Journal of Geographical Information Science*, *12*(7), 699–714. doi:10.1080/136588198241617

Clawson, M., & Stewart, C. L. (1965). *Land use information. A critical survey of U.S. statistics including possibilities for greater uniformity*. Baltimore, MD: The Johns Hopkins Press for Resources for the Future, Inc.

Clergue, B., Amiaud, B., Pervanchon, F., Lasserre-Joulin, F., & Plantureux, S. (2005). Biodiversity: Function and assessment in agricultural areas, a review. *Agronomy for Sustainable Development*, *25*, 1–15. doi:10.1051/agro:2004049

CLUE. (2008). *Land use and land cover change model.* Retrieved February 1, 2009, from http://www.cluemodel.nl/

Cochard, R., & Agosti, D. (2008). Putting ant-acacia mutualisms to the fire. *Science*, *319*, 759–761. doi:10.1126/science.319.5871.1759d

Cochard, R., Ranamukharachchi, S. L., Shivakoti, G., Shipin, O., & Edwards, P. J. (2008). The 2004 tsunami in Aceh and Southern Thailand: A review on coastal ecosystems, wave hazards and vulnerability. *Perspectives in Plant Ecology, Evolution and Systematics*, *10*, 3–40. doi:10.1016/j.ppees.2007.11.001

Cohen, W. B., Maiersperger, T. K., Yang, Z. Q., Gower, S. T., Turner, D. P., & Ritts, W. D. (2003). Comparisons of land cover and LAI estimates derived from ETM plus and MODIS for four sites in North America: A quality assessment of 2000/2001 provisional MODIS products. *Remote Sensing of Environment*, *88*(3), 233–255. doi:10.1016/j.rse.2003.06.006

Cohen, W. B., Maiersperger, T. K., Gower, S. T., & Turner, D. P. (2003). An improved strategy for regression of biophysical variables and Landsat ETM+ data. *Remote Sensing of Environment*, *84*(4), 561–571. doi:10.1016/S0034-4257(02)00173-6

Colman, R. A., & Power, S. B. (2010). Atmospheric radiative feedbacks associated with transient climate change and climate variability. *Climate Dynamics*, *34*(7-8), 919–933. doi:10.1007/s00382-009-0541-8

Conservation International. (2007). *Biodiversity hotspots. Mesoamerica.* Retrieved from http://www.biodiversityhotspots.org/xp/hotspots/mesoamerica/Pages/default.aspx

Convention on Biological Diversity. (1992). *Text of the convention on biological diversity*. Retrieved April 15, 2010, from http://www.cbd.int/convention/convention.shtml

Coppin, P., Jonckheere, I., Nackaerts, K., Muys, B., & Lambin, E. (2004). Digital change detection methods in ecosystem monitoring: A review. *International Journal of Remote Sensing*, *25*(9), 1565–1596. doi:10.1080/0143116031000101675

Corlett, R. T. (2009). *The ecology of tropical East Asia*. Oxford, UK: Oxford University Press.

Corsi, F., De Leeuw, J., & Skidmore, A. K. (2000). Modeling species distribution with GIS. In Boitani, L., & Fuller, T. K. (Eds.), *Research techniques in animal ecology: Controversies and consequences* (pp. 389–434). New York, NY: Colombia University Press.

Corves, C., & Place, C. J. (1994). Mapping the reliability of satellite-derived land cover maps - an example from central Brazilian Amazon Basin. *International Journal of Remote Sensing, 15*, 1283–1294. doi:10.1080/01431169408954161

Council of Europe. UNEP, ECNC. (1996). *The Pan-European biological and landscape diversity strategy: A vision for Europe's natural heritage*. Strasbourg, Austria: Council of Europe Publishing.

Cowling, S. A., Betts, R. A., Cox, P. M., Ettwein, V. J., Jones, C. D., Maslin, M. A., & Spall, S. A. (2004). Contrasting simulated past and future responses of the Amazonian forest to atmospheric change. *Philosophical transactions of the Royal Society B, 359*, 539-547.

Cox, P. A., & Elmquist, T. (2000). Pollinator extinction in the Pacific Islands. *Conservation Biology, 14*(5), 1237–1239. doi:10.1046/j.1523-1739.2000.00017.x

Cox, P. M., Betts, R. A., Collins, M., Harris, P. P., Huntingford, C., & Jones, C. D. (2004). Amazonian forest dieback under climate-carbon cycle projections for the 21st century. *Theoretical and Applied Climatology, 78*, 137–156. doi:10.1007/s00704-004-0049-4

Cramer, W., Bondeau, A., Schaphoff, S., Lucht, W., Smith, B., & Sitch, S. (2004). Tropical forests and the global carbon cycle: impacts of atmospheric carbon dioxide, climate change and rate of deforestation. *Philosophical Transactions of the Royal Society of London Series B, 359*, 331–343. doi:10.1098/rstb.2003.1428

Cramer, W., Mesquita, R. C. G., & Williamson, G. B. (2007). Forest fragmentation differentially affects seed dispersal of large and small-seeded tropical trees. *Biological Conservation, 137*, 415–423. doi:10.1016/j.biocon.2007.02.019

Crisp, M. D., Laffan, S., Linder, H. P., & Monro, A. (2001). Endemism in the Australian flora. *Journal of Biogeography, 28*, 183–198. doi:10.1046/j.1365-2699.2001.00524.x

Cropper, M., Griffiths, C., & Mani, M. (1996). *Roads, population pressures and deforestation in Thailand, 1976-1989*. New York, NY: The World Bank.

Cropper, M., Griffiths, C. W., & Mani, M. (1997). *Roads, population pressures, and deforestation in Thailand, 1976-89*. (Policy Research Working Paper 1726). World Bank, Policy Research Department, Washington, D.C.

Crossland, M. D., Perkins, W. C., & Wynne, B. E. (1995). Spatial decision support systems: An overview of technology and a test efficiency. *Decision Support Systems, 14*(3), 219–235. doi:10.1016/0167-9236(94)00018-N

Cruz, R. B. (2006). Land use information system for local government: The case of Naga City, Philippines. *Systemics. Cybernetics and Informatics, 4*, 67–74.

CSIRO. (2009). *Climate Change and the 2009 Bushfires.* ([Victorian Bushfires Royal Commission). Commonwealth Scientific and Industrial Research Organisation, Canberra.]. *CSIRO Submission, 091345*, 2009.

Cuesta-Camacho, F., Ganzenmuller, A., Peralvo, M., Novoa, J., & Riofrio, G. (2006). *Predicting species' niche distribution shifts and biodiversity change within climate change scenarios: A regional assessment for bird and plant species in the Northern Tropical Andes*. Unpublished report, Biodiversity Monitoring Program (EcoCiencia), Peru.

Cuesta-Camocho, F., Ganzenmuller, A., Peralvo, M., Novoa, J., & Riofrio, G. (2006). *Predicting species' niche distribution shifts and biodiversity change within climate change scenarios: A regional assessment for bird and plant species in the Northern Tropical Andes*. EcoCiencia, Peru: Biodiversity Monitoring Program.

Cumming, G. S. (2007). Global biodiversity scenarios and landscape ecology. *Landscape Ecology, 22*, 671–685. doi:10.1007/s10980-006-9057-3

Curran, L. M., Caniago, I., Paoli, G. D., Astianti, D., Kusneti, M., & Leighton, M. (1999). Impact of El Niño and logging on canopy tree recruitment in Borneo. *Science, 286*, 2184–2188. doi:10.1126/science.286.5447.2184

Currie, D. J. (1991). Energy and large-scale patterns of animal and plant species richness. *American Naturalist, 137*, 27–49. doi:10.1086/285144

Cutter, S. L., & Renwick, W. H. (1999). *Exploitation, conservation, preservation: A geographic perspective on natural resource use*. New Jersey: John Wiley & Sons, Inc.

Dale, V. H., Joyce, L. A., McNulty, S., & Neilson, R. P. (2001). Climate change and forest disturbances. *Bioscience*, *51*(9), 723–734. doi:10.1641/0006-3568(2001)051[0723:CCAFD]2.0.CO;2

Dallmeyer, A., Claussen, M., & Otto, J. (2010). Contribution of oceanic and vegetation feedbacks to Holocene climate change in monsoonal Asia. *Climate of the Past*, *6*(2), 195–218. doi:10.5194/cp-6-195-2010

DANE. (2007). Colombia: Departamento Administrativo Nacional de Estadística, Indicadores de producción agropecuaria. Retrieved from http://www.dane.gov.co/index.php? option=com_content&task=category&sectionid=18&id=41&Itemid=152

DANE. (2008). Producto interno Bruto. Segundo trimestre de 2008 Base 2000. Departamento Nacional de estadística. Retrieved from http://www.dane.gov.co/files/investigaciones/boletines/pib/presen_PIB_IItrim08.pdf

Danielson, F., Beukema, H., Burgess, N. D., Parish, F., Brühl, C. A., & Donald, P. F. (2008). Biofuel plantations on forested lands: Double jeopardy for biodiversity and climate. *Conservation Biology*, *23*(2), 348–358. doi:10.1111/j.1523-1739.2008.01096.x

Dasgupta, S. Laplante. B., Meisner, C., Wheeler, D., & Yan, J. (2007). *The impact of sea level rise on developing countries: A comparative analysis*. World Bank Policy Research Working Paper 4136, February 2007.

David, M. S., Davis, F. W., & Hollander, A. D. (1994). Hierarchical representation of species distribution for biological survey and monitoring. In Goodchild, M., Parks, B. O., & Steyaert, L. (Eds.), *Environmental modeling: Progress and research issues* (pp. 445–449). Fort Collins, CO: GIS World Books.

Davis, S. D., Heywood, V. H., & Hamilton, A. C. (1995). *Centres of plant diversity 2*. Cambridge, U.K.: WWF and IUCN Publications Unit.

Davis, S., Heywood, V. H., & Hamilton (Eds.). (1997). *The Americas*. Centres of plant diversity (IUCN.). Gland.

De Aranzabal, I., Schmitz, M. F., Aguilera, P., & Pineda, F. D. (2008). Modelling of landscape changes derived from the dynamics of socio-ecological systems. A case of study in a semiarid Mediterranean landscape. *Ecological Indicators*, *8*(5), 672–685. doi:10.1016/j.ecolind.2007.11.003

De Fries, R. S., Hansen, M. C., & Townshend, J. R. G. (2000). Global continuous fields of vegetation characteristics: A linear mixture model applied to multi-year 8km AVHRR data. *International Journal of Remote Sensing*, *21*, 1389–1414. doi:10.1080/014311600210236

De Groot, R. S., Wilson, M. A., & Boumans, R. M. J. (2002). A typology for the classification, description and valuation of ecosystem functions, goods and services. *Ecological Economics*, *41*, 393–408. doi:10.1016/S0921-8009(02)00089-7

De Silva, A. L. C., & De Costa, W. A. J. M. (2004). Varietal variation in growth, physiology and yield of sugarcane under two contrasting water regimes. *Tropical Agricultural Research*, *16*, 1–12.

De Vries, B., Bollen, J., Bouwman, L., Den Elzen, M., Janssen, M., & Kreileman, E. (2000). Greenhouse gas emissions in an equity-environment- and service-oriented world: An IMAGE-based scenario for the 21st Century. *Technological Forecasting and Social Change*, *63*, 137–174. doi:10.1016/S0040-1625(99)00109-2

De Wit, A., & Mücher, C. A. (2009). *Satellite-derived trends in phenology over Europe: Real trends or algorithmic effects*. Poster presented at the General Assembly 2009 of the European Geosciences Union (EGU), from 19 – 24 April 2009, Vienna, Austria.

Defourny, P., Schouten, L., Bartalev, S., Bontemps, S., Cacetta, P., de Wit, A. J. W., et al. Arino, O. (2009). *Accuracy assessment of a 300 m global land cover map: The GlobCover experience*. 33rd International Symposium on Remote Sensing of Environment (ISRSE), May 4-8.2009, Stresa, Italy

DeFries, R., Bounoua, L., & Collatz, G. J. (2002). Human modification of the landscape and surface climate in the next fifty years. *Global Change Biology*, *8*, 438–458. doi:10.1046/j.1365-2486.2002.00483.x

DeFries, R., Hansen, M., & Townshend, J. G. R. (1995). Global discrimination of land cover types from metrics derived from AVHRR Pathfinder data. *International Journal of Remote Sensing*, *19*, 3141–3168.

DeFries, R. S., Foley, J. A., & Asner, G. P. (2004). Land-use choices: Balancing human needs and ecosystem function. *Frontiers in Ecology and the Environment*, *2*, 249–257. doi:10.1890/1540-9295(2004)002[0249:LCBHNA]2.0.CO;2

Delang, C. O. (2002). Deforestation in northern Thailand: The result of Hmong farming practices or Thai development strategies? *Society & Natural Resources*, *15*, 483–501. doi:10.1080/08941920290069137

Delin, A. E., & Andren, H. (2004). Effects of habitat fragmentation on Eurasian red squired (*Sciurus vulgaris*) in a forest landscape. *Landscape Ecology*, *14*, 62–72.

Dennis, R. L. H., & Hardy, P. B. (1999). Targeting squares for survey: Predicting species richness and incidence of species for a butterfly atlas. *Global Ecology and Biogeography*, *8*(6), 443–454. doi:10.1046/j.1365-2699.1999.00148.x

Densham, P., & Rushton, G. (1987). Decision support systems for locational planning. In Golledge, R. G., & Timmermans, H. (Eds.), *Behavioral modeling in geography and planning* (pp. 56–90). New York, NY: Croom Heelm.

Densham, P. J. (1991). Spatial decision support systems. In Maguire, D. J., Goodchild, M. F., & Rhind, D. W. (Eds.), *Geographical Information Systems: Principles and applications* (pp. 403–421). Harlow, Essex, UK: Longman.

Densham, P. J., & Goodchild, M. F. (1989). Spatial decision support systems: A research agenda. *Processing of GIS/LIS'89* (pp. 707-716). Orlando, FL.

Department of Local Administration. (2007). *Population census in Thailand from 1994 to 2007*. Retrieved March 2, 2008, from http://www.dopa.go.th/ hpstat9/people2.htm

Diane, M. (2002). *Community approaches to watershed management.* Centre for Rural Studies and Enrichment. St. Peter's College, Muenster. SK. Retrieved from http://www.saskriverbasin.ca/Conference/2002/Presentations/Diane%20 Martz.ppt

Didham, R. K., Hammond, P. M., Lawton, J. H., Eggleton, P., & Stork, N. E. (1998). Beetle species responses to tropical forest fragmentation. *Ecological Monographs*, *68*, 295–323. doi:10.1890/0012-9615(1998)068[0295:BSRTTF]2.0.CO;2

Diemont, W. H., Nabuurs, G. J., Rieley, J. O., & Rijksen, H. D. (1997). Climate change and management of tropical peatlands as a carbon reservoir. In Rieley, J. O., & Page, S. E. (Eds.), *Tropical peatlands* (pp. 363–368). Cardigan, UK: Samara Publishing Limited.

Dinerstein, E. (1997). *A framework for identifying high priority areas and actions for the conservation of tigers in the wild.* World Wildlife Fund-US.

Dixon, J., Gulliver, A., & Gibbon, D. (2001). *Farming systems and poverty*. Rome and Washington DC: FAO and World Bank.

Dobson, A. P. (1996). *Conservation and biodiversity*. New York, NY: Scientific American Library.

Dole, M. (1965). The natural history of oxygen. *The Journal of General Physiology*, *49*, 5–27. doi:10.1085/jgp.49.1.5

Donovan, T. M., & Flather, C. H. (2002). Relationships among North American songbird trends, habitat fragmentation and landscape occupancy. *Ecological Applications*, *12*, 364–374.

Dormann, C. F. (2007). Promising the future? Global change projections of species distributions. *Basic and Applied Ecology*, *8*, 387–397. doi:10.1016/j.baae.2006.11.001

Drake, D. R., & Hunt, T. L. (2009). Invasive rodents on islands: Integrating historical and contemporary ecology. *Biological Invasions*, *11*(7), 1483–1487. doi:10.1007/s10530-008-9392-1

Dufrêne, M., & Legendre, P. (1997). Species assemblages and indicator species: The need for a flexible asymmetrical approach. *Ecological Monographs*, *67*, 345–366.

Dullinger, S., Kleinbauer, I., Peterseil, J., Smolik, M., & Essl, F. (2009). Niche based distribution modelling of an invasive alien plant: Effects of population status, propagule pressure and invasion history. *Biological Invasions*, *11*(10), 2401–2414. doi:10.1007/s10530-009-9424-5

Dumanski, J., & Pieri, C. (2000). Land quality indicators: Research plan. *Agriculture Ecosystems & Environment*, *81*, 93–102. doi:10.1016/S0167-8809(00)00183-3

Dunn, P. O., & Winkler, D. W. (1999). Climatic change has affected breeding date of tree swallows throughout North America. *Proceedings. Biological Sciences*, *266*, 2487–2490. doi:10.1098/rspb.1999.0950

Duro, D. C., Coops, N. C., Wulder, M. A., & Han, T. (2007). Development of a large area biodiversity monitoring system driven by remote sensing. *Progress in Physical Geography*, *31*(3), 235–260. doi:10.1177/0309133307079054

Dymond, C. C., Mladenoff, D. J., & Radeloff, V. C. (2002). Phenological differences in Tasseled Cap indices improve deciduous forest classification. *Remote Sensing of Environment*, *80*, 460–472. doi:10.1016/S0034-4257(01)00324-8

Eastman, J. R., Jin, W., Kyem, A. K., & Toledano, J. (1995). Raster procedures for multi-criteria/multi-objective decisions. *Photogrammetric Engineering and Remote Sensing*, *61*(5), 539–547.

ECNC, IUCN and Council of Europe. (2004). *Pan-European biological and landscape diversity strategy*. Council for the European Biological and Landscape Diversity Strategy. Third Intergovernmental Conference 'Biodiversity in Europe'. Follow up of the Kyiv Biodiversity Resolution: Pan-European Ecological Network Action Plan Proposal. STRA-CO (2004) 3c rev. Retrieved October 25, 2009, from http://www.peblds.org

Edwards, J. L., Lane, M. A., & Nielsen, E. S. (2000). Interoperability of biodiversity databases: Biodiversity information on every desktop. *Science*, *289*, 2312–2314. doi:10.1126/science.289.5488.2312

EEA. (2005). *The European environment - state and outlook 2005*. Copenhagen.

EEA. (2007). *Halting the loss of biodiversity by 2010: Proposal for a first set of indicators to monitor progress in Europe*. (EEA Technical report No 11/2007). ISSN 1725–2237, EEA, Copenhagen.

Eickhout, B., Den Elzen, M., & Kreileman, E. (2004). *The atmosphere-ocean system in IMAGE 2.2. Report 481508017*. Bilthoven, The Netherlands: National Institute for Public Health and the Environment.

Eickhout, B., Van Meijl, H., Tabeau, A., & Van Rheenen, R. (2007). Economic and ecological consequences of four European land use scenarios. *Land Use Policy*, *24*(3), 562–575. doi:10.1016/j.landusepol.2006.01.004

Elith, J., Graham, C. H., Anderson, R. P., Dudik, M., Ferrier, S., & Guisan, A. (2006). Novel methods improve prediction of species' distributions from occurrence data. *Ecography*, *29*, 129–151. doi:10.1111/j.2006.0906-7590.04596.x

Elith, J. (2002). Quantitative methods for modeling species habitat: Comparative performance and an application to Australian plants. In Ferson, S., & Burgman, M. (Eds.), *Quantitative methods for conservation biology* (pp. 39–58). New York, NY: Springer-Verlag.

Elton, C. (1927). *Animal ecology*. London, UK: Sedgwick and Jackson.

Emch, M., James, W., Quinn, M. P., & Alexander, M. (2005). Forest cover change in the Toledo district, Belize from 1975 to 1999: A remote sensing approach. *The Professional Geographer*, *57*(2), 256–267. doi:10.1111/j.0033-0124.2005.476_1.x

Encyclopedia of Earth. (2007). *Land-use and land-cover change. Washington DC*. Retrieved March 20, 2010, from http://www.eoearth.org/article/Land-use_and_land-cover_change

Engler, R., Guisan, A., & Rechsteiner, L. (2004). An improved approach for predicting the distribution or rare and endangered species from occurrence and pseudo-absence data. *Journal of Applied Ecology*, *41*, 263–274. doi:10.1111/j.0021-8901.2004.00881.x

Englund, E. J. (1993). Spatial simulation: Environmental applications. In Goodchild, M. F., Steyaert, L. T., & Parks, B. O. (Eds.), *GIS and environmental modeling: Process and research issues* (pp. 432–437). Fort Collins, CO: GIS World, Inc.

ESRI, Inc. (1996). *Using Arc View GIS*. Redlands, CA: ESRI, Inc.

Eswaran, H., Lal, R., & Reich, P. F. (2001). Land degradation: An overview. In E. M. Bridges, I. D. Hannam, L. D. Oldeman, F. W. T. Pening de Vries, S. J. Scherr & S. Sompatpanit (Eds.). *Responses to land degradation. Proceedings of the 2nd International Conference on Land Degradation and Desertification*, Khon Kaen, Thailand.

EU Council Directive. (1992). *Natura 2000: Network on the conservation of natural habitats and of wild fauna. COUNCIL DIRECTIVE 92/43/EEC (1) of 21 May 1992*. The Council of the European Communities.

European Commission. (1992). *Development of a framework for the evaluation of policy options to deal with the greenhouse effect*. Brussels: Commission of the European Community, Directorate General for Environment, Nuclear Safety and Civil Protection.

Eva, H., de Miranda, E., Di Bella, C., Gond, V., Huber, O., Sgrenzaroli, M., et al. (2003). The Land Cover Map for South America in the Year 2000. European Commision Joint Research Centre. Retrieved from http://www-gem.jrc.it/glc2000

Evans, D. (2006). The habitats of the European union habitats directive. *Proceedings of the Royal Irish Academy-Section B, Biological Environments, 106*(3), 167-173.

DMA. (1992). Digital chart of the world. Defence Mapping Agency. In Fairfax, V., & Duellman, W. E. (Eds.), *Patterns of distribution of amphibians: A global perspective*. Baltimore, MD: John Hopkins University Press.

Faith, D. P., Ferrier, S., & Williams, K. J. (2008). Getting biodiversity intactness indices right: Ensuring that biodiversity reflects diversity. *Global Change Biology, 14*, 207–221. doi:10.1111/j.1365-2486.2007.01500.x

Falcucci, A., Maiorano, L., & Boitani, L. (2007). Changes in land-use/land-cover patterns in Italy and their implications for biodiversity conservation. *Landscape Ecology, 22*, 617–631. doi:10.1007/s10980-006-9056-4

FAO. (1979). *FAO-UNESCO soil map of the world: Southeast Asia*. Paris: Food and Agricultural Organization of the United Nations.

FAO. (1997). *Africover land cover classification. Environment and Natural Resources Service (SDRN), Food and Agriculture Organization of the United Nations*. Rome: FAO.

FAO. (1997). *Estimating biomass and biomass change of tropical forests: A primer. (Forestry Paper No. 134)*. Rome: FAO.

FAO. (2000). *On definitions of forest and forest change*. Rome, Italy: Food and Agriculture Organisation of the United Nations.

FAO. (2002). *TerraStat: Global land resources GIS models and databases for poverty and food insecurity mapping*. Land and Water Digital Media Services.

FAO. (2002). *World agriculture: Towards 2015/2030. Summary report*. Rome: Food and Agriculture Organization.

FAO. (2006). *Global forest resources assessment 2005. Progress towards sustainable forest management. FAO Forestry Paper* (p. 320). Rome: FAO.

FAO. (2006). *Global forest resources assessment 2005. (FAO Forestry Paper No. 147)*. Rome: UN Food and Agriculture Organization.

FAO. (2006). *Global forest resources assessments 2005. Progress towards sustainable forest management. FAO forestry paper*. Rome: FAO.

FAO. (2007). *State of the world's forests*. Rome, Italy: Food and Agriculture Organisation of the United Nations.

FAO. (2010). *Global forest resources assessment. Food and Agriculture Organization of the United Nations*. Rome: FAO.

FAO. (1999). *Agricultural biodiversity*. Multifunctional Character of Agriculture and Land Conference. Background paper 1. Maastricht Netherlands.

FAO. (2003). *TERRSTAT: Land resource potential and constraints statistics at country and regional level*. Retrieved May 15, 2010, from http://www.fao.org/ag/agl/agll/terrastat/

FAO. (2004). *FAO—The Africover Initiative*. Food and Agricultural Organization of the United Nations. Retrieved March 25, 2010, from http://www.africover.org/africover_initiative.htm

FAO. (2005). *Properties and management of drylands*, Retrieved June 12, 2006, from http://www.fao.org/ag/agl/agll/drylands/assessment.htm

FAO. *(2005)*. Working paper 18: Global forest resources assessment update 2005, specification of national reporting tables for FRA 2005. *Rome*

Fedra, K. (1993). GIS and environmental modeling. In Goodchild, M. F., Parks, B. O., & Steyaert, L. T. (Eds.), *Environmental modeling with GIS* (pp. 35–50). New York, NY: Oxford University Press.

Fedra, K. (1994). Model-based environment information and decision support systems. In Hilty, J. (Ed.), *Informatik für den Umweltschutz* (pp. 37–58). Metropolis Verlag, Marburg.

Feeley, K. J., & Silman, M. R. (2010). Land-use and climate change effects on population size and extinction risk of Andean plants. *Global change biology, Accepted article*. doi:10.1111/j.1365-2486.2010.02197.x

Feike, A. D., Sarah, E. H., Peter, B. R., & Johannes, M. H. K. (2005). Divergent effects of elevated CO_2, N fertilization, and plant diversity on soil C and N dynamics in a grassland field experiment. *Plant and Soil*, *272*, 41–52. doi:10.1007/s11104-004-3848-6

Fensham, R. J., Fairfax, R. J., & Cannell, R. J. (1994). The invasion of Lantana camara L. in Forty Mile Scrub National Park, North Queensland. *Austral Ecology*, *19*(3), 297–305. doi:10.1111/j.1442-9993.1994.tb00493.x

Feranec, J., Hazeu, G., Christensen, S., & Jaffrain, G. (2007). Corine land cover change detection in Europe (case studies ofThe Netherlands and Slovakia). *Land Use Policy*, *24*(1), 234–247. doi:10.1016/j.landusepol.2006.02.002

Ferrier, S., Manion, G., Eilth, J., & Richardson, K. (2007). Using generalized dissimilarity modelleing to analyse and predict patterns of beta diversity in regional biodiversity assessment. *Diversity & Distributions*, *13*, 252–264. doi:10.1111/j.1472-4642.2007.00341.x

Fielding, A. H., & Bell, J. F. (1997). A review of methods for the assessment of prediction errors in conservation presence/absence models. *Environmental Conservation*, *24*, 38–49. doi:10.1017/S0376892997000088

Fisher, P. F. (1991). Modeling soil map-unit inclusions by Monte Carlo simulation. *International Journal of Geographical Information Systems*, *5*(2), 193–208. doi:10.1080/02693799108927843

Fisher, P. F., Comber, A. J., & Wadsworth, R. A. (2005). Land use and land cover: Contradiction or complement. In Fisher, P., & Unwin, D. (Eds.), *Re-Presenting GIS* (pp. 85–98). Chichester, UK: Wiley Press.

Fitz, H. C., DeBellevue, E. B., Costanza, R., Boumans, R., Maxwell, T., Wainger, L., & Sklar, F. H. (1996). Development of a general ecosystem model (GEM) for a range of scales and ecosystems. *Ecological Modelling*, *88*, 263–297. doi:10.1016/0304-3800(95)00112-3

Fitzherbert, E. B., Struebig, M. J., Morel, A., Danielson, F., Brühl, C. A., Donald, P. F., & Phalan, B. (2008). How will oil palm expansion affect biodiversity? *Trends in Ecology & Evolution*, *23*(10), 538–545. doi:10.1016/j.tree.2008.06.012

Floren, A., & Linsenmair, E. (2001). The influence of anthropogenic disturbances on the structure of arboreal arthropod communities. *Plant Ecology*, *153*, 153–167. doi:10.1023/A:1017510312462

Fontaine, C., Dajoz, I., Meriguet, J., & Loreau, M. (2006). Functional diversity of plant—pollinator interaction webs enhances the persistence of plant communities. *PLoS Biology*, *4*(1), e1. doi:10.1371/journal.pbio.0040001

Food and Agriculture Organization of the United Nations. (2010). *Global forest assessment 2010*. Rome: Food and Agriculture Organization of the United Nations.

Foody, G. M., Boyd, D. S., & Curran, P. J. (1996). Relations between tropical forest biophysical properties and data acquired in AVHRR channels 1± 5. *International Journal of Remote Sensing*, *17*, 1341–1355. doi:10.1080/01431169608948707

Foody, G. M., Sargent, I. M. J., Atkinson, P. M., & Williams, J. W. (2004). Thematic labelling from hyperspectral remotely sensed imagery: Trade-offs in image properties. *International Journal of Remote Sensing*, *25*(12), 2337–2363. doi:10.1080/01431160310001654969

Foottit, R. G., & Adler, P. H. (Eds.). (2009). *Insect biodiversity*. Oxford, UK: Blackwell Publishing. doi:10.1002/9781444308211

Foreman, R. T. T. (1995). Some general principles of landscape and regional ecology. *Landscape Ecology*, *10*(3), 133–142. doi:10.1007/BF00133027

Foreman, R. T. T., & Gordon, M. (1986). *Landscape ecology*. New York, NY: John Wiley & Sons.

Forman, R. T. T. (1995). *Land mosaics: The ecology of landscapes and regions*. Cambridge, UK: Cambridge University Press.

Forseth, I. N. Jr, & Innis, A. F. (2004). Kudzu (*Pueraria montana*): History, physiology, and ecology combine to make a major ecosystem threat. *Critical Reviews in Plant Sciences, 23*(5), 401–413. doi:10.1080/07352680490505150

Förster, M., Frick, A., Walentowski, H., & Kleinschmit, B. (2008). Approaches to utilising QuickBird data for the monitoring of NATURA 2000 habitats. *Community Ecology, 9*(2), 155–168. doi:10.1556/ComEc.9.2008.2.4

Foster, P. (2001). The potential negative impacts of global climate change on tropical montane cloud forests. *Earth-Science Reviews, 55*(1-2), 73–106. doi:10.1016/S0012-8252(01)00056-3

Fotheringham, S., & Rogers, P. (1994). *Spatial analysis and GIS*. London, UK: Taylor & Francis.

Fox, J., & Vogler, J. B. (2005). Land-use and land-cover change in montane mainland Southeast Asia. *Environmental Management, 36*(3), 394–403. doi:10.1007/s00267-003-0288-7

Fox, J. M. (1991). Spatial information resource management in Asia: A review of institutional issues. *International Journal of Geographical Information Systems, 5*(1), 59–72. doi:10.1080/02693799108927831

Francl, K. E., Hayhoe, K., Saunders, M., & Maurer, E. P. (in press). Ecosystem adaptation to climate change: Small mammal migration pathways in the Great Lakes states. *Journal of Great Lakes Research*. doi: 10.1018.jglr.2009.09.007.

Franklin, J. (2009). *Mapping species distributions: Spatial inference and prediction*. Cambridge, UK: Cambridge University Press.

Fraser, R. H., Abuelgasim, A., & Latifovic, R. (2005). A method for detecting large-scale forest cover change using coarse spatial resolution imagery. *Remote Sensing of Environment, 95*, 414–427. doi:10.1016/j.rse.2004.12.014

Friedl, M. A., McIver, D. K., Hodges, J. C. F., Zhang, X. Y., Muchoney, D., & Strahler, A. H. (2002). Global land cover mapping from MODIS: Algorithms and early results. *Remote Sensing of Environment, 83*(1-2), 287–302. doi:10.1016/S0034-4257(02)00078-0

Fritz, S., & See, L. (2008). Identifying and quantifying uncertainty and spatial disagreement in the comparison of Global Land Cover for different applications. *Global Change Biology, 14*, 1057–1075. doi: 10.1111/j.1365-2486.2007.01519.x

Fukushima, M., Kanzaki, M., Hara, M., Ohkubo, T., Preechapanya, P., & Chocharoen, C. (2008). Secondary forest succession after the cessation of swidden cultivation in the montane forest area in northern Thailand. *Forest Ecology and Management, 255*(5-6), 1994–2006. doi:10.1016/j.foreco.2007.12.022

Gardiner, J. L. (1994). Sustainable development for river catchments. *Journal of the Institution of Water and Environmental Management, 8*, 308–319. doi:10.1111/j.1747-6593.1994.tb01109.x

Gash, J. H. C., & Nobre, C. A. (1997). Climatic effects of Amazonian deforestation: Some results from ABRACOS. *Bulletin of the American Meteorological Society, 78*, 823–830. doi:10.1175/1520-0477(1997)078<0823:CEOADS>2.0.CO;2

Gaston, K. J. (Ed.). (1996). *Biodiversity: A biology of numbers and difference*. Oxford, UK: Blackwell Science.

Gaston, K. J. (2000). Global patterns in biodiversity. *Nature, 405*, 220–227. doi:10.1038/35012228

Gaston, K. J. (2010). Biodiversity. In Sodhi, N. S., & Ehrlich, P. R. (Eds.), *Conservation Biology for all* (pp. 27–44). Cambridge, UK: Oxford University Press. doi:10.1093/acprof:oso/9780199554232.003.0003

Geertman, S., & Stillwell, J. (2002). *Planning support systems in practice: Advances in spatial science*. New York, NY: Springer Publishers.

GEF. (1999). *Report of the STAP Expert Group Workshop on Land Degradation*. (GEF/C.14/Inf. 15). Global Environment Facility.

Gehrig-Fasel, J., Guisan, A., & Zimmermann, N. E. (2007). Tree line shifts in the Swiss Alps: Climate change or land abandonment? *Journal of Vegetation Science, 18*, 571–582. doi:10.1111/j.1654-1103.2007.tb02571.x

Geist, H. J., & Lambin, E. F. (2002). Proximate causes and underlying driving force of tropical deforestation. *Bioscience, 52*(2), 143–150. doi:10.1641/0006-3568(2002)052[0143:PCAUDF]2.0.CO;2

GEO. (2005). *The g=Global Earth Observation System of Systems (GEOSS)—10-year implementation plan and reference document*. Retrieved on May 12, 2010, from http://earthobservations.org

GEO-BON. (2008). *The geo biodiversity observation network: Concept document*. Postdam, Germany: Groups on Earth Observation.

Gerard, F., Petit, S., Smith, G., Thomson, A., Brown, N., & Manchester, S. (2010). Huitu, Tuominen, S., Köhler, R., Olschofsky, K., Ziese, H., Kolar, J., Sustera, J., Luque, S., Pino, J., Pons, X., Roda, F., Roscher, M., Feranec, J., 2010. Land cover change in Europe between 1950 to 2000 determined employing aerial photography. *Progress in Physical Geography*, *34*(2), 183–205. doi:10.1177/0309133309360141

Gereti, E. J., Wolanski, E., & Chiombola, E. A. T. (2003). *Assessment of the environmental, social and economic impacts on the Serengeti ecosystem of the developments in the Mara River Catchment in Kenya*. Dar Es Salaam: Tanzania National Parks.

Ghazoul, J. (2005). Pollen and seed dispersal among dispersed plants. *Biological Reviews of the Cambridge Philosophical Society*, *80*, 413–443. doi:10.1017/S1464793105006731

Ghazoul, J., Liston, K. A., & Boyle, T. J. B. (1998). Disturbance-induced density-dependent seed set in *Shorea siamensis* (Dipterocarpaceae), a tropical forest tree. *Journal of Ecology*, *86*, 462–473. doi:10.1046/j.1365-2745.1998.00270.x

Ghazoul, J., & Sheil, D. (2010). *Tropical rainforest: Ecology, diversity and conservation*. New York, NY: Oxford University Press.

Ghosh, A., & Rushton, G. (1986). *Spatial analysis and location-allocation models*. New York, NY: Van Nostrand Reinhold.

Giesen, W. (2004). *Causes of peatswamp forest degradation in Berbak NP, Indonesia, and recommendations for restoration. International Agricultural Centre (IAC)*. Netherlands: Arcadis Euroconsult Arcadis Euroconsult, Wageningen University.

Gilman, E. L., Ellison, J., Duke, N. C., & Field, C. (2008). Threats to mangroves from climate change and adaptation options, a review. *Aquatic Botany*, *89*(2), 237–250. doi:10.1016/j.aquabot.2007.12.009

Gilruth, P. T., Marsh, S. E., & Itami, R. (1995). A dynamic spatial model of shifting cultivation in the highlands of Guinea, West Africa. *Ecological Modelling*, *79*(1-3), 179–197. doi:10.1016/0304-3800(93)E0145-S

Gitay, H., Suárez, A., Watson, R. T., & Dokken, D. J. (Eds.). (2002). *Climate change and biodiversity*. Geneva: IPCC Technical Paper V. Intergovernmental Panel on Climate Change.

Givnish, T. J. (1999). On the causes of gradients in tropical tree diversity. *Journal of Ecology*, *87*, 193–210. doi:10.1046/j.1365-2745.1999.00333.x

Global Land Cover Network (GLCN). (2000). *Global land cover 2000*. Retrieved March 15, 2010, from www.glcn.org

GLOBCOVER. (2008). *Globcover products description validation report I2.1*. European Space Agency. Retrieved from http://www.esa.int/esaEO/ SEMGSY2IU7E_index_0.html.

GLOBIO Consortium. (2010). *What is GLOBIO?* Retrieved from http://www.globio.info/what-is-globio

GLOBIO. (2007). *Modelling human impacts on biodiversity*. Retrieved February 1, 2009, from http://www.globio.info/

GLP. (2005). *Global land project science plan and implementation strategy*. (IGBP Report No. 53/IHDP Report No. 19). IGBP Secretariat, Stockholm. Retrieved March 27, 2010, from http://www.globallandproject.org/Documents/report_53.pdf

GMS-EOC. (2008). *Subregional environmental performance assessment (EPA) report. National performance assessment and subregional strategic environment framework for the greater Mekong Subregion. TA No. 6069*. Bangkok, Thailand: GMS Environmental Operations Center.

Gnanavelrajah, N., Shrestha, R. P., Schmidt-Vogt, D., & Samarakoon, L. (2008). Assessment of carbon stock and soil carbon management in agricultural landuses in Thailand. *Land Degradation and Development*, *19*(3), 242–256. doi:10.1002/ldr.838

Gnanavelrajah, N. (2007). *Sustainable land-use options for enhancing carbon sequestration, plant diversity and productivity: A case of Khlong Yai Sub-watershed, Thailand*. Unpublished PhD dissertation, Asian Institute of Technology, Bangkok.

Godfray, C. (2002). Challenges for taxonomy. *Nature, 417*, 17–19. doi:10.1038/417017a

Goldewijk, K. (2001). Estimating global land use change over the past 300 years: The HYDE database. *Global Biogeochemical Cycles, 15*, 417–434. doi:10.1029/1999GB001232

Gonçalves, P., Carrão, H., Pinheiro, A., & Caetano, M. (2006). Land cover classification with Support Vector Machine Applied to MODIS imagery. In Marçal, A. (Ed.), *Global developments in environmental earth observation from space* (pp. 517–526). Portugal: MillPress.

Goodchild, M. F. (1994). Integrating GIS and remote sensing for vegetation analysis and modeling: Methodological issues. *Journal of Vegetation Science, 5*(5), 615–626. doi:10.2307/3235878

Gopal, S., Friedl, M. A., McIver, D. K., Hodges, J. C. F., Zhang, X. Y., & Muchnoy, D. (2002). Global land cover mapping from MODIS: Algorithms and early results. *Remote Sensing of Environment, 83*, 287–302. doi:10.1016/S0034-4257(02)00078-0

Gordon, L. J., Peterson, G. D., & Bennett, E. M. (2008). Agricultural modifications of hydrological flows create ecological surprises. *Trends in Ecology & Evolution, 23*, 211–219..doi:10.1016/j.tree.2007.11.011

Gordon, J. (1993). Ecosystem management: An idiosyncratic overview. In Aplet, G. H., Johnson, N., Olson, J. T., & Sample, V. A. (Eds.), *Defining sustainable forestry* (pp. 240–247). Washington, DC: Island Press.

Gotelli, N. J. (2001). *A primer of ecology*. Sunderland, MA: Sinauer Associates.

Graham, C. H., Elith, J., Hijmans, R. J., Guisan, A., Peterson, A. T., & Loiselle, B. A. (2008). The influence of spatial errors in species occurrence data used in distribution models. *Journal of Applied Ecology, 45*, 239–247. doi:10.1111/j.1365-2664.2007.01408.x

Grainger, A. (1980). The state of the world's tropical forests. *The Ecologist, 10*, 6–54.

Grainger, A., Boucher, D. H., Frumhoff, P. C., Laurance, W. F., Lovejoy, T., & McNeely, J. (2009). Biodiversity and REDD at Copenhagen. *Current Biology, 19*, R974–R976. doi:10.1016/j.cub.2009.10.001

Grainger, A., Rose, S., Trisurat, Y., & Brockelman, W. (1995). A GIS approach to mapping spatial variation in tropical rain forest biodiversity. In T. J. B. Boyle & B. Boontawee (Eds.), *Proceedings of the IUFRO Symposium on Measuring and Monitoring Biodiversity in Tropical and Temperate Forests* (pp. 335-354). Chiang Mai, Thailand: Royal Forest Department.

Grambine, R. E. (1994). What is ecosystem management? *Conservation Biology, 8*, 27–38. doi:10.1046/j.1523-1739.1994.08010027.x

Green, P. T., Lake, P. S., & O'Dowd, D. J. (2004). Resistance of island rainforest to invasion by alien plants: Influence of microhabitat and herbivory on seedling performance. *Biological Invasions, 6*(1), 1–9. doi:10.1023/B:BINV.0000010144.12808.cb

Green, R., Collingham, Y., Willis, S., Gregory, R., Willis, S. G., & Gregory, R. D. (2008). Performance of climate envelope models in retrodicting recent changes in bird population size from observed climatic change. *Biology Letters, 4*, 599–602. doi:10.1098/rsbl.2008.0052

Grigg, D. (1965). The logic of regional systems. *Annals of the Association of American Geographers. Association of American Geographers, 55*(3), 465–991. doi:10.1111/j.1467-8306.1965.tb00529.x

Groom, G., Mücher, C. A., Ihse, M., & Wrbka, T. (2006). Remote sensing in landscape ecology: Experiences and perspectives in a European context. *Landscape Ecology, 21*, 391–408. doi:10.1007/s10980-004-4212-1

Groom, G. (2005). Methodological review of existing classifications. In D. M. Wascher (Ed.), *European landscape character areas–typology, cartography and indicators for the assessment of sustainable landscapes* (pp. 32-45). Final ELCAI project report, Landscape Europe.

Groombridge, B., & Jenkins, M. D. (2002). *World atlas of biodiversity. Earth's living resources in the 21st century*. Berkeley, CA: University of California Press.

Gross, J. E., Goetz, S. J., & Cihlar, J. (2009). Application of remote sensing to parks and protected area monitoring: Introduction to the special issue. *Remote Sensing of Environment, 113*, 1343–1345. doi:10.1016/j.rse.2008.12.013

Gruijter, J. J., Bierkens, M. F. P., Brus, D. J., & Knotters, M. (2006). *Sampling for natural resource monitoring*. Berlin/ Heidelberg, Germany: Springer-Verlag.

Guimarães, P. R. Jr, Galetti, M., & Jordano, P. (2008). Seed dispersal anachronisms: Rethinking the fruits extinct megafauna ate. *PLoS ONE, 3*(3), e1745.

Guisan, A., & Thuiller, W. (2005). Predicting species distribution: Offering more than simple habitat models. *Ecology Letters, 8*(9), 993–1009. doi:10.1111/j.1461-0248.2005.00792.x

Guisan, A., & Zimmermann, N. E. (2000). Predictive habitat distribution models in ecology. *Ecological Modelling, 135*(2-3), 147–186. doi:10.1016/S0304-3800(00)00354-9

Gulmon, S. L., Chiariello, N. R., Mooney, H. A., & Chu, C. C. (1983). Phenology and resource use in three co-occuring grassland annuals. *Oecologia, 58*, 33–42. doi:10.1007/BF00384539

Gupta, B.N. and Biswas Sas (1997). Biodiversity characterization at landscape level using satellite remote sensing. Paper presented in a workshop, " Biodiversity characterization using remote sensing " project of National Remote sensing agency, Hyderabad, sponsored by department of space and biotechnology, Govt. of India organized by NRSA at Hyderabad, India.

Gurevitch, J., Scheiner, S. M., & Fox, G. A. (2006). *The ecology of plants*. Massachusetts: Sinauer Associates, Inc.

Gustafson, E. J., & Parker, G. R. (1994). Using an index of habitat patch proximity for landscape design. *Landscape and Urban Planning, 29*, 117–130. doi:10.1016/0169-2046(94)90022-1

Gutman, G., Janetos, A., Justice, C., Moran, E., Mustard, J., & Rindfuss, R. … Turner, II B. J. (2004). *Land change science: Observing, monitoring, and understanding trajectories of change on the Earth's surface*. New York, NY: Kluwer Academic Publishers.

Hadi, A., Inubushi, K., & Furukawa, Y., Pur nomo, E., Rasmadi, M., & Tsurata, H. (2005). Greenhouse gas emissions from tropical peatlands of Kalimantan, Indonesia. *Nutrient Cycling in Agroecosystems, 71*, 73–80. doi:10.1007/s10705-004-0380-2

Haest, B., Kempeneers, P., Cheung-Wai Chan, J., Vanden Borre, J., Kooistra, L., Mücher, C. A., & Deronde, B. (2009). *Spectral comparison of Chris/Proba and AHS hyperspectral imagery in the framework of NATURA 2000 habitat monitoring*. 6th EARSeL SIG IS workshop "Imaging Spectroscopy: Innovative tool for scientific and commercial environmental applications" Tel-Aviv, Israel, 16-19 March 2009.

Haila, Y., & Kouki, J. (1994). The phenomenon of biodiversity in conservation biology. In Gaston, K. J. (Ed.), *Biodiversity: A Biology of numbers and difference*. Oxford, UK: Blackwell Science.

Haines-Young, R., Green, D. R., & Cousins, S. H. (1993). *Landscape ecology and GIS*. London, UK: Taylor and Francis.

Haines-Young, R. (2009). Land use and biodiversity relationships. *Land Use Policy, 26S*, S178–S186.. doi:10.1016/j.landusepol.2009.08.009

Hall, C. A. S., Tian, H., Qi, Y., Pontius, G., & Cornell, J. (1995). Modelling spatial and temporal patterns of tropical land use change. *Journal of Biogeography, 22*(4/5), 753–757. doi:10.2307/2845977

Hall, R. K., Watkins, R. L., Heggem, D. T., Jones, K. B., Kaufmann, P. R., Moore, S. B., & Gregory, S. J. (2009). Quantifying structural physical habitat attributes using LIDAR and hyperspectral imagery. *Environmental Monitoring and Assessment, 159*(1-4), 63–83. doi:10.1007/s10661-008-0613-y

Hall, J. B. (1984). Juniperus excela in Africa: A biogeographical study of an Afromontane tree. *Journal of Biogeography, 11*, 47–61. doi:10.2307/2844775

Hannah, L., Lohse, D., Hutchinson, C., Carr, J. L., & Lankerani, A. (1994). A preliminary inventory of human disturbance of world ecosystems. *Ambio, 23*(4/5), 246–250.

Hansen, A. J., deFries, R. S., & Turner, W. (2004). Land use change and biodiversity: A synchapter of rates consequences during the period of satellite imagery. In G. Gutman, A. C. Janetos, C. O. Justice, E. F. Moran, J. F. Mustard, R. R. Rindfuss,... M. A. Cochrane (Eds.), *Land change science: Observing, monitoring and understanding trajectories of change on the Earth's surface* (vol. 6) (pp. 277-300). Dordrecht, The Netherlands: Kluwer Academic Publishers.

Hanson, J., Sato, M., Ruedy, R., Lo, K., Lea, D. W., & Medina-Elizade, M. (2006). Global temperature change. *Proceedings of the National Academy of Sciences of the United States of America, 103*(39), 14288–14293. doi:10.1073/pnas.0606291103

Hardie, I. W., & Parks, P. J. (1997). Land use with heterogeneous land quality: An application of an area-base model. *American Journal of Agricultural Economics, 79*(2), 299–310. doi:10.2307/1244131

Harris, L. D. (1984). *The fragmented forest: Island biogeography theory and the preservation of biotic diversity.* Chicago, IL: University of Chicago Press.

Hastie, T., & Tibshirani, R. (1990). *Generalized additive models*. London, UK: Chapman & Hall.

Hay, G. J., Dubé, P., Bouchard, A., & Marceau, D. J. (2002). A scale-space primer for exploring and quantifying complex landscapes. *Ecological Modelling, 153*(1-2), 27–49. doi:10.1016/S0304-3800(01)00500-2

Hellmann, F., & Verburg, P. H. (2010). Impact assessment of the European biofuel directive on land use and biodiversity. *Journal of Environmental Management, 91*, 1389–1396. doi:10.1016/j.jenvman.2010.02.022

Herkt, M. (2007). *Modelling habitat suitability to predict the potential distribution of Erhard's Wall Lizard Podarcis erhardii on Crete*. The Netherlands: ITC.

Herold, M., Woodcock, C. E., Loveland, T. R., Townshend, J., Brady, M., Steenmans, C., & Schmullius, C. C. (2008). Land-cover observations as part of a global earth observation system of systems (GEOSS): Progress, activities, and prospects. *IEEE Systems Journal, 2*(3), 414–423. doi:10.1109/JSYST.2008.925983

Herold, M. (2000). *UN Global land cover network: An international framework for standardized development of land cover data.* Global Terrestrial Observing System (GTOS) of the United Nations.

Hewitson, B. (2003). Developing perturbations for climate change impact assessments. *Eos, Transactions, American Geophysical Union, 84*, 337–348. doi:10.1029/2003EO350001

Hewlett, J. D., & Nutter, W. L. (1969). *An outline of forest hydrology. School of Forest Resources*. Georgia: University of Georgia Press.

Heywood, V. H., & Bates, I. (1995). Introduction. In Heywood, V. H., & Watson, R. T. (Eds.), *Global biodiversity assessment*. Cambridge, UK: Cambridge University Press.

Higgins, P. A. T. (2007). Biodiversity loss under existing land use and climate change: an illustration using northern South America. *Global Ecology and Biogeography, 16*, 197–204. doi:10.1111/j.1466-8238.2006.00278.x

Hijmans, R. J., Cameron, S. E., Parra, J. L., Jones, P. G., & Jarvis, A. (2005). Very high resolution interpolated climate surfaces for global land areas. *International Journal of Climatology, 25*, 1965–1978. doi:10.1002/joc.1276

Hilderink, H. B. M. (2001). *World population in transition: An integrated regional modeling framework*. Unpublished Thela thesis, Rozenberg, Amsterdam.

Hill, J. K. (1999). Butterfly spatial distribution and habitat requirements in a tropical forest: impacts of selective logging. *Journal of Applied Ecology, 36*, 564–572. doi:10.1046/j.1365-2664.1999.00424.x

Hill, J. K., Hamer, K. C., Lace, L. A., & Banham, W. M. T. (1995). Effects of selective logging on tropical forest butterflies on Buru, Indonesia. *Journal of Applied Ecology, 32*, 754–760. doi:10.2307/2404815

Hirsch, P. (1987). Deforestation and development in Thailand and Singapore. *The Journal of Tropical Geography, 8*(2), 129–138. doi:10.1111/j.1467-9493.1987.tb00190.x

Hirzel, A., Hausser, H., Chessel, J. D., & Perrin, N. (2002). Ecological-niche factor analysis: How to compute habitat-suitability maps without absence data. *Ecology, 83*(7), 2027–2036. doi:10.1890/0012-9658(2002)083[2027:ENFAHT]2.0.CO;2

Hoffman, W. A., Schroeder, W., & Jackson, R. B. (2003). Regional feedbacks among fire, climate, and tropical deforestation. *Journal of Geophysical Research*, *108*, 4721. doi:10.1029/2003JD003494

Holmgren, P. K., Holmgren, N. H., & Barnett, L. C. (1990). *Index Herbariorum*. New York: International Association of Plant Taxonomists.

Homer, C. G., Huang, C., Yang, L., Wylie, B., & Coan, M. (2004). Development of a 2001 national land-cover database for the United States. *Photogrammetric Engineering and Remote Sensing*, *70*(7), 829–840.

Hook, J., Susan, N., & Robyn, J. (2003). *Social atlas of the Lower Mekong Basin*. Phnom Penh: Mekong River Commission.

Hooper, D. U., Chapin, F. S., Ewel, J. J., Hector, A., Inchausti, P., & Lavorel, S. (2005). Effects of biodiversity on ecosystem functioning: A consensus of current knowledge. *Ecological Monographs*, *75*, 3–5. doi:10.1890/04-0922

Hooper, D. U. (1998). The role of complementarity and competition in ecosystem responses to variation in plant diversity. *Ecology*, *79*(2), 704–719. doi:10.1890/0012-9658(1998)079[0704:TROCAC]2.0.CO;2

Hooper, D. U., & Vitousek, P. M. (1998). Effect of plant composition and diversity on nutrient cycling. *Ecological Monographs*, *68*, 121–149. doi:10.1890/0012-9615(1998)068[0121:EOPCAD]2.0.CO;2

Horne, R., & Hickey, J. (2006). Ecological sensitivity of Australian rainforests to selective logging. *Austral Ecology*, *16*(1), 119–129. doi:10.1111/j.1442-9993.1991.tb01487.x

Hosmer, D. W., & Lemeshow, S. (2000). *Applied logistic regression*. New York, NY: John Wiley & Sons, Inc. doi:10.1002/0471722146

Houlahan, J. E., Findlay, C. S., Schmidt, B. R., Meyer, A. H., & Kuzmin, S. L. (2000). Quantitative evidence for global amphibian population declines. *Nature*, *404*, 752–755. doi:10.1038/35008052

Hubbell, S. (2001). *The Unified Neutral Theory of Biodiversity and Biogeography*. Princeton University Press.

Hughes, J. B., Round, P. D., & Woodruff, S. D. (2003). The Indochinese-Sundaic faunal transition at the Isthmus of Kra: An analysis of resident forest bird species distributions. *Journal of Biogeography*, *30*, 569–580. doi:10.1046/j.1365-2699.2003.00847.x

Hulme, M. (2010). *Why we disagree about climate change. Understanding controversy, inaction, and opportunity*. Cambridge, UK: Cambridge University Press.

Hurkmans, R. T. W. L., Terink, W., Uijlenhoet, R., Moors, E. J., Troch, P. A., & Verburg, P. H. (2009). Effects of land use changes on streamflow generation in the Rhine basin. *Water Resources Research*, *45*, W06405.. doi:10.1029/2008WR007574

Hurst, P. (1990). *Rainforest politics: Ecological destruction in Southeast Asia*. London, UK: Zed Books Ltd.

Hutchinson, M. F. (1995). Interpolating mean rainfall using thin plate smoothing splines. *International Journal of GIS*, *9*, 305–403.

Hutchinson, M. F. (2000). *ANUSPLIN, ver 4.1. user guide. Centre for Resource and Environmental Studies*. Canberra, Australia: The Australian National University.

Ibáñez, I., Clarck, J. S., Dietze, M. C., Feeley, K. J., Hersh, M., & LaDeau, S. (2006). Predicting biodiversity change: outside the climate envelope, beyond the species–area curve. *Ecology*, *87*(8), 1896–1906. doi:10.1890/0012-9658(2006)87[1896:PBCOTC]2.0.CO;2

Inghe, O. (2001). The Swedish landscape monitoring programme: Current status and prospects for the near future. In Groom, G., & Reed, T. (Eds.), *Strategic landscape monitoring for the Nordic countries (TemaNord 2001: 523)* (pp. 61–67). Copenhagen: Nordic Council of Ministers.

International Geosphere Biosphere Programme. (2010). *IGBP projects*. Retrieved March 22, 2010, from http://www.igbp.net

Inubushi, K., Furukawa, Y., & Hadi, A., Pur nomo, E., & Tsurata, H. (2003). Seasonal changes of CO_2, CH_4 and N_2O fluxes in relation to land-use change in tropical peatlands located in coastal area of South Kalimantan. *Chemosphere*, *52*, 603–608. doi:10.1016/S0045-6535(03)00242-X

IPCC. (2000). *IPCC special report on emission scenarios: Summary for policymakers*. Nairobi: Intergovernmental Panel on Climate Change. United Nations Environment Programme and World Meteorological Organization.

IPCC. (2000). *Special report on emissions scenarios - a special report of working group III of the Intergovernmental Panel on Climate Change*. Cambridge, UK: Cambridge University Press.

IPCC (2007). *Climate Change 2007: Synthesis Report*. Contribution of Working Groups I, II and III to the Fourth Assessment Report of the Intergovernmental Panel on Climate Change [Core Writing Team, Pachauri, R.K and Reisinger, A. (eds.)]. IPCC, Geneva, Switzerland, 104 pp.

IPCC. (2007). *Climate change 2007 – The physical science basis: Contribution of working group 1 to the fourth assessment report of the IPCC*. Cambridge, UK: Cambridge University Press.

IPCC. (2007). *Climate change 2007: Impacts, adaptation, and vulnerability. The fourth assessment report of the Intergovernmental Panel of Climate Change*. Cambridge, UK: Cambridge University Press.

IPCC. (2007). *Observations: Oceanic climate change and sea level. Working Group 1: The Physical Science Basis*. Geneva, Switzerland: Intergovernmental Panel on Climate Change.

Iskander, D. T. (1999). Amphibian declines monitoring in the Leuser Management Unit, Aceh, North Sumatra, Indonesia. *Froglog, 34*, 2.

IUCN. (1980). *World conservation strategy*. International Union for the Conservation of Nature and Natural Resources, Gland.

IUCN. (1991). *Caring for the Earth*. The World Conservation Union, Gland.

IUCN. (2003). *2003 IUCN red list of threatened species*. Retrieved May 10, 2009, from http://www.redlist.org

IUCN. (2009). *The IUCN red list of threatened species*. Gland, Switzerland: The World Conservation Union (IUCN).

IUCN. 2004. *The Durban action plan. March 2004*. Retrieved April 8, 2008 from http://www.iucn.org/themes/wcpa/ wpc2003/english/outputs/durban/daplan.html

Jackson, M. T., Ford-Lloyd, B. V., & Parry, M. L. (Eds.). (1990). *Climatic change and plant genetic resources*. London, UK: Belhaven Press.

Jackson, M. T., & Ford-Lloyd, B. V. (1990). Plant genetic resources - a perspective. In Jackson, M. T., Ford-Lloyd, B. V., & Parry, M. L. (Eds.), *Climatic change and plant genetic resources* (pp. 1–17). London, UK: Belhaven Press.

Jacobs, M. (1962). Reliquiae Kerrianae. *Blumea, 11*, 427–493.

Jain, S. (2007). Technical note: Use of IKONOS satellite data to identify informal settlements in Dehradun, India. *International Journal of Remote Sensing, 28*, 3227–2333. doi:10.1080/01431160600705122

Jarvis, A., Castano, S., Hyman, G., Gebhardt, S., Guevara, E., Castro, M., et al. (2006). TNC Threats Assessment Version 1.1. CIAT. Retrieved from http://conserveonline.org/workspaces/ersm.pilots/pilot/SACRThreats/view.html

Jayakumar, S., & Arockiasmy, D. I. (2003). Landuse/landcover mapping and change detection in part of Eastern Ghats of Tamil Nadu using remote sensing and GIS. *Journal of Indian Society of Remote Sensing, 31*, 251–260. doi:10.1007/BF03007345

Jenkins, C. N., & Pimm, S. L. (2003). How big is the global weed patch? *Annals of the Missouri Botanical Garden, 90*, 172–178. doi:10.2307/3298581

Jermar, M. K. (1987). *Water resources and management*. Amsterdam, The Netherlands: Elsevier Science, Publishing Company Inc.

Jeschke, J. M., & Strayer, D. L. (2008). Usefulness of bioclimatic models for studying climate change and invasive species. *Annals of the New York Academy of Sciences, 1134*, 1–24. doi:10.1196/annals.1439.002

Johnes, P. J. (1996). Evaluation and management of the impact of land use change on the nitrogen and phosphorus load delivered to surface waters: the export coefficient modelling approach. *Journal of Hydrology (Amsterdam), 183*, 323–349. doi:10.1016/0022-1694(95)02951-6

Johns, A. G. (1986). Effects of selective logging on the behavioural ecology of West Malaysian primates. *Ecology, 67*, 684–694. doi:10.2307/1937692

Johns, A. G. (1997). *Timber production & biodiversity: Conservation in tropical forests*. Cambridge, UK: Cambridge University Press. doi:10.1017/CBO9780511525827

Johns, A. G., & Skorupa, J. P. (1987). Responses of rainforest primates to habitat disturbances: A review. *International Journal of Primatology*, *8*, 157–191. doi:10.1007/BF02735162

Johnston, C. A., Watson, T., & Wolter, P. T. (2007). Sixty-three years of land alteration in Erie Township. *Journal of Great Lakes Research*, *33*, 253–268. doi:10.3394/0380-1330(2007)33[253:SYOLAI]2.0.CO;2

Jolliffe, I. T. (2002). *Mathematics*. New York, NY: Springer-Verlag.

Jones-Walters, L. (2007). Pan-European ecological networks. *Journal for Nature Conservation*, *15*(4), 262–264. doi:10.1016/j.jnc.2007.10.001

Jongman, R. H. G., Bunce, R. G. H., Metzger, M. J., Mücher, C. A., Howard, D. C., & Mateus, V. L. (2006). Objectives and applications of a statistical environmental stratification of Europe. *Landscape Ecology*, *21*(3), 409–419. doi:10.1007/s10980-005-6428-0

Jongman, R. H. G., & Bunce, R. G. H. (2000). Landscape classification, scales and biodiversity in Europe. In Mander, U., & Jongman, R. H. G. (Eds.), *Consequences of land use changes: Advances in ecological sciences* (pp. 11–38). Southampton/ Boston, MA: Wessex Institute of Technology Press.

Jongman, R. H. G. (Ed.). (1996). Ecological and landscape consequences of land use change in Europe. *Proceedings of the First ECNC Seminar on L and Use Change and its Ecological Consequences,* Tilburg, the Netherlands, 16–18 February 1995. ECNC Publication Series on Man and Nature, Volume 2, November 1996.

Joshi, P. K., Roy, P. S., Singh, S., Agrawal, S., & Yadav, D. (2006). Vegetation cover mapping in India using multi-temporal IRS Wide Field Sensor (WiFS) data. *Remote Sensing of Environment*, *103*(2), 190–202. doi:10.1016/j.rse.2006.04.010

Joshi, P. K., Gupta, B., & Roy, P. S. (2008). Spectral evaluation of vegetation features using multi-satellite sensor system (Terra ASTER, Landsat ETM+ and IRS 1D LISS III) in man-made and natural landscape. *Sensor Review*, *28*(1), 52–61. doi:10.1108/02602280810850035

Josse, C., Navarro, G., Comer, P., Evans, R., Faber-Langendoen, D., & Fellows, M. (2003). *Ecological systems of Latin America and the Caribbean. A working classification of terrestrial systems*. NatureServe.

Josse, C., Cuesta, F., Barrena, V., Cabrera, E., Chacón-Moreno, E., Ferreira, W., et al. (2009). *Ecosistemas de los Andes del Norte y Centro. Bolivia, Colombia, Ecuador, Perú y Venezuela* (Secretaría General de la Comunidad Andina, Programa Regional ECOBONA – Intercooperation, CONDESAN-Proyecto Páramo Andino, Programa Bio-Andes, EcoCiencia, NatureServe, IAvH, LTA-UNALM, ICAE-ULA, CDC-UNALM, RUMBOL SRL.). Lima.

Justice, C. O., Vermote, E., Townshend, J. R. G., Defries, R., Roy, D. P., & Hall, D. K. (1998). The Moderate Resolution Spectroradiometer (MODIS): Land remote sensing for global change research. *IEEE Transactions on Geoscience and Remote Sensing*, *36*, 1228–1249. doi:10.1109/36.701075

Kahn, J., & McDonald, J. (1994). International debt and deforestation. In K. Brown. & D.W. Pearce (Eds.), *The causes of tropical deforestation*, (pp. 55-106). Berkeley, CA: University of California Press.

Kanowski, J., Catterall, C. P., & Wardell-Johnson, G. W. (2005). Consequences of broadscale timber plantations for biodiversity in cleared rainforest landscapes of tropical and subtropical Australia. *Forest Ecology and Management*, *208*(1-3), 359–372. doi:10.1016/j.foreco.2005.01.018

Kapos, V. (1989). Effects of isolation on the water status of forest patches in the Brazilian Amazon. *Journal of Tropical Ecology*, *5*, 173–185. doi:10.1017/S0266467400003448

Kapos, V., Ravilious, C., Campbell, A., Dickson, B., Gibbs, H. K., & Hansen, M. C. … Trumper, K. C. (2008). *Carbon and biodiversity, a demonstration atlas*. UNEP World Conservation Monitoring Centre, Cambridge.

Kasperson, R. E., Kasperson, J. X., & Turner, B. L. II. (1999). Risk and criticality: Trajectories of regional environmental degradation. *Ambio*, *28*(6), 562–568.

Keen, P. G. W., & Scott-Morton, M. S. (1978). *Decision support systems: An organizational perspective*. Reading, MA: Addison-Wesley.

Keith, D. A., Akçakaya, H. R., Thuiller, W., Midgley, G. F., Pearson, R. G., & Phillips, S. G. (2008). Predicting extinction risks under climate change: Coupling stochastic population models with dynamic bioclimatic habitat models. *Biology Letters*, *4*, 560–563. doi:10.1098/rsbl.2008.0049

Keramitsoglou, I., Kontoes, C., Sifakis, N., Mitchley, J., & Xofis, P. (2005). Kernel based re-classification of Earth observation data for fine scale habitat mapping. *Journal for Nature Conservation*, *13*(2-3), 91–99. doi:10.1016/j.jnc.2005.02.004

Kerber, A. G., & Schutt, J. B. (1986). Utility of AVHRR channel 3 and 4 in land-cover mapping. *Photogrammetric Engineering and Remote Sensing*, *52*, 1877–1883.

Kikkawa, J. (2006). Reforestation and biodiversity in the Asia-Pacific region. In Suzuki, K., Ishii, K., Sakurai, S., & Sasaki, S. (Eds.), *Plantation technology in tropical forest science* (pp. 247–263). Tokyo, Japan: Springer. doi:10.1007/4-431-28054-5_25

Kim, J. (2004). Regime interplay: The case of biodiversity and climate change. *Global Environmental Change*, *14*, 315–324. doi:10.1016/j.gloenvcha.2004.04.001

Kimmins, J. P. (2004). *Forest ecology: A foundation for sustainable forest management and environmental ethics in forestry* (3rd ed.). British Columbia, Canada: Prentice Hall.

Kirtland, D., Gaydos, L., Clarke, K., DeCola, L., Acevedo, W., & Bell, C. (1994). An analysis of human-induced land transformations in the San Francisco Bay/Sacramento area. *World Resource Review*, *6*(2), 206–217.

Kitcher, P. (2010). The climate change debates. *Science*, *328*(5983), 1230–1234. doi:10.1126/science.1189312

Klein Goldewijk, K., & Ramankutty, N. (2004). Land cover change over the last three centuries due to human activities: The availability of new global data sets. *GeoJournal*, *61*(4), 335–344. doi:10.1007/s10708-004-5050-z

Knight, J. F., Lunetta, R. L., Ediriwickrema, J., & Khorram, S. (2006). Regional scale land-cover characterization using MODIS-NDVI 250 m multi-temporal imagery: A phenology based approach. *GIScience and Remote Sensing*, *43*, 1–23. doi:10.2747/1548-1603.43.1.1

Kobler, A., Džeroski, S., & Keramitsoglou, I. (2006). Habitat mapping using machine learning-extended kernel-based reclassification of an Ikonos satellite image. *Ecological Modelling*, *191*(1), 83–95. doi:10.1016/j.ecolmodel.2005.08.002

Koh, L. P., & Ghazoul, J. (2008). Biofuels, biodiversity, and people: Understanding the conflicts and finding opportunities. *Biological Conservation*, *141*(10), 2450–2460. doi:10.1016/j.biocon.2008.08.005

Köhler, R., Olschofsky, K., & Gerard, F. (Eds.). (2006). *Land cover change in Europe from the 1950s to 2000. Aerial photo interpretation and derived statistics from 59 samples distributed across Europe. Hamburg, Germany: University of Hamburg*. Germany: World Forestry Institute.

Kok, K., Verburg, P. H., & Veldkamp, T. (2007). Integrated assessment of the land system: The future of land use. *Land Use Policy*, *24*(3), 517–520. doi:10.1016/j.landusepol.2006.04.007

Kok, K., & Veldkamp, A. (2001). Evaluating impact of spatial scales on land use pattern analysis in Central America. *Agriculture Ecosystems & Environment*, *85*, 205–221. doi:10.1016/S0167-8809(01)00185-2

Kok, K., & Winograd, M. (2002). Modelling land-use change for Central America, with special reference to the impact of hurricane Mitch. *Ecological Modelling*, *149*, 53–69. doi:10.1016/S0304-3800(01)00514-2

Koomen, E., Rietveld, P., & De Nijs, T. (2008b). Modelling land-use change for spatial planning support [Editorial]. *The Annals of Regional Science*, *42*, 1–10. doi:10.1007/s00168-007-0155-1

Koomen, E., Loonen, W., & Hilferink, M. (2008a). Climate-change adaptations in land-use planning: A scenario-based approach. In Bernard, L., Friis-Christensen, A., & Pundt, H. (Eds.), *The European information society: Taking geoinformation science one step further* (pp. 261–282). Berlin, Germany: Springer.

Korner, C. (2009). Response of humid tropical trees to rising CO_2. *Annual Review of Ecology Evolution and Systematics*, *40*, 61–79. doi:10.1146/annurev.ecolsys.110308.120217

Korner, C. (1999). *Alpine plant life*. Berlin, Germany: Springer.

Kram, T., & Stehfest, E. (2006) The IMAGE model: History, current status and prospects. In A. F. Bouwman, T. Kram & K. Klein Goldewijk (Eds.), *Integrated modelling of global environmental change. An overview of IMAGE 2.4* (pp. 7-24). Netherlands Environmental Assessment Agency (MNP), Bilthoven, the Netherlands.

Krhoda, G. O. (1988). The impact of resource utilization on the hydrology of the Mau Hills Forest in Kenya. *Mountain Research and Development*, *8*, 193–200. doi:10.2307/3673447

Krishtalka, L., & Humphrey, P. S. (2000). Can natural history museums capture the future? *Bioscience*, *50*, 611–617. doi:10.1641/0006-3568(2000)050[0611:CNHMCT]2.0.CO;2

Krishtalka, L., Peterson, A. T., Vieglais, D. A., Beach, J. H., & Wiley, E. O. (2002). The green Internet: A tool for conservation science. In Levitt, J. N. (Ed.), *Conservation in the Internet age: Strategic threats and opportunities* (pp. 143–164). Washington, DC: Island Press.

Krutilla, K., Hyde, W. F., & Barnes, D. (1995). Peri-urban deforestation in developing countries. *Forest Ecology and Management*, *74*(2), 181–195. doi:10.1016/0378-1127(94)03474-B

Küchler, M., Ecker, K., Feldmeyer-Christe, E., Graf, U., Küchler, H., & Waser, L. T. (2004). Combining remotely sensed spectral data and digital surface models for fine-scale modelling of mire ecosystems. *Community Ecology*, *5*(1), 55–68. doi:10.1556/ComEc.5.2004.1.6

Kummer, D. M. (1991). *Deforestation in the postwar Philippines*. The University of Chicago and London.

Kundzewicz, Z. W., Mata, L. J., Arnell, N. W., Döll, P., Jimenez, B., & Miller, K. (2008). The implications of projected climate change for freshwater resources and their management. *Hydrological Sciences Journal*, *53*, 3–10.

Kuniyasu, M. (2002). Environments and people of Sumatran peat swamp forest II: Distribution of villages interactions between people and forests. *Southeast Asian Studies*, *40*(1), 87–108.

Kunwar, P., Kachchwaha, T. S., Kumar, A., Aggarwal, A. K., Singh, A. N., & Mendiratta, N. (2010)... *Current Science*, *98*(2).

Kunwar, P., & Kachhwaha, T. S. (2001). Transformation of landuse-landcover for optimal utilization of natural resources on sustainable basis in Mandawara Block, Lalitpur District, Uttar Pradesh using remote sensing and GIS techniques. In Singh, S. B., & Singh Rana, P. B. (Eds.), *National Geographical Society of India, BHU* (pp. 151–156). Varanasi: Environment and Development.

La Sorte, F. A., & Jetz, W. (2010). Avian distribution under climate change: Towards improved projections. *The Journal of Experimental Biology*, *213*, 862–869. doi:10.1242/jeb.038356

Lake, J. C., & Leishman, M. R. (2004). Invasion success of exotic plants in natural ecosystems: The role of disturbance, plant attributes and freedom from herbivores. *Biological Conservation*, *117*(2), 215–226. doi:10.1016/S0006-3207(03)00294-5

Lal, R., Kimble, J. M., Follet, R. F., & Cole, C. V. (1998). *The potential of US croplands sequester carbon and mitigate the green house effect*. Chelsea, MI: Ann Arbor Press.

Lambert, F. R., & Collar, N. J. (2002). The future of Sundaic lowland forest birds: Long-term effects of commercial logging and fragmentation. *Forktail*, *18*, 127–146.

Lambin, E. F., Turner, B. L. II, Geist, H. J., Agbola, S. B., Angelsen, A., & Bruce, J. W. (2001). The causes of land-use and land-cover change – moving beyond the myths. *Global Environmental Change: Human and Policy Dimensions*, *11*, 261–269.

Lambin, E. F., & Geist, H. (Eds.). (2006). *Land-use and land-cover change: Local processes and global impacts*. New York, NY: Springer.

Lambin, E. F., Geist, H., & Lepers, E. (2003). Dynamics of land use and cover change in tropical regions. *Annual Review of Environment and Resources*, *28*, 205–241. doi:10.1146/annurev.energy.28.050302.105459

Lambin, E. F., Turner, B. L. II, Geist, H., Agbola, S., Angelsen, A., & Bruce, J. W. (2001). The causes of land use and land cover change: Moving beyond the myths. *Global Environmental Change*, *11*(4), 261–269. doi:10.1016/S0959-3780(01)00007-3

Lambin, E. F., Geist, J., & Lepers, E. (2003). Dynamics of land-use and land-cover change in tropical regions. *Annual Review of Environment and Resources*, *28*, 205–241. doi:10.1146/annurev.energy.28.050302.105459

Lambin, E., Turner, B. L., Geist, H., Agbola, S. B., Angelsen, A., & Bruce, J. W. (2001). The causes of land-use and land-cover change: moving beyond the myths. *Global Environmental Change*, *11*, 261–269. doi:10.1016/S0959-3780(01)00007-3

Lambin, E. F., Turner, B. L., Geist, H. J., Agbola, S. B., Angelsen, A., & Bruce, J. W. (2001). The causes of land-use and land-cover change: Moving beyond the myths. *Global Environmental Change*, *11*(4), 261–269. doi:10.1016/S0959-3780(01)00007-3

Lambin, E. F., Baulies, X., Bockstael, N., Fischer, G., Krug, T., & Leemans, R. … Vogel, C. (1999). *Land-use and cover change: Implementation strategy*. (IGBP Report No. 48/IHDP Report No. 10). IGBP, Stockholm, 125 pp.

Land Development Department. (2001). *Soil erosion map*. Land Development Department, Ministry of Agriculture and Co-operatives, Bangkok, Thailand.

Land Development Department. (2003). *Annual statistics report year 2003*. Land Development Department, Ministry of Agriculture and Co-operatives, Bangkok, Thailand.

Landis, J. D. (1996). Imagining land use futures: Applying the California urban futures model. *Journal of the American Planning Association. American Planning Association*, *61*(4), 438. doi:10.1080/01944369508975656

Landis, J. D., Monzon, J. P., Reilly, M., & Cogan, C. (1998). *Development and pilot application of the California Urban and Biodiversity Analysis (CURBA) Model*. Paper presented at the 1998 ESRI International User Conference, October 7–9. Retrieved from http://www.esri.com/library/userco…oc98/PROCEED/TO600/PAP571/P571.HTM

Larsen, K. (1988). Botany in Aarhus 1963–1988. The first 25 years of the Botanical Institute, University of Aarhus. *Reports from the Botanical Institute. University of Aarhus*, *17*, 1–92.

Larsen, K., & Warncke, E. (1966). Report on the flora of Thailand project – I. Expedition. *The Natural History Bulletin of the Siam Society*, *21*, 251–262.

Larsen, K. (1979). Exploration of the flora of Thailand. In Larsen, K., & Holm-Nielsen, L. B. (Eds.), *Tropical botany* (pp. 125–333). London, UK: Academic Press.

Laurance, W. F. (1997). Biomass collapse in Amazonian forest fragments. *Science*, *278*, 1117–1118. doi:10.1126/science.278.5340.1117

Laurance, W. F. (2008). Global warming and amphibian extinctions in eastern Australia. *Austral Ecology*, *33*, 1–9. doi:10.1111/j.1442-9993.2007.01812.x

Laurance, W. F. (2010). Habitat destruction: Death by a thousand cuts. In Sodhi, N. S., & Ehrlich, P. R. (Eds.), *Conservation Biology for all* (pp. 73–87). New York, NY: Oxford University Press. doi:10.1093/acprof:oso/9780199554232.003.0005

Laurance, W. F., Nascimiento, H. E. M., Laurance, S. G., Andrade, A., Ribeiro, J. E. L. S., & Giraldo, J. P. … D'Angelo, S. (2006). Rapid decay of tree-community composition in Amazonian forest fragments. *Proceedings of the National Academy of Sciences of the United States of America*, *103*(50), 19010-19014.

Lavorel, S., McIntyre, S., Landsberg, J., & Forbes, T. D. A. (1997). Plant functional classifications: From general groups to specific groups based on response to disturbance. *Trends in Ecology & Evolution*, *12*(12), 474–478. doi:10.1016/S0169-5347(97)01219-6

Le Houerou, H. N. (1984). Rain-use efficiency: A unifying concept in arid-land ecology. *Journal of Arid Environments*, *7*, 213–247.

Leadley, P., Pereira, H. M., Alkemade, R., Fernandez-Manjarrés, J. F., Proenca, V., Scharlemann, J. P. W., & Walpole, M. J. (2010). *Biodiversity scenarios: Projections of 21st century change in biodiversity and associated ecosystem services, Technical series no. 50*. Montreal: Secretariat of the Convention on Biological Diversity.

Leadley, P., Pereira, H. M., Alkemade, R., Fernandez-Manjarrés, J. F., Proenca, V., Scharlemann, J. P. W., & Walpole, M. (2010). *Biodiversity scenarios: Projections of 21st century change of biodiversity and associated ecosystem services*. Secretariat of the Convention on biological Diversity, Montreal.

Lee, J. T., Elton, M. J., & Thompson, S. (1999). The role of GIS in landscape assessment: using land use based criteria for an area of the Chiltern Hills Area of outstanding natural beauty. *Land Use Policy*, *16*, 23–32. doi:10.1016/S0264-8377(98)00033-7

Lee, Y., & Nelder, J. A. (1996). Heirarchical generalized linear models (with discussion). *Journal of the Royal Statistical Society. Series B. Methodological*, *58*, 619–678.

Lee, P., & Olson, C. E. (2004). *Landscape changes projection using remote sensing and geographic information system*. Retrieved January 27, 2009, from http://wagner.zo.ntu.edu.tw/Download/Landscape%20changes%20projection%20by%20remote%20sensing.pdf

Leemans, R., Eickhout, B. J., Strengers, B., Bouwman, A. F., & Schaeffer, M. (2002). The consequences for the terrestrial carbon cycle of uncertainties inland use, climate and vegetation responses in the IPCC SRES scenarios. *Science in China*, *43*, 1–15.

Leemans, R., & Eickhout, B. (2004). Another reason for concern: Regional and global impact of ecosystems for different levels of climate change. *Global Environmental Change Part A*, *14*, 219–228. doi:10.1016/j.gloenvcha.2004.04.009

Leemans, R. (1999). Modelling for species and habitats: New opportunities for problem solving. *The Science of the Total Environment*, *240*, 51–73. doi:10.1016/S0048-9697(99)00320-4

Leemans, R., Gaston, K. J., van Jaarsveld, A. S., Dixon, J., Harrison, J., & Cheatle, M. E. (2007). *International review of the GLOBIO model version 3. (MNP Report No 555050002/2007)*. Bilthoven: Netherlands Environmental Assessment Agency.

Lehner, P., & Doll, P. (2004). *Global lakes and wetlands database*.

Lengyel, S., Déri, E., Varga, Z., Horváh, R., Tóthmérész, B., & Henry, P. Y. (2008). Habitat monitoring in Europe: A description of current practices. *Biodiversity and Conservation*, *17*(14), 3327–3339. doi:10.1007/s10531-008-9395-3

Leopoldo, P. R., Franken, W. K., & Villa Nova, N. A. (1995). Real evapotranspiration and transpiration through a tropical rain forest in central Amazonia as estimated by the water balance method. *Forest Ecology and Management*, *73*(1-3), 185–195. doi:10.1016/0378-1127(94)03487-H

Lewis, S. L., Lopez-Gonzalez, G., Sonke, B., Affum-Baffoe, K., Baker, T. R., & Ojo, L. O. (2009). Increasing carbon storage in intact African tropical forests. *Nature*, *457*, 1003–U3. doi:10.1038/nature07771

Lezama-López, M. (2007). *El Índice de Capital Natural como instrumento de análisis de pérdida de biodiversidad en Nicaragua*. Technical Report. Netherlands Environmental Assessment Agency. Retrieved from http://www.globio.info/assessments-with-globio

Li, Y., & Wilcove, D. S. (2005). Threats to vertebrate species in China and the United States. *Bioscience*, *55*, 147–153. doi:10.1641/0006-3568(2005)055[0147:TTVSIC]2.0.CO;2

Lillesand, T. M., Kiefer, R. W., & Chipman, J. W. (2007). *Remote sensing and image interpretation* (5th ed.). New Delhi, India: Wiley India Pvt. Ltd.

Lillesand, T. M., Kiefer, R. W., & Chipman, J. W. (2008). *Remote sensing and image interpretation* (6th ed.). New York, NY: Wiley.

Limin, S. H., & Putir, P. E. (2000). The massive exploitation of peat swamp forest potentially has not successfully increased the local people's prosperity in Central Kalimantan. In T. Iwakuma, T. Inoue, T. Kohyama, M. Osaki, H. Simbolon, H. Tachibana, H.... K. Yabe (Eds.), *Proceedings of the International Symposium on: Tropical Peatlands. Bogor, Indonesia, 22-24 November 1999* (pp. 491-498). Japan: Sapporo Editors.

Lin, W., Zhang, L., Du, D., Yang, L., Lin, H., Zhang, Y., & Li, J. (2009). Quantification of land use/land cover changes in Pearl River Delta and its impact on regional climate in summer using numerical modeling. *Regional Environmental Change*, *9*(2), 75–82. doi:10.1007/s10113-008-0057-5

Liu, C., Berry, P. M., Dawson, T. P., & Pearson, R. G. (2005). Selecting tresholds of occurrence in the prediction of species distributions. *Ecography*, *28*, 385–393. doi:10.1111/j.0906-7590.2005.03957.x

Loh, J., Green, R. E., Ricketts, T., Lamoreux, J., Jenkins, M., Kapos, V., & Randers, J. (2005). The living plant Index: Using species population time series to track trends in biodiversity. *Philosophical Transactions of the Royal Society of London. Series B, Biological Sciences*, *360*, 286-295.

Loi, N. K. *(2002).* Effect of land use/land cover changes and practices on sediment contribution to The Tri An Reservoir of Dong Nai Watershed, Vietnam. *Unpublished M.Sc. thesis, Graduate School, Kasetsart University, Bangkok, Thailand.*

Loi, N. K. (2005). *Decision support system (DSS) for sustainable watershed management in Dong Nai watershed – Vietnam.* Unpublished Ph.D thesis, The Graduate School, Kasetsart University, Bangkok, Thailand.

Loiselle, B. A., Howell, C. A., Graham, C. H., Goerck, J. M., Brooks, T., Smith, K. G., & Williams, P. H. (2003). Avoiding pitfalls of using species distribution models in conservation planning. *Conservation Biology*, *17*, 1591–1600. doi:10.1111/j.1523-1739.2003.00233.x

Loiselle, B. A., Jorgensen, P. M., Consiglio, T., Jimenez, I., Blake, J. G., Lohmann, L. G., & Montiel, O. M. (2008). Predicting species distributions from herbarium collections: Does climate bias in collection sampling influence model outcomes? *Journal of Biogeography*, *35*, 105–116.

Lomolinol, M. V. (1982). Species-area and species-distance relationships of terrestrial mammals in the Thousand Island Region. *Journal Oecologia*, *54*, 1432–1939.

Louette, G., Maes, D., Alkemade, J. R. M., Boitani, L., de Knegt, B., & Eggers, J. (2010). BioScore–cost-effective assessment of policy impact on biodiversity using species sensitivity scores. *Journal of Nature Conservation*, *18*, 142–148. doi:10.1016/j.jnc.2009.08.002

Lovejoy, T. E., & Hannah, L. (Eds.). (2005). *Climate change and biodiversity*. New Haven, CT: Yale University Press.

Lovejoy, T. E. (1997). Biodiversity: What is it? In Reaka-Kudld, M. L., Wilson, D. E., & Wilson, E. O. (Eds.), *Biodiversity II: Understanding and protecting our biological resources* (pp. 7–14). Washington, DC: Joseph Henry Press.

Lovejoy, T. E. (2010). Climate change. In Sodhi, N. S., & Ehrlich, P. R. (Eds.), *Conservation Biology for all* (pp. 153–161). New York, NY: Oxford University Press. doi:10.1093/acprof:oso/9780199554232.003.0009

Loveland, T. R., Reed, B. C., Brown, J. F., Ohlen, D. O., Zhu, Z., Yang, L., & Merchant, J. W. (2000). Development of a global land cover characteristics database and IGBP DISCover from 1-km AVHRR data. *International Journal of Remote Sensing*, *21*(6-7), 1303–1330. doi:10.1080/014311600210191

Lubchenco, J., Olson, A. M., Brubaker, L. B., Carpenter, S. R., Holland, M. M., & Hubbell, S. P. (1991). The sustainable biosphere initiative: An ecological research agenda: A report from the Ecological Society of America. *Journal of Ecology*, *72*(2), 371–412. doi:10.2307/2937183

Lujten, J., Miles, L., & Cherrington, E. (2006). *Land use change modelling for three scenarios for the MAR region*. Technical Report. UNEP World Conservation Monitoring Center.

MA. (2005). *Millennium ecosystem assessment. Ecosystems and human well-being: Scenarios* (*Vol. 2*). Washington, DC: Island Press.

MacDonald, G. (2003). *Biogeography: Introduction to space, time and life*. New York, NY: John Wiley & Sons, Inc.

Mace, G. M., Cramer, W., Diaz, S., Faith, D. P., Larigauderie, A., & La Prestre, P. (2010). Biodiversity targets after 2010. *Current Opinion in Environmental Sustainability*, *2*, 3–8. doi:10.1016/j.cosust.2010.03.003

MacKinnon, J. (1997). *Protected areas systems review of the Indo-Malayan realm*. ABS/WCMW/World Bank, Canterbury.

Maffi, L., & Woodley, E. (2010). *Biocultural diversity conservation. A global sourcebook*. London, UK: Earthscan Ltd.

Magguran, A. E. (1998). *Ecological diversity and its measurement*. Princeton, NJ: Princeton University Press.

Maguire, D. I. (1991). An overview and definitions of GIS. In Maguire, D. J., Goodchild, M. F., & Rhind, D. W. (Eds.), *Geographical information systems* (pp. 9–20). London, UK: Longman.

Magurran, A. E. (2003). *Measuring biological diversity*. Oxford, UK: Blackwell Publishing.

Maidment, D. R. (1993). GIS and hydrologic modeling. In Goodchild, M. F., Parks, B. O., & Steyaert, L. T. (Eds.), *Environmental modeling with GIS*. New York, NY: Oxford University Press.

Majer, J. D., & Beeston, G. (1996). The biodiversity integrity index: An illustration using ants in Western Australia. *Conservation Biology, 10*(1), 65–73. doi:10.1046/j.1523-1739.1996.10010065.x

Malcolm, J. R., Liu, C., Neilson, R. P., Hansen, L., & Hannah, L. (2006). Global warming and extinctions of endemic species from biodiversity hotspots. *Conservation Biology, 20*, 538–548. doi:10.1111/j.1523-1739.2006.00364.x

Malczewski, J. (1999). *GIS and multi-criteria decision analysis*. New York, NY: John Wiley & Sons, Inc.

Malhi, Y., & Wright, J. (2004). Spatial patterns and recent trends in the climate of tropical rainforest regions. *Philosophical Transactions of the Royal Society of London Series B, 359*, 311–329. doi:10.1098/rstb.2003.1433

Malhi, Y., Roberts, J. T., Betts, R. A., Killeen, T. J., Li, W., & Nobre, C. A. (2008). Climate change, deforestation and the fate of the Amazon. *Science, 319*, 169–172. doi:10.1126/science.1146961

Malhi, Y., Aragão, L. E. O. C., Galbraith, D., Huntingford, C., Fisher, R., & Zelazowski, P. … Meir, P. (2009). Exploring the likelihood and mechanism of a climate-change-induced dieback of the Amazon rainforest. *Proceedings of the National Academy of Sciences of the United States of America, 106*(49), 20610-20615.

Mander, Ü., Mitchley, J., Xofis, P., Keramitsoglou, I., & Bock, M. (2005). Earth observation methods for habitat mapping and spatial indicators for nature conservation in Europe. *Journal for Nature Conservation, 13*(2-3), 69–73. doi:10.1016/j.jnc.2005.04.001

Manel, S., Williams, H. C., & Ormerod, S. J. (2001). Evaluating presence—absence models in ecology: The need to account for prevalence. *Journal of Applied Ecology, 38*, 921–931. doi:10.1046/j.1365-2664.2001.00647.x

Manifold System. (2008). *Manifold® System 8.0 Universal Edition*. Nevada: Manifold Net Ltd.

Margules, C. R., & Sarkar, S. (2007). *Systematic conservation planning*. Cambridge, UK: Cambridge University Press.

Marod, D. (2003). *Management of the Pha Taem protected forest complex to promote cooperation for trans-boundary biodiversity conservation between Thailand, Cambodia and Laos (Phase I): Forest ecology final report*. Bangkok: Faculty of Forestry, Kasetsart University.

Martin-Smith, K. M. (1998). Effects of disturbance caused by selective timber extraction on fish communities in Sabah, Malaysia. *Environmental Biology of Fishes, 53*, 155–167. doi:10.1023/A:1007496424730

Mather, P. M. (1976). *Computation methods of multivariate analysis in physical geography*. Chichester, West Sussex, UK: Wiley.

Mathur, V. B., Raza, R., Lal, P., Agarwal, M. K., & Singsit, S. (2003). Biodiversity characterization in Middle Goriganga Valley, Askot Wildlife Sanctuary, Uttarakhand State. *Technical report submitted to Indian Institute of Remote Sensing, Dehradun*. Retrieved July 7, 2010 from http://www.wii.gov.in/publications/researchreports/2005/diversity&rarityinfloralwh.pdf

Matthews, R., Gilbert, N., Roach, A., Polhill, J. G., & Gotts, N. M. (2007). Agent-based land-use models: A review of applications. *Landscape Ecology, 22*(10), 1447–1459. doi:10.1007/s10980-007-9135-1

Matthews, E. (1983). Global vegetation and land use: New high resolution data bases for limited studies. *Journal of Climate and Applied Meteorology, 22*, 474–487. doi:10.1175/1520-0450(1983)022<0474:GVALUN>2.0.CO;2

Mattison, E. H. A., & Norris, K. (2005). Bridging the gaps between agricultural policy, land-use and biodiversity. *Trends in Ecology & Evolution, 20*(11), 610–616.. doi:10.1016/j.tree.2005.08.011

Maxwell, J. F. (2004). A synopsis of the vegetation of Thailand. *The Natural History Journal of Chulalongkorn University*, *4*, 19–29.

Mayaux, P., Eva, H., Gallego, J., Strahler, A. H., Herold, M., & Agrawal, S. (2006). Validation of the global land cover 2000 map. *IEEE Transactions in Geoscience Remote*, *44*(7), 1728–1737. doi:10.1109/TGRS.2006.864370

McArthur, R. H., & McArthur, J. W. (1961). On bird species diversity. *Ecology*, *42*, 594–598. doi:10.2307/1932254

McCarthy, J. J., Canziani, O. F., Leary, N. A., Dokken, D. J., & White, K. S. (Eds.). (2001). *Climate change 2001. Working group II: Impacts, adaptation and vulnerability. Intergovernmental Panel on Climate Change*. Cambridge, UK: Cambridge University Press.

McCune, B., & Grace, J. B. (2002). *Analysis of ecological communities*. Gleneden Beach, OR: MjM Software Design.

McGarigal, K., & Marks, B. (1995). *FRAGSTATS: Spatial pattern analysis program for quantifying landscape structure*. (Gen. Tech. Rep. PNW-GTR-351). Portland, USA.

McKee, J. K., Sciullli, P. W., Fooce, C. D., & Waite, T. A. (2003). Forecasting global biodiversity threats associated with human population growth. *Biological Conservation*, *115*, 161–164. doi:10.1016/S0006-3207(03)00099-5

McLaughlin, J. F., Hellmann, J. J., Boggs, C. L., & Ehrlich, P. R. (2002). Climate change hastens population extinctions. *Proceedings of the National Academy of Sciences of the United States of America*, *99*, 6070–6074. doi:10.1073/pnas.052131199

McNeely, J. A. (1998). *Economics and biological diversity: Developing and using economic incentives to conserve biological resources*. Gland, Switzerland: IUCN.

McNeely, J. A., Miller, K. R., Reid, W. V., Mittermeier, R. W., & Werner, T. B. (1990). *Conserving the world's biological diversity*. International Union for the Conservation of Nature, Gland.

McNeely, J. A. (1998). *Economics and biological diversity: Developing and using economic incentives to conserve biological resources*. Gland, Switzerland: IUCN.

McPherson, J. M., Jetz, W., & Rogers, D. J. (2004). The effects of species' range sizes on the accuracy of distribution models: Ecological phenomenon or statistical artefact? *Journal of Applied Ecology*, *41*, 811–823. doi:10.1111/j.0021-8901.2004.00943.x

McRoberts, E., & Tomppo, E. (2007). Remote sensing support for national forest inventories. *Remote Sensing of Environment*, *110*(4), 412–419. doi:10.1016/j.rse.2006.09.034

MEA. (2005). *Millennium Ecosystem Assessment: Ecosystems and human well-being*. Washington, DC: Island Press.

Meeus, J. H. A. (1995). Pan-European landscapes. *Landscape and Urban Planning*, *31*(1-3), 57–79. doi:10.1016/0169-2046(94)01036-8

Meijaard, E., & Sheil, D. (2008). A logged forest in Borneo is better than none at all. *Nature*, *446*, 976.

Meijl, Hv., van Rheenen, T., Tabeau, A., & Eickhout, B. (2006). The impact of different policy environments on agricultural land use in Europe. *Agriculture Ecosystems & Environment*, *114*, 21–38. doi:10.1016/j.agee.2005.11.006

Millennium Ecosystem Assessment (MEA). (2002). *Ecosystems and human well-being*. Retrieved from http://www.millenniumassessment.org

Mertens, B., & Lambin, E. (1997). Spatial modelling of deforestation in southern Cameroon: Spatial disaggregation of diverse deforestation processes. *Applied Geography (Sevenoaks, England)*, *17*(2), 143–162. doi:10.1016/S0143-6228(97)00032-5

Mertz, O., Ravnborg, H. M., Lövei, G., Nielsen, I., & Konijnendijk, C. C. (2007). Ecosystem services and biodiversity in developing countries. *Biodiversity and Conservation*, *16*, 2729–2737..doi:10.1007/s10531-007-9216-0

Metcalfe, I., Smith, J. M. B., Morwood, M., & Davidson, I. (Eds.). (2002). *Faunal and floral migrations and evolution in SE Asia-Australasia*. New York, NY: Oceania Publication.

Metz, B., Davidson, O., Bosch, P., Dave, R., & Meyer, L. (Eds.). (2007). *Climate change 2007. Working group III: mitigation. Intergovernmental Panel on Climate Change*. Cambridge, UK: Cambridge University Press.

mFood and Agriculture Organization of the United Nations. (2005). *Global forest assessment.* Retrieved March 10, 2008, from http://www.fao.org/forestry/site/fra/24690/en

Michalski, F., & Peres, C. A. (2005). Anthropogenic determinants of primate and carnivore local extinctions in a fragmented forest landscape of southern Amazonia. *Biological Conservation, 124*, 383–396. doi:10.1016/j.biocon.2005.01.045

Mickelson, J. G., Civco, D. L., & Silander, J. A. Jr. (1998). Delineating forest canopy species in the Northeastern United States using multi-temporal TM imagery. *Photogrammetric Engineering and Remote Sensing, 64*(9), 891–904.

Middleton, D. J. (2003). Progress on the flora of Thailand. *Telopea, 10*, 33–42.

Milanova, E. V., & Kushlin, A. V. (Eds.). (1993). *World map of present day landscapes. An explanatory note.* Department of World Physical Geography and Geoecology, Moscow State University, in collaboration with UNEP.

Miles, L., Grainger, A., & Phillips, O. (2004). Impact of global climate change on tropical forest biodiversity in Amazonia. *Global Ecology and Biogeography, 13*, 553–565. doi:10.1111/j.1466-822X.2004.00105.x

Miles, L., & Kapos, V. (2008). Reducing greenhouse gas emissions from deforestation and forest degradation, global land-use implications. *Science, 320*, 1454–1455. doi:10.1126/science.1155358

Miles, L. (2002). *The impact of global climate change on tropical forest biodiversity in Amazonia.* Unpublished doctoral dissertation, University of Leeds, Leeds.

Millennium Ecosystem Assessment-MA. (2005). *Ecosystems and human well-being: synthesis note: The designation of areas undergoing land-cover change resulting from degradation in drylands has been omitted.* Washington, DC: Island Press.

Miller, G. H., Mangan, J., Pollard, D., Thompson, S., Felzer, B., & Magee, J. (2005). Sensitivity of the Australian monsoon to insolation and vegetation: implications for human impact on continental moisture balance. *Geology, 33*, 65–68. doi:10.1130/G21033.1

Miller, M. (1994). Large African herbivores, bruchid beetles and their interactions with Acacia seeds. *Oecologia, 97*(2), 265–270. doi:10.1007/BF00323159

Miller, R. L. (Ed.). (1994). *Mapping the diversity of nature.* London, UK: Chapman & Hall.

Miller-Rushing, A., & Primack, R. B. (2008). Global warming and flowering times in Thoreau's Concord: A community perspective. *Ecology, 89*, 332–341. doi:10.1890/07-0068.1

Mittermeier, R., & Mittermeier, C. (2005). *Megadiversity: Earth's Biologically Wealthiest Nations.* Mexico: Cemex.

MNP. (2004). *Quality and the future. Sustainability outlook (summary).* Bilthoven, The Netherlands: Netherlands Environmental Assessment Agency.

MOF RI. (2007). Statistik kehutanan [Departemen Kehutanan Indonesia] [In Indonesia]. *Indonesia*, 2007.

Moller, M. S., & Blaschke, T. (2005). *Monitoring LU/LC dynamics in the urban-rural fringe.* Paper presented at the Anais XII Simpósio Brasileiro de Sensoriamento Remoto, Goiânia, Brasil, 16-21 abril 2005, INPE, (pp. 3821-3828).

Mooney, H. A. (2010). The ecosystem-service chain and the biological diversity crisis. *Philosophical Transactions of the Royal Society B, 365*, 31–39. doi:10.1098/rstb.2009.0223

Moran, E. F., Skole, D. L., & Turner, B. L. II. (2004). The development of the international land use and land cover change (LCLUC) initiative. In Gutman, G., Janetos, A. C., Justice, C. O., & Moran, E. F. (Eds.), *Land change science: Observing, monitoring and understanding trajectories of change on the Earth's surface* (pp. 1–16). Dordrecht, The Netherlands: Kluwer Academic Publishers.

Morin, X., Viner, D., & Chuine, I. (2008). Tree species range shifts at a continental scale: New predictive insights from a process-based model. *Journal of Ecology, 96*, 784–794. doi:10.1111/j.1365-2745.2008.01369.x

Mortenson, H. S., Dupont, Y. L., & Olesen, J. M. (2008). A snake paradise: Disturbance of plant reproduction following extirpation of bird flower-visitors on Guam. *Biological Conservation, 141*(8), 2146–2154. doi:10.1016/j.biocon.2008.06.014

Mücher, C. A., Hennekens, S. M., Bunce, R. G. H., Schaminée, J. H. J., & Schaepman, M. E. (2009). Modelling the spatial distribution of Natura 2000 habitats across Europe. *Landscape and Urban Planning*, *92*(2), 148–159. doi:10.1016/j.landurbplan.2009.04.003

Mücher, C. A., Klijn, J. A., Wascher, D. M., & Schaminée, J. H. J. (2010). A new European Landscape Classification (LANMAP): A transparent, flexible and user-oriented methodology to distinguish landscapes. *Ecological Indicators*, *10*, 87–103. doi:10.1016/j.ecolind.2009.03.018

Mücher, C. A., Steinnocher, K. T., Kressler, F. P., & Heunks, C. (2000). Land cover characterization and change detection for environmental monitoring of pan-Europe. *International Journal of Remote Sensing*, *21*(6-7), 1159–1181. doi:10.1080/014311600210128

Mücher, C. A., & Wascher, D. M. (2007). European landscape characteracterization. In Pedroli, B., van Doorn, A., de Blust, G., Paracchini, M. L., Wascher, D., & Bunce, F. (Eds.), *Europe's living landscapes* (pp. 37–43). KNNV Publishing.

Mücher, C. A. (2009). *Geo-spatial modelling and monitoring of European landscapes and habitats using remote sensing and field surveys*. PhD thesis, Wageningen University, Wageningen, The Netherlands. ISBN 978-90-8585-453-1, 269 pp.

Mücher, C. A., Champeaux, J. L., Steinnocher, K. T., Griguolo, S., Wester, K., Heunks, C., et al. Nieuwenhuis, G. J. A. (2001). *Development of a consistent methodology to derive land cover information on a European scale from remote sensing for environmental monitoring: The PELCOM report*. Alterra-rapport 178, Wageningen, the Netherlands.

Mücher, C. A., Hazeu, G. W., Swetnam, R., Pino, J., Halada, L., & Gerard, F. (2008). Historic land cover changes at Natura 2000 sites and their associated landscapes across Europe. *Proceedings 28th EARSeL Symposium: Remote Sensing for a Changing Europe* (pp. 226-231). Istanbul, Turkey.

Mücher, C. A., Hennekens, S. M., Bunce, R. G. H., & Schaminée, J. H. J. (2005). Spatial identification of European habitats to support the design and implementation of a Pan-European Ecological Network. In D. McCollin & J. I. Jackson (Eds.), *Planning, people and practice. The landscape ecology of sustainable landscapes* (pp. 217-225). 13th Annual IALE(UK) Conference held at the University of Northampton.

Mücher, C. A., Vos, C. C., Renetzeder, C., Wrbka, T., Kiers, M. A., van Eupen, M., & Bugter, R. J. F. (2007). *The application of satellite imagery to identify landscape structure. Proceedings of the IALE 2007 World Congress* (pp. 590-591). Wageningen, the Netherlands.

Munday, M., & Munday, G. (1992). The climate of south–west Ecuador. In Best, B. (Ed.), *The threatened forests of south-west Ecuador* (pp. 7–59). Leeds: Biosphere Publications.

Mundia, C. N., & Aniya, M. (2005). Analysis of land use/cover changes and urban expansion of Nairobi city using remote sensing and GIS. *International Journal of Remote Sensing*, *26*(13), 2831–2849. doi:10.1080/01431160500117865

Murphy, P. G., & Lugo, A. (1995). Dry forests of Central America and the Caribbean. In Bullock, S., Mooney, H. A., & Medina, E. (Eds.), *Seasonally Dry Tropical Forests* (pp. 9–34). Cambridge University Press. doi:10.1017/CBO9780511753398.002

Murthy, M. S. R., Giriraj, A., & Dutt, C. B. S. (2003). Geoinformatics for biodiversity assessment. *Biology Letters*, *40*(2), 75–100.

Myers, N., Mittermeier, R. A., Mittermeier, C. G., da Fonseca, G. A. B., & Kent, J. (2000). Biodiversity hotspots for conservation priorities. *Nature*, *403*, 853–858. doi:10.1038/35002501

Myers, N., Mittermeier, R., Mittermeier, C., Fonseca, G., & Kent, J. (2000). Biodiversity hotspots for conservation priorities. *Nature*, *403*, 853–858. doi:10.1038/35002501

Myers, N. (1988). Threatened biotas: Hot spots in tropical forests. *The Environmentalist*, *8*, 1–20. doi:10.1007/BF02240252

Myers, N., Mittermeier, R. A., Mittermeier, C. G., Fonseca, G. A. B., & Kent, J. (2000). Biodiversity hotspots for conservation priorities. *Nature*, *403*, 853–858. doi:10.1038/35002501

Naeem, S., & Li, S. (1997). Biodiversity enhances ecosystem reliability. *Nature*, *390*, 507–509. doi:10.1038/37348

Nagendra, H. (2001). Review article. Using remote sensing to assess biodiversity. *International Journal of Remote Sensing*, *22*(12), 2377–2400. doi:10.1080/01431160117096

Naidoo, R., Balmford, A., Costanza, R., Fisher, B., & Green, R. E. Lehner. B.,... Ricketts, T. H. (2008). Global mapping of ecosystem services and conservation priorities. *Proceedings of the National Academy of Sciences USA*, *105*, 9495-9500.

Nakicenovic, N., Alcamo, J., Davis, G., De Vries, B., Fenhann, J., & Gaffin, S. (2000). *Special report on emissions scenarios. IPCC Special Reports*. Cambridge, UK: Cambridge University Press.

Nascimiento, H. E. M., Laurance, W. F., Condit, R., Laurance, S., D'Angelo, S., & Andrade, A. C. (2005). Demographic and life-history correlates for Amazonian trees. *Journal of Vegetation Science*, *16*, 625–634. doi:10.1111/j.1654-1103.2005.tb02405.x

National Economic and Social Development Board-NESDB. (2008). *The 10th national economic and social development plan (2008-2011)*. Bangkok: Prime Minister Office.

National Research Council (NRC). (2001). *Compensating for wetland losses under the Clean Water Act*. Washington, DC: National Academy Press.

National Resources Inventory (NRCS). (2001). *Percent change in cropland area* (pp. 1982-1997), USDA-NRCS-RID. Retrieved January 15, 2010, from http://www.nrcs.usda.gov/technical/land/index.html

National Resources Inventory (NRCS). (2007). *Change in average annual soil erosion by wind and water on cropland and CPR land*, (pp. 1982-1992). USDA-NRCS-RID. Retrieved January 15, 2010, from http://www.nrcs.usda.gov/technical/land/index.html

Naughton-Treves, L., Mena, J. L., Treves, A., Alvarez, N., & Radeloff, V. C. (2003). Wildlife Survival Beyond Park Boundaries: the Impact of Slash-and-Burn Agriculture and Hunting on Mammals in Tambopata, Peru. *Conservation Biology*, *17*(4), 1106–1117. doi:10.1046/j.1523-1739.2003.02045.x

Navalgund, R. R., Jayaraman, V., & Roy, P. S. (2007). Remote sensing applications: An overview. *Current Science*, *93*(12), 1747–1766.

NEEA. (2007). *The International Biodiversity Project: Understanding biodiversity, ecosystem services and poverty in order to support policy makers*. The Netherlands Environmental Assessment Agency. PBL website. Retrieved November 15, 2007, from http://www.pbl.nl/en/publications/2007/WorldwideBiodiversityLossAnd-Poverty_HowDoTheyRelate.html

Neitsch, S. L., Arnold, J. G., Kiniry, J. R., Srinivasan, R., & Williams, J. R. (2002). *Soil and water assessment tool: User's manual* (Version 2000). (GSWRL Report 02-02, BRC Report 2-06). Temple, Texas, USA.

Nelleman, C. (Ed.). (2004). *The fall of the water*. Arenda, Norway: United Nations Environmental Programme – GRID.

Nellemann, C., Vistness, I., Ahlenius, H., Rekacewicz, P., Kaltenborn, B. P., & Magomedova, M. … Furuhovde, T. (2003). Environment and security, 2050 scenarios. In R. O. Rasmussen & N. E. Koroleva (Eds.), *Social and environmental impacts in the North* (pp. 129-148). Dordrecht, The Netherlands: Kluwer Academic Publishers.

Nelson, B. W., Ferreira, C. A. C., da Silva, M. F., & Kawasaki, M. L. (1990). Endemism centers, refugia and botanical collection density in Brazilian Amazonia. *Nature*, *345*, 714–716. doi:10.1038/345714a0

NEMA. (2001). *State of the environment report for Uganda*. Kampala: National Environment Management Authority, Ministry of Water Lands and Environment.

NERC/ESRC/DFID. (2008). *Ecosystem services for poverty alleviation: Strengthening research capacity in developing countries*. Swindon: Natural Environment Research Council.

Neter, J., Wasserman, W., Nachtsheim, C. J., & Kutner, M. H. (1996). *Applied linear regression models* (3rd ed.). Chicago, IL: Irwin.

Ng, P. K. L., Tay, J. B., & Lim, K. K. P. (1994). Diversity and conservation of blackwater fishes in Peninsular Malaysia, particularly in the North Selangor peat swamp forest. *Hydrobiologia, 285*, 203–218. doi:10.1007/BF00005667

Nichol, J. (1997). Bioclimatic impacts of the 1994 smoke haze event in Southeast Asia. *Atmospheric Environment, 31*(8), 1209–1219. doi:10.1016/S1352-2310(96)00260-9

Nichol, J., & Wang, M. S. (2007). Remote sensing of urban vegetation life from by spectral mixture analysis of high resolution IKONOS satellite images. *International Journal of Remote Sensing, 28*, 985–1000. doi:10.1080/01431160600784176

Nichols, E., Larsen, T., Spector, S., Davis, A. L., Escobar, F., Favila, M., & Vulinec, K. (2007). Global dung beetle response to tropical forest modification and fragmentation: A quantitative literature review and meta-analysis. *Biological Conservation, 137*, 1–19. doi:10.1016/j.biocon.2007.01.023

Nieder, J., Prospera, J., & Michaloud, G. (2001). Epiphytes and their contribution to canopy diversity. *Plant Ecology, 153*, 51–63. doi:10.1023/A:1017517119305

Nogué, S., Rull, V., & Vegas-Vilarrúbia, T. (2009). Modeling biodiversity loss by global warming on Pantepui, northern South America: projected upward migration and potential habitat loss. *Climatic Change, 94*, 77–85. doi:10.1007/s10584-009-9554-x

Norse, E. A., & McManus, R. E. (1980). *Ecology and living resources biological diversity.* In Council on Environmental Quality, 11th Annual Report, 31-80. Council on Environmental Quality, Washington, DC.

Noss, R. F. (1990). Indicators for monitoring biodiversity: A hierarchical approach. *Conservation Biology, 4*(4), 355–364. doi:10.1111/j.1523-1739.1990.tb00309.x

Noss, R. F. (1990). Indicators for monitoring biodiversity: A hierarchical approach. *Conservation Biology, 4*(4), 355–364. doi:10.1111/j.1523-1739.1990.tb00309.x

NRC (National Research Council). (2003). *NEON: Addressing the nation's environmental challenges.* Washington, DC: National Academies Press.

NRC. (2006). *Net primary productivity.* Retrieved June 2, 2006, from http://ccrs.nrcan.gc.ca/optic/coarse/beps/npp_e.php

O'Dowd, D. J., Green, P. T., & Lake, P. S. (2003). Invasional meltdown on an oceanic island. *Ecology Letters, 6*(9), 812–817. doi:10.1046/j.1461-0248.2003.00512.x

O'Dowd, D. J., & Lake, P. S. (2009). Red crabs in rain forest, Christmas Island: Removal and relocation of leaf-fall. *Journal of Tropical Ecology, 5*, 337–348. doi:10.1017/S0266467400003746

O'Rourke, E. (2006). Biodiversity and land use change on the Causse Méjan, France. *Biodiversity and Conservation, 15*, 2611–2626..doi:10.1007/s10531-005-5402-0

Oberhauser, U. (1997). Secondary forest regeneration beneath pine (*Pinus kesiya*) plantations in the northern Thai highlands: A chronosequence study. *Forest Ecology and Management, 99*(1-2), 171–183. doi:10.1016/S0378-1127(97)00203-X

O'Callaghan, J. R. (1995). NELUP: An introduction. *Journal of Environmental Planning and Management, 38*(1), 5–20. doi:10.1080/09640569513084

Ochoa-Gaona, S. (2001). Traditional land-use systems and patterns of forest fragmentation in the highlands of Chiapas, Mexico. *Environmental Management, 27*, 571–586. doi:10.1007/s002670010171

OECD. (2008). *Environmental outlook to 2030.* Paris, France: Organisation for Economic Cooperation and Development.

OECD. (2008). OECD Environmental. *Outlook to*, 2030.

Office of Agricultural Economics. (2007). *Agricultural statistics of Thailand 2004.* Ministry of Agriculture and Co-operatives, Bangkok, Thailand.

Office of Environmental Policy and Planning. (1997). *Thailand policy and perspective plan for enhancement and conservation of national environmental quality, 1997-2016.* Bangkok, Thailand: Ministry of Science, Technology and Environment.

Office of Natural Resources and Environmental Policy and Planning. (2006). *The state of the environment for the year 2005-2006.* Bangkok: Office of Natural Resource and Environmental Policy and Planning, Ministry of Natural Resources and Environment.

Office of Technology Assessment (OTA). (1987). *Technologies to maintain biological diversity. (OTA-F-330)*. Washington, DC: Government Printing Office.

Ogawa, S., Rikimaru, A., Masumoto, T., Tanji, H., & Maharaxay, M. (2005). Landuse and water management of agriculture in Mekong River basin. *Journal of Agricultural Meteorology*, *60*(5), 371–374.

Oglethorpe, D. R., & O'Callaghan, J. R. (1995). Farmlevel economic modelling within a river catchment decision support system. *Journal of Environmental Planning and Management*, *38*(1), 93–106. doi:10.1080/09640569513138

Oldeman, L. R., & van Lynden, G. W. J. (1998). Revisiting the GLASOD methodology. In Lal, R., Blum, W. H., Valentine, C., & Stewart, B. A. (Eds.), *Methods for assessment of soil degradation* (pp. 423–440). Boca Raton, FL: CRC Press.

Oliver, I., Pik, A., Britton, D., Dangerfield, J. M., Colwell, R. K., & Beattie, A. J. (2000). Virtual biodiversity assessment systems. *Bioscience*, *50*, 441–449. doi:10.1641/0006-3568(2000)050[0441:VBAS]2.0.CO;2

Olson, D. M., Dinerstein, E., Wikramanayake, E. D., Burgess, N. D., Powell, G. V. N., & Underwood, E. C. (2001). Te rrestrial Ecoregions of the Wo rld: A New Map of Life on Earth. *Bioscience*, *51*(11), 933–938. doi:10.1641/0006-3568(2001)051[0933:TEOTWA]2.0.CO;2

Ongsomwang, S. (1995). *Fundamental of GIS*. Bangkok: Forest Resources Analysis Division, Forestry Academic Office, Royal Forest Department.

Opdam, P., Verboom, J., & Pouwels, R. (2003). Landscape cohesion: An index for the conservation potential of landscapes for biodiversity. *Landscape Ecology*, *18*, 113–126. doi:10.1023/A:1024429715253

Openshaw, S., & Whitehead, P. (1985). A Monte Carlo simulation approach to solving multi-criteria optimization problems related to plan making, evaluation, and monitoring in local planning. *Environment and Planning*, *12*(4), 321–334.

Openshaw, S. (1991). Developing appropriate spatial analysis method for GIS. In Maguire, D. J., Goodchild, M. F., & Rhind, D. W. (Eds.), *Geographical information systems* (pp. 389–402). London, UK: Longman.

Osenberg, C. W., Sarnelle, O., Cooper, S. D., & Holt, R. D. (1999). Resolving ecological questions through meta-analysis: Goals, metrics and models. *Ecology*, *80*, 1105–1117. doi:10.1890/0012-9658(1999)080[1105:REQTMA]2.0.CO;2

Ostermann, O. P. (1998). The need for management of nature conservation sites designated under Natura 2000. *Journal of Applied Ecology*, *35*(6), 968–973. doi:10.1111/j.1365-2664.1998.tb00016.x

Ounon, T. (1990). The way to GIS development in academic institute. *Proceedings of the Conference Decumbent Entitled Application of Remote Sensing Information and GIS for Development and Management of Natural Resource* (pp. 10-17). Division of Natural Resource Assessment by Satellite, National Research Council, Bangkok.

Overmars, K. P., Verburg, P. H., & Veldkamp, T. (2007). Comparison of a deductive and an inductive approach to specify land suitability in a spatially explicit land-use model. *Land Use Policy*, *24*(3), 584–599. doi:10.1016/j.landusepol.2005.09.008

Padmawathe, R., Qureshi, Q., & Rawat, G. S. (2004). Effects of selective logging on vascular epiphyte diversity in a moist lowland forest of Eastern Himalaya, India. *Biological Conservation*, *119*(1), 81–92. doi:10.1016/j.biocon.2003.10.024

Page, S. E., Siegert, F., Rieley, J. O., Boehm, H. V., Jaya, A., & Limin, S. (2002). The amount of carbon released from peat and forest fires in Indonesia during 1997. *Nature*, *420*, 61–65. doi:10.1038/nature01131

Page, S. E., Wüst, R. A. J., Weiss, D., Rieley, J. O., Shotyk, W., & Limin, S. (2004). A record of late Pleistocene and Holocene carbon accumulation and climate change from an equatorial peat bog (Kalimantan, Indonesia): Implications for past, present and future carbon dynamics. *Journal of Quaternary Science*, *19*, 625–635. doi:10.1002/jqs.884

Palmer, C. (2010). REDD+: Property rights and liability. *Science*, *328*, 1105. doi:10.1126/science.328.5982.1105-a

Palmer, T. M., Stanton, M. L., Young, T. P., Goheen, J. R., Pringle, R. M., & Karban, R. (2008). Breakdown of an ant-plant mutualism follows the loss of large herbivores from an African savanna. *Science*, *319*, 192–195. doi:10.1126/science.1151579

Palo, M. (1994). Population and deforestation. In Brown, K., & Pearce, D. W. (Eds.), *The causes of tropical deforestation* (pp. 55–106). London, UK: UCL Press.

Panayotou, T., & Sungsuwan, S. (1989). *An economic study of the causes of tropical deforestation: The case of northeast Thailand.* (Development Discussion Paper No. 284). Harvard University Institute of Economic Development, Harvard University, Massachusetts, USA.

Papadimitriou, F. (2009). Modelling spatial landscape complexity using the Levenshtein algorithm. *Ecological Informatics*, *4*(1), 48–55. doi:10.1016/j.ecoinf.2009.01.001

Parker, D. C., Manson, S. M., Janssen, M. A., Hoffman, M., & Deadman, P. (2001). *Multi-agent systems for the simulation of land-use and land-cover change: A review*. Indiana University, Retrieved April 20, 2010, from http://www.csiss.org/events/other/agent-based/.../maslucc_overview.pdf

Parks, B. O., Fornwall, M. D., & Quinn, J. F. (Eds.). (2004). *Proceedings from NBII: First NBII Biodiversity Modeling Workshop: Results and Recommendations.* U.S. Geological Survey, Center for Bioinformatics, Denver, CO.

Parmesan, C. (2006). Ecological and evolutionary responses to recent climate change. *Annual Review of Ecology Evolution and Systematics*, *37*, 637–669. doi:10.1146/annurev.ecolsys.37.091305.110100

Parmesan, C., Ryrholm, N., Stefanescu, C., Hill, J. K., Thomas, C. D., & Descimon, H. (1999). Poleward shifts in geographic ranges of butterfly species associated with regional warming. *Nature*, *399*, 579–583. doi:10.1038/21181

Parnell, J., Simpson, D. A., Moat, J., Kirkup, D. W., Chantaranothai, P., & Boyce, P. C., & contributors. (2003). Plant collecting spread and densities: Their potential impact on biogeographical studies in Thailand. *Journal of Biogeography*, *30*, 193–209. doi:10.1046/j.1365-2699.2003.00828.x

Parnell, J. A. N., Simpson, D. A., Moat, J., Kirkup, D. W., Chantaranothai, P., & Boyce, P. C. (2003). Plant collecting spread and densities: Their potential impact on biogeographical studies in Thailand. *Journal of Biogeography*, *30*, 193–209. doi:10.1046/j.1365-2699.2003.00828.x

Parnell, J. A. N. (2000). The conservation of biodiversity: Aspects of Ireland's role in the study of tropical plant diversity with particular reference to the study of the flora of Thailand and *Syzygium*. In Rushton, B. (Ed.), *Biodiversity: The Irish dimension* (pp. 205–216). Dublin, Ireland: Royal Irish Academy Special Publication.

Parolo, G., & Rossi, G. (2008). Upward migration of vascular plants following a climate warming trend in the Alps. *Basic and Applied Ecology*, *9*(2), 100–107. doi:10.1016/j.baae.2007.01.005

Parr, T. W., Ferretti, M., Simpson, I. C., Forsius, M., & Kovács-Láng, E. (2002). Towards a long-term integrated monitoring programme in Europe: Network design in theory and practice. *Environmental Monitoring and Assessment*, *78*(3), 253–290. doi:10.1023/A:1019934919140

Parry, M. L., Canziani, O. F., Palutikof, J. P., Van Der Linden, P. J., & Hanson, C. E. (Eds.). (2007). *Climate change 2007. Working group II: Impacts, adaptation and vulnerability*. Cambridge, UK: Cambridge University Press/ Intergovernmental Panel on Climate Change.

Pastor, J., & Wolter, P. T. (2002). Mapping and modeling forest change in a Boreal Landscape. In *Proceeding NASA LCLUC*, Science Team Meeting Washington D.C. 19–21. November.

Patta, S. (2004). *Application of stochastic dowscaling techniques to global climate model data for regional climate prediction.* Unpublished M.Sc. thesis, Louisiana State University and Agricultural and Mechanical College, Louisiana.

Pattanavibool, A., Dearden, P., & Kutintara, U. (2004). Habitat fragmentation in north Thailand: A case study. *Bird Conservation International*, *14*, S13–S22. doi:10.1017/S0959270905000195

Patton, D. R. (1992). *Wildlife-habitat relationships in forested ecosystems*. Oregon, USA: Timber Press Inc.

PBL. (2010). *Rethinking Global Biodiversity Strategies, Exploring structural changes in production and consumption to reduce biodiversity loss.* PBL-Netherlands Environmental Assessment agency, Bilthoven.

Pearce, D. W. (2001). The economic value of forest ecosystems. *Ecosystem Health*, *7*(4), 284–296. doi:10.1046/j.1526-0992.2001.01037.x

Pearson, R. G., Dawson, T. P., & Liu, C. (2004). Modelling species distributions in Britain: A hierarchical integration of climate and land-cover data. *Ecography*, *27*(3), 285–298. doi:10.1111/j.0906-7590.2004.03740.x

Peh, K. S. H. (2007). Potential effects of climate change on altitudinal distributions in tropical birds. *The Condor*, *109*, 437–440. doi:10.1650/0010-5422(2007)109[437:PEOCCO]2.0.CO;2

Pekkarinen, A., Reithmaier, L., & Strobl, P. (2009). Pan-European forest/non-forest mapping with Landsat ETM+ and CORINE Land Cover 2000 data. *ISPRS Journal of Photogrammetry and Remote Sensing*, *64*(2), 171–183. doi:10.1016/j.isprsjprs.2008.09.004

Perdigão, V., & Annoni, A. (1997). *Technical and methodological guide for updating CORINE land cover database*. Joint Research Centre. Ispra.

Pereira, H. M., Leadley, P. W., Proença, V., Alkemade, R., Scharlemann, J. P. W., & Fernandez-Manjarrés, J. F. (2010). Scenarios for Global Biodiversity in the 21st Century. *Science*, *330*, 1496–1501. doi:10.1126/science.1196624

Pereira, H. M., & Cooper, H. D. (2006). Towards the global monitoring of biodiversity change. *Trends in Ecology & Evolution*, *21*(3), 123–129. doi:10.1016/j.tree.2005.10.015

Peres, C. A. (2010). Overexploitation. In Sodhi, N. S., & Ehrlich, P. R. (Eds.), *Conservation Biology for all* (pp. 107–130). Oxford, UK: Oxford University Press. doi:10.1093/acprof:oso/9780199554232.003.0007

Pérez-Soba, M., Verburg, P. H., & Koomen, E. (2010). *Land use modelling-implementation: Preserving and enhancing the environmental benefits of land-use services. Final report to the European Commission, DG Environment*. Wageningen: Alterra Wageningen UR/ Geodan Next/ Object Vision/ BIOS/ LEI and PBL.

Perrings, C. (2006). Ecological economics after the millennium assessment. *International Journal of Ecological Economics and Statistics*, *6*(FO6), 8–22.

Perry, G. L. W., & Enright, N. J. (2002). Spatial modelling of landscape composition and pattern in a maquis-forest complex, Mont Do, New Caledonia. *Ecological Modelling*, *152*(2-3), 279–302. doi:10.1016/S0304-3800(02)00004-2

Peterseil, J., Wrbka, T., Plutzar, C., Schmitzberger, I., Kiss, A., & Szerencsits, E. (2004). Evaluating the ecological sustainability of Austrian agricultural landscapes-the SINUS approach. *Land Use Policy*, *21*(3), 307–320. doi:10.1016/j.landusepol.2003.10.011

Peterson, A. T., Ortega-Huerrta, M. A., Bartley, J., Sanchez-Cordero, V., & Soberon, J., V., Buddemeier, R. H., & Stockwell, D. R. B. (2002). Future projections for Mexican faunas under global climate change scenarios. *Nature*, *416*, 626–629. doi:10.1038/416626a

Peterson, A. T. (2001). Predicting species' geographical distributions based on ecological niche modeling. *The Condor*, *103*, 599–605. doi:10.1650/0010-5422(2001)103[0599:PSGDBO]2.0.CO;2

Peterson, A. T., Papes, M., & Eaton, M. (2007). Transferability and model evaluation in ecological niche modeling: A comparison of GARP and Maxent. *Ecography*, *30*, 550–560.

Phat, N. K., Knorr, W., & Kim, S. (2004). Appropriate measures for conservation of terrestrial carbon stocks – analysis of trends of forest management in Southeast Asia. *Forest Ecology and Management*, *191*, 283–299. doi:10.1016/j.foreco.2003.12.019

Phelps, J., Webb, E. L., & Agrawal, A. (2010). Does REDD+ threaten to recentralize forest governance? *Science*, *328*, 312–313. doi:10.1126/science.1187774

Phillips, V. D. (1998). Peatswamp ecology and sustainable development in Borneo. *Biodiversity and Conservation*, *7*, 651–671. doi:10.1023/A:1008808519096

Phillips, S. J., Andersion, R. P., & Schapire, R. E. (2006). Maximum entropy modeling of species geographic distributions. *Ecological Modelling*, *190*, 231–239. doi:10.1016/j.ecolmodel.2005.03.026

Phillips, S. J., & Dudik, M. (2008). Modeling of species distributions with Maxent: New extensions and a comprehensive evaluation. *Ecography*, *31*, 161–175. doi:10.1111/j.0906-7590.2008.5203.x

Pimm, S. L. (2001). *The world according to Pimm: A scientist audits the Earth*. New York, NY: McGraw Hill.

Pimm, S. L. (2007). Climate disruption and biodiversity. *Current Biology*, *183*, R117–R119.

Pimm, S. L., & Jenkins, C. N. (2010). Extinctions and the practice of preventing them. In Sodhi, N. S., & Ehrlich, P. R. (Eds.), *Conservation Biology for all* (pp. 181–198). Oxford, UK: Oxford University Press. doi:10.1093/acprof:oso/9780199554232.003.0011

Pontius, R. G., Boersma, W., Castella, J.-C., Clarke, K., de Nijs, T., Dietzel, C., & Verburg, P. H. (2008). Comparing the input, output, and validation maps for several models of land change. *The Annals of Regional Science*, *42*, 11–37. doi:10.1007/s00168-007-0138-2

Pontius, R. G. Jr, Huffaker, D., & Denman, K. (2004). Useful techniques of validation for spatially explicit land-change models. *Ecological Modelling*, *179*, 445–461.. doi:10.1016/j.ecolmodel.2004.05.010

Pontius, R. G. Jr, & Schneider, L. C. (2001). Land-cover change model validation by an ROC method for the Ipswich watershed, Massachusetts, USA. *Agriculture Ecosystems & Environment*, *85*, 239–248. doi:10.1016/S0167-8809(01)00187-6

Porro, R., Benitez, S., Rubiano, J., Mulligan, M., Naranjo, L. G., & Jarvis, A. (2008). *Securing biostability in the Amazon/Andes: A situation analysis. Report to NERC/ESRC/DFID*. London, UK: Kings College, University of London.

Poschlod, P., Bakker, J. P., & Kahmen, S. (2005). Changing land use and its impact on biodiversity. *Basic and Applied Ecology*, *6*, 93–98..doi:10.1016/j.baae.2004.12.001

Potschin, M. B., & Haines-Young, R. H. (2006). Landscapes and sustainability. *Landscape and Urban Planning*, *75*(3-4), 155–161. doi:10.1016/j.landurbplan.2005.03.006

Potting, J., & Bakkes, J. (Eds.). (2004). *The GEO-3 scenarios 2002-2032: Quantification and analysis of environmental impacts*. Report UNEP/DEWA/RS.03-4 and RIVM 402001022, Division of Early Warning and Assessment (DEWA), United Nations Environment Programme (UNEP) / National Institute for Public Health and the Environment, Nairobi / Bilthoven.

Poulsen, J. R., Clark, C. J., Mavah, G., & Elkan, P. W. (2009). Bushmeat supply and consumption in a tropical logging concession in Northern Congo. *Conservation Biology*, *23*(6), 1597–1608. doi:10.1111/j.1523-1739.2009.01251.x

Pounds, J. A., Bustamante, M. R., Coloma, L. A., Consuegra, J. A., Fogden, M. P. L., & Foster, P. N. (2006). Widespread amphibian extinctions from epidemic disease driven by global warming. *Nature*, *439*, 161–167. doi:10.1038/nature04246

Pounds, J. A., Fodgen, M. P. L., & Campbell, J. H. (1999). Biological response to climate change on a tropical mountain. *Nature*, *398*, 611–615. doi:10.1038/19297

Power, M. E., Tilman, D., Estes, J. A., Menge, B. A., Bond, W. J., & Mills, L. S. (1996). Challenges in the quest for keystones. *Bioscience*, *46*, 609–620. doi:10.2307/1312990

Power, J. F., & Follett, R. F. (1987). Monocultures: Advantages, limitations, and alternatives. *Scientific American*, *256*(3), 78–86. doi:10.1038/scientificamerican0387-78

Prance, G. T., Beentje, H. J. D., & Johns, R. (2000). The tropical flora remains undercollected. *Annals of the Missouri Botanical Garden*, *87*, 67–71. doi:10.2307/2666209

Prasetyo, L. B., Wibowo, S. A., Kartodihardjo, H., & Tonny, F., Haryanto, Sonaji, R., & Setiawan, Y. (2008). Land use and land-cover changes of conservation area during transition to regional autonomy: Case study of Balairaja Wildlife Reserve in Riau Province, Indonesia. *Tropics*, *17*(2), 99–108. doi:10.3759/tropics.17.99

Prentice, I. C., Cramer, W., Harrison, S., Leemans, R., Monserud, R. A., & Solomon, A. M. (1992). A global biome model based on plant physiology and dominance, soil properties and climate. *Journal of Biogeography*, *19*, 117–134. doi:10.2307/2845499

Prentice, C., Cramer, W., Harrison, S. P., Leemans, R., Monserud, R. A., & Solomon, A. M. (1992). A global biome model based on plant physiology and dominance, soil properties and climate. *Journal of Biogeography*, *19*, 117–134. doi:10.2307/2845499

Priess, J. A., & Schaldach, R. (2008). Integrated models of the land system: A review of modelling approaches on the regional to global scale. *Living Reviews in Landscape Research*, *2*. Retrieved January 12, 2010, from http://www.livingreviews.org/lrlr-2008-1.

Priess, J. A., & Schaldach, R. (2008). Integrated models of the land system: A review of modelling approaches on the regional to global scale. *Living Reviews in Landscape Research, 2*. Retrieved from http://www.livingreviews.org/lrlr-2008-1

Prince, S. D., Brown de Colstoun, E., & Kravitz, L. L. (1998). Evidence from rain-use efficiency does not indicate extensive Sahelian desertification. *Global Change Biology, 4*, 359–379. doi:10.1046/j.1365-2486.1998.00158.x

Pringle, R. M. (2008). Elephants as agents of habitat creation for small vertebrates at the patch scale. *Ecology, 89*(1), 26–33. doi:10.1890/07-0776.1

Prins, H. T. P., & van der Jeugd, H. (1993). Herbivore population crashes and woodland structure in East Africa. *Journal of Ecology, 81*, 305–314. doi:10.2307/2261500

Prommakul, P. (2003). *Habitat utilization and prey of the tiger (Panthera tigris) in the Eastern Thung Yai Naresuan Wildlife Sanctuary*. Unpublished M.S. thesis (in Thai). Kasetsart University, BangkoK, Thailand.

Prydatko, V., & Kolomytsev, G. (2008). Climate and biodiversity changes by 2050 in GLOBIO Ukraine region. *Ukrainian-American Environmental Association Newsletter, 7*(5), 34.

Prydatko, V., Pristinska, G., Panina, O., & Vasylkivsky, B. (2006). *Textbook on collection and processing of the information for national reports of Ukraine on implementation of the Convention on Biological Diversity*. Kyiv, Ukraine: EcoPravo.

Prydatko, V. I., Kolomytsev, G. O., Burda, R. I., & Chumachenko, S. M. (2008). *Landscape ecology: Textbook on application of pressure-based biodiversity modelling for national and regional educational purposes. Book 1 & 2*. Kyiv, Ukraine: NAU.

Prydatko, V. I. (2005). Indication and indicators: Development and use of it for purpose to evaluate state of agrobiodiversity in Ukraine. In Sozinov, O. O. (Eds.), *Agrobiodiversity of Ukraine: Theory, methodology, indicators, examples. Book 1* (pp. 94–113). Kyiv, Ukraine: Nichlava.

Prydatko, V. (2000). *Biodiversity and bioresources of Ukraine: Review of SoE publications*. (1992-1998). *Re-evaluation of trends* (1966-1999). *The environment and resources*. Kyiv: ERRIU. ISBN 966-95141-1-6

Raabová, J., Münzbergová, Z., & Fischer, M. (2007). Ecological rather than geographic or genetic distance affects local adaptation of the rare perennial herb, Aster amellus. *Biological Conservation, 139*, 348–357. doi:10.1016/j.biocon.2007.07.007

Raes, N., Roos, M. C., Slik, J. W. F., van Loon, E. E., & Ter Steege, H. (2009). Botanical richness and endemicity patterns of Borneo derived from species distribution models. *Ecography, 32*, 180–192. doi:10.1111/j.1600-0587.2009.05800.x

Raes, N., & Ter Steege, H. (2007). A null-model for significance testing of presence-only species distribution models. *Ecography, 30*, 727–736. doi:10.1111/j.2007.0906-7590.05041.x

Raes, N., & Van Welzen, P. C. (2009). The demarcation and internal division of Flora Malesiana: 1857 – Present. *Blumea, 54*, 6–8. doi:10.3767/000651909X475888

Raes, N. (2009). *Borneo: A quantitative analysis of botanical richness, endemicity and floristic regions based on herbarium records*. Unpublished PhD thesis, National Herbarium of the Netherlands, Leiden University Branch, Leiden.

Rahman, M. H., Pal, S. K., & Alam, F. (1992). Effect of nitrogen, phosphorus, potassium, sulphur, zinc and manganese nutrients on yield and sucrose content of sugarcane (Saccharum officinarum) in floodplain soils of Bangladesh. *Indian Journal of Agricultural Sciences, 62*, 450–455.

Rajan, K. S., & Shibasaki, R. (1997). Estimation of agricultural productivity and its application to modelling the expansion of agricultural land in Thailand. *Journal of Agricultural Meteorology, 52*(5), 815–818.

Ramachandra, T. V., & Kumar, U. (2004). Geographic resources decision support system for land use, land cover dynamics analysis. In *Proceedings of the FOSS/GRASS Users Conference*, Bangkok, Thailand

Ramankutty, N., & Foley, J. A. (1999). Estimating historical changes in global land cover: Croplands from 1700 to 1992. *Global Biogeochemical Cycles, 13*(4), 997–1028. doi:10.1029/1999GB900046

Ramankutty, N., Foley, J. A., & Olejniczak, N. J. (2002). People on the land: Changes in global population and croplands during the 20th century. *Ambio, 31*, 251–257.

Raskin, P. D., & Kemp-Benedict, E. (2002). *Global Environmental Outlook Scenario Framework. Background Paper for UNEP's Third Global Environmental Outlook Report* (p. 102). Stockholm Environment Institute–Boston Center.

Reddy, C. S., Babar, S., Sudha, K., & Raju, V. S. (2008). Vegetation cover mapping and landscape level disturbance gradient analysis in warangal district, Andhra Pradesh, India using satellite remote sensing and GIS. *Space Research Journal, 1*, 29–38. doi:10.3923/srj.2008.29.38

Reddy, S., & Davalos, L. M. (2003). Geographical sampling bias and its implications for conservation priorities in Africa. *Journal of Biogeography, 30*, 1719–1727. doi:10.1046/j.1365-2699.2003.00946.x

Redford, K. H., & Richter, B. D. (1999). Conservation of biodiversity in a world of use. *Conservation Biology, 13*(6), 1246–1256. doi:10.1046/j.1523-1739.1999.97463.x

Reese, H. M., Lillesand, T. M., Nagel, D. E., Stewart, J. S., Goldmann, R. A., & Simmons, T. E. (2002). Statewide land cover derived from multiseasonal Landsat TM data—a retrospective of the WISCLAND project. *Remote Sensing of Environment, 82*(2), 3224–3237. doi:10.1016/S0034-4257(02)00039-1

Reidsma, P., Tekelenburg, T., van den Berg, M., & Alkemade, R. (2006). Impacts of land use change on biodiversity: An assessment of agricultural biodiversity in the European Union. *Agriculture Ecosystems & Environment, 114*, 86–102. doi:10.1016/j.agee.2005.11.026

Reis, S. (2008). Analyzing land use/land cover changes using remote sensing and GIS in Rize. *North-East Turkey Sensors, 8*(10), 6188–6202.

Renetzeder, C., Schindler, S., Peterseil, J., Prinz, M. A., Mücher, C. A., & Wrbka, T. (2010). Can we measure ecological sustainability? Landscape pattern as an indicator for naturalness and land use intensity at regional, national and European level. *Ecological Indicators, 10*, 39–48. doi:10.1016/j.ecolind.2009.03.017

Renetzeder, C., van Eupen, M., Mücher, C. A., & Wrbka, T. (2008). A spatial regional reference framework for sustainability assessment in Europe. In Helming, K., Perez-Soba, M., & Tabbush, P. (Eds.), *Sustainability impact assessment of land use changes* (pp. 249–268). Berlin/ Heidelberg, Germany & New York, NY: Springer. doi:10.1007/978-3-540-78648-1_13

Repetto, R., & Gillis, M. (1988). *Public policies and misuse of forest resources*. Cambridge, UK: Cambridge University Press. doi:10.1017/CBO9780511601125

Republic of Kenya. (2004). *Report of the Commission of Inquiry into illegal/ irregular allocation of public land.* Nairobi, Kenya: Government Printer.

Rerkasem, B., & Rerkasem, K. (2000). Agrodiversity for in situ conservation of Thailand's native rice germplasm. *CMU Journal, 1*(2), 15–23.

Revenga, C., Campbell, I., Abell, R., de Villiers, P. & Bryer, M. (2005). Prospects for monitoring freshwater ecosystems towards the 2010 targets. *Philosophical Transactions of the Royal Society B Biological Sciences, 360*(1454), 397-413.

RFD/ITTO. (2002). *Management of the Pha Taem protected forest complex to promote cooperation for trans-boundary biodiversity conservation between Thailand, Cambodia and Laos (Phase I)*. Bangkok: International Tropical Timber Organization and Royal Forest Department.

RFD/ITTO. (2004). *Management of the Emerald Triangle protected forest complex to promote cooperation for trans-boundary biodiversity conservation between Thailand, Cambodia and Laos (Phase II)*. Bangkok: International Tropical Timber Organization and Royal Forest Department.

Rieley, J. O., Page, S. E., & Shepherd, P. A. (1997). Tropical bog forests of South East Asia. In Stoneman, L. P. R. E., & Ingram, H. A. P. (Eds.), *Conserving peatlands* (pp. 35–41). Wallingford, UK: CAB International.

Rienks, W. A. (Ed.). (2008). *The future of rural Europe*. Wageningen, The Netherlands: Wageningen University Research and Netherlands Environmental Assessment Agency.

Riginos, C., & Grace, J. B. (2008). Savanna tree density, herbivores, and the herbaceous community: Bottom-up versus top-down effects. *Ecology, 89*, 2228–2238. doi:10.1890/07-1250.1

Ripple, W. J., & Larson, E. J. (2000). Historic aspen recruitment, elk, and wolves in northern Yellowstone National Park, USA. *Biological Conservation, 95*(3), 361–370. doi:10.1016/S0006-3207(00)00014-8

Rivas-Martínez, S., Sánchez-Mata, D., & Costa, M. (1999). North American boreal and western temperate forest vegetation. Retrieved from http://www.ucm.es/info/cif/book/namerica2/namerica_02_1.htm

Roberts, D. A., Keller, M., & Soares, J. V. (2003). Studies of land-cover, land-use and biophysical of vegetation in the large scale biosphere atmosphere experiment in Amazonia. *Remote Sensing of Environment*, *87*, 377–388. doi:10.1016/j.rse.2003.08.012

Rooij, W. van., & Tekelenburg, T. (2007). Land use modelling focusing on the impact on biodiversity. Retrieved October 20, 2009, from www.fao.org/forestry/foris/ppt/outlook2020/land-use-modelling.pdf

Rose, S., & Grainger, A. (2002). Multivariate mapping of spatial variation in biodiversity in Peruvian Amazonia. *Diversity & Distributions*, *9*, 237–250. doi:10.1046/j.1472-4642.2003.00019.x

Roshier, D. A., Whetton, P., Allan, R. J., & Robertson, A. I. (2001). Distribution and persistence of temporary wetland habitats in arid Australia in relation to climate. *Austral Ecology*, *26*, 371–384. doi:10.1046/j.1442-9993.2001.01122.x

Rotmans, J. (1990). *IMAGE. An integrated model to assess the greenhouse effect*. Dordrecht, The Netherlands: Kluwer Academic Publishers.

Rounsevell, M. D. A., Berry, P. M., & Harrison, P. A. (2006). Future environmental change impacts on rural land use and biodiversity: A synthesis of the ACCELERATES project. *Environmental Science & Policy*, *9*, 93–100.. doi:10.1016/j.envsci.2005.11.001

Rounsevell, M. D. A., Reginster, I., Araújo, M. B., Carter, T. R., Dendoncker, N., & Ewert, F. (2006). A coherent set of future land use change scenarios in Europe. *Agriculture Ecosystems & Environment*, *114*, 57–68..doi:10.1016/j.agee.2005.11.027

Roy, P. S., & Giriraj, A. (2008). Land use and land cover analysis in Indian context. *Journal of Applied Sciences*, *8*(8), 1346–1353. doi:10.3923/jas.2008.1346.1353

Roy, P. S., & Tomar, S. (2000). Biodiversity characterization at landscape level using geospatial modelling technique. *Biological Conservation*, *95*, 95–109. doi:10.1016/S0006-3207(99)00151-2

Royal Forest Department. (2007). *Forestry statistics year 2006*. Bangkok, Thailand: Ministry of Natural Resources and Environment.

Royal Thai Survey Department. (2002). *Topographic map scale 1: 50,000*. Ministry of Defense, Bangkok, Thailand.

Rubinstein, R. Y. (1981). *Simulation and the Monte Carlo method*. New York, NY: Wiley.

Rubio, J. L., & Bochet, E. (1998). Desertification indicators as diagnosis criteria for desertification risk assessment in Europe. *Journal of Arid Environments*, *39*, 113–120. doi:10.1006/jare.1998.0402

Rudel, T. K., Coomes, O. T., Moran, E., Achard, F., Angelsen, A., Xu, J. C., & Lambin, E. (2005). Forest transitions: Towards a global understanding of land use change. *Global Environmental Change*, *15*, 23–31. doi:10.1016/j.gloenvcha.2004.11.001

Runhaar, J., & De Haes, H. A. (1994). The use of site factors as ecosystem classification characteristics. In Klijn, F. (Ed.), *Ecosystem classification for environmental management* (pp. 139–172). Dordrecht, The Netherlands: Kluwer.

Sadler, B., & Verheem, R. (1996). *Strategic environmental assessment: Status, challenges and future directions*. The Netherlands: Ministry of Housing, Spatial Planning and the Environment.

Saipunkaew, W., Wolseley, P. A., Chimonides, P. J., & Boonpragob, K. (2007). Epiphytic macrolichens as indicators of environmental alteration in northern Thailand. *Environmental Pollution*, *146*, 366–376. doi:10.1016/j.envpol.2006.03.044

Sala, O. E., Chapin, F. S. III, Armesto, J. J., Berlow, E., Bloomfield, J., Irzo, R.... Wall, D. H. (2000). Global biodiversity scenarios for the year 2100. *Science*, *287*, 1770–1774. doi:10.1126/science.287.5459.1770

Samuels, R., Rimmer, A., Hartmann, A., Krichak, S. & Alpert, P. (2010). *Change impacts on Jordan River flow: Downscaling application from a regional climate model*. American Meterological Society. doi: 10.1175/2010JHM1177.1.

Sanderson, E. W., Jaiteh, M., Levy, M. A., Redford, K. H., Wannebo, A. V., & Woolmer, G. (2002). The human footprint and the last of the wild. *Bioscience*, *52*(10), 891–904. doi:10.1641/0006-3568(2002)052[0891:THFATL]2.0.CO;2

Sangha, K. K., & Jalota, R. J. (2005). Value of ecological services of exotic Eucalyptus tereticornis and native albergia sissoo tree plantations of North-Western India. *Conservation & Society*, *3*(1), 92–105.

Santilli, M., Moutinho, P., Schwartzman, S., Nepstad, D., Curran, L., & Nobre, C. (2005). Tropical deforestation and the Kyoto Protocol. *Climatic Change*, *71*, 267–276. doi:10.1007/s10584-005-8074-6

Santisuk, T., Smitinand, T., Hoamuangkaew, W., Ashton, P., Sohmer, S. H., & Vincent, J. R. (Eds.). (1991). *Plants for our future: Botanical research and conservation needs in Thailand*. Bangkok, Thailand: Royal Forest Department Thailand.

Santisuk, T. (1988). *An account of the vegetation of northern Thailand. Geological Research, 5*. Stuttgart, Germany: Franz Steiner Verlag.

Santisuk, T., & Larsen, S. (Eds.), *Flora of Thailand* (*Vol. 6*). Bangkok, Thailand: The Diamond Printing Co.

Satterlund, D. R., & Adams, P. W. (1992). *Wildland watershed management* (2nd ed.). New York, NY: John Wiley & Sons, Inc.

Saunders, D. A., Hobbs, R. J., & Margules, C. R. (1991). Biological consequences of ecosystem fragmentation: A review. *Conservation Biology*, *5*(1), 18–32. doi:10.1111/j.1523-1739.1991.tb00384.x

Sayer, J., Harcourt, C., & Collins, N. M. (1992). *Conservation atlas of tropical forests: Africa*. London, UK: Macmillan.

Sayre, R., Roca, E., Sedaghatkish, G., Young, B., Keel, S., Roca, R., & Sheppard, S. (2000). *Nature in focus: Rapid ecological assessment*. Washington D.C. and California: The Nature Conservancy, Island Press.

sCBD & MNP. (2007). *Cross-roads of life on earth—exploring means to meet the 2010 Biodiversity target*. Solution-oriented scenarios for global biodiversity outlook 2. Technical Series no 31. Secretariat of the Convention on Biological Diversity, Montreal.

sCBD (2010) *Global Biodiversity Outlook 3*. Secretariat of the Convention on Biological Diversity Montréal, 94 pages.

sCBD. (2006). *Global biodiversity outlook 2*. Secretariat of the Convention on Biological Diversity, Montreal.

sCBD., & MNP. (2007). *Cross-roads of life on Earth - exploring means to meet the 2010 biodiversity target. Solution-oriented scenarios for Global Biodiversity Outlook 2*. In Technical Series no 31. Secretariat of the Convention on Biological Diversity, Montreal.

Scepan, J., Menz, G., & Hansen, M. C. (1999). The DISCover validation image interpretation process. *Photogrammetry Engineering and Remote Sing*, *65*(9), 1075–1081.

Schaepman, M. E., Malenovsky, Z., Mücher, C. A., Kooistra, L., & Thullier, W. (2007). Bridging scaling gaps for the assessment of biodiversity from space. In G. E. O. Secretariat (Ed.), *The full picture. A publication for the GEO Ministerial Summit, Earth Observation for Sustainable Growth and Development* (pp. 258-161). Cape Town, Geneva (CH) and Tudor Rose.

Schaepman-Strub, G., Limpens, J., Menken, M., Bartholomeus, H. M., & Schaepman, M. E. (2009). Towards spatial assessment of carbon sequestration in peatlands: Spectroscopy based estimation of fractional cover of three plant functional types. *Biogeosciences*, *6*(2), 275–284. doi:10.5194/bg-6-275-2009

Schaepman-Strub, G., Schaepman, M. E., Martonchik, J., & Schaaf, C. (2006). What's in a satellite albedo product? *International Geoscience and Remote Sensing* (IGARSS), 2848-2851.

Schaminée, J. H. J., Hennekens, S. M., & Ozinga, W. A. (2007). Use of the ecological information system SynBioSys for the analysis of large datasets. *Journal of Vegetation Science*, *18*, 463–470. doi:10.1111/j.1654-1103.2007.tb02560.x

Schlesinger, M. E., Malyshev, S., Rozanov, E. V., Yang, F., Andronova, N. G., & De Vries, B. (2000). Geographical distributions of temperature change for scenarios of greenhouse gas and sulphur dioxide emissions. *Technological Forecasting and Social Change*, *65*, 167–193. doi:10.1016/S0040-1625(99)00114-6

Schmid, B., Joshi, J., & Schlapfer, F. (2001). Empirical evidence for biodiversity-ecosystem functioning relationships. In Kinzing, A. P., Pacala, S. W., & Tilman, D. (Eds.), *The functional consequences of biodiversity: Empirical progress and theoretical extension. Monographs in Population Biology* (pp. 120–150). Princeton, NJ: Princeton University.

Schmidt-Vogt, D. (1999). *Swidden farming and fallow vegetation in northern Thailand. Geological Research, 8*. Stuttgart, Germany: Franz Steiner Verlag.

Schmiegelow, F. K. A., & Monkkonen, M. (2002). Habitat loss and fragmentation in dynamic landscape: Avian perspectives from the boreal forest. *Ecological Applications, 12*, 375–389.

Scholes, R. J., & Biggs, R. (2005). A biodiversity intactness index. *Nature, 434*, 45–49. doi:10.1038/nature03289

Schuur, E. A. G., Bockheim, J., Canadell, J. G., Euskirchen, E., Field, C. B., & Goryachkin, S. V. (2008). Vulnerability of permafrost carbon to climate change: Implications for the global carbon cycle. *Bioscience, 58*(8), 701–714. doi:10.1641/B580807

Schwartz, M. W., Brigham, C. A., Hoeksema, J. D., Lyons, K. G., Mills, H., & van Mantgem, P. J. (2000). Linking biodiversity to ecosystem function: Implications for conservation biology. *Oecologia, 122*(3), 297–305. doi:10.1007/s004420050035

Scientific modelling. (n.d.). *Wikipedia*. Retrieved March 27, 2010, from http://en.wikipedia.org/wiki/Scientific_modelling

SCS. (1972). *Hydrology sect. 4, SCS national engineering handbook*. Washington, DC: USDA-SCS.

Secretariat of the Convention on Biological Diversity. (2003). *Interlinakages between biological diversity and climate change. Advise on the integration of biodiversity considerations into the implementation of the United Nations Framework Convention on Climate Change and Its Kyoto protocol* (CBD Technical Series no. 10). Montreal.

Secretariat of the Convention on Biological Diversity (sCBD). (2006). *Convention on biological diversity: Global biodiversity outlook 2*. Montreal.

Secretariat of the Convention on Biological Diversity (sCBD). (2010). *Global biodiversity outlook 3 – executive summary.* Montreal.

Secretariat of the Convention on Biological Diversity. (2006). *Global biodiversity outlook 2*. Montreal.

Secretariat of the Convention on Biological Diversity & Netherlands Environmental Assessment Agency. (2007). *Cross-roads of planet Earth's life. Exploring means to meet the 2010 biodiversity target*. Montreal.

Secretariat of the Convention on Biological Diversity. (2003). *Interlinakages between biological diversity and climate change. Advise on the integration of biodiversity considerations into the implementation of the United Nations Framework Convention on Climate Change and Its Kyoto protocol.* Montreal.

Secretariat of the Convention on Biological Diversity. (2006). *Global biodiversity outlook* 2. Montreal.

Sedano, F., Gong, P., & Ferrão, M. (2005). Land cover assessment with MODIS imagery in southern African Miombo ecosystems. *Remote Sensing of Environment, 98*, 429–441. doi:10.1016/j.rse.2005.08.009

Seitzinger, S., Mayorga, E., Bouwman, A. F., Kroeze, C., Beusen, A. H. W., & Billen, G. (2010). Global river nutrient export: A scenario analysis of past and future trends. *Global Biogeochemical Cycles, 24*, GB0A08.. doi:10.1029/2009GB003587

Seitzinger, S. P., Harrison, J. A., Dumont, E., Beusen, A. H. W., & Bouwman, A. F. (2005). Sources and delivery of carbon, nitrogen, and phosphorus to the coastal zone: An overview of Global NEWS models and their application. *Global Biogeochemical Cycles, 19*, GB4S. doi:10.1029/2004GB002453

Sekercioglu, C. H. (2006). Conservation ecology: Area trumps mobility in fragment bird extinctions. *Current Biology, 17*, R283–R286. doi:10.1016/j.cub.2007.02.019

Sekercioglu, C. H. (2010). Ecosystem functions and services. In Sodhi, N. S., & Ehrlich, P. R. (Eds.), *Conservation Biology for all* (pp. 45–72). Oxford, UK: Oxford University Press. doi:10.1093/acprof:oso/9780199554232.003.0004

Sekercioglu, C. H., Ehrlich, P. R., Daily, G. C., & Aygen, D. Goehring, D., & Sandi, R. F. (2002). Disappearance of insectivorous birds from tropical fragments. *Proceedings of the National Academy of Sciences of the United States of America, 99*, 263-267.

Sharma, P. K., Lahkar, B. P., Ghosh, S., Rabha, A., Das, J. P., & Nath, N. K. (2008). Land use and land cover change and future implication analysis in Manas National Park, India using multitemporal satellite data. *Current Science*, *95*(2), 223–227.

Sharma, K. D. (1998). The hydrological indicators of desertification. *Journal of Arid Environments*, *39*, 121–132. doi:10.1006/jare.1998.0403

Shearer, A. W. (2005). Approaching scenario-based studies: Three perceptions about the future and considerations for landscape planning. *Environment and Planning. B, Planning & Design*, *32*(1), 67–87..doi:10.1068/b3116

Shi, W. Z., Ehlers, M., & Molenaar, M. (2005). Uncertainties in integrated remote sensing and GIS. *International Journal of Remote Sensing*, *26*(14), 2911–2915. doi:10.1080/01431160500197537

Shrestha, R. P., Eiumnoh, A., & Baimoung, S. (1996). *Soil erosion assessment and its policy implications: A case study of RS and GIS applications in Uthai Thani, Thailand*. Retrieved January 28, 2010, from http://www.gisdevelopment.net/aars/acrs/1996/ss/ss1004.shtml

Sierra, R. (1999). Traditional resource-use systems and tropical deforestation in a multiethnic region in North-west Ecuador. *Environmental Conservation*, *26*(2), 136–145. doi:10.1017/S0376892999000181

Sierra, R., & Stallings, J. (1998). The Dynamics and Social Organization of Tropical Deforestation in Northwest Ecuador, 1983-1995. *Human Ecology*, *26*(1), 135–161. doi:10.1023/A:1018753018631

Sigman, D. M., & Boyle, E. A. (2000). Glacial/interglacial variations in atmospheric carbon dioxide. *Nature*, *407*, 859–869. doi:10.1038/35038000

Simberloff, D. (2010). Invasive species. In Sodhi, N. S., & Ehrlich, P. R. (Eds.), *Conservation Biology for all* (pp. 131–148). Oxford, UK: Oxford University Press. doi:10.1093/acprof:oso/9780199554232.003.0008

Simpson, G. G. (1964). Species density of North American recent mammals. *Systematic Zoology*, *13*, 57–73. doi:10.2307/2411825

Simpson, E. H. (1949). Measurement of diversity. *Nature*, *163*, 688. doi:10.1038/163688a0

Sitch, S., Huntingford, C., Gedney, N., Levy, P. E., Lomas, M., & Piao, S. L. (2008). Evaluation of the terrestrial carbon cycle, future plant geography and climate-carbon cycle feedbacks using fice dynamic Global Vegetation Models (DGVMs). *Global Change Biology*, *95*, 2015–2039. doi:10.1111/j.1365-2486.2008.01626.x

Skidmore, A. K., Toxopeus, A. G., De Bie, C. A. J. M., Corsi, F., & Venus, V. (2006). *Herpetological species mapping for the Mediterranean*. Unpublished manuscript presented at the ISPRS mid-term Symposium 2006 Remote Sensing: From pixels to processes. ITC, Enschede, the Netherlands.

Slootweg, R., Kolhoff, A., Verheem, R., & Höft, R. (2006). *Biodiversity in EIA and SEA*. Background document to CBD Decision VIII/28: Voluntary Guidelines on Biodiversity-Inclusive Impact Assessment. The Netherlands.

Smith, A. P., & Quin, D. G. (1996). Patterns and causes of extinction and decline in Australian conilurine rodents. *Biological Conservation*, *77*, 243–267. doi:10.1016/0006-3207(96)00002-X

Smitinand, T., & Larsen, K. (1966). The flora of Thailand project. Report on the work in 1966. *The Natural History Bulletin of the Siam Society*, *21*, 341.

Smitinand, T. (1958). The genus Dipterocarpus Gaertn.f. in Thailand. *Thai Forest Bulletin (Botany)*, *4*, 1–50.

Sobero, J., & Peterson, A. T. (2004). Biodiversity informatics: Managing and applying primary biodiversity data. *Philosophical Transactions of the Royal Society of London. Series B, Biological Sciences*, *359*, 689–698. doi:10.1098/rstb.2003.1439

Sodhi, N. S., & Brook, B. W. (2006). *Southeast Asian biodiversity in crisis*. Cambridge, UK: Cambridge University Press.

Sodhi, N. S., Brook, B. W., & Bradshaw, C. J. A. (2007). *Tropical conservation biology*. Singapore: Blackwell Publishing.

Sodhi, N. S., Koh, L. P., Brook, B. W., & Ng, P. L. K. (2004). Southeast Asian biodiversity: An impending disaster. *Trends in Ecology & Evolution, 19*(12), 654–660. doi:10.1016/j.tree.2004.09.006

Solomon, S., Plattner, G., Knutti, R., & Friedlingstein, P. (2009). Irreversible climate change due to carbon dioxide emissions. *Proceedings of the National Academy of Sciences of the United States, 106*(6), 1704–1709. doi:10.1073/pnas.0812721106

Sommer, M., Gerke, H. H., & Deumlich, D. (2008). Modelling soil landscape genesis - a time split approach for hummocky agricultural landscapes. *Geoderma, 145*(3-4), 480–493. doi:10.1016/j.geoderma.2008.01.012

Somporn, S. (1995). *Land use/land cover change detection in the Chiang Mai Area using Landsat TM.* Department Of Geography and Earth Sciences- A Report.

Somura, H., Hoffman, D., Arnold, J. G., Takeda, I., & Mori, Y. (2009). *Application of the SWAT Model to the Hii River Basin, Shimane Prefecture, Japan. Soil and Water Assessment Tool (SWAT): Global Applications.* World Association of Soil and Water Conservation. Special Pub. No.4.

Sonna, P. (2003). *The status and distribution of Eld's deer cervus eldi siamensis in Preah Vihear Province. Wildlife protection Office*. Phnom Phen, Cambodia: Forest Administration.

Sorensen, K. W. (1993). Indonesian peat swamp forests, and their role as a carbon sink. *Chemosphere, 27*(6), 1065–1082. doi:10.1016/0045-6535(93)90068-G

Soule, M. E., & Kohm, K. A. (Eds.). (1989). *Research priorities for conservation biology*. Washington, DC: Island Press.

Southworth, F., Dale, V. H., & O'Neill, R. V. (1991). Contrasting patterns of land use in Rondonia, Brazil: Simulating the effects on carbon release. *International Social Science Journal, 130*, 681–698.

Sozinov, O. O., Prydatko, V. I., Tarariko, O. H., & Shtepa, Y. N. (2005a). *Agrobiodiversity of Ukraine: Theory, methodology, indicators, examples. Book 1 &2*. Kyiv, Ukraine: Nichlava.

Sozinov, O., & Prydatko, V. (2006). *Basic report on the implementation of the Convention on Biological Diversity in Ukraine.* Retrieved December 12, 2006, from http://www.undp.org.ua/i/files/Bio_base.pdf

Spangenberg, J. H. (2007). Biodiversity pressure and the driving force behind. *Ecological Economics, 61*, 146–158.. doi:10.1016/j.ecolecon.2006.02.021

Spellenberg, I. F. (2005). *Monitoring ecological change* (2nd ed.). Cambridge, UK: Cambridge University Press. doi:10.1017/CBO9780511614699

Srivastava, S. K., & Gupta, R. D. (2003). *Monitoring of changes in land use/ land cover using multi- sensor satellite data.* Map India Conference 2003.

Srivastavata, S., Singh, T. P., Singh, H., Kushwaha, S. P. S., & Roy, P. S. (2002). Large scale deforestation in Sonitpur district of Assam. *Current Science, 82*(12), 1479–1484.

Stanners, D., & Bordeaux, P. (Eds.). (1995). *Europe's environment; The Dobříš assessment.* Luxembourg: European Environmental Agency, Copenhagen. Office for Official Publications of the European Communities.

Stefanov, W. L., Ramsey, M. S., & Christensen, P. R. (2001). Monitoring urban land cover change: An expert system approach to land cover classification of semiarid to arid urban centers. *Remote Sensing of Environment, 77*, 173–185. doi:10.1016/S0034-4257(01)00204-8

Stein, B. R., & Wieczorek, J. (2004). Mammals of the world: MaNIS as an example of data integration in a distributed network environment. *Biodiversity Informatics, 1*, 14–22.

Steinmann, K., Linder, H. P., & Zimmermann, N. E. (2009). Modelling plant species richness using functional groups. *Ecological Modelling, 220*(7), 962–967. doi:10.1016/j.ecolmodel.2009.01.006

Stewart-Cox, B. (1995). *Wild Thailand*. Cambridge, MA: MIT Press.

Steyaert, L. T., & Goodchild, M. F. (1994). Integrating geographic information systems and environmental simulation models: A status review. In Michener, W. K., Brunt, J. W., & Stafford, S. G. (Eds.), *Environmental information management and analysis: Ecosystem to global scales* (pp. 333–355). London, UK: Taylor & Francis.

Stibig, H. J., Belward, A. S., Roy, P. S., Rosalina-Wasrin, U., Agrawal, S., & Joshi, P. K., Hildanus, Beuchle, R., Fritz, S., Mubareka, S., & Giri, C. (2007). A land-cover map for South and Southeast Asia derived from SPOT-vegetation data. *Journal of Biogeography*, *34*, 625–637. doi:10.1111/j.1365-2699.2006.01637.x

Still, C. J., Foster, P. N., & Schneider, S. H. (1999). Simulating the effects of climate change on tropical montane cloud forests. *Nature*, *398*, 608–610. doi:10.1038/19293

Stocking, M. (2002). Diversity: A new strategic direction for soil conservation. Sustainable utilization of global soil and water resources. *Proceedings of the 12th International Soil Conservation Conference* (pp. 53-58). Beijing, China: Tsinghua University Press.

Stomph, T. J., Mücher, C. A., & Fresco, L. O. (1997). Environmental impact of land use: A new basis for analysis. *The Land*, *1*(2), 129–142.

Stoms, D. M., & Estes, J. E. (1993). A remote sensing research agenda for mapping and monitoring biodiversity. *International Journal of Remote Sensing*, *14*, 1839–1860. doi:10.1080/01431169308954007

Stoms, D. M. (1992). Effects of habitat map generalization in biodiversity assessment. *Photogrammetric Engineering and Remote Sensing*, *58*(11), 1587–1591.

Stouffer, P. C., & Bierregaard, R. O. (1995). Use of Amazonian forest fragments by understorey insectivorous birds. *Ecology*, *76*, 2429–2445. doi:10.2307/2265818

Stracey, C. M., & Pimm, S. L. (2009). Testing island biogeography theory with visitation rates of birds to British islands. *Journal of Biogeography*, *36*, 1532–1539. doi:10.1111/j.1365-2699.2009.02090.x

Strahler, A. H., Woodcok, C. E., & Smith, J. A. (1986). On the nature of models in remote sensing. *Remote Sensing of Environment*, *30*(2), 121–139. doi:10.1016/0034-4257(86)90018-0

Strand, H., Höft, R., Strittholt, J., Miles, L., Horning, N., Fosnight, E., & Turner, W. (Eds.). (2007). *Sourcebook on remote sensing and biodiversity indicators*. Secretariat of the Convention on Biological Diversity, Montreal, Technical Series no. 32.

Su, L., Chopping, M. J., Rango, A., Martonchik, J. V., & Peters, D. P. C. (2007). Differentiation of semi-arid vegetation types based on multi-angular observations from MISR and MODIS. *International Journal of Remote Sensing*, *28*(6), 1419–1424. doi:10.1080/01431160601085995

Sunderlin & Resosudarmo. (1996). *Rates and causes of deforestation in Indonesia: Towards a resolution of the ambiguities*. CIFOR.

Swallow, S. K., Talukdar, P., & Wear, D. N. (1997). Spatial and temporal specialization in forest ecosystem management under sole ownership. *American Journal of Agricultural Economics*, *79*, 311–326. doi:10.2307/1244132

Swallow, B. M. (2006). Forests, flowers or flamingos. *ETFRN News*, *45*, 52.

Swallow, B. M., Kallesoe, M. F., Iftikhar, U. A., Van Noordwijk, M., Bracer, C., & Scherr, S. J. (2009). Compensation and rewards for environmental services in the developing world: Framing pan-tropical analysis and comparison. *Ecology and Society*, *14*(2), 26.

Swets, J. A. (1986). Measuring the accuracy of diagnostic systems. *Science*, *240*, 1285–1293. doi:10.1126/science.3287615

Symeonakis, E., & Drake, N. (2004). Monitoring desertification and land degradation over sub-Saharan Africa. *International Journal of Remote Sensing*, *25*, 573–592. doi:10.1080/0143116031000095998

Tabeau, A., Eickhout, B., & van Meijl, H. (2006). *Endogenous agricultural land supply: Estimation and implementation in the GTAP model*. Ninth Annual Conference on Global Economic Analysis, June 2006, Addis Ababa, Ethiopia.

Takacs, D. (1996). *Philosophies of paradise*. Retrieved May 16, 2010, from http://www.dhushara.com/book/diversit/restor/takacs.htm#anchor370984

Takakai, F., Morishita, T., Hashidoko, Y., Darung, U., Kuramochi, K., & Dohong, S. (2006). Effects of agricultural land-use change and forest fire on N_2O emission from tropical peatlands, Central Kalimantan, Indonesia. *Soil Science and Plant Nutrition*, *52*, 662–674. doi:10.1111/j.1747-0765.2006.00084.x

Tallis, H., Kareiva, P., Marvier, M., & Chang, A. (2008). An ecosystem services framework to support both practical conservation and economic development. *Proceedings of the National Academy of Sciences of the United States of America, 105*, 9457–9464. doi:10.1073/pnas.0705797105

Tan, K. C., Lim, H. S., & Matjafri, M. Z. (2009). Landsat data to evaluate urban expansion and determine land use/ land cover changes in Penang Island, Malaysia. *Environmental Earth Sciences*. doi. 10.1007/s12665-009-0286-z

Tekelenburg, A., Prydatko, V., Alkemade, J. R. M., Schaub, D., Luhmann, E., & Meijer, J. R. (2003). *Assessment of wild biodiversity in agricultural land use. First design and perspectives of a pressure-based Global Biodiversity Model.* In 6th Annual Ukraine's ESRI User Conference Geoinformation Technologies in the Management of Territorial Development. (pp. 184-196). Simpheropol, Ukraine: TNU. Retrieved from http://www.ulrmc.org.ua/publication/envmanag/globio%20rivm%20ulrmc_ru.pdf

Ten Brink, B. J. E., & Tekelenburg, T. (2002). *Biodiversity: How much is left*? The Natural Capital Index framework (NCI). (RIVM report 402001014). Bilthoven.

Terborgh, J., Feeley, K., Silman, M., Nunez, P., & Balukjian, B. (2006). Vegetation dynamics of predator-free land-bridge islands. *Journal of Ecology, 94*, 253–263. doi:10.1111/j.1365-2745.2006.01106.x

Thanapakpawin, P., Richey, J., Thomas, D., Rodda, S., Campbell, B., & Logsdon, M. (2006). Effects of land-use change on the hydrologic regimes of the Mae Chaem river basin, NW Thailand. *Journal of Hydrology (Amsterdam), 334*(1-2), 215–230. doi:10.1016/j.jhydrol.2006.10.012

Thanh, S. V. (2008). *Biodiversity assessment and modeling: Review and potential application in Vietnam.* Paper presented at International Symposium on Geoinformatics for Spatial Infrastructure Development in Earth and Allied Sciences 2008. Retrieved from http://www.globio.info/assessments-with-globio/national-sub-regional-assessments

Thapa, G. B., & Weber, K. E. (1990). Actors and factors of deforestation in tropical Asia. *Environmental Conservation, 17*(1), 19–27. doi:10.1017/S0376892900017252

Thenkabail, P. S. (2004). Inter-sensor relationships between IKONOS and Landsat-7 ETM+ NDVI data in three ecoregions of Africa. *International Journal of Remote Sensing, 25*(2), 389–408. doi:10.1080/01431160310001148 42

Thenkabail, P. S., Hall, J., Lin, T., Ashton, M. S., Harris, D., & Enclona, E. A. (2003). Detecting floristic structure and pattern across topographic and moisture gradients in a mixed species Central African forest using IKONOS and Landsat-7 ETM+ images. *International Journal of Applied Earth Observation and Geoinformation, 4*, 255–270. doi:10.1016/S0303-2434(03)00006-0

Theobald, M. R., Dragosits, U., Place, C. J., Smith, J. U., Sozanska, M., & Brown, L. (2004). Modelling nitrogen fluxes at the landscape scale. *Water Air and Soil Pollution Focus, 4*(6), 135–142. doi:10.1007/s11267-004-3023-3

Theobald, D. M. (2005). *GIS concepts and ArcGIS methods* (2nd ed.). Colorado: Colorado State University.

Thomas, C. D., Cameron, A., Green, R. E., Bakkenes, M., Beaumont, L. J., & Collingham, Y. C. (2004). Extinction risk from climate change. *Nature, 427*(8), 145–148. doi:10.1038/nature02121

Thomas, J. W., & Huke, S. (1996). The forest service approach to healthy ecosystems. *Journal of Forestry, 94*(8), 14–18.

Thomas, R. H., & Huggett, R. J. (1980). *Modelling in geography: Mathematical approach.* London, UK: Harper & Row.

Thrupp, L. (1998). *Cultivating diversity: Agrobiodiversity and food security.* Washington, DC: World Resources Institute.

Thuiller, W., Araujo, M. B., & Lavorel, S. (2004). Do we need land-cover data to model species distributions in Europe? *Journal of Biogeography, 31*, 353–361. doi:10.1046/j.0305-0270.2003.00991.x

Thuiller, W., Broennimann, O., Hughes, G., Alkemade, J. R. M., Midgley, G. F., & Corsi, F. (2006). Vulnerability of African mammals to anthropogenic climate change under conservative land transformation assumptions. *Global Change Biology, 12*, 424–440. doi:10.1111/j.1365-2486.2006.01115.x

Thuiller, W., Alberta, C., Araujo, M. B., Berry, P. M., Cabeza, M., & Guisane, A. (2008). Predicting global change impacts on plant species distributions. *Perspectives in Plant Ecology, Evolution and Systematics*, *9*, 137–152. doi:10.1016/j.ppees.2007.09.004

Tilman, D., Reich, P. B., Knops, J., Wedin, D., Mielke, T., & Lehman, C. (2008). Diversity and productivity in a long-term grassland experiment. *Science*, *294*, 843–845. doi:10.1126/science.1060391

Tilman, D. (1996). Biodiversity: Population versus ecosystem stability. *Ecology*, *77*, 350–363. doi:10.2307/2265614

Tilman, D., Knops, J., Wedin, D., Reich, P., Ritchie, M., & Siemann, E. (1997). The influence of functional diversity and composition on ecosystem process. *Science*, *277*, 1300–1302. doi:10.1126/science.277.5330.1300

Tilman, D. (1997). Biodiversity and ecosystem functioning. In Daily, G. C. (Ed.), *Nature's services: Societal dependence on natural ecosystems* (pp. 93–112). Washington, DC: Island Press.

Tinker, B. (2000). *Report of the third session of the IMAGE advisory board. Report 481508014*. Bilthoven, The Netherlands: National Institute of Public Health and the Environment.

Tobler, M. W., Cochard, R., & Edwards, P. J. (2003). The impact of cattle ranching on large-scale vegetation patterns in a coastal savanna in Tanzania. *Journal of Applied Ecology*, *40*, 430–444. doi:10.1046/j.1365-2664.2003.00816.x

Tognelli, M. F., Roig-Junent, S. A., Marvald, A. E., Flores, G. E., & Lobo, J. M. (2009). An evaluation of methods for modeling distribution of Patagonian insects. *Revista Chilena de Historia Natural (Valparaiso, Chile)*, *82*, 347–360. doi:10.4067/S0716-078X2009000300003

Tomlinson, R. F. (1985). *An introduction to geographic information system*. A paper presented at the U.N. Seminar on the Role of Surveying. Mapping and Charting in Country Development Programming Aylmer, Quebec.

Townsend, P. A., & Cohoon, K. P. (1999). Sensitivity of distributional prediction algorithms to geographic data completeness. *Ecological Modelling*, *117*(1), 159–164. doi:10.1016/S0304-3800(99)00023-X

Townshend, J., Justice, C., Li, W., Gurney, C., & McManus, J. (1991). Global land cover classification by remote sensing—present capabilities and future possibilities. *Remote Sensing of Environment*, *35*(2-3), 243–255. doi:10.1016/0034-4257(91)90016-Y

Toxopeus, A. G., Skidmore, A. K., de Bie, C. K., Venus, V., Marquez, J. R. G., & Pablos, N. S. … Lymbarakis, P. (2007). BIOFRAG-herpetological biodiversity and habitat fragmentation in the Mediterranean Basin. *First Mediterranean Herpetological Congress (CMH1), 16-20 April 2007, Marrakesh-Morocco.*

Trisurat, Y. (2007). Applying gap analysis and a comparison index to assess protected areas in Thailand. *Environmental Management*, *39*, 235–245. doi:10.1007/s00267-005-0355-3

Trisurat, Y., Alkemade, R., & Arets, E. (2009). Projecting forest tree distributions and adaptation to climate change in northern Thailand. *Journal of Ecology and Natural Environment*, *1*(3), 55–63.

Trisurat, Y., Alkemade, R., & Verburg, P. (2010). Projecting land-use change and its consequences for biodiversity in Northern Thailand. *Environmental Management*, *45*, 626–639. doi:10.1007/s00267-010-9438-x

Trisurat, Y. (2004). *GIS database and its applications in ecosystem management*. Bangkok: Western Forest Complex Ecosystem Management Project, National Park, Wildlife and Plant Conservation.

Trisurat, Y. (2009). Application of geo-informatics for transboundary biodiversity conservation of the Pha Taem Protected Forest. *International Journal of Terrestrial Observation*, *1*(2), 17–29.

Trisurat, Y., Alkemade, R., & Arets, E. (2009). Projecting forest tree distributions and daptation to climate change in northern Thailand. *Journal of Ecology and Natural Environment*, *1*(3), 55–63.

Trisurat, Y., Alkeman, R., & Verburg, P. H. (2010). Projecting land-use change and its consequences for biodiversity in Northern Thailand. *Environmental Management*, *45*, 626–639. doi:10.1007/s00267-010-9438-x

Trisurat, Y., Pattanavibool, A., Gale, A. G., & Reed, D. (2010). Improving the viability of large mammal populations using landscape indices for conservation planning. *Wildlife Research, 36*, 401–412. doi:10.1071/WR09110

Trisurat, Y. (1999). *Land-use changes inside and around Srinakharin and Erawan National Parks: Fnal report submitted to the Royal Forest Department*. Bangkok, Thailand: Environ Planning, Inc.

Trisurat, Y. (2007). Applying gap analysis and a comparison index to assess protected areas in Thailand. *Environmental Management, 39*(2), 235–245. doi:10.1007/s00267-005-0355-3

Trisurat, Y., Alkemade, R., & Arets, E. (2009). Projecting forest tree distributions and adaptation to climate change in northern Thailand. *Journal of Ecology and Natural Environment, 1*, 55–63.

Trisurat, Y., Alkemade, R., & Verburg, P. H. (2010). Projecting land-use change and its consequences for biodiversity in Northern Thailand. *Environmental Management, 45*, 626–639. doi:10.1007/s00267-010-9438-x

Trisurat, Y. (2007). Applying gap analysis and a comparison index to assess protected areas in Thailand. *Environmental Management, 39*, 235–245. doi:10.1007/s00267-005-0355-3

Trisurat, Y. Bhumpakphan, N., Dechyosdee, U., Kanchanasakha, B., & Tanhikorn, S. (2010). Identifying priority areas for biodiversity conservation in northern Thailand: Land use changes and species modeling approaches. In V. Rescigno & S. Maletta (Eds.), Biodiversity hotspots. (pp. 81-103). New York, NY: Nova Science Publishers.

Trisurat, Y., Suraphabmiatree, S., & Saengnil, S. (2010). *Plant species vulnerability to climate change during 2002-2100*. Unpublished report submitted to the National Research Council of Thailand, Bangkok.

Trivedi, M. R., Morecroft, M. D., Berry, P. M., & Dawson, T. P. (2008). Potential effects of climate change on plant communities in three Montane nature reserves in Scotland, UK. *Biological Conservation, 141*(6), 1665–1675. doi:10.1016/j.biocon.2008.04.008

Turner, I. M., & Corlett, R. T. (1996). The conservation value of small, isolated fragments of lowland tropical rain forest. *Trends in Ecology & Evolution, 11*, 330–333. doi:10.1016/0169-5347(96)10046-X

Turner, M. G. (2004). Landscape ecology: The effect of pattern of process. *Annual Review of Ecology and Systematics, 20*, 171–197. doi:10.1146/annurev.es.20.110189.001131

Turner, D. P., Cohen, W. B., Kennedy, R. E., Fassnacht, K. S., & Briggs, J. M. (1999). Relationships between leaf area index and Landsat TM spectral vegetation indices across three temperate zone sites. *Remote Sensing of Environment, 70*(1), 52–68. doi:10.1016/S0034-4257(99)00057-7

Turner, M. G. (1989). Landscape ecology: The effect of pattern on process. *Annual Review of Ecology and Systematics, 20*(1), 171–197. doi:10.1146/annurev.es.20.110189.001131

Turner, W. R., Brandon, K., Brooks, T. M., Costanza, R., Da Fonseca, G. A. B., & Portela, R. (2007). Global conservation of biodiversity and ecosystem services. *Bioscience, 57*, 868–873. doi:10.1641/B571009

Turner, I. M., & Corlett, R. T. (1996). The conservation value of small isolated fragments of lowland tropical rain forest. *Trends in Ecology & Evolution, 11*, 330–333. doi:10.1016/0169-5347(96)10046-X

Turpin, N., Dupraz, P., Thenail, C., Joannon, A., Baudry, J., Herviou, S., & Verburg, P. (2009). Shaping the landscape: Agricultural policies and local biodiversity schemes. *Land Use Policy, 26*, 273–283..doi:10.1016/j.landusepol.2008.03.004

U.S. Department of Interior. (1980). *Habitat as a basis for environmental assessment. ESM 101*. Washington, DC: Fish and Wildlife Service, Division of Ecological Service.

Uhl, C., & Viera, I. C. G. (1989). Ecological impacts of selective logging in the Brazilian Amazon: A case study from the Paragominas region of the state of Para. *Biotropica, 21*, 98–106. doi:10.2307/2388700

ULRMC, Ukrainian Land and Resource Management Center. (2010a). *EEBIO searchable service: Maps, species based models, habitats, pressures*. Retrieved April 6, 2010, from http://www.ulrmc.org.ua/services/eebio/is/index.asp?lang=EN

ULRMC, Ukrainian Land and Resource Management Center. (2010b). *BINU searchable list of agrobiodiversity indicators*. Retrieved April 6, 2010, from www.ulrmc.org.ua/services/binu/is/index.asp?lang=EN

UN. (2000). *Millennium declaration*. New York, NY: United Nations.

UN ESCAP. (2010). *Statistical yearbook for Asia and the Pacific 2009*. Retrieved June 8, 2010, from http://www.unescap.org/stat/data/syb2009/

UNDP. (2009). *Human development report (2009)*. New York, NY: Palgrave Macmillan/UN Development Programme.

UNEP. (2002). *GEO-3. Past, present and future perspectives*. London, UK: Earthscan.

UNEP. (2006). *Global deserts outlook. Devision of early Warning and Assessment*. Nairobi, Kenya: United Nations Environmental Programme.

UNEP. (2007). *Global outlook for ice and snow*. Nairobi, Kenya: United Nations Environment Programme.

UNEP. (2007). *Global environmental outlook 4*. Nairobi, Kenya: Environment for Development. United Nations Environmental Programme.

UNEP. (2009). *Integrated assessment: Mainstreaming sustainability into policymaking. A guidance manual*. Nairobi, Kenya: United Nations Environmental Programme.

UNEP. (2001). *GLOBIO. Global methodology for mapping human impacts on the biosphere*. In: Report UNEP/DEWA/TR 25. United Nations Environmental Programme, Nairobi.

UNEP. (2004). *Decisions adopted by the Conference of the Parties to the Convention on Biological Diversity at its seventh meeting* (UNEP/CBD/COP/7/21/Part 2), Decision VII/30 (CBD 2004).Retrieved from http://www.biodiv.org/decisions/

UNEP/CBD. (2006). *Decision adopted by the conference of the parties to the convention on biological diversity at its eighth meeting VIII/28. Impact assessment: Voluntary guidelines on biodiversity-inclusive impact assessment*. Curitiba, Brazil, 20-31 March 2006.

UNEP/RIVM. (2004). *The GEO-3 Scenarios 2002-2032: Quantification and analysis of environmental impacts*. In: Report UNEP/DEWA/RS03-04; RIVM 402001022. Division of Early Warning and Assessment. DEWA-UNEP. / National Institute for Public Health and the Environment. RIVM., Nairobi, Kenya; Bilthoven, The Netherlands.

UNFCCC. (2009). *Revised negotiating text. Ad hoc Working Group on long-term cooperative action under the convention. FCCC/AWGLCA/ 2009/INF.1*. Bonn: Framework Convention on Climate Change.

United Nation Environmental Program. (2001). UNEP/DEWA/TR.01-3. In C. Nellemann, L. Kullerud, I. Vistnes, B., C. Forbes, E. Husby, G. P. Kofinas, B. P. Kaltenborn, et al (Eds.), *GLOBIO. Global methodology for mapping human impacts on the biosphere.*

United Nations. (2002). (*UN*). New York: Report of the World Summit on Sustainable Development.

United Nations Environmental Programme. (1991). *Fourth revised draft convention on biodiversity*. Nairobi, Kenya.

United Nations Environmental Programme. (2001). *GLOBIO: Global methodology for mapping human impacts on the biosphere*. Report UNEP/DEWA/TR 25. United Nations Environmental Programme, Nairobi.

United Nations. (1993). *Multilateral Convention on biological diversity*. United Nation Treaty Series. 1760, I 31619.

United Nations. (2010) *Goal 7: Ensure environmental sustainability*. Retrieved from http://www.un.org/millenniumgoals/environ.shtml

United Nations. (UN). (2005). *The millennium development goals report 2005*. New York.

University of Tokyo. (2005) *Rainfall data*. GAME-T2 Data Center. Retrieved September 16, 2005, from http://hydro.iis.u-tokyo.ac.jp/GAME-T/GAIN-T/updates.html

UNODC. (2008). Colombia Coca Cultivation Survey. United Nations Office on Drugs and Crime (UNODC) and Government of Colombia. Retrieved from http://www.unodc.org/unodc/en/crop-monitoring/index.html

Urban, D. L., O'Neill, R. V., & Shugart, H. H. (1987). Landscape ecology. *Bioscience*, *37*, 119–127. doi:10.2307/1310366

Urrutia, R., & Vuille, M. (2009). Climate change projections for the tropical Andes using a regional climate model: Temperature and precipitation simulations for the end of the 21st century. *Journal of Geophysical Research*, *14*, D02108. doi:10.1029/2008JD011021

USGS. (2004). *Shuttle radar topography mission.* United States Geological Survey. Retrieved October 20, 2005, from ftp://ftp.glcf.umiacs.umd.edu/glcf/srtm

Usher, A. D. (2009). *Thai forestry. A critical history*. Chiang Mai, Thailand: Silkworm Books.

Ustin, S., Gitelson, A. A., Jacquemoud, S., Schaepman, M. E., Asner, G., Gamon, J. A., & Zarco-Tejada, P. (2009). Retrieval of foliar information about plant pigment systems from high resolution spectroscopy. *Remote Sensing of Environment*, *113*(1), S67–S77. doi:10.1016/j.rse.2008.10.019

Vallan, D., Andreone, F., Raherisoa, V. H., & Dolch, R. (2004). Does selective wood exploitation affect amphibian diversity? The case of An'Ala, a tropical rainforest in eastern Madagascar. *Oryx*, *38*, 410–417. doi:10.1017/S003060530400078X

Vallega, A. (2002). The regional approach to the ocean, the ocean regions, and ocean regionalization – A post-modern dilemma. *Ocean and Coastal Management*, *45*, 753. doi:10.1016/S0964-5691(02)00104-7

Van Aardenne, J. A., Dentener, F. J., Olivier, J. G. J., Klein Goldewijk, C. G. M., & Lelieveld, J. (2001). A 1 x 1 degree resolution dataset of historical anthropogenic tracé gas emissions for the period 1890-1990. *Global Biogeochemical Cycles*, *15*, 909–928. doi:10.1029/2000GB001265

van Andel, T. H., Zangger, E., & Demitrack, A. (1990). Land use and soil erosion in prehistoric and historical Greece. *Journal of Field Archaeology*, *17*(4), 379–396. doi:10.2307/530002

Van Dam, R., Gitay, H., & Finlayson, M. (2002). *Climate change and wetlands: Impact and mitigation.* Ramsar Draft, COP8 paper.

Van der Heijden, R. B. J. (2007). *Characterization of European landscapes and analysis of their dynamics. Internal CGI Report.* Wageningen, the Netherlands: Alterra.

Van der Sluijs, J. P., Craye, M., Funtowicz, S., Kloprogge, P., Ravetz, J., & Risbey, J. (2005). Combining quantitative and qualitative measures of uncertainty in model based environmental assessment: The NUSAP system. *Risk Analysis*, *25*(2), 481–492. doi:10.1111/j.1539-6924.2005.00604.x

van der Werf, G. R., Morton, D. C., DeFries, R. S., Olivier, J. G. J., Kasibhatla, P. S., & Jackson, R. B. (2009). CO_2 emissions from forest loss. *Nature Geoscience*, *2*, 737–738. doi:10.1038/ngeo671

Van Drecht, G., & Bouwman, A. F. (2009). Global nitrogen and phosphate in urban wastewater for the period 1970 to 2050. *Global Biogeochemical Cycles*, *23*(3).

Van Duuren, L., Eggink, G. J., Kalkhovan, J., Notenboom, J., van Strien, A. J., & Wortelboer, R. (Eds.). (2003). *Natuurcompendium 2003.* Natuur in cijfers. CBS (Voorburg), MNP Bilthoven en Wageningen. Retrieved from November 20, 2009, from http://www.pbl.nl/nl/publicaties/mnp/2003/Natuurcompendium_2003.html, ISBN 906960101X

van Houtan, K. S., Pimm, S. L., Halley, J. M., Bierregaard, R. O., & Lovejoy, T. E. (2007). Dispersal of Amazonian birds in continuous and fragmented forest. *Ecology Letters*, *10*(3), 219–229. doi:10.1111/j.1461-0248.2007.01004.x

van Mantgem, P. J., & Stephenson, N. L. (2007). Apparent climatically induced increase of tree mortality rates in a temperate forest. *Ecology Letters*, *10*(10), 909–916. doi:10.1111/j.1461-0248.2007.01080.x

Van Opstal, A. (1999). *The architecture of the Pan-European Ecological Network: Suggestions for concept and criteria.* Discussion report on behalf of the committee of experts of the Pan European Ecological Network. Report IKC Natuurbeheer Nr.37, Wageningen, The Netherlands.

van Rooij, W. (2006). *Manual: Using GLOBIO-3 for determining the current level of biodiversity on a national/regional scale for the course*. Enschede: Regional and National Biodiversity Modelling and Analysis. ITC.

van Schaik, C. P., Monk, K. A., & Robertson, J. M. Y. (2001). Dramatic decline in orangutan numbers in the Leuser Ecosystem, Northern Sumatra. *Oryx*, *35*, 14–25.

Van Steenis, C. G. G. J. (1950). The delimitation of Malaysia and its main plant geographical divisions. In C. G. G. J. van Steenis (Ed.), *Flora Malesiana* Ser. 1, 1. (pp. lxx—lxxv). Noordhoff-Kolff n.v., Djakarta.

van Vuuren, D., Sala, O., & Pereira, H. M. (2006). The future of vascular plant diversity under four global scenarios. *Ecology and Society*, *11*, 25.

Van Vuuren, D. P., Den Elzen, M., Lucas, P., Eickhout, B., Strengers, B., & Van Ruijven, B. (2007). Stabilizing greenhouse gas concentrations at low levels: An assessment of reduction strategies and costs. *Climatic Change*, *81*, 119–159. doi:10.1007/s10584-006-9172-9

Vancutsem, C., Pekel, J. F., Evrard, C., Malaisse, F., & Defourny, P. (2009). Mapping and characterizing the vegetation types of the Democratic Republic of Congo using SPOT vegetation time series. *International Journal of Applied Earth Observation and Geoinformation*, *11*(1), 62–76. doi:10.1016/j.jag.2008.08.001

Veldkamp, A., & Lambin, E. F. (2001). Predicting land-use change. *Agriculture Ecosystems & Environment*, *85*, 1–6. doi:10.1016/S0167-8809(01)00199-2

Veldkamp, A., Kok, K., de Koning, G. H. J., Verburg, P. H., Priess, J., & Bergsma, A. R. (2001). The need for multi-scale approaches in spatial specific land use change modelling. *Environmental Modeling and Assessment*, *6*, 111–121. doi:10.1023/A:1011572301150

Veldkamp, A., & Lambin, E. F. (2001). Editorial: Predicting land-use change. *Agriculture Ecosystems & Environment*, *85*, 1–6. doi:10.1016/S0167-8809(01)00199-2

Veliz, C., Tovar, A., Tovar, C., Regal, F., & Vásquez, P. (2008). ¿Qué áreas conservar en nuestras Zonas Áridas? Seleccionando sitios prioritarios para la conservación en la Ecorregión Desierto de Sechura - Perú. *Zonas Áridas*, *12*(1), 36–59.

Verboom, J., Alkemade, R., Klijn, J., Metzger, M. J., & Reijnen, R. (2007). Combining biodiversity modelling with political and economic development scenarios for 25 EU countries. *Ecological Economics*, *62*, 267–276. doi:10.1016/j.ecolecon.2006.04.009

Verboom, J., Alkemade, R., Klijn, J., Metzger, M. J., & Reijnen, R. (2007). Combining biodiversity modeling with political and economic development scenarios for 25 EU countries. *Ecological Economics*, *62*, 267–276.. doi:10.1016/j.ecolecon.2006.04.009

Verburg, P., Eickhout, B., & van Meijl, H. (2008). A multi-scale, multi-model approach for analyzing the future dynamics of European land use. *The Annals of Regional Science*, *42*, 57–77. doi:10.1007/s00168-007-0136-4

Verburg, P. H., & Veldkamp, A. (2004). Projecting land use transitions at forest fringes in the Philippines at two spatial scales. *Landscape Ecology*, *19*, 77–98. doi:10.1023/B:LAND.0000018370.57457.58

Verburg, P., & Overmars, K. (2009). Combining top-down and bottom-up dynamics in land use modelling: Exploring the future of abandoned farmlands in Europe with the Dyna-CLUE model. *Landscape Ecology*, *24*, 1167–1181. doi:10.1007/s10980-009-9355-7

Verburg, P. H., Overmars, K. P., Huigen, M. G. A., de Groot, W. T., & Veldkamp, A. (2006). Analysis of the effects of land use change on protected areas in the Philippines. *Applied Geography (Sevenoaks, England)*, *26*, 153–173. doi:10.1016/j.apgeog.2005.11.005

Verburg, P. H., Soepboer, W., Limpiada, R., Espaldon, M. V. O., Sharifa, M., & Veldkamp, A. (2002). Land use change modelling at the regional scale: The CLUE-S model. *Environmental Management*, *30*, 391–405. doi:10.1007/s00267-002-2630-x

Verburg, P. H., van de Steeg, J., Veldkamp, A., & Willemen, L. (2009). From land cover change to land function dynamics: A major challenge to improve land characterization. *Journal of Environmental Management*, *90*, 1327–1335. doi:10.1016/j.jenvman.2008.08.005

Verburg, P. H., & Veldkamp, A. (2004). Projecting land use transitions at forest fringes in the Philippines at two spatial scales. *Landscape Ecology*, *19*, 77–98. doi:10.1023/B:LAND.0000018370.57457.58

Verburg, P. H., Veldkamp, A., de Koning, G. H. J., Kok, K., & Bouma, J. (1999). A spatial explicit allocation procedure for modelling the pattern of land use change based upon actual land use. *Ecological Modelling*, *116*, 45–61. doi:10.1016/S0304-3800(98)00156-2

Verburg, P. H., Soepboer, W., Limpiada, R., Espaldon, M. V. O., Sharifa, M., & Veldkamp, A. (2002). Land use change modelling at the regional scale: The CLUE-S model. *Environmental Management*, *30*, 391–405. doi:10.1007/s00267-002-2630-x

Verburg, P., Eickhout, B., & van Meijl, H. (2008). A multi-scale, multi-model approach for analyzing the future dynamics of European land use. *The Annals of Regional Science*, *42*, 57–77. doi:10.1007/s00168-007-0136-4

Verburg, P. H., Schot, P., Dijst, M. J., & Veldkamp, A. (2004). Land use change modelling: Current practice and research priorities. *GeoJournal*, *61*(4), 309–324. doi:10.1007/s10708-004-4946-y

Verburg, P. H., Soepboer, W., Limpiada, R., Espaldon, M. V. O., Sharifa, M., & Veldkamp, A. (2002). Land-use change modelling at the regional scale: The CLUE-S model. *Environmental Management*, *30*(3), 391–405. doi:10.1007/s00267-002-2630-x

Verburg, P. H., & Veldkamp, A. (2004). Projecting land use transitions at forest fringes in the Philippines at two spatial scales. *Landscape Ecology*, *19*(1), 77–98. doi:10.1023/B:LAND.0000018370.57457.58

Verburg, P., Soepboer, W., Veldkamp, A., Limpiada, R., Espaldon, V., & Mastura, S. S. A. (2002). Modeling the spatial dynamics of regional land use: the CLUE-S model. *Environmental Management*, *30*(3), 391–405. doi:10.1007/s00267-002-2630-x

Verburg, P., & Veldkamp, A. (2004). Projecting land use transitions at forest fringes in the Philippines at two spatial scales. *Landscape Ecology*, *19*, 77–98. doi:10.1023/B:LAND.0000018370.57457.58

Verburg, P. H. (2006). Simulating feedbacks in land use and land cover change models. *Landscape Ecology*, *21*, 1171–1183..doi:10.1007/s10980-006-0029-4

Verburg, P. H., Schulp, C. J. E., Witte, N., & Veldkamp, A. (2006). Downscaling of land use change scenarios to assess the dynamics of European landscapes. *Agriculture Ecosystems & Environment*, *114*, 39–56..doi:10.1016/j.agee.2005.11.024

Verburg, P. H., Soepboer, W., Veldkamp, A., Limpiada, R., Espaldon, V., & Mastura, S. S. A. (2002). Modelling the spatial dynamics of regional land use: The CLUE-S model. *Environmental Management*, *30*(3), 391–405. doi:10.1007/s00267-002-2630-x

Verburg, P. H., Veldkamp, A., & Bouma, J. (1999). Land use under condition of high population pressure: The case of Java. *Global Environmental Change*, *9*, 303–312. doi:10.1016/S0959-3780(99)00175-2

Verburg, P., & Overmars, K. (2009). Combining top-down and bottom-up dynamics in land use modeling: Exploring the future of abandoned farmlands in Europe with the Dyna-CLUE model. *Landscape Ecology*, *24*, 1167–1181. doi:10.1007/s10980-009-9355-7

Verburg, P., & Overmars, K. (2009). *Combining top-down and bottom-up dynamics in land use modeling: Exploring the future of abandoned farmlands in Europe with the Dyna-CLUE model*. Landsc Ecol. Retrieved August 15, 2009, from http://dx.doi.org/10.1007/s10980-009-9355-7

Verma, R. K., & Totey, N. G. (1999). Biological diversity, medicinal potential of ground flora and improvement in soil quality under plantations raised on degraded bhata land. *Advances in Forestry Research in India*, *20*, 37–69.

Vié, J.-C., Hilton-Taylor, C., Pollock, C., Ragle, J., Smart, J., Stuart, S. N., & Tong, R. (2008). The IUCN red list: a key conservation tool. In Vié, J.-C., Hilton-Taylor, C., & Stuart, S. N. (Eds.), *The 2008 review of the IUCN red lst of threatened species*. International Union for the Conservation of Nature and Natural Resources, Gland.

Vitousek, P. M., Mooney, H. A., Lubchenco, J., & Melillo, J. M. (1997). Human domination of Earth's ecosystems. *Science*, *277*, 494–499. doi:10.1126/science.277.5325.494

Vogelmann, J. E., Howard, S. M., Yang, L., Larson, C. R., Wylie, C. R., & Van Driel, N. (2001). Completion of the 1990s national land cover data set for the Conterminous United States from Landsat Thematic map-per data and ancillary data sources. *Photogrammetric Engineering and Remote Sensing*, *67*, 6650–6662.

Voinov, A., Costanza, R., Wainger, L., Boumans, R., Villa, F., Maxwell, T., & Voinov, H. (1999). Patuxent landscape model: Integrated ecological economic modelling of a watershed. *Environmental Modelling & Software*, *14*, 473–491. doi:10.1016/S1364-8152(98)00092-9

Volney, W. J. A., & Fleming, R. A. (2000). Climate change impacts of boreal forest insects. *Agriculture Ecosystems & Environment*, *82*, 283–294. doi:10.1016/S0167-8809(00)00232-2

Vorosmarty, C. J., Green, P., Salisbury, J., & Lammers, R. B. (2000). Global water resources: Vulnerability from climate change and population growth. *Science*, *289*(5477), 284–288. doi:10.1126/science.289.5477.284

Vors, L. S., & Boyce, M. S. (2009). Global declines of caribou and reindeer. *Global Change Biology*, *15*, 2626–2633. doi:10.1111/j.1365-2486.2009.01974.x

Vos, W., & Stortelder, A. H. F. (1992). *Vanishing Tuscan landscapes: Landscape ecology of a Submediterranean-Montane area (Solano Basin, Tuscany, Italy)*. Wageningen, The Netherlands: Pudoc Scientific Publishers.

Wackernagel, M., Schulz, N. B., Deumling, D., Callejas Linares, A., Jenkins, M., Kapos, V., et al. Randers, J. (2002). Tracking the ecological overshoot of the human economy. *Proceedings of the National Academy of Sciences, 99*, 9266-9271.

Wadsworth, R. A., & Treweek, J. (1999). *Geographical information systems for ecology: An introduction*. Harlow: Addison Wesley Longman.

Walker, B. (1995). Conserving biological diversity through ecosystem resilience. *Conservation Biology*, *9*(4), 747–752. doi:10.1046/j.1523-1739.1995.09040747.x

Walsh, S., Messina, J., Crews-Meyer, K., Bilsborrow, R., & Pan, W. (2002). Characterizing and modeling patterns of deforestation and agricultural extensification in the Ecuadorian Amazon. In Crews-Meyer, K., & Walsh, S. (Eds.), *Linking people, place and policy: a GIS science approach* (pp. 187–214). Boston: Kluwer Academic Publishers.

Walther, G. (2003). Plants in a warmer world. *Perspectives in Plant Ecology, Evolution and Systematics*, *6*(3), 169–185. doi:10.1078/1433-8319-00076

Walther, G., Post, E., Convey, P., Menzel, A., Parmesan, C., & Beebee, T. J. C. (2002). Ecological responses to recent climate change. *Nature*, *416*, 389–395. doi:10.1038/416389a

Wang, Y., Mitchell, B. R., Nugranad-Marzilli, J., Bonynge, G., Zhou, Y., & Shriver, G. (2009). Remote sensing of land-cover change and landscape context of the National Parks: A case study of the Northeast Temperate Network. *Remote Sensing of Environment*..doi:10.1016/j.rse.2008.09.017

Wassenaar, T., Gerber, P., Verburg, P. H., Rosales, M., Ibrahim, M., & Steinfeld, H. (2007). Projecting land use changes in the Neotropics: The geography of pasture expansion into forest. *Global Environmental Change*, *17*, 86–104. doi:10.1016/j.gloenvcha.2006.03.007

Wataru, K. (2001). *Aqua/Terra MODIS bi-weekly composite database (Asia)*. Institute of Industrial Science, University of Tokyo. Retrieved October 20, 2005, from http://webmodis.iis.u-tokyo.ac.jp/Asia/

Watcharakitti, S. (1976). *Tropical forest land-use evolution in northern Thailand. Forest Research Bulletin, 44*. Bangkok, Thailand: Faculty of Forestry, Kasetsart University.

Watson, R. T., Noble, I. R., Bolin, B., Ravindrath, N. H., Verardo, D. J., & Dokken, D. J. (Eds.). (2000). *Land use, land-use change and forestry*. Cambridge, UK: Cambridge University Press.

Wear, D. N., Liu, R., Foreman, J. M., & Sheffield, R. M. (1999). The effects of population growth on timber management and inventories in Virginia. *Forest Ecology and Management*, *118*, 107–115. doi:10.1016/S0378-1127(98)00491-5

Wear, D. N., Apt, R., & Mangold, R. (1998). People, space, time: Factors that will govern forest sustainability. *Proceedings of the 63rd North American Wildlife and Natural Resources Conference* (pp. 348–361). Wildlife Management Institute, Washington, D.C.

Webster, P. J., Holland, G. J., Curry, J. A., & Chang, H.-R. (2005). Changes in tropical cyclone number, duration, and intensity in a warming environment. *Science*, *309*(5742), 1844–1846. doi:10.1126/science.1116448

WEFCOM (Western Forest Complex for Ecosystem Management Project). (2004). *GIS database and its application for ecosystem management. Final Technical Report*. Bangkok: Royal Forest Department.

Weijters, M. J., Janse, J. H., Alkemade, R., & Verhoeven, J. T. A. (2009). Quantifying the effect of catchment land-use and water nutrient concentrations on freshwater river and stream biodiversity. *Aquatic Conservation: Marine and Freshwater Ecosystems*, *19*, 104–112. doi:10.1002/aqc.989

Welton, L. J., Siler, C. D., Bennet, D., Diesmos, A., Duya, M. R., & Dugay, R.... Brown, R. M. (2010). A spectacular new Philippine monitor lizard reveals a hidden biogeographic boundary and a novel flagship species for conservation. *Biology Letters*. doi:.doi:10.1098/rsbl.2010.0119

Westhoek, H. J., van den Berg, M., & Bakkes, J. A. (2006). Scenario development to explore the future of Europe's rural areas. *Agriculture Ecosystems & Environment*, *114*, 7–20. doi:10.1016/j.agee.2005.11.005

White, M. A., de Beurs, K. M., Didan, K., Inouye, D. W., Richardson, A. D., & Jensen, O. P. (2009). Intercomparison, interpretation, and assessment of spring phenology in North America estimated from remote sensing for 1982-2006. *Global Change Biology*, *15*(10), 2335–2359. doi:10.1111/j.1365-2486.2009.01910.x

White, M. A., & Nemani, R. R. (2006). Real-time monitoring and short-term forecasting of land surface phenology. *Remote Sensing of Environment*, *104*(1), 43–49. doi:10.1016/j.rse.2006.04.014

Whitemore, T. C., & Sayer, J. A. (1992). *Tropical deforestation and species extinction*. IUCN, Gland.

Whitmore, T. C. (1995). Comparing South East Asian and other tropical rainforests. In Primack, R. E., & Lovejoy, T. E. (Eds.), *Ecology, conservation and management of South East Asian rainforests* (pp. 5–15). New Haven, CT: Yale University Press.

Whittaker, R. H. (1975). *Communities and ecosystems*. New York, NY: Macmillan.

Whitten, T., Damanik, S. J., Anwar, J., & Hisyam, N. (1997). *The ecology of Sumatra*. Jakarta, Indonesia: Periplus.

Wich, S. A., Singleton, I., Utami-Atmoko, S. S., Geurts, M. L., Rijksen, H. D., & van Schaik, C. P. (2003). The status of the Sumatran orangutan *Pongo abelii*: An update. *Oryx*, *37*(1), 49–54. doi:10.1017/S0030605303000115

Wikipedia (n.d.). *Decline in amphibian populations*. Retrieved July 11, 2010, from http://en.wikipedia.org/wiki/Decline_in_amphibian_populations

Wikramanayake, E. D., Dinerstein, E., Loucks, C. J., Olson, D. M., Morrison, J., & Lamoreux, J. ... Hedao, P. (2002). *Terrestrial ecoregions of the Indo-Pacific: A conservation assessment*. Washington, DC: Covelo and London: Island Press

Wilby, R. L., Dawson, R., & Barrow, E. M. (2002). SDSM: A decision support tool for the assessment of regional climate change assessments. *Environmental Modelling & Software*, *17*, 145–157. doi:10.1016/S1364-8152(01)00060-3

Wilby, R. L., & Harris, I. (2006). *A framework for assessing uncertainties in climate change impacts: Low-flow scenarios for the River Thames, UK*. doi: 10.1029/2005WR004065

Wildlife Conservation Division and Forestry Research Center. (1997). *Application of remote sensing and GIS for monitoring forest land use in Huai Kha Khaeng Wildlife Sanctuary: Final report*. Bangkok, Thailand: Office of Natural Resources Conservation, Royal Forest Department/Faculty of Forestry, Kasetsart University.

Wildlife Conservation Division and Forestry Research Center. (1999). *Application of remote sensing and GIS for monitoring forest land use in Khao-Ang-Runai Wildlife Sanctuary: Final report*. Bangkok, Thailand: Office of Natural Resources Conservation, Royal Forest Department/Faculty of Forestry, Kasetsart University.

Wildlife Conservation Division and Khon Kaen University. (2000). *Application of remote sensing and GIS for monitoring forest land use in Phu Luan Wildlife Sanctuary: Final report*. Bangkok, Thailand: Office of Natural Resources Conservation, Royal Forest Department/Faculty of Forestry, Kasetsart University.

Williams, W. D. (1998). Dryland wetlands. In McComb, A. J., & Davis, J. A. (Eds.), *Wetlands for the future*. Glen Osmond, Australia: Gleneagles Publishing.

Wilson, E. O. (Ed.). (1988). *Biodiversity*. Washington: National Academy Press.

Wilson, A. G., Coelho, J. D., MacGill, S. M., & Williams, H. C. W. L. (1981). *Optimization in location and transport analysis*. Chichester, West Sussex, UK: Wiley.

Wischmeier, W. H., & Smith, D. D. (1978). *Predicting rainfall erosion losses: A guide to conservation planning*. U.S. Department of Agriculture, Agricultural handbook No 537.

Wisz, M. S., Hijmans, R. J., Li, J., Peterson, A. T., Graham, C. H., & Guisan, A. (2008). Effects of sample size on the performance of species distribution models. *Diversity & Distributions, 14*, 763–773. doi:10.1111/j.1472-4642.2008.00482.x

Wisz, M. S., & Guisan, A. (2009). Do pseudo-absence selection strategies influence species distribution models and their predictions? An information-theoretic approach based on simulated data. *BMZ Ecology, 9*, 8. Retrieved February 20, 2009 from http://www.biomedcentral.com/1472-6785/9/8

Wolter, P. T., & White, M. A. (2002). Recent forest cover type transitions and landscape structural changes in northeast Minnesota. *Landscape Ecology, 17*, 133–155. doi:10.1023/A:1016522509857

Wood, E. C., Lewis, J. E., Gray Tappan, G., & Lietzow, R. W. (1997). *The development of a land cover change model for southern Senegal*. Paper presented at the Land Use Modeling Workshop, June 1997, Sponsored by the USGS and NCGIA. Retrieved June 5, 2010, from http://ncgia.ncgia.ucsb.edu/conf/landuse97/papers/wood eric/pecdoc.html

Woodruff, D. S. (2003). Neogene marine transgressions, palaeogeography and biogeographic transitions on the Thai-Malay Peninsula. *Journal of Biogeography, 30*, 551–567. doi:10.1046/j.1365-2699.2003.00846.x

World Conservation Union. (2010). *About biodiversity. The biodiversity challenge.* Retrieved from http://www.iucn.org/what/tpas/biodiversity/about/

World Wildlife Fund. (1998). *Living planet report 1998*. Gland, Switzerland: World Wildlife Fund.

Wösten, J. H. M., Ismail, A. B., & van Wijk, A. L. M. (1997). Peat subsidence and its practical implications: A case study in Malaysia. *Geoderma, 78*, 25–36. doi:10.1016/S0016-7061(97)00013-X

Wrbka, T., Erb, K. H., Schulz, N. B., Peterseil, J., Hahn, C., & Haberl, H. (2004). Linking pattern and process in cultural landscapes. An empirical study based on spatially explicit indicators. *Land Use Policy, 21*(3), 289–306. doi:10.1016/j.landusepol.2003.10.012

WRI. (2002). *Drylands, people, and ecosystem goods and services: A Web-based geospatial analysis*. World Resource Institute. Retrieved June 11, 2006, from http://www.wri.org

WRI. IUCN, & UNEP. (1992). *Global biodiversity strategy: A policy-makers' guide*. A report produced in consultation with FAO and UNESCO. Washington, DC.

Wright, J. A., DiNicola, A., & Gaitan, E. (2000). Latin American forest plantations: Opportunities for carbon sequestration, economic development, and financial returns. *Journal of Forestry, 98*(9), 20–23.

Wright, S. J. (2002). Plant diversity in tropical forests: A review of mechanisms of species coexistence. *Oecologia, 130*, 1–14.

Wright, S. J., Muller-Landau, H. C., & Schipper, J. (2009). The future of tropical species on a warmer planet. *Conservation Biology, 23*(6), 1418–1426. doi:10.1111/j.1523-1739.2009.01337.x

Wu, J., & David, J. L. (2002). A spatially explicit hierarchical approach to modeling complex ecological systems: Theory and applications. *Ecological Modelling, 153*(1-2), 7–26. doi:10.1016/S0304-3800(01)00499-9

Wu, J., & Marceau, D. (2002). Modeling complex ecological systems: An introduction. *Ecological Modelling, 153*(1-2), 1–6. doi:10.1016/S0304-3800(01)00498-7

Xian, G., & Crane, M. (2005). Assessments of urban growth in the Tampa Bay watershed using remote sensing data. *Remote Sensing of Environment, 97*, 203–215. doi:10.1016/j.rse.2005.04.017

Xian, G., Homer, C., & Fry, J. (2009). Updating the 2001 National Land Cover Database land cover classification to 2006 by using Landsat imagery change detection methods. *Remote Sensing of Environment, 113*(6), 1133–1147. doi:10.1016/j.rse.2009.02.004

Xiubin, H., Tang, K., & Zhang, X. (2004). Soil erosion dynamics on the Chinese Loess Plateau in the last 10'000 years. *Mountain Research and Development*, *24*(4), 342–347. doi:10.1659/0276-4741(2004)024[0342:SEDOTC]2.0.CO;2

Yahner, R. H. (1988). Changes in wildlife communities near edges. *Conservation Biology*, *2*(4), 333–339.. doi:10.1111/j.1523-1739.1988.tb00197.x

Yahner, R. H., & Mahan, C. G. (1997). Effects of logging roads on depredation of artificial ground nests in a forested landscape. *Wildlife Society Bulletin*, *25*(1), 158–162.

Yang, L., Homer, C. G., Hegge, K., Huang, C., Wylie, B., & Reed, B. (2001). A Landsat 7 scene selection strategy for a National Land Cover Database. In *Proceedings IEEE 2001 International Geoscience and Remote Sensing Symposium* Sydney, Australia.

Yatsyk, A. V., & Shenchuk, V. Y. (2006). *Encyclopedia on water industry, nature management, nature procreation, sustainable development*. Kyiv, Ukraine: Geneza.

Yemefack, M., Bijker, W., & Jong, S. M. (2006). Investigating relationships between Landsat-7 ETM+ data and spatial segregation of LU/LC types under shifting agriculture in southern Cameroon. *International Journal of Applied Earth Observation and Geoinformation*, *8*(2), 2234–2256. doi:10.1016/j.jag.2005.08.003

Yost, A. C., Petersen, S. L., Gregg, M., & Miller, R. (2008). Predictive modeling and mapping sage grouse (Centrocercus urophasianus) nesting habitat using maximum entropy and a long-term dataset from Southern Oregon. *Ecological Informatics*, *3*, 375–386. doi:10.1016/j.ecoinf.2008.08.004

Young, J., Watt, A., Nowicki, P., Alard, D., Clitherow, J., & Henle, K. (2005). Towards sustainable land use: Identifying and managing the conflicts between human activities and biodiversity conservation in Europe. *Biodiversity and Conservation*, *14*, 1641–1661..doi:10.1007/s10531-004-0536-z

Yu, X., Gillespie, A., Burns, W., & Taplin, R. (Eds.). (2006). *Climate change in the South Pacific: Impacts and responses in Australia, New Zealand, and small island states*. Dordrecht, Netherlands: Springer.

Zachos, J., Pagani, M., Sloan, L., Thomas, E., & Billups, K. (2001). Trends, rhythms, and aberrations in global climate 65 Ma to present. *Science*, *292*(5517), 686–693. doi:10.1126/science.1059412

Zanette, L. (2000). Fragment size and the demography of an area-sensitive songbird. *Journal of Animal Ecology*, *69*, 458–470. doi:10.1046/j.1365-2656.2000.00408.x

Zarin, D. J., Huijun, G., & Enu-kwesi, L. (2002). Guidelines on the assessment of plant species diversity in agricultural landscapes. In Brookfield, H., Padoch, C., Parsons, H., & Stocking, M. (Eds.), *Cultivating biodiversity: Understanding, analyzing and using agricultural diversity*. London, UK: ITDG Publishing.

Zhan, X., DeFries, R. S., Townshend, J. R. G., Dimiceli, C., Hansen, M., Huang, C., & Sohlberg, R. (2000). The 250m global land cover change product from the moderate resolution imaging spectroradiometer of NASA's Earth Observing System. *International Journal of Remote Sensing*, *21*(6-7), 1433–1460. doi:10.1080/014311600210254

Zhang, X., Drake, N. A., Wainwright, J., & Mulligan, N. (1999). Comparison of slope estimates from low resolution DEMs: Scaling issues and fractal method for their solution. *Earth Surface Processes and Landforms*, *24*, 763–779. doi:10.1002/(SICI)1096-9837(199908)24:9<763::AID-ESP9>3.0.CO;2-J

Zhang, Y., Uusivuori, J., & Kuuluvainen, J. (2000). Econometric analysis of the causes of forestland use/cover change in Hainan, China. *Canadian Journal of Forest Research*, *30*, 1913–1921. doi:10.1139/cjfr-30-12-1913

Zheng, N., & Yoon, J. (2009). Expansion of the world's deserts due to vegetation-albedo feedback under global warming. *Geophysical Research Letters*, *36*(L17401).

Zhou, C., Wei, X., Zhou, G., Yan, J., Wang, X., & Wang, C. (2008). Impacts of a large-scale reforestation program on carbon storage dynamics in Guangdong, China. *Forest Ecology and Management*, *255*(3-4), 847–854. doi:10.1016/j.foreco.2007.09.081

Zimmermann, N. E., Edwards, T. C. Jr, Moisen, G. G., Frescino, T. S., & Blackard, J. A. (2007). Remote sensing-based predictors improve distribution models of rare, early successional and broadleaf tree species in Utah. *Journal of Applied Ecology*, *44*(5), 1057–1067. doi:10.1111/j.1365-2664.2007.01348.x

Zimmermann, N. E., & Kienast, F. (1999). Predictive mapping of alpine grasslands in Switzerland: Species versus community approach. *Journal of Vegetation Science*, *10*(4), 469–482. doi:10.2307/3237182

Zimmermann, N. E., Washington-Allen, R. A., Ramsey, R. D., Schaepman, M. E., Mathys, L., Kötz, B., et al. Edwards, T. C. (2007). Modern remote sensing for environmental monitoring of landscape states and trajectories. In F. Kienast, O. Wildi & S. Ghosh (Eds.), *A changing world: Challenges for landscape research* (pp.65 – 91). Dordrecht, The Netherlands: Springer.

Zsuzsanna, D., Bartholy, J., Pongracz, R., & Barcza, Z. (2005). Analysis of land-use/land-cover change in the Carpathian region based on remote sensing techniques. *Physics and Chemistry of the Earth*, *30*, 109–115.

ZVS. (2010). *Zürcher Vogelatlas*. Retrieved August 9th, 2010, from http://www.birdlife-zuerich.ch/vogelfinder/atlas-nach-vogelarten.html

About the Contributors

Rob Alkemade is a senior researcher at the Netherlands Environmental Assessment Agency (PBL). He obtained his PhD at Wageningen University in the role of nematodes in coastal ecosystems. He developed and applied models for assessing the effects of environmental change on biodiversity, first at the National Institute for the Environment and Public Health (RIVM) and later at PBL. He has a wide experience in biodiversity assessment and scenario analysis at the global level and contributed to the Millennium Ecosystem Assessment, Global Environmental Outlooks and Global Biodiversity Outlooks. For this purpose, he developed the GLOBIO3 model. He is a visiting scientist at Wageningen University doing research on the relationship between biodiversity and ecosystem services.

Rajendra P Shrestha received his PhD in natural resources management and currently is an Associate Professor at the Asian Institute of Technology, Thailand. His areas of research interests include land use and land cover change focusing on land change/degradation-human interface for policy support in the context of climate change. He also has interest in livelihood studies and food security in relation to land use. He has extensively published on these topics in southeast and south Asia. His research collaboration has been with several organizations, FAO, UNDP, UNEP, IUCN, WAC and the universities in the region. Previously, he has worked as lecturer and agriculture officer in Nepal. He was also a Senior Programme Officer at the United Nations Environment Programme, Bangkok, a Visiting Scientist at Nihon University of Japan, and Roskilde University of Denmark.

Yongyut Trisurat is an Associate Professor of Forestry at Kasetsart University in Bangkok, Thailand. He received PhD in natural resources management and conservation from the Asian Institute of Technology (AIT) in Thailand. He was a Research Fellow at the Institute of Geography, Freie University Berlin in 1995, a Fulbright Visiting Scholar affiliated with University of Hawaii and the East-West Center in 2005, and a Visiting Researcher at AIT in 2009. He has been active in the area of protected areas, biodiversity conservation, landscape ecology and GIS for over 15 years and has been a frequent contributor to several international agencies (e.g., ITTO, IUCN, ADB, CIDA, DANCED/DANIDA, WWF). His current research involves biodiversity conservation and climate change. In addition, he has published a number of peer-reviewed papers and book chapters on these subjects.

* * *

Carlos Alberto Arnillas is a research fellow of the Conservation Data Centre of the National Agrarian University of Lima. He received his bachelor degree in Biology in the same university. His research is focused on landscape ecology, with emphasis on conservation planning and climate change impact on biodiversity. Currently, he is part of an international team researching climate change impact on tropical Andes.

Peter C. Boyce is a visiting lecturer, School of Biological Sciences, Universiti Sains Malaysia. Previously he held a BRT Research Associate post for two years. When not teaching in Penang he is based in Kuching, Sarawak. Awards include a Silver Engler Medal from the International Association of Plant Taxonomists (1996) and the Henry Allan Gleason Award, New York Botanical Gardens (2000). Research interests centre on the Araceae of tropical Asia, Hanguana, speciation dynamics in everwet and perhumid Sunda, and morphological adaptations in specific ecological niches, notably rheophytic plants. Current research foci include taxonomy and systematics of *Homalomena*, the Schismatoglottideae, and Nephythyrideae, and of *Hanguana*.

Caroline Byrne received her B.A. degree in Natural Sciences and Ph.D. on the Systematics of the Thai Clusiaceae and Hypericaceae at Trinity College, Dublin, Ireland. Following the completion of her Ph.D. in 2009, she was a research assistant at Trinity College on the Interactive Flora of the Burren Project for 6 months. At present, she is preparing and finishing papers for publication.

Kongkanda Chayamarit received her B.Sc. in biology at Kasetsart University, as well as her M.Sc. in Botany. She obtained her doctorate in Plant Systematics from the Faculty of Science of the University of Tokyo (Japan). From 1979 to 1984, she was plant taxonomist at the Forest Herbarium in Bangkok, till 2005 she worked there as curator, followed by the position of Director until 2008. In 2009, she became director of the Botanical Garden Organization of Thailand and is, therein, in charge of Queen Sirikit Botanic Garden. She is the production manager and motor behind the Flora of Thailand project.

Roland Cochard is Assistant Professor (since 2009) at the Asian Institute of Technology near Bangkok, Thailand. He received his Bachelor in Environmental Science (with Honours) in 1999 from James Cook University in North Queensland, Australia, and his PhD in 2004 from the Institute of Integrative Biology at the Swiss Federal Institute of Technology (ETH) in Zurich, Switzerland. During and after his PhD (2000-2005) he conducted research on African savanna vegetation dynamics for ETH and the GTZ. In 2006, he conducted a survey of tsunami affected coastal ecosystems in Thailand and Indonesia (collaboration of ETH, AIT and ZIL), before he was involved in a bird atlas project (in 2007) and served as an advisor (in 2008) to Zurich Financial Services in a Country Risk Assessment Project. He is currently conducting research on biodiversity and conservation, savanna and rainforest vegetation dynamics, invasive species management, ecological restoration, and climate change and sustainability issues.

Charlotte Couch is a botanical researcher at the Royal Botanic Gardens, Kew. She received a BSc. (Hons) from University of Wales, Aberystwyth and subsequently a MSc in Plant Diversity (Biodiversity and Conservation) from the University of Reading. She has recently worked on conservation assessments for Cyperaceae species from Thailand and on the Interactive Key for Flora Malesiana.

Tom Curtis is a plant taxonomist, ecologist and horticulturalist. His doctoral research was on dactylorchids in Ireland and Europe, and he has over 36 years field experience in orchids and Ireland's wild plants. He was co-author of *The Irish Red Data Book: 1 Vascular Plants*, *The Orchids of Ireland* and co-editor of *Ireland and the Water Framework Directive*. He has published extensively on the flora of Ireland and its coastal ecology. He formerly worked in the research branch of the National Parks and Wildlife Service. Since 2002, he has worked as an ecological consultant on projects as diverse as the Water Framework Directive, the rare plants and montane flora of County Wicklow and on a fen restoration project with BirdWatch Ireland. Currently, he is a Research Associate of the Botany School, Trinity College, Dublin, Adjunct Lecturer in Botany and Plant Science in the National University of Ireland, Galway, and Chairman of Genetic Heritage Ireland.

Soejatmi Dransfield is a plant taxonomist specializing in bamboos, who gained her first degree in Plant Taxonomy from Academy of Agriculture, Ciawi, Bogor, Indonesia. She began her botanical career as a staff member of Herbarium Bogoriense, Indonesia, and gained her PhD from Reading University, UK, in 1975 with her thesis 'The revision of Cymbopogon (Gramineae)'. After she moved to UK in 1978, she continued her research on bamboo taxonomy including the generic delimitation of the Old World tropical bamboos. She is currently Honorary Research Fellow at the Royal Botanic Gardens, Kew, UK, writing the account of bamboos from Malesia, Thailand and Madagascar.

Hans-Joachim Esser is Curator and Research Scientist at the herbarium of the Botanische Staatssammlung München. He received his Diploma and Doctorate in Biology at the University of Hamburg, Germany. He was Mercer Fellow at the Arnold Arboretum, Harvard University, USA, in 2000-2002. He worked as Postdoc and Visiting Researcher at Trinity College Dublin, Ireland, the Rijksherbarium Leiden, Netherlands, and the University of Utrecht, Netherlands. He worked at the Forest Herbarium Bangkok with a grant of the Thai Biodiversity Research and Training Program (BRT). He has been specializing in Systematic Botany for 20 years. He contributed to various floras of tropical areas of Asia and South America; currently he is member of the Editorial Board of the Flora of Thailand.

Gustavo Galindo is currently working for the Instituto de Hidrología, Meteorología y Estudios Ambientales (Ideam) in Colombia. He received his B.Sc. in Biology from the Universidad de los Andes and has postgraduate studies in GIS and remote sensing from CIAF and the Universidad Distrital Francisco José de Caldas. Gustavo has more than 10 years of experience in spatial analysis in the areas of biodiversity conservation, landscape ecology and ecosystem mapping; he worked for the Instituto de Investigación Alexander von Humboldt (IAvH), for more than 5 years where he received the support to do this research. His work is centered on biomass estimation of tropical forests in the frame of REDD.

Alan Grainger is Senior Lecturer in the School of Geography, University of Leeds, which he joined in 1992. He has undertaken research into modelling and monitoring tropical deforestation since 1980, gaining his D.Phil. at the University of Oxford for building the world's first global simulation model of long-term trends in tropical forest resources. For the past 20 years, he has also modelled the role of tropical forests in global climate change and the impacts of the latter on biodiversity. His interests also extend to sustainable development, desertification, and the analysis of forest policy and institutions.

Jan Janse has some 25 years of experience in modelling of aquatic ecosystems. He studied biology and environmental sciences at Utrecht University and specialized in freshwater systems. He worked at several institutions like a regional water board, the Research Institute for Nature Management, Wageningen University, the National Institute of Public Health and the Environment, and now the Netherlands Environmental Assessment Agency, on (policy-oriented) research and advisory projects in the fields of water quality, biodiversity, and water management. He graduated at Wageningen University on the Ph.D. thesis 'Model studies on the eutrophication of shallow lakes and ditches.' These models linking human impact to tipping points in aquatic systems are nationally and internationally acknowledged. He is currently involved in the development and application of aquatic models in a global context.

P.K. Joshi has held the positions of Associate Professor and Head of Department of Natural Resources at TERI University, New Delhi, India. He is trained originally as an environmentalist, and then as an ecologist, developing skills in remote sensing and GIS with a firm scientific research basis. Prior to joining TERI University, Dr Joshi spent a decade with the Indian Space Research Organization (ISRO) on secondment from the Indian Institute of Remote Sensing (IIRS), Dehradun, an internationally-renowned institution in the field of RS and GIS. His research has been recognized by the Indian Academy of Sciences (INSA) and NASI (National Academy of Sciences India (NASI) through the award of their highly prestigious Young Scientist Medal (2006) and Young Scientist Platinum Jubilee Award (2009) respectively and many others of similar kind. He is widely published, has experience of the successful supervision of graduate research students at PhD and masters levels, and, in addition to his BSc (Hons), MSc in Environmental Sciences, Post-grad Diploma in Marketing and a PhD, recently (2005) obtained a masters degree in Sustainable Development (Climate Change). His current research involves landscape analysis, climate change, and natural resource assessment using Geo-informatics.

Aung Pyeh Khant is a Geo-informatics Scientist working with Assoc. Prof. Dr. Nitin Kumar Tripathi of the Asian Institute of Technology (AIT) in Bangkok, Thailand. He obtained M.S. in Remote Sensing & Geographic Information Systems from AIT in 2002. His research interests are on biodiversity monitoring and geo-informatics.

Eric Koomen is assistant professor at the Department of Spatial Economics of the Vrije Universiteit Amsterdam. He holds a Ph.D. in 'Spatial analysis in support of physical planning.' This dissertation combined economic topics (valuation of open space, urban development, rural vitality) with earth science related issues (water management, flood-risk assessment) and combinations thereof (agricultural land-use change, open-space preservation). His current research interests include land-use change analysis and climate adaptation. He is a tutor on GIS and environmental impact assessment and European aspects of GI in the UNIGIS MSc programme and responsible for the courses on 'Land-use change' and 'Assessing the landscape' in the Earth and Economics programme. Eric, furthermore, works part-time at the Geodan Next Company where he informs regional authorities about likely spatial developments, their potential impacts and possible policy alternatives.

Grygoriy Kolomytsev, is a Lead engineer of the I.I.Schmalhausen Institute of Zoology of National Academy of Sciences of Ukraine (IZ NASU). Since 2010, he is a PhD student at the Taras Shevchenko National University, Kyiv. In 2007, he participated in IT-training on pressure-based-biodiversity modeling, MSA, and GLM application at the Faculty of Geo-Information Science and Earth Observation (ITC) at the University of Twente (the Netherlands). Since 2006, he holds an M.S. in Biology & Zoology from the Taras Shevchenko National University, Kyiv, Ukraine.

Tom Kram is a programme manager for integrated assessment modeling at the Netherlands Environmental Assessment Agency (PBL). He earned a M.Sc. degree in Electrical Engineering and Operations Research from Technical University Delft, specializing in economics of electric power production. His core responsibilities include the development and application of the IMAGE modeling framework, working with national and international research partnerships. He has contributed to IPCC in a variety of functions, including Lead Author of the 2nd Assessment Report and the Special Report on Emissions Scenarios (SRES). Currently he is a member of IPCC-TGICA, a task group set up to support data and scenario information for impact and climate analysis. His current research focuses on the role of land-use in as pivot for climate change impacts, adaptation, and mitigation (e.g. bio-energy, forestry options) in close conjunction with providing other ecological goods and services for human development (food, water, biodiversity, etc.).

Jan Peter Lesschen is researcher at Alterra in Wageningen (Netherlands), which is part of Wageningen University and Research centre. He has a MSc. degree in soil science from Wageningen University and obtained his PhD degree at the University of Amsterdam on the study of multi-scale interactions between soil, vegetation, and erosion in Southeast Spain. He is currently working in EU and Dutch funded projects on greenhouse gas emissions, land use change, bio-energy, nutrient management, and regional scale modeling. Furthermore, he is responsible for the development of the MITERRA model, which assesses effects and interactions of policies and measures in agriculture on GHG emissions, nitrogen fluxes and soil carbon stocks at regional level for the EU-27 and at local level for the Netherlands.

Nguyen Kim Loi is a Lecturer at Department of Applied Geo-infomatics, Nong Lam University (NLU) in Ho Chi Minh City, Vietnam. He received B.S. degree in Forest Resources from Nong Lam University and the M.Sc. and Ph.D. in Watershed Management and Environmental from the Kasetsart University (KU) in Thailand. Dr. Nguyen Kim Loi has extensive experience with watershed and environmental management, GIS, and land use planning issues. He is expert in GIS application and related spatial techniques for watershed modeling, land use mapping, soil erosion control, and climate change. His current research involves GIS, Soil and Waters Assessment Tool (SWAT) model, and climate change.

Denisse McLean R. has a B.Sc. in Socioeconomic Development and Environment from the Panamerican Agricultural University, Zamorano in Honduras. She works as a research assistant for the Biodiversity Modeling Project of the Regional Biodiversity Institute (IRBIO). She was responsible for the national biodiversity assessments for Honduras and Nicaragua and for the integration of models into the regional assessment for Central America. She also worked in the design of a handbook on biodiversity modeling on the national scale with GLOBIO3 methodology for Spanish speaking audience. Currently, she is working on the Central American model results validation with countries' Biodiversity Technical Committees, on the transfer of outputs to environmental authorities and on developing other biodiversity modeling proposals for the region.

Conor Meade is a Lecturer in Ecology at the National University of Ireland Maynooth. A BSc graduate in Biology from University College Dublin, he completed a PhD in Plant Systematics at the University of Dublin, Trinity College in 2001. He joined the National University of Ireland as a postdoctoral researcher in 2001 and was appointed University Lecturer in 2006. His research interests include angiosperm systematics (especially the Annonaceae), gene-flow and hybridization in plant populations, and plant biogeography in Europe and Southeast Asia.

David Middleton is a tropical botanist at the Royal Botanic Garden Edinburgh, Scotland. He received his BSc and PhD degrees in Botany from Aberdeen University. He has furthered his research on the taxonomy of Southeast Asian plants at Trinity College Dublin, Ireland, the Rijksherbarium Leiden, Netherlands, the Arnold Arboretum of Harvard University, USA and the Royal Botanic Garden Edinburgh. He has contributed accounts of the Apocynaceae to the Flora of Thailand, Flora Malesiana, the Tree Flora of Sabah and Sarawak, the Flora of Peninsular Malaysia, the Flore du Cambodge, du Laos et du Vietnam, and is a coauthor on the Flora of China account. He currently specialises in research on the Gesneriaceae of Southeast Asia and is the editor of the Edinburgh Journal of Botany.

Justin Moat has been employed for 18 years at The Royal Botanic Gardens, Kew with the initial remit of setting up a GIS unit, which he currently heads. Justin develops and manages the GIS unit, projects and related research, especially webmapping, vegetation mapping and conservation assessments.

Muthama Muasya is a Senior Lecturer at University of Cape Town (South Africa). He holds BSc and MPhil degrees from Moi University (Kenya), PhD from University of Reading (UK), and postdoctoral stints at Royal Botanic Gardens Kew (UK), Rutgers University (USA) and KU Leuven (Belgium). He teaches courses in Biodiversity and Evolutionary Biology and does primary research in Angiosperm Systematics. He has broad interests on the taxonomy, biogeography, and phylogenetics of the cosmopolitan monocotyledonous family Cyperaceae, the evolution of the Cape flora, diversity and use of wetland plants, and the origin, diversity, and utilization of the African savanna biome.

Ir C.A. (Sander) Mücher is head of the team Earth Observation at Alterra, which is part of Wageningen University and Research Centre (WUR). He is a senior researcher in Remote Sensing & GIS with a background in Tropical Crop Science, with specialisations in Rural Surveys & Land Ecology, Soil Science and Geo-Information. His research activities at Alterra started in 1993, as a project coordinator of various studies funded by the National Remote Sensing Programme (NRSP). In 1997, he started as a project coordinator of the EU-FP4 project PELCOM which aimed at land use monitoring with low resolution satellite data for environmental applications. He is involved in many European research projects in which the integration of RS and GIS with ecological knowledge plays an important role. Most recent EU projects are ECOCHANGE, which aims to assess and forecast biodiversity and ecosystem changes in Europe, and EBONE, which aims at an integrated biodiversity observing system in space in time.

John Parnell, currently Head of the School of Natural Sciences, is Professor of Systematic Botany and Curator of the Herbarium in Trinity College Dublin, Ireland. He obtained both his B.Sc. in Botany and Ph.D. from the University of Aberdeen, Scotland and was then appointed to a Lectureship in Trinity College. His research in higher plant systematics, especially plant taxonomy and floristics, is deliberately split between and uniquely links Europe and Thailand. This split allows for better understand of the variation patterns and biogeographic patterns of tropical taxa, usually known from only a few individuals, by studying population scale variation and biogeography in Ireland.

Colin Pendry is a Researcher and Editor of the Flora of Nepal at the Royal Botanic Garden Edinburgh. He received a BSc in Biological Sciences from the University of Edinburgh and PhD in Tropical Ecology from the University of Stirling. He was a Royal Society Research Fellow at Trinity College Dublin from 1994-1996, and in 1997, was a visiting lecturer at Khon Kaen University. He has extensive field experience in the UK, SE Asia, Latin America, and Nepal, and has taxonomic expertise in the Polygalaceae of Thailand and Indochina and Latin American Polygonaceae. He has published on the historical biogeography of Latin American seasonally dry forests and the ecology of SE Asian rainforests.

Manuel Peralvo is a geographer currently working as an associate researcher at CONDESAN in Quito, Ecuador. He received a MA from the Department of Geography and the Environment at the University of Texas at Austin and is a PhD candidate at the same institution. His main area of research is focused in human-environment relationships with emphasis in the use of environmental models to support decision making processes. Currently, Manuel is working in different projects in the Andean region aimed at characterizing and supporting adaptation mechanisms to the combined effects of climate change, land use, and land cover change. Other researches initiatives are related to the generation of environmental information to support REDD mechanisms and the analysis of the impacts of environmental changes on the structure and function of Andean social and environmental systems.

Nannapat Pattharahirantricin is a researcher of the Forest Herbarium, Department of National Parks, Wildlife and Plant Conservation. She received the Master degree in Forest Biology from Faculty of Forestry, Kasetsart University in Thailand. She has been working on some genera in Euphorbiaceae and Malvaceae for the Flora of Thailand treatments. She is now responsible for the productions of Thai Forest Bulletin (Botany), an international botanical journal and the Flora of Thailand publications, and also working as the Forest Herbarium curator assistant.

Rachun Pooma is a researcher of the Forest Herbarium, Department of National Parks, Wildlife and Plant Conservation. He received the Ph.D. in Botany from Kasetsart University in Thailand. He has been working on plant taxonomy, especially in Dipterocarpaceae and Burseraceae for Flora of Thailand Project, and has been surveying and collecting plants though out the country. He is now a curator of the Forest Herbarium.

Lilik Budi Prasetyo is Associate Professor at Department of Forest Resources Conservation and Ecotourism, Forestry Faculty of Bogor Agricultural University (IPB), Indonesia. He received B.S. degree in Faculty of Agriculture of IPB and the Master degree in the Department of environmental sciences, Tsukuba University. He completed his PhD degree in the same University in Forest Management at the Institute Agriculture & Forestry. He has visited some institution such as the Tokyo University, Japan, National Institute for Agricultural and Environmental Sciences, Tsukuba Japan, and Viikki Tropical Forest Research Institute of Helsinki University as visiting researcher. Most of his research is on the application of Remote Sensing and Geographical Information System in the field of Landscape Ecology.

Neena Priyanka is a Doctoral Research Fellow at TERI University, New Delhi, India. She earned her B.Sc. degree in Botany (Hons) from Delhi University, and the Master's degrees in Environmental Studies from TERI University. She was a visiting researcher to Kyushu Institute of Technology (KIT), Fukuoka, Japan to carry out studies on Urban Heat Islands (UHI) and a faculty guest in alliance with Prof. P. K. Joshi at Guru Govind Indraprastha and TERI University. Her work mostly focuses on remote sensing, GIS ad spatial modeling in the context of biodiversity conservation and natural resources management with some empirical studies on the endangered *Olive ridley* sea turtle habitat assessment, *Simarouba glauca* modeling for livelihood adaptation, *Lantana camara* invasion potential, to provide scientific basis to decision makers for species conservation and management arena. Her current PhD research involves invasive species modeling in context of climate change and anthropogenic disturbances.

Vasyl Prydatko is a Senior Specialist at the Ukrainian Land and Resource Management Center (ULRMC), NGO, which objectives include applying RS, digital mapping, GIS, and other IT data to support rendering public and private sector decisions, both in Ukraine and in the region. He worked as the Associated Professor at the National University of Life and Environmental Sciences of Ukraine (2007-2009), Senior Scientist at the environmental institute of the National Security and Defense Council of Ukraine (1999-2001), and Head of the Department of Ministry of the Environmental Protection of Ukraine (1993-1999). In ULRMC, he coordinates and manages international IT-projects (USAID, UNDP, GEF, NEAA, PBL). Vasyl began his carrier as biology scientist at the Wrangel Island Reserve and participated in scientific expeditions in the Arctic (1978-1988). He holds a Ph.D. in Biology from Schmalhausen Institute of Zoology NASU and an M.S. in Biology & Zoology from the Taras Shevchenko National University, Kyiv, Ukraine.

Niels Raes is a Postdoctoral Research Fellow at the NCB Naturalis. His main research interests concern macroecological patterns of biodiversity and biogeography derived from species distribution models and the predicted impacts of global climate change on these patterns.

Wilbert van Rooij works as a senior consultant at the non-profit organisation Aidenvironment in Amsterdam, the Netherlands. He did his Master's in tropical forest management at the Wageningen Agricultural University and worked for several years in Ethiopia and Malaysia as a forestry, GIS, and Remote Sensing specialist. From 2006-2010, Wilbert specialised in biodiversity modelling at the Netherlands Environmental Assessment Organization (PBL) and joined Aidenvironment in 2010. He developed a modelling training manual and organized several training courses mainly in tropical regions with participants from over 20 countries. Currently, he is involved with the integration of land use and biodiversity modelling with strategic environmental assessment projects in Vietnam and Papua.

David A. Simpson is Assistant Keeper for Systematics in the Herbarium, Library, Art and Archives at the Royal Botanic Gardens, Kew. He graduated from the University of Wales in 1977 with an Honours degree in Botany and Forestry. This was followed by a MSc in Pure and Applied Plant Taxonomy from the University of Reading in 1978 and a PhD at the University of Lancaster in 1983. His research focuses on the taxonomy and systematics of Sedges (Cyperaceae), Grasses (Poaceae) and related families worldwide. He has published seven books and over 150 papers. He is Editor in Chief of Kew Bulletin and a member of the Flora of Thailand and Flora of China Editorial Boards.

Marta Pérez-Soba is senior researcher at the Centre for Geo-Information at Alterra Wageningen University Research (the Netherlands). She received her degree as Agricultural Engineer from the Polytechnic University of Madrid (Spain) and has a PhD on environmental impacts on forest ecosystems (Groningen University, The Netherlands, 1995). She has been active in the topics of eco-toxicology, landscape ecology, and GIS for over 20 years and contributes as researcher or coordinator of projects for several European organisations (e.g. European Commission, European Environment Agency, ESPON). Her current research involves impact assessment of land use change, sustainable development, and future regional developments in the European countryside.

George Staples is Senior Researcher in the Singapore Botanic Gardens, a post he has held since 2007. He earned B.A. and M.Sc. degrees from Florida Atlantic University and A.M. and Ph.D. degrees from Harvard University. For 19 years, he was Botanist at the Bishop Museum, Honolulu, Hawaii, where he authored three books, including a major new identification manual for tropical cultivated plants. Current research interests include taxonomy and systematics of Convolvulaceae, Asian floristics, invasive species biology, and economic and useful plants. He has studied the Thai flora for 25 years and contributed botanical specimen data to the chapter on Thai phytogeography in this book.

Somran Suddee is a Senior Scientific Researcher at The Forest Herbarium (BKF), Bangkok, Thailand. He received his B.Sc. degree in Forestry from Kasetsart University, his M.Sc. in Botany from Chulalongkorn University, and his Ph.D. in Plant Taxonomy from Trinity College, University of Dublin, Ireland (in collaboration with the Royal Botanic Gardens, Kew, London). He is currently working on the families Labiatae and Orchidaceae for the Flora of Thailand Project. He is a member of the Plant Taxonomy committee in the Royal Thai Institute.

Sarawood Sungkaew is now a lecturer in the Forest Biology Department, Faculty of Forestry, Kasetsart University. He gained B.S. and M.S. degrees in Forestry from Kasetsart University, and Ph.D. in Plant Taxonomy and Systematics from Trinity College, University of Dublin, Ireland. He is one of the collaborators of Bamboo Phylogeny Group, an international team of researchers with expertise in bamboo systematics and dedicated to producing a robust phylogeny of the woody bamboos. His areas of research interests are forest plant diversity, forest plant ecology, and taxonomy and systematics of bamboos.

Atchara Teerawatananon is an official researcher in the Natural History Museum, Thailand. She obtained her B.S. degree in Agriculture, M.S. degree in Botany, both from Kasetsart University, and her Ph.D. in Plant Taxonomy and Systematics from Trinity College, University of Dublin, Ireland. Her research area involves museum management, plant diversity, grass taxonomy, and systematics.

Anna Trias-Blasi has been recently appointed Bulbous Monocot Systematics & Conservation researcher at the Royal Botanic Gardens Kew in the UK. She received a *Licenciatura en Biología* (equivalent to a B.S. degree) in Biology from the *Universitat de Girona* in Spain, a M.Sc. in the Biodiversity and Taxonomy of Plants from the University of Edinburgh and the Royal Botanic Garden Edinburgh in the UK, and a Ph.D. entitled *Systematics of the Thai Vitaceae* from Trinity College Dublin in Ireland. She was a Postdoctoral Researcher at the Royal Botanic Garden Edinburgh in 2010. Her research involves plant systematics, taxonomy, biogeography, and conservation.

Nguyen Dieu Trinh is Official at the Ministry of Planning and Investment of Vietnam. She got the Master Degree in Environmental Economics Management at Hanoi Economics University. Her experience is the involvement in the planning process where environment and climate change issues are taken into account for sustainable socio-economic development strategies/plans. She is also active in working with development partners/donors across the globe for international knowledge transfer, experiences sharing, and policy update at all levels. Her daily job is either doing research or integrating research results into planning and contributing to the environmentally friendly investment policy-making process.

Nitin Kumar Tripathi is Associate Professor of Remote Sensing and GIS at Asian Institute of Technology in Bangkok, Thailand. He received B.Tech. degree in Civil Engineering from National Institute of Technology, India, and the M.Tech. and Ph.D. in Geoinformatics from the Indian Institute of Technology (IIT) in India. He was a Visiting Outstanding Researcher in Osaka City University in 2008. He has been active in the area of remote sensing applications to protected areas, biodiversity conservation, and GIS for over 20 years and has been a frequent contributor to several international agencies (e.g., DANIDA/ NACA/ MPEDA/ SIDA/AIT/UNEP). His current research involves biodiversity conservation, climate change, and green house gas mapping using remote sensing.

Carolina Tovar is a research fellow of the Conservation Data Centre of the National Agrarian University of Lima. She received an MSc degree in Conservation of Forestry Resources from the National Agrarian University, Lima, Peru, and a second MSc in Biological Sciences from the University of Amsterdam. Her research is mainly related to landscape ecology, species distribution modeling, and land use/cover change. She has been involved in conservation planning for the last 8 years, collaborating with local actors and national and international research centers related to the tropical Andes and Amazonia. She is currently a PhD student at the University of Oxford, on the integration of long term ecological analysis in conservation issues.

Albertus G. Toxopeus is an Associate Professor at the Department of Natural Resources (NRS) at the Faculty of Geo-Information Science and Earth Observation at the University Twente (UT) in Enschede, The Netherlands. He received B.S. and the M.Sc.degree in Biology from the University of Groningen (RUG), and Ph.D. in Natural Resources Management and Conservation from the University of Amsterdam in The Netherlands. He has been active in the area of protected areas, biodiversity conservation, RS and GIS for over 20 years and has been a frequent contributor to several international agencies (e.g., UNESCO, FAO, IUCN, KWS, MICOA). His current research involves biodiversity, conservation, and climate change.

Peter H. Verburg is a professor and the Head of the Department Spatial Analysis and Decision Support of the Institute for Environmental Studies at VU University Amsterdam, the Netherlands. Peter obtained his PhD at Wageningen University in the field of land use modeling in the Asian region. Peter is a geographer specialized in the integrated analysis of land use change at multiple spatial and temporal scales. As part of his activities, he has developed the CLUE model that has been used for land use change modeling in a wide range of scenario studies across the globe. Peter has published over 80 peer-reviewed papers in the fields of geography, landscape ecology, agricultural and environmental science.

Peter C. van Welzen is Professor in Tropical Plant Biogeography at Leiden University (The Netherlands) and works on the Malesian Euphorbiaceae in the Netherlands Centre for Biodiversity Naturalis. He received his B.Sc., M.Sc. and Ph.D. in Biology at Leiden. He is an active contributor in several Asian flora projects (e.g., Flora Malesiana, Flora of Thailand) and combines alpha-taxonomy with phylogenetic and biogeographic research. Peter is board member of the Flora of Thailand Project.

Chandra Irawadi Wijaya is Graduate Student in Information Technology for Natural Resources Management at Bogor Agricultural University in Indonesia. He received his Bachelor Degree in Forestry from Bogor Agricultural University. He was an exchange research student at Division of Spatial Information Science, University of Tsukuba, Japan in 2009/2010. He worked at Tropenbos International Indonesia Programme in 2009 as GIS Specialist and Center for International Forestry Research (CIFOR) during 2005 - 2007 as GIS Consultant. Currently, he works at World Agroforestry (ICRAF) as Research Assistant. His current research involves land use change study, conservation, environmental services, and GIS.

Paul Wilkin has been Lilioid & Alismatid Monocots Team Leader in the Herbarium at the Royal Botanic Gardens, Kew since 2002. He received a B.A. Degree in Natural Sciences from Cambridge University and a M.Sc. and PhD in plant systematics from the University of Reading. His main research focus is the systematics, ethnobotany, sustainable use and conservation of Dioscoreales, the yams and their allies. This programme is underpinned by baseline surveys and inventories in Dioscoreaceae, especially in Madagascar and Thailand. He is the Contribution Editor of Thai Forest Bulletin (Botany) has been active in imaging and databasing Kew's Monocot herbarium holdings and in developing e-Taxonomy. Other monocot taxa under systematic study include Dracaenoids (currently *Dracaena* and *Sansevieria* (Asparagceae), *Gagea* and *Erythronium (*Liliaceae*), Sternbergia* (Amaryllidaceae) and *Tigridieae* (Iridaceae) of Bolivia.

Index

D

E

I

J

K

L

M

N

P

R

S

T

U

V

W

www.ingramcontent.com/pod-product-compliance
Lightning Source LLC
LaVergne TN
LVHW062136200926
848484LV00007B/1822